Applying Nanotechnology for Environmental Sustainability

Sung Hee Joo
University of Miami, USA

A volume in the Advances in Environmental Engineering and Green Technologies (AEEGT) Book Series

www.igi-global.com

Published in the United States of America by
IGI Global
Information Science Reference (an imprint of IGI Global)
701 E. Chocolate Avenue
Hershey PA, USA 17033
Tel: 717-533-8845
Fax: 717-533-8661
E-mail: cust@igi-global.com
Web site: http://www.igi-global.com

Copyright © 2017 by IGI Global. All rights reserved. No part of this publication may be reproduced, stored or distributed in any form or by any means, electronic or mechanical, including photocopying, without written permission from the publisher. Product or company names used in this set are for identification purposes only. Inclusion of the names of the products or companies does not indicate a claim of ownership by IGI Global of the trademark or registered trademark.

Library of Congress Cataloging-in-Publication Data

Names: Joo, Sung Hee, editor.
Title: Applying nanotechnology for environmental sustainability / Sung Hee
   Joo, editor.
Description: Hershey, PA : Information Science Reference, 2017. | Includes
   bibliographical references and index.
Identifiers: LCCN 2016017815| ISBN 9781522505853 (hardcover) | ISBN
   9781522505860 (ebook)
Subjects: LCSH: Nanostructured materials--Environmental aspects. |
   Environmental protection--Materials. | Nanostructured
   materials--Industrial applications. | Micropollutants. | Nanofiltration. |
   Sustainable engineering.
Classification: LCC TD196.N36 A67 2017 | DDC 620.1/150286--dc23 LC record available at https://lccn.loc.gov/2016017815

This book is published in the IGI Global book series Advances in Environmental Engineering and Green Technologies (AEEGT) (ISSN: 2326-9162; eISSN: 2326-9170)

British Cataloguing in Publication Data
A Cataloguing in Publication record for this book is available from the British Library.

All work contributed to this book is new, previously-unpublished material. The views expressed in this book are those of the authors, but not necessarily of the publisher.

For electronic access to this publication, please contact: eresources@igi-global.com.

Advances in Environmental Engineering and Green Technologies (AEEGT) Book Series

ISSN: 2326-9162
EISSN: 2326-9170

MISSION

Growing awareness and an increased focus on environmental issues such as climate change, energy use, and loss of non-renewable resources have brought about a greater need for research that provides potential solutions to these problems. Research in environmental science and engineering continues to play a vital role in uncovering new opportunities for a "green" future.

The **Advances in Environmental Engineering and Green Technologies (AEEGT)** book series is a mouthpiece for research in all aspects of environmental science, earth science, and green initiatives. This series supports the ongoing research in this field through publishing books that discuss topics within environmental engineering or that deal with the interdisciplinary field of green technologies.

COVERAGE
- Green Technology
- Green Transportation
- Sustainable Communities
- Cleantech
- Air Quality
- Water Supply and Treatment
- Industrial Waste Management and Minimization
- Pollution Management
- Renewable Energy
- Biofilters and Biofiltration

IGI Global is currently accepting manuscripts for publication within this series. To submit a proposal for a volume in this series, please contact our Acquisition Editors at Acquisitions@igi-global.com or visit: http://www.igi-global.com/publish/.

The Advances in Environmental Engineering and Green Technologies (AEEGT) Book Series (ISSN 2326-9162) is published by IGI Global, 701 E. Chocolate Avenue, Hershey, PA 17033-1240, USA, www.igi-global.com. This series is composed of titles available for purchase individually; each title is edited to be contextually exclusive from any other title within the series. For pricing and ordering information please visit http://www.igi-global.com/book-series/advances-environmental-engineering-green-technologies/73679. Postmaster: Send all address changes to above address. Copyright © 2017 IGI Global. All rights, including translation in other languages reserved by the publisher. No part of this series may be reproduced or used in any form or by any means – graphics, electronic, or mechanical, including photocopying, recording, taping, or information and retrieval systems – without written permission from the publisher, except for non commercial, educational use, including classroom teaching purposes. The views expressed in this series are those of the authors, but not necessarily of IGI Global.

Titles in this Series
For a list of additional titles in this series, please visit: www.igi-global.com

Smart Cities as a Solution for Reducing Urban Waste and Pollution
Goh Bee Hua (National University of Singapore, Singapore)
Information Science Reference • copyright 2016 • 362pp • H/C (ISBN: 9781522503026) • US $190.00 (our price)

Food Science, Production, and Engineering in Contemporary Economies
Andrei Jean-Vasile (Petroleum-Gas University of Ploiesti, Romania)
Information Science Reference • copyright 2016 • 473pp • H/C (ISBN: 9781522503415) • US $240.00 (our price)

Smart Grid as a Solution for Renewable and Efficient Energy
Ayaz Ahmad (COMSATS Institute of Information Technology, Pakistan) and Naveed Ul Hassan (LUMS School of Science & Engineering, Pakistan)
Information Science Reference • copyright 2016 • 415pp • H/C (ISBN: 9781522500728) • US $220.00 (our price)

Biologically-Inspired Energy Harvesting through Wireless Sensor Technologies
Vasaki Ponnusamy (Universiti Tunku Abdul Rahman, Malaysia) Noor Zaman (King Faisal University, Saudi Arabia) Tang Jung Low (Universiti Teknologi Petronas, Malaysia) and Anang Hudaya Muhamad Amin (Multimedia University, Malaysia)
Information Science Reference • copyright 2016 • 318pp • H/C (ISBN: 9781466697928) • US $200.00 (our price)

Handbook of Research on Waste Management Techniques for Sustainability
Ulas Akkucuk (Bogazici University, Turkey)
Information Science Reference • copyright 2016 • 438pp • H/C (ISBN: 9781466697232) • US $240.00 (our price)

Control and Treatment of Landfill Leachate for Sanitary Waste Disposal
Hamidi Abdul Aziz (Universiti Sains Malaysia, Malaysia) and Salem Abu Amr (Universiti Sains Malaysia, Malaysia)
Information Science Reference • copyright 2016 • 459pp • H/C (ISBN: 9781466696105) • US $220.00 (our price)

Toxicity and Waste Management Using Bioremediation
Ashok K. Rathoure (Vardan Environet Guargaon, India) and Vinod K. Dhatwalia (Uttaranchal University, India)
Engineering Science Reference • copyright 2016 • 421pp • H/C (ISBN: 9781466697348) • US $205.00 (our price)

Handbook of Research on Climate Change Impact on Health and Environmental Sustainability
Soumyananda Dinda (Sidho-Kanho-Birsha University, India)
Information Science Reference • copyright 2016 • 711pp • H/C (ISBN: 9781466688148) • US $325.00 (our price)

www.igi-global.com

701 E. Chocolate Ave., Hershey, PA 17033
Order online at www.igi-global.com or call 717-533-8845 x100
To place a standing order for titles released in this series, contact: cust@igi-global.com
Mon-Fri 8:00 am - 5:00 pm (est) or fax 24 hours a day 717-533-8661

List of Reviewers

Tatiana Andreani, *University of Porto & CITAB-University of Trás-os-Montes and Alto Douro, Portugal*
Jiyeol Bae, *School of Environmental Science and Engineering, GIST, Korea*
César Barbero, *Universidad Nacional de Río Cuarto, Argentina*
Emrah Celik, *University of Miami, USA*
Hyeon-gyu Choi, *School of Environmental Science and Engineering, GIST, Korea*
Ahmed Emam, *National Research Centre, Egypt*
Jie Fu, *Georgia Institution of Technology, USA*
Krishna Giri, *Rain Forest Research Institute, India*
Bing Han, *Auburn University, USA*
Yang Hu, *University of Waterloo, Canada*
Wen Liu, *Auburn University, USA*
Ahmed Mansour, *Cairo University, Egypt*
Gaurav Mishra, *Rain Forest Research Institute, India*
Kyle Moor, *Dept. of Chemical and Environmental Engineering, USA*
Verónica Nogueira, *University of Porto, Portugal*
Kathryn Nunnelley, *University of Virginia, USA*
Allan Philippe, *University Koblenz-Landau, Germany*
Nicolas Rongione, *University of Miami, USA*
James Smith, *University of Virginia, USA*
Samuel Snow, *Michigan State University, USA*
Moon Son, *School of Environmental Science and Engineering, GIST, Korea*
Chunming Su, *Environmental Protection Agency, USA*
Narendra Sura, *R.V. College of Engineering, India*
Irshad Wani, *J&K Government, India*
Edith Yslas, *Universidad Nacional de Río Cuarto, Argentina*

Table of Contents

Preface xviii

Acknowledgment xxvii

Chapter 1
Evaluation of Currently Available Techniques for Studying Colloids in Environmental Media:
Introduction to Environmental Nanometrology 1
    Allan Philippe, University Koblenz-Landau, Germany

Chapter 2
Nanotechnology for Filtration-Based Point-of-Use Water Treatment: A Review of Current
Understanding 27
    Kathryn Gwenyth Nunnelley, University of Virginia, USA
    James A Smith, University of Virginia, USA

Chapter 3
Nanotechnology in Engineered Membranes: Innovative Membrane Material for Water-Energy
Nexus 50
    Heechul Choi, Gwangju Institute of Science and Technology (GIST), South Korea
    Moon Son, Gwangju Institute of Science and Techonology (GIST), South Korea
    Jiyeol Bae, Gwangju Institute of Science and Techonology (GIST), South Korea
    Hyeon-gyu Choi, Korea Research Institute of Chemical Technology (KRICT), South Korea

Chapter 4
Removal of Emerging Contaminants from Water and Wastewater Using Nanofiltration
Technology 72
    Yang Hu, University of Waterloo, Canada
    Yue Peng, Georgia Institute of Technology, USA
    Wen Liu, Auburn University, USA
    Dongye Zhao, Auburn University, USA
    Jie Fu, Georgia Institute of Technology, USA

Chapter 5
Long-Term Performance Evaluation of Groundwater Chlorinated Solvents Remediation Using Nanoscale Emulsified Zerovalent Iron at a Superfund Site 92

    Chunming Su, United States Environmental Protection Agency, USA
    Robert W. Puls, United States Environmental Protection Agency, USA (ret.)
    Thomas A. Krug, Geosyntec Consultants Inc., Canada
    Mark T. Watling, Geosyntec Consultants Inc., Canada
    Suzanne K. O'Hara, Geosyntec Consultants Inc., Canada
    Jacqueline W. Quinn, NASA Kennedy Space Center, USA
    Nancy E. Ruiz, US Navy, USA

Chapter 6
In-Situ Oxidative Degradation of Emerging Contaminants in Soil and Groundwater Using a New Class of Stabilized MnO_2 Nanoparticles 112

    Bing Han, Auburn University, USA
    Wen Liu, Auburn University, USA
    Dongye Zhao, Auburn University, USA

Chapter 7
Light Sensitized Disinfection with Fullerene 137

    Kyle Moor, Yale University, USA
    Samuel Snow, Michigan State University, USA
    Jaehong Kim, Yale University, USA

Chapter 8
Nanotechnology Applications for Sustainable Crop Production 164

    Gaurav Mishra, Rain Forest Research Institute, India
    Shailesh Pandey, Rain Forest Research Institute, India
    Antara Dutta, Rain Forest Research Institute, India
    Krishna Giri, Rain Forest Research Institute, India

Chapter 9
Developments in Antibacterial Disinfection Techniques: Applications of Nanotechnology 185

    Nicolas Augustus Rongione, University of Miami, USA
    Scott Alan Floerke, University of Miami, USA
    Emrah Celik, University of Miami, USA

Chapter 10
Assessment of Advanced Biological Solid Waste Treatment Technologies for Sustainability 204

    Duygu Yasar, University of Miami, USA
    Nurcin Celik, University of Miami, USA

Chapter 11
Hybrid Nanostructures: Synthesis and Physicochemical Characterizations of Plasmonic Nanocomposites 231

    Ahmed Nabile Emam, National Research Centre, Egypt
    Ahmed Sadek Mansour, Cairo University, Egypt
    Emad Girgis, National Research Centre, Egypt
    Mona Bakr Mohamed, Cairo University, Egypt

Chapter 12
Hybrid Plasmonic Nanostructures: Environmental Impact and Applications 276
    Ahmed Nabile Emam, National Research Centre, Egypt
    Ahmed Sadek Mansour, Cairo University, Egypt
    Emad Girgis, National Research Centre, Egypt
    Mona Bakr Mohamed, Cairo University, Egypt

Chapter 13
Ecotoxicity and Toxicity of Nanomaterials with Potential for Wastewater Treatment Applications 294
    Verónica Inês Jesus Oliveira Nogueira, University of Porto, Portugal
    Ana Gavina, University of Porto, Portugal
    Sirine Bouguerra, Engineering School of Sfax, Tunisia
    Tatiana Andreani, University of Porto, Portugal & CITAB-University of Trás-os-Montes and Alto Douro, Portugal
    Isabel Lopes, University of Aveiro, Portugal
    Teresa Rocha-Santos, University of Aveiro, Portugal
    Ruth Pereira, University of Porto, Portugal

Chapter 14
Ecotoxicity Effects of Nanomaterials on Aquatic Organisms: Nanotoxicology of Materials on Aquatic Organisms 330
    César A Barbero, Universidad Nacional de Río Cuarto, Argentina
    Edith Inés Yslas, Universidad Nacional de Río Cuarto, Argentina

Chapter 15
Copper and Copper Nanoparticles Induced Hematological Changes in a Freshwater Fish *Labeo rohita* – A Comparative Study: Copper and Copper Nanoparticle Toxicity to Fish 352
    Kaliappan Krishnapriya, Bharathiar University, India
    Mathan Ramesh, Bharathiar University, India

Chapter 16
Control of Perishable Goods in Cold Logistic Chains by Bionanosensors 376
    David Bogataj, Universidad Politécnica de Cartagena, Spain
    Damjana Drobne, University of Ljubljana, Slovenia

Chapter 17
Understanding Toxicity of Nanomaterials in Biological Systems 403
    Irshad Ahmad Wani, Jamia Millia Islamia, India
    Tokeer Ahmad, Jamia Millia Islamia, India

Compilation of References 428

About the Contributors 547

Index 555

Detailed Table of Contents

Preface xviii

Acknowledgment xxvii

Chapter 1
Evaluation of Currently Available Techniques for Studying Colloids in Environmental Media:
Introduction to Environmental Nanometrology 1
    Allan Philippe, University Koblenz-Landau, Germany

Engineered nanoparticles are emerging pollutants with poorly known environmental fate and impact. Studying the fate of engineered colloids in the environment is highly challenging due to the complexity of their possible interactions with environmental components and to the need of dedicated analytical methods. Many relevant processes like e.g. agglomeration and dissolution can be studied by monitoring the size of colloids. Techniques dedicated to the determination of the size of colloids in environmental media are thus required. Such techniques should remain accurate at low concentrations and be specific, widely matrix independent and free of artefact due to sample preparation. This chapter aims at evaluating and comparing systematically the currently used tools (e.g. microscopy, light scattering, particle counters) for sizing colloids considering these requirements. As an example of a highly promising solution, the current development of separation techniques coupled to (single particle) ICP-MS is described in more details.

Chapter 2
Nanotechnology for Filtration-Based Point-of-Use Water Treatment: A Review of Current
Understanding 27
    Kathryn Gwenyth Nunnelley, University of Virginia, USA
    James A Smith, University of Virginia, USA

With significant infrastructure investments required for centralized water treatment, in home treatment technologies, known as point-of-use, have become a popular solution in the developing world. This review discusses current filtration-based point-of-use water treatment technologies in three major categories: ceramics, papers and textiles. Each of these categories has used silver for added antimicrobial effectiveness. Ceramics have had the most development and market infiltration, while filter papers are a new development. Textiles show promise for future research as a cheap, socially acceptable, and effective method. Also, a new method of silver incorporation in ceramics is explored.

Chapter 3
Nanotechnology in Engineered Membranes: Innovative Membrane Material for Water-Energy Nexus 50

Heechul Choi, Gwangju Institute of Science and Technology (GIST), South Korea
Moon Son, Gwangju Institute of Science and Techonology (GIST), South Korea
Jiyeol Bae, Gwangju Institute of Science and Techonology (GIST), South Korea
Hyeon-gyu Choi, Korea Research Institute of Chemical Technology (KRICT), South Korea

The membrane processes have received extensive attention as comprehensive and interdisciplinary approaches for water-energy nexus. Nanotechnology has induced significant attention in improving membrane performances to mitigate global water and energy scarcity because of its unique characteristics and simple application for membrane fabrication. Nano-sized materials could provide highly enhanced characteristics to a membrane material, resulting in excellent performance enhancement, such as permeability, selectivity, and fouling resistance, of membrane. Carbon Nanotube (CNT), a widely utilized or studied nanomaterial in membrane science, is discussed in this chapter with a focus on its state-of-the-art applications and future prospects. Electrospun nanofiber, which is one of the feasible nano-structured membrane materials, is also discussed as a promising material for water-energy nexus. Therefore, this chapter also describes its application cases and its potential as an innovative membrane for water-energy nexus.

Chapter 4
Removal of Emerging Contaminants from Water and Wastewater Using Nanofiltration Technology............ 72

Yang Hu, University of Waterloo, Canada
Yue Peng, Georgia Institute of Technology, USA
Wen Liu, Auburn University, USA
Dongye Zhao, Auburn University, USA
Jie Fu, Georgia Institute of Technology, USA

Conventional water/wastewater treatment methods are incapable of removing the majority of Emerging Contaminants (ECs) and a large amount of them and their metabolites are ultimately released to the aquatic environment or drinking water distribution networks. Recently, nanofiltration, a high pressure membrane filtration process, has shown to be superior to other conventional filtration methods, in terms of effluent quality, easy operation and maintenance procedures, low cost, and small required operational space. This chapter provides a comprehensive overview of the most relevant works available in literature reporting the use of nanofiltration for the removal of emerging contaminants from water and wastewater. The fundamental knowledge of nanofiltration such as separation mechanisms, characterization of nanofiltration membranes, and predictive modeling has also been introduced. The literature review has shown that nanofiltration is a promising tool to treat ECs in environmental cleaning and water purification processes.

Chapter 5

Long-Term Performance Evaluation of Groundwater Chlorinated Solvents Remediation Using
Nanoscale Emulsified Zerovalent Iron at a Superfund Site 92

    Chunming Su, United States Environmental Protection Agency, USA
    Robert W. Puls, United States Environmental Protection Agency, USA (ret.)
    Thomas A. Krug, Geosyntec Consultants Inc., Canada
    Mark T. Watling, Geosyntec Consultants Inc., Canada
    Suzanne K. O'Hara, Geosyntec Consultants Inc., Canada
    Jacqueline W. Quinn, NASA Kennedy Space Center, USA
    Nancy E. Ruiz, US Navy, USA

This chapter addresses a case study of long-term assessment of a field application of environmental nanotechnology. Dense Non-Aqueous Phase Liquid (DNAPL) contaminants such as Tetrachloroethene (PCE) and Trichloroethene (TCE) are a type of recalcitrant compounds commonly found at contaminated sites. Recent research has focused on their remediation using environmental nanotechnology in which nanomaterials such as nanoscale Emulsified Zerovalent Iron (EZVI) are added to the subsurface environment to enhance contaminant degradation. The chapter finds that the main limitations of the EZVI technology are difficulty in effectively distributing the viscous EZVI to all areas impacted with DNAPL; potential decrease in hydraulic conductivity due to iron corrosion products buildup or biofouling; potential to adversely impact secondary groundwater quality through mobilization of metals and production of sulfides or methane; injection of EZVI may displace DNAPL away from the injection point; and repeated injections may be required to completely destroy the contaminants.

Chapter 6

In-Situ Oxidative Degradation of Emerging Contaminants in Soil and Groundwater Using a New
Class of Stabilized MnO_2 Nanoparticles 112

    Bing Han, Auburn University, USA
    Wen Liu, Auburn University, USA
    Dongye Zhao, Auburn University, USA

Emerging Organic Contaminants (EOCs) such as steroidal estrogen hormones are of growing concern in recent years, as trace concentrations of these hormones can cause adverse effects on the environmental and human health. While these hormones have been widely detected in soil and groundwater, effective technology has been lacking for in-situ degradation of these contaminants. This chapter illustrates a new class of stabilized MnO_2 nanoparticles and a new in-situ technology for oxidative degradation of EOCs in soil and groundwater. The stabilized nanoparticles were prepared using a low-cost, food-grade Carboxymethyl Cellulose (CMC) as a stabilizer. The nanoparticles were then characterized and tested for their effectiveness for degradation of both aqueous and soil-sorbed E2 (17β-estradiol). Column tests confirmed the effectiveness of the nanoparticles for in-situ remediation of soil sorbed E2. The nanoparticle treatment decreased both water leachable and soil-sorbed E2, offering a useful alternative for in-situ remediation of EOCs in the subsurface.

Chapter 7
Light Sensitized Disinfection with Fullerene... 137
    Kyle Moor, Yale University, USA
    Samuel Snow, Michigan State University, USA
    Jaehong Kim, Yale University, USA

Fullerene has drawn wide interest across many fields due to its favorable electronic and optical properties, which has spurred its use in a myriad of applications. One of the hallmark properties of fullerene is its ability to act as a photosensitizer and efficiently generate $^1O_2$, a form of Reactive Oxygen Species (ROS), upon visible irradiation when dispersed in organic solvents. However, the application of fullerene in environmental systems has been somewhat limited due to fullerene's poor solubility in water, which causes individual fullerene molecules to aggregate and form large colloidal species, quenching much of fullerene's $^1O_2$ production. This is unfortunate given that $^1O_2$ provides many advantages as an oxidant compared to ROS produced from typical advanced oxidation processes, such as OH radicals, due to $^1O_2$'s greater chemical selectivity and its ability to remain unaffected by the presence of background water constituents, such as natural organic matter and carbonate. Hence, fullerene materials may hold great potential for the oxidation and disinfection of complex waters. Herein, we chronicle the advances that have been made to propel fullerene materials towards use in emerging water disinfection technologies. Two approaches to overcome fullerene aggregation and the subsequent loss of $^1O_2$ production in aqueous systems are herein outlined: 1) addition of hydrophilic functionality to fullerene's cage, creating highly photoactive colloidal fullerenes; and 2) covalent attachment of fullerene to solid supports, which physically prevents fullerene aggregation and allows efficient 1O2 photo-generation. An emphasis is placed on the inactivation of MS2 bacteriophage, a model for human enteric viruses, highlighting the potential of fullerene materials for light-activated disinfection technologies.

Chapter 8
Nanotechnology Applications for Sustainable Crop Production 164
    Gaurav Mishra, Rain Forest Research Institute, India
    Shailesh Pandey, Rain Forest Research Institute, India
    Antara Dutta, Rain Forest Research Institute, India
    Krishna Giri, Rain Forest Research Institute, India

Innovations in nanotechnology revolutionized the world in all sectors, including medicine, biotechnology, electronics, material science, energy sectors etc. In fact, the existing literature also points towards the potential application of nanotechnology in global food production and alarming situation of food scarcity. With the advancement of nanotechnology, use of nanofertilizers for yield enhancement, nanopesticides for insect pests and disease control and nanosensors to monitor soil and plant health becomes one of the most fascinating and promising lines of investigation to achieve sustainability. Furthermore, nanobiotechnology application enables gene transfer to expand the genetic base of crop varieties against different biotic and abiotic stresses. Some lacunas which may resist commercialization of nanotechnology are cost, industrial setup, and public attitude towards the health and food safety issues. Sincere attempts are required to answer these questions and develop suitable strategies to solve any problems, we might encounter in near future.

Chapter 9
Developments in Antibacterial Disinfection Techniques: Applications of Nanotechnology 185
Nicolas Augustus Rongione, University of Miami, USA
Scott Alan Floerke, University of Miami, USA
Emrah Celik, University of Miami, USA

One of the most daunting challenges facing nations today is controlling the spread of increasingly lethal bacteria. Today, a handful of bacteria can no longer be treated with traditional antibiotics and show antibacterial resistance. In this regard, nanotechnology possesses tremendous potential for the development of novel tools which help prevent and combat the spread of unwanted microorganisms. These tools can provide unique solutions for the challenges of the traditional disinfection methods, such as increased antibacterial activity, cost reduction, biocompatibility and personalized treatment. Despite its great potential, nanotechnology remains in its infancy and continued research efforts are required to achieve its full potential. In this chapter, traditional methods and their associated limitations are reviewed for their efficacy against microbial spread, and potential solutions in nanotechnology are described. A review of the state of the art disinfection techniques using nanotechnology is presented, and promising new areas in the field are discussed.

Chapter 10
Assessment of Advanced Biological Solid Waste Treatment Technologies for Sustainability 204
Duygu Yasar, University of Miami, USA
Nurcin Celik, University of Miami, USA

53.8% of annually generated US Municipal Solid Waste was discarded in landfills by 2012. However, landfills fail to provide a sustainable solution to manage the waste. The State of Florida has responded to the need of establishing sustainable SWM systems by setting an ambitious 75% recycling goal to be achieved by 2020. To this end, Advanced Biological Treatment (ABT) and Thermal Treatment (ATT) of municipal solid waste premise a sustainable solution to manage the waste as it drastically reduces the volume of waste discarded in landfills and produces biogas that can be used to generate energy. In this chapter, ABT and ATT technologies are analyzed; and their advantages and disadvantages are examined from a sustainability perspective. A comprehensive top-to-bottom assessment of ABT technologies is provided for Florida using Analytic Hierarchy Process based on the collected subject matter expert rankings.

Chapter 11
Hybrid Nanostructures: Synthesis and Physicochemical Characterizations of Plasmonic
Nanocomposites 231
Ahmed Nabile Emam, National Research Centre, Egypt
Ahmed Sadek Mansour, Cairo University, Egypt
Emad Girgis, National Research Centre, Egypt
Mona Bakr Mohamed, Cairo University, Egypt

The recent extensive interest of nanostructure materials associated with their unique properties is motivated to develop new hybrid nanocomposites that couple two nano-components together in the form of Core/Shell, nanoalloys, and doped nanostructures. Hybrid nanostructure provides another opportunity for tuning the physical and chemical properties at the nanoscale. This opens the door for the discovery of new properties and potential for more applications. This chapter is devoted to present, and discuss the recent advances and progress relevance for Plasmonic hybrid nanocomposites. In addition, literature reviewed on different attempts to obtain high quality plasmonic nanocomposites via chemical routes, and their physico-chemical aspects for this class of novel nanomaterials. The authors presented their recent published work regarding Plasmonic hybrid nanostructure regarding plasmonic-semiconductor, plasmonic magnetic and plasmonic graphene nanocomposites.

Chapter 12
Hybrid Plasmonic Nanostructures: Environmental Impact and Applications 276
    Ahmed Nabile Emam, National Research Centre, Egypt
    Ahmed Sadek Mansour, Cairo University, Egypt
    Emad Girgis, National Research Centre, Egypt
    Mona Bakr Mohamed, Cairo University, Egypt

Plasmonic hybrid nanostructure including Semiconductor-metallic nanoparticles, and graphene-plasmonic nanocomposites have great potential to be used as photocatalyst for hydrogen production and for photodegradation of organic waste. Also, they are potential candidate as active materials in photovoltaic devices. Plasmonic-magnetic nanocomposites could be used in photothermal therapy and biomedical imaging. This chapter will focus on the environmental impact of these materials and their in-vitro and in-vivo toxicity. In addition, the applications of these hybrid nanostructures in energy and environment will be discussed in details.

Chapter 13
Ecotoxicity and Toxicity of Nanomaterials with Potential for Wastewater Treatment
Applications 294
    Verónica Inês Jesus Oliveira Nogueira, University of Porto, Portugal
    Ana Gavina, University of Porto, Portugal
    Sirine Bouguerra, Engineering School of Sfax, Tunisia
    Tatiana Andreani, University of Porto, Portugal & CITAB-University of Trás-os-Montes and
        Alto Douro, Portugal
    Isabel Lopes, University of Aveiro, Portugal
    Teresa Rocha-Santos, University of Aveiro, Portugal
    Ruth Pereira, University of Porto, Portugal

Nanotechnology holds the promise of develop new processes for wastewater treatment. However, it is important to understand what the possible impacts on the environment of NMs. This study joins all the information available about the toxicity and ecotoxicity of NMs to human cell lines and to terrestrial and aquatic biota. Terrestrial species seems more protected, since effects are being recorded for concentrations higher than those that could be expected in the environment. The soil matrix is apparently trapping and filtering NMs. Further studies should focus more on indirect effects in biological communities rather than only on effects at the individual level. Aquatic biota, mainly from freshwater ecosystems, seemed to be at higher risk, since dose effect concentrations recorded were remarkable lower, at least for some NMs. The toxic effects recorded on different culture lines, also give rise to serious concerns regarding the potential effects on human health. However, few data exists about environmental concentrations to support the calculation of risks to ecosystems and humans.

Chapter 14
Ecotoxicity Effects of Nanomaterials on Aquatic Organisms: Nanotoxicology of Materials on
Aquatic Organisms... 330
    César A Barbero, Universidad Nacional de Río Cuarto, Argentina
    Edith Inés Yslas, Universidad Nacional de Río Cuarto, Argentina

The increasing production and use of engineered nanomaterials raise concerns about inadvertent exposure and the potential for adverse effects on the aquatic environment. The aim of this chapter is focused on studies of nanotoxicity in different models of aquatic organisms and their impact. Moreover, the chapter provides an overview of nanoparticles, their applications, and the potential nanoparticle-induced toxicity in aquatic organisms. The topics discussed in this chapter are the physicochemical characteristic of nanomaterials (size, aggregation, morphology, surface charge, reactivity, dissolution, etc.) and their influence on toxicity. Further, the text discusses the direct effect of nanomaterials on development stage (embryonic and adult) in aquatic organisms, the mechanism of action as well as the toxicity data of nanomaterials in different species.f action as well as the toxicity data of nanomaterials in different species.

Chapter 15
Copper and Copper Nanoparticles Induced Hematological Changes in a Freshwater Fish *Labeo rohita* – A Comparative Study: Copper and Copper Nanoparticle Toxicity to Fish........................... 352
    Kaliappan Krishnapriya, Bharathiar University, India
    Mathan Ramesh, Bharathiar University, India

In the present study, fish Labeo rohita were exposed to 20, 50 and 100 µg/L of both Cu NPs and copper sulphate ($CuSO_4$, bulk copper) for 24 h and hematological profiles were estimated. A significant ($P<0.01$) change in the hemoglobin (Hb), hematocrit (Hct), white blood cells (WBC) and Mean Corpuscular Volume (MCV) levels were observed in all the three concentrations of both bulk and Cu NPs treated fish when compared to control groups. However a non significant change in red blood cells (RBC) (20 and 50 µg/L Cu NPs) and mean corpuscular hemoglobin (MCH) (20 and 50 µg/L bulk Cu) were observed. The alteration in Mean Corpuscular Hemoglobin Concentration (MCHC) value was found to be non significant both in bulk and Cu NPs treated fish. The alterations of these parameters can be used as a potential indicator to examine the health of fish in aquatic ecosystem contaminated with metal and metal based nanoparticles.

Chapter 16
Control of Perishable Goods in Cold Logistic Chains by Bionanosensors.. 376
    David Bogataj, Universidad Politécnica de Cartagena, Spain
    Damjana Drobne, University of Ljubljana, Slovenia

Nanotechnology can contribute to food security in supply chains of agri production-consumption systems. The unique properties of nanoparticles have stimulated the increasing interest in their application as biosensing. Biosensing devices are designed for the biological recognition of events and signal transduction. Many types of nanoparticles can be used as biosensors, but gold nanoparticles have sparked most interest. In the work presented here, we will address the problem of fruit and vegetable decay and rotting during transportation and storage, which could be easily generalized also onto post-harvest loss prevention in general. During the process of rotting, different compounds, including different gasses, are released into the environment. The application of sensitive bionanosensors in the storage/transport containers can detect any changes due to fruit and vegetable decay and transduce the signal. The goal of this is to reduce the logistics cost for this items. Therefore, our approach requires a multidisciplinary and an interdisciplinary approach in science and technology. The cold supply chain is namely a science, a technology and a process which combines applied bio-nanotechnology, innovations in the industrial engineering of cooling processes including sensors for temperature and humidity measurements, transportation, and applied mathematics. It is a science, since it requires the understanding of chemical and biological processes linked to perishability and the systems theory which enables the developing of a theoretical framework for the control of systems with perturbed time-lags. Secondly, it is a technology developed in engineering which relies on the physical means to assure appropriate temperature conditions along the CSC and, thirdly, it is also a process, since a series of tasks must be performed to prepare, store, and transport the cargo as well as monitor the temperature and humidity of sensitive cargo and give proper feedback control, as it will be outlined in this chapter. Therefore, we shall discuss how to break the silos of separated knowledge to build an interdisciplinary and multidisciplinary science of post-harvest loss prevention. Considering the sensors as floating activity cells, modelled as floating nodes, in a graph of such a system, an extended Material Requirement Planning (MRP) theory will be described which will make it possible to determine the optimal feedback control in post-harvest loss prevention, based on bionanosensors. Therefore, we present also a model how to use nanotechnology from the packaging facility to the final retail. Any changes in time, distance, humidity or temperature in the chain could cause the Net Present Value (NPV) of the activities and their added value in the supply chain to be perturbed, as presented in the subchapter. In this chapter we give the answers to the questions, how to measure the effects of some perturbations in a supply chain on the stability of perishable agricultural goods in such systems and how nanotechnology can contribute with the appropriate packaging and control which preserves the required level of quality and quantity of the product at the final delivery. The presented model will not include multicriteria optimization but will stay at the NPV approach. But the annuity stream achieved by improved sensing and feedback control could be easily combined with environmental and medical/health criteria. An interdisciplinary perspective of industrial engineering and management demonstrates how the development of creative ideas born in separate research fields can be liaised into an innovative design of smart control devices and their installation in trucks and warehouses. These innovative technologies could contribute to an increase in the NPV of activities in the supply chains of perishable goods in general.

Chapter 17
Understanding Toxicity of Nanomaterials in Biological Systems 403
    Irshad Ahmad Wani, Jamia Millia Islamia, India
    Tokeer Ahmad, Jamia Millia Islamia, India

Nanotechnology is a growing applied science having considerable global socioeconomic value. Nanoscale materials are casting their impact on almost all industries and all areas of society. A wide range of engineered nanoscale products has emerged with widespread applications in fields such as energy, medicine, electronics, plastics, energy and aerospace etc. While the market for nanotechnology products will have grown over one trillion US dollars by 2015, the presence of these material is likely to increase leading to increasing likelihood of exposure. The direct use of nanomaterials in humans for medical and cosmetic purposes dictates vigorous safety assessment of toxicity. Therefore this book chapter provides the detailed toxicity assessment of various types of nanomaterials.

Compilation of References 428

About the Contributors 547

Index 555

Preface

The journey to the invention of nanotechnology began several decades ago. Nanomaterials (NMs) have been applied for multiple purposes and are commonly found in many consumer products. Applications of nanotechnology also include the analysis of physicochemical properties of materials, degradation of contaminants, textile production, biomedicine, renewable energy, environmental systems, health care, electronics, and food agriculture. The technology involving nanomaterials is expected to grow tremendously, yet research studies on environmental safety, toxicity, transport, transformation, removal pathways, and dissipation kinetics under environmental parameters have been limited.

Moreover, despite nano-foods emergence in early 2000 after which it became a focus of interest in discussions about the sustainability of nanotechnology, research on nanotechnology applications in food production remains outside the mainstream, with debates ongoing about the regulation of NMs. The lack of long-term safety data, assessments limited to a case-by-case basis, the lack of evidence regarding human risk, and the lack of analytical tools to evaluate the safety of nanomaterials used in consumer products need to be addressed in nanotechnology applications. The remediation of contaminants in environmental media, technical feasibility, cost-effectiveness, and potential hazards to the environment and human also need to be addressed, even if NMs offer versatile properties to transform various types of contaminants. The physicochemical properties such as size, shape, surface charge, dissolution, and surface area can affect the interactions between nanomaterials and cells, animals, humans and the environment and consequently may control the ultimate impact of nanomaterials on health and the environment

This book covers a wide range of aspects in the applications of nanomaterials, beginning with "Introduction to Environmental Nanotechnology," "Nanotechnology for Water Treatment," "Nanotechnology in Engineered Membranes," "Removal of Emerging Contaminants," "Assessment of Long-Term Performance of Nanotechnology for Groundwater Remediation," "Disinfection Uses," "Applications of Nanotechnology in Agriculture," "Hybrid Nanostructures," "Toxicity of Nanomaterials," and "Nanomaterials in Biological Systems."

A brief summary of each chapter in the book is presented here.

Chapter 1: Evaluation of Currently Available Techniques for Studying Colloids in Environmental Media: Introduction to Environmental Nanometrology

The chapter describes analytical methods and techniques to monitor the release of nanomaterials and to determine the size of colloids in environmental media. One of primary issues in evaluating nanoparticles in environmental media is attributed to the lack of analytical techniques to accurately analyze low concentrations. The chapter contributes to understanding and reviewing the currently used tools.

Preface

Abstract: Engineered nanoparticles are emerging pollutants with poorly known environmental fate and impact. Studying the fate of engineered colloids in the environment is highly challenging due to the complexity of their possible interactions with environmental components and to the need of dedicated analytical methods. Many relevant processes like agglomeration and dissolution can be studied by monitoring the size of colloids. Techniques dedicated to the determination of the size of colloids in environmental media are thus required. Such techniques should remain accurate at low concentrations and be specific, widely matrix independent, and free of artefact due to sample preparation. This chapter aims at evaluating and comparing systematically the currently used tools (e.g., microscopy, light scattering, particle counters) for sizing colloids considering these requirements. As an example of a highly promising solution, the current development of separation techniques coupled to (single particle) ICP-MS is described in more details.

Chapter 2: Nanotechnology for Filtration-Based Point-of-Use Water Treatment: A Review of Current Understanding

The chapter reviews current filtration-based, point-of-use (POU) water treatment technologies, especially by the addition of silver since silver has an antimicrobial property. Contributions are made to suggest new methods for centralized water treatment and future research directions in POU technologies.

Abstract: With significant infrastructure investments required for centralized water treatment, in-home treatment technologies, known as point-of-use, have become a popular solution in the developing world. This review discusses current filtration-based, point-of-use water treatment technologies in three major categories: ceramics, papers, and textiles. Each of these categories has used silver for added antimicrobial effectiveness. Ceramics have had the most development and market infiltration, while filter papers are a new development. Textiles show promise for future research as a cheap, socially acceptable, and effective method. Also, a new method of silver incorporation in ceramics is explored.

Chapter 3: Nanotechnology in engineered membranes: Innovative membrane material for water-energy nexus

Integration of nanotechnology in membrane processes is relatively unexplored. As such, the chapter discusses ways to improve membrane performances with nanotechnology in terms of permeability, selectivity, and fouling resistance. Feasible nano-structured membrane materials as a novel solution for water-energy nexus are discussed along with several cases for the applications.

Abstract: The membrane processes have received extensive attention as comprehensive and interdisciplinary approaches for water-energy nexus. Nanotechnology has induced significant attention in improving membrane performances to mitigate global water and energy scarcity because of its unique characteristics and simple application for membrane fabrication. Nano-sized materials could provide highly enhanced characteristics to a membrane material, resulting in excellent performance enhancement, such as permeability, selectivity, and fouling resistance, of the membrane. Carbon nanotube (CNT), a widely utilized or studied nanomaterial in membrane science, is discussed in this chapter with a focus on its state-of-the-art applications and future prospects. Electrospun nanofiber, which is one of the feasible nano-structured membrane materials, is also discussed as a promising material for water-energy nexus. Therefore, this chapter also describes its application cases and its potential as an innovative membrane for water-energy nexus.

Chapter 4: Removal of Emerging Contaminants from Water and Wastewater using Nanofiltration Technology

Emerging contaminants of concern have been problematic due to lack of suitable treatment and detection at trace concentrations. This chapter introduces one of promising membrane technologies, nanofiltration, for treating emerging contaminants and contributes to review the applications of nanofiltration for environmental cleanup.

Abstract: Conventional water/wastewater treatment methods are incapable of removing the majority of emerging contaminants (ECs) and a large amount of them and their metabolites are ultimately released to the aquatic environment or drinking water distribution networks. Recently, nanofiltration (NF), a high-pressure membrane filtration process, has been shown to be superior to other conventional filtration methods in terms of effluent quality and easy operation and maintenance procedures, low cost, and small required operational space. This chapter provides a comprehensive overview of the most relevant works available in literature reporting the use of nanofiltration for the removal of emerging contaminants from water and wastewater. The fundamental knowledge of nanofiltration such as separation mechanisms, characterization of NF membranes, and predictive modeling has also been introduced. The literature review has shown that nanofiltration is a promising tool to treat ECs in environmental cleaning and water purification processes.

Chapter 5: Long-term Performance Evaluation of Groundwater Chlorinated Solvents Remediation Using Nanoscale Emulsified Zerovalent Iron at a Superfund Site

The chapter presents a case study of field performance in applying nanoscale zero-valent iron (nZVI) for remediation of groundwater plumes contaminated with chlorinated organic solvents. There has been a lack of long-term data in the fulfillment of applying nanomaterials for environmental remediation. Contribution through assessment and evaluation of data obtained in the field application is accomplished by this chapter.

Abstract: This chapter addresses a case study of the long-term assessment of a field application of environmental nanotechnology. Dense non-aqueous phase liquid (DNAPL) contaminants such as tetrachloroethene (PCE) and trichloroethene (TCE) are a type of recalcitrant compound commonly found at contaminated sites. Recent research has focused on their remediation using environmental nanotechnology in which nanomaterials such as nanoscale emulsified zerovalent iron (EZVI) are added to the subsurface environment to enhance contaminant degradation. Such a nano-remediation approach may be mostly applicable to the source zone where the contaminant mass is the greatest and source removal is a critical step in controlling the further spreading of the groundwater plume. While NZVI shows promise in both laboratory and field tests, limited information is available about the long-term effectiveness of nano-remediation because previous field tests are mostly less than two years. Here an update is provided for a six-year performance evaluation of EZVI for treating PCE and its daughter products at a Superfund site at Parris Island, South Carolina, USA.

Preface

Chapter 6: In-situ Oxidative Degradation of Emerging Contaminants in Soil and Groundwater Using a New Class of Stabilized MnO_2 Nanoparticles

The chapter provides recent research studies in applying novel nanomaterials (stabilized MnO_2 nanoparticles) for treating emerging organic contaminants (EOCs), particularly steroidal estrogen hormones. Their effectiveness for degradation of both aqueous and soil-sorbed E2 and characterization are presented, contributing to a new addition to existing knowledges on nanomaterials used for environmental remediation.

Abstract: Emerging organic contaminants (EOCs) such as steroidal estrogen hormones are of growing concern in recent years, as trace concentrations of these hormones can cause adverse effects on the environmental and human health. While these hormones have been widely detected in soil and groundwater, effective technology has been lacking for in-situ degradation of these contaminants. This chapter illustrates a new class of stabilized MnO_2 nanoparticles and a new in-situ technology for oxidative degradation of EOCs in soil and groundwater. The stabilized nanoparticles were prepared using a low-cost, food-grade carboxymethyl cellulose (CMC) as a stabilizer. The nanoparticles were then characterized and tested for their effectiveness for degradation of both aqueous and soil-sorbed E2 (17β-estradiol). Column tests confirmed the effectiveness of the nanoparticles for in-situ remediation of soil sorbed E2. The nanoparticle treatment decreased both water leachable and soil-sorbed E2, offering a useful alternative for in-situ remediation of EOCs in the subsurface.

Chapter 7: Light-Sensitized Disinfection with Fullerene

The chapter illustrates how fullerene is applied to environmental systems, especially for the oxidation and disinfection of complex waters. The chapter contributes to advance existing knowledge on fullerene and offers new methods for overcoming fullerene aggregation, presenting promising future applications to disinfection technologies.

Abstract: Fullerene has drawn wide interest across many fields due to its favorable electronic and optical properties, which have spurred its use in a myriad of applications. One of the hallmark properties of fullerene is its ability to act as a photosensitizer and efficiently generate $^1O_2$, a form of reactive oxygen species (ROS), upon visible irradiation when dispersed in organic solvents. However, the application of fullerene in environmental systems has been somewhat limited due to fullerene's poor solubility in water, which causes individual fullerene molecules to aggregate and form large colloidal species, quenching much of fullerene's $^1O_2$ production. Two approaches to overcome fullerene aggregation and the subsequent loss of $^1O_2$ production in aqueous systems are herein outlined: 1) addition of hydrophilic functionality to fullerene's cage, creating highly photoactive colloidal fullerenes and 2) covalent attachment of fullerene to solid supports, which physically prevents fullerene aggregation and allows efficient $^1O_2$ photo-generation. An emphasis is placed on the inactivation of MS2 bacteriophage, a model for human enteric viruses, highlighting the potential of fullerene materials for light-activated disinfection technologies.

Chapter 8: Nanotechnology Applications for Sustainable Crop Production

The chapter focuses on applications of nanotechnology in agriculture. Nanomaterials in food industry have been debated in terms of long-term safety and impact on public health. The chapter contributes to one of the hottest topics in this field by presenting various potential applications of nanomaterials and issues on the cost and industrial setup and addresses public attitudes towards the health and food safety.

Abstract: Innovations in nanotechnology revolutionized the world in all sectors, including medicine, biotechnology, electronics, material science, energy sectors, etc. In fact, the existing literature also points towards the potential application of nanotechnology in global food production and alarming situation of food scarcity. With the advancement of nanotechnology, use of nanofertilizers for yield enhancement, nanopesticides for insect pests and disease control, and nanosensors to monitor soil and plant health becomes one of the most fascinating and promising lines of investigation to achieve sustainability. Furthermore, nanobiotechnology application enables gene transfer to expand the genetic base of crop varieties against different biotic and abiotic stresses. Some lacunas that may resist commercialization of nanotechnology are cost, industrial setup, and public attitude towards the health and food safety issues. Sincere attempts are required to answer these questions and develop suitable strategies to solve any problems we might encounter in near future.

Chapter 9: Developments in Antibacterial Disinfection Techniques: Applications of Nanotechnology

The chapter contributes to the field of nanomedicine and presents one of the most challenging and unexplored issues by offering novel solutions with applications of nanotechnology to disinfect lethal bacteria.

Abstract: One of the most daunting challenges facing nations today is controlling the spread of increasingly lethal bacteria. Today, a handful of bacteria can no longer be treated with traditional antibiotics and show antibacterial resistance. In this regard, nanotechnology possesses tremendous potential for the development of novel tools that help prevent and combat the spread of unwanted microorganisms. These tools can provide unique solutions for the challenges of the traditional disinfection methods, such as increased antibacterial activity, cost reduction, biocompatibility, and personalized treatment. Despite its great potential, nanotechnology remains in its infancy and continued research efforts are required to achieve its full potential. In this chapter, traditional methods and their associated limitations are reviewed for their efficacy against microbial spread, and potential solutions in nanotechnology are described. A review of the state-of-the-art disinfection techniques using nanotechnology is presented, and promising new areas in the field are discussed.

Chapter 10: Assessment of Advanced Biological Solid Waste Treatment Technologies for Sustainability

The chapter describes evaluation of sustainable solid waste treatment technologies by comparing two technologies (ABT and ATT) in terms of their sustainability. Particular contribution is made to reviewing the application of nanotechnology for waste management and a comprehensive top-to-bottom assessment of ABT technologies using analytic hierarchy process.

Abstract: Annually, 53.8% of U.S. municipal solid waste was discarded in landfills by 2012. However, landfills fail to provide a sustainable solution to managing the waste. The state of Florida has responded to the need for establishing sustainable SWM systems by setting an ambitious 75% recycling goal to be achieved by 2020. To this end, advanced biological treatment (ABT) and thermal treatment (ATT) of municipal solid waste premise a sustainable solution to manage the waste as it drastically reduces the volume of waste discarded in landfills and produces biogas that can be used to generate energy. In this chapter, ABT and ATT technologies are analyzed, and their advantages and disadvantages are examined

Preface

from a sustainability perspective. A comprehensive top-to-bottom assessment of ABT technologies is provided for Florida using the Analytic Hierarchy Process based on the collected subject matter expert rankings.

Chapter 11: Hybrid Nanostructures: Synthesis and Physicochemical Characterizations of Plasmonic Nanocomposites

This chapter introduces hybrid nanostructure, which provides another opportunity for controlling physicochemical properties at the nanoscale. Particular contribution is made to the recent advances and progress relevance for Plasmonic hybrid nanocomposites.

Abstract: The recent extensive interest of nanostructure materials associated with their unique properties is motivated to develop new hybrid nanocomposites that couple two nano-components together in the form of core/shell, nanoalloys, and doped nanostructures. Hybrid nanostructures provide another opportunity for tuning the physical and chemical properties at the nanoscale. This opens the door for the discovery of new properties and potential for more applications. This chapter is devoted to the present situation and discusses the recent advances and progress relevance for plasmonic hybrid nanocomposites. In addition, literature is reviewed on different attempts to obtain high-quality plasmonic nanocomposites via chemical routes, and their physico-chemical aspects for this class of novel nanomaterials. The authors presented their recently published work regarding plasmonic hybrid nanostructures, plasmonic-semiconductor, and plasmonic magnetic and plasmonic graphene nanocomposites.

Chapter 12: Hybrid Plasmonic Nanostructures: Environmental Impact and Applications

The chapter illustrates the applications of plasmonic hybrid nanostructure as photocatalyst for hydrogen production and for photodegradation of organic waste and discusses other potential promising applications for photothermal therapy and biomedical imaging, along with environmental impact and their toxicity.

Abstract: Plasmonic hybrid nanostructures, such as semiconductor-metallic nanoparticles, and graphene-plasmonic nanocomposites, have great potential to be used as photocatalyst for hydrogen production and for photodegradation of organic waste. Also, they are a potential candidate as active materials in photovoltaic devices. Plasmonic-magnetic nanocomposites could be used in photothermal therapy and biomedical imaging. This chapter will focus on the environmental impact of these materials and their in-vitro and in-vivo toxicity. In addition, the applications of these hybrid nanostructures in energy and environment will be discussed in details.

Chapter 13: Ecotoxicity and toxicity of nanomaterials with potential for wastewater treatment applications

This chapter contributes to the ecotoxicity of nanomaterials (NMs) by reviewing all available information on the potential impact, especially the toxicity of NMs to human cell lines and environmental biota. Significant contributions are made to unveiling serious concerns regarding the potential effects on human health.

Abstract: Nanotechnology holds the promise of develop new processes for wastewater treatment. However, it is important to understand what the possible impacts on the environment of NMs. This study joins all the information available about the toxicity and ecotoxicity of NMs to human cell lines and to terrestrial and aquatic biota. Terrestrial species seem more protected since the effects are being recorded for concentrations higher than those that could be expected in the environment. The soil matrix is apparently trapping and filtering NMs. Further studies should focus more on indirect effects in biological communities rather than only on effects at the individual level. Aquatic biota, mainly from freshwater ecosystems, seemed to be at higher risk since dose effect concentrations recorded were remarkable lower, at least for some NMs. The toxic effects recorded on different culture lines also give rise to serious concerns regarding the potential effects on human health. However, few data exist about environmental concentrations to support the calculation of risks to ecosystems and humans.

Chapter 14: Ecotoxicity effects of nanomaterials on aquatic organisms: Nanotoxicology of materials on aquatic organisms

This chapter devotes and contributes to the ecotoxicity effects of nanomaterials, especially by assessing the potential toxicity of NMs on different models of aquatic organisms and their impact.

Abstract: The increasing production and use of engineered nanomaterials raise concerns about inadvertent exposure and the potential for adverse effects on the aquatic environment. The aim of this chapter is focused on studies of nanotoxicity in different models of aquatic organisms and their impact. Moreover, the chapter provides an overview of nanoparticles, their applications, and the potential nanoparticle-induced toxicity in aquatic organisms. The topics discussed in this chapter are the physicochemical characteristics of nanomaterials (size, aggregation, morphology, surface charge, reactivity, dissolution, etc.) and their influence on toxicity. Further, the text discusses the direct effect of nanomaterials on development stage (embryonic and adult) in aquatic organisms, the mechanism of action, as well as the toxicity data of nanomaterials in different species and the toxicity data of nanomaterials in different species.

Chapter 15: Copper and Copper Nanoparticles Induced Hematological Changes in a Freshwater Fish Labeo Rohita. A Comparative Study: Copper and Copper Nanoparticle Toxicity to Fish

This chapter presents research findings on toxicity comparison between copper and
copper nanoparticles upon exposure to a freshwater fish (Labeo Rohita) and contributes to evaluation of aquatic ecosystems contaminated with metal and metal-based nanoparticles.

Abstract: In the present study, fish Labeo rohita were exposed to 20, 50, and 100 µg/L of both Cu NPs and copper sulphate ($CuSO_4$, bulk copper) for 24 h and hematological profiles were estimated. A significant ($P< 0.01$) change in the hemoglobin (Hb), hematocrit (Hct), white blood cells (WBC) and mean corpuscular volume (MCV) levels were observed in all the three concentrations of both bulk and Cu NPs treated fish when compared to control groups. However, a non-significant change in red blood cells (RBC) (20 and 50 µg/ Cu NPs) and mean corpuscular hemoglobin (MCH) (20 and 50 µg/ bulk Cu) were observed. The alteration in mean corpuscular hemoglobin concentration (MCHC) value was found to be non-significant both in bulk and Cu NP treated fish. The alterations of these parameters can be used as a potential indicator to examine the health of fish in aquatic ecosystem contaminated with metal and metal based nanoparticles.

Chapter 16: Control of Perishable Goods in Cold Logistic Chains by Bionanosensors

The chapter contributes to applications of nanotechnology in biosensing devices and discusses a multidisciplinary and an interdisciplinary approach in science and technology. Discussion is also provided

Preface

on how to break the silos of separated knowledge to build an interdisciplinary and multidisciplinary science of post-harvest loss prevention.

Abstract: Nanotechnology can contribute to food security in supply chains of agriproduction-consumption systems. The unique properties of nanoparticles have stimulated the increasing interest in their application as biosensing. Biosensing devices are designed for the biological recognition of events and signal transduction. In the work presented here, we will address the problem of fruit and vegetable decay and *rotting* during transportation and storage, which could be easily generalized also onto post-harvest loss prevention in general. During the process of rotting, different compounds, including different gasses, are released into the environment. The application of sensitive bionanosensors in the storage/transport containers can detect any changes due to fruit and vegetable decay and transduce the signal. *The goal of this is to reduce the logistics cost for this items. Therefore, our approach requires a multidisciplinary and an interdisciplinary approach in science and technology. The cold supply chain is namely a science, a technology and a process which combines applied bio-nanotechnology, innovations in the industrial engineering of cooling processes including sensors for temperature and humidity measurements, transportation, and applied mathematics.*

Chapter 17: Understanding Toxicity of Nanomaterials in Biological Systems

The chapter assesses various types of nanomaterials and provides safety assessment of toxicity, especially in biological systems, contributing to the nanotoxicity field.

Abstract: Nanotechnology is a growing applied science having considerable global socioeconomic value. Nanoscale materials are casting their impact on almost all industries and all areas of society. A wide range of engineered nanoscale products has emerged with widespread applications in fields such as energy, medicine, electronics, plastics, energy and aerospace, etc. While the market for nanotechnology products will exceed over one trillion U.S. dollars by 2015, the presence of these materials is likely to increase leading to increasing likelihood of exposure. The direct use of nanomaterials in humans for medical and cosmetic purposes dictates vigorous safety assessment of toxicity. Therefore, this chapter provides the detailed toxicity assessment of various types of nanomaterials.

Overall, the book describes and explains in detail both fundamental scientific concepts and advanced cutting-edge technologies in nanoscience, as well as to provide an educational reference for practitioners working in this and related fields. Thus, the book offers readers a unique combination of scientific and technological knowledge. It serves as an essential reference for all levels in science, technology, and engineering. Fundamental and applied research are presented throughout the text, and recent studies are also offered along with critical reviews. The overall objective of the book is to provide both fundamental and advanced knowledge for practitioners, students, consultants, regulators, engineers, and scientists as well as entry-level professionals in the increasingly important field of nanotechnology. Offering a wide range of information and practical experience in nanotechnology, this book will benefit students and professionals in areas such as environmental science/engineering, material science/engineering, medicine, biology, toxicity, and chemical engineering/science. The wide-ranging material included in this book is expected to have a positive impact on both current practice and future research design.

A core value of the book is its combination of basic knowledge with the presentation of key current issues and research findings. This book adds significant contributions to existing subject matters by addressing the most up-to-date issues, critical reviews in the subject, field studies, presenting relatively unexplored aspects in nanotoxicity, characterizations of nanomaterials, and applications of nanomaterials in membrane filtration, biomedicine, and food production. The target audience includes research scientists and engineers, industry consultants, regulators, practitioners, students (both undergraduate and graduate), and also entry-level professionals including those with non-major educational backgrounds. For this audience, the book will be an essential reference, providing fundamental information for practice and research.

Acknowledgment

This book was not possible without all the contributors who served as the authors. Production of a book often requires a significant investment of time to ensure both the quality and quantity of contents of the particular theme. As such, the editor greatly appreciates the reviewers who have contributed constructive comments and suggestions to enhance the quality of all chapters for the book. All the feedback, suggestions and recommendations were valuable and strengthened the contents of the new fields in nano-research and applications.

The editor acknowledges all the authors in the book and particular acknowledgements go to the following fellows. Dr. Chunming Su (U.S. EPA), Dr. Jaehong Kim (Yale University), Dr. Don Zhao (Auburn University), Dr. James Smith (University of Virginia), Kathryn Nunnelley (University of Virginia), Bing Han, Wen Liu, Kyle Moor, Samuel Snow (Michigan State University), Moon Son (GIST), Dr. Heecheol Choi (GIST), Jiyeol Bae, Hyeongue Choi, Dr. Verónica Nogueira (University of Porto), Dr. Tatiana Andreani, Ahmed Emam, Dr. César Barbero (Universidad Nacional de Río Cuarto), Yang Hu (University of Waterloo), Jie Fu (Georgia Institution of Technology), Dr. Allan Philippe (University Koblenz-Landau), and Dr. Edith Yslas (Universidad Nacional de Río Cuarto).

The editorial assistant, Janine Haughton, is also grateful for the assistance of all the contributors.

Finally, I would like to greatly acknowledge my parents for their never-ending support and love. This book is especially dedicated to my parents.

Thank you to all of you.

Chapter 1
Evaluation of Currently Available Techniques for Studying Colloids in Environmental Media:
Introduction to Environmental Nanometrology

Allan Philippe
University Koblenz-Landau, Germany

ABSTRACT

Engineered nanoparticles are emerging pollutants with poorly known environmental fate and impact. Studying the fate of engineered colloids in the environment is highly challenging due to the complexity of their possible interactions with environmental components and to the need of dedicated analytical methods. Many relevant processes like e.g. agglomeration and dissolution can be studied by monitoring the size of colloids. Techniques dedicated to the determination of the size of colloids in environmental media are thus required. Such techniques should remain accurate at low concentrations and be specific, widely matrix independent and free of artefact due to sample preparation. This chapter aims at evaluating and comparing systematically the currently used tools (e.g. microscopy, light scattering, particle counters) for sizing colloids considering these requirements. As an example of a highly promising solution, the current development of separation techniques coupled to (single particle) ICP-MS is described in more details.

DOI: 10.4018/978-1-5225-0585-3.ch001

INTRODUCTION

Numerous studies have addressed the fate of engineered colloids in the environment due to the strong need for societal and environmental risk assessment (Batley, Kirby, & McLaughlin, 2012; Nowack et al., 2012). Furthermore, colloids are increasingly used in water purification or soil remediation processes (Karn, Kuiken, & Otto, 2009), where prediction of the long-term effects of their application requires understanding their fate in those media. This implies monitoring their most relevant properties being: size, shape, mass, density, crystalline phase, charge, and elemental composition. For instance, understanding agglomeration requires being able to distinguish between agglomerates formed by different mechanisms such as reaction or diffusion limited agglomerates, flocs and hetero-agglomerates, which were all observed under simulated environmental conditions and are characterized by different structures and shapes (Hotze, Phenrat, & Lowry, 2010; Allan Philippe & Schaumann, 2014b). Thus, distinguishing them requires information on their mass, size, shape and elemental composition. Over the years, various techniques were developed for determining these parameters (Hassellöv, Readman, Ranville, & Tiede, 2008; Jimenez, Gomez, Bolea, Laborda, & Castillo, 2011; López-Serrano, Olivas, Landaluze, & Cámara, 2014; Simonet & Valcárcel, 2009; Singh, Stephan, Westerhoff, Carlander, & Duncan, 2014; K. Tiede et al., 2008).

This chapter attempts to systematically evaluate and compare the most used analytical methods for determining the size and the shape of colloids in environmental media and propose a systematic approach for future selection or development of analytical techniques. Only methods with a broad applicability are addressed. Some techniques like UV-visible-, Raman-, IR-, EPR-, NMR-spectroscopy, HPLC, ESI-, MALDI-, TOF-mass spectrometry, can be applied to some specific samples (K. Tiede et al., 2008) but lack of versatility and, therefore, are not considered here. For further comprehensive reviews on nano-analytical techniques, the reader is invited to refer to published reviews (Burleson, Driessen, & Penn, 2005; Fedotov, Vanifatova, Shkinev, & Spivakov, 2011; Hassellöv et al., 2008; Jimenez et al., 2011; López-Serrano et al., 2014; Simonet & Valcárcel, 2009; Singh et al., 2014; K. Tiede et al., 2008; Frank von der Kammer et al., 2012). This chapter focuses on aqueous samples; methods dedicated to pedogenic or aerial particles are beyond the scope of this work. The largest part of the discussions presented in this chapter was already addressed in author's PhD work (Allan Philippe, 2015).

The following section described the specific challenges raised by environmental samples. In the next sections, common techniques are described, evaluated and compared on the basis of these criteria. Considering this current analytical toolbox for colloids, the remaining challenges and needs for the development of new techniques are described followed by an example of a recent answer to this needs: hydrodynamic chromatography coupled to single particle ICP-MS.

REQUIREMENTS FOR ENVIRONMENTAL SAMPLES

Detection Limit

During the second half of the XXth century, biologists and polymer scientists leaded the analytics of colloids. Environmental scientists focused mostly on the characterization of natural colloids (Buffle, Wilkinson, Stoll, Filella, & Zhang, 1998). Natural colloids are mainly composed of organic matter, iron oxides, and aluminosilicates with concentrations in the mg L^{-1} range (J. R. Lead & Wilkinson, 2006). The

Evaluation of Currently Available Techniques for Studying Colloids in Environmental Media

development of commercial nanomaterials in the last 20 years raised new challenges for environmental analysts since they combine the challenges of working with colloids and working at low concentrations. Indeed, recent efforts in modelling concentrations of engineered colloids in various environmental compartments suggest that only (ultra)trace amount of engineered colloids are present in surface waters with realistic concentrations most probably below 0.1 mg L^{-1} (H. H. Liu & Cohen, 2014). Therefore, analytical techniques for sizing and quantifying engineered colloids should remain accurate in the µg L^{-1} range. This is often challenging as, in this range, particle number concentrations become extremely small in terms of moles of analytes. For instance, a realistic mass concentration for engineered TiO_2 spherical nanoparticles in surface water would be 0.1 µg L^{-1} (H. H. Liu & Cohen, 2014) corresponding to 60 ng L^{-1} of Ti and 45 000 particles with a diameter of 100 nm in one milliliter natural water corresponding to less than 0.1 attomole of individual analytes (here one particle). This is in the same range than the limit of detection of most advanced mass spectrometers (Watson & Sparkman, 2007). This pinpoints the difficulties encountered at such concentration ranges and the strong limitations of using analytical tools dedicated to molecules for characterizing colloids.

Sample pre-concentration procedures involving, for instance, centrifugation, filtration, cloud-point extraction or separation using magnetic or electric field are helpful in some cases (Baalousha & Lead, 2012; Fabricius, Duester, Meermann, & Ternes, 2013). However, they are rarely applicable on environmental samples due to high natural colloidal background and preparation artefacts. Instruments capable of detecting and counting individual particles represent a more promising solution. In such cases, the detection limit can also refer to the smallest particle size detectable by the system.

Matrix Effects and Selectivity

Since environmental media contain high amount of diverse inorganic and organic compounds, an analytical technique should remain accurate in their presence. For instance, surface water contains inorganic ions (e.g. Na^+, K^+, Ca^{2+}, Mg^{2+}, Cl^-, NO_3^-, HCO_3^-), which usually do not affect the sizing of particles, unless the respective technique is highly sensitive to viscosity, density or refractive index. Dissolved organic matter (DOM) can disturb the sizing of particles when large molecular agglomerates form in the presence of divalent cations (Caceci & Billon, 1990; Maurice & Namjesnik-Dejanovic, 1999; Tipping, 2002) or when DOM coating forms on the particles and thus modify surface properties (e.g. charge, chemical affinity) (Allan Philippe & Schaumann, 2014b). However, DOM is generally not of a major concern for the sizing of particles. Matrix effects can be sometime quantified and the results corrected using calibrants measured in the same media as the analytes. However, the high variability of environmental samples makes this approach, at best, tedious. Separation techniques are generally robust upon matrix effects compared to other methods as analytes are separated from their ionic or molecular background before being detected and measured.

Natural colloids are the main challenge for detecting and sizing colloids in environmental matrices. Indeed, their concentrations can reach several mg L^{-1} in natural waters (Wigginton, Haus, & Hochella Jr, 2007), which is much higher than the expected concentrations for engineered colloids. Furthermore, distinguishing between natural and artificial colloids is a challenge for most techniques, especially if their size ranges overlay. At present, element or even isotope specific detection systems are the most practical tools for facing this challenge. This is one of the reasons for the recent development of separation methods coupled with ICP-MS detection systems.

Sample Perturbation

For molecular analytes, an extraction procedure is often required prior to analysis. However, colloids can be sensitive to both chemical and physical processes. For instance, the drying step necessary for typical electron microscopy analysis can induce agglomeration of the analytes and thus modify the particle morphology to be determined (Burleson et al., 2005). This has motivated the development of environmental scanning electron microscopy, that can be used under low vacuum and with an atmosphere saturated in water and, thus, limits drying artefacts, under some extend (K. Tiede et al., 2008). In addition, non-perturbing sample preparations for electron microscopy using, for instance, ultracentrifugation or embedment in resins have been reported for natural samples (Perret et al., 1991). Hence, sample perturbation for a given technique can be limited by developing appropriate preparation methods. However, such developments are usually tedious and have to be validated for each type of sample and media, especially for quantitative measurements.

Separation techniques require usually less sample preparation and, therefore, are advantageous compared to techniques requiring sample drying. However, the effect of common preparation steps such as dilution in the carrier media and filtration should be investigated. In situ methods are the only fully non-perturbing type of analysis. Such methods are unfortunately often highly sensitive to matrix components, especially to natural colloids.

Cost/Time Efficiency

Most reviews about nanometrology do not address cost or time efficiency issues. Although these criteria do not concern directly the scientific aspect of the research, this is often a crucial point for laboratory decision makers and for the planning of large scale experiments. For instance, the running of ten parallel mesocosm experiments can easily afford the measurement of hundreds of samples per week. It is obviously not realistic to consider a systematic characterization of these samples using electron microscopy or atomic force microscopy. Furthermore, most laboratories cannot easily access expensive techniques such as transmission electron microscope. A readily available, time efficient and affordable technique will often be preferred to more powerful but expensive devices.

Comparing accessible techniques, the measurement duration is an important factor as analytes alteration increases with time, if colloids are in a metastable state. For instance, typical measurement durations are between 20 and 45 minutes with flow field flow fractionation (AF4) (Bolea, Jiménez-Lamana, Laborda, & Castillo, 2011; Giddings, 1993; F von der Kammer, Baborowski, & Friese, 2005). Considering the necessity to replicate each measurement, the characterization of 100 suspensions requiring the same measurement conditions should take at least one week of full-time measurement while same amount of samples could be analyzed in 33 hours with hydrodynamic or size exclusion chromatography (McHugh & Brenner, 1984). A high sample throughput is only possible with fast and automatized techniques. Furthermore, one should always consider the time required for the method development.

OVERVIEW ON ANALYTICAL TECHNIQUES

Classification of Sizing Techniques

Analytical techniques can be classified based on their measurement principle such as light scattering or size exclusion. A more useful approach is to cluster techniques using their analytical outputs (measured parameters, weighting used in the obtained distribution). From this point of view, four categories clearly appear: imaging techniques (microscopy), particle counting techniques, light scattering based techniques, and separation techniques. While filtration, centrifugation, size exclusion chromatography, Coulter counter and electron microscopy were the most used techniques in the past century, the recent need for lower concentration and size detection limits stimulated the (further) development of ICP-MS detection systems, NTA, and AF4, for instance.

Individual detection and analysis of particles is probably the best approach for environmental samples. However, individual detection of nanoparticle is challenging. Imaging techniques such as electron microscopy (EM) and atomic force microscopy (AFM) are the most efficient representatives of this category for dry samples (Burleson et al., 2005). However, determining absolute concentrations requires high care by preparing samples and analyzing images. For liquid samples, Coulter counter (CC) and, more recently, nanoparticle tracking analysis (NTA), laser induced breakdown detection (LIBD), and single particle inductively coupled plasma mass spectrometry (SP-ICP-MS) were developed for detecting particles individually (Carr et al., 2009; Hassellöv et al., 2008; Simonet & Valcárcel, 2009; K. Tiede et al., 2008). These techniques can be designated as counting methods since a number based size/volume distribution can be obtained from the data. Furthermore, only counting techniques can be used for determining particle number concentration without calibration.

The most important in situ techniques for liquid samples are based on light scattering (Finsy, 1994). Dynamic light scattering (DLS), multi-angle light scattering (MALS) and ζ-potential measurements rely on the analysis of scattered light intensity signal from the whole sample and determine overall average values only. Data analysis is based on complex theories implying several hypotheses on the sample (Finsy, 1994).

Separation techniques isolate the analyte based on one or several properties that are quantified using appropriate calibration or theoretical models. Most common separation techniques are asymmetrical flow-field flow fractionation (AF4), sedimentation field flow fractionation (Sed-FFF), size exclusion chromatography (SEC), analytical ultracentrifugation (AUC), hydrodynamic chromatography (HDC), and capillary electrophoresis (CE) (Fedotov et al., 2011; Lespes & Gigault, 2011; McHugh & Brenner, 1984). The analytical outputs depend on the separation mechanism and the detection system.

Depending on the sizing principle and the data evaluation, the "size" can be represented by the core diameter (D_{Core}), the effective diameter (D_{Eff}), the hydrodynamic diameter (D_{Hydr}) or the gyration diameter (D_{Gyr}) (Hassellöv et al., 2008; Hunter, 2001; Striegel & Brewer, 2012). Definitions for these terms can be found in the index. Figure 1 summarizes, without being exhaustive, the different parameters and the corresponding techniques used for measuring them. The above mentioned groups of techniques are described in further details in the next sections.

Figure 1. Schematic summary of the analytical inputs (parameters that need to be known for measuring) and outputs (parameters that are determined) of common analytical techniques for colloids characterization. Separation techniques are considered for themselves, without additional specific information from the detector.

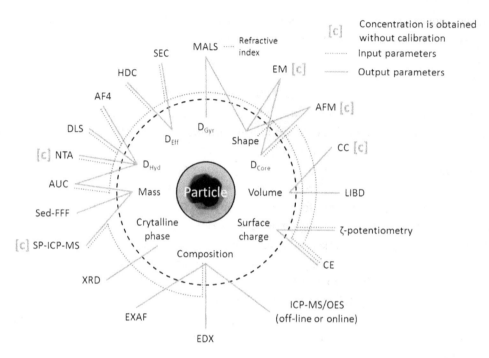

Imaging Techniques

Transmission and scanning electron microscopes are the most used electron optical systems for the characterization of colloids. For high resolution microscopes, analyzing and counting individual particles of a few nanometers or larger is possible by direct visualization. Thus, number size distribution can be obtained directly after image processing. However, it is often difficult to observe materials with low electron density (e.g. organic coating). Therefore, only the dense core of particles can be observed, unless an appropriate staining method is developed (Burleson et al., 2005). Since such systems have to operate under high vacuum for maintaining an acceptable resolution, samples have to be dried. Drying can induce crystallization of salts, agglomeration of colloids or molecules and alteration of some structure such as gels (Burleson et al., 2005).

Sample preparation methods minimizing artefacts were developed for environmental samples (Burleson et al., 2005; Perret et al., 1991; Wilkinson, Balnois, Leppard, & Buffle, 1999; Wilkinson et al., 1999). For instance, heteroagglomerates of natural organic matter and iron oxides could be observed using TEM in lake water after preconcentration and staining steps (Wilkinson et al., 1999). Nonetheless, these methods have to be optimized for each type of sample and are not completely free of artifacts. SEM devices operating under low vacuum have been developed for the study of wet samples (Burleson et al., 2005; Paul, Jamie, Harrison, Jones, & Stoll, 2005); unfortunately with a much lower resolution than traditional SEM.

AFM can measure the interaction forces between a tip and the sample by rastering the tip over the sample surface (Burleson et al., 2005). Force measurements and topographic images with vertical and lateral resolutions in the sub-nanometer range can be obtained. AFM is useful for imaging size, shape, and topography of colloids with or without solvent and, with modified AFM tips, measuring the forces of interaction between colloids as a function of solution chemistry (Burleson et al., 2005). Therefore, AFM has been used for studying various natural colloids (Baalousha & Lead, 2007; J. Lead, Muirhead, & Gibson, 2005; Plaschke, Römer, & Kim, 2002). Analytes have to be fixed on a flat surface for imaging, which makes the sample preparation tedious for environmental samples (Baalousha & Lead, 2012). AFM, like most microscopy techniques, is highly time-consuming.

Light Scattering Techniques

Light scattering techniques are in situ and measurements are straightforward with modern instruments. While a sample is illuminated by a light source, the intensity of the scattered light is monitored at one angle and over time with DLS and at several angles simultaneously with MALS (Finsy, 1994). Some instruments can perform both types of analysis simultaneously. In DLS, the complex temporal variations in intensity due to the Brownian motion of the particles are auto-correlated in order to extract a diffusion coefficient distribution by using dedicated algorithms (Finsy, 1994). As for NTA, the diffusion coefficient distribution can be transformed into size distribution using Stokes-Einstein equation. The resulting size distribution is intensity weighted. The intensity being proportional to the sixth power of particle size, the size distribution is weighted by the fifth power of particle size (Finsy, 1994). Therefore, the average hydrodynamic diameter calculated for polydisperse suspensions can be much larger than the number average obtained with microscopy techniques. In other words, for highly polydisperse samples, the largest particles may be considerably overrepresented. Limit of detection for the concentration depends on particle size and material and is usually in the mg L^{-1} range for most type of colloids, whereas the smallest measurable particle size can reach several nanometers for the most performant instruments. Since particle shape strongly influences the scattered intensity, it needs to be provided for interpreting DLS data (Lin et al., 1990). DLS is often used for determining aggregation kinetic parameters in environmental matrices, as the effect of concentration can be easily extracted in most of those studies (Metreveli, Philippe, & Schaumann, 2014).

Using MALS and knowing the refractive index of the particle, it is possible to determine an average D_{Gyr}. Indications on the particle shape such as fractal dimension can also be obtained (Bushell & Amal, 2000). Based on similar principles, small angle X-ray scattering and small angle neutron scattering can be used for particles in the range of some nanometers or for macromolecules (Gilbert, Lu, & Kim, 2007; King & Jarvie, 2012).

Particle Counting Techniques

Particle counting techniques are highly promising for characterizing colloidal suspensions since they combine the advantage of operating in the aqueous phase with individual particle detection. A number based size distribution is obtained directly from the data. Depending on the detection method, different parameters can be collected for each particle. Due to the detection mode, limits of detection for concentrations can be lower than 1 ng L^{-1} and sample concentration has to be optimized for avoiding the detection of several particles at once. However, the limit of detection for size depends strongly on the detector and is usually higher than 10 nm.

Coulter counter is one of the oldest types of particle counter. Suspended particles are forced to move through a narrow hole or tube by the mean of an electric field. If the size of one particle is on the same order of magnitude than the dimensions of the hole, the overall current will be significantly affected by the path of this particle and a spike signal will be measured by monitoring the current intensity over time. Spike intensity is related to the volume of the particle. The main drawback of this method is the high detection limit for the size (usually around 100 nm). However, future devices with improved materials and channels geometries may become available in the next years (Ito, Sun, & Crooks, 2003; Zhe, Jagtiani, Dutta, Hu, & Carletta, 2007).

Ultramicroscopy, since recently also called nanotracking analysis (NTA), is a special mode of optical microscopy where the liquid sample is illuminated by a laser. The light scattered by one particle can be observed with a classical darkfield optical microscope. The diffusion coefficients of each individual particle can be extracted by analyzing their Brownian motion using appropriate data evaluation software (Carr et al., 2009). Stokes-Einstein equation is then used for converting diffusion coefficients into hydrodynamic diameters resulting in a hydrodynamic diameter number based distribution. Particle concentration often has to be optimized (Filipe, Hawe, & Jiskoot, 2010). The size limit for the detection depends on the refractive index of the particle material and is higher than 10 nm (Filipe et al., 2010). Since Stokes-Einstein equation includes a shape factor, the shape of the particle has to be known (Hunter, 2001).

SP-ICP-MS has been developed recently (Mitrano et al., 2011; Tuoriniemi, Cornelis, & Hasellöv, 2012) for the detection and quantification of highly diluted heavy metal based nanoparticles in complex matrices (limit of detection in the ng L^{-1} range for Ag, Au, Ce, Cd, etc.). Liquid samples are pumped into a nebulizer that produces droplets of some micrometers. The droplet components are vaporized, atomized and ionized in argon plasma. Produced ions are guided into a quadrupole mass-spectrometer measuring the amount of ions having a defined mass over the time (Figure 2).

SP-ICP-MS principle is that, at low concentrations, a suspension can be injected sufficiently slowly into the detector for detecting particles individually. Indeed, a properly tuned instrument is sensitive enough for detecting an ion cloud produced by the atomization of one single particle (Figure 3). The intensity of the obtained spike signal is directly proportional to the amount of atoms of the monitored element contained in one particle. If calibrants of the monitored element are available, an elemental mass can be calculated for each detected particle (Pace et al., 2012).

Figure 2. Simplified principle of inductively coupled plasma mass spectrometry (ICP-MS)

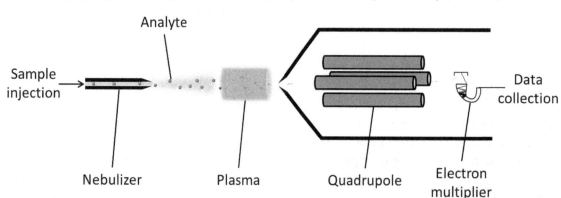

Figure 3. Illustration of SP-ICP-MS detection principle. Left side: a pulse of 50 nm large gold nanoparticles dispersion was injected into the ICP-MS detector. Right side: the same dispersion was diluted 1:500 before injection. On the right graph signal spike represent single particle detection events.

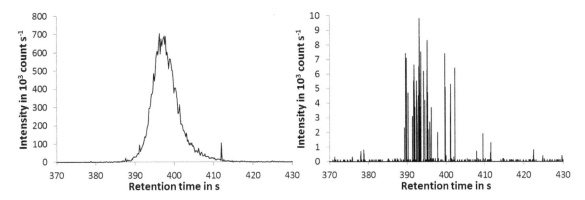

Separation Methods

For complex samples, separation of analytes from the medium is generally advantageous, especially if samples contain different types of colloids. Centrifugation and filtration can be used for preparing sample (Fabricius et al., 2013; K. Tiede et al., 2008). However, these methods have to be validated by a size analysis. Therefore, combining separation and size analysis online is highly interesting. Most separation mechanisms are based on particle size, mass or surface charge. One major advantage of separation techniques is the possibility to combine several detectors online for collecting additional information on the analyte.

Analytical ultracentrifugation (AUC) is a disk centrifuge equipped with a detector (usually turbidity or X-ray detectors) for monitoring the analyte sedimentation under the centrifugal field (Schuch & Wohlleben, 2010; Wohlleben, 2012). With this technique, a mass distribution is obtained provided that the density of the analytes is known. If this is not the case, the sample can be measured after dilution in pure water and in deuterated water in order to determine the absolute mass distribution of the suspension (Schuch & Wohlleben, 2010). Particle shape influences sedimentation rate and should be known. One main disadvantage of AUC is that coupling with high performance detectors such as ICP-MS is technically not yet possible. As traditional detectors used with AUC are often not selective and not appropriate for detecting trace amount of colloids, AUC is rarely used for environmental samples and is more dedicated to routine quality control in the industry (Wohlleben, 2012).

Sedimentation field flow fractionation is another method for determining the mass distribution of particles dispersion. A centrifugal force is applied on a flat tube where eluent is pumped at a defined flow velocity. The analytes are injected into this flow and pushed by gravity to the external walls of the tube. Since the flow rate is reduced near to the wall, heavy particles will be slower than lighter particles by the combined effect of the gravity field and the flow profile (Giddings, 1993). This method can be coupled to various detectors including ICP-MS (Dubascoux et al., 2010), whereas it is limited to particles that sediment readily under fair centrifugal field. A minimal size of 50 nm is generally required (Fedotov et al., 2011).

Hydrodynamic chromatography can be used to separate particles by their size. Size separation occurs when particles flow in a pipe with a small internal diameter. In this case large particles cannot sample the whole parabolic flow velocity profile as they center of mass cannot approach near to the wall, whereas small particles can access to these low velocity regions. Large particles will, therefore tend to elute earlier than small particles. Using an appropriate calibration method, it is possible to determine the effective diameter (geometric diameter for a sphere) of unknown samples. HDC is simple and robust but less efficient than other separation methods (McHugh & Brenner, 1984; Striegel & Brewer, 2012; Karen Tiede et al., 2009). Particles with diameters in the size range 5-3000 nm can be separated using commercially available columns. Further details related to HDC about separation mechanism and coupling techniques can be found later in this chapter.

HDC can be carried out on packed monodisperse spheres columns. If the spheres used for the packing are porous with pores size in the same range as the particles to analyze, size exclusion occurs and large particles elute faster than small particles, since the latter are delayed in the pores. Various SEC columns are commercially available. However they generally have relatively narrow separation ranges and are limited to the separation of particles smaller than 100 nm. While SEC is limited for particles, it is often the method of choice for studying natural or artificial macromolecules. Effective diameters can be determined with HDC and SEC using size calibration.

Hydrodynamic diameter distribution can be determined with AF4, in which the analytes are eluted through a flat channel. A crossflow stream, entering and exiting through permeable walls, drives components toward a membrane (Giddings, 1993). The pore size of the membrane determines the lower size limit for separation. Particles with high diffusion coefficients have access to a larger flow profile and hence elute faster than particles with lower diffusion coefficients. Particles from one nanometer to several micrometers can be separated. Diffusion coefficients can be determined from retention times using physical models or size calibration curves (Gray et al., 2012). Stokes-Einstein equation can be used to convert diffusion coefficients into hydrodynamic diameters. AF4 has become very popular in environmental sciences due to its large size range, high resolution and mild elution conditions (Baalousha, Stolpe, & Lead, 2011). For instance, AF4 was used to determine the concentration of silver and TiO_2 nanoparticles in antimicrobial consumer products and sunscreens, respectively (Cascio et al., 2015; Nischwitz & Goenaga-Infante, 2012).

Sed-FFF, AF4, SEC, and HDC were mostly combined with UV-visible, differential refractometry, and fluorescence detectors. Direct information on the size or the nature of particles cannot be obtained using these detectors and, therefore, they are used as unspecific detection systems for determining retention times. Detectors capable of determining particle properties in an online modus make possible to carry out multidimensional characterization. For instance, coupling of separation techniques with DLS and MALS detectors was used to characterize particle shape (Brewer & Striegel, 2011; F von der Kammer et al., 2005). ICP-MS detectors become increasingly common in environmental sciences as it is isotope specific and has a detection limit in the range of ng L^{-1} for most elements (Lespes & Gigault, 2011). Thus, ICP-MS has been coupled to SEC (Jimenez et al., 2011), HDC (Karen Tiede et al., 2009) and AF4 (Dubascoux et al., 2010). Furthermore, SP-ICP-MS has been used as a detector with HDC (Pergantis, Jones-Lepp, & Heithmar, 2012). SP-ICP-MS is especially interesting as it cumulates the advantages of a particle counter detector with the selectivity of ICP-MS. Combined to information obtained from the retention time particle size, mass, and number concentration can be obtained in a single run (Pergantis et al., 2012). Further discussions concerning this coupling can be found later in this chapter.

Comparison of Sizing Techniques

Comparing different sizing techniques is difficult, since the analytical outputs may intrinsically differ and the performance depends on the used instrument, the experimental design and sometime the analyte and the matrix. For instance, size and polydispersity can influence the measurement duration and the size resolution of AF4 (Baalousha et al., 2011). The choice of the most appropriate technique thus depends on the type of analyte, the expected concentrations, the matrix components and the exact parameters to be determined. Nonetheless, it is possible to compare the common techniques used for sizing colloids based on the most relevant criteria for environmental samples. Such criteria can be quantified using a numerical indicator based on estimated average performances.

For instance, the size ranges can be easily described using the lowest particle size that can be detected by the techniques and the broadness of the range itself. Imaging and light scattering techniques have generally a very broad size range in contrast to counting techniques (Figure 4). However, a broad size range does not ensure accurate measurements of highly polydisperse samples. For instance, the complex method optimization required for AF4, which depends on the particles type and size, makes it practically impossible to determine precisely the size of each fraction in a highly polydisperse sample in one single run (Baalousha et al., 2011). Concerning DLS, the light scattering signal from larger particles masks the signal from smaller particles resulting in a monodisperse distribution, which does not reflect the actual polydispersity.

Measurement duration can also be easily quantified (Figure 5). Imaging techniques are clearly more time-consuming than other methods due to the time required for preparing samples and the tedious image production and analysis, especially when few engineered particles have to be found in a complex matrix. In such cases, it can be difficult to image enough particles (typically hundreds) for determining a representative size distribution. On the other hand, chromatographic, spectroscopic and counting methods are generally much faster and can be easily automatized.

Figure 4. Size ranges of the most used sizing techniques for colloids in environmental samples (Lespes & Gigault, 2011; K. Tiede et al., 2008)

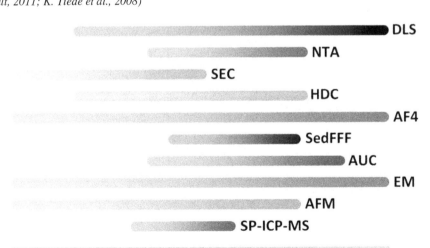

Figure 5. Typical measurement duration (for one replicate) of the most used sizing techniques for colloids in environmental sciences

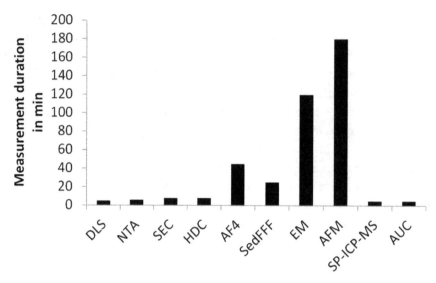

Table 1. Qualitative criteria used for the evaluation of sizing techniques. For comparison purpose, limits of detection (LOD) were calculated for silver particles. Separation techniques are considered coupled with ICP-MS if this coupling has been reported. Information about individual techniques was collected in dedicated review articles and from personal experience.

Grade in Normalized Qualitative Units	0	2.5	4	5	7	7.5	8	10
Minimal Size in Nm	100	50		20-30		5		1
Size Range Broadness in Nm	<100	300		1000		3000		>10 000
Sample Perturbation	Drying/ extraction			Dilution/ filtration				In situ/ direct injection
Matrix Sensitivity	Influenced by polys-dispersity or impurities			Influenced by other colloids of the same size				No matrix influence
LOD in ng L^{-1}	100 000		100		10			1
Measurement Duration in Min	>60	45	45	25		8		<8

(Hasselöv et al., 2008; Lespes & Gigault, 2011; Simonet & Valcárcel, 2009; K. Tiede et al., 2008)

Considering criteria that cannot be easily quantified, an arbitrary grade can be deserved to each technique as proposed in the Table 1. The correspondence system is purely arbitrary. The proposed system is based on typical challenges reported in the literature for environmental samples and was optimized

Evaluation of Currently Available Techniques for Studying Colloids in Environmental Media

for differentiating performances of evaluated techniques. For instance, matrix sensitivity is considered the highest when few larger particles or impurities can compromise the final result. Then, a significant improvement is met, when measurement accuracy stands the presence of natural particles having similar sizes than the analytes. This is the case for chromatographic techniques where target particles are separated from the natural background and then detected using element/isotope specific detector (e.g. ICP-MS).

Concerning the sample perturbation that can occur during the pre-treatment or during the analysis, the ideal situation is in situ measurement where the sample is analyzed without any alteration as for light scattering techniques. More frequently, a dilution and/or filtration step is required. In this case, the preparation steps have to be optimized in order to reduce loss or alteration of the analyte. However, these preparation steps are usually less problematic than extraction of the analyte or drying of the sample, since strong alterations of the colloids state are expectable and careful method validation is required for each type of sample.

The graphical representation of this evaluation makes possible to visually compare several techniques (figure 6). While NTA, HDC, AF4, and SP-ICP-MS are balanced considering the chosen parameters, DLS,

Figure 6. a-b. Evaluation of the fulfillment of the environmental samples requirements by most used sizing techniques. Details about the evaluation are provided in Table 1.

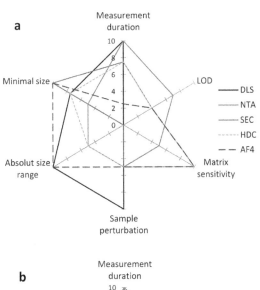

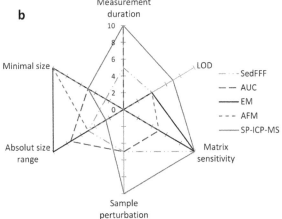

SEC, AUC, and AFM have strong drawbacks. Depending of the investigated question, some parameters may be of more importance. In cases where sample disturbance is critical and where natural colloids with the same size range than the analyte are expected, SP-ICP-MS will be the method of choice provided the analytes can be detected using this method. In cases where expected concentrations and matrix are not problematic, DLS can be used with a fair confidence. However, a combination of complementary techniques is required for unknown environmental samples.

For instance, combining AF4 or HDC with SP-ICP-MS would have the advantage to cover the whole size range, while maintaining limit of detection in the ng L^{-1} range. Indeed, the lower size range (< 30 nm) in which particles are generally numerous but too small to be detected individually could be covered with a separation technique using a size calibration approach and the larger particles could be detected and quantified by SP-ICP-MS even at low number concentration. Another powerful combination is EM with NTA or SP-ICP-MS. Indeed sample perturbation (e.g. aggregation) can be revealed since sample preparation is minimal for NTA and SP-ICP-MS in contrast to EM. In order to make the preceding discussion more concrete, the next chapter summarizes the recent development of one sizing technique that takes advantage of such a powerful combination: HDC-SP-ICP-MS.

CASE STUDY: HDC-SP-ICP-MS FOR ENVIRONMENTAL SAMPLES

Theory of HDC

H. Small reported first the use of HDC as an analytical technique for sizing macromolecules (Small, 1974) based on a theory developed for explaining the elution of protein samples in narrow tubes (DiMarzio & Guttman, 1970). HDC can be performed using a capillary or a column packed with monodisperse microspheres. Interestingly, the model used for the capillary can be applied straightforwardly to the case of a packed column (McHugh & Brenner, 1984). In a cylindrical tube, a Poiseuille flow transports particles at different velocities. Due to steric exclusion, large particles can access only to the central region of the tube where the average velocity is the highest due to the parabolic shape of the flow (Figure 7). Smaller particles can access to the region near to the wall; their average velocity is therefore lower. Thus, large particles will tend to elute faster than small particles and all particles will elute faster than or as fast as the eluent.

Figure 7. Separation mechanism in HDC

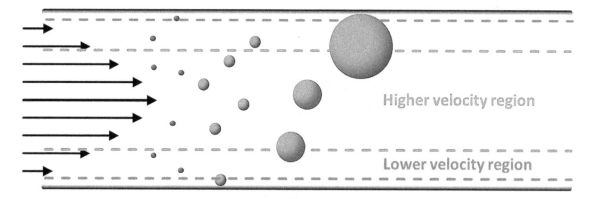

The model is based on the following hypotheses (Striegel & Brewer, 2012):

- Flow is laminar.
- Particles efficiently sample the whole velocity profile. This is generally true for colloidal particles, provided that the elution time is long enough.
- Affinity or collision of the particles with themselves or the wall are negligible.
- Particles are spherical or similar to spheres.
- Packed columns can be modelled by parallel tubes.

The retention factor of an analyte (R_F) can be defined by the ratio of the average retention time of the particles (τ_p) measured and the average retention time of a marker (τ_0). R_F is related to the reduced diameter $\bar{a}$ of the particle (ratio between the diameter of the particle and the diameter of the fictitious tube) as shown in the following formula:

$$R_F = \frac{\tau_p}{\tau_0} = \left(1 + 2\bar{a} - \bar{a}^2\right)^{-1} \tag{1}$$

Hence, retention factor depends only on the particle size for a given type of column/capillary and is independent of the elution conditions. This simple model successfully described the elution of various spherical particles (Chun, Park, & Kim, 1990; McHugh & Brenner, 1984; Stoisits, Poehlein, & Vanderhoff, 1976; Tijssen, Bos, & Van Kreveld, 1986). The wide independency of the retention factor on the flow velocity, temperature, eluent composition (as far as electrostatic forces are negligible), particle composition and surface coating has been demonstrated experimentally (A Philippe, Gangloff, Rakcheev, & Schaumann, 2014; Allan Philippe & Schaumann, 2014a).

The size-based selectivity can be used to quantify the intrinsic capability of a technique to separate analytes according to their size and is calculated using the following formula (Lespes & Gigault, 2011):

$$S = \left|\frac{d\left(\log \tau_p\right)}{d\left(\log d\right)}\right| \tag{2}$$

where τ_p is determined experimentally using standard suspensions with known diameter d. This parameter can be used for comparing performances of different size separation techniques as shown in the Table 2. HDC is clearly less selective than other techniques. However, its low performance is compensated by its high flexibility, simplicity and robustness towards challenging matrices (Allan Philippe & Schaumann, 2014a; Karen Tiede et al., 2009, 2010).

First Applications to Environmental Samples

The high flexibility of HDC is reflected in its numerous applications in biology (Brough, Hillman, & Perry, 1981; L. Liu et al., 2013; Roman et al., 2015), polymer science (Chenal, Rieger, Philippe, & Bouteiller, 2014; Langhorst et al., 1986; Striegel, 2012) and, more recently, in environmental sciences

Table 2. Size selectivity ranges for the main separation techniques used for particle characterization

Technique	Size Range (nm)	Optimal Selectivity	Usual Selectivity
Size exclusion chromatography	0.1-100	0.2	0.05-0.15
Packed beads hydrodynamic chromatography	5 to few 1000	0.1	0.02-0.1
Capillary hydrodynamic chromatography	10 to few 1000	0.2	0.05-0.15
Flow-FFF	1 to 50 000	1	0.5-1
Sedimentation-FFF	50 to 50 000	3	0.5-2.5

Adapted from Lespes and Gigault (Lespes & Gigault, 2011)

(Karen Tiede et al., 2009). Some criticisms based on observed artefacts measurements (affinity effects for some coated particles) have been raised (Gray et al., 2012). However, that problem can be minimized by optimizing the eluent composition (A Philippe et al., 2014). First applications of HDC to environmental samples aimed at detecting silver nanoparticles in sludge from waste water treatment plant (Karen Tiede et al., 2010), River Rhine water (Metreveli et al., 2014), and drinking water (Pergantis et al., 2012).

During the pioneer's time, the most used detectors were dedicated to the detection of polymers. Still today, the most common detectors used with HDC are UV-visible, optical density detectors, and differential refractometer (DR) (Penlidis, Hamielec, & MacGregor, 1983; Striegel & Brewer, 2012).

However, ICP-MS is clearly the detector of choice for environmental samples (Lespes & Gigault, 2011), since its isotope specificity can be used for distinguishing between natural and artificial colloids, if their respective elemental composition/isotopic ratios differ. Detection limits are element dependent and are in the µg L^{-1} range for most common elements (e.g. Ti, Fe, Al) and in the ng L^{-1} for rare elements (e.g. Au, Ag, Ce) (Karen Tiede et al., 2009, 2010), which fairly cover the expected range of engineered colloids in natural media.

The quantity of ions produced as particles pass through the plasma is supposed to be independent of their size, since particles smaller than 500 nm are completely atomized in the plasma (Dubascoux et al., 2010). Thus, it is assumed that the number of counts registered by the detector is proportional to the number of atoms passing through the detector and thereafter to the total mass of particles. Thus, an HDC chromatogram can be interpreted as the elemental mass-weighted distribution of the injected sample. This is a major advantage of mass spectrometer detectors compared to classical detectors, for which a complex analysis of the signal is required and implies hypotheses on the nature of the analytes.

Coupling with SP-ICP-MS

SP-ICP-MS used as a detector with HDC combined the advantages of a counting method and elemental specificity (Pergantis et al., 2012). As explained above, the mass of unknown particles can be determined on a single particle basis using SP-ICP-MS. When combined to HDC, mass and effective diameter of each detected particle can be determined by using SP-ICP-MS, provided that appropriate calibrants are available (Figure 8). This approach is highly promising as it is possible to compute the mass weighted and the number weighted size distribution from the data of a single measurement. Furthermore, large particles present in small amount can be detected in single particle mode, whereas small particles that may not be accessible by SP-ICP-MS due to its intrinsic limit of detection can be detected in plain mode

Figure 8. Illustration of chromatogram interpretation for HDC-SP-ICP-MS

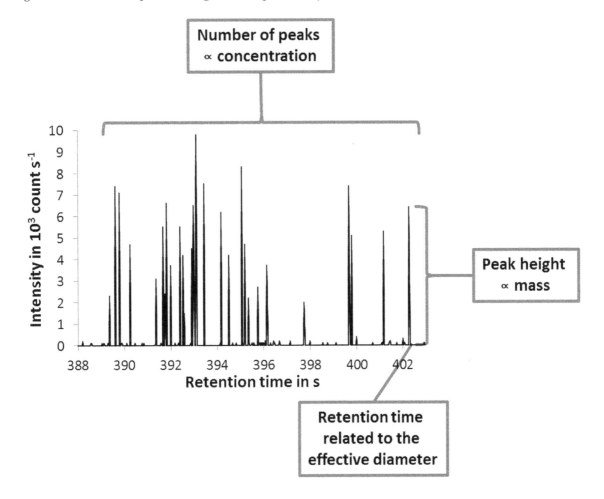

using HDC and calibration curve (retention factors over size). Therefore, mixtures of large and small particles with similar mass concentration can be accurately characterized using HDC-SP-ICP-MS. Characterization of silver and gold nanoparticles and monitoring of their dissolution in human blood and river water samples, respectively, was achieved using this technique (Proulx & Wilkinson, 2014; Roman et al., 2015).

Particle shape analysis is another highly promising application of HDC-SP-ICP-MS. Two parameters are measured per particle: the intensity of the spike, which is proportional to the mass of the particle, and the retention time/factor related to the effective diameter. Thus, it is possible to build a two dimensional distribution graph with these parameters (Figure 9). For instance, a fictitious diameter can be calculated from the mass assuming a spherical shape and be compared with the diameter obtained from the HDC. For spherical particles, these two diameters should be equal, while for cylinders or fractal agglomerates they can strongly differ (A Philippe et al., 2014; Rakcheev, Philippe, & Schaumann, 2013). By analyzing such 2D-distributions, it is possible to evaluate the particle shape. For instance, fractal agglomerates of gold nanoparticles and rode-like silver nanoparticles could be distinguished from spherical standard particles (A Philippe et al., 2014; Rakcheev et al., 2013).

Figure 9. An example of 2-D plot (effective diameter measured using the retention time over the core diameter calculated from the spike height) that can be obtained using HDC-SP-ICP-MS. Each dot represents one detected particle. Red dots denote spherical standard gold nanoparticles (100 nm nominal diameter), while blue dots denote fractal agglomerate of 10 nm large gold nanoparticles.

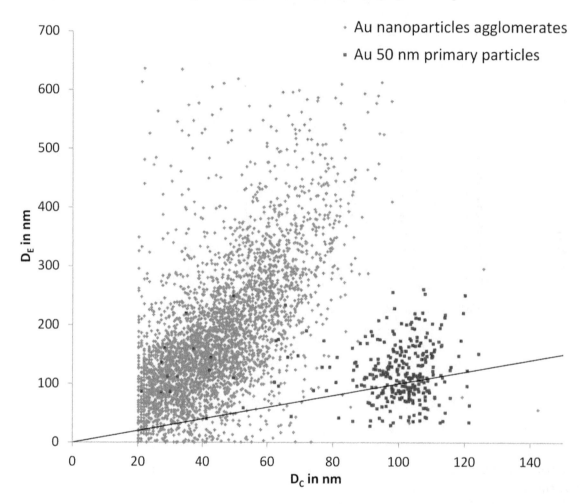

The main limitation of this approach is that the elution behavior of non-spherical particles is more complex than for hard spheres and still requires additional efforts for being implemented in the data analysis (A Philippe et al., 2014). Another limitation, especially for unknown sample, is that only one element can be monitored at once implying that, at least, the main element of the particles to be measured has to be known. Recent technical advances like e.g. the development of fast ICP-MS that can produce several data points per ions-cloud offer interesting possibilities to study complex unknown particles containing several elements such as heteroagglomerates (Montaño, Badiei, Bazargan, & Ranville, 2014).

Another limitation that will be minimized by technical improvement is that the theoretical minimal size that can be detected in single particle mode is above 40 nm for relevant elements such as Al, Fe, Ti, and Zn (Lee et al., 2014). This implies that SP-ICP-MS can be applied to a limited number of particles containing heavy metals (typically: Au, Ag, Pt, Ce etc.). This size limit can even become higher if a relatively high concentration of dissolved species of the monitored element is present. Indeed, the signal produced by the background dissolved species can overlap the spikes produced by small particles. Some efforts were made to minimize this effect by using statistical tools for analyzing data produced by the detector (Tuoriniemi et al., 2012).

CONCLUSION

In order to characterize engineered nanoparticles in environmental media at realistic concentrations, a combination of techniques is clearly required, since no one of the currently available techniques can cover all the requirements of environmental samples. Combination of mass and particle number based techniques coupled off- or on-line is currently the most promising approach for quantification and characterization of engineered colloids in environmental media, as demonstrated by recent advances in the development of HDC-SP-ICP-MS. Therefore, further efforts are required for developing routine data analysis methods and improve detection limits of detectors (e.g. ICP-MS) in order to be able to characterize all of the types of engineered colloids that could be found in the aquatic environment in a near future. Without doubts, new coupling techniques such as AF4-SP-ICP-MS and HDC-SP-ICP-TOF-MS (Borovinskaya, Gschwind, Hattendorf, Tanner, & Günther, 2014) will be developed in the next years and will extend the capability of the current analytical toolbox dedicated to the characterization of engineered colloids. However, many challenges remain such as the characterization of non-spherical particles, hetero-agglomerates or flocs. Furthermore, techniques for detection and characterization of non-fluorescent organic nanomaterials in natural matrices (e.g. nanoplastic in seawater) still have to be developed.

REFERENCES

Baalousha, M., & Lead, J. (2007). Characterization of natural aquatic colloids (< 5 nm) by flow-field flow fractionation and atomic force microscopy. *Environmental Science & Technology, 41*(4), 1111–1117. doi:10.1021/es061766n PMID:17593707

Baalousha, M., & Lead, J. (2012). Rationalizing nanomaterial sizes measured by atomic force microscopy, flow field-flow fractionation, and dynamic light scattering: Sample preparation, polydispersity, and particle structure. *Environmental Science & Technology, 46*(11), 6134–6142. doi:10.1021/es301167x PMID:22594655

Baalousha, M., Stolpe, B., & Lead, J. (2011). Flow field-flow fractionation for the analysis and characterization of natural colloids and manufactured nanoparticles in environmental systems: A critical review. *Journal of Chromatography. A, 1218*(27), 4078–4103. doi:10.1016/j.chroma.2011.04.063 PMID:21621214

Batley, G. E., Kirby, J. K., & McLaughlin, M. J. (2012). Fate and risks of nanomaterials in aquatic and terrestrial environments. *Accounts of Chemical Research, 46*(3), 854–862. doi:10.1021/ar2003368 PMID:22759090

Bolea, E., Jiménez-Lamana, J., Laborda, F., & Castillo, J. (2011). Size characterization and quantification of silver nanoparticles by asymmetric flow field-flow fractionation coupled with inductively coupled plasma mass spectrometry. *Analytical and Bioanalytical Chemistry, 401*(9), 2723–2732. doi:10.1007/s00216-011-5201-2 PMID:21750882

Borovinskaya, O., Gschwind, S., Hattendorf, B., Tanner, M., & Gunther, D. (2014). Simultaneous mass quantification of nanoparticles of different composition in a mixture by microdroplet generator-ICPTOFMS. *Analytical Chemistry*, *86*(16), 8142–8148. doi:10.1021/ac501150c PMID:25014784

Brewer, A. K., & Striegel, A. M. (2011). Characterizing the size, shape, and compactness of a polydisperse prolate ellipsoidal particle via quadruple-detector hydrodynamic chromatography. *Analyst (London)*, *136*(3), 515–519. doi:10.1039/C0AN00738B PMID:21109889

Brough, A., Hillman, D., & Perry, R. (1981). Capillary hydrodynamic chromatography-an investigation into operational characteristics. *Journal of Chromatography. A*, *208*(2), 175–182. doi:10.1016/S0021-9673(00)81929-9

Buffle, J., Wilkinson, K. J., Stoll, S., Filella, M., & Zhang, J. (1998). A generalized description of aquatic colloidal interactions: The three-colloidal component approach. *Environmental Science & Technology*, *32*(19), 2887–2899. doi:10.1021/es980217h

Burleson, D. J., Driessen, M. D., & Penn, R. L. (2005). On the characterization of environmental nanoparticles. *Journal of Environmental Science and Health. Part A*, *39*(10), 2707–2753.

Bushell, G., & Amal, R. (2000). Measurement of fractal aggregates of polydisperse particles using small-angle light scattering. *Journal of Colloid and Interface Science*, *221*(2), 186–194. doi:10.1006/jcis.1999.6532 PMID:10631019

Caceci, M. S., & Billon, A. (1990). Evidence for large organic scatterers (50-200 nm diameter) in humic acid samples. *Organic Geochemistry*, *15*(3), 335–350. doi:10.1016/0146-6380(90)90011-N

Carr, R., Hole, P., Malloy, A., Nelson, P., Wright, M., & Smith, J. (2009). Applications of nanoparticle tracking analysis in nanoparticle research-a mini-review. *European Journal of Parenteral & Pharmaceutical Sciences*, *14*(2), 45–50.

Cascio, C., Geiss, O., Franchini, F., Ojea-Jimenez, I., Rossi, F., Gilliland, D., & Calzolai, L. (2015). Detection, quantification and derivation of number size distribution of silver nanoparticles in antimicrobial consumer products. *Journal of Analytical Atomic Spectrometry*, *30*(6), 1255–1265. doi:10.1039/C4JA00410H

Chenal, M., Rieger, J., Philippe, A., & Bouteiller, L. (2014). High Yield Preparation of All-Organic Raspberry-like Particles by Heterocoagulation via Hydrogen Bonding Interaction. *Polymer*, *55*(16), 3516–3524. doi:10.1016/j.polymer.2014.05.057

Chun, M. S., Park, O. O., & Kim, J. K. (1990). Flow and dynamic behavior of dilute polymer solutions in hydrodynamic chromatography. *Korean Journal of Chemical Engineering*, *7*(2), 126–137. doi:10.1007/BF02705057

DiMarzio, E., & Guttman, C. (1970). Separation by flow. *Macromolecules*, *3*(2), 131–146. doi:10.1021/ma60014a005

Dubascoux, S., Le Hecho, I., Hasellöv, M., Von Der Kammer, F., Potin Gautier, M., & Lespes, G. (2010). Field-flow fractionation and inductively coupled plasma mass spectrometer coupling: History, development and applications. *Journal of Analytical Atomic Spectrometry*, *25*(5), 613–623. doi:10.1039/b927500b

Fabricius, A.-L., Duester, L., Meermann, B., & Ternes, T. A. (2013). ICP-MS-based characterization of inorganic nanoparticles—sample preparation and off-line fractionation strategies. *Analytical and Bioanalytical Chemistry*, 1–13. PMID:24292431

Fedotov, P. S., Vanifatova, N. G., Shkinev, V. M., & Spivakov, B. Y. (2011). Fractionation and characterization of nano-and microparticles in liquid media. *Analytical and Bioanalytical Chemistry*, *400*(6), 1787–1804. doi:10.1007/s00216-011-4704-1 PMID:21318253

Filipe, V., Hawe, A., & Jiskoot, W. (2010). Critical evaluation of Nanoparticle Tracking Analysis (NTA) by NanoSight for the measurement of nanoparticles and protein aggregates. *Pharmaceutical Research*, *27*(5), 796–810. doi:10.1007/s11095-010-0073-2 PMID:20204471

Finsy, R. (1994). Particle sizing by quasi-elastic light scattering. *Advances in Colloid and Interface Science*, *52*, 79–143. doi:10.1016/0001-8686(94)80041-3

Giddings, J. C. (1993). Field-flow fractionation: analysis of macromolecular, colloidal, and particulate materials. *Science, 260*(5113), 1456–1465.

Gilbert, B., Lu, G., & Kim, C. S. (2007). Stable cluster formation in aqueous suspensions of iron oxyhydroxide nanoparticles. *Journal of Colloid and Interface Science*, *313*(1), 152–159. doi:10.1016/j.jcis.2007.04.038 PMID:17511999

Gray, E. P., Bruton, T. A., Higgins, C. P., Halden, R. U., Westerhoff, P., & Ranville, J. F. (2012). Analysis of gold nanoparticle mixtures: A comparison of hydrodynamic chromatography (HDC) and asymmetrical flow field-flow fractionation (AF4) coupled to ICP-MS. *Journal of Analytical Atomic Spectrometry*, *27*(9), 1532–1539. doi:10.1039/c2ja30069a

Hassellöv, M., Readman, J. W., Ranville, J. F., & Tiede, K. (2008). Nanoparticle analysis and characterization methodologies in environmental risk assessment of engineered nanoparticles. *Ecotoxicology*, *17*(5), 344–361.

Hotze, E. M., Phenrat, T., & Lowry, G. V. (2010). Nanoparticle aggregation: Challenges to understanding transport and reactivity in the environment. *Journal of Environmental Quality*, *39*(6), 1909–1924. doi:10.2134/jeq2009.0462 PMID:21284288

Hunter, R. J. (2001). *Foundations of Colloid Science* (2nd ed.). Oxford University Press.

Ito, T., Sun, L., & Crooks, R. M. (2003). Simultaneous determination of the size and surface charge of individual nanoparticles using a carbon nanotube-based Coulter counter. *Analytical Chemistry*, *75*(10), 2399–2406. doi:10.1021/ac034072v PMID:12918983

Jimenez, M., Gomez, M., Bolea, E., Laborda, F., & Castillo, J. (2011). An approach to the natural and engineered nanoparticles analysis in the environment by inductively coupled plasma mass spectrometry. *International Journal of Mass Spectrometry*, *307*(1), 99–104. doi:10.1016/j.ijms.2011.03.015

Karn, B., Kuiken, T., & Otto, M. (2009). Nanotechnology and in situ remediation: A review of the benefits and potential risks. *Environmental Health Perspectives*, *117*(12), 1813. doi:10.1289/ehp.0900793 PMID:20049198

King, S. M., & Jarvie, H. P. (2012). Exploring how organic matter controls structural transformations in natural aquatic nanocolloidal dispersions. *Environmental Science & Technology*, *46*(13), 6959–6967. doi:10.1021/es2034087 PMID:22260303

Langhorst, M. A., Stanley, F. W. Jr, Cutie, S. S., Sugarman, J. H., Wilson, L. R., Hoagland, D. A., & Prud'homme, R. K. (1986). Determination of nonionic and partially hydrolyzed polyacrylamide molecular weight distributions using hydrodynamic chromatography. *Analytical Chemistry*, *58*(11), 2242–2247. doi:10.1021/ac00124a027

Lead, J., Muirhead, D., & Gibson, C. (2005). Characterization of freshwater natural aquatic colloids by atomic force microscopy (AFM). *Environmental Science & Technology*, *39*(18), 6930–6936. doi:10.1021/es050386j PMID:16201613

Lead, J. R., & Wilkinson, K. J. (2006). Aquatic colloids and nanoparticles: Current knowledge and future trends. *Environmental Chemistry*, *3*(3), 159–171. doi:10.1071/EN06025

Lee, S., Bi, X., Reed, R. B., Ranville, J. F., Herckes, P., & Westerhoff, P. (2014). Nanoparticle size detection limits by single particle ICP-MS for 40 elements. *Environmental Science & Technology*, *48*(17), 10291–10300. doi:10.1021/es502422v PMID:25122540

Lespes, G., & Gigault, J. (2011). Hyphenated analytical techniques for multidimensional characterisation of submicron particles: A review. *Analytica Chimica Acta*, *692*(1), 26–41. doi:10.1016/j.aca.2011.02.052 PMID:21501709

Lin, M., Lindsay, H., Weitz, D., Klein, R., Ball, R., & Meakin, P. (1990). Universal diffusion-limited colloid aggregation. *Journal of Physics Condensed Matter*, *2*(13), 3093–3113. doi:10.1088/0953-8984/2/13/019

Liu, H. H., & Cohen, Y. (2014). Multimedia environmental distribution of engineered nanomaterials. *Environmental Science & Technology*, *48*(6), 3281–3292. doi:10.1021/es405132z PMID:24548277

Liu, L., Veerappan, V., Pu, Q., Cheng, C., Wang, X., Lu, L., & Guo, G. et al. (2013). High-Resolution Hydrodynamic Chromatographic Separation of Large DNA Using Narrow, Bare Open Capillaries: A Rapid and Economical Alternative Technology to Pulsed-Field Gel Electrophoresis? *Analytical Chemistry*, *86*(1), 729–736. doi:10.1021/ac403190a PMID:24274685

López-Serrano, A., Olivas, R. M., Landaluze, J. S., & Cámara, C. (2014). Nanoparticles: A global vision. Characterization, separation, and quantification methods. Potential environmental and health impact. *Analytical Methods*, *6*(1), 38–56. doi:10.1039/C3AY40517F

Maurice, P., & Namjesnik-Dejanovic, K. (1999). Aggregate structures of sorbed humic substances observed in aqueous solution. *Environmental Science & Technology*, *33*(9), 1538–1541. doi:10.1021/es981113+

McHugh, A. J., & Brenner, H. (1984). Particle size measurement using chromatography. *Critical Reviews in Analytical Chemistry*, *15*(1), 63–117. doi:10.1080/10408348408085429

Metreveli, G., Philippe, A., & Schaumann, G. E. (2014). Disaggregation of silver nanoparticle homoaggregates in a river water matrix. *The Science of the Total Environment*. PMID:25433382

Mitrano, D. M., Lesher, E. K., Bednar, A., Monserud, J., Higgins, C. P., & Ranville, J. F. (2011). Detecting nanoparticulate silver using single-particle inductively coupled plasma-mass spectrometry. *Environmental Toxicology and Chemistry*, *31*(1), 115–121. doi:10.1002/etc.719 PMID:22012920

Montaño, M. D., Badiei, H. R., Bazargan, S. & Ranville, J. (2014). Improvements in the detection and characterization of engineered nanoparticles using spICP-MS with microsecond dwell times. *Environmental Science: Nano*.

Nischwitz, V., & Goenaga-Infante, H. (2012). Improved sample preparation and quality control for the characterisation of titanium dioxide nanoparticles in sunscreens using flow field flow fractionation online with inductively coupled plasma mass spectrometry. *Journal of Analytical Atomic Spectrometry*, *27*(7), 1084–1092. doi:10.1039/c2ja10387g

Nowack, B., Ranville, J. F., Diamond, S., Gallego-Urrea, J. A., Metcalfe, C., Rose, J., & Klaine, S. J. et al. (2012). Potential scenarios for nanomaterial release and subsequent alteration in the environment. *Environmental Toxicology and Chemistry*, *31*(1), 50–59. doi:10.1002/etc.726 PMID:22038832

Pace, H. E., Rogers, N. J., Jarolimek, C., Coleman, V. A., Gray, E. P., Higgins, C. P., & Ranville, J. F. (2012). Single Particle Inductively Coupled Plasma-Mass Spectrometry: A Performance Evaluation and Method Comparison in the Determination of Nanoparticle Size. *Environmental Science & Technology*, *46*(22), 12272–12280. doi:10.1021/es301787d PMID:22780106

Paul, S., Jamie, R., Harrison, R. M., Jones, I. P., & Stoll, S. (2005). Characterization of humic substances by environmental scanning electron microscopy. *Environmental Science & Technology*, *39*(7), 1962–1966. doi:10.1021/es0489543 PMID:15871224

Penlidis, A., Hamielec, A., & MacGregor, J. (1983). Hydrodynamic and Size Exclusion Chromatography of Particle Suspensions-An Update. *Journal of Liquid Chromatography*, *6*(sup002S2), 179–217. doi:10.1080/01483918308062874

Pergantis, S. A., Jones-Lepp, T. L., & Heithmar, E. M. (2012). Hydrodynamic chromatography online with single particle-inductively coupled plasma mass spectrometry for ultratrace detection of metal-containing nanoparticles. *Analytical Chemistry*, *84*(15), 6454–6462. doi:10.1021/ac300302j PMID:22804728

Perret, D., Leppard, G. G., Müller, M., Belzile, N., De Vitre, R., & Buffle, J. (1991). Electron microscopy of aquatic colloids: Non-perturbing preparation of specimens in the field. *Water Research*, *25*(11), 1333–1343. doi:10.1016/0043-1354(91)90111-3

Philippe, A. (2015). *Hydrodynamic Chromatography for Studying Interactions between Colloids and Dissolved Organic Matter in the Environment*. Academic Press.

Philippe, A., Gangloff, M., Rakcheev, D., & Schaumann, G. (2014). Evaluation of hydrodynamic chromatography coupled with inductively coupled plasma mass spectrometry detector for analysis of colloids in environmental media-effects of colloid composition, coating and shape. *Analytical Methods*, *6*(21), 8722–8728. doi:10.1039/C4AY01567C

Philippe, A., & Schaumann, G. E. (2014a). Evaluation of Hydrodynamic Chromatography Coupled with UV-Visible, Fluorescence and Inductively Coupled Plasma Mass Spectrometry Detectors for Sizing and Quantifying Colloids in Environmental Media. *PLoS ONE*, *9*(2), e90559. doi:10.1371/journal.pone.0090559 PMID:24587393

Philippe, A., & Schaumann, G. E. (2014b). Interactions of dissolved organic matter with natural and engineered inorganic colloids: A review. *Environmental Science & Technology*, *48*(16), 8946–8962. doi:10.1021/es502342r PMID:25082801

Plaschke, M., Römer, J., & Kim, J. I. (2002). Characterization of Gorleben groundwater colloids by atomic force microscopy. *Environmental Science & Technology*, *36*(21), 4483–4488. doi:10.1021/es0255148 PMID:12433155

Proulx, K., & Wilkinson, K. (2014). Separation, detection and characterization of engineered nanoparticles in natural waters using hydrodynamic chromatography and multi-method detection (light scattering, analytical ultracentrifugation and single particle ICP-MS). *Environmental Chemistry*, *11*(4), 392. doi:10.1071/EN13232

Rakcheev, D., Philippe, A., & Schaumann, G. E. (2013). Hydrodynamic chromatography coupled with single particle-inductively coupled plasma mass spectrometry for investigating nanoparticles agglomerates. *Analytical Chemistry*, *85*(22), 10643–10647. doi:10.1021/ac4019395 PMID:24156639

Roman, M., Rigo, C., Castillo-Michel, H., Munivrana, I., Vindigni, V., Mivceti'c, I., & Cairns, W. R. et al. (2015). Hydrodynamic chromatography coupled to single-particle ICP-MS for the simultaneous characterization of AgNPs and determination of dissolved Ag in plasma and blood of burn patients. *Analytical and Bioanalytical Chemistry*, 1–16. PMID:26396079

Schuch, H., & Wohleben, W. (2010). Measurement of Particle Size Distribution of Polymer Latexes (L. M. Gugliotta & J. R. Veda, Eds.). Academic Press.

Simonet, B. M., & Valcárcel, M. (2009). Monitoring nanoparticles in the environment. *Analytical and Bioanalytical Chemistry*, *393*(1), 17–21. doi:10.1007/s00216-008-2484-z PMID:18974979

Singh, G., Stephan, C., Westerhoff, P., Carlander, D., & Duncan, T. V. (2014). Measurement methods to detect, characterize, and quantify engineered nanomaterials in foods. *Comprehensive Reviews in Food Science and Food Safety*, *13*(4), 693–704. doi:10.1111/1541-4337.12078

Small, H. (1974). Hydrodynamic chromatography a technique for size analysis of colloidal particles. *Journal of Colloid and Interface Science*, *48*(1), 147–161. doi:10.1016/0021-9797(74)90337-3

Stoisits, R. F., Poehlein, G. W., & Vanderhoff, J. W. (1976). Mathematical modeling of hydrodynamic chromatography. *Journal of Colloid and Interface Science*, *57*(2), 337–344. doi:10.1016/0021-9797(76)90208-3

Striegel, A. M. (2012). Hydrodynamic chromatography: Packed columns, multiple detectors, and microcapillaries. *Analytical and Bioanalytical Chemistry*, *402*(1), 1–5. doi:10.1007/s00216-011-5334-3 PMID:21901463

Striegel, A. M., & Brewer, A. K. (2012). Hydrodynamic Chromatography. *Annual Review of Analytical Chemistry*, *5*(1), 15–34. doi:10.1146/annurev-anchem-062011-143107 PMID:22708902

Tiede, K., Boxall, A., Tear, S., Lewis, J., David, H., & Hassellov, M. (2008). Detection and characterization of engineered nanoparticles in food and the environment-a review. *Food Additives and Contaminants*, *25*(07), 795–821. doi:10.1080/02652030802007553 PMID:18569000

Tiede, K., Boxall, A. B., Tiede, D., Tear, S. P., David, H., & Lewis, J. (2009). A robust size-characterisation methodology for studying nanoparticle behaviour in "real"environmental samples, using hydrodynamic chromatography coupled to ICP-MS. *Journal of Analytical Atomic Spectrometry*, *24*(7), 964–972. doi:10.1039/b822409a

Tiede, K., Boxall, A. B., Wang, X., Gore, D., Tiede, D., Baxter, M., & Lewis, J. et al. (2010). Application of hydrodynamic chromatography-ICP-MS to investigate the fate of silver nanoparticles in activated sludge. *Journal of Analytical Atomic Spectrometry*, *25*(7), 1149–1154. doi:10.1039/b926029c

Tijssen, R., Bos, J., & Van Kreveld, M. E. (1986). Hydrodynamic chromatography of macromolecules in open microcapillary tubes. *Analytical Chemistry*, *58*(14), 3036–3044. doi:10.1021/ac00127a030

Tipping, E. (2002). *Cation binding by humic substances*. Cambridge University Press. doi:10.1017/CBO9780511535598

Tuoriniemi, J., Cornelis, G., & Hassellov, M. (2012). Size discrimination and detection capabilities of single-particle ICPMS for environmental analysis of silver nanoparticles. *Analytical Chemistry*, *84*(9), 3965–3972. doi:10.1021/ac203005r PMID:22483433

Von der Kammer, F., Baborowski, M., & Friese, K. (2005). Field-flow fractionation coupled to multi-angle laser light scattering detectors: Applicability and analytical benefits for the analysis of environmental colloids. *Analytica Chimica Acta*, *552*(1), 166–174. doi:10.1016/j.aca.2005.07.049

Von der Kammer, F., Ferguson, P. L., Holden, P. A., Masion, A., Rogers, K. R., Klaine, S. J., & Unrine, J. M. et al. (2012). Analysis of engineered nanomaterials in complex matrices (environment and biota): General considerations and conceptual case studies. *Environmental Toxicology and Chemistry*, *31*(1), 32–49. doi:10.1002/etc.723 PMID:22021021

Watson, J. T., & Sparkman, O. D. (2007). *Introduction to mass spectrometry: instrumentation, applications, and strategies for data interpretation*. John Wiley & Sons. doi:10.1002/9780470516898

Wigginton, N. S., Haus, K. L., & Hochella, M. F. Jr. (2007). Aquatic environmental nanoparticles. *Journal of Environmental Monitoring*, *9*(12), 1306–1316. doi:10.1039/b712709j PMID:18049768

Wilkinson, K. J., Balnois, E., Leppard, G. G., & Buffle, J. (1999). Characteristic features of the major components of freshwater colloidal organic matter revealed by transmission electron and atomic force microscopy. *Colloids and Surfaces. A, Physicochemical and Engineering Aspects*, *155*(2-3), 287–310. doi:10.1016/S0927-7757(98)00874-7

Wohlleben, W. (2012). Validity range of centrifuges for the regulation of nanomaterials: From classification to as-tested coronas. *Journal of Nanoparticle Research*, *14*(12), 1–18. doi:10.1007/s11051-012-1300-z PMID:23239934

Zhe, J., Jagtiani, A., Dutta, P., Hu, J., & Carletta, J. (2007). A micromachined high throughput Coulter counter for bioparticle detection and counting. *Journal of Micromechanics and Microengineering*, *17*(2), 304–313. doi:10.1088/0960-1317/17/2/017

KEY TERMS AND DEFINITIONS

Core Diameter: Diameter of the solid core of a particle. For multilayer particles, the definition can be operationally based on the detection system. In such cases, the core corresponds to the extension of the most external layer that can be detected.

Effective Diameter: Geometric diameter of the particle (largest distance between two points belonging to the particle).

Fractal Agglomerates: Assembly of several particles attached to each other and having a fractal geometry at a given scale (e.g. agglomerate average size), at least. The fractal dimension characterizes the geometry of fractal object.

Gyration Diameter: The root mean square distance from the center of gravity of the points belonging to the particle.

Hydrodynamic Chromatography: Chromatographic process where the separation process is based solely on the geometric size (effective diameter) of the particles. Larger particles elute faster than smaller particles.

Hydrodynamic Diameter: Diameter calculated from the diffusion coefficient using the Stokes-Einstein equation.

ICP-MS: Inductively coupled plasma mass spectrometry, the analyte (as a liquid or a gas) is introduced into a plasma chamber were it is fully atomized and ionized. The produced ions are then introduced into a mass spectrometer (e.g. quadrupole, sector field, TOF) for being selectively detected.

SP-ICP-MS: Special mode of ICP-MS in which particles are introduced one by one into the detector.

Chapter 2
Nanotechnology for Filtration-Based Point-of-Use Water Treatment:
A Review of Current Understanding

Kathryn Gwenyth Nunnelley
University of Virginia, USA

James A Smith
University of Virginia, USA

ABSTRACT

With significant infrastructure investments required for centralized water treatment, in home treatment technologies, known as point-of-use, have become a popular solution in the developing world. This review discusses current filtration-based point-of-use water treatment technologies in three major categories: ceramics, papers and textiles. Each of these categories has used silver for added antimicrobial effectiveness. Ceramics have had the most development and market infiltration, while filter papers are a new development. Textiles show promise for future research as a cheap, socially acceptable, and effective method. Also, a new method of silver incorporation in ceramics is explored.

INTRODUCTION

Approximately 1 billion people worldwide do not have access an "improved water supply", which is defined by the World Health Organization (WHO) as "a household connection or access to a public standpipe, a protected well or spring, or a source of rainwater collection"(Cosgrove & Rijsberman, 2000). This definition also requires that at least 20 liters per person per day are available within 1 km of a person's home. Because this definition does not refer to water quality, reliability of service, or even cost, it is estimated that an additional 2-3 billion people also have unsafe water supplies, at least for part of the time (Hutton & Haller, 2004). Studies have also shown that numbers reported by the Joint Monitoring

DOI: 10.4018/978-1-5225-0585-3.ch002

Programme of the proportion of drinking water coming from improved sources are likely substantial overestimations when water quality data is included (Bain et al., 2012). Access as well as quality is key.

The dangers of consuming unsafe water are substantial, particularly for children. The WHO estimates that consuming unsafe water causes the deaths of over 4 million people per year, with more than 1.5 million of these deaths being children under the age of 5. This is about equal to having 10 jumbo jets crashing every day with 90% of the passengers being children (Dillingham & Guerrant, 2004). Not included in this figure are the added health burdens on children who experience cognitive impairment and growth stunting as a result of gastrointestinal infections caused by consumption of water with pathogenic organisms like Shigella, pathogenic strains of *Escherichia coli*, *Vibrio cholerae*, and *Cryptosporidium parvum (Dillingham & Guerrant, 2004)*. People living with AIDS are particularly susceptible to infections from waterborne pathogens because of their weakened immune systems (Dillingham et al., 2006). In many parts of sub-Saharan Africa, there is an unfortunate confluence of low-quality drinking water and high rates of HIV infection. Providing safe water to our global population has been identified as one of the Grand Challenges of Engineering by the National Academy of Engineering.

In most cities and suburban areas of the developed world, water is treated at centralized water-treatment plants and delivered directly to homes through distribution systems with chlorine residual. About 3-4 billion people receive this high level of service, which generally prevents gastrointestinal infections by waterborne pathogens. This service is usually not available in the developing world or even in some suburban and rural areas of the developed world. These regions cannot economically sustain centralized water treatment plants that meet the full demands of the population. A centralized water treatment facility is a very large investment not feasible for many communities. The treatment plants plus the distribution system is not practical in developing rural areas due to the financial and structural requirements for safe treatment and delivery of water. The WHO has suggested that one possible solution is to take a more decentralized approach to water treatment, wherein people treat their water in their households immediately before consumption. This idea, often referred to as point-of-use (POU) water treatment, has the potential to significantly improve the microbial quality of household water and reduce the risk of diarrheal disease and death, particularly among children (Thomas Clasen, Nadakatti, & Menon, 2006). Table 1 shows a summary of some point-of-use systems currently in use and their effectiveness against *Escherichia coli* (*E. coli*).

Ideally, local markets can ideally drive the proliferation of these technologies without reliance on government interventions and/or subsidies. Peter-Varbanets, et al. divides current point-of-use treatments being used into three general categories: heat and UV-based systems (boiling with fuel, solar radiation, SODIS, combined action of heat and solar UV, and UV lamps), chemical treatment methods (coagulation, flocculation and precipitation, adsorption, ion exchange and chemical disinfection), and physical removal processes (sedimentation or settling, filtration, including membranes, ceramic and fiber filters, and granular media filters, including sand filters and aeration).

Rural areas of developing countries have been unable to implement centralized drinking water treatment because it is expensive. This leads to using untreated natural water sources (rivers, lakes, groundwater or rain) (Peter-Varbanets, Zurbrugg, Swartz, & Pronk, 2009). Without centralized treatment and direct piping into the home, there are many steps before consumption including collection from the source, storage, contamination, and treatment (J. E. Mellor, Smith, Learmonth, Netshandama, & Dillingham, 2012). There are three types of conventional methods for treatment including physical, chemical, and biological (Praveena & Aris, 2015). Common practices for point of use water treatment not utilizing filtration include boiling, chlorination, or solar disinfection. Membrane systems for use in the home

Table 1. A summary of E. coli removal efficiency in some POU technologies that use silver nanoparticles

Type of Material	Removal Efficiency (LRV)	Removal Efficiency (%)	Silver Concentration (mg/L)	References
Ceramic				
Ceramic materials (clay from Indonesia)	3		10	(Rayner et al., 2013)
Ceramic materials (clay from Tanzania)	2.8		10	(Praveena & Aris, 2015)
Ceramic materials (clay from Nicaragua)	2.5		11	(Praveena & Aris, 2015)
Ceramic materials (clay mixed with sawdust)		92	.02	(Kallman, Vinka A. Oyanedel-Craver, & and James A. Smith, 2011)
Ceramic filter (40%soil, 10%flour, 50%grog)		97.8-100	<.1	(V. A. Oyanedel-Craver & J. A. Smith, 2008)
Fiber				
Synthetic fiber		100	10	(S. Chen, Liu, & Zeng, 2005)
Natural fiber pure cellulose paper	8.7		.1	(T. A. Dankovich & Gray, 2011)

(Praveena & Aris, 2015)

have become increasingly popular, but still remain too expensive for the poorest of countries. Microbial pollution is the most threatening issue with water supplies, and fecal contamination of a water supply leads to water-borne diseases. Boiling of water is the most common form of treatment in developing countries, but is not as effective as hoped. A study in semi-urban India found 40.4% of stored drinking water of those who reported boiling their water were contaminated with fecal coliform (Thomas Clasen et al., 2008). In addition, 25.1% had more than 100 fecal coliforms (FC) per 100 mL, while the WHO standard is 0 FC/100mL) (Thomas Clasen et al., 2008). Boiling also creates a high-energy demand and poses the additional problem of deforestation when wood fires are used to boil the water. Assuming that households boiled 6 liters per day, the estimated monthly fuel cost for boiling water in semi-urban India was US $1.50 for households using liquid petroleum gas and US $3.21 for those using wood (Thomas Clasen et al., 2008). The cost is higher for those using wood as fuel because of the opportunity cost of lost time earning money that is instead spent gathering wood. Places like Bangladesh also lack access to the fuel necessary to boil water (A. Huq et al., 1996). Chlorine has been tested as an in-home option and has proven effective, but unpopular; household users complain about the taste and have reported a refusal to drink the water since the smell is similar to that of household bleach (Kirchhoff & McClelland, 1985). Solar disinfection has been proven as effective but requires sunlight and a large supply of plastic bottles while only treating a limited amount of water at one time (D. S. Lantagne, Quick, & Mintz, 2006). When there is limited water quantity, common solutions include groundwater wells and rainwater harvesting. However, groundwater wells are not always well planned and can become polluted, while temporal and spatial variations in rainfall can make rainwater harvesting unreliable in the absence of large-volume storage (Peter-Varbanets et al., 2009).

BACKGROUND

High surface area to volume ratio leads to better bactericidal activity, making nanomaterials excellent adsorbents (Praveena & Aris, 2015). This chapter will focus on one particular nanomaterial, silver. Silver has been used for disinfection for millennia and in wound treatments since the 1960s (LeOuay & Stellacci, 2015). Silver has bactericidal activity in its oxidized state, Ag^+ (Karel, Koparal, & Kaynak, 2015). Silver nanoparticles are between 1 and 100 nm in size (T. A. Dankovich, 2012).

Growth studies of microbial cultures have been performed with both silver and copper nanoparticles. One study showed that silver and copper nanoparticles have great promise as antimicrobial agents against *E. coli, B. subtilis,* and *S. aureus (Ruparelia, Chatterjee, Duttagupta, & Mukherji, 2008). E. coli* is a common choice for testing the effectiveness of a technology since it is a gram-negative, rod-shaped bacterium, 1-2 µm long and 0.1-0.5 µm in diameter, and a common water microbe with strains that cause gastrointestinal infections (Brown, Chai, Wang, & Sobsey, 2012). Silver nanoparticles perform better as an antimicrobial agent against *E. coli* and *S. aureus* than copper. Copper nanoparticles form an oxide layer and demonstrate better antimicrobial activity towards *B. subtilis* than silver, possibly due to a greater affinity to the surface-active groups. Ruparelia et al. admits that the antimicrobial mechanism of the silver and copper nanoparticles is not yet fully established, and a combination of silver and copper nanoparticles may produce a more complete bactericidal effect against mixed bacterial populations that would be more common in actual use (Ruparelia et al., 2008).

Silver is usually in concentrations of approximately 0.1 mg/kg in aquatic systems at subnanomolar levels. Higher silver concentrations in aquatic environments often come from waste streams of commercial or industrial processes. Silver in nature can be monovalent (Ag(I)) or metallic (Ag(0)). Studies have shown that free or hydrated Ag(I) exhibits toxic effects on a variety of aquatic organisms ranging from zooplankton to rainbow trout (Herrin, Andren, & Armstrong, 2001). Several studies have demonstrated that silver ions are selectively toxic for prokaryotic microorganisms, with little effect on eukaryotic cells (Park & Jang, 2003). However, the exact reasons for the antimicrobial ability of the nanoparticles are still unknown. Among all antimicrobial nanomaterials, nanosilver is probably the most widely used (Li et al., 2008). Silver's antimicrobial property is related to the amount and the rate at which it is released. Metallic state silver is inert, but silver will react with moisture in the skin or fluid in a wound to become ionized. Ionized silver is highly reactive, binding to bacterial DNA and RNA by denaturing. The nanoparticles bind to the cell membrane, a vital point of protection, and accumulate to causing irreparable damage (Pradeep & Anshup, 2009). This inhibits a bacterial cell from replication. In binding to tissue proteins, the silver brings structural changes to the cell wall and nuclear membranes causing the cell to distort and die (Rai, Yadav, & Gade, 2009). It has also been found that if the antimicrobial and the bacterial cell (negative charge) have the same surface charge, there is repulsion and less contact (Pradeep & Anshup, 2009). Oxygen presence in a system with bulk silver provides complete destructive oxidation of microorganisms (Pradeep & Anshup, 2009).

Health concerns are the chief barrier to the implementation of new technologies that use silver nanoparticles. Silver release from a treatment material is affected by the water's pH, temperature and chemical composition. Just 60 milligrams can be toxic to humans, while 1.3 to 6.2 grams is lethal (D. S. Lantagne, 2001). The WHO sets a limit to ensure that these technologies do not cause harm, stating that up to .1 mg/L can be tolerated without risk. The United States Environmental Protection Agency sets a maximum silver concentration for drinking water at .1 mg/L (Quang et al., 2013). The retention rate of silver in humans is only 0-10 percent, and silver is mainly stored in the liver and skin. The WHO set a guideline of 10 grams of silver in a lifetime (D. S. Lantagne, 2001). Potential human risks of silver consumption at high levels include high blood pressure, kidney damage, gastrointestinal irritation, cancer and neurological damage, in addition to DNA damage (Praveena & Aris, 2015). The silver levels after treatment must be within a safe range for human consumption. Risks to environment and human health are a rising concern. The technologies discussed in the chapter release silver at levels below the drinking water standard, meaning they are safe and may cause a substantial improvement in water quality and human health, particularly in resource-limited settings. Some of the ceramic technologies are made with silver nitrate instead of silver nanoparticles and thus release silver ions instead of nanoparticles.. There is a strong understanding of the health effects of silver in ion form, but the effects of silver in the nano form are less understood.

CURRENT POINT-OF-USE WATER TREATMENT TECHNOLOGIES

Herein, we discuss current point-of-use water treatment practices that use silver nanoparticles in porous media. The three main categories of treatment type that will be addressed are ceramic, paper and textile. These technologies us e filtration, which benefit from removal of turbidity and are gravity fed, which saves on energy costs (Albert, Luoto, & Levine, 2010). In addition, we present new data on a novel "silver-nanopatch" fabrication technique as applied to ceramic filters.

Ceramic

Silver has been imbedded in clay and can be designed to control silver release kinetics (Karel et al., 2015). Silver imbedded clay can also be used with textiles by soaking a textile then oven drying (Kasuga et al., 2011). Although this method has not been used for water, it has been used for air purifiers, and the clay has been used in plates to treat drinking water (Kasuga et al., 2011). Fired clay makes a ceramic that can be used for ceramic filtration which has been shown to have high effectiveness against bacteria and protozoa (D. S. Lantagne et al., 2006). Potters for Peace is a US-based nongovernmental organization (NGO) that has been teaching communities to produce and manufacture ceramic pots that filter water (D. S. Lantagne, 2001). The pots are made from water, clay and a combustible material like sawdust and shaped like a pot. The kiln is typically fired at 900 °C for 8 hours, below the melting point of silver (Praveena & Aris, 2015; Ren & Smith, 2013). During firing, the clay forms a ceramic and the sawdust combusts, leaving behind porosity for water flow. Some filter factories then perform a pressure test by submerging the filter to the rim in water for 10 seconds and to see if any wet spots appear on the interior of the filter pot. A wet spot would suggest a macropore that in turn may limit filter performance for pathogen removal (Rayner, 2009). Filters are then soaked and subjected to a flow test. If the filters pass a flow rate test they are then painted or dipped in colloidal silver solution that provides antimicrobial effects (Rayner, 2009). Water containing a colloidal silver has been filtered through as a way to embed the silver in the filters, but has been proven less efficient (D. S. Lantagne, 2001). Also, painting on silver nitrate is less efficient due to high washout, and silver nanoparticles are recommended (Rayner et al., 2013). Mixing in silver nanoparticles prior to firing has shown higher retention of silver in the ceramic and potentially better long term use when compared to the paint on or dipped application method which release silver at high levels quickly during early use (Ren & Smith, 2013). This has not been tested for bacterial disinfection nor has it been field tested, but it could be a promising alternative. The ceramic sits inside a 20-30 liter plastic bucket with a spigot. Figure 1 is a diagram of a completed ceramic water filter (Brown & Sobsey, 2010). These can be manufactured locally and pots are a common storage container for water already, so there is no break to the social norm (D. S. Lantagne et al., 2006). The filtering of water is gravity-fed and would save on energy cost. The clay-to-sawdust ratio by volume of 47% has proven best in optimizing flow rate without losing coliform removal (D. Lantagne et al., 2010). Also, the type of combustible used has been shown to make a difference even if the same size and amount is used, like sawdust versus coffee or rice husk (D. Lantagne et al., 2010).

The silver-impregnated ceramic water filters have shown to be both sustainable and socially acceptable as well as effective with physical filtration and chemical treatment both contributing to pathogen removal (Abebe et al., 2014). The ceramic filter without silver has been shown to successfully remove $E.$ coli but the addition of silver improves the water quality of the effluent (Kallman et al., 2011). Porosimetry of the filters has shown that the pores of the ceramic are too big to remove bacteria by only size exclusion-- there are several other mechanisms like depth straining, particle bridging inertial impaction, interception, diffusion and electrostatic attraction that help with capturing bacteria (Matthies et al., 2015). The filter has also been shown to be a preferred option by a study of 400 households in Kenya. Households were given the option of three different technologies, PUR (chlorine-based), Waterguard (chlorine-based) and a ceramic water filter. The households had a two-month trial with each, and then were asked their preference again at the conclusion of the trial. The ceramic water filter was preferred both before and after (Albert et al., 2010).

The method of application of silver does not appear to be a factor affecting disinfection efficiency (D. Lantagne et al., 2010; V. Oyanedel-Craver & J. A. Smith, 2008). Instead the mass of colloidal silver that remains after the saturation period determines effectiveness. The bacterial removal may be helped by the ceramic pores being filled by silver instead of the bacteria in transport. Testing has shown between 97.8 and 100% removal of $E.$ coli (V. Oyanedel-Craver & J. A. Smith, 2008). It is also important that the silver release remains below the regulatory limit for silver in drinking water, which is 0.1 mg/L. This level was not exceeded for any of the filters tested after passing water through the filter for about 200 min (V. Oyanedel-Craver & J. A. Smith, 2008). The shape of the filter, flat bottom versus round bottom, also has no effect (D. Lantagne et al., 2010).

Ceramic candles are often used in a gravity-fed, ceramic water filter system. The set up involves two 20 L plastic buckets stacked on top each other with the silver impregnated ceramic candles vertical allowing their effluent to flow to the bottom bucket and be dispensed out a spigot. Each candle can treat 20000 L depending on water quality (THOMAS Clasen, Parra, Boisson, & Collin, 2005). This technique has been associated with 70% lower diarrheal rates, and 92% of users report "feeling better" after using the filter (T. F. Clasen, Brown, Collin, Suntura, & Cairncross, 2004). Ceramic candles have shown that after 16 months, only 48.7% are still operating properly, and 54% of those produce coliform free water (T Clasen & Boisson, 2006). The main problem with this technology is breakage. The ceramic filters are popular with the end user, but with so many parts, any failure commonly leads to discontinued use because of difficulty to repair (T Clasen & Boisson, 2006). It has been found that the treatments are not always done correctly by the user (D. S. Lantagne & Clasen, 2012). This lowers the effectiveness of the technology and therefore does not provide recipients with the safe drinking water they should be receiving. Proper education on use and maintenance of the treatment system is vital (D. S. Lantagne, 2001). Proper maintenance of ceramic water filters can reduce early childhood diarrhea by an extra 45% (J. Mellor, Abebe, Ehdaie, Dillingham, & Smith, 2014). Often the new technology may be provided

Figure 1. Schematic of the ceramic water filter produced by Resource Development International—Cambodia (courtesy of Mickey Sampson). The complete filter unit consists of a lid (A), ceramic filter element (B) in a plastic container (C) that gathers treated water and dispenses through a tap at the base (D).
Brown & Sobsey, 2010.

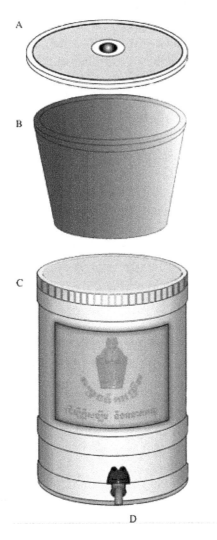

to a developing country, but with English directions. Keeping methods simple and similar to current practices could improve the utilization phase of treatment processes (D. S. Lantagne & Clasen, 2012).

The benefits of ceramic filtration include its proven effectiveness against bacteria and protozoa, its simplicity, its long-term use, and its low cost due to the ability to locally produce. One study of the ceramic filter reported a 90-99% removal of MS2, a viral pathogen (Brown & Sobsey, 2010), but conversely it has also been claimed that there is a minimal efficiency against viruses (Matthies et al., 2015). But with that, there is a risk of recontamination in the lower reservoir and need for education for optimal use (T. F. Clasen et al., 2004). Ceramic filters provide health improvements, cost-effectiveness and are environmentally friendly, making them sustainable and a better option than a centralized treatment system in the developing world (Ren, Colosi, & Smith, 2013). Unfortunately, ceramic filters have shown a 2% decrease in likelihood of use after each month of ownership with the most common reason being the filter breakage (Brown, Proum, & Sobsey, 2009), and 67% of the discontinued use is because of breakage (Sobsey, Stauber, Casanova, Brown, & Elliott, 2008).

As for human health effects, a study has shown that the use of silver-impregnated ceramic filters reduced diarrhea rates by 49% across age groups and sexes (Brown, Sobsey, & Loomis, 2008), and by 80% among HIV-positive users (Abebe et al., 2014). This is important since individuals who are immunocompromised can be particularly susceptible to gastrointestinal infections from waterborne pathogens (Abebe et al., 2014). There has also been epidemiological research on ceramic filter use and reduction of the hepatitis A virus (T. F. Clasen et al., 2004). Ceramic water filters have been found to reduce early childhood diarrhea by 41% over two years (J. Mellor et al., 2014), but the effectiveness is reduced after 3 years (J. Mellor et al., 2014).

Other ceramic technologies have been developed that do not use filtration. Silver-coated ceramic beads can be used in water, and has been tested by pouring through columns rather than a batch method (Han, Lee, Lee, Uzawa, & Park, 2005; Lin et al., 2013). Recently, a novel method of metallic silver formation in a porous ceramic was reported by Ehdaie et al. (2014). These researchers combined clay, sawdust, water, and silver nitrate, pressed the mixture into the form of a ceramic tablet, and fired the tablet in a kiln at 900 °C. During firing, the clay formed a ceramic, the sawdust combusts, and the ionic silver is reduced to metallic silver, forming metallic silver "nanopatches" with an average diameter of 2.71 nm as seen in Figure 2. The ceramic tablet can be added to a 5-15-L household water storage container

Figure 2. Transmission electron microscopy image of silver nanopatches in the center of a ceramic tablet using scanning electron microscopy mode. The scale bar represents 20 nm in length
(Ehdaie et al., 2014).

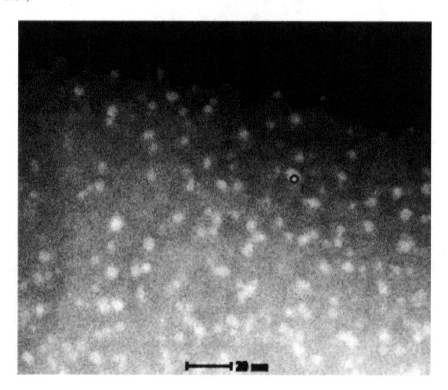

and it passively releases ionic silver for the inactivation of *E. coli* and other gram-negative waterborne pathogenic bacteria. A 3-log reduction (99.9% removal) was shown for *E. coli* in a 10 L water source after 8 hours. This technology shows promise as a successful stand-alone technology for household water treatment for very low cost, or in combination with other filter or boiling treatments, where recontamination is a common issue (Ehdaie, Krause, & Smith, 2014). Similarly, colloidal silver has been coated on a ceramic brick for suspension in large drinking water storage tanks (Nover, McKenzie, Joshi, & Fleenor, 2013). This provides continuous, low-maintenance disinfection quickly (Nover et al., 2013), but the silver levels may become too high and require attention to residence time and size of the water container (Nover et al., 2013).

In a recent investigation in 2015, we conducted a study on incorporating this new method of silver application to ceramic filters. This study was done at the PureMadi Mukondeni Pottery Cooperative in Limpopo Province, South Africa. Currently, ceramic filters are being made by mixing sawdust, water and clay. The mixture is pressed into the shape of a filter and then fired. After that, a silver nanoparticle suspension is painted on the inside and the outside of the filters. With the help of local potters, filters were made using a new method. The new method of incorporating silver is mixing clay, sawdust and water with $AgNO_3$. During the firing process the Ag^+ reduces to Ag^0 forming small nanopatches, like in Figure 1. Filters were made with 0.3 g Ag (the same silver mass used in the paint on method) and 1.0 g Ag.

Filters were tested at the University of Venda, in Thoyondou, South Africa. Controls (filters fabricated by painting on silver nanoparticles), filters with 0.3 g of Ag, and filters with 1.0 g Ag were compared over 12 days for coliform bacteria removal efficiency. For testing, 4L of challenge water from a local stream was poured into the top reservoir of each filter. This was allowed to percolate and gather in the lower reservoir for 12 hours before collection. Samples were then tested for total coliform and *E. coli* removal using membrane filtration. To calculate results, the log of the effluent concentration (C) divided by influent concentration (C_0). Results are displayed in Figure 3. The error bars represent one standard deviation above and below the mean and show there is no statistical difference between using the paint on method and the new silver nitrate method of incorporating prior to firing.

The method is less expensive and easier to manufacture as the painting step is omitted and $AgNO_3$ is much less expensive than silver nanoparticles. Potentially, even more silver in the form of silver nitrate for be added to improve the performance of the filters without increasing filter cost. On the downside, filters that do not pass quality control tests (pressure or flow rate tests) will waste the silver that has been applied to them. Filters that are manufactured conventionally will not receive silver nanoparticle application until passing quality control tests. Further research needs to be done to expand this data set, but these results show promise for an improvement to the technology.

Paper

Paper is mostly made up of cellulose, the most abundant natural polymer. It is readily available, inexpensive, renewable, and biodegradable as well as having a great ability for metal ion sorption (Tang, Meng, Lu, & Zhu, 2009) (T. A. Dankovich, 2012). Paper is a cellulosic non-woven material like some fabrics (Ngo, Li, Simon, & Garnier, 2011). Silver nanoparticles have been formed on cellulose in bandages, clothing, food packaging, and much more. This proves a potential use of paper for water treatment (T. A. Dankovich, 2012). The most common way of synthesizing silver nanoparticles on cellulosic material is *in situ* chemical reduction of metal ions, with a reducing agent commonly being sodium borohydride (T. A. Dankovich, 2012; Tang et al., 2009; Zeng et al., 2007). Dankovich (2012) has described a method

Figure 3. Plot of the average log removal of total coliform bacteria and E. coli by conventional filters manufactured by painting on 0.3 g of silver nanoparticles (Control) and filters fired after mixing with silver nitrate (0.3 and 1.0 g as Ag). During firing, the silver nitrate is reduced to produce silver nanopatches throughout the porous ceramic media.

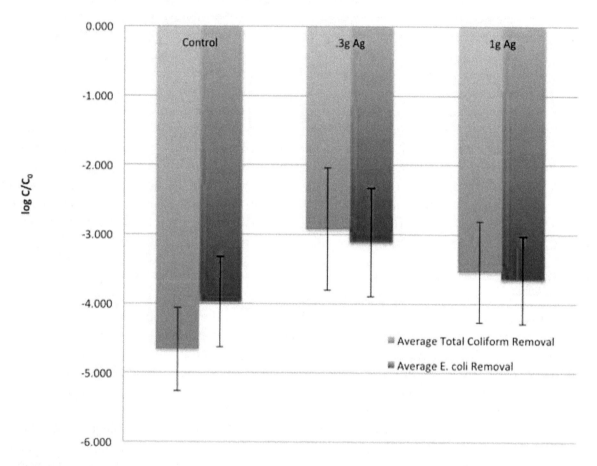

of creating silver nanoparticle (AgNP) paper by immersing blotting paper sheets (6.5 cm by 6.5 cm) in 20 mL of silver nitrate solutions at concentrations from 0 to 100 mM for 30 minutes. The sheets are then rinsed with ethanol to remove excess silver nitrate that was not absorbed into the blotter. To form nanoparticles, the paper is placed in aqueous $NaBH_4$ solutions (ranging from 1:1 to 10:1 molar ratio, $NaBH_4/AgNO_3$) for 15 min. Lastly, the paper is soaked in water for 60 min, and dried in an oven at 60° C for 2 to 3 hours (T. A. Dankovich & Gray, 2011).

A microware irradiation method for *in situ* preparation of silver nanoparticles in paper has also been developed. Using a microwave, or a conventional oven, a silver nanoparticle covered paper can be produced in a "green" way. Blotting papers are submerged in aqueous solutions of silver nitrate and glucose (T. Dankovich, Clinch, Weinronk, Dillingham, & Smith, 2014). The papers are then heated in either a microwave or a conventional oven before being soaked in water to remove excess reagents (T. Dankovich et al., 2014). Glucose functions as the reducing agent with higher concentrations leading to smaller, more uniform nanoparticles and a faster reaction time. The average size of the resulting nanoparticles can be measured with TEM images to be 5.5 nm (T. A. Dankovich, 2012). Figure 4 shows scanning

Figure 4. SEM image of AgNP paper with 2.5 mg Ag g⁻¹ paper, (a) 35 000× and (b) 60 000× magnification (T. A. Dankovich, 2014).

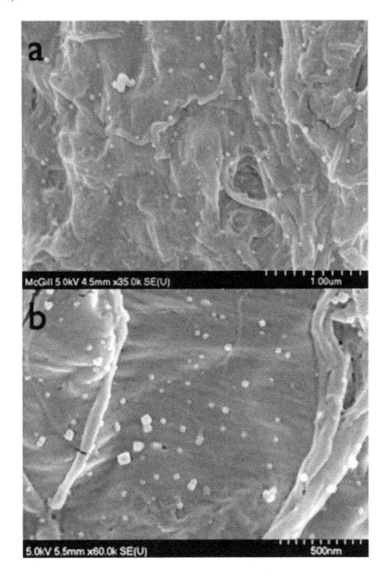

electron microscope (SEM) images of AgNP on paper (T. A. Dankovich, 2014). This way of producing the papers is more benign than the use of sodium borohydride as the reducing agent, while being just as effective (T. Dankovich et al., 2014).

Silver nanoparticles have the largest antibacterial effects with the smallest particle sizes, with average diameters under 10 nm being most effective. Results showed the AgNP paper was effective with *E. coli* bacteria suspensions poured through the sheets. After passage through the paper, effluent bacteria was isolated and used for plate counts, showing log 7.6 (±1.3) and log 3.4 (±0.9) reductions of viable *E. coli* and *E. faecalis* bacteria, respectively. Bacteria were reduced by log 0.95 (± 0.5), probably because some bacteria remained on the fiber surfaces in the blotting paper. Graphite furnace atomic absorption was used to measure the silver content in the effluent, which measured 0.0475 (±0.0177) ppm meeting the USEPA guideline for drinking water of less than 0.1 ppm (T. A. Dankovich & Gray, 2011).

Gottesman et al. (2010) describe a way of depositing silver nanoparticles onto paper by ultrasonication. The paper is held immersed in a solution of silver nitrate, ethanol, ethylene glycol and water which is then irradiated with a high intensity ultrasonic horn.. These papers have been shown to have minimal loss of silver from the coated surface, which is important for the longevity of the material. They were also shown to be effective against both gram-negative and gram-positive bacteria. This shows promise for water treatment and the avoidance of cross contamination (Gottesman et al., 2010).

Silver nanoparticles have been imbedded in paper by first immobilizing initiators on a paper's surface. Then a surface-initiated atom transfer radical polymerization technique has been developed to graft poly(tert-butyl acrylate) brushes to cellulose filters. Then the brushes are transferred into poly(acrylic acid) by hydrolyzation. Silver nitrate is then reduced to Ag^+ *in situ* by reducing agent sodium borohydride. These filters display good antibacterial ability against *E. coli* (Tang et al., 2009).

Textile

Silver nanoparticles are commonly used in clothing, especially socks, as an antimicrobial agent (Benn & Westerhoff, 2008). Antimicrobial silver is also popular for improving dressing for burn wounds. An ethanol/water medium with silver nitrate uses cation exchange between silver and sodium from sodium carboxymethyl cotton gauze (Parikh et al., 2005). The use of textiles for filtration is not a new idea. Different types of sari material have been used in Bangladesh as an inexpensive and easily available option for point-of-use water filtration (Colwell et al., 2003; A. Huq et al., 1996). Shesh, a traditional textile of North Africa, and more modern textiles like denim or shower curtains have also been tested for water filtration performance (Saidani-Scott, Tierney, & Sánchez-Silva, 2009). Textiles are very effective at removing turbidity when used for filtration (Tammisetti, 2010). The waterborne pathogen *Vibrio cholerae* often attaches to something larger like zooplankton, particularly copepods (Colwell et al., 2003; A. Huq et al., 1996). This makes for easier removal by fabric filtration. The larger particles are more likely to get filtered out via fabric filtration, while smaller organisms could pass straight through the holes in the fabric. Folding of the fabric is encouraged to increase efficiency so more levels of fabric are used for filtration (Colwell et al., 2003). The simple filtration technique of pouring water through a fabric folded at least four times has been shown to effectively remove zooplankton, most phytoplankton and particulates bigger than 20 micrometers (Colwell et al., 2003). Most importantly, follow up studies have shown that the method is accepted and sustainable by villagers in Bangladesh and also reduced the incidence of cholera by 48% (Anwar Huq et al., 2010). Gram-negative bacterium *E. coli* and Gram-positive bacterium *Staphylococcus aureus* are commonly used for antimicrobial activity with fabric testing and fungus *C. albicans* is also used frequently (Ilic et al., 2010; Ilic et al., 2009; Maneerung, Tokura, & Rujiravanit, 2007; Mecha & Pillay, 2014). Textiles show this promise because without alteration they remove larger suspensions and reduce turbidity.

Radetić (2013) offers a comprehensive overview of nanoparticle-embedded textiles. The quick push for silver nanoparticles in textiles led to a growing industry with lack of insight into the possible environmental effects. Cotton seems to be the original textile of choice, especially pre-dyed or printed. The natural fiber allows for silver application to be commonly done without any prep work. Cotton is also a common fiber that is easily available and widely used. Like paper, sodium borohydride and glucose are common reducing agents used for the formation of silver nanoparticles (Maneerung et al., 2007; Reda M. El-Shishtawy, 2011). Silk has also be used with the reduction of silver nitrate with reducing agents like hydrazine and glucose (Gulrajani, Gupta, Periyasamy, & Muthu, 2008). Wool is also a common fiber, but not as well explored for AgNP applications. It has been shown to have antimicrobial abilities, and its main benefit is said to be the bright brilliant colors it produces (Radetic, 2013). Silver nanoparticle wool composites have been made by the reduction of silver ions in solution by trisodium citrate in solution with merino wool fibers or fabrics (Kelly & Johnston, 2011). This process creates a bind between the silver metal nanoparticles and the amino acids of the keratin protein in the wool fibers with the reducing agent functioning as the linker (Kelly & Johnston, 2011). Cotton fabrics also do not need pre-treatment to be able to load silver nanoparticles efficiently, even though 3-aminopropyltriethoxysilane has been used for surface modification with improved results (Guo et al., 2013; Ilic et al., 2009; Liu, Lee, Norsten, & Chong, 2013), but, the ability of cotton textiles to retain moisture leads to an excellent environment for growing microorganisms (C.-Y. Chen & Chiang, 2008). Because of surface charges, inorganic particles and polymeric materials, like manmade textiles, demonstrate repelling between silver and the fiber at the interface; this requires a modification be made to the surface, but this commonly washes away with laundering leading to leaching (Dastjerdi, Montazer, & Shahsavan, 2009). Chelating monomers, like glycidyl methacrylate-iminodiacetic acid, can be used to graft onto cotton fibers to increase the amount of Ag^+ sorbed. The Ag^+ can then be reduced by UV radiation to form silver nanoparticles on the surface (C.-Y. Chen & Chiang, 2008). Polyester and polyamide offer two synthetic options, and polypropylene fibers (PP) are widely used in the manufacturing of medical and hygienic textile materials where a high level of antimicrobial activity is required (surgical masks, diapers, hygienic bands, etc.) (Radetic, 2013). Figure 5 show untreated PP fiber as well as a PP fiber loaded with silver nanoparticles. Activated carbon fibers have high adsorption capacities and can be impregnated with silver. Because of the high adsorption capacity and large surface area, bacteria will breed on the activated carbon filter (Yoon, Byeon, Park, & Hwang, 2008). Silver-deposited activated carbon filters have been shown to be effective for air filters, but also have a low life expectancy with greatly decreasing performance quickly when used for water filtration (LePape et al., 2002, 2004; Yoon et al., 2008). When designing a method to incorporate silver nanoparticles with textiles, it is key to control the chemistry of the solution including the pH, ionic strength, and surface complexation (Yin, Walker, Chen, & Yang, 2014). A crosslinkable polysiloxane layer can be used to permanently modify the surface of polyester with a polymeric layer on the surface that is resistant to light, heat, chemical and microbial attacks making it safer during washing (Dastjerdi et al., 2009; Dastjerdi, Montazer, & Shahsavan, 2010).

Another way of synthesizing silver nanoparticles in textiles involves using Tollen's reagent; silver nitrate is combined with NaOH to form Ag_2O powder. An ammonia solution is added and sonicated to produce an aqueous solution of Tollen's reagent, a complex ion, $[Ag(NH_3)_2]^+$. Cotton fabric is soaked in a boiling bath of Tollen's reagent solution then washed with deionized water and dried at room temperature forming silver nanoparticles (Montazer, Alimohammadi, Shamei, & Rahimi, 2012). Using electrostatic interactions, cellulose can be modified to be positively charged by using an epoxy substitution reaction on the abundant hydroxyl groups on cellulose to graft ammonium ions. This now positively charged cellulose fiber surface can be involved in adsorption of either negatively charged metal complex ions followed by reduction or with metal nanoparticles capped with negative citrate ions (Dong & Hinestroza, 2009).

In efforts to take a more "green approach", natural extracts from *Eucalyptus citriodora* and *Ficus bengalenis* are used in production of silver nanoparticles. Silver nanoparticles are produced on cotton fibers by an aqueous silver nitrate solution with fresh leaf extract. The leaf broth reduces Ag^+ to silver nanoparticles through the interactions of functional groups present in the leaf extracts (Ravindra, Mohan, Reddy, & Raju, 2010). This does not require any harsh chemicals or dangerous byproducts in the production process. *Curcuma longa* tuber powder and extract have also been used for their water-soluble

Figure 5. SEM image of untreated PP fiber and PP fiber loaded with AgNP (Radetic, 2013).

organics to reduce silver ions from aqueous silver nitrate to nano-sized silver particles on cotton cloth. The extract works better than the powder since there is more of the reducing agent to form the nanoparticles (Sathishkumar, Sneha, & Yun, 2010).

Low-temperature radio frequency plasma has been used to activate the surface of fibers for improved binding ability of colloidal silver nanoparticles onto polyester fabrics. The plasma treatment reduces the number of coatings necessary for the same level of antibacterial ability in half. These fabrics have been tested for antimicrobial ability as well as laundering ability and have proven to be an improved method. The antibacterial ability was tested against the common Gram-negative bacterium *E. coli* and Gram-positive bacterium *S. aureus* (Ilic et al., 2010).

Nanotechnology for Filtration-Based Point-of-Use Water Treatment

As with papers, the cellulose structure of fabric can be used to soak the fabric in silver nitrate solution followed by either physical (heat/UV) or chemical (sodium borohydride) reduction (Fernández et al., 2009). Microwaves are a way of delivering irradiation and heat after soaking with a reducing agent, like sodium citrate (Guo et al., 2013). Mecha and Pillay uses a modified version of the Dankovich method described in the previous section on woven fabric for microfiltration membranes (T. A. Dankovich, 2014). This is a point-of-use filter that is gravity fed and has coated membranes that are hydrophilic and have high water permeability ($p<0.05$). Filtration of turbid water (40–700 NTU) produced clear permeate (<1 NTU) and treatment of water spiked with bacteria (2500– 77,000 CFU/100 mL *E. coli*) showed the removal efficiency of coated membranes was 100% (Mecha & Pillay, 2014).

It has been reported that silver-nanoparticle-coated textile fabrics possess antibacterial activity against *S. aureus*. Silver nanocomposite fibers have been made containing silver nanoparticles incorporated inside the fabric but scanning electron microscopy has concluded that the silver nanoparticles incorporated in the sheath part of fabrics possessed significantly better antibacterial ability compared to the fabrics with silver nanoparticles in the core. The same results came from silver nanoparticles on polyester nonwovens (Rai et al., 2009). Aqueous silver nitrate has been electrospun with polymers to create a fiber that is then reduce with UV irradiation to create silver nanoparticle fibers (Lala et al., 2007). These fibers could then be woven into a textile for use in filtration. An interesting application method has also been developed for applying silver to TiO_2 thin film. An aqueous solution of 1 M silver nitrate ($AgNO_3$) is mixed with ethanol and a 0.6 mL aliquot of the solution is uniformly cast on the TiO_2 thin film. The film is then irradiated with UV light for 1 min, to deposit Ag nanoparticles in the porous film by photocatalytic reduction of Ag^+. The resulting $Ag-TiO_2$ film is thoroughly rinsed with DI water and dried by blowing air with an electric duster (Naoi et al., 2004). Schoen et al., presents an interesting textile technology that combines several mechanisms into one process. It is a "textile based multiscale device for the high speed electrical sterilization of water using silver nanowires, carbon nanotubes, and cotton" (Schoen et al., 2010). This gravity fed device can operate at 100,000 L/(h m^2) while inactivating >98% of bacteria with only seconds of total incubation time. An electrical mechanism rather than size exclusion, allows for the flow to be fast but effective with a very high surface area and large electric field with silver nanowire (Schoen et al., 2010).

CONCLUSION AND FUTURE RESEARCH DIRECTIONS

Increasing the market penetration of point-of-use (POU) technologies will require continued development of inexpensive, easy to use, socially acceptable technologies that do not change the taste of water. These design efforts must be complimented by education and training. Current filtration technologies possess many of the requisite design criteria. Each of the technologies introduced in the chapter have their own strengths and weaknesses. Ceramics filters are the most widely used technology and have the most market penetration to date, but the fragile nature of ceramics makes brakeage a concern. Fabric filters can successfully remove copepods, reducing cholera, but their effectiveness against other pathogens is unknown. Adding silver to fabrics and evaluating performance and social acceptance would be a significant advancement. Paper filters are the least developed technology but may be effective once the proper way of incorporating the technology into household point-of-use is determined. The new method of silver incorporation by the reduction of silver nitrate in papers and ceramics also shows promise as a way to avoid harsh chemicals and the potential health risks of silver nanoparticles.

Moving forward, important work should also be done to understand the anti-bacterial activity of silver against all classes of bacteria (Pradeep & Anshup, 2009). New POU technologies, in addition to being effective, need to be culturally appropriate to encourage use and motivate the purchasing of the treatment system (D. S. Lantagne et al., 2006). An optimized POU technology should be easy to use, low maintenance, independent of energy and chemicals and have a low cost (Peter-Varbanets et al., 2009). Finding a technology that satisfies all these requirements is challenging. Other nanomaterials have proven effective against other waterborne pathogens and the potential for combination of silver with another metal could improve technologies (Ruparelia et al., 2008). Long term effectiveness is important, making the antifouling ability necessary (Praveena & Aris, 2015). Expanding production of, marketing and distributing any technology will also require research and knowledge of underlying economic factors (Sobsey et al., 2008). It has also been suggested that the introduction of multiple interventions to prevent the transmission of disease is the most effective way to improve the quality of all of the interventions (J. E. Mellor et al., 2012). Solving one point of exposure will not maximize effect. Multiple interventions to improve knowledge and treatments will have the most effect.

REFERENCES

Abebe, L. S., Smith, J. A., Narkiewicz, S., Oyanedel-Craver, V., Conaway, M., Singo, A., & Dillingham, R. et al. (2014). Ceramic water filters impregnated with silver nanoparticles as a point-of-use water-treatment intervention for HIV-positive individuals in Limpopo Province, South Africa: A pilot study of technological performance and human health benefits. *Journal of Water and Health*, *12*(2), 288–300. doi:10.2166/wh.2013.185 PMID:24937223

Albert, J., Luoto, J., & Levine, D. (2010). End-User Preferences for and Performance of Competing POU Water Treatment Technologies among the Rural Poor of Kenya. *Environmental Science & Technology*, *44*(12), 4426–4432. doi:10.1021/es1000566 PMID:20446726

Bain, R. E., Gundry, S. W., Wright, J. A., Yang, H., Pedley, S., & Bartram, J. K. (2012). Accounting for water quality in monitoring access to safe drinking-water as part of the Millennium Development Goals: Lessons from five countries. *Bulletin of the World Health Organization*, *90*(3), 228–235. doi:10.2471/BLT.11.094284 PMID:22461718

Benn, T. M., & Westerhoff, P. (2008). Nanoparticle Silver Released into Water from Commercially Available Sock Fabrics. *Environmental Science & Technology*, *42*(11), 41334139. doi:10.1021/es7032718 PMID:18589977

Brown, J., Chai, R., Wang, A., & Sobsey, M. D. (2012). Microbiological Effectiveness of Mineral Pot Filters in Cambodia. *Environmental Science & Technology*, *46*(21), 12055–12061. doi:10.1021/es3027852 PMID:23030639

Brown, J., Proum, S., & Sobsey, M. D. (2009). Sustained use of a household-scale water filtration device in rural Cambodia. *Journal of Water and Health*, *7*(3), 404–412. doi:10.2166/wh.2009.085 PMID:19491492

Brown, J., & Sobsey, M. D. (2010). Microbiological effectiveness of locally produced ceramic filters for drinking water treatment in Cambodia. *Journal of Water and Health*, *8*(1), 1–11. doi:10.2166/wh.2009.007 PMID:20009242

Brown, J., Sobsey, M. D., & Loomis, D. (2008). Local drinking water filters reduce diarrheal disease in Cambodia: A randomized, controlled trial of the ceramic water purifier. *The American Journal of Tropical Medicine and Hygiene, 79*(3), 394–400. PMID:18784232

Chen, C.-Y., & Chiang, C.-L. (2008). Preparation of cotton fibers with antibacterial silver nanoparticles. *Matterials Letters,* (62), 3607-3609.

Chen, S., Liu, J., & Zeng, H. (2005). Structure and antibacterial activity of silver-supporting activated carbon fibers. *Journal of Materials Science, 40*(23), 6223–6231. doi:10.1007/s10853-005-3149-3

Clasen, T., & Boisson, S. (2006). Household-Based Ceramic Water Filters for the Treatment of Drinking Water in Disaster Response: An Assessment of a Pilot Programme in the Dominican Republic. *Water Practice & Technology, 1*(2). doi:10.2166/wpt.2006031

Clasen, T., McLaughlin, C., Nayaar, N., Boisson, S., Gupta, R., Desai, D., & Shah, N. (2008). Microbiological Effectiveness and Cost of Disinfecting Water by Boiling in Semi-urban India. *The American Society of Tropical Medicine and Hygiene, 79*(3), 407–413. PMID:18784234

Clasen, T., Nadakatti, S., & Menon, S. (2006). Microbiological performance of a water treatment unit designed for household use in developing countries. *Tropical Medicine & International Health, 11*(9), 1399–1405. doi:10.1111/j.1365-3156.2006.01699.x PMID:16930262

Clasen, T., Parra, G. G., Boisson, S., & Collin, S. (2005). Household-Based Ceramic Water Filters for the Prevention of Diarrhea: A Randomized, Controlled Trial of a Pilot Program in Columbia. *The American Society of Tropical Medicine and Hygiene, 73*(4), 790–795. PMID:16222027

Clasen, T. F., Brown, J., Collin, S., Suntura, O., & Cairncross, S. (2004). Reducing diarrhea through the use of household-based ceramic water filters: A randomized, controlled trial in rural Bolivia. *The American Journal of Tropical Medicine and Hygiene, 70*(6), 651–657. PMID:15211008

Colwell, R. R., Huq, A., Islam, M. S., Aziz, K. M. A., Yunus, M., Khan, N. H., & Russek-Cohen, E. et al. (2003). Reduction of cholera in Bangladeshi villages by simple filtration. *Proceedings of the National Academy of Sciences of the United States of America, 100*(3), 1051–1055. doi:10.1073/pnas.0237386100 PMID:12529505

Cosgrove, W. J., & Rijsberman, F. R. (2000). *World Water Vision: Making Water Everybody's Business.* London: Earthscan Publications.

Dankovich, T., Clinch, C., Weinronk, H., Dillingham, R., & Smith, J. A. (2014). (Manuscript submitted for publication). Inactivation of bacteria from contaminated streams in Limpopo, South Africa by silver- or copper-nanoparticle paper filters. *Water Research.*

Dankovich, T. A. (2012). *Bactericidal Paper Containing Silver Nanoparticles for Water Treatment.* (Doctor of Philosophy Thesis). McGill University.

Dankovich, T. A. (2014). Microwave-assisted incorporation of silver nanoparticles in paper for point-of-use water purification. *Royal Society of Chemistry, 1,* 367–378. PMID:25400935

Dankovich, T. A., & Gray, D. G. (2011). Bactericidal Paper Impregnated with Silver Nanoparticles for Point-of-Use Water Treatment. *Environmental Science & Technology*, *45*(5), 1992–1998. doi:10.1021/es103302t PMID:21314116

Dastjerdi, R., Montazer, M., & Shahsavan, S. (2009). A new method to stabilize nanoparticles on textile surfaces. *Colloids and Surfaces. A, Physicochemical and Engineering Aspects*, *345*(1-3), 202–210. doi:10.1016/j.colsurfa.2009.05.007

Dastjerdi, R., Montazer, M., & Shahsavan, S. (2010). A novel technique for producing durable multi-functional textiles using nanocomposite coating. *Colloids and Surfaces. B, Biointerfaces*, *81*(1), 32–41. doi:10.1016/j.colsurfb.2010.06.023 PMID:20675103

Dillingham, R., & Guerrant, R. L. (2004). Childhood stunting: Measuring and stemming the staggering costs of inadequate water and sanitation. *Lancet*, *363*(9403), 94–95. doi:10.1016/S0140-6736(03)15307-X PMID:14726158

Dillingham, R., Pinkerton, R., Leger, P., Severe, P., Pape, J., & Fitzgerald, D. (2006). *Mortality in Haitian patients treated with antiretroviral therapy (ART) in a community setting*. Paper presented at the 13th Conference on Retroviruses and Opportunistic Infections, Denver, CO.

Dong, H., & Hinestroza, J. P. (2009). Metal Nanoparticles on Natural Cellulose Fibers: Electrostatic Assembly and In Situ Synthesis. *Applied Materials & Interfaces*, *1*(4), 797–803. doi:10.1021/am800225j PMID:20356004

Ehdaie, B., Krause, C., & Smith, J. A. (2014). Porous Ceramic Tablet Embedded with Silver Nanopatches for Low- Cost Point-of-Use Water Purification. *Environmental Science & Technology*, *48*(23), 13901–13908. doi:10.1021/es503534c PMID:25387099

Fernández, A., Soriano, E., López-Carballo, G., Picouet, P., Lloret, E., Gavara, R., & Hernández-Muñoz, P. (2009). Preservation of aseptic conditions in absorbent pads by using silver nanotechnology. *Food Research International*, *42*(8), 1105–1112. doi:10.1016/j.foodres.2009.05.009

Gottesman, R., Shukla, S., Perkas, N., Solovyov, L. A., Nitzan, Y., & Gedanken, A. (2010). Sonochemical Coating of Paper by Microbiocidal Silver Nanoparticles. *Langmuir*, *27*(2), 720–726. doi:10.1021/la103401z PMID:21155556

Gulrajani, M. L., Gupta, D., Periyasamy, S., & Muthu, S. G. (2008). Preparation and Application of Silver Nanoparticles on Silk for Imparting Antimicrobial Properties. *Journal of Applied Polymer Science*, *108*(1), 614–623. doi:10.1002/app.27584

Guo, R., Li, Y., Lan, J., Jiang, S., Liu, T., & Yan, W. (2013). Microwave-Assisted Synthesis of Silver Nanoparticles on Cotton Fabric Modified with 3-Aminopropyltrimethoxysilane. *Journal of Applied Polymer Science*. doi:10.1002/app.39636

Han, D.-W., Lee, M. S., Lee, M. H., Uzawa, M., & Park, J.-C. (2005). The use of silver-coated ceramic beads for sterilization of Sphingomonas sp. in drinking mineral water. *World Journal of Microbiology & Biotechnology*, *21*(6-7), 921–924. doi:10.1007/s11274-004-6721-0

Herrin, R. T., Andren, A. W., & Armstrong, D. E. (2001). Determination of Silver Speciation in Natural Waters. 1. Laboratory Tests of Chelex-100 Chelating Resin as a Competing Ligand. *Environmental Science & Technology, 35*(10), 1953–1958. doi:10.1021/es001509x PMID:11393973

Huq, A., Xu, B., Chowdhury, M. A. R., Islam, M. S., Montilla, R., & Colwell, R. R. (1996). A Simple Filtration Method To Remove Plankton-Associated Vibrio cholerae in Raw Water Supplies in Developing Countries. *Applied and Environmental Microbiology, 62*(7), 2508–2512. PMID:8779590

Huq, A., Yunus, M., Sohel, S. S., Bhuiya, A., Emch, M., Luby, S. P., & Colwell, R. R. et al. (2010). Simple Sari Cloth Filtration of Water Is Sustainable and Continues To Protect Villagers from Cholera in Matlab, Bangladesh. *mBio, 1*(1), e00034-10. doi:10.1128/mBio.00034-10 PMID:20689750

Hutton, G., & Haller, L. (2004). *Evaluation of the costs and benefits of water and sanitation improvements at the global level*. Geneva: World Health Organization.

Ilic, V., Šaponjić, Z., Vodnik, V., Lazović, S., Dimitrijević, S., Jovančić, P., & Radetić, M. et al. (2010). Bactericidal Efficiency of Silver Nanoparticles Deposited onto Radio Frequency Plasma Pretreated Polyester Fabrics. *Industrial & Engineering Chemistry Research, 49*(16), 7287–7293. doi:10.1021/ie1001313

Ilic, V., Šaponjić, Z., Vodnik, V., Potkonjak, B., Jovančić, P., Nedeljković, J., & Radetić, M. (2009). The influence of silver content on antimicrobial activity and color of cotton fabrics functionalized with Ag nanoparticles. *Carbohydrate Polymers, 78*(3), 564–569. doi:10.1016/j.carbpol.2009.05.015

Kallman, E. N., Vinka A. Oyanedel-Craver, A. M. A., & and James A. Smith, M. A. (2011). Ceramic Filters Impregnated with Silver Nanoparticles for Point-of-Use Water Treatment in Rural Guatemala. *Journal of Enviromental Engineering, 137*, 407-415. doi: .000033010.1061/(ASCE)EE.1943-7870

Karel, F. B., Koparal, A. S., & Kaynak, E. (2015). Development of Silver Ion Doped Antibacterial Clays and Investigation of Their Antibacterial Activity. *Advances in Materials Science and Engineering, 2015*, 1–6. doi:10.1155/2015/409078

Kasuga, E., Kawakami, Y., Matsumoto, T., Hidaka, E., Oana, K., Ogiwara, N., & Honda, T. et al. (2011). Bactericidal activities of woven cotton and nonwoven polypropylene fabrics coated with hydroxyapatite-binding silver/titanium dioxide ceramic nanocomposite "Earth-plus". *International Journal of Nanomedicine, 6*, 1937–1943. PMID:21931489

Kelly, F. M., & Johnston, J. H. (2011). Colored and Functional Silver Nanoparticle Wool Fiber Composites. *Applied Materials & Interfaces, 3*(4), 1083–1092. doi:10.1021/am101224v PMID:21381777

Kirchhoff, L. V., McClelland, K. E., Pinho, M. D. C., Araujo, J. G., De Sousa, M. A., & Guerrant, R. L. (1985). Feasibility and efficacy of in-home water chlorination in rural North-eastern Brazil. *The Journal of Hygiene, 94*(2), 173–180. doi:10.1017/S0022172400061374 PMID:2985691

Lala, N. L., Ramaseshan, R., Bojun, L., Sundarrajan, S., Barhate, R. S., Ying-jun, L., & Ramakrishna, S. (2007). Fabrication of Nanofibers With Antimicrobial Functionality Used as Filters: Protection Against Bacterial Contaminants. *Biotechnology and Bioengineering, 97*(6), 1357–1365. doi:10.1002/bit.21351 PMID:17274060

Lantagne, D., Klarman, M., Mayer, A., Preston, K., Napotnik, J., & Jellison, K. (2010). Effect of production variables on microbiological removal in locally-produced ceramic filters for household water treatment. *International Journal of Environmental Health Research, 20*(3), 171–187. doi:10.1080/09603120903440665 PMID:20162486

Lantagne, D. S., & Clasen, T. F. (2001). Investigation of the Potters for Peace Colloidal Silver Impregnated Ceramic Filter Report 2: Field Investigations. *Environmental Science & Technology, 46*(20), 11352–11360. doi:10.1021/es301842u PMID:22963031

Lantagne, D. S., Quick, R., & Mintz, E. D. (2006). *Household Water Treatment and Safe Storage Options in Developing Coutries: A Review of Current Implementation Practices*. Academic Press.

LeOuay, B., & Stellacci, F. (2015). Antibacterial activity of silver nanoparticles: A surface science insight. *Nano Today, 10*(3), 339–354. doi:10.1016/j.nantod.2015.04.002

LePape, H., Solano-Serena, F., Contini, P., Devillers, C., Maftah, A., & Leprat, P. (2002). Evaluation of the anti-microbial properties of an activated carbon fibre supporting silver using a dynamic method. *Carbon, 40*(15), 2947–2954. doi:10.1016/S0008-6223(02)00246-4

LePape, H., Solano-Serena, F., Contini, P., Devillers, C., Maftah, A., & Leprat, P. (2004). Involvement of reactive oxygen species in the bactericidal activity of activated carbon fibre supporting silver: Bactericidal activity of ACF(Ag) mediated by ROS. *Journal of Inorganic Biochemistry, 98*(6), 1054–1060. doi:10.1016/j.jinorgbio.2004.02.025 PMID:15149815

Li, Q., Mahendra, S., Lyon, D. Y., Brunet, L., Liga, M. V., Li, D., & Alvarez, P. J. J. (2008). Antimicrobial nanomaterials for water disinfection and microbial control: Potential applications and implications. *Water Research, 42*(18), 4591–4602. doi:10.1016/j.watres.2008.08.015 PMID:18804836

Lin, S., Huang, R., Cheng, Y., Liu, J., Lau, B. L. T., & Wiesner, M. R. (2013). Silver nanoparticle-alginate composite beads for point-of-use drinking water disinfection. *Water Research, 47*(12), 3959–3965. doi:10.1016/j.watres.2012.09.005 PMID:23036278

Liu, H., Lee, Y.-Y., Norsten, T. B., & Chong, K. (2013). In situ formation of anti-bacterial silver nanoparticles on cotton textiles. *Journal of Industrial Textiles*. doi:10.1177/1528083713481833

Maneerung, T., Tokura, S., & Rujiravanit, R. (2007). (in press). Impregnation of silver nanoparticles into bacterial cellulose for antimicrobial wound dressing. *Carbohydrate Polymers*.

Matthies, K., Bitter, H., Deobald, N., Heinle, M., Diedel, R., Obstaand, U., & Brenner-Weissa, G. (2015). Morphology, composition and performance of a ceramic filter for household water treatment in Indonesia. *Water Practice & Technology, 10*(2), 361–370. doi:10.2166/wpt.2015.044

Mecha, C. A., & Pillay, V. L. (2014). Development and evaluation of woven fabric microfiltration membranes impregnated with silver nanoparticles for potable water treatment. *Journal of Membrane Science, 458*, 149–156. doi:10.1016/j.memsci.2014.02.001

Mellor, J., Abebe, L., Ehdaie, B., Dillingham, R., & Smith, J. (2014). Modeling the sustainability of a ceramic water filter intervention. *Water Research, 49*, 286–299. doi:10.1016/j.watres.2013.11.035 PMID:24355289

Mellor, J. E., Smith, J. A., Learmonth, G. P., Netshandama, V. O., & Dillingham, R. A. (2012). Modeling the Complexities of Water, Hygiene, and Health in Limpopo Province, South Africa. *Environmental Science & Technology, 46*(24), 13512–13520. doi:10.1021/es3038966 PMID:23186073

Montazer, M., Alimohammadi, F., Shamei, A., & Rahimi, M. K. (2012). In situ synthesis of nano silver on cotton using Tollens' reagent. *Carbohydrate Polymers, 87*(2), 1706–1712. doi:10.1016/j.carbpol.2011.09.079

Ngo, Y. H., Li, D., Simon, G. P., & Garnier, G. (2011). Paper surfaces functionalized by nanoparticles. *Advances in Colloid and Interface Science, 163*(1), 23–38. doi:10.1016/j.cis.2011.01.004 PMID:21324427

Nover, D. M., McKenzie, E. R., Joshi, G., & Fleenor, W. E. (2013). Assessment of colloidal silver impregnated ceramic bricks for small-scale drinking water treatment applications. *International Journal for Service Learning, 8*(1), 18–35.

Oyanedel-Craver, V., & Smith, J. A. (2008). Sustainable colloidal-silver-impregnated ceramic filter for point-of-use water treatment. *Environmental Science & Technology, 42*(3), 927–933. doi:10.1021/es071268u PMID:18323124

Parikh, D. V., Fink, T., Rajasekharan, K., Sachinvala, N. D., Sawhney, A. P. S., Calamari, T. A., & Parikh, A. D. (2005). Antimicrobial Silver/Sodium Carboxymethyl Cotton Dressings for Burn Wounds. *Textile Research Journal, 75*(2), 134–138. doi:10.1177/004051750507500208

Park, S.-J., & Jang, Y.-S. (2003). Preparation and Characterization of Activated Carbon Fibers Supported with Silver Metal for Antibacterial Behavior. *Journal of Colloid and Interface Science, 261*(2), 238–243. doi:10.1016/S0021-9797(03)00083-3 PMID:16256528

Peter-Varbanets, M., Zurbrugg, C., Swartz, C., & Pronk, W. (2009). Decentralized systems for potable water and the potential of membrane technology. *Water Research, 43*(2), 245–265. doi:10.1016/j.watres.2008.10.030 PMID:19010511

Pradeep, T., & Anshup, . (2009). Noble metal nanoparticles for water purification: A critical review. *Thin Solid Films, 517*(24), 6441–6478. doi:10.1016/j.tsf.2009.03.195

Praveena, S. M., & Aris, A. Z. (2015). *Application of Low-Cost Materials Coated with Silver Nanoparticle as Water Filter in Escherichia coli Removal.* Water Qual Expo Health; doi:10.1007/s12403-015-0167-5

Quang, D. V., Sarawade, P. B., Jeon, S. J., Kim, S. H., Kim, J.-K., Chai, Y. G., & Kim, H. T. (2013). Effective water disinfection using silver nanoparticle containing silica beads. *Applied Surface Science, 266*, 280–287. doi:10.1016/j.apsusc.2012.11.168

Radetic, M. (2013). Functionalization of textile materials with silver nanoparticles. *Journal of Materials Science, 48*(1), 95–107. doi:10.1007/s10853-012-6677-7

Rai, M., Yadav, A., & Gade, A. (2009). Silver nanoparticles as a new generation of antimicrobials. *Biotechnology Advances, 27*(1), 76–83. doi:10.1016/j.biotechadv.2008.09.002 PMID:18854209

Ravindra, S., Mohan, Y. M., Reddy, N. N., & Raju, K. M. (2010). Fabrication of antibacterial cotton fibres loaded with silver nanoparticles via "Green Approach". *Colloids and Surfaces. A, Physicochemical and Engineering Aspects, 367*(1-3), 31–40. doi:10.1016/j.colsurfa.2010.06.013

Rayner, J. (2009). *Current Practices in Manufacturing of Ceramic Pot Filters for Water Treatment. (Master of Science)*. Loughborough University.

Rayner, J., Zhang, H., Schubert, J., Lennon, P., Lantagne, D., & Oyanedel-Craver, V. (2013). Laboratory Investigation into the Effect of Silver Application on the Bacterial Removal Efficacy of Filter Material for Use on Locally Produced Ceramic Water Filters for Household Drinking Water Treatment. *ACS Sustainabe Chemical Engineering, 1*, 737–745. doi:10.1021/sc400068p

Reda, M. (2011). In situ production of silver nanoparticle on cotton fabric and its antimicrobial evaluation. *Cellulose (London, England), 18*(1), 75–82. doi:10.1007/s10570-010-9455-1

Ren, D., Colosi, L. M., & Smith, J. A. (2013). Evaluating the Sustainability of Ceramic Filters for Point-of-Use Drinking Water Treatment. *Environmental Science & Technology, 47*(19), 11206–11213. doi:10.1021/es4026084 PMID:23991752

Ren, D., & Smith, J. A. (2013). Retention and Transport of Silver Nanoparticles in a Ceramic Porous Medium Used for Point-of-Use Water Treatment. *Environmental Science & Technology, 47*(8), 3825–3832. doi:10.1021/es4000752 PMID:23496137

Ruparelia, J., Chatterjee, A., Duttagupta, S., & Mukherji, S. (2008). Strain specificity in antimicrobial activity of silver and copper nanoparticles. *Acta Biomaterialia, 4*(3), 707–716. doi:10.1016/j.actbio.2007.11.006 PMID:18248860

Saidani-Scott, H., Tierney, M., & Sánchez-Silva, F. (2009). Experimental Study of Water Filtering Using Textiles as in Traditional Methods. *Applied Mechanics and Materials, 15*, 15–20. doi:10.4028/www.scientific.net/AMM.15.15

Sathishkumar, M., Sneha, K., & Yun, Y.-S. (2010). Immobilization of silver nanoparticles synthesized using Curcuma longa tuber powder and extract on cotton cloth for bactericidal activity. *Bioresource Technology, 101*(20), 7958–7965. doi:10.1016/j.biortech.2010.05.051 PMID:20541399

Schoen, D. T., Schoen, A. P., Hu, L., Kim, H. S., Heilshorn, S. C., & Cui, Y. (2010). High Speed Water Sterilization Using One-Dimensional Nanostructures. *Nano Letters, 10*(9), 3628–3632. doi:10.1021/nl101944e PMID:20726518

Sobsey, M. D., Stauber, C. E., Casanova, L. M., Brown, J. M., & Elliott, M. A. (2008). Point of Use Household Drinking Water Filtration: A Practical, Effective Solution for Providing Sustained Access to Safe Drinking Water in the Developing World. *Environmental Science & Technology, 42*(12), 4261–4267. doi:10.1021/es702746n PMID:18605542

Tammisetti, R. (2010). *Research on the Effectiveness of Using Cloth as a Filter to Remove Turbidity from Water*. Academic Press.

Tang, S., Meng, X., Lu, H., & Zhu, S. (2009). PVP-assisted sonoelectrochemical growth of silver nanostructures with various shapes. *Materials Chemistry and Physics, 116*(2-3), 464–468. doi:10.1016/j.matchemphys.2009.04.004

Yin, T., Walker, H. W., Chen, D., & Yang, Q. (2014). Influence of pH and ionic strength on the deposition of silver nanoparticles on microfiltration membranes. *Journal of Membrane Science, 449*, 9–14. doi:10.1016/j.memsci.2013.08.020

Yoon, K. Y., Byeon, J. H., Park, C. W., & Hwang, J. (2008). Antimicrobial Effect of Silver Particles on Bacterial Contamination of Activated Carbon Fibers. *Environmental Science & Technology, 42*(4), 1251–1255. doi:10.1021/es0720199 PMID:18351101

Zeng, F., Hou, C., Wu, S., Liu, X., Tong, Z., & Yu, S. (2007). Silver nanoparticles directly formed on natural macroporous matrix and their anti-microbial activities. *Nanotechnology, 18*(5), 055605. doi:10.1088/0957-4484/18/5/055605

Chapter 3
Nanotechnology in Engineered Membranes:
Innovative Membrane Material for Water–Energy Nexus

Heechul Choi
Gwangju Institute of Science and Technology (GIST), South Korea

Jiyeol Bae
Gwangju Institute of Science and Techonology (GIST), South Korea

Moon Son
Gwangju Institute of Science and Techonology (GIST), South Korea

Hyeon-gyu Choi
Korea Research Institute of Chemical Technology (KRICT), South Korea

ABSTRACT

The membrane processes have received extensive attention as comprehensive and interdisciplinary approaches for water-energy nexus. Nanotechnology has induced significant attention in improving membrane performances to mitigate global water and energy scarcity because of its unique characteristics and simple application for membrane fabrication. Nano-sized materials could provide highly enhanced characteristics to a membrane material, resulting in excellent performance enhancement, such as permeability, selectivity, and fouling resistance, of membrane. Carbon Nanotube (CNT), a widely utilized or studied nanomaterial in membrane science, is discussed in this chapter with a focus on its state-of-the-art applications and future prospects. Electrospun nanofiber, which is one of the feasible nano-structured membrane materials, is also discussed as a promising material for water-energy nexus. Therefore, this chapter also describes its application cases and its potential as an innovative membrane for water-energy nexus.

INTRODUCTION

Water and energy are critical, mutually dependent resources. The production of energy requires large volumes of water, and a water infrastructure requires large amounts of energy. Therefore, the nexus between water and energy is a clear agenda for sustainable development. Recently, membrane processes have received considerable attention as comprehensive and interdisciplinary approaches for the

DOI: 10.4018/978-1-5225-0585-3.ch003

water-energy nexus. This process is considered one of the most realistic and reliable technical solutions to address water/energy challenges. Microfiltration (MF), ultrafiltration (UF), reverse osmosis (RO), forward osmosis (FO), and pressure-retarded osmosis (PRO) in particular have been actively utilized not only in academia but in the industrial field. A membrane material fabrication technology could be a core technology of a membrane process for the water-energy nexus, which requires particularly excellent characteristics and a performance beyond those of the common membrane materials.

Nanotechnology has induced significant attention in improving membrane properties and performances because of its unique characteristics and simple application for membrane fabrication. Nano-sized materials could provide highly enhanced characteristics to a membrane material, resulting in excellent membrane performance enhancements, such as permselectivity, fouling resistance, and mechanical properties. Various nanomaterials, such as carbon nanotubes (CNTs), graphene oxide, titania, alumina, silica, and silver, have been widely studied to fabricate advanced functional membranes due to their unique physico-chemical properties. Despite its potentials, critical issues, such as low stability, leaching problem and self-aggregation, are still challenges that need to be overcome.

CNT, a commonly utilized or studied nanomaterial in membrane science, is deeply discussed in this chapter with a focus on its state-of-the-art applications, implications, and future prospects. Nanotechnology for a membrane includes not only nanomaterials but also nano-structured membrane material. Electrospun nanofiber is a feasible nano-structured membrane material with a one-dimensional structure. The distinct interconnected nanostructure of a nanofiber from nonwoven media can provide unique properties. Therefore, electrospun nanofiber has been actively applied in water filtration membranes. This chapter also describes its application cases and its potential as a new type of membrane for water treatment and energy harvesting processes. This chapter provides not only the advantages of these innovative membranes but also their limitations or challenges.

NANOMATERIAL-EMBEDDED POLYMER MEMBRANES (NANO-ENHANCED MEMBRANES)

The aims of membrane-based processes can be classified into two main categories: water production and energy generation. Fresh water can be obtained from waste water, industry water, and seawater via the MF, UF, RO, and FO processes. On the other hand, different combinations of salinity gradients, such as seawater and river water or seawater brine and seawater, lead to harvesting energy via the PRO process. Currently, polymer-based membranes are widely applied in both academic studies and industrial applications. It is worth mentioning that the convergence of nanotechnology and membrane technology can mitigate global water and energy scarcity by maximizing membrane performance (Figure 1).

A nano-enhanced membrane (NeM), which has been investigated since the 2000s and whose term was proposed for the first time in 2013 (Buonomenna, 2013), is defined as a functionalized membrane with discrete nanoparticles or nanotubes. Over the past decade, nanotechnology has led to a next-generation membrane process in water treatment and has enabled excellent membrane performances, such as high permeability, selectivity, and fouling resistance. NeMs are classified into three categories: inorganic, organic, and bio-inspired NeMs. A commercial ceramic membrane coated with reactive or functionalized nanomaterials is categorized as inorganic NeMs. In organic NeMs, a nanomaterial is blended into a polymer solution, and a nanocomposite membrane is fabricated via phase inversion and interfacial

Figure 1. Membrane processes for water and energy generation.

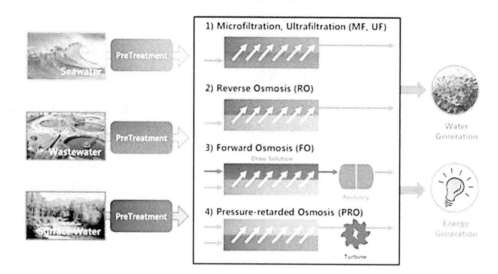

polymerization. Bio-inspired NeMs are a new concept of a membrane structure based on a biomimetic. Recently, vertically aligned nanotubes and aquaporin-based membranes have been proposed.

Nanomaterials can enhance membrane properties, such as permeability, selectivity, fouling resistance, and durability, as listed in Table 1. CNTs, graphene oxide, titania, alumina, silica, and silver are widely used as inorganic fillers for membrane fabrication due to their unique physico-chemical properties.

Among these nanomaterials, CNTs are currently receiving considerable attention because they have been proven to increase the water flux, rejection, and fouling resistance of membranes simultaneously (Evrim Celik, Hosik Park, et al., 2011; Evrim Celik, Lei Liu, et al., 2011). CNTs are described as seamless "rolled-up" graphene sheets with both sides capped. More specifically, CNTs are coaxial cylindrical carbonaceous nanomaterials with a dimension of a nano-meter in diameter and a micro-meter in length. Generally, CNTs are divided into single and multi-walled nanostructures depending on the number of walls the CNTs are composed of (Baughman, Zakhidov, & de Heer, 2002). Currently, chemical vapor

Table 1. Nano-materials' effects on polymeric membrane properties

Nano-Materials	Reported Features
Carbon nanotubes	Hydrophilicity enhanced, Water flux enhanced, Fouling decreased
Graphene oxide	Water flux enhanced, Anti-biofouling enhanced, Mechanical strength enhanced
Titania	Hydrophilicity enhanced, Water flux enhanced, Anti-fouling enhanced by UV
Alumina	Hydrophilicity enhanced, Water flux enhanced, Anti-fouling enhanced, Mechanical strength enhanced
Silica	Water flux enhanced by more pores on membrane surface
Silver	Bio-fouling decreased, Released out of membrane
Iron oxides	High rejection when combined with ozonation

(Evrim Celik, Hosik Park, HyeongyuChoi, & Heechul Choi, 2011; Evrim Celik, Lei Liu, & Heechul Choi, 2011; Hyeon-gyu Choi et al., 2015; J. Lee et al., 2013; Lei Liu et al., 2014; Moon Son et al., 2016; Moon Son et al., 2015; M. M. Pendergast & Hoek, 2011)

deposition (CVD), arc-discharge, and laser-ablation prevail over other techniques for the synthesis of nanotubes (Dai, 2002).

Recently, thin-film composite (TFC) membranes prepared by interfacial polymerization have evolved as another branch of a high-selectivity membrane. A negatively charged and ultra-dense polyamide (PA) active layer is one representative TFC membrane (Misdan, Lau, & Ismail, 2012). Nanomaterials can be added into the PA active layer or support layer of the TFC membrane, the thin-film nanocomposite (TFN) membrane, based on their application fields. The nanomaterials are mainly embedded into the active layer of the TFN membrane when feed water first meets the active layer. On the other hand, it is incorporated into the support layer when TFN is applied to the osmotic driven membrane because most of the nanomaterials can enhance the hydrophilicity of the polymer. In most cases of the TFN membrane, nanomaterials use less than 10 wt% of the total polymer solution (Table 2).

According to the International Renewable Energy Agency (IRENA), the total technical potential for the salinity gradient power is estimated to be around 647 gigawatts (GW) globally, which is equivalent to 23% of the electricity consumed in 2011 (R. Kempener & Neumann, 2014). Unlike other renewable energy sources, such as solar and wind sources, this plant is not affected by climate change and can be operated 24 hours per day. In addition, the plant does not cause noise pollution and only requires a compact space for operation. Therefore, a membrane-based PRO plant could generate promising renewable and emission-free energy and thus contribute to eco-friendly power production (C. Klaysom, T.Y. Cath, T. Depuydt, & Vankelecom, 2013). Moreover, the efficiency of the second law of thermodynamics

Table 2. Summary of the TFN membranes formed

Membrane Type	Used Amount/wt%	Membrane Properties Altered and Novel Characteristics Shown with Particle Additions
PA-CNTs/PES (Moon Son et al., 2015)	0.5	Increased hydrophilicity Increased water permeability by 44% without sacrificing selectivity Increased alginate fouling resistance by 300%
Graphene oxide/PA-PSf (Hu & Mi, 2013)	2.8	Increased water flux Decreased monovalent and divalent salt ion rejection
Ag-zeolite/PA-PSf (B.-H. J. Mary Laura Lind, Arun Subramani, Xiaofei Huang, Eric M.V. Hoek 2009)	0.4	Increased hydrophilicity Increased water permeability by 66% Decreased propensity for biofouling
Titania/PA-PES (H. S. Lee et al., 2008)	1-9	Decreased permeability and increased rejection at low particle additions Increased permeability and decreased salt rejection above 5 wt%
PA-Titania/PSf (Emadzadeh, Lau, Matsuura, Rahbari-Sisakht, & Ismail, 2014)	1	Increased water flux by over 400% Increased hydrophilicity Increased reverse solute flux
Zeolite/PA-PSf (Jeong et al., 2007)	0.4	Increased hydrophilicity and surface charge Decreased surface roughness Increased water permeability by 80%
Zeolite/PA-PSf (Lind, Suk, Nguyen, & Hoek, 2010)	0.2	Increased hydrophilicity Increased water permeability Increased salt rejection in RO testing
nAg/PA-(Psf+CNT) (E.-S. Kim et al., 2012)	10 (nAg)	Increased antibacterial and antifouling properties Increased hydrophilicity Increased water permeability

increases to 20%, and the input power decreases by 38% by introducing the PRO system as an energy recovery device for a reverse osmosis desalination plant (Sharqawy, Zubair, & Lienhard V, 2011). The TFC membranes are much preferred among the various membranes used for the PRO process due to their high water permeability and selectivity (Chalida Klaysom, Tazhi Y. Cath, Tom Depuydt, & Vankelecom, 2013; Ngai Yin Yip et al., 2011). To develop a high-performance TFC membrane, various nanomaterials, such as CNTs, zeolite, silver, titania (TiO_2), and silica (SiO_2), have been investigated due to their unique functionalities (e.g., hydrophilicity, antimicrobial functionality, and mechanical property) (Amini, Jahanshahi, & Rahimpour, 2013; Charles-François de Lannoy, David Jassby, Katie Gloe, Alexander D. Gordon, & Wiesner, 2013; Emadzadeh, Lau, & Ismail, 2013; Emadzadeh, Lau, Matsuura, Ismail, & Rahbari-Sisakht, 2014; Hai Huang, Xinying Qu, Hang Dong, Lin Zhang, & Chen, 2013; Hai Huang, Xinying Qu, Xiaosheng Ji, et al., 2013; Hee Joong Kim et al., 2014; Jadav, Aswal, & Singh, 2010; Jadav & Singh, 2009; Jeong et al., 2007; Junwoo Park et al., 2010; Kim, Hwang, Gamal El-Din, & Liu, 2012; Lin Zhang, Guo-Zhong Shi, Shi Qiu, Li-Hua Cheng, & Chen, 2011; Mary Laura Lind, Byeong-Heon Jeong, Arun Subramani, Xiaofei Huang, & Hoek, 2009; Mary Laura Lind, Daniel Eumine Suk, The-Vinh Nguyen, & Hoek, 2010; Moon Son et al., 2015; M. T. M. Pendergast, Nygaard, Ghosh, & Hoek, 2010; Rajaeian, Rahimpour, Tade, & Liu, 2013; Xiaoxiao Song, Zhaoyang Liu, & Sun, 2013; Yin, Kim, Yang, & Deng, 2012; Zhao et al., 2014). The CNTs have been shown to be one of the most efficient inorganic nanomaterials used to increase membrane performance in terms of permeability, selectivity, and fouling resistance in the PRO process. Therefore, combining nanotechnology and membrane technology contributes to generating renewable, alternative, and sustainable energy, which has been recommended as a fundamental solution to global energy scarcity.

Challenging Considerations for NeM Application

Despite its potential, critical considerations have been remained for field applications of NeMs. Nanomaterial tends to be aggregated in the membrane due to its stability, which hinders the homogeneous dispersion of nanomaterial in the membrane matrix. Nanomaterial aggregates enlarge microvoids in the membrane and thus cause limited improvement in the membrane's permeability and antifouling performance or even defects in the membrane (Zhu, Wang, Jiang, & Jin, 2014). Further, it also weakens chemical bonding formation between a nanomaterial and a polymer monomer, which disturbs the immobilization of nanomaterial in the membrane, thus causing nanomaterial to release out of the membrane (Chen et al., 2010). Because it has been reported that nanomaterial is more toxic in nature than that in the bulk-phase, released nanomaterial to nature is carefully considered (Nazarenus et al., 2014; Rivera-Gil et al., 2013; Soenen, Parak, Rejman, & Manshian, 2015). Additionally, the leaching issue is related to the scale-up of NeMs for field application. As discussed above, the nanomaterial aggregate in the membrane might change the membrane's structure. This trend could become greater when the membrane's size is enlarged.

Although nanomaterial manufacturing costs are decreasing rapidly, it is still expensive, which relates to increases in membrane fabrication costs (Y.-W. Huang, Wu, & Aronstam, 2010). This demonstrates that the optimized nanomaterial concentration could be a critical factor governing membrane fabrication costs.

The proposed issues should be carefully considered or studied for field applications of NeMs in water treatment.

CNTS-POLYMER COMPOSITE MEMBRANE

CNTs have been widely investigated in the environmental field (Das et al., 2014). Due to their large surface area-to-volume ratios, they are considered promising adsorbents for pollutant removal (Rao, Lu, & Su, 2007); their remarkable mechanical strength, negative charge, and hydrophilicity after modification render them an ideal component of nanocomposites (Coleman, Khan, Blau, & Gun'ko, 2006; Evrim Celik, Hosik Park, et al., 2011; Evrim Celik, Lei Liu, et al., 2011; Lei Liu et al., 2014). Since the recent important findings on water molecules' ultra-fast transport in CNTs' inner hole in a lab-scale experiment and molecular dynamics simulation, additional attempts have been made to achieve vertically aligned CNT (VA-CNT) membranes with several orders of magnitude of enhanced water flux (Corry, 2007; Fornasiero et al., 2008; Joseph & Aluru, 2008; Melechko et al., 2005; Noy et al., 2007; Striolo, 2006; Verweij, Schillo, & Li, 2007; Whitby & Quirke, 2007); however, the restricted experimental conditions and the size of the synthesized membrane via the CVD method must still be resolved in terms of the processability and scaling up of VA-CNT membranes (Ahn et al., 2012). In addition, challenges in pore size and diameter along with the ion rejection of VA-CNT membranes further undermine their applicability in seawater desalination (Kar, Bindal, & Tewari, 2012). For these reasons, some researchers have shifted their focus from VA-CNT to CNT-based composite membrane fabrication. For this approach, the purification and modification of randomly entangled raw CNTs are the prerequisites for the practical applications of CNTs to obtain satisfactory dispersion and compatibility with polymers (Byrne & Gun'ko, 2010; Eder, 2010; Karousis, Tagmatarchis, & Tasis, 2010; S. W. Kim et al., 2012; Majumder & Ajayan, 2010; Sahoo, Rana, Cho, Li, & Chan, 2010; Spitalsky, Tasis, Papagelis, & Galiotis, 2010; Xie, Mai, & Zhou, 2005). The amorphous carbon impurities and metallic catalysts were removed during these physical or chemical modification steps, and the CNTs gained charged functional groups and open ends (Karousis et al., 2010). The incorporation of functionalized CNTs (f-CNTs) in polymer matrix induced hydrophilicity, membrane pore morphology, and surface charges could improve the membrane's performance in terms of permeability, selectivity, and fouling resistance, as previously discussed (Figure 2) (Ahn et al., 2012).

Historically, polymeric membranes have dominated the water treatment membrane market for decades, and the cellulose acetate (CA) membrane could be classified as the first-generation membrane (K. P.

Figure 2. Schematics of CNT-based membranes: (a) VA-CNT (Hinds et al., 2004), (b) TFN with a CNT in the PA layer (Chan et al., 2013), and (c) TFN with a CNT in the support layer membrane (Moon Son et al., 2015).

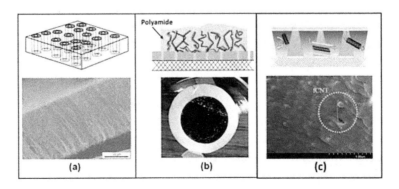

Lee, Arnot, & Mattia, 2011). The change of flux and the rejection by the carboxylic f-CNT incorporated CA membrane have been investigated, indicating that f-CNTs acted as a water passage, which led to an improved permeation rate and a slightly decreased rejection (El Badawi, Ramadan, Esawi, & El-Morsi, 2014).

For the recently developed TFN membrane, which is mainly used for seawater desalination or saline energy harvesting, incorporations of f-CNTs into PA cause the TFN membranes to possess superior properties in terms of high flux, rejection, and fouling resistance. These features are attributed to the f-CNTs' functions in the PA active layer. Recently, it was reported that the f-CNTs' blended support layer can improve membrane water flux, rejection, and alginate fouling simultaneously due to the enhanced hydrophilicity, pore properties, and negatively charged surface (Moon Son et al., 2015). The research on f-CNTs embedded in the TFN membrane illustrated that the flux increase basically resulted from three aspects: 1) the membrane's hydrophilicity increased through the hydrophilic f-CNTs addition, 2) the macrovoids formed between the f-CNTs and the polymer favored the water permeation, and 3) f-CNTs served as the ultra-fast transport channel to facilitate water molecules (H. J. Kim, K. Choi, et al., 2014; Zhao et al., 2014). It is worth mentioning that the flux enhancement is not pronounced for f-CNT nanocomposites compared to VA-CNTs because the f-CNTs are embedded in the PA or in the support polymer layer rather than having an aligned orientation, but the advantages of f-CNTs lead the way for the application of the f-CNT membrane in the real field.

The improved membrane rejection was caused by two mechanisms: 1) the steric repulsion generated by the tips of the f-CNTs excluded the target compound's penetration; 2) a more negatively charged TFN membrane induced by f-CNTs caused a high electrostatic repulsion between the membrane's surface and the negatively charged target compounds (Chan et al., 2013). It is interesting to note that most of the f-CNTs' TFN membrane showed an insignificant "trade-off" between the flux and ion rejection in the RO process for seawater desalination.

The surface properties enable the TFN membrane to adsorb less organic foulants through hydration and electrostatic forces, which generates the anti-organic fouling properties of the TFN membrane in seawater desalination (Zhao et al., 2014). The presence of f-CNTs in the TFN membrane was also found to have antimicrobial properties toward bacteria (H. J. Kim, Y. Baek, et al., 2014; Tiraferri, Vecitis, & Elimelech, 2011). Biofouling is one of the most critical considerations in membrane processes. A biofilm is formed on the membrane's surface by variable bacteria growth, which is difficult to control. High-molecular weight compounds in particular are secreted by bacteria in the biofilm, called extracellular polymeric substances (EPSs). An EPS is mostly composed of polysaccharides and proteins, constituting 50–90% of the organic matter in the biofilm (Donlan & Costerton, 2002). Alginate has been most commonly studied as a surrogate of an EPS in the biofilm. The f-CNTs allowed for a lower water flux decline, demonstrating a smaller alginate fouling layer formation on the membrane's surface. As described, the f-CNTs allow for a negatively charged membrane surface characteristic. It has been reported that the electrostatic repulsive force is dominant in the foulant-membrane interaction (Hyeongyu Choi et al., 2015).

The research on CNTs embedded in the TFN membrane for energy harvesting via the PRO process is in its initial stage, and related research is scarce (Miao Tian, Wang, Goh, Liao, & Fane, 2015); however, CNTs embedded in the TFN membrane have huge potential because feed and draw solutions face both sides of the membrane in the PRO process. Figure 3 illustrates the development of a CNTs-polymer composite membrane in a time scale (Mauter & Elimelech, 2008).

Figure 3. The development of f-CNTs -polymer composites (Mauter & Elimelech, 2008)

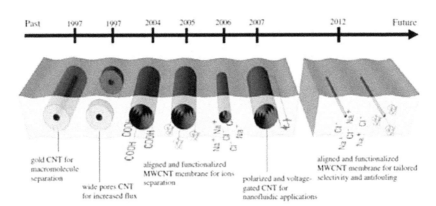

Challenges for CNTs-Polymer Composite Membrane Application

The strong attracting forces between CNTs account for their tightly bundled structure (Thess et al., 1996). Moreover, this results in low dispersion in polar and apolar solvents (Mauter & Elimelech, 2008). It is critical to control the attraction forces of the CNTs, hence different types of surface modifications have been actively studied (Banerjee, Hemraj-Benny, & Wong, 2005; Holzinger et al., 2001; Jin et al., 2007; Sun, Fu, Lin, & Huang, 2002). An interesting study was reported in 1998, dealing with the accelerated transport of water molecules through nanoscale channels of CNTs' inner-sidewalls (J. Liu et al., 1998). Vertically aligned nanotube membranes have been developed recently; however, the aligned CNTs membrane was developed only in limited area so far due to the tiny size of membranes. The scale-up of this membrane size might be one of the challenging areas of research. Finally, CNTs need to be post-treated to control for defects, metal catalyst contamination, and physical heterogeneities (Chiang, Brinson, Smalley, Margrave, & Hauge, 2001; Shelimov, Esenaliev, Rinzler, Huffman, & Smalley, 1998; Shi et al., 1999). Therefore, highly purified CNT synthesis on a large scale should be developed for commercial or field applications of the CNTs-polymer composite membrane (Plata, Gschwend, & Reddy, 2008).

Those challenges have been actively investigated around the world and will be addressed in the near future.

ELECTROSPUN NANOFIBER MEMBRANES

Nanotechnology for membranes not only includes nanomaterials, but also nano-structured membrane materials. An electrospun nanofiber is one of the most feasible nano-structured membrane materials with a one-dimensional structure. The distinct interconnected nanostructure of the nanofiber from the nonwoven media can provide unique properties for various separation engineering processes. Therefore, electrospun nanofibers have been actively applied in water filtration membranes.

In recent years, the electrospinning technique has been actively utilized in nanofiber fabrication studies. Unlike mechanical fiber spinning techniques, electrospinning is a widely used technology for electrostatic fiber formation, which utilizes electrical forces to produce polymer fibers with diameters ranging from 2 nm to several micrometers using polymer solutions of both natural and synthetic polymers (Doshi

& Reneker, 1995). The fiber formation mechanisms are as follows. A polymer solution or a polymer melt is charged in a high electric potential that produces an electrical field between the spinneret and the collector. When the electric force acting on the surface of a liquid overcomes the surface tension, a charged liquid jet is pulled from the liquid surface and moves straight toward a grounded collector (Yarin, Koombhongse, & Reneker, 2001). This process of electrospinning has gained considerable attention in the last decade not only due to its versatility in spinning a wide variety of polymeric fibers, but also due to its ability to consistently produce fibers in the submicron range, which is otherwise difficult to achieve using standard mechanical fiber-spinning techniques (Ma, Kotaki, Inai, & Ramakrishna, 2005). This process offers unique capabilities for producing novel natural nanofibers and fabrics with controllable pore structures (He, Wan, & Yu, 2005). With smaller pores and a higher surface area than regular fibers, electrospun fibers have been successfully applied in various fields. Spun nanofibers also offer several advantages, such as an extremely high surface-to-volume ratio, tunable porosity, malleability to conform to a wide variety of sizes and shapes, and the ability to control the nanofiber composition to achieve the desired results from its properties and functionality. Because of these advantages, electrospun nanofibers have been widely investigated in the past several years for their use in water purification processes.

A high porosity, interconnectivity, microscale interstitial space, and surface-to-volume ratio indicate that nonwoven electrospun nanofiber meshes are an excellent material for membrane preparation, particularly in environmental engineering applications (Figure 4.) (Ramakrishna et al., 2006). Electrospun nanofibers can form an effective size exclusion membrane for particulate removal from wastewater. Polymeric nanofibers have been used in filtration applications for more than a decade. For filtration, the channels and structural elements of a filter must be matched to the scale of the particles or droplets that are to be captured in the filter. Thus, the unique properties of electrospun membranes consisting of nano-sized-diameter fibers can be used as an advantage. It has been shown that electrospinning meets the challenge of providing solutions for the removal of undesirable particles in submicron ranges. In fact, electrospun nanofibers have been used in air filtration applications more often than in water treatment processes due to their deficient mechanical properties for water treatment applications. Therefore, nonwoven fabrics, such as polyethylene terephthalate (PET), have been used in many studies as a support layer to augment the mechanical strength of the electrospun nanofiber membrane. Particle removal from an aqueous solution by a nanofiber membrane has been studied by many researchers. Gopal et al. reported that polyvinylidene fluoride (PVDF) electrospun membranes were successful in rejecting more than 90% of the micro-particles from the solution (Gopal et al., 2006). Barhate et al. investigated the structural and transport properties of an electrospun membrane in relation to the processing parameters to determine the distribution, deposition, and orientation of nanofibers in nanofibrous filtering media. Their results demonstrate that control over the pore size distribution can be achieved by coordinating the drawing and collection rates (Barhate, Loong, & Ramakrishna, 2006). Moreover, the polyacrylonitrile (PAN) nanofiber/PET nonwoven fabric MF membranes performed significantly better than the commercial MF membrane of the same mean pore size (0.22 μm) with a two to three times higher flux (~1.5 $L/m^2 \cdot h$). The nanofibrous MF filter could maintain a very high rejection ratio of micro-particles and bacteria (LRV = 6) (R. Wang, Liu, Li, Hsiao, & Chu, 2012). Chemically cross-linked poly(vinyl alcohol) (PVA) nanofiber membranes by glutaraldehyde showed a 3–7 times higher pure water flux than the commercial membrane (Millipore GSWP 0.22 mm). The nanofibrous PVA membranes with an average thickness of 20 μm could successfully reject more than 98% of the polycarboxylate microsphere particles with a diameter of 0.209 ± 0.011 μm and still maintain a 1.5–6 times higher permeate flux than that of the Millipore GSWP 0.22 μm membrane (Y. Liu, Wang, Ma, Hsiao, & Chu, 2013).

Figure 4. An electrospun polysulfone membrane: (a) surface, (b) cross-section, and (c) magnified cross-section images

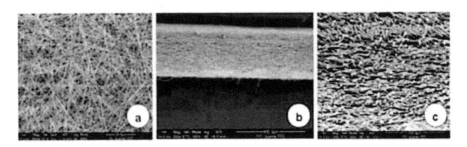

Furthermore, in UF and nanofiltration processes, applications of electrospun membranes have been successfully demonstrated. Wang et al. reported that a novel high flux filtration medium consisting of a three-tier composite structure, i.e., a nonporous hydrophilic nanocomposite coating top layer, an electrospun nanofibrous substrate midlayer, and a conventional nonwoven microfibrous support, exhibited a high flux rate (up to 330 L/m^2·h at the feed pressure of 100 psi) and an excellent total organic solute rejection rate (99.8%) without appreciable fouling (X. Wang et al., 2005). An electrospun PAN nanofibrous scaffold was used as a mid-layer support by Yoon et al. in a new type of a high-flux thin film nanofibrous composite (TFNC) membrane for nanofiltration (NF) applications. The results indicate that the TFNC membranes exhibited over a 2.4 times greater permeate flux than the TFC membranes with the same chemical compositions while maintaining the same rejection rate (ca. 98%) (Yoon, Hsiao, & Chu, 2009).

The filtration mechanisms of contaminants through electrospun nanofiber membranes are based on specific interactions and are controlled by tuning the properties of the membranes. The removal of pollutants by synthesized electrospun nanofiber materials may be primarily responsible through various interactions, such as adsorption and oxidation. The removal ability is dependent on the surface properties of the membranes. Therefore, functionalized electrospun nanofiber membranes are very effective for contaminant removal in water. The removal of various contaminants from aqueous solutions through different interactions, such as adsorption and oxidation, by a functionalized electrospun nanofiber membrane has been studied by many researchers. Various functionalized electrospun nanofibers and their effects are listed in Table 3.

Various polymer electrospun nanofibers have been functionalized by nanomaterials and organic functional groups. Nanofibers coated with nanomaterials, such as silver, copper, and zinc, show antibacterial effects, and polymer nanofibers functionalized by nano-sized metal oxides, such as TiO_2 and CuO, show catalytic activity. Moreover, additional functionalization can enhance different physical and chemical properties of polymer nanofibers.

The electrospun nanofiber membranes have a much higher porosity due to their unique interconnected fiber structure compared with the traditional polymer membrane synthesized by phase inversion. Thus far, nanofiber membranes have been used as a scaffold-like porous support layer to fabricate a new type of TFC membrane for the engineered osmosis process. It was reported that the support nanofiber membrane could facilitate the mixing of accumulated solutions with a diluted feed stream and overcome the internal concentration polarization (ICP) problem as well as increase the potential of the electro-spun

Table 3. Various functionalized electrospun nanofibers

Polymer	Functionalization	Effects
Poly lactide/chitosan (De Faria, Perreault, Shaulsky, Arias Chavez, & Elimelech, 2015)	Graphene oxide decorated /silver nano-particles	Antibacterial effect
Teapolyphenols/polystyrene (Z. Liu, Yan, Miao, Huang, & Liu, 2015)	Silver nano-particle	Catalytic activities Antibacterial activity
PU (Pant et al., 2014)	Silver-doped fly ash	Enhanced absorption capacity
PVA (Destaye, Lin, & Lee, 2013)	Silver-nano particle	Mechanically strong Antibacterial effect
PVA/PDMAPAAm (Yun, Jin, Lee, & Kim, 2010)	Protonation of amine Titanium dioxide particle	pH sensitive Enhanced absorption capacity
PVP (S. S. Lee, Bai, Liu, & Sun, 2013)	TiO2/CuO	Photocatalytic activities
PVP (Quirós, Borges, Boltes, Rodea-Palomares, & Rosal, 2015)	Silver, copper, and zinc nanoparticles	Antibacterial effect
PVP (M. J. Nalbandian et al., 2015)	Ag and TiO_2	Photocatalytic activities Antibacterial effect
Bismuth vanadate (Michael J. Nalbandian et al., 2015)	Ag and Au	VL-activated photocatalysis
PAN, PSu (L. Huang, Manickam, & McCutcheon, 2013)	Exposed DMF vapor	Improve mechanical strength

nanofiber-supported TFC membranes for osmosis process applications due to the superior porosity and pore interconnectivity of the electrospun nanofiber membranes. Various types of electrospun nanofiber-supported TFC membranes for the osmosis process are listed in Table 4.

Polyamide has been widely used as an active layer, and various polymer nanofibers, such as PAN, PVDF, and PVA, have been used as porous support layers. Fabricated electrospun nanofiber-supported TFC membranes have been applied in osmosis-based desalination and energy harvesting processes. Electrospun nanofiber-supported TFC membranes increased the water flux by 200 to 300% and reduced the salt passage by 90% (Bui & McCutcheon, 2013); however, electrospun nanofiber-based FO/PRO

Table 4. Various types of electrospun nanofiber supported TFC membranes for the osmosis process

Polymer	Active Layer	Applications
PAN+CA (Bui & McCutcheon, 2013)	PA	FO, PRO
PVDF (M. Tian, Qiu, Liao, Chou, & Wang, 2013)	PA	FO
PA (Liwei Huang & McCutcheon, 2014)	PA	FO, PRO
PAN (Bui & McCutcheon, 2014)	PA	PRO
PVA/PET (E. L. Tian et al., 2014)	PA	FO
PAN (Kaur, Sundarrajan, Rana, Matsuura, & Ramakrishna, 2012)	PA	NF
PVDF (Park et al., 2012)	PEI (cross-linked)	NF

membranes are still in the early stage of development. Since electrospun nanofiber membrane fabrication is not strictly constrained by the materials used in the phase inversion, it provides an opportunity to explore the applicability of different types of polymers for making electrospun nanofiber membranes, which can be used as the substrate for FO/PRO membrane preparation (M. Tian et al., 2013).

Challenges of Applying Electrospun Nanofiber as Water Treatment Membranes

Despite the huge potential of electrospun nanofiber as water treatment membranes, critical drawbacks remain for field applications of electrospun nanofiber membrane (ENM) due to their mechanical, geometrical properties.

One of the drawbacks of using electrospun nanofibers in liquid filtration applications is that they are mechanically unstable compared to cast membranes made from the same polymer. ENMs have demonstrated very poor mechanical strength because the fibers do not adhere to each other and can be easily disintegrated to form a cotton-like fluffy structure. Therefore, many researchers have used the support layer to apply ENMs to the water treatment process. In addition, due to their geometrical properties, membrane cleaning by reverse flow is very difficult. When high-pressure reverse fluid is introduced at the bottom of an ENM, the ENM becomes delaminated and possibly unusable. Another drawback of ENMs is their weakness for membrane fouling. Generally, when a membrane's surface is quite rough, membrane fouling is more likely. ENMs have a very rough surface due to their nonwoven structure, so their real field applicability is reduced. Maeng reported that significant short-term fouling occurred when MBR was used with an electrospun nanofibrous membrane (H.-C. Kim et al., 2014). Lastly, compared to cast membranes, ENMs are very difficult to modulate. ENMs were fabricated as a flat form; thus, it was necessary to develop a modulation technique that is suitable for real field applications of ENMs. The proposed challenges should be carefully considered or addressed for their field applications in water treatment.

CONCLUSION

Nanotechnology is gaining momentum as a solution to resolve global water and energy shortages, which are currently major issues. The technical progress of conventional polymer-based membranes is stagnant, and biomimetic membrane technology, considered a next-generation membrane, is still insufficiently developed. Therefore, the convergence of nanotechnology and membrane technology has been proposed as a potentially powerful and alternative solution to lead recent membrane technology research for the water-energy nexus.

To realize the application and commercialization of a nanotechnology-based membrane, there are some obstacles in replacing conventional membranes: 1) the prevention of the nanomaterials from leaching from the membranes' matrix for long-term operation regarding potential risks to the environment, 2) the investigation of the membranes' durability, particularly for electrospun nanofiber membranes, 3) the development of standard membrane fabrication, and 4) the evaluation of the operating costs and the optimization of nanotechnology-based membrane processes.

REFERENCES

Ahn, C. H., Baek, Y., Lee, C., Kim, S. O., Kim, S., Lee, S., & Yoon, J. et al. (2012). Carbon nanotube-based membranes: Fabrication and application to desalination. *Journal of Industrial and Engineering Chemistry, 18*(5), 1551–1559. doi:10.1016/j.jiec.2012.04.005

Amini, M., Jahanshahi, M., & Rahimpour, A. (2013). Synthesis of novel thin film nanocomposite (TFN) forward osmosis membranes using functionalized multi-walled carbon nanotubes. *Journal of Membrane Science, 435*(0), 233–241. doi:10.1016/j.memsci.2013.01.041

Banerjee, S., Hemraj-Benny, T., & Wong, S. S. (2005). Covalent Surface Chemistry of Single-Walled Carbon Nanotubes. *Advanced Materials, 17*(1), 17–29. doi:10.1002/adma.200401340

Barhate, R. S., Loong, C. K., & Ramakrishna, S. (2006). Preparation and characterization of nanofibrous filtering media. *Journal of Membrane Science, 283*(1-2), 209–218. doi:10.1016/j.memsci.2006.06.030

Baughman, R. H., Zakhidov, A. A., & de Heer, W. A. (2002). Carbon Nanotubes--the Route Toward Applications. *Science, 297*(5582), 787–792. doi:10.1126/science.1060928 PMID:12161643

Bui, N. N., & McCutcheon, J. R. (2013). Hydrophilic nanofibers as new supports for thin film composite membranes for engineered osmosis. *Environmental Science & Technology, 47*(3), 1761–1769. doi:10.1021/es304215g PMID:23234259

Bui, N. N., & McCutcheon, J. R. (2014). Nanofiber supported thin-film composite membrane for pressure-retarded osmosis. *Environmental Science & Technology, 48*(7), 4129–4136. doi:10.1021/es4037012 PMID:24387600

Buonomenna, M. G. (2013). Nano-enhanced reverse osmosis membranes. *Desalination, 314*, 73–88. doi:10.1016/j.desal.2013.01.006

Byrne, M. T., & Gun'ko, Y. K. (2010). Recent Advances in Research on Carbon Nanotube–Polymer Composites. *Advanced Materials, 22*(15), 1672–1688. doi:10.1002/adma.200901545 PMID:20496401

Celik, E., Liu, L., & Choi, H. (2011). Protein fouling behavior of carbon nanotube/polyethersulfone composite membranes during water filtration. *Water Research, 45*(16), 5287–5294. doi:10.1016/j.watres.2011.07.036 PMID:21862096

Celik, E., Park, H., Choi, H., & Choi, H. (2011). Carbon nanotube blended polyethersulfone membranes for fouling control in water treatment. *Water Research, 45*(1), 274–282. doi:10.1016/j.watres.2010.07.060 PMID:20716459

Chan, W.-F., Chen, H.-y., Surapathi, A., Taylor, M. G., Shao, X., Marand, E., & Johnson, J. K. (2013). Zwitterion Functionalized Carbon Nanotube/Polyamide Nanocomposite Membranes for Water Desalination. *ACS Nano, 7*(6), 5308–5319. doi:10.1021/nn4011494 PMID:23705642

Chen, W., Su, Y., Zhang, L., Shi, Q., Peng, J., & Jiang, Z. (2010). In situ generated silica nanoparticles as pore-forming agent for enhanced permeability of cellulose acetate membranes. *Journal of Membrane Science, 348*(1–2), 75–83. doi:10.1016/j.memsci.2009.10.042

Chiang, I. W., Brinson, B. E., Smalley, R. E., Margrave, J. L., & Hauge, R. H. (2001). Purification and Characterization of Single-Wall Carbon Nanotubes. *The Journal of Physical Chemistry B*, *105*(6), 1157–1161. doi:10.1021/jp003453z

Choi, H.-, Son, M., Yoon, S. H., Celik, E., Kang, S., Park, H., & Choi, H et al.. (2015). Alginate fouling reduction of functionalized carbon nanotube blended cellulose acetate membrane in forward osmosis. *Chemosphere*, *136*, 204–210. doi:10.1016/j.chemosphere.2015.05.003 PMID:26022283

Coleman, J. N., Khan, U., Blau, W. J., & Gun'ko, Y. K. (2006). Small but strong: A review of the mechanical properties of carbon nanotube-polymer composites. *Carbon*, *44*(9), 1624–1652. doi:10.1016/j.carbon.2006.02.038

Corry, B. (2007). Designing Carbon Nanotube Membranes for Efficient Water Desalination. *The Journal of Physical Chemistry B*, *112*(5), 1427–1434. doi:10.1021/jp709845u PMID:18163610

Dai, H. (2002). Carbon Nanotubes: Synthesis, Integration, and Properties. *Accounts of Chemical Research*, *35*(12), 1035–1044. doi:10.1021/ar0101640 PMID:12484791

Das, R., Abd Hamid, S. B., Ali, M. E., Ismail, A. F., Annuar, M. S. M., & Ramakrishna, S. (2014). Multifunctional carbon nanotubes in water treatment: The present, past and future. *Desalination*, *354*, 160–179. doi:10.1016/j.desal.2014.09.032

De Faria, A. F., Perreault, F., Shaulsky, E., Arias Chavez, L. H., & Elimelech, M. (2015). Antimicrobial Electrospun Biopolymer Nanofiber Mats Functionalized with Graphene Oxide-Silver Nanocomposites. *ACS Applied Materials & Interfaces*, *7*(23), 12751–12759. doi:10.1021/acsami.5b01639 PMID:25980639

de Lannoy, C.-F., Jassby, D., Gloe, K., Gordon, A. D., & Wiesner, M. R. (2013). Aquatic Biofouling Prevention by Electrically Charged Nanocomposite Polymer Thin Film Membranes. *Environmental Science & Technology*, *47*(6), 2760–2768. doi:10.1021/es3045168 PMID:23413920

Destaye, A. G., Lin, C. K., & Lee, C. K. (2013). Glutaraldehyde vapor cross-linked nanofibrous PVA mat with in situ formed silver nanoparticles. *ACS Applied Materials & Interfaces*, *5*(11), 4745–4752. doi:10.1021/am401730x PMID:23668250

Donlan, R. M., & Costerton, J. W. (2002). Biofilms: Survival mechanisms of clinically relevant microorganisms. *Clinical Microbiology Reviews*, *15*(2), 167–193. doi:10.1128/CMR.15.2.167-193.2002 PMID:11932229

Doshi, J., & Reneker, D. H. (1995). Electrospinning process and applications of electrospun fibers. *Journal of Electrostatics*, *35*(2-3), 151–160. doi:10.1016/0304-3886(95)00041-8

Eder, D. (2010). Carbon Nanotube–Inorganic Hybrids. *Chemical Reviews*, *110*(3), 1348–1385. doi:10.1021/cr800433k PMID:20108978

El Badawi, N., Ramadan, A. R., Esawi, A. M. K., & El-Morsi, M. (2014). Novel carbon nanotube–cellulose acetate nanocomposite membranes for water filtration applications. *Desalination*, *344*(0), 79–85. doi:10.1016/j.desal.2014.03.005

Emadzadeh, D., Lau, W. J., & Ismail, A. F. (2013). Synthesis of thin film nanocomposite forward osmosis membrane with enhancement in water flux without sacrificing salt rejection. *Desalination, 330*(0), 90–99. doi:10.1016/j.desal.2013.10.003

Emadzadeh, D., Lau, W. J., Matsuura, T., Ismail, A. F., & Rahbari-Sisakht, M. (2014). Synthesis and characterization of thin film nanocomposite forward osmosis membrane with hydrophilic nanocomposite support to reduce internal concentration polarization. *Journal of Membrane Science, 449*(0), 74–85. doi:10.1016/j.memsci.2013.08.014

Emadzadeh, D., Lau, W. J., Matsuura, T., Rahbari-Sisakht, M., & Ismail, A. F. (2014). A novel thin film composite forward osmosis membrane prepared from PSf–TiO2 nanocomposite substrate for water desalination. *Chemical Engineering Journal, 237*, 70–80. doi:10.1016/j.cej.2013.09.081

Fornasiero, F., Park, H. G., Holt, J. K., Stadermann, M., Grigoropoulos, C. P., Noy, A., & Bakajin, O. (2008). Ion exclusion by sub-2-nm carbon nanotube pores. *Proceedings of the National Academy of Sciences of the United States of America, 105*(45), 17250–17255. doi:10.1073/pnas.0710437105 PMID:18539773

Gopal, R., Kaur, S., Ma, Z., Chan, C., Ramakrishna, S., & Matsuura, T. (2006). Electrospun nanofibrous filtration membrane. *Journal of Membrane Science, 281*(1–2), 581–586. doi:10.1016/j.memsci.2006.04.026

He, J. H., Wan, Y. Q., & Yu, J. Y. (2005). Scaling law in electrospinning: Relationship between electric current and solution flow rate. *Polymer, 46*(8), 2799–2801. doi:10.1016/j.polymer.2005.01.065

Hinds, B. J., Chopra, N., Rantell, T., Andrews, R., Gavalas, V., & Bachas, L. G. (2004). Aligned Multiwalled Carbon Nanotube Membranes. *Science, 303*(5654), 62–65. doi:10.1126/science.1092048 PMID:14645855

Holzinger, M., Vostrowsky, O., Hirsch, A., Hennrich, F., Kappes, M., Weiss, R., & Jellen, F. (2001). Sidewall Functionalization of Carbon Nanotubes. *Angewandte Chemie International Edition, 40*(21), 4002–4005. doi:10.1002/1521-3773(20011105)40:21<4002::AID-ANIE4002>3.0.CO;2-8 PMID:12404474

Hu, M., & Mi, B. (2013). Enabling graphene oxide nanosheets as water separation membranes. *Environmental Science & Technology, 47*(8), 3715–3723. doi:10.1021/es400571g PMID:23488812

Huang, H., Qu, X., Dong, H., Zhang, L., & Chen, H. (2013). Role of NaA zeolites in the interfacial polymerization process towards a polyamide nanocomposite reverse osmosis membrane. *RSC Advances, 3*(22), 8203–8207. doi:10.1039/c3ra40960k

Huang, H., Qu, X., Ji, X., Gao, X., Zhang, L., Chen, H., & Hou, L. (2013). Acid and multivalent ion resistance of thin film nanocomposite RO membranes loaded with silicalite-1 nanozeolites. *Journal of Materials Chemistry A, 1*(37), 11343–11349. doi:10.1039/c3ta12199b

Huang, L., Manickam, S. S., & McCutcheon, J. R. (2013). Increasing strength of electrospun nanofiber membranes for water filtration using solvent vapor. *Journal of Membrane Science, 436*, 213–220. doi:10.1016/j.memsci.2012.12.037

Huang, L., & McCutcheon, J. R. (2014). Hydrophilic nylon 6,6 nanofibers supported thin film composite membranes for engineered osmosis. *Journal of Membrane Science, 457*, 162–169. doi:10.1016/j.memsci.2014.01.040

Huang, Y.-W., Wu, C.-, & Aronstam, R. S. (2010). Toxicity of Transition Metal Oxide Nanoparticles: Recent Insights from in vitro Studies. *Materials (Basel)*, *3*(10), 4842–4859. doi:10.3390/ma3104842

Jadav, G. L., Aswal, V. K., & Singh, P. S. (2010). SANS study to probe nanoparticle dispersion in nanocomposite membranes of aromatic polyamide and functionalized silica nanoparticles. *Journal of Colloid and Interface Science*, *351*(1), 304–314. doi:10.1016/j.jcis.2010.07.028 PMID:20701923

Jadav, G. L., & Singh, P. S. (2009). Synthesis of novel silica-polyamide nanocomposite membrane with enhanced properties. *Journal of Membrane Science*, *328*(1–2), 257–267. doi:10.1016/j.memsci.2008.12.014

Jeong, B.-H., Hoek, E. M. V., Yan, Y., Subramani, A., Huang, X., Hurwitz, G., & Jawor, A. et al. (2007). Interfacial polymerization of thin film nanocomposites: A new concept for reverse osmosis membranes. *Journal of Membrane Science*, *294*(1–2), 1–7. doi:10.1016/j.memsci.2007.02.025

Jin, H., Jeng, E. S., Heller, D. A., Jena, P. V., Kirmse, R., Langowski, J., & Strano, M. S. (2007). Divalent Ion and Thermally Induced DNA Conformational Polymorphism on Single-walled Carbon Nanotubes. *Macromolecules*, *40*(18), 6731–6739. doi:10.1021/ma070608t

Joseph, S., & Aluru, N. R. (2008). Why Are Carbon Nanotubes Fast Transporters of Water? *Nano Letters*, *8*(2), 452–458. doi:10.1021/nl072385q PMID:18189436

Kar, S., Bindal, R. C., & Tewari, P. K. (2012). Carbon nanotube membranes for desalination and water purification: Challenges and opportunities. *Nano Today*, *7*(5), 385–389. doi:10.1016/j.nantod.2012.09.002

Karousis, N., Tagmatarchis, N., & Tasis, D. (2010). Current Progress on the Chemical Modification of Carbon Nanotubes. *Chemical Reviews*, *110*(9), 5366–5397. doi:10.1021/cr100018g PMID:20545303

Kaur, S., Sundarrajan, S., Rana, D., Matsuura, T., & Ramakrishna, S. (2012). Influence of electrospun fiber size on the separation efficiency of thin film nanofiltration composite membrane. *Journal of Membrane Science*, *392-393*, 101–111. doi:10.1016/j.memsci.2011.12.005

Kempener, R., & Neumann, F. (2014). *Salinity gradient energy*. IRENA.

Kim, E.-S., Hwang, G., Gamal El-Din, M., & Liu, Y. (2012). Development of nanosilver and multi-walled carbon nanotubes thin-film nanocomposite membrane for enhanced water treatment. *Journal of Membrane Science*, *394–395*(0), 37–48. doi:10.1016/j.memsci.2011.11.041

Kim, H.-C., Choi, B. G., Noh, J., Song, K. G., Lee, S.-, & Maeng, S. K. (2014). Electrospun nanofibrous PVDF–PMMA MF membrane in laboratory and pilot-scale study treating wastewater from Seoul Zoo. *Desalination*, *346*, 107–114. doi:10.1016/j.desal.2014.05.005

Kim, H. J., Baek, Y., Choi, K., Kim, D.-G., Kang, H., Choi, Y.-S., & Lee, J.-C. et al. (2014). The improvement of antibiofouling properties of a reverse osmosis membrane by oxidized CNTs. *RSC Advances*, *4*(62), 32802–32810. doi:10.1039/C4RA06489E

Kim, H. J., Choi, K., Baek, Y., Kim, D.-G., Shim, J., Yoon, J., & Lee, J.-C. (2014). High-Performance Reverse Osmosis CNT/Polyamide Nanocomposite Membrane by Controlled Interfacial Interactions. *ACS Applied Materials & Interfaces*, *6*(4), 2819–2829. doi:10.1021/am405398f PMID:24467487

Kim, H. J., Choi, K., Baek, Y., Kim, D.-G., Shim, J., Yoon, J., & Lee, J.-C. (2014). High-Performance Reverse Osmosis CNT/Polyamide Nanocomposite Membrane by Controlled Interfacial Interactions. *ACS Applied Materials & Interfaces*, *6*(4), 2819–2829. doi:10.1021/am405398f PMID:24467487

Kim, S. W., Kim, T., Kim, Y. S., Choi, H. S., Lim, H. J., Yang, S. J., & Park, C. R. (2012). Surface modifications for the effective dispersion of carbon nanotubes in solvents and polymers. *Carbon*, *50*(1), 3–33. doi:10.1016/j.carbon.2011.08.011

Klaysom, , Cath, , & Depuydt, , & Vankelecom, I. F. J. (2013). Forward and pressure retarded osmosis: Potential solutions for global challenges in energy and water supply. *Chemical Society Reviews*, *42*(6959). PMID:23778699

Klaysom, C., Cath, T. Y., Depuydt, T., & Vankelecom, I. F. J. (2013). Forward and pressure retarded osmosis: Potential solutions for global challenges in energy and water supply. *Chemical Society Reviews*, 42. PMID:23778699

Lee, H. S., Im, S. J., Kim, J. H., Kim, H. J., Kim, J. P., & Min, B. R. (2008). Polyamide thin-film nanofiltration membranes containing TiO_2 nanoparticles. *Desalination*, *219*(1–3), 48–56. doi:10.1016/j.desal.2007.06.003

Lee, J., Chae, H.-R., Won, Y. J., Lee, K., Lee, C.-H., Lee, H. H., & Lee, J.- et al.. (2013). Graphene oxide nanoplatelets composite membrane with hydrophilic and antifouling properties for wastewater treatment. *Journal of Membrane Science*, *448*, 223–230. doi:10.1016/j.memsci.2013.08.017

Lee, K. P., Arnot, T. C., & Mattia, D. (2011). A review of reverse osmosis membrane materials for desalination—Development to date and future potential. *Journal of Membrane Science*, *370*(1-2), 1–22. doi:10.1016/j.memsci.2010.12.036

Lee, S. S., Bai, H., Liu, Z., & Sun, D. D. (2013). Novel-structured electrospun TiO2/CuO composite nanofibers for high efficient photocatalytic cogeneration of clean water and energy from dye wastewater. *Water Research*, *47*(12), 4059–4073. doi:10.1016/j.watres.2012.12.044 PMID:23541306

Lind, M. L., Eumine Suk, D., Nguyen, T.-V., & Hoek, E. M. V. (2010). Tailoring the Structure of Thin Film Nanocomposite Membranes to Achieve Seawater RO Membrane Performance. *Environmental Science & Technology*, *44*(21), 8230–8235. doi:10.1021/es101569p PMID:20942398

Lind, M. L., Jeong, B.-H., Subramani, A., Huang, X., & Hoek, E. M. V. (2009). Effect of mobile cation on zeolite-polyamide thin film nanocomposite membranes. *Journal of Materials Research*, *24*(5), 1624–1631. doi:10.1557/jmr.2009.0189

Lind, M. L., Suk, D. E., Nguyen, T.-V., & Hoek, E. M. V. (2010). Tailoring the Structure of Thin Film Nanocomposite Membranes to Achieve Seawater RO Membrane Performance. *Environmental Science & Technology*, *44*(21), 8230–8235. doi:10.1021/es101569p PMID:20942398

Liu, J., Rinzler, A. G., Dai, H., Hafner, J. H., Bradley, R. K., Boul, P. J., & Smalley, R. E. et al. (1998). Fullerene Pipes. *Science*, *280*(5367), 1253–1256. doi:10.1126/science.280.5367.1253 PMID:9596576

Liu, L., Son, M., Park, H., Celik, E., Bhattacharjee, C., & Choi, H. (2014). Efficacy of CNT-bound polyelectrolyte membrane by spray-assisted layer-by-layer (LbL) technique on water purification. *RSC Advances*, *4*(62), 32858–32865. doi:10.1039/C4RA05272B

Liu, Y., Wang, R., Ma, H., Hsiao, B. S., & Chu, B. (2013). High-flux microfiltration filters based on electrospun polyvinylalcohol nanofibrous membranes. *Polymer (United Kingdom)*, *54*(2), 548–556. doi:10.1016/j.polymer.2012.11.064

Liu, Z., Yan, J., Miao, Y. E., Huang, Y., & Liu, T. (2015). Catalytic and antibacterial activities of green-synthesized silver nanoparticles on electrospun polystyrene nanofiber membranes using tea polyphenols. *Composites. Part B, Engineering*, *79*, 217–223. doi:10.1016/j.compositesb.2015.04.037

Ma, Z., Kotaki, M., Inai, R., & Ramakrishna, S. (2005). Potential of nanofiber matrix as tissue-engineering scaffolds. *Tissue Engineering*, *11*(1-2), 101–109. doi:10.1089/ten.2005.11.101 PMID:15738665

Majumder, M., & Ajayan, P. M. (2010). Carbon Nanotube Membranes: A New Frontier in Membrane Science. In D. Enrico & G. Lidietta (Eds.), *Comprehensive Membrane Science and Engineering* (pp. 291–310). Oxford: Elsevier. doi:10.1016/B978-0-08-093250-7.00038-4

Mary Laura Lind, B.-H. J. (2009). Effect of mobile cation on zeolite-polyamide thin film nanocomposite membranes. *Journal of Materials Research*, *24*(5), 1624–1631. doi:10.1557/jmr.2009.0189

Mauter, M. S., & Elimelech, M. (2008). Environmental Applications of Carbon-Based Nanomaterials. *Environmental Science & Technology*, *42*(16), 5843–5859. doi:10.1021/es8006904 PMID:18767635

Melechko, A. V., Merkulov, V. I., McKnight, T. E., Guillorn, M. A., Klein, K. L., Lowndes, D. H., & Simpson, M. L. (2005). Vertically aligned carbon nanofibers and related structures: Controlled synthesis and directed assembly. *Journal of Applied Physics*, *97*(4), 041301. doi:10.1063/1.1857591

Misdan, N., Lau, W. J., & Ismail, A. F. (2012). Seawater Reverse Osmosis (SWRO) desalination by thin-film composite membrane—Current development, challenges and future prospects. *Desalination*, *287*(0), 228–237. doi:10.1016/j.desal.2011.11.001

Nalbandian, M. J., Zhang, M., Sanchez, J., Choa, Y.-H., Cwiertny, D. M., & Myung, N. V. (2015). Synthesis and optimization of BiVO4 and co-catalyzed BiVO4 nanofibers for visible light-activated photocatalytic degradation of aquatic micropollutants. *Journal of Molecular Catalysis A Chemical*, *404–405*, 18–26. doi:10.1016/j.molcata.2015.04.003

Nalbandian, M. J., Zhang, M., Sanchez, J., Kim, S., Choa, Y. H., Cwiertny, D. M., & Myung, N. V. (2015). Synthesis and optimization of Ag-TiO$_2$ composite nanofibers for photocatalytic treatment of impaired water sources. *Journal of Hazardous Materials*, *299*, 141–148. doi:10.1016/j.jhazmat.2015.05.053 PMID:26101968

Nazarenus, M., Zhang, Q., Soliman, M. G., del Pino, P., Pelaz, B., Carregal-Romero, S., & Parak, W. J. et al. (2014). In vitro interaction of colloidal nanoparticles with mammalian cells: What have we learned thus far? *Beilstein Journal of Nanotechnology*, *5*, 1477–1490. doi:10.3762/bjnano.5.161 PMID:25247131

Noy, A., Park, H. G., Fornasiero, F., Holt, J. K., Grigoropoulos, C. P., & Bakajin, O. (2007). Nanofluidics in carbon nanotubes. *Nano Today*, *2*(6), 22–29. doi:10.1016/S1748-0132(07)70170-6

Pant, H. R., Kim, H. J., Joshi, M. K., Pant, B., Park, C. H., Kim, J. I., & Kim, C. S. et al. (2014). One-step fabrication of multifunctional composite polyurethane spider-web-like nanofibrous membrane for water purification. *Journal of Hazardous Materials*, *264*, 25–33. doi:10.1016/j.jhazmat.2013.10.066 PMID:24269971

Park, J., Choi, W., Kim, S. H., Chun, B. H., Bang, J., & Lee, K. B. (2010). Enhancement of Chlorine Resistance in Carbon Nanotube Based Nanocomposite Reverse Osmosis Membranes. *Desalination and Water Treatment*, *15*(1-3), 198–204. doi:10.5004/dwt.2010.1686

Park, S. J., Cheedrala, R. K., Diallo, M. S., Kim, C., Kim, I. S., & Goddard, W. A. (2012). Nanofiltration membranes based on polyvinylidene fluoride nanofibrous scaffolds and crosslinked polyethyleneimine networks. *Journal of Nanoparticle Research*, *14*(7), 884. doi:10.1007/s11051-012-0884-7

Pendergast, M. M., & Hoek, E. M. V. (2011). A review of water treatment membrane nanotechnologies. *Energy & Environmental Science*, *4*.

Pendergast, M. T. M., Nygaard, J. M., Ghosh, A. K., & Hoek, E. M. V. (2010). Using nanocomposite materials technology to understand and control reverse osmosis membrane compaction. *Desalination*, *261*(3), 255–263. doi:10.1016/j.desal.2010.06.008

Plata, D. L., Gschwend, P. M., & Reddy, C. M. (2008). Industrially synthesized single-walled carbon nanotubes: Compositional data for users, environmental risk assessments, and source apportionment. *Nanotechnology*, *19*(18), 185706. doi:10.1088/0957-4484/19/18/185706 PMID:21825702

Quirós, J., Borges, J. P., Boltes, K., Rodea-Palomares, I., & Rosal, R. (2015). Antimicrobial electrospun silver-, copper- and zinc-doped polyvinylpyrrolidone nanofibers. *Journal of Hazardous Materials*, *299*, 298–305. doi:10.1016/j.jhazmat.2015.06.028 PMID:26142159

Rajaeian, B., Rahimpour, A., Tade, M. O., & Liu, S. (2013). Fabrication and characterization of polyamide thin film nanocomposite (TFN) nanofiltration membrane impregnated with TiO2 nanoparticles. *Desalination*, *313*(0), 176–188. doi:10.1016/j.desal.2012.12.012

Ramakrishna, S., Fujihara, K., Teo, W. E., Yong, T., Ma, Z., & Ramaseshan, R. (2006). Electrospun nanofibers: Solving global issues. *Materials Today*, *9*(3), 40–50. doi:10.1016/S1369-7021(06)71389-X

Rao, G. P., Lu, C., & Su, F. (2007). Sorption of divalent metal ions from aqueous solution by carbon nanotubes: A review. *Separation and Purification Technology*, *58*(1), 224–231. doi:10.1016/j.seppur.2006.12.006

Rivera-Gil, P., Jimenez De Aberasturi, D., Wulf, V., Pelaz, B., Del Pino, P., Zhao, Y., & Parak, W. J. et al. (2013). The Challenge To Relate the Physicochemical Properties of Colloidal Nanoparticles to Their Cytotoxicity. *Accounts of Chemical Research*, *46*(3), 743–749. doi:10.1021/ar300039j PMID:22786674

Sahoo, N. G., Rana, S., Cho, J. W., Li, L., & Chan, S. H. (2010). Polymer nanocomposites based on functionalized carbon nanotubes. *Progress in Polymer Science*, *35*(7), 837–867. doi:10.1016/j.progpolymsci.2010.03.002

Sharqawy, M. H., Zubair, S. M., & Lienhard, V. J. H. (2011). Second law analysis of reverse osmosis desalination plants: An alternative design using pressure retarded osmosis. *Energy, 36*(11), 6617–6626. doi:10.1016/j.energy.2011.08.056

Shelimov, K. B., Esenaliev, R. O., Rinzler, A. G., Huffman, C. B., & Smalley, R. E. (1998). Purification of single-wall carbon nanotubes by ultrasonically assisted filtration. *Chemical Physics Letters, 282*(5–6), 429–434. doi:10.1016/S0009-2614(97)01265-7

Shi, Z., Lian, Y., Liao, F., Zhou, X., Gu, Z., Zhang, Y., & Iijima, S. (1999). Purification of single-wall carbon nanotubes. *Solid State Communications, 112*(1), 35–37. doi:10.1016/S0038-1098(99)00278-1

Soenen, S. J., Parak, W. J., Rejman, J., & Manshian, B. (2015). (Intra)Cellular Stability of Inorganic Nanoparticles: Effects on Cytotoxicity, Particle Functionality, and Biomedical Applications. *Chemical Reviews, 115*(5), 2109–2135. doi:10.1021/cr400714j PMID:25757742

Son, M., Hyeon-gyu, C., Liu, L., Celik, E., & Park, H. (2015). Efficacy of carbon nanotube positioning in the polyethersulfone support layer on the performance of thin-film composite membrane for desalination. *Chemical Engineering Journal, 266*, 376–384. doi:10.1016/j.cej.2014.12.108

Son, M., Park, H., Liu, L., Choi, H., Kim, J. H., & Choi, H. (2016). Thin-film nanocomposite membrane with CNT positioning in support layer for energy harvesting from saline water. *Chemical Engineering Journal, 284*, 68–77. doi:10.1016/j.cej.2015.08.134

Song, X., Liu, Z., & Sun, D. D. (2013). Energy recovery from concentrated seawater brine by thin-film nanofiber composite pressure retarded osmosis membranes with high power density. *Energy & Environmental Science, 6*(4), 1199–1210. doi:10.1039/c3ee23349a

Spitalsky, Z., Tasis, D., Papagelis, K., & Galiotis, C. (2010). Carbon nanotube-polymer composites: Chemistry, processing, mechanical and electrical properties. *Progress in Polymer Science, 35*(3), 357–401. doi:10.1016/j.progpolymsci.2009.09.003

Striolo, A. (2006). The Mechanism of Water Diffusion in Narrow Carbon Nanotubes. *Nano Letters, 6*(4), 633–639. doi:10.1021/nl052254u PMID:16608257

Sun, Y.-P., Fu, K., Lin, Y., & Huang, W. (2002). Functionalized Carbon Nanotubes: Properties and Applications. *Accounts of Chemical Research, 35*(12), 1096–1104. doi:10.1021/ar010160v PMID:12484798

Thess, A., Lee, R., Nikolaev, P., Dai, H., Petit, P., Robert, J., & Smalley, R. E. et al. (1996). Crystalline Ropes of Metallic Carbon Nanotubes. *Science, 273*(5274), 483–487. doi:10.1126/science.273.5274.483 PMID:8662534

Tian, E. L., Zhou, H., Ren, Y. W., mirza, Z., Wang, X. Z., & Xiong, S. W. (2014). Novel design of hydrophobic/hydrophilic interpenetrating network composite nanofibers for the support layer of forward osmosis membrane. *Desalination, 347*, 207–214. doi:10.1016/j.desal.2014.05.043

Tian, M., Qiu, C., Liao, Y., Chou, S., & Wang, R. (2013). Preparation of polyamide thin film composite forward osmosis membranes using electrospun polyvinylidene fluoride (PVDF) nanofibers as substrates. *Separation and Purification Technology, 118*, 727–736. doi:10.1016/j.seppur.2013.08.021

Tian, M., Wang, R., Goh, K., Liao, Y., & Fane, A. G. (2015). Synthesis and characterization of high-performance novel thin film nanocomposite PRO membranes with tiered nanofiber support reinforced by functionalized carbon nanotubes. *Journal of Membrane Science, 486*, 151–160. doi:10.1016/j.memsci.2015.03.054

Tiraferri, A., Vecitis, C. D., & Elimelech, M. (2011). Covalent Binding of Single-Walled Carbon Nanotubes to Polyamide Membranes for Antimicrobial Surface Properties. *ACS Applied Materials & Interfaces, 3*(8), 2869–2877. doi:10.1021/am200536p PMID:21714565

Verweij, H., Schillo, M. C., & Li, J. (2007). Fast Mass Transport Through Carbon Nanotube Membranes. *Small, 3*(12), 1996–2004. doi:10.1002/smll.200700368 PMID:18022891

Wang, R., Liu, Y., Li, B., Hsiao, B. S., & Chu, B. (2012). Electrospun nanofibrous membranes for high flux microfiltration. *Journal of Membrane Science, 392-393*, 167–174. doi:10.1016/j.memsci.2011.12.019

Wang, X., Chen, X., Yoon, K., Fang, D., Hsiao, B. S., & Chu, B. (2005). High flux filtration medium based on nanofibrous substrate with hydrophilic nanocomposite coating. *Environmental Science & Technology, 39*(19), 7684–7691. doi:10.1021/es050512j PMID:16245845

Whitby, M., & Quirke, N. (2007). Fluid flow in carbon nanotubes and nanopipes. *Nat Nano, 2*(2), 87–94. doi:10.1038/nnano.2006.175 PMID:18654225

Xie, X.-L., Mai, Y.-W., & Zhou, X.-P. (2005). Dispersion and alignment of carbon nanotubes in polymer matrix: A review. *Materials Science and Engineering R Reports, 49*(4), 89–112. doi:10.1016/j.mser.2005.04.002

Yarin, A. L., Koombhongse, S., & Reneker, D. H. (2001). Taylor cone and jetting from liquid droplets in electrospinning of nanofibers. *Journal of Applied Physics, 90*(9), 4836–4846. doi:10.1063/1.1408260

Yin, J., Kim, E.-S., Yang, J., & Deng, B. (2012). Fabrication of a novel thin-film nanocomposite (TFN) membrane containing MCM-41 silica nanoparticles (NPs) for water purification. *Journal of Membrane Science, 423–424*(0), 238–246. doi:10.1016/j.memsci.2012.08.020

Yip, N. Y., Tiraferri, A., Phillip, W. A., Schiffman, J. D., Hoover, L. A., Kim, Y. C., & Elimelech, M.Ngai Yin Yip. (2011). Thin-Film Composite Pressure Retarded Osmosis Membranes for Sustainable Power Generation from Salinity Gradients. *Environmental Science & Technology, 45*(10), 4360–4369. doi:10.1021/es104325z PMID:21491936

Yoon, K., Hsiao, B. S., & Chu, B. (2009). High flux nanofiltration membranes based on interfacially polymerized polyamide barrier layer on polyacrylonitrile nanofibrous scaffolds. *Journal of Membrane Science, 326*(2), 484–492. doi:10.1016/j.memsci.2008.10.023

Yun, J., Jin, D., Lee, Y. S., & Kim, H. I. (2010). Photocatalytic treatment of acidic waste water by electrospun composite nanofibers of pH-sensitive hydrogel and TiO2. *Materials Letters, 64*(22), 2431–2434. doi:10.1016/j.matlet.2010.08.001

Zhang, L., Shi, G.-Z., Qiu, S., Cheng, L.-H., & Chen, H.-L. (2011). Preparation of high-flux thin film nanocomposite reverse osmosis membranes by incorporating functionalized multi-walled carbon nanotubes. *Desalination and Water Treatment, 34*(1-3), 19–24. doi:10.5004/dwt.2011.2801

Zhao, H., Qiu, S., Wu, L., Zhang, L., Chen, H., & Gao, C. (2014). Improving the performance of polyamide reverse osmosis membrane by incorporation of modified multi-walled carbon nanotubes. *Journal of Membrane Science, 450*(0), 249–256. doi:10.1016/j.memsci.2013.09.014

Zhu, Y., Wang, D., Jiang, L., & Jin, J. (2014). Recent progress in developing advanced membranes for emulsified oil/water separation. *NPG Asia Mater, 6*(5), e101. doi:10.1038/am.2014.23

KEY TERMS AND DEFINITIONS

Electrospinning: This technique forms micro-or nano-scale fine fibers from an organic or an inorganic precursor solution using an electrical charge. The characteristics of the electrospun fiber are quite similar to those of conventional fiber fabricated by dry spinning or electrospraying.

Forward Osmosis (FO): The process, which is based on the natural phenomenon of water molecule transport across a semi-permeable membrane, using the osmotic pressure difference between feed and permeate solutions as a driving force. A highly concentrated solution, a "draw" solution, is required to allow water molecules to pass through the membrane from a feed solution to a draw solution. Unlike reverse osmosis, the FO process does not require external hydraulic pressure to operate.

Microfiltration (MF): One of the conventional water treatment filtration processes, operated under pressure through a membrane of small pore size (0.1–10 μm; larger pores than for ultrafiltration). Generally, particles and bacteria are removed by the MF process.

Nano-Enhanced Membrane (NeM): A functionalized membrane with discrete nanoparticles or nanotubes.

Nanomaterials: Materials of which a single unit is between 1 and 1000 nanometers in size, but usually 1–100 nm, including nanoparticles and nanotubes.

Pressure-Retarded Osmosis (PRO): The membrane-based osmotic driven process to generate electric power. A permeated solution is pressurized by water passed through a membrane using osmotic pressure difference between feed and permeate solution, which is a similar concept to forward osmosis, but PRO requires hydraulic pressure to pressurize a permeate solution, not greater than the osmotic pressure difference. Typically, the salinity gradient has been utilized as an osmotic driving force, for example, seawater and river water.

Reverse Osmosis (RO): A water purifying process that uses a semipermeable membrane to remove particles, micro solutes, molecules, and ions from feed solution. In reverse osmosis, an applied hydraulic pressure is necessary to overcome osmotic pressure. The pore size of a reverse osmosis membrane is in the range of less than 1 nm on a nonporous scale, which allows water molecule permeation only.

Ultrafiltration (UF): A water-purifying filtration process, operated under pressure through a membrane of small pore size nanometers to hundreds of nanometers, which is smaller than that of microfiltration. Generally, colloids, virus, proteins, and macromolecules are removed using the UF process. This separation process is used in industry and research for purifying and concentrating macromolecular (10^3 - 10^6 Da) solutions.

Chapter 4
Removal of Emerging Contaminants from Water and Wastewater Using Nanofiltration Technology

Yang Hu
University of Waterloo, Canada

Wen Liu
Auburn University, USA

Yue Peng
Georgia Institute of Technology, USA

Dongye Zhao
Auburn University, USA

Jie Fu
Georgia Institute of Technology, USA

ABSTRACT

Conventional water/wastewater treatment methods are incapable of removing the majority of Emerging Contaminants (ECs) and a large amount of them and their metabolites are ultimately released to the aquatic environment or drinking water distribution networks. Recently, nanofiltration, a high pressure membrane filtration process, has shown to be superior to other conventional filtration methods, in terms of effluent quality, easy operation and maintenance procedures, low cost, and small required operational space. This chapter provides a comprehensive overview of the most relevant works available in literature reporting the use of nanofiltration for the removal of emerging contaminants from water and wastewater. The fundamental knowledge of nanofiltration such as separation mechanisms, characterization of nanofiltration membranes, and predictive modeling has also been introduced. The literature review has shown that nanofiltration is a promising tool to treat ECs in environmental cleaning and water purification processes.

1. INTRODUCTION

Emerging contaminants (ECs) can be broadly defined as any synthetic or naturally occurring chemicals but cause known or suspected adverse ecological and(or) human health effects. This ever-increasing contaminants pose potential environmental and health threat to the living organism (Bolong, Ismail,

DOI: 10.4018/978-1-5225-0585-3.ch004

Salim, & Matsuura, 2009). In some cases, due to low concentration emerging contaminants have likely existed for a long time, which may not be detected until new analytical methods are developed. Hence, some ECs are not necessarily new chemicals, which are from municipal, agricultural, and industrial wastewater sources and pathways (Petrović, Gonzalez, & Barceló, 2003).

There are two reasons why emerging contaminants are of continued concern for the health and safety of consuming public. The first one is the trace level of most ECs with concentration at μg/L or ng/L, and broad range of physiochemical characteristics, which make both detection and elimination extremely difficult (Richardson & Ternes, 2005); And the second one is the adverse health and environmental effects of ECs even at a low concentration. Some intermediate metabolites or transforming products of ECs (especially for endocrine disrupting chemicals) exhibit biologically active effect as well (Diamanti-Kandarakis et al., 2009).

The main issue of ECs is nonexistence of limiting regulations, especially for new compounds, by-products, pharmaceuticals as related to the water and wastewater treatment industry. The first Contaminant Candidate List (CCL) was created in 1998 by United States Environmental Protection Agency (US EPA) for contaminants that are currently not subject to any proposed or promulgated national primary drinking water regulations. Even through many efforts have been devoted, there is limited number of contaminants with Maximum Contaminant Level (MCL) regulated by US EPA.

The main groups of ECs are described in Table 1.

As most ECs are small organic molecules, most conventional water treatment processes exhibit ineffective removal of emerging contaminants. Given the variety of emerging contaminants, advanced water treatment methods should be suitable for this purpose. The ability to reject small organic contaminants makes nanofiltration membranes almost a nature choice for emerging contaminants. Compared with

Table 1. Main group of emerging contaminants, definitions and examples

Group	Definition	Examples	Environmental Risks	Comments
Pesticides	Any substance or mixture of substances intended for preventing, destroying, repelling, or mitigating any pest (by US EPA)	Atrazine, Diuron, Alachlor, Diazinon	Probable human carcinogen, endocrine disrupting potential	Dominant group in the list of persistent organic pollutants (Berg, Hagmeyer, & Gimbel, 1997)
Disinfection by-products (DBPs)	The byproducts formed by the reaction between disinfectants and naturally-occurring materials	Trihalomethanes, Haloacetic acids	Increased risk of cancer and liver, kidney, or central nervous system problems	
Endocrine disrupting chemicals (EDCs)	The chemicals which can interfere with body's endocrine system	Estrone, Estriol, Testosterone, Progesterone	Developmental, reproductive, neurological, and immune effects in both humans and wildlife	Certain fish and wildlife are easy to be affected.
Pharmaceutically Active Compounds (PhACs)	Pharmaceutical residues and their metabolites	Ibuprofen, Diclofenac, Diatrizoateb	Inherent potential for a wide range of physiological effects	Potential for induction of proliferation of antibiotic resistance (Van Wyk, 2015)

Figure 1. Filtration membrane spectrum
(www.axeonwater.com).

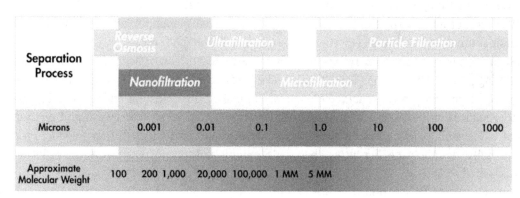

other membrane system (e.g., microfiltration, ultrafiltration and reverse osmosis), nanofiltration show remarkable potential for the removal of small organic contaminants (Figure 1), since the molecular weight cut-off values of most nanofiltration membranes are in the range of 200 to 300 Da, which coincides with the molecular weight of many ECs of concern.

Nanofiltration, defined as process between ultrafiltration and reverse osmosis, however is a controversial technology. The origin of nanofiltration can be traced back to 1960s, even before the term of nanofiltration was coined at the second half of 1980s (Loeb, Titelman, Korngold, & Freiman, 1997). In the beginning, the nanofiltration was categorized as open reverse osmosis or loose ultrafiltration. By the early 1970s, a full series of first-generation cellulose acetate membrane has been developed by Loeb and Sourirajan (Strathmann, Scheible, & Baker, 1971). The cellulose acetate was the standard material for nanofiltration until the composite nanofiltration was developed (Pepper, 1988). This breakthrough significantly enhanced the performance and popularity of nanofiltration, making nanofiltration membranes widely available. In the last decade, majority of studies on nanofiltration moved to chemically stable nanofiltration membranes. Many pH/solvent stable membranes were developed and commercially available (Musale & Kumar, 2000; Toh, Lim, & Livingston, 2007).

Nanofiltration membranes have been shown to be able to remove turbidity, microorganisms and hardness, as well as a fraction of the dissolved salts, which is similar with reverse osmosis. The initial application of nanofiltration in water industry was pretreatment of desalination process with significantly lower operating pressure, which brings energy-efficient advantages. However it is the high rejection of charged or uncharged organic compounds that makes nanofiltration membranes the suitable option to remove emerging contaminants. Even though the reverse osmosis membrane shows higher rejection of low molecular weight compounds, the lower operation pressure brings nanofiltration excellent efficiency. The use of NF to replace RO shows reduced energy costs and reduced volumes of toxic pollutants for the water treatment. For a typical nanofiltration membrane, the rejection of a divalent ion is above 95%, while the rejection of a monovalent ion can be anywhere between 20 and 80% with 600 - 100 PSI working pressure. For example, NF270 is a commercially available nanofiltration membrane with pure water permeability of 13.5 (L m^{-2} h^{-1} bar^{-1}) (Schaep, Vandecasteele, Mohammad, & Bowen, 2001).

In this chapter, the fundamental of nanofiltration will be introduced in second section, while the application of emerging contaminant removal will be discussed in third and fourth section.

2. FUNDAMENTAL OF NANOFILTRATION

As discussed in the first section, nanofiltration as a pressure-driven membrane has been studied for over 40 years. In this section, the fundamental of nanofiltration will be focused, and the material, the preparation, characterization and predictive models of the nanofiltration will be covered as well.

2.1. Materials and Preparation of Nanofiltration Membranes

There are three major groups of nanofiltration: asymmetric membranes, composite membranes and ceramic membranes. The first two belongs to the polymeric membranes. But the asymmetric membranes can be prepared via one step, the phase inversion method. The classical cellulose acetate membranes are the most important member of this group. Composite membranes are composed of porous nonselective supporting layer coated with a thin-film layer. This multilayer approach offers much more flexibility and control of the synthesis, which makes thin film composite nanofiltration more competitive over asymmetric membrane (Matsuura, 2001; Petersen, 1993; Song, Liu, & Sun, 2005). The membranes based on ceramics are the hot spot especially in the last decade. It has shown higher resistance against organic solvent and high pressure (Benhui, 1996).

2.1.1 Phase Inversion

Phase inversion is the process of controlled polymer transformation from a liquid into a solid state. Immersion precipitation is definitely the most popular method to prepare asymmetric membranes (Mulder, 1996). Since the casting solution of polymer and solvent is immersed into a non-solvent bath (coagulation bath) in typical immersion precipitation, the morphology and the performance of the membrane depends on thermodynamics as well as kinetics. Therefore the resulting membranes can be controlled by the composition of the casting solution, the type of polymer, coagulation medium and post treatment (if any) (Kawakami, Mikawa, & Nagaoka, 1996; Paulsen, Shojaie, & Krantz, 1994).

Even through the asymmetric membranes have been studied for decades, the general quantitative model for formation of asymmetric membrane by immersion precipitation is still under investigation. Based on the previous researches, some qualitative conclusions can be drawn as follows (Schäfer, Fane, & Waite, 2005):

- More charged group in polymer can lead to higher hydrophilic polymer, which would increase water flux of the membranes and alter the salt rejection (Dias, Rosa, & de Pinho, 1998; Stamatialis, Dias, & de Pinho, 1999).
- Increased polymer concentration of casting solution can lead to increased thickness and lower fluxes (Mulder, 1996).
- High interaction between the non-solvent and solvent often produces high porosity of the membrane by addition of additives and increased temperature of coagulation bath (in certain range) (Yeow, Liu, & Li, 2004).

2.1.2 Interfacial Polymerization

The interfacial polymerization has become a very successful and useful method to synthesize the composite nanofiltration membrane, since the concept of interfacial polymerization was first introduced in 1966 (Zimmermann, 1966). Due to better performance of composite membranes over asymmetric ones, many progresses have been made on the preparation of novel composite membranes in the last decade. Among the commonly used monomers, the cross-linked aromatic polyamide composite membrane produced by interfacial polymerization of phenylenediamine (MPD) and trimesoyl chloride (TMC) is still most successful in the present.

Compared with phase inversion, interfacial polymerization brings us more flexible optimization, which makes the research on interfacial polymerization more systemic. The hydrophilicity of polymer is still an important factor on membrane performance. Triamine 3,5-diamino-N-(4-aminophenyl) benzamide (DABA) monomer was used with m-phenylenediamine (MPD) and trimesoyl chloride (TMC) to improve the water flux of membranes, by more hydrophilic (giving rise to lower contact angles), smoother and thinner surface (H. Wang, Li, Zhang, & Zhang, 2010). Membranes with high resistance against chlorine (one of the common disinfectants used in water and wastewater treatment) were also prepared with 1,3-cyclohexanebis(methylamine) (CHMA) in water with TMC in hexane (Buch, Mohan, & Reddy, 2008). Besides the researches in polymer, many studies were performed on porous supports of composite membranes. By using novel support polymers, several thermally/solvent stable composite membranes were synthesized. For example, a thermal stable composite membrane by polymerization of piperazine (PIP) and TMC on poly (phthalazinone ether amide) (PPEA) membrane can function effectively at 80 °C (C. Wu, Zhang, Yang, & Jian, 2009).

Another important development in composite membrane is the addition of nanoscale particles into film of membranes, which is named as thin film nanocomposite (TFN) membrane. By incorporating of titanium oxide (TiO_2), this hybrid thin film composite show photocatalytically destructive capability on microorganisms, which can be used to solve biofouling problem (Kim, Kwak, Sohn, & Park, 2003). Surface modification of TiO_2 can also increase water flux up to 2-fold compared with the pure polyamide membrane with negligible rejection loss (Rajaeian, Rahimpour, Tade, & Liu, 2013). Multiwall carbon nanotubes (MWNTs)/polyester thin film nanocomposite membranes showed superior high water flux due to the high hydrophilic property (H. Wu, Tang, & Wu, 2010). Similar research with Metal–organic frameworks to yield high flux membrane (3-fold increase in permeance) was performed on polyamide (PA) thin film (Sorribas, Gorgojo, Téllez, Coronas, & Livingston, 2013).

2.1.3 Ceramic Membranes

Ceramic based membranes are a unique class of nanofiltration membranes due to inorganic components. Ceramic membrane shows higher chemical, structural and thermal stability over polymeric membranes. Even though it only holds a small membrane market share, many progresses of ceramic membranes were made, especially in solvent-resistant nanofiltration membrane area (Hendrix & Vankelecom, 2013). Similar to polymeric membranes, support materials are still required for membranes based on ceramic. The difference is mesoporous intermediate layer is usually chosen as support. Al, Si, Ti or Zr oxides can be used to build ceramic membranes (Guizard, Ayral, & Julbe, 2002).

Figure 2. Flow diagram for the preparation of microporous or mesoporous membranes via sol–gel techniques
(Hong & Elimelech, 1997).

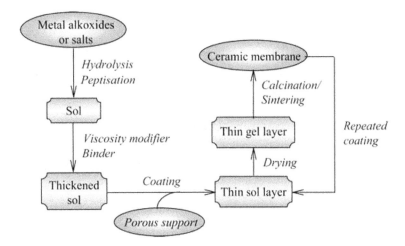

The general procedure to prepare active layer of ceramic membrane is sol-gel method, which involves converting colloidal solution into a gelatinous substance (Hsieh, 1996). The general procedure for ceramic membranes is shown in Figure 2.

2.2 Characterization of Nanofiltration Membrane

In this section, characterization for nanofiltration membranes will be divided into performance parameters, morphology parameters, as well as charge parameters to discuss (Artuğ & Hapke, 2006).

For performance, retention of the membrane and permeability coefficient are the most important parameters. By definition, the retention can be calculated with Equation 1:

$$R = \frac{c_f - c_p}{c_f} \tag{1}$$

in which c_f represents the concentration of the feed and c_p represents the concentration of permeate. The permeability is characterized by permeability coefficient (L_p), which is calculated by the slope of plotting the water flux as a function of the pressure.

For morphology, pore size, porosity, surface roughness and hydrophobicity are important parameters. Gas adsorption is usually the first choice to study the pore characteristics of solid materials (Schull, 1948), but for nanofiltration membrane with nanometer range pores (for example, the average pore diameter of NF90 is only 0.68 nm) this method is not accurate any more. Same issue occurs when permporometry is used to measure small pores, which produces same inaccurate results with gas adsorption (Cao, Meijernik, Brinkman, & Burggraaf, 1993). Microscopic techniques, especially atomic force microscopy (AFM) are very suitable for nanofiltration membranes. High resolution images of many commercial membranes were achieved by AFM. Even though several emerging technics or methods were applied to measure the pore size distribution, AFM and SEM are still preferred for the measurement (Otero et al., 2008).

Hydrophobicity is another essential morphology parameter, which is often characterized by contact angle (defined in Equation 2):

$$\cos\theta = \frac{\gamma_{SV} - \gamma_{SL}}{\gamma_{LV}} \qquad (2)$$

in which γ_{SV} and γ_{LV} are the surface tensions of the solid and the liquid with the vapor of the liquid and γ_{SL} is the tension of the solid-liquid interface.

The charge of the membrane can be obtained by measuring the streaming potential, and zeta potential.

2.3 Modeling of Nanofiltration Membrane

It has been 50 years since the nanofiltration concept was established in 1960s. But still a universal quantitative model for the description and prediction of nanofiltration performance has not been founded, though many theories have been established to describe the fundamental. Nanofiltration is a very complex process, because the rejection event occurs on a scale of the order of nanometer, which is very close to atomic dimensions. This would force us to consider the whole process with microscopic concept, which explains why more models of nanofiltration are transplanted from reserve osmosis. The filtration process involves several mechanisms, like steric interaction, charge interaction, adsorption and sorption-diffusion. Complex interactions between solutes, solvents and membranes make the nanofiltration modeling extremely difficult.

This section does not intend to include exhaustive physical equations, and the following will instead focus on the status of each modeling type. From a general point of view, modeling can be founded by experimental phenomena (empirical) or physical mechanism.

The empirical model, Spiegler-Kedem model, was initially founded at 1960s based on the irreversible thermodynamics and all the parameters and coefficients were determined experimentally (Kedem & Katchalsky, 1963). The composite membranes were considered as a "black box", which meant all the transport processes were not include in the model. This model was successfully applied in reverse osmosis membranes afterwards (Spiegler & Kedem, 1966). A few following researches confirmed the model was only valid under certain conditions, but the main drawback is lack of considering the variety of rejection and the charge effect of the membrane. The concentration dependency on rejection and the electrostatic exclusion are considered in the model developed later in 1989 and 1992 (Perry & Linder, 1989; Schirg & Widmer, 1992). Even though new parameters were included, the application of extended Spiegler-Kedem model was limited due to poor predictive ability.

The other models are based on transport-mechanistic modeling. These physical models take into account the transport mechanisms occurring in the membrane, which can be divided into two branches: space-charged (SC) model (Morrison Jr & Osterle, 1965) and Teorell-Meyer-Sievers (TMS) model (Teorell, 1953). In SC models, the pores on the membrane are considered as charged capillaries. Ions are considered as point charges, which are defined by Poisson-Boltzmann equation; while ion transport is defined by extended Nernst-Planck equation. With respect of TMS models, the radial distribution is neglected. The following researches confirmed two models to show good agreement for capillaries having smaller radius and lower surface charge density, which is the case of nanofiltration (X.-L. Wang, Tsuru, Nakao, & Kimura, 1995). Many modified models have been proposed recently based on the similar assumption, which shows a rapid growing realization of physical models.

By limited parameters and soluble mathematic calculations, well-founded models should serve for two purposes: description and prediction. Unfortunately, despite of many accomplishes in modeling of nanofiltration, the current modeling and simulation are more descriptive than predictive. Attempt to combine all current knowledge for a predictive model led to the conclusion that much depended on fitting parameters and not on physically relevant parameters (Straatsma, Bargeman, Van der Horst, & Wesselingh, 2002). For now empirical model for certain membrane is much reliable to predict membrane performance. Hence further research is still needed.

3. REMOVAL OF EMERGING CONTAMINANTS WITH NANOFILTRATION

During the last decade, nanofiltration has become more and more popular in drinking water treatment and wastewater treatment, not only for conventional water softening, but also for the removal of micro-pollutants, especially organic emerging contaminants. In this section, the removal of different types of emerging contaminants as well as the removal mechanisms will be discussed.

3.1. Removal of Emerging Contaminates

3.1.1 Pesticides

The use of pesticides to control unwanted pests dates back hundreds of years. Pesticides became a severe issue since the hundreds of synthetic organic chemicals were used without regulatory. Some of them can cause immediate toxic effects or act as potential carcinogen with long-term exposure, while some of them show superior stability in natural environment, such as DDT (dichlorodiphenyltrichloroethane). The use of nanofiltration for the removal of pesticides has been investigated by many researchers. To achieve

Table 2. Retention of selected pesticides using nanofiltration membranes

Compound	Membrane	Feed Concentration	Retention (%)	Reference
Atrazine	NF45	300 µg/L	91.8	(Van der Bruggen, Schaep, Maes, Wilms, & Vandecasteele, 1998)
	NF90	20 mg/L	95	(Ahmad, Tan, & Shukor, 2008)
	UTC20	5 mg/l	93	(Zhang, Van der Bruggen, Chen, Braeken, & Vandecasteele, 2004)
Simazine	NF70	300 µg/L	88.5	(Van der Bruggen et al., 1998)
	UTC20	5 mg/L	83	(Zhang et al., 2004)
Diuron	NF70	300 µg/L	85.9	(Van der Bruggen et al., 1998)
Isoproturon	NF70	300 µg/L	90.3	(Van der Bruggen et al., 1998)
	ESNA	149 µg/L	98.3	(Plakas, Karabelas, Wintgens, & Melin, 2006)
Isoxathion	NTR-729HF	n/a	99.84	(Kiso, Sugiura, Kitao, & Nishimura, 2001)

best removal outcome, optimized membrane properties, additives during filtration and different operation conditions have been tried. Table 2 shows high retention of selected pesticides with nanofiltration.

Ghaemi and coworkers performed a series of solid study about removal of nitroaromatic pesticides with modified nanofiltration membranes. They first investigated the effect of sodium dodecyl sulfate (SDS) anionic surfactant as additives in cellulose acetate (CA) nanofiltration membrane preparation. AFM measurements showed addition of SDS decreased thickness of the top layer and performance evaluation indicated higher pure water flux and rejection compared to CA membranes. The rejection of 3,5-Dinitrosalicylic acid (DNSA) and p-nitrophenol (PNP) reached 97.5% and 90% respectively under optimized condition (Ghaemi, Madaeni, Alizadeh, Daraei, Vatanpour, et al., 2012). Then they continued to study the effect of organic acids (ascorbic acid, citric acid and malic acid) as addictive in polysulfone nanofiltration membrane. Following the same strategy, retention of pesticides, 2,4-Dinitrophenol (DNP) and p-nitrophenol (PNP) reached 99% and 90% respectively with increasing hydrophilicity and water flux of the membrane (Ghaemi, Madaeni, Alizadeh, Daraei, Badieh, et al., 2012). Also the effects of addition of amphiphilic fatty acids (palmitic, oleic, and linoleic acids) were investigated on morphology, performance of cellulose acetate nanofiltration membranes and retention of nitroaromatic pesticides. Fabricated membranes presented higher efficiency in retention after the addition of palmitic acid with five pesticides as target solutes (Ghaemi, Madaeni, Alizadeh, Rajabi, et al., 2012). Košutić obtained good results for the removal of arsenic and pesticides from natural ground water. The NF270 nanofiltration membranes exhibited superior permeation rate values, which were more than five times higher than the corresponding values of the reverse osmosis CPA2 membrane. This can be explained by the charge exclusion mechanism during filtration process (Košutić, Furač, Sipos, & Kunst, 2005).

Interestingly some studies indicated dissolved natural organic matter (DOM) could enhance atrazine rejection in most cases and the enhancement was dependent upon the concentrations of the DOM and the atrazine and the ionic strength (Devitt, Ducellier, Cote, & Wiesner, 1998). Some following studies showed the presence of humic acid, sulfates, and chlorides can improve the elimination of certain pesticides (Košutić et al., 2005). But other contrasting results showed DOM can decrease or have negligible effect on micropollutant removal (Devitt et al., 1998). The charge exclusion is found to be the prevailing mechanism of small model compounds retention by the negatively charged nanofiltration membranes. The size exclusion mechanism also is observed to be important but not sufficient for the unionized small organic molecules rejections (Lin, Chiang, & Chang, 2007).

Besides the pesticide itself, the potential of nanofiltration membranes for treatment of groundwater polluted with some of these key transformation products was assessed. It turned out the nanofiltration membranes were capable of rejecting the regular pesticides, but did not give satisfactory rejections of the transformation products due to decreasing molecular size. Filtration system needed to be modified in order to obtain better performance (Madsen & Søgaard, 2014).

3.1.2 Disinfection By-Products (DBPs)

Many DBPs pose potential health risks due to carcinogenic effect with long-term exposure (Waller, Swan, DeLorenze, & Hopkins, 1998). In the past 40 years since the trihalomethanes (THMs) were identified as DBPs in drinking water, significant research efforts have been directed toward increasing our understanding of DBP formation and removal (Richardson & Postigo, 2012; Waller et al., 1998). Several publications have shown good rejection of DBP with nanofiltration. Vedat reported the removal efficiencies of THM by nanofiltration technique with commercial NF200 and DS5 membranes. With synthesized fee water, feed concentration and operating pressure show little effect on rejection, while the brominated THMs with higher molecular weight show higher removal efficiency compared with other species of THM (Uyak, Koyuncu, Oktem, Cakmakci, & Toroz, 2008).

Besides removing the DBPs directly, removing the DBP precursors is another promising strategy, like removal of bromide and other organic matters before disinfection to substantially lower the formation of DBP (Kimbrough & Suffet, 2002). Nanofiltration can lead to a high rejection of NOM and a high reduction of THM precursors in water, which can be efficient to reduce THM formation potential (De la Rubia, Rodríguez, León, & Prats, 2008). Commercial nanofiltration membranes, NF70 and NF270 were utilized to remove small THM precursors as well. High removals of THM precursors were observed at high pH values (Lin et al., 2007). Generally this strategy is more effective than directly removing DBPs as the DBP precursors are of large molecular weight.

Due to variety of DBPs, even though most NF studies reported promising DBP and DBP precursor rejections (Blau, Taylor, Morris, & Mulford, 1992; Jarusutthirak, Mattaraj, & Jiraratananon, 2007; Lee & Lee, 2007), removal pattern of individual species should be considered as well (Chellam, Sharma, Shetty, & Wei, 2008).

3.1.3 Pharmaceutically Active Compounds (PhACs)

Pharmaceutically active compounds (PhACs) present an increasingly large group of emerging contaminants. As pharmaceuticals are designed to be biologically active and chemically stable, more than 83 pharmaceuticals have been detected all over the world (Baena-Nogueras, Pintado-Herrera, González-Mazo, & Lara-Martín, 2015). Pharmaceuticals as well as their metabolites can influence none-target organisms with acute or chronic effect. Conventional sewage treatment shows unsatisfactory removal of PhACs, which is at concentration range of µg/L or lower (Heberer, 2002). Many researchers have reported almost 100% retention of PhACs with nanofiltration membranes in their studies. For example, two commercial composite membranes (SR2 and SR3) show 75–95% and 95–100% removal efficiency of five model PhACs with different physicochemical properties (i.e., cephalexin, tetracycline, acetaminophen, indomethacin and amoxicillin) (Zazouli, Susanto, Nasseri, & Ulbricht, 2009). Even in the full-scale drinking water treatment plant, nanofiltration was reported to successfully remove >85% of 12 PhACs (Radjenović, Petrović, Ventura, & Barceló, 2008).

3.1.4 Endocrine Disrupting Chemicals (EDCs)

Steroid hormones such as estrone are prominent in EDCs due to high endocrine-disrupting potency (Colborn, vom Saal, & Soto, 1993). Given the low concentration yet adverse effects in aquatic environment, EDCs have attracted significant interests recently (Andersen, Siegrist, Halling-Sørensen, & Ternes, 2003; Vos et al., 2000). However a definitive list of EDCs does not exist, as most chemicals in commerce have not been tested for endocrine toxicity. Nanofiltration membranes are demonstrated to have fair removal efficiencies for EDCs as well. Removal of estrone and 17β-estradiol with nanofiltration are summarized in Table 3. Numerous studies were also performed with other hormones as reference EDCs. Koyuncu et al. investigated the rejections of estradiol and testosterone (by 64% and 62%, respectively) by NF200 membrane (Koyuncu, Arikan, Wiesner, & Rice, 2008). Pereira et al. utilized the same NF200 membrane and obtained similar results. NF200 membrane rejected the selected hormones with efficiencies higher than 71% (except estriol with only 38% removal) directly from surface water (Pereira, Galinha,

Table 3. Retention of selected EDCs using nanofiltration membranes

Compound	Membrane	Feed Concentration	Retention (%)	Reference
Estrone	TFC-SR2	100 ng/L	13	(Agenson, Oh, Kikuta, & Urase, 2003; L. Nghiem, Manis, Soldenhoff, & Schäfer, 2004; L. D. Nghiem, Schäfer, & Elimelech, 2004; Weber, Gallenkemper, Melin, Dott, & Hollender, 2004)
	TFC-5	100 ng/L	76	
	X-20	100 ng/L	87	
	NF90	100 ng/L	89	
	NF270	100 ng/L	85	
	UTC60	50 ng/L	80	
	NTR7250	50 ng/L	57	
	PES10	100 ng/L	40	
17β-estradiol	TFC-SR2	100 ng/L	21	
	TFC-5	100 ng/L	82	
	NF90	100 ng/L	86	
	NF270	100 ng/L	85	
	UTC60	50 ng/L	72	
	NTR7250	50 ng/L	58	
	PES10	100 ng/L	50	

Crespo, Matos, & Crespo, 2012). Unlike the almost complete removal of other emerging contaminants, nanofiltration didn't work that well in EDCs removal, even the molecular weight of target EDCs was lower than molecular weight cut-off of the membrane in several studies (Sandra Sanches et al., 2012; Semião & Schäfer, 2013).

Even though, in the last decade many researches show promising result of EC revmoval. Most of studies were performed in lab scale without using "real" water. However, regardless of many breakthroughs in materials or configuration of the NF membrane itself, the potential of nanofiltration in industrial application is still underdeveloped.

3.2. Removal Mechanisms of Nanofiltration for Emerging Contaminants

As discussed in the last section, nanofiltration shows generally high retention of emerging contaminants and varying performance for different target micropollutants. Since the molecular weight cut-off values of nanofiltration membranes fall in the same range as the molecular weight of most emerging contaminants, the overall good rejection can be predicted. The reason for variations will be discussed in this section.

Size exclusion is the simplified, yet useful retention model, which basically depends on the pore size of membranes and the physical size of contaminants. In this model, average pore sizes and estimated diameter of solutes are the two key parameters. For the diameter of solute molecule, the most easily estimated method is molecular weight, and predictions of retention coefficient have been made based on molecular weight. Molecular weight cut-off, which is defined as the molecular weight of a solute with retention of 90%, is used as the primary parameter of commercial nanofiltration membranes. As molecular weight is not dimensional parameter, the hydrodynamic volume was tested and provided better prediction than the molecular weight (Meireles, Bessieres, Rogissart, Aimar, & Sanchez, 1995). In order to take structure parameter into account, Stoke radius was considered as better fit, regarding the molecules as spheres (Fei & Bart, 2001). However, some other methods indicate the molecules are not spherical, and the molecular width and length are calculated as STERIMOL parameters (Kiso et al., 2001). Computer aided methods have been introduced as well, random energetic optimization by software can produce molecular size directly. Recently researchers tend to choose multiple size parameters in order to produce more complicated and accurate predictions, like distribution coefficient, molecular weight, and molar volume, etc. (S Sanches, Galinha, Crespo, Pereira, & Crespo, 2013). In addition, operation conditions, like pH, ionic strength, and concentration of contaminants, can influence actual sizes of the contaminants and the pores (Childress & Elimelech, 2000).

The other models are involving the interactions between contaminants and membranes, which includes electrostatic interaction, physical adsorption, and hydrogen bonding (Braeken et al., 2005; L. D. Nghiem et al., 2004). Estrone is a typical example that size exclusion mechanism is not the dominant factor for retention, due to hydrogen bond and strong hydrophobic interaction with membranes. The accumulation of contaminants could lower the performance of membranes, which explains relatively low retention of estrone with nanofiltration (Braeken et al., 2005; D. Nghiem, Schäfer, & Waite, 2002).

Considering the complexity of nanofiltration process, these mechanisms always work together, even in many cases one mechanism can be a dominant factor, which is difficult to predict.

4. PLANT DESIGN

Even though major studies on nanofiltration are fundamental researches, like novel membrane development and retention mechanism, plant design of nanofiltration with respect to emerging contaminant removal is very crucial in practice.

In water treatment industry, all treatment technics should work together, including nanofiltration. And the pre-treatment and post-treatment may vary based on certain plant. Nanofiltration membranes are usually applied near the end of the long line of water treatment process, while they are used as pretreatment for desalination. For the removal of trace organic contaminants, pretreatment of the feed is extremely important for nanofiltration, which is the main principle in nanofiltration design. Besides to avoid fouling, mechanical damage from solids or oxidization from disinfection can lead to severe effect on performance. Due to the pore size in nanometer range, soluble inorganic and dissolved organic components are the major foulant species (Speth, Gusses, & Summers, 2000). Even though fouling can be minimized, but not prevented, by adjustment of the parameters of filtration process (like module configurations, flow, pressure, pH and temperature) (Maartens, Swart, & Jacobs, 1999), a series of pretreatments are still necessary. In general, the pretreatment for nanofiltration can be designed based on same requirement of reverse osmosis.

In the water treatment, nanofiltration is usually not the final step; the post-treatment may be disinfection and/or pH adjustment. UV-disinfection is used as post-treatment more than chlorination disinfection method, which is installed as a back-up.

5. CONCLUSION AND PROSPECTS

In this chapter, the removal of emerging contaminants with nanofiltration in water and wastewater treatment was discussed. Emerging contaminants, as micropollutants with low concentration, should be rejected efficiently by nanofiltration based on the fundamental understanding of filtration mechanism. Extensive studies in various scale confirmed this repeatedly with different model compounds, nanofiltration membranes. The retention mechanisms and factors affecting the membrane's retention were discussed in many researches. Computer aided analysis also improved the modeling and prediction of the nanofiltration. High retention (complete removal in some cases) of almost all emerging contaminant groups indicates a promising potential of nanofiltration in this area.

On the other hand, these findings are most obtained from laboratory-scale researches, and direct application in water treatment industry is lacking. For the practical application, the pretreatment requirement, module design and post-treatment options are important research topics associating with nanofiltration technology, and urgently need investigations.

The implementation of nanofiltration in the industry is a success story even with several challenges in practice: 1) higher energy consumption than ultrafiltration; 2) pre-treatment is needed; 3) nanofiltration membranes are a little more expensive than reverse osmosis membranes. Although it is unlikely to reach the total area of global, installed membrane for either reverse osmosis or ultrafiltration, it can be expected that large steps forward will be made in the soon future.

REFERENCES

Agenson, K., Oh, J., Kikuta, T., & Urase, T. (2003). Rejection mechanisms of plastic additives and natural hormones in drinking water treated by nanofiltration. *Water Supply*, *3*(5), 311–319.

Ahmad, A., Tan, L., & Shukor, S. A. (2008). Dimethoate and atrazine retention from aqueous solution by nanofiltration membranes. *Journal of Hazardous Materials*, *151*(1), 71–77. doi:10.1016/j.jhazmat.2007.05.047 PMID:17587496

Andersen, H., Siegrist, H., Halling-Sørensen, B., & Ternes, T. A. (2003). Fate of estrogens in a municipal sewage treatment plant. *Environmental Science & Technology*, *37*(18), 4021–4026. doi:10.1021/es026192a PMID:14524430

Artuğ, G., & Hapke, J. (2006). Characterization of nanofiltration membranes by their morphology, charge and filtration performance parameters. *Desalination*, *200*(1), 178–180. doi:10.1016/j.desal.2006.03.287

Baena-Nogueras, R. M., Pintado-Herrera, M. G., González-Mazo, E., & Lara-Martín, P. A. (2015). *Determination of Pharmaceuticals in Coastal Systems Using Solid Phase Extraction (SPE) Followed by Ultra Performance Liquid Chromatography–tandem Mass Spectrometry*. UPLC-MS/MS.

Benhui, S. Y. S. (1996). Preparation and Application of Nanofiltration Membrane. *Petrochemical Design*, *3*, 009.

Berg, P., Hagmeyer, G., & Gimbel, R. (1997). Removal of pesticides and other micropollutants by nanofiltration. *Desalination*, *113*(2), 205–208. doi:10.1016/S0011-9164(97)00130-6

Blau, T. J., Taylor, J. S., Morris, K. E., & Mulford, L. (1992). DBP control by nanofiltration: Cost and performance. *Journal - American Water Works Association, 84*(12), 104–116.

Bolong, N., Ismail, A., Salim, M. R., & Matsuura, T. (2009). A review of the effects of emerging contaminants in wastewater and options for their removal. *Desalination, 239*(1), 229–246. doi:10.1016/j.desal.2008.03.020

Braeken, L., Ramaekers, R., Zhang, Y., Maes, G., Van der Bruggen, B., & Vandecasteele, C. (2005). Influence of hydrophobicity on retention in nanofiltration of aqueous solutions containing organic compounds. *Journal of Membrane Science, 252*(1), 195–203. doi:10.1016/j.memsci.2004.12.017

Buch, P., Mohan, D. J., & Reddy, A. (2008). Preparation, characterization and chlorine stability of aromatic–cycloaliphatic polyamide thin film composite membranes. *Journal of Membrane Science, 309*(1), 36–44. doi:10.1016/j.memsci.2007.10.004

Cao, G., Meijernik, J., Brinkman, H., & Burggraaf, A. (1993). Permporometry study on the size distribution of active pores in porous ceramic membranes. *Journal of Membrane Science, 83*(2), 221–235. doi:10.1016/0376-7388(93)85269-3

Chellam, S., Sharma, R. R., Shetty, G. R., & Wei, Y. (2008). Nanofiltration of pretreated Lake Houston water: Disinfection by-product speciation, relationships, and control. *Separation and Purification Technology, 64*(2), 160–169. doi:10.1016/j.seppur.2008.09.007

Childress, A. E., & Elimelech, M. (2000). Relating nanofiltration membrane performance to membrane charge (electrokinetic) characteristics. *Environmental Science & Technology, 34*(17), 3710–3716. doi:10.1021/es0008620

Colborn, T., vom Saal, F. S., & Soto, A. M. (1993). Developmental effects of endocrine-disrupting chemicals in wildlife and humans. *Environmental Health Perspectives, 101*(5), 378–384. doi:10.1289/ehp.93101378 PMID:8080506

De la Rubia, A., Rodríguez, M., León, V. M., & Prats, D. (2008). Removal of natural organic matter and THM formation potential by ultra-and nanofiltration of surface water. *Water Research, 42*(3), 714–722. doi:10.1016/j.watres.2007.07.049 PMID:17765283

Devitt, E., Ducellier, F., Cote, P., & Wiesner, M. (1998). Effects of natural organic matter and the raw water matrix on the rejection of atrazine by pressure-driven membranes. *Water Research, 32*(9), 2563–2568. doi:10.1016/S0043-1354(98)00043-8

Diamanti-Kandarakis, E., Bourguignon, J.-P., Giudice, L. C., Hauser, R., Prins, G. S., Soto, A. M., & Gore, A. C. et al. (2009). Endocrine-disrupting chemicals: An Endocrine Society scientific statement. *Endocrine Reviews, 30*(4), 293–342. doi:10.1210/er.2009-0002 PMID:19502515

Dias, C. R., Rosa, M. J., & de Pinho, M. N. (1998). Structure of water in asymmetric cellulose ester membranes—and ATR-FTIR study. *Journal of Membrane Science, 138*(2), 259–267. doi:10.1016/S0376-7388(97)00226-3

Fei, W., & Bart, H.-J. (2001). Predicting diffusivities in liquids by the group contribution method. *Chemical Engineering and Processing: Process Intensification*, *40*(6), 531–535. doi:10.1016/S0255-2701(00)00151-3

Ghaemi, N., Madaeni, S. S., Alizadeh, A., Daraei, P., Badieh, M. M. S., Falsafi, M., & Vatanpour, V. (2012). Fabrication and modification of polysulfone nanofiltration membrane using organic acids: Morphology, characterization and performance in removal of xenobiotics. *Separation and Purification Technology*, *96*, 214–228. doi:10.1016/j.seppur.2012.06.008

Ghaemi, N., Madaeni, S. S., Alizadeh, A., Daraei, P., Vatanpour, V., & Falsafi, M. (2012). Fabrication of cellulose acetate/sodium dodecyl sulfate nanofiltration membrane: Characterization and performance in rejection of pesticides. *Desalination*, *290*, 99–106. doi:10.1016/j.desal.2012.01.013

Ghaemi, N., Madaeni, S. S., Alizadeh, A., Rajabi, H., Daraei, P., & Falsafi, M. (2012). Effect of fatty acids on the structure and performance of cellulose acetate nanofiltration membranes in retention of nitroaromatic pesticides. *Desalination*, *301*, 26–41. doi:10.1016/j.desal.2012.06.008

Guizard, C., Ayral, A., & Julbe, A. (2002). Potentiality of organic solvents filtration with ceramic membranes. A comparison with polymer membranes. *Desalination*, *147*(1), 275–280. doi:10.1016/S0011-9164(02)00552-0

Heberer, T. (2002). Occurrence, fate, and removal of pharmaceutical residues in the aquatic environment: A review of recent research data. *Toxicology Letters*, *131*(1), 5–17. doi:10.1016/S0378-4274(02)00041-3 PMID:11988354

Hendrix, K., & Vankelecom, I. F. (2013). Solvent-Resistant Nanofiltration Membranes. Encyclopedia of Membrane Science and Technology.

Hong, S., & Elimelech, M. (1997). Chemical and physical aspects of natural organic matter (NOM) fouling of nanofiltration membranes. *Journal of Membrane Science*, *132*(2), 159–181. doi:10.1016/S0376-7388(97)00060-4

Hsieh, H. (1996). *Inorganic membranes for separation and reaction*. Elsevier.

Jarusutthirak, C., Mattaraj, S., & Jiraratananon, R. (2007). Factors affecting nanofiltration performances in natural organic matter rejection and flux decline. *Separation and Purification Technology*, *58*(1), 68–75. doi:10.1016/j.seppur.2007.07.010

Kawakami, H., Mikawa, M., & Nagaoka, S. (1996). Gas permeability and selectivity through asymmetric polyimide membranes. *Journal of Applied Polymer Science*, *62*(7), 965–971. doi:10.1002/(SICI)1097-4628(19961114)62:7<965::AID-APP2>3.0.CO;2-Q

Kedem, O., & Katchalsky, A. (1963). Permeability of composite membranes. Part 1.—Electric current, volume flow and flow of solute through membranes. *Transactions of the Faraday Society*, *59*, 1918–1930. doi:10.1039/tf9635901918

Kim, S. H., Kwak, S.-Y., Sohn, B.-H., & Park, T. H. (2003). Design of TiO 2 nanoparticle self-assembled aromatic polyamide thin-film-composite (TFC) membrane as an approach to solve biofouling problem. *Journal of Membrane Science*, *211*(1), 157–165. doi:10.1016/S0376-7388(02)00418-0

Kimbrough, D. E., & Suffet, I. (2002). Electrochemical removal of bromide and reduction of THM formation potential in drinking water. *Water Research, 36*(19), 4902–4906. doi:10.1016/S0043-1354(02)00210-5 PMID:12448534

Kiso, Y., Sugiura, Y., Kitao, T., & Nishimura, K. (2001). Effects of hydrophobicity and molecular size on rejection of aromatic pesticides with nanofiltration membranes. *Journal of Membrane Science, 192*(1), 1–10. doi:10.1016/S0376-7388(01)00411-2

Košutić, K., Furač, L., Sipos, L., & Kunst, B. (2005). Removal of arsenic and pesticides from drinking water by nanofiltration membranes. *Separation and Purification Technology, 42*(2), 137–144. doi:10.1016/j.seppur.2004.07.003

Koyuncu, I., Arikan, O. A., Wiesner, M. R., & Rice, C. (2008). Removal of hormones and antibiotics by nanofiltration membranes. *Journal of Membrane Science, 309*(1), 94–101. doi:10.1016/j.memsci.2007.10.010

Lee, S., & Lee, C.-H. (2007). Effect of membrane properties and pretreatment on flux and NOM rejection in surface water nanofiltration. *Separation and Purification Technology, 56*(1), 1–8. doi:10.1016/j.seppur.2007.01.007

Lin, Y.-L., Chiang, P.-C., & Chang, E.-E. (2007). Removal of small trihalomethane precursors from aqueous solution by nanofiltration. *Journal of Hazardous Materials, 146*(1), 20–29. doi:10.1016/j.jhazmat.2006.11.050 PMID:17212977

Loeb, S., Titelman, L., Korngold, E., & Freiman, J. (1997). Effect of porous support fabric on osmosis through a Loeb-Sourirajan type asymmetric membrane. *Journal of Membrane Science, 129*(2), 243–249. doi:10.1016/S0376-7388(96)00354-7

Maartens, A., Swart, P., & Jacobs, E. (1999). Feed-water pretreatment: Methods to reduce membrane fouling by natural organic matter. *Journal of Membrane Science, 163*(1), 51–62. doi:10.1016/S0376-7388(99)00155-6

Madsen, H. T., & Søgaard, E. G. (2014). Applicability and modelling of nanofiltration and reverse osmosis for remediation of groundwater polluted with pesticides and pesticide transformation products. *Separation and Purification Technology, 125*, 111–119. doi:10.1016/j.seppur.2014.01.038

Matsuura, T. (2001). Progress in membrane science and technology for seawater desalination—a review. *Desalination, 134*(1), 47–54. doi:10.1016/S0011-9164(01)00114-X

Meireles, M., Bessieres, A., Rogissart, I., Aimar, P., & Sanchez, V. (1995). An appropriate molecular size parameter for porous membranes calibration. *Journal of Membrane Science, 103*(1), 105–115. doi:10.1016/0376-7388(94)00311-L

Morrison, F. Jr, & Osterle, J. (1965). Electrokinetic energy conversion in ultrafine capillaries. *The Journal of Chemical Physics, 43*(6), 2111–2115. doi:10.1063/1.1697081

Mulder, M. (1996). *Basic principles of membrane technology*. Springer Science & Business Media. doi:10.1007/978-94-009-1766-8

Musale, D. A., & Kumar, A. (2000). Solvent and pH resistance of surface crosslinked chitosan/poly (acrylonitrile) composite nanofiltration membranes. *Journal of Applied Polymer Science, 77*(8), 1782–1793. doi:10.1002/1097-4628(20000822)77:8<1782::AID-APP15>3.0.CO;2-5

Nghiem, D., Schäfer, A., & Waite, T. (2002). *Adsorption of estrone on nanofiltration and reverse osmosis membranes in water and wastewater treatment.* Academic Press.

Nghiem, L., Manis, A., Soldenhoff, K., & Schäfer, A. (2004). Estrogenic hormone removal from wastewater using NF/RO membranes. *Journal of Membrane Science, 242*(1), 37–45. doi:10.1016/j.memsci.2003.12.034

Nghiem, L. D., Schäfer, A. I., & Elimelech, M. (2004). Removal of natural hormones by nanofiltration membranes: Measurement, modeling, and mechanisms. *Environmental Science & Technology, 38*(6), 1888–1896. doi:10.1021/es034952r PMID:15074703

Otero, J., Mazarrasa, O., Villasante, J., Silva, V., Prádanos, P., Calvo, J., & Hernández, A. (2008). Three independent ways to obtain information on pore size distributions of nanofiltration membranes. *Journal of Membrane Science, 309*(1), 17–27. doi:10.1016/j.memsci.2007.09.065

Paulsen, F. G., Shojaie, S. S., & Krantz, W. B. (1994). Effect of evaporation step on macrovoid formation in wet-cast polymeric membranes. *Journal of Membrane Science, 91*(3), 265–282. doi:10.1016/0376-7388(94)80088-X

Pepper, D. (1988). RO-fractionation membranes. *Desalination, 70*(1), 89–93. doi:10.1016/0011-9164(88)85046-X

Pereira, V. J., Galinha, J., Crespo, M. T. B., Matos, C. T., & Crespo, J. G. (2012). Integration of nanofiltration, UV photolysis, and advanced oxidation processes for the removal of hormones from surface water sources. *Separation and Purification Technology, 95*, 89–96. doi:10.1016/j.seppur.2012.04.013

Perry, M., & Linder, C. (1989). Intermediate reverse osmosis ultrafiltration (RO UF) membranes for concentration and desalting of low molecular weight organic solutes. *Desalination, 71*(3), 233–245. doi:10.1016/0011-9164(89)85026-X

Petersen, R. J. (1993). Composite reverse osmosis and nanofiltration membranes. *Journal of Membrane Science, 83*(1), 81–150. doi:10.1016/0376-7388(93)80014-O

Petrović, M., Gonzalez, S., & Barceló, D. (2003). Analysis and removal of emerging contaminants in wastewater and drinking water. *TrAC Trends in Analytical Chemistry, 22*(10), 685–696. doi:10.1016/S0165-9936(03)01105-1

Plakas, K., Karabelas, A., Wintgens, T., & Melin, T. (2006). A study of selected herbicides retention by nanofiltration membranes—the role of organic fouling. *Journal of Membrane Science, 284*(1), 291–300. doi:10.1016/j.memsci.2006.07.054

Radjenović, J., Petrović, M., Ventura, F., & Barceló, D. (2008). Rejection of pharmaceuticals in nanofiltration and reverse osmosis membrane drinking water treatment. *Water Research, 42*(14), 3601–3610. doi:10.1016/j.watres.2008.05.020 PMID:18656225

Rajaeian, B., Rahimpour, A., Tade, M. O., & Liu, S. (2013). Fabrication and characterization of polyamide thin film nanocomposite (TFN) nanofiltration membrane impregnated with TiO 2 nanoparticles. *Desalination*, *313*, 176–188. doi:10.1016/j.desal.2012.12.012

Richardson, S. D., & Postigo, C. (2012). *Drinking water disinfection by-products. In Emerging organic contaminants and human health* (pp. 93–137). Springer.

Richardson, S. D., & Ternes, T. A. (2005). Water analysis: Emerging contaminants and current issues. *Analytical Chemistry*, *77*(12), 3807–3838. doi:10.1021/ac058022x PMID:15952758

Sanches, S., Galinha, C., Crespo, M. B., Pereira, V., & Crespo, J. (2013). Assessment of phenomena underlying the removal of micropollutants during water treatment by nanofiltration using multivariate statistical analysis. *Separation and Purification Technology*, *118*, 377–386. doi:10.1016/j.seppur.2013.07.020

Sanches, S., Penetra, A., Rodrigues, A., Ferreira, E., Cardoso, V. V., Benoliel, M. J., & Crespo, J. G. et al. (2012). Nanofiltration of hormones and pesticides in different real drinking water sources. *Separation and Purification Technology*, *94*, 44–53. doi:10.1016/j.seppur.2012.04.003

Schaep, J., Vandecasteele, C., Mohammad, A. W., & Bowen, W. R. (2001). Modelling the retention of ionic components for different nanofiltration membranes. *Separation and Purification Technology*, *22*(1-2), 169–179. doi:10.1016/S1383-5866(00)00163-5

Schäfer, A. I., Fane, A. G., & Waite, T. D. (2005). *Nanofiltration: principles and applications*. Elsevier.

Schirg, P., & Widmer, F. (1992). Characterisation of nanofiltration membranes for the separation of aqueous dye-salt solutions. *Desalination*, *89*(1), 89–107. doi:10.1016/0011-9164(92)80154-2

Schull, C. (1948). The determination of pore size distribution from gas adsorption data. *Journal of the American Chemical Society*, *70*(4), 1405–1410. doi:10.1021/ja01184a034

Semião, A. J., & Schäfer, A. I. (2013). Removal of adsorbing estrogenic micropollutants by nanofiltration membranes. Part A—Experimental evidence. *Journal of Membrane Science*, *431*, 244–256. doi:10.1016/j.memsci.2012.11.080

Song, Y., Liu, F., & Sun, B. (2005). Preparation, characterization, and application of thin film composite nanofiltration membranes. *Journal of Applied Polymer Science*, *95*(5), 1251–1261. doi:10.1002/app.21338

Sorribas, S., Gorgojo, P., Téllez, C., Coronas, J., & Livingston, A. G. (2013). High flux thin film nanocomposite membranes based on metal–organic frameworks for organic solvent nanofiltration. *Journal of the American Chemical Society*, *135*(40), 15201–15208. doi:10.1021/ja407665w PMID:24044635

Speth, T. F., Gusses, A. M., & Summers, R. S. (2000). Evaluation of nanofiltration pretreatments for flux loss control. *Desalination*, *130*(1), 31–44. doi:10.1016/S0011-9164(00)00072-2

Spiegler, K., & Kedem, O. (1966). Thermodynamics of hyperfiltration (reverse osmosis): Criteria for efficient membranes. *Desalination*, *1*(4), 311–326. doi:10.1016/S0011-9164(00)80018-1

Stamatialis, D. F., Dias, C. R., & de Pinho, M. N. (1999). Atomic force microscopy of dense and asymmetric cellulose-based membranes. *Journal of Membrane Science*, *160*(2), 235–242. doi:10.1016/S0376-7388(99)00089-7

Straatsma, J., Bargeman, G., Van der Horst, H., & Wesselingh, J. (2002). Can nanofiltration be fully predicted by a model? *Journal of Membrane Science, 198*(2), 273–284. doi:10.1016/S0376-7388(01)00669-X

Strathmann, H., Scheible, P., & Baker, R. (1971). A rationale for the preparation of Loeb-Sourirajan-type cellulose acetate membranes. *Journal of Applied Polymer Science, 15*(4), 811–828. doi:10.1002/app.1971.070150404

Teorell, T. (1953). Transport processes and electrical phenomena in ionic membranes. *Progress in Biophysics and Biophysical Chemistry, 3*, 305–369.

Toh, Y. S., Lim, F., & Livingston, A. (2007). Polymeric membranes for nanofiltration in polar aprotic solvents. *Journal of Membrane Science, 301*(1), 3–10.

Uyak, V., Koyuncu, I., Oktem, I., Cakmakci, M., & Toroz, I. (2008). Removal of trihalomethanes from drinking water by nanofiltration membranes. *Journal of Hazardous Materials, 152*(2), 789–794. doi:10.1016/j.jhazmat.2007.07.082 PMID:17768007

Van der Bruggen, B., Schaep, J., Maes, W., Wilms, D., & Vandecasteele, C. (1998). Nanofiltration as a treatment method for the removal of pesticides from ground waters. *Desalination, 117*(1), 139–147. doi:10.1016/S0011-9164(98)00081-2

Van Wyk, H. (2015). Antibiotic resistance[review]. *SA Pharmaceutical Journal, 82*(3), 20–23. PMID:26415379

Vos, J. G., Dybing, E., Greim, H. A., Ladefoged, O., Lambré, C., Tarazona, J. V., & Vethaak, A. D. et al. (2000). Health effects of endocrine-disrupting chemicals on wildlife, with special reference to the European situation. *CRC Critical Reviews in Toxicology, 30*(1), 71–133. doi:10.1080/10408440091159176 PMID:10680769

Waller, K., Swan, S. H., DeLorenze, G., & Hopkins, B. (1998). Trihalomethanes in drinking water and spontaneous abortion. *Epidemiology (Cambridge, Mass.), 9*(2), 134–140. doi:10.1097/00001648-199803000-00006 PMID:9504280

Wang, H., Li, L., Zhang, X., & Zhang, S. (2010). Polyamide thin-film composite membranes prepared from a novel triamine 3, 5-diamino-N-(4-aminophenyl)-benzamide monomer and m-phenylenediamine. *Journal of Membrane Science, 353*(1), 78–84. doi:10.1016/j.memsci.2010.02.033

Wang, X.-L., Tsuru, T., Nakao, S.-i., & Kimura, S. (1995). Electrolyte transport through nanofiltration membranes by the space-charge model and the comparison with Teorell-Meyer-Sievers model. *Journal of Membrane Science, 103*(1), 117–133. doi:10.1016/0376-7388(94)00317-R

Weber, S., Gallenkemper, M., Melin, T., Dott, W., & Hollender, J. (2004). Efficiency of nanofiltration for the elimination of steroids from water. *Water Science and Technology, 50*(5), 9–14. PMID:15497823

Wu, C., Zhang, S., Yang, D., & Jian, X. (2009). Preparation, characterization and application of a novel thermal stable composite nanofiltration membrane. *Journal of Membrane Science, 326*(2), 429–434. doi:10.1016/j.memsci.2008.10.033

Wu, H., Tang, B., & Wu, P. (2010). MWNTs/polyester thin film nanocomposite membrane: An approach to overcome the trade-off effect between permeability and selectivity. *The Journal of Physical Chemistry C*, *114*(39), 16395–16400. doi:10.1021/jp107280m

Yeow, M., Liu, Y., & Li, K. (2004). Morphological study of poly (vinylidene fluoride) asymmetric membranes: Effects of the solvent, additive, and dope temperature. *Journal of Applied Polymer Science*, *92*(3), 1782–1789. doi:10.1002/app.20141

Zazouli, M. A., Susanto, H., Nasseri, S., & Ulbricht, M. (2009). Influences of solution chemistry and polymeric natural organic matter on the removal of aquatic pharmaceutical residuals by nanofiltration. *Water Research*, *43*(13), 3270–3280. doi:10.1016/j.watres.2009.04.038 PMID:19520413

Zhang, Y., Van der Bruggen, B., Chen, G., Braeken, L., & Vandecasteele, C. (2004). Removal of pesticides by nanofiltration: Effect of the water matrix. *Separation and Purification Technology*, *38*(2), 163–172. doi:10.1016/j.seppur.2003.11.003

Zimmermann, R. (1966). Condensation Polymers: By Interfacial and Solution Methods. Von PW Morgan. John Wiley & Sons, New York London-Sydney 1965. 1. Aufl., XVIII, 561 S., zahlr. Abb., mehrere Tab., geb.£ 9.10.–. *Angewandte Chemie*, *78*(16), 787–787. doi:10.1002/ange.19660781632

Chapter 5
Long-Term Performance Evaluation of Groundwater Chlorinated Solvents Remediation Using Nanoscale Emulsified Zerovalent Iron at a Superfund Site

Chunming Su
United States Environmental Protection Agency, USA

Robert W. Puls
United States Environmental Protection Agency, USA (ret.)

Thomas A. Krug
Geosyntec Consultants Inc., Canada

Mark T. Watling
Geosyntec Consultants Inc., Canada

Suzanne K. O'Hara
Geosyntec Consultants Inc., Canada

Jacqueline W. Quinn
NASA Kennedy Space Center, USA

Nancy E. Ruiz
US Navy, USA

ABSTRACT

This chapter addresses a case study of long-term assessment of a field application of environmental nanotechnology. Dense Non-Aqueous Phase Liquid (DNAPL) contaminants such as Tetrachloroethene (PCE) and Trichloroethene (TCE) are a type of recalcitrant compounds commonly found at contaminated sites. Recent research has focused on their remediation using environmental nanotechnology in which nanomaterials such as nanoscale Emulsified Zerovalent Iron (EZVI) are added to the subsurface environment to enhance contaminant degradation. Such nanoremediation approach may be mostly applicable to the source zone where the contaminant mass is the greatest and source removal is a critical step in controlling the further spreading of the groundwater plume. Compared to micro-scale and granular

DOI: 10.4018/978-1-5225-0585-3.ch005

counterparts, NZVI exhibits greater degradation rates due to its greater surface area and reactivity from its faster corrosion. While NZVI shows promise in both laboratory and field tests, limited information is available about the long-term effectiveness of nanoremediation because previous field tests are mostly less than two years. Here an update is provided for a six-year performance evaluation of EZVI for treating PCE and its daughter products at a Superfund site at Parris Island, South Carolina, USA. The field test consisted of two side-by-side treatment plots to remedy a shallow PCE source zone (less than 6 m below ground surface) using pneumatic injection and direct injection, separately in October 2006. For the pneumatic injections, a two-step injection procedure was used. First, the formation was fluidized by the injection of nitrogen gas alone, followed by injection of the EZVI with nitrogen gas as the carrier. In the pneumatic injection plot, 2,180 liters of EZVI containing 225 kg of iron (Toda RNIP-10DS), 856 kg of corn oil, and 22.5 kg of surfactant were injected to remedy an estimated 38 kg of chlorinated volatile compounds (CVOC)s. Direct injections were performed using a direct push rig. In the direct injection plot, 572 liters of EZVI were injected to treat an estimated 0.155 kg of CVOCs. Visual inspection of collected soil cores before and after EZVI injections shows that the travel distance of EZVI was dependent on the method of delivery with pneumatic injection achieving a greater distance of 2.1 m than did direct injection reaching a distance of 0.89 m. Significant decreases in PCE and TCE concentrations were observed in downgradient wells with corresponding increases in degradation products including significant increases in ethene. In the pneumatic injection plot, there were significant reductions in the downgradient groundwater mass flux values for chlorinated ethenes (>58%) and a significant increase in the mass flux of ethene (628%). There were significant reductions in total CVOCs mass (78%), which was less than an estimated 86% decrease in total CVOCs made at 2.5 years due to variations in soil cores collected for CVOCs extraction and determination; an estimated reduction of 23% (vs.63% at 2.5 years) in the sorbed and dissolved phases and 95% (vs. 93% at 2.5 years) reduction in the PCE DNAPL mass. Significant increases in dissolved sulfide, volatile fatty acids (VFA), and total organic carbon (TOC) were observed and dissolved sulfate and pH decreased in many monitoring wells. The apparent effective destruction of CVOC was accomplished by a combination of abiotic dechlorination by nanoiron and biological reductive dechlorination stimulated by the oil in the emulsion. No adverse effects of EZVI were observed for the microbes. In contrast, populations of dehalococcoides showed an increase up to 10,000 fold after EZVI injection. The dechlorination reactions were sustained for the six-year period from a single EZVI delivery. Repeated EZVI injections four to six years apart may be cost-effective to more completely remove the source zone contaminant mass. Overall, the advantages of the EZVI technology include an effective "one-two punch" of rapid abiotic dechlorination followed by a sustained biodegradation; contaminants are destroyed rather than transferred to another medium; ability to treat both DNAPL source zones and dissolved-phase contaminants to contain plume migration; ability to deliver reactants to targeted zones not readily accessible by conventional permeable reactive barriers; and potential for lower overall costs relative to alternative technologies such as groundwater pump-and-treat with high operation and maintenance costs or thermal technologies with high capital costs. The main limitations of the EZVI technology are difficulty in effectively distributing the viscous EZVI to all areas impacted with DNAPL; potential decrease in hydraulic conductivity due to iron corrosion products buildup or biofouling; potential to adversely impact secondary groundwater quality through mobilization of metals and production of sulfides or methane; injection of EZVI may displace DNAPL away from the injection point; and repeated injections may be required to completely destroy the contaminants.

INTRODUCTION

Chlorinated volatile organic compounds (CVOCs) such as tetrachloroethene (PCE) and trichloroethene (TCE) are dense non-aqueous phase liquid (DNAPL) contaminants widely found in groundwater at Superfund sites in the U.S. They have very low maximum contaminant levels (MCL) for drinking water set up by the US Environmental Protection Agency (EPA) (0.005 mg L^{-1} for both PCE and TCE). There is an urgent need to develop and verify cost-effective methods for groundwater remediation. During the past decade, a novel approach called nanoremediation has gained support from both environmental research and application communities. Nanoremediation is an environmental nanotechnology that uses nanomaterials for environmental cleanup. It has the potential to decrease the overall costs of site remediation, reduce cleanup time, eliminate excavation and disposal of contaminated soil, and reduce contaminant concentration, and it can be done *in situ* (Karn et al., 2009). Of the nanomaterials explored for remediation, nanoscale zerovalent iron (NZVI) has been the most widely used in both laboratory and field studies for groundwater and hazardous waste treatment (Zhang, 2003; Li et al., 2006; Zhan et al., 2011; Tang and Lo, 2013; Guo et al., 2015). A pioneer field test showed potential applications of the nanoscale Fe/Pd bimetallic particles for treating groundwater chlorinated ethenes such as TCE (Elliott and Zhang, 2001). A latter field study conducted in Taiwan confirmed the effectiveness of Pd/Fe bimetallic particles for dechlorination of vinyl chloride (Wei et al., 2009). Other field studies have tested uncoated NZVI (Lacina et al., 2015) and a variety of stabilized and composite NZVI nanomaterials such as palladium-catalyzed and polymer-coated NZVI (Henn and Waddill, 2006) and carboxymethyl cellulose stabilized NZVI (Bennett et al., 2010; He et al., 2010) for groundwater remediation of chlorinated solvents with promising results.

Recent progress has been made in several key areas that has deepened our understanding of the merits and uncertainties of NZVI-based remediation applications. These areas include the materials chemistry of NZVI in its simple and modified forms, the NZVI reactivity with a wide spectrum of contaminants in addition to the well-documented chlorinated solvents, methods to enhance the colloidal stability and transport properties of NZVI in porous media, and the effects of NZVI amendment on the biogeochemical environment (Yan et al., 2013, O'Carroll et al., 2013). Nevertheless, concerns about the safety (the long-term fate, transformation, and ecotoxicity of NZVI in environmental systems) and efficiency of the NZVI technology have limited its applications (Crane and Scott, 2012). The extent and type of NZVI technology applications differ between Europe and the USA (Mueller et al., 2012). Europeans have been more conservative and only three full-scale remediations with NZVI had been carried out up to year 2012, while NZVI has become an established treatment method in the USA. Bimetallic particles and emulsified NZVI, which are extensively used in the USA, have not yet been applied in Europe. Economic constraints and the precautionary attitude in Europe raise questions regarding whether NZVI is a cost-effective method for aquifer remediation. Challenges to the commercialization of NZVI include mainly non-technical aspects such as the possibility of a public backlash, the fact that the technology is largely unknown to consultants, governments and site owners as well as the lack of long-term experiences (Mueller et al., 2012). A more recent review shows that nanotechnology is more effective for removing emerging contaminants and treatment cost of some nanotechnology is comparable to that of conventional methods (Adeleye et al., 2016).

A special formulation of ZVI, namely emulsified zerovalent iron (EZVI) was developed using a surfactant, biodegradable vegetable oil, water, and nanoscale or microscale zero valent iron (ZVI) particles in an emulsion (Geiger et al., 2003; Quinn et al., 2005). The emulsion droplets enhance contact between the ZVI and the DNAPL. The ZVI provides rapid abiotic degradation of the DNAPL constituents and the oil provides an immediate sequestration of the DNAPL constituents as well as a long-term electron donor source to enhance further biodegradation. The EZVI was field tested and monitored for up to two years to treat source-zone CVOCs dominated by TCE at a site in the Cape Canaveral Air Force Station, FL (Quinn et al., 2005; O'Hara et al., 2006). Results showed that EZVI injection into a source area containing TCE using pressure pulse technology had decreased the mass flux of dissolved phase TCE from the DNAPL source zone, and decreased the amount of free-phase DNAPL mass over time. Most of the field tests involving NZVI have been monitored in a short-term fashion (several months to two years) and long-term monitoring is lacking. There is a continuing need for long-term assessments of groundwater remedial systems and some efforts have been made to meet such needs. One example is a field test of permeable reactive barrier of ZVI for treating groundwater hexavalent chromium [Cr(VI)] and TCE that had been continually monitored for 14 years (Wilkin et al., 2014) and is still being monitored today.

Built on previous EZVI success, Su et al. (2012, 2013) further tested EZVI injection using pneumatic fracturing injection and direct injection methods for treating source-zone PCE contamination at Parris Island, South Carolina, USA, and reported a two-and-a-half-year performance evaluation. Here we present an update on the treatment efficiency after six years of operation. Such long-term performance evaluation will help fill the knowledge gaps of such technology and help design cost-effective strategies for site remediation and closure.

BACKGROUND

Groundwater contamination to depths of approximately 5.8 m below ground surface (bgs) by PCE occurred from a 1994 spill at the Marine Corps Recruit Depot (MCRD), Parris Island, South Carolina, superfund site (Vroblesky et al., 2011; Su et al., 2012). A PCE DNAPL source area was delineated in field investigations conducted in 2005 and 2006 (Figure 1a). Groundwater table is about 1.3 m bgs with small seasonal variations. Groundwater flow follows a northwest-southeast path with very slow velocity at about 0.046–0.055 m day^{-1}. Containment and treatment of the source zone was critical in controlling the migration of the contaminant plume. EZVI technology was chosen due to its previous success at the National Aeronautical and Space Administration (NASA)'s Launch Complex 34 (LC34) located on the 45th Space Wing's Cape Canaveral Air Force Station, FL (Quinn et al., 2005; O'Hara, et al., 2006). EZVI was injected using pressure pulse technology in that study resulting in uneven distribution of the injected EZVI although a decrease in TCE mass flux and TCE DNAPL mass were achieved. Following that study NASA tested four techniques (pneumatic fracturing, direct injection, hydraulic fracturing, pressure pulsing) for separately delivering 379 liters of EZVI to depths of 4.9 to 5.8 m below ground surface (bgs) in an open field near the original injection location. NASA concluded that pneumatic injection and direct injection methods allowed for more controlled injection without loss of EZVI above or below the target depth interval. In the present study, an additional field test was conducted to provide more field demonstration data to improve the EZVI delivery approach and to confirm and validate the technology for widespread use for DNAPL source zone treatment.

Two types of performance monitoring wells (six fully screened and seven multilevel monitoring wells) were installed in two test plots at the site in June 2006 (Figure 1b). Five 5.79-m deep and 5.08-cm diameter monitoring wells (PMW-2, PMW-3, PMW-4, PMW-5, and PMW-6) were installed, screened from 1.22 to 5.79 m bgs in the pneumatic injection plot where an estimated 38 kg of CVOCs was pres-

Figure 1. Left panel (a) shows locations of soil cores (green triangles) and temporary wells (orange circles) collected in June 2005 and June 2006 to further evaluate contaminant distribution. The demolished building is shown by the green area. Right panel (b) shows locations of fully screened monitoring wells (blue circles) and multilevel monitoring wells (quartered circles) installed in June 2006 to target the source areas shaded by pink color.

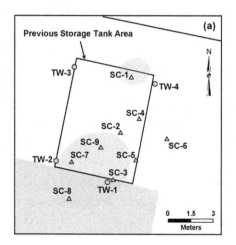 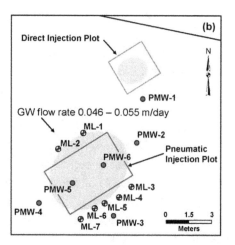

ent. Additionally, seven multilevel monitoring wells (1.27-cm diameter) were installed, with each well screened at seven depth intervals (mid-screen depths at 1.22, 1.98, 2.74, 3.51, 4.27, 5.03, and 5.79 m bgs for ML3, ML4, ML5, and ML6; mid-screen depth at 1.07, 1.83, 2.59, 3.35, 4.11, 4.88, and 5.64 m bgs for ML2 and ML7). In the direct injection plot where an estimated 155 g of CVOCs were present, one 4.11-m deep and 5.08-cm diameter monitoring well (PMW-1) was installed, screened from 1.07 to 4.11 m bgs.

Baseline soil core and groundwater sampling and analyses were conducted before EZVI injection. Eight soil cores (SC-1 through SC-8, Figure. 1a) were collected from ground surface to a depth of 6.10 m bgs at the site during the June 2005 site investigation to evaluate the CVOC mass distribution following the procedures as described by Quinn et al. (2005). An additional soil core (SC-9) was collected from within the pneumatic injection test plot in June 2006 to complete the baseline CVOC mass evaluation. Soil cores were sub-sampled in 0.61-m long sections and CVOCs were extracted and preserved on-site from each section using methanol. The 200 to 300 g wet soil sample was collected and placed in a pre-weighed 500 mL glass container. After capping, the container was reweighed to determine the wet weight of the soil. The container was then filled with 250 mL of reagent grade methanol. The container was then weighed a third time to determine the weight of the methanol added. After the container was filled with methanol and the soil sample, it was agitated by hand for approximately 10 minutes. After shaking, the container was re-weighed to ensure that no methanol was lost during the agitation period. The container was then placed upright and suspended soil matter was allowed to settle for at least 30 minutes. Following the settling period, methanol extract was decanted into glass VOA vials using disposable pipettes. The vials were then capped and stored in a cooler on ice at 4°C until they were shipped on ice to an analytical laboratory. The dry weight of each of the soil samples was determined by sending a subset of the core to the laboratory for moisture content analysis. The final concentrations of VOCs were calculated per the dry weight (105 °C dry basis) of the soil.

Test plot dimensions of 4.57 m wide, 3.05 m long, and 3.66 m in vertical thickness (1.83 to 5.49 m bgs) were used for the VOC mass estimate calculations in the Pneumatic Injection test plot (these are the same test plot dimensions used for VOC mass estimate calculations during the Environmental Security Technology Certification Program or ESTCP demonstration). PCE DNAPL mass was estimated from threshold PCE soil concentrations using equation (1) below to evaluate the presence of DNAPL:

$$C_t = C_{water} (K_d \rho_b + n)/\rho_b \tag{1}$$

where:

C_t = maximum PCE concentration in the dissolved and adsorbed phases (mg Kg^{-1})
C_{water} = PCE Solubility; 240 mg L^{-1} at 20°C
ρ_b = bulk density of soil (g cm^{-3}): clay/silty clay = 0.98; sand = 1.55; peat = 0.3 (from soil core SC-9)
n = porosity (unitless): clay/silty clay = 0.625; sand = 0.281; peat = 0.844 (from soil core SC-9)
K_d = partitioning coefficient of PCE in soil (mg Kg^{-1}) = $K_{oc} f_{oc}$
K_{oc} = organic carbon partition coefficient for PCE (L Kg^{-1}): clay/silty clay = 447; sand = 355; peat = 631 (Montgomery, 2000)
f_{oc} = fraction organic carbon (unitless): clay/silty clay = 0.045; sand = 0.00088; peat = 0.25 (from soil core SC-9)

Measurements of soil porosity, bulk density, and fraction of organic carbon were performed on select soil samples from soil core SC-9 to further refine CVOC mass estimates.

CVOC and ethene mass flux estimation was made using groundwater concentrations in the upgradient (ML-1 and ML-2) and downgradient (ML-3 through ML-7) multilevel well transects by assuming an effective cross sectional area (perpendicular to groundwater flow) for each multilevel well screen. CVOC and ethene mass flux was also made using the integral pump tests (IPTs) performed downgradient of the pneumatic injection test plot in well PMW-3 prior to EZVI injection in October 2006, March 2009, and at the end of the performance monitoring period in September 2012.

Groundwater samples were analyzed off-site within the limits of sample holding times for CVOCs [PCE, TCE, cis-1,2-dichloroethene (cDCE), trans-1,2-dichloroethene (tDCE), 1,1-Dichloroethene, and vinyl chloride (VC)], dissolved hydrocarbon gases (DHGs) (acetylene, methane, ethene, and ethane), and volatile fatty acids (VFAs) (acetic, CH_3-COOH; butyric, CH_3-CH_2-CH_2-COOH; lactic, CH_3-CHOH-COOH; propionic, CH_3-CH_2-COOH; and pyruvic acids, CH_3-CO-COOH). For CVOCs, groundwater samples (preserved by acidification using concentrated H_2SO_4 to pH less than 2 and cooled to 4 °C on ice) were analyzed by EPA Method 8260B using gas chromatography-mass spectrometry. For DHGs, groundwater samples were measured by gas chromatography (Kampbell and Vandegrift, 1998). For VFA analysis, groundwater samples were collected in 250-mL amber glass jars and preserved by adding concentrated H_3PO_3 to achieve pH 2 and refrigerated to 4 °C. Samples were filtered through a 0.45-μm membrane and a subsample of 50-mL aliquot was injected directly into the high-pressure liquid chromatography (HPLC), separated by a chromatography column, and the eluting compounds were measured with a UV detector. Total organic carbon (TOC) and total inorganic carbon (TIC) in groundwater samples were determined using a carbon analyzer (Dohrmann DC-80). Dissolved anions (chloride, sulfate, and fluoride) were measured using capillary ion electrophoresis (CE, Waters Quanta 4000E). Samples were collected, filtered with a 0.45-μm filter (Gelman Aquaprep) and acidified to pH less than 2 with HNO_3 (Optima,

Fisher Scientific) for total dissolved metals plus Si and S. Analyte concentrations were determined using inductively coupled plasma-optical emission spectrometry (ICP-OES) (PerkineElmer Optima 3300DV). Analytical quality assurance was practiced during sample collection, storage, and analyses. Detailed information about detection limits, recovery rates, and standard materials was provided in the standard operation procedures for each analyte, for example, method detection limit was 0.17 mg L^{-1} for PCE, 0.20 mg L^{-1} for ethene, 0.39 mg L^{-1} for acetic acid, and 0.263 mg L^{-1} for chloride. The percentage of recovery ranged generally between 95% and 105%. For evaluating reproducibility of groundwater analysis, one replicate groundwater sample was collected and analyzed for every 10 samples.

The EZVI was synthesized on-site by combining 10% by weight NZVI (RNIP-10DS from Toda Kogyo Corporation, 35–140 nm, and a composition of 66% Fe^0 and 34% Fe_3O_4), 51% tap water, 38% corn oil and 1% surfactant (Span 85). The EZVI injections were carried out between October 14 and 18, 2006. Detailed injections were described previously (Su et al., 2012, 2013). Briefly, in the pneumatic injection plot, 2180 liters of EZVI containing 225 kg of iron (Toda RNIP-10DS), 856 kg of corn oil, and 22.5 kg of surfactant were injected to remedy an estimated 38 kg of CVOCs. In the direct injection plot, 572 liters of EZVI were injected to treat an estimated 0.155 kg of CVOCs. The CVOC total mass change was estimated from pro- and post-demonstration soil cores. Daylighting of injected EZVI occurred at several locations but the total amount of daylight was less than 10%.

Groundwater sampling and analyses after EZVI injection were conducted in November/December 2006; January, March, and July 2007; January and July 2008; March 2009; September 2010; and October 2012. For each sampling event, groundwater samples were collected from select fully screened and multilevel monitoring wells and analyzed for either some or all of the parameters initially tested during baseline sampling activities. All of the monitoring wells were sampled for the October 2012, last event.

In October 2012, four post-demonstration soil cores was also collected in the pneumatic injection test plot from locations adjacent to the baseline soil cores (less than 0.3 m away), and cores were analyzed for CVOC concentrations to determine post-demonstration CVOC mass estimates. As was done during baseline sampling, post-demonstration CVOC concentrations from wells PMW-5 and PMW-6 were used to calculate estimates of dissolved phase CVOC mass in the southwestern and northeastern halves of the pneumatic injection test plot, respectively and the estimate of the sorbed and DNAPL mass was determined using the soil cores.

EZVI PERFORMANCE EVALUATION

Concentrations of CVOCs and Ethene in Relation to Geochemical Parameters

Baseline data show that PCE and *c*DCE were the predominant CVOCs in downgradient well PMW-3 in the pneumatic injection plot (Figure 2a). The presence of *c*DCE indicates reductive dechlorination of PCE through natural attenuation processes had been occurring. After EZVI injection, concentrations of PCE and TCE temporarily increased but quickly decreased; whereas, *c*DCE showed an increase then decrease after 21 months (Figure 2a). Concentrations of VC were mostly lower than those of *c*DCE. Concentration of *c*DCE was the lowest four years after EZVI injection then it bounced up after that. The rebound of *c*DCE may be a result of near exhaustion of added carbon sources for microbial dechlorination since the NZVI had corroded completely about three months after injection (Su et al., 2013). This also indicates that one single EZVI injection was able to maintain the activity of dechlorination for as long as four years. Dissolved ethene and methane also increased after injection (Figure 2b). Dissolved DHGs were in the order: methane > ethene > ethane > acetylene (not shown but less than 0.12 mg L^{-1}). Eh values showed steady decrease from about 150 mV to -45 mV then rebounded after four years; whereas, pH values showed a minimum of 5.46 five months after injection and then rebounded to 5.82 thirty months after injection followed by a slight decrease afterwards (Figure 2c). Dissolved Fe(II) was variable and sulfide concentrations showed a spike three months following injection and then slowly decreased (Figure 2d). Dissolved sulfate concentrations decreased after injection likely as a result of microbial reduction (data not shown). TOC showed a spike after EZVI injection (data not shown) likely due to organics released by the breakdown of the corn oil and surfactant components of the EZVI. DO levels were between 0 and 1.5 mg L^{-1}.

Figure 2. Temporal changes in monitoring well PMW-3 (1.22–5.79 m bgs, downgradient of the pneumatic injection test plot) of VCOC (a), DHG (b), Eh and pH (c), and dissolved ferrous iron and sulfide (d) before and after EZVI injection.

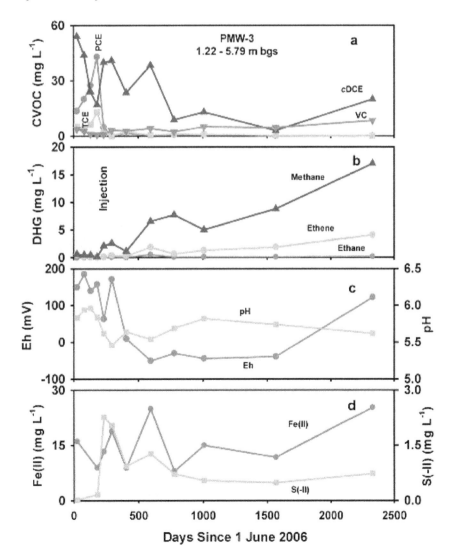

Prior to EZVI injection, cDCE and PCE showed the highest CVOCs concentrations in the upgradient multilevel monitoring well ML2-3 in the pneumatic injection plot, which was similar to other upgradient wells (Figure 3a). Natural attenuation of PCE was occurring by the presence of PCE degradation products of TCE, cDCE, and vinyl chloride (VC). After injection, concentrations of PCE and TCE decreased to low levels, and cDCE peaked then decreased before rebounding; whereas, VC increased over time to reach steady state (Figure 3a). The DHGs especially ethene increased dramatically after injection (Figure 3b) indicating enhanced dechlorination of CVOCs upon EZVI addition.

Figure 3. Temporal changes in monitoring well ML2-3 (2.59 m bgs, upgradient of the pneumatic injection test plot) of VCOC (a), DHG (b), Eh and pH (c), and dissolved ferrous iron and sulfide (d) before and after EZVI injection.

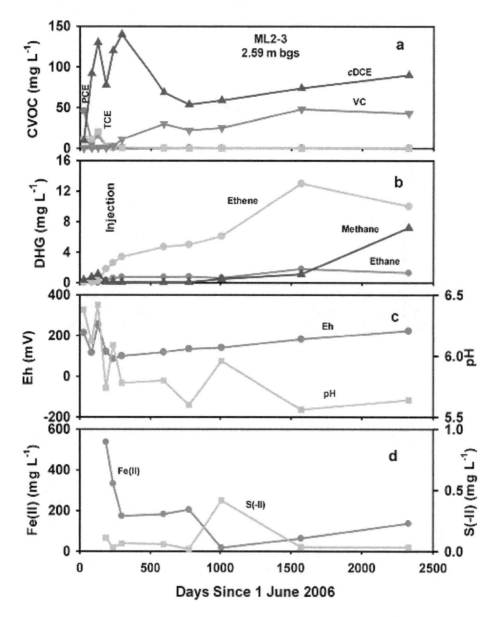

Slightly different trends were observed in the downgradient well ML7-5. Concentrations of cDCE increased after injection, reached a plateau from year three to year four, and finally declined thereafter (Figure 4a). Dissolved DHGs increased after injection and were in the order: methane > ethene > ethane > acetylene (not shown but less than 0.12 mg L^{-1} (Figure 4b). Greater decreases in CVOCs were observed in downgradient well PMW-1 in the direct injection plot (data not shown) and PMW-3 in the pneumatic injection plot. Redox potential in the upgradient well ML2-3 decreased right after EZVI injection (Figure 3c); whereas, Eh values in downgradient well ML7-5 showed a slower (time lag) but smaller decrease over time (Figure 4c). The pH values decreased from 6.4 to the lowest value of 5.6 for ML2-3, and to the lowest value of 5.0 for ML7-5 (Figure 4c) before rebounding back.

Figure 4. Temporal changes in monitoring well ML7-5 (4.11 m bgs, downgradient of the pneumatic injection test plot in the direct injection test plot) of CVOC (a), DHG (b), Eh and pH (c), and dissolved ferrous iron and sulfide (d) before and after EZVI injection.

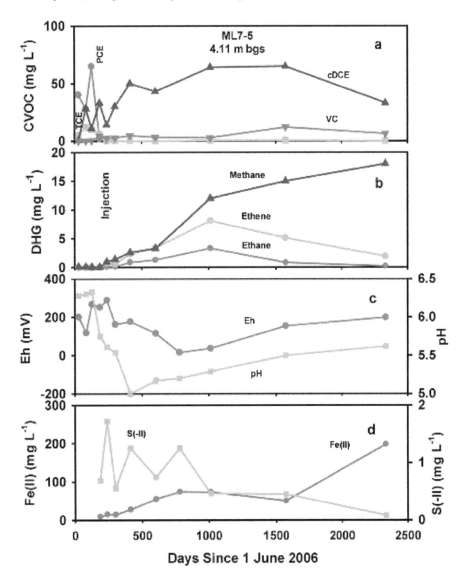

Dissolved Fe(II) concentrations was estimated to be from 2 to 21 mg/L from total dissolved Fe measurement assuming all the dissolved Fe is ferrous iron in monitoring well ML2-3. Dissolved Fe(II) reached 538 mg L^{-1} after injection then decreased over time (Figure 3d). Dissolved sulfide was low possibly due to precipitation with Fe(II) to form iron sulfides. Dissolved Fe(II) increased over time after EZVI injection reaching 200 mg L^{-1} at year six. Changes in geochemical parameters generally agreed with other field studies (for example, Wei et al., 2010).

Significance of pH Changes

The pH is a master variable in biogeochemistry that is controlled by a combination of multiple reactions and vice versa. The aquifer at the site is slightly acidic before EZVI injection. It became even more acidic after injection. Such decrease in pH after injection could be a result of several biogeochemical processes simultaneously taking place. The fermentation of organic compounds in the corn oil produces VFAs that generates H$^+$ ions. PCE undergoes abiotic dechlorination by NZVI to produce ethene, magnetite, HCl, and hydrogen gas as indicated in Eq. (2).

$$3C_2Cl_4 + 12Fe^0 + 16H_2O \rightarrow 3C_2H_4 + 4Fe_3O_4 + 12 HCl + 4H_2 (g) \qquad (2)$$

X-ray diffraction results for the iron corrosions products collected from purging water in monitoring wells and from soil cores showed that magnetite was a major NZVI corrosion product identified in this field study (Su et al., 2013) although other minerals such as green rust and siderite cannot be ruled out. Bacterial reductive dechlorination using H$_2$ gas (McCarty, 1997; Smidt and de Vos, 2004) derived from iron corrosion could also generate HCl as shown in Eq. (3).

$$C_2Cl_4 + 4H_2 (g) \rightarrow C_2H_4 + 4HCl \qquad (3)$$

The VFAs such as acetic acid can be used by microbes in biological dechlorination to produce HCl and CO$_2$ (gas) that lower the groundwater pH (Borden, 2007) as shown in Eq. (4).

$$C_2Cl_4 + CH_3\text{-}COOH + 2H_2O \rightarrow C_2H_4 + 4HCl + 2CO_2 (g) \qquad (4)$$

The pH decrease and concentrations of chloride ions increase in monitoring wells are additional evidence of the active dechlorination of CVOCs taking place in groundwter at the site due to EZVI injection. An optimal pH of 4.9 was reported for abiotic TCE degradation by zerovalent iron (Chen et al., 2001); whereas, the optimum pH values of 6.8 to 7.6 were reported for the maximum microbial reductive dechlorination (Zhuang and Pavlostathis, 1995). The pH values observed in the source area were mostly above pH 4.9 and below pH 6.8. Nevertheless, microbial reductive degradation was not prohibited under the lower-than-optimum pH conditions in the source area. There was no need to artificially adjust the pH of the aquifer for the optimal performance of EZVI at this site. The pH values in most monitoring wells showed an increasing trend between 2.5 and six years after injection due to soil buffering effect as shown in an example for ML7-5 (Figure 4c). This rebound in pH should provide a more favorable environment for the dechlorinating microbes to do their work over long periods of time after EZVI injection.

Dissolved VFAs

Typical distribution of dissolved VFAs is shown for downgradient multilevel cluster ML-7 wells at three depths (Figure 5). VFAs were negligible before EZVI injection and were dramatically increased for three of the five VFAs measured. Dissolved propionic acid, acetic acid, and butyric acid were the dominant acids with lactic acid and pyruvic acid showing only trace amounts. The acid concentrations were highest at the shallow depth than at the bottom depth, probably because the lighter corn oil (relative to water) from disintegrated EZVI emulsions tended to migrate (float) towards the top portion of the saturated subsurface where it was used most by microbes as a food source. Corn oil as a component in the EZVI formulation seemed to supply a variety of electron donors such as VFAs that persisted in the source area subsurface over a prolonged period of time up to six years due to the very slow groundwater velocity (0.046 – 0.055 m day^{-1}). An earlier study showed similar VFA distribution trends when emulsified soybean oil was used in a study to enhance anaerobic bioremediation of a TCE source area (Borden

Figure 5. Temporal changes in dissolved volatile fatty acids in monitoring well ML7-3 (2.59 m bgs) (a), ML7-5 (4.11 m bgs) (b), and ML7-7 (5.64 m bgs) (c) downgradient of the pneumatic injection test plot before and after EZVI injection.

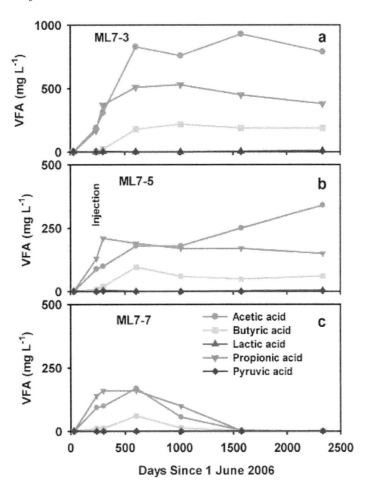

et al., 2007). In that study, no NZVI was used. Although vegetable oil itself is a good carbon source for the dechlorinating bacteria, vegetable oil alone may not be as effective as a combined vegetable oil and NZVI. NZVI helps a rapid establishment of reducing conditions that is favorable to the microbial dechlorination. In addition, NZVI provides additional iron that can be transformed to form iron sulfides. Iron sulfides can abiotically degrade CVOCs (He et al., 2015).

Mass Flux

Results of pre-injection and post-demonstration mass flux estimates calculated using CVOC concentrations from the five multilevel wells ML-3 through ML-7 on the downgradient side of the test plot revealed significant reductions in mass flux of the parent compounds PCE and TCE, and of the degradation product cDCE (data not shown). In contrast, the mass flux of the degradation products VC and ethene increased significantly over the test period. The increase in ethene mass flux indicates that the reduction in PCE, TCE, and cDCE concentrations were due to degradation and not just displacement of water or DNAPL out of the plot.

Total CVOC Mass Reduction

It was estimated from the baseline groundwater and soil sampling events that the total mass of target CVOCs in the pneumatic injection test plot was approximately 38 kg, of which roughly 29 kg (or ~76%) is attributed to PCE DNAPL (Table 1). Post-demonstration sampling results were used to calculate a CVOC total mass of approximately 8.3 kg (Table 1) in the pneumatic injection test plot, of which roughly 1.4 kg (or ~17%) is attributed to PCE DNAPL. Injection of EZVI into the pneumatic injection test plot resulted in approximately 95% reduction in the estimated mass of PCE DNAPL, and approximately 78% reduction in the total mass of target CVOCs. This is lower than an estimated 86% reduction in the total mass of target CVOCs made two and a half years after EZVI injection as reported by Su et al. (2012).

Table 1. Pre- (year 2006) and post-demonstration (year 2012) CVOC mass estimates in the pneumatic injection plot

Media	VOC	Pre-Injection Mass (g)			Post-demonstration Mass (g)		
		Sorbed/Dissolved	DNAPL	Total	Sorbed/Dissolved	DNAPL	Total
Soil	PCE	2,760	29,028	31,788	3,116	1,384	4,500
	TCE	1,317	0	1,317	672	0	672
	Cis-DCE	1,254	0	1,254	1,542	0	1,542
	VC	2,214	0	2,214	204	0	204
Groundwater	PCE	577	0	577	48	0	48
	TCE	267	0	267	50	0	50
	Cis-DCE	588	0	588	1,226	0	1,226
	VC	12	0	12	103	0	103
Total Mass (g)		8,990	29,028	38,018	6,962	1,384	8,346
% Reduction					23%	95%	78%

Variability of soil cores collected for CVOCs extraction and determination may have contributed to the difference. Nevertheless such reduction is significant from a single EZVI injection event. Repeated EZVI injections could have been used to more completely degrade CVOCs in the source zone but this was not an objective of this field test. Because a single injection was able to maintain active removal of CVOCs for at least four years (Figure 2a), repeated injections may be arranged four years apart to achieve a more cost-effective injection schedule for complete source remediation.

PROCESSES OF DECHLORINATION IN THE SOURCE AREA

Mechanisms and pathways of dechlorination in the source area involve several processes that were likely responsible for the accelerated CVOC degradation in the present study. This field test further shows that the EZVI provided multiple reactions for reduction in CVOC mass including rapid abiotic dechlorination resulting from the NZVI and a slower biological dechlorination resulting from the presence of the vegetable oil that was sustained for a more prolonged period of time; however, it is difficult to assess how much contributions of CVOC degradation are from abiotic and biotic pathways. The injected NZVI creates reducing conditions that result in immediate abiotic dechlorination of CVOCs and these conditions are consistent with the reduced redox required for biological reductive dechlorination processes. Anaerobic and aerobic corrosions of NZVI also causes a small increase in pH (Eqs. (5) and (6)) that can be beneficial in buffering the production of acid that can be associated with biological processes such as fermentation producing VFAs and other biological reactions (e.g., Eqs. (3) and (4)).

$$Fe^0 + 2H_2O \rightarrow Fe^{2+} + H_2 + 2OH^- \text{ anaerobic corrosion} \tag{5}$$

$$2Fe^0 + 2H_2O + O_2 \rightarrow 2Fe^{2+} + 4OH^- \text{ aerobic corrosion} \tag{6}$$

The influence of the NZVI on the redox conditions in the bulk fluid in the vicinity of the injected EZVI will be less significant than if the NZVI was injected in the absence of the emulsion but as the corn oil and surfactant components of the EZVI breakdown and the NZVI is no longer shielded within the emulsion, the beneficial impact of the NZVI will become more significant. In general, there is strong indication that biological dechlorination is significant at both test plots. This generalization is supported by: (1) presence of daughter products of PCE (TCE, cDCE, and VC) at elevated concentrations prior to EZVI injection; and (2) cDCE and VC as major dechlorination products after EZVI injection. Similar biological processes also have been shown in a previous field test using Pd-catalyzed and polymer-coated nanoiron suspension to treat TCE, cDCE, and VC (Henn and Waddill, 2006). Microbial analysis was performed using soil cores collected from the source area, and in agreement with another study in the southern groundwater plume area (Vroblesky et al., 2009), the results revealed the presence of microbes (*Dehalococcoides* sp. and *Dehalobacter* sp.) known to degrade chlorinated solvents. A variety of microbes can degrade PCE and TCE, but only microbes related to the genus *Dehalococcoides* are known to degrade cDCE and VC to ethene (He et al., 2003 a, b; Schmidt and de Vos, 2004). Complete sequential dechlorination of PCE to ethene has been observed for *Dehalococcoides etheneogenes* (Maymo-Gatell et al., 1997). Microcosm studies show no deleterious effect on total bacterial abundance from NZVI (Kirschling et al., 2010). There is no evidence of adverse impact to microbial communities after EZVI injection; in contrast, population of dehalococcoides was increased from less than 10^3 cells per g of fresh soil in the saturated zone to up to 10^7 cells per g of fresh soil with a 10,000 fold increase after injection in this study, in agreement with an earlier field study (Kirschling et al., 2010). It is likely that a collection of dechlorinators are involved in the degradation of PCE and its chlorinated daughter products in the source area.

ADVANTAGES AND DISADVANTAGES OF THE EZVI TECHNOLOGY

EZVI holds promise in remediating sites cost effectively and addressing challenging site conditions, such as the presence of DNAPL. Based on the previous and present work, the following general statements can be made with regard to the pros and cons, or strengths and weaknesses of the EZVI technology. The main advantages of the EZVI technology over other treatment technologies (pump-and-treat, thermal, conventional PRBs) include:

- An effective "one-two punch" of rapid abiotic degradation followed by slower biological degradation;
- Contaminants are destroyed rather than transferred to another medium;
- Ability to treat both DNAPL source zones and dissolved-phase contaminants to contain plume migration;
- Due to the enhanced reactivity and fast corrosion potential of NZVI, EZVI is most suitable for treating the source-zone DNAPL where the greatest economic benefits can be made;
- Ability to deliver reactants to targeted zones not readily accessible by conventional PRBs; and
- Potential for lower overall costs relative to alternative technologies such as groundwater pump-and-treat with high operation and maintenance costs or thermal technologies with high capital costs.

The main disadvantages of the EZVI technology are:

- Difficulty in effectively distributing the viscous EZVI to all areas impacted with DNAPL;
- Potential decrease in hydraulic conductivity due to iron oxide buildup or biofouling;
- Potential to adversely impact secondary groundwater quality through mobilization of metals and production of sulfides or methane if excess electron donor, in the form of the vegetable oil, is added;
- Injection of EZVI may displace DNAPL away from the injection point; and
- Repeated injections of EZVI may be required to completely destroy the contaminants.

FUTURE RESEARCH DIRECTIONS

It is still a challenge to deliver NZVI to a targeted subsurface zone in a uniform fashion. EZVI injection may be more applicable to deeper aquifers than the shallow aquifers that may help minimize daylighting. Research is needed to explore more effective injection technologies to minimize preferential flow and daylighting at sites with varying lithological and hydrogeological conditions. Formulation of NZVI to make it injectable while maintaining its reactivity can be further improved. Injected nanoiron was shown to transform to iron oxides (with greater particle size) mostly within three months (Su et al., 2013). Fate and transport of NZVI in the subsurface during expanded periods up to many years are still poorly understood. Corrosion products of NZVI such as green rusts, iron sulfides, and magnetite may offer additional benefits for CVOC degradation – such abiotic pathways and their relative contributions to the overall contaminant contamination need to be evaluated in detail in both laboratory and field studies. Being consistent with other field studies, the results of the current field applications with respect to contaminant reduction are promising, and no major adverse impacts on the environment have been observed so far. It is thus expected that these tests will contribute to promoting the technology in other geographical places such as Europe (Mueller et al., 2012).

CONCLUSION

Positive results have been obtained from a six-year field test using a single EZVI injection event for source zone remediation at a Superfund site contaminated with DNAPL PCE. Addition of EZVI resulted in more reducing conditions that stimulated dechlorinating bacteria (dehalococcoides); there is no evidence of adverse effect to the microbial communities. High dissolved VFA concentrations were sustained in the shallow monitoring wells even after six years of EZVI injection, indicating the excellent sustainability of the EZVI technology at this site where groundwater flow is slow. There were significant reductions in the downgradient groundwater mass flux. There were significant reduction in total VOC and DNAPL. A combination of abiotic (from NZVI and its corrosion products) and biotic (from bacteria) pathways lead to contaminant degradation through dechlorination reactions. EZVI technology can be successfully applied to treat source zone DNAPL.

ACKNOWLEDGMENT

This project is a collaboration among the United States Environmental Protection Agency through its Office of Research and Development, Geosyntec Consultants Inc., NASA, and the Naval Facilities Engineering Service Center. Funding was provided by the Environmental Security Technology Certification Program (ESTCP) (project ER-0431) and the U.S. EPA. Although the research described in this article has been funded partly by the U.S.EPA, it has not been subjected to the Agency's peer and administrative review and, therefore, does not necessarily reflect the views of the Agency, and no official endorsement should be inferred. Mention of trade names or commercial products does not constitute endorsement or recommendation for use. We are grateful to the following individuals and organizations: Deborah Schnell, Cornel Plebani, and their team of Pneumatic Fracturing, Inc. (Alpha, NJ) for high-pressure pneumatic injection of EZVI, Andrew Thornton and Corey Gamwell of Vironex Environmental Field Service (Golden, CO) for direct push injection of EZVI, Drs. Cherie Geiger and Christian Clausen of the University of Central Florida for assisting with on-site preparation of EZVI, Mr. Tim Harrington of MCRD at Parris Island and Ms. Bridget Toews (Independent Contractor) for providing logistical support, Messrs. Justin Groves, Brad Scroggins, Ken Jewell, Russell Neil, Tim Lankford, and Pat Clark of EPA, and Steve Randall of Geosyntec for field support, Ms. Lynda Callaway and Kristie Hargrove and Mr. Mark White of EPA for TOC/TIC and anions analysis, Messrs. Steve Markham and Andrew

Greenwood and Ms. Sandra Saye of Shaw Environmental & Infrastructure, Inc. for metals analysis, and Columbia Analytical Services, In. (Rochester, NY) for VOC, DHG, and VFA analysis, and TestAmerica (Knoxville, TN) for soils CVOC analysis.

REFERENCES

Adeleye, A. S., Conway, J. R., Garner, K., Huang, Y., Su, Y., & Keller, A. A. (2016). Engineered nanomaterials for water treatment and remediation: Costs, benefits, and applicability. *Chemical Engineering Journal, 286*, 640–662. doi:10.1016/j.cej.2015.10.105

Bennett, P., He, F., Zhao, D., Aiken, B., & Feldman, L. (2010). In situ testing of metallic iron nanoparticle mobility and reactivity in a shallow granular aquifer. *Journal of Contaminant Hydrology, 116*(1-4), 35–46. doi:10.1016/j.jconhyd.2010.05.006 PMID:20542350

Borden, R. C. (2007). Concurrent bioremediation of perchlorate and 1,1,1-trichloeoethane in an emulsified oil barrier. *Journal of Contaminant Hydrology, 94*(1-2), 13–33. doi:10.1016/j.jconhyd.2007.06.002 PMID:17614158

Borden, R. C., Beckwith, W. J., Lieberman, M. T., Akladiss, N., & Hill, S. R. (2007). Enhanced anaerobic bioremediation of a TCE source at the Tarheel Army Missile Plant using EOS. *Remediation Journal, 17*(3), 5–19. doi:10.1002/rem.20130

Chen, J.-L., Al-Abed, S. R., Ryan, J. A., & Li, Z. (2001). Effects of pH on dechlorination of trichloroethylene by zero-valent iron. *Journal of Hazardous Materials, B83*(3), 243–254. doi:10.1016/S0304-3894(01)00193-5 PMID:11348735

Crane, R. A., & Scott, T. B. (2012). Nanoscale zero-valent iron: Future prospects for an emerging water treatment technology. *Journal of Hazardous Materials, 211-212*, 112–125. doi:10.1016/j.jhazmat.2011.11.073 PMID:22305041

Elliott, D. W., & Zhang, W.-X. (2001). Field assessment of nanoscale bimetallic particles for groundwater treatment. *Environmental Science & Technology, 35*(24), 4922–4926. doi:10.1021/es0108584 PMID:11775172

Geiger, C. L., Clausen, C. A., Brooks, K., Clausen, C., Huntley, C., Filipek, L., & Major, D. et al. (2003). Nanoscale and microscale iron emulsions for treating DNAPL. *ACS Symposium Series. American Chemical Society, 837*, 132–140. doi:10.1021/bk-2002-0837.ch009

Guo, X., Zhao, Y., Qiu, Y., & Shi, X. (2015). Zero-valent iron nanoparticle-supported composite materials for environmental remediation applications. *Current Nanoscience, 11*(6), 748–759. doi:10.2174/1573413711666150430223749

He, F., Zhao, D., & Paul, C. (2010). Field assessment of carboxymethyl cellulose stabilized iron nanoparticles for in situ destruction of chlorinated solvents in source zones. *Water Research, 44*(7), 2360–2370. doi:10.1016/j.watres.2009.12.041 PMID:20106501

He, J., Ritalahti, K. M., Aiello, M. R., & Löffler, F. E. (2003a). Complete detoxification of vinyl chloride by an anaerobic enrichment culture and identification of the reductively dechlorinating population as a *Dehalococcoides* species. *Applied and Environmental Microbiology, 69*(2), 996–100. doi:10.1128/AEM.69.2.996-1003.2003 PMID:12571022

He, J., Ritalahti, K. M., Yang, K., Koenigsberg, S. S., & Löffler, F. E. (2003b). Detoxification of vinyl chloride to ethene coupled to growth of an anaerobic bacterium. *Nature, 424*(6944), 62–65. doi:10.1038/nature01717 PMID:12840758

He, Y. T., Wilson, J. T., Su, C., & Wilkin, R. T. (2015). Review of abiotic degradation of chlorinated solvents by reactive iron minerals. *Ground Water Monitoring and Remediation, 35*, 57–75.

Henn, K. W., & Waddill, D. W. (2006). Utilization of nanoscale zero-valent iron for source remediation - A case study. *Remediation Journal, 16*(2), 57–77. doi:10.1002/rem.20081

Kampbell, D. H., & Vandegrift, S. A. (1998). Analysis of dissolved methane, ethane, and ethylene in ground water by a standard gas chromatographic technique. *Journal of Chromatographic Science, 36*(5), 253–256. doi:10.1093/chromsci/36.5.253 PMID:9599433

Karn, B., Kuiken, T., & Otto, M. (2009). Nanotechnology and in situ remediation: A review of the benefits and potential risks. *Environmental Health Perspectives, 117*(12), 1823–1831. doi:10.1289/ehp.0900793 PMID:20049198

Kirschling, T. L., Gregory, K. B., Minkley, E. G. Jr, Lowry, G. V., & Tilton, R. D. (2010). Impact of nanoscale zero valent iron on geochemistry and microbial populations in trichloroethylene contaminated aquifer materials. *Environmental Science & Technology, 44*(9), 3474–3480. doi:10.1021/es903744f PMID:20350000

Lacina, P., Dvorak, V., Vodickova, E., Barson, P., Kalivoda, J., & Goold, S. (2015). The Application of nano-sized zero-valent iron for in situ remediation of chlorinated ethylenes in groundwater: A field case study. *Water Environment Research, 87*(4), 326–333. doi:10.2175/106143015X14212658613596 PMID:26462077

Li, X.-Q., Elliott, D. W., & Zhang, W.-X. (2006). Zero-valent iron nanoparticles for abatement of environmental pollutants: Materials and engineering aspects. *Critical Reviews in Solid State and Material Sciences, 31*(4), 111–122. doi:10.1080/10408430601057611

Maymo-Gatell, X., Chien, Y., Gossett, J. M., & Zender, S. H. (1997). Isolation of a bacterium that reductively dechlorinates tetrachloroethene to ethene. *Science, 276*(5318), 1568–1571. doi:10.1126/science.276.5318.1568 PMID:9171062

McCarty, P. L. (1997). Breathing with chlorinated solvents. *Science, 276*(5318), 1521–1522. doi:10.1126/science.276.5318.1521 PMID:9190688

Montgomery, J. H. (2000). *Groundwater Chemicals* (3rd ed.). Boca Raton, FL: Lewis Publishers and CRC Press.

Mueller, N. C., Bruns, B. J., Černík, M., Rissing, P., Rickerby, D., & Nowack, B. (2012). Application of nanoscale zero valent iron (NZVI) for groundwater remediation in Europe. *Environmental Science and Pollution Research International, 19*(2), 550–558. doi:10.1007/s11356-011-0576-3 PMID:21850484

O'Carroll, D., Sleep, B., Krol, M., Boparai, H., & Kocur, C. (2013). Nanoscale zero valent iron and bimetallic particles for contaminated site remediation. *Advances in Water Resources, 51*, 104–122. doi:10.1016/j.advwatres.2012.02.005

O'Hara, S., Krug, T., Quinn, J., Clausen, C., & Geiger, C. (2006). Field and laboratory evaluation of the treatment of DNAPL source zones using emulsified zero-valent iron. *Remediation Journal, 16*(2), 35–56. doi:10.1002/rem.20080

Quinn, J., Geiger, C., Clausen, C., Brooks, K., Coon, C., O'Hara, S., & Holdsworth, T. et al. (2005). Field demonstration of DNAPL dehalogenation using emulsified zero-valent iron. *Environmental Science & Technology, 39*(5), 1309–1318. doi:10.1021/es0490018 PMID:15787371

Smidt, H., & de Vos, W. M. (2004). Anaerobic microbial dehalogenation. *Annual Review of Microbiology, 58*(1), 43–73. doi:10.1146/annurev.micro.58.030603.123600 PMID:15487929

Su, C., Puls, R. W., Krug, T. A., Watling, M. T., O'Hara, S. K., Quinn, J. W., & Ruiz, N. E. (2012). A two and half-year-performance evaluation of a field test on treatment of source zone tetrachloroethene and its chlorinated daughter products using emulsified zero valent iron nanoparticles. *Water Research, 46*(16), 5071–5084. doi:10.1016/j.watres.2012.06.051 PMID:22868086

Su, C., Puls, R. W., Krug, T. A., Watling, M. T., O'Hara, S. K., Quinn, J. W., & Ruiz, N. E. (2013). Travel distance and transformation of injected emulsified zero valent iron nanoparticles in the subsurface during two and half years. *Water Research, 47*(12), 4095–4106. doi:10.1016/j.watres.2012.12.042 PMID:23562563

Tang, S. C. N., & Lo, I. M. C. (2013). Magnetic nanoparticles: Essential factors for sustainable environmental applications. *Water Research, 47*(8), 2613–2632. doi:10.1016/j.watres.2013.02.039 PMID:23515106

Vroblesky, D. A., Petkewich, M. D., Lowery, M. A., & Landmeyer, J. E. (2011). Sewer as a source and sink of chlorinated-solvent groundwater contamination, Marine Corps Recruit Depot, Parris Island, South Carolina. *Ground Water Monitoring and Remediation, 31*(4), 63–69. doi:10.1111/j.1745-6592.2011.01349.x

Wei, Y.-T., Wu, S.-C., Chou, C.-M., Che, C.-H., Tsai, S.-M., & Lien, H.-L. (2010). Influence of nanoscale zero-valent iron on geochemical properties of groundwater and vinyl chloride degradation: A field case study. *Water Research, 44*(1), 131–140. doi:10.1016/j.watres.2009.09.012 PMID:19800096

Wilkin, R. T., Lee, T. R., McNeil, M. S., Su, C., & Adair, C. (2014). Fourteen-year assessment of a permeable reactive barrier for treatment of hexavalent chromium and trichloroethylene. In R. Naidu & V. Birke (Eds.), *Permeable Reactive Barrier: Sustainable Groundwater Remediation* (pp. 99–107). CRC Press.

Yan, W., Lien, H.-L., Koel, B. E., & Zhang, W.-X. (2013). Iron nanoparticles for environmental cleanup: Recent developments and future outlook. *Environmental Science: Processes & Impacts, 15*, 63–77. PMID:24592428

Zhan, J., Kolesnichenko, I., Sunkara, B., He, J., McPherson, G. L., Piringer, G., & John, V. T. (2011). Multifunctional iron – carbon nanocomposits through an aerosol-based process for the in situ remediation of chlorinated hydrocarbons. *Environmental Science & Technology, 45*(5), 1949–1954. doi:10.1021/es103493e PMID:21299241

Zhang, W.-X. (2003). Nanoscale iron particles for environmental remediation: An overview. *Journal of Nanoparticle Research, 5*(3/4), 323–332. doi:10.1023/A:1025520116015

Zhuang, P., & Pavlostathis, S. G. (1995). Effect of temperature, pH, and electron donor on the microbial reductive dechlorination of chloroalkenes. *Chemosphere, 31*(6), 3537–3548. doi:10.1016/0045-6535(95)00204-L

KEY TERMS AND DEFINITIONS

Abiotic Dechlorination: Dechlorination by inorganic reactants such as elemental iron and sulfides.

Biotic Dechlorination: Dechlorination by microbes such as *Dehalococcoides* sp. and *Dehalobacter* sp.

Chlorinated Volatile Organic Compounds: Compounds that contain chlorine atoms and are volatile. Examples include tetrachloroethene or perchloroethene (PCE) and trichloroethene (TCE).

Dense Non-Aqueous Phase Liquid: A denser-than-water NAPL, i.e. a liquid that is both denser than water and is immiscible in or does not dissolve in water.

Groundwater: Liquid water that seeps through the soil or rocks underground.

Groundwater Remediation: The process that is used to treat polluted groundwater by removing the pollutants or converting them into harmless products.

Mass Flux: The rate of mass flow per unit area.

Maximum Contaminant Levels: Standards that are set by the United States Environmental Protection Agency (EPA) for drinking water quality. An MCL is the legal threshold limit on the amount of a substance that is allowed in public water systems under the Safe Drinking Water Act.

Non-Aqueous Phase Liquids or NAPLs: Liquid solution contaminants that do not dissolve in or easily mix with the water. Examples are oil, gasoline, and petroleum products.

Nanoremediation: An environmental nanotechnology that uses nanomaterials for environmental cleanup.

National Priorities List (NPL): The list of national priorities among the known releases or threatened releases of hazardous substances, pollutants, or contaminants throughout the United States and its territories. The NPL is intended primarily to guide the EPA in determining which sites warrant further investigation.

Superfund Site: An uncontrolled or abandoned place where hazardous waste is located, possibly affecting local ecosystems or people. Sites are listed on the National Priorities List (NPL) upon completion of Hazard Ranking System (HRS) screening, public solicitation of comments about the proposed site, and after all comments have been addressed.

Chapter 6
In-Situ Oxidative Degradation of Emerging Contaminants in Soil and Groundwater Using a New Class of Stabilized MnO$_2$ Nanoparticles

Bing Han
Auburn University, USA

Wen Liu
Auburn University, USA

Dongye Zhao
Auburn University, USA

ABSTRACT

Emerging Organic Contaminants (EOCs) such as steroidal estrogen hormones are of growing concern in recent years, as trace concentrations of these hormones can cause adverse effects on the environmental and human health. While these hormones have been widely detected in soil and groundwater, effective technology has been lacking for in-situ degradation of these contaminants. This chapter illustrates a new class of stabilized MnO$_2$ nanoparticles and a new in-situ technology for oxidative degradation of EOCs in soil and groundwater. The stabilized nanoparticles were prepared using a low-cost, food-grade Carboxymethyl Cellulose (CMC) as a stabilizer. The nanoparticles were then characterized and tested for their effectiveness for degradation of both aqueous and soil-sorbed E2 (17β-estradiol). Column tests confirmed the effectiveness of the nanoparticles for in-situ remediation of soil sorbed E2. The nanoparticle treatment decreased both water leachable and soil-sorbed E2, offering a useful alternative for in-situ remediation of EOCs in the subsurface.

DOI: 10.4018/978-1-5225-0585-3.ch006

1. INTRODUCTION AND BACKGROUND

EOCs in Environment

With the development of more powerful analytical tools, our ability to detect contaminants in water at the trace levels (ng/L-μg/L) has been greatly improved. As a result, a host of emerging organic contaminants (EOCs) have been revealed in water bodies (Pal, Gin, Lin, & Reinhard, 2010). Over the last decade or so, increasing attention has been placed on various endocrine disrupting chemicals (EDCs) including pharmaceuticals and personal care products. Typically, these EOCs are physiological toxic but remain unregulated or are undergoing the regulatory process (Rivera-Utrilla, Sanchez-Polo, Ferro-Garcia, Prados-Joya, & Ocampo-Perez, 2013). EDCs can interfere with hormonal and homeostatic system functions. Steroidal estrogens are one class of EDCs, of particular environmental concern are endogenous estrone (E1), 17β-estradiol (E2), and estriol (E3) released by human and wildlife, as well as synthetic 17α-ethinylestradiol (EE2), which has been used in almost all oral contraceptive pills. When released into the environment, these hormones may cause a wide range of adverse effects including abnormal development, reproductive disorders, sexual disorders as well as cancers in wildlife and humans (Colborn, Saal, & Soto, 1993; Diamanti-Kandarakis et al., 2009). Of the estrogen hormones, E2 has been a major concern for its physiological effects at lower concentrations than other steroid hormones (Shore & Shemesh, 2003). For instance, the male Japanese medaka was found to produce female specific proteins when exposed to E2 at 5 ng/L (Tabata et al., 2001). E2 has been detected in aqueous environment at concentrations above 10 ng/L, which is above the lowest observable effect level (LOEL) for some fish and plants (Shore & Shemesh, 2003). A resent U.S. Geological Survey detected various reproductive hormones in 139 streams sampled across 30 states, of which E2 was found in 10% of the 139 streams at a maximum concentration of 0.1 μg/L (Kolpin et al., 2002). The analysis of 112 samples in 59 groundwater sites in Austria detected E2 in more than 50% of the samples with a maximum concentration of 0.79 ng/L (Hohenblum et al., 2004).

Estrogens in the environment originate from humans, livestock, wildlife, and pharmaceuticals. Municipal wastewater discharge is considered one of the major sources of hormones in the aquatic system as current wastewater treatment processes could not completely remove these hormones (Baronti et al., 2000; Rosenfeldt & Linden, 2004; Ternes et al., 1999). Animal manure represents another major source of estrogens. Large amounts of estrogen and testosterone can leach from manure piles, and besides, land application of animal wastes contributes significant amounts of hormones to soil (Finlay-Moore, Hartel, & Cabrera, 2000; Shore & Shemesh, 2003).

The mobility of estrogens is generally low and they are easily bound to soil. However, low concentrations of estrogens were reported to breakthrough in porous media and release into groundwater, which are high enough to cause harmful effects (Hohenblum, Gans, Moche, Scharf, & Lorbeer, 2004; Kolodziej, Harter, & Sedlak, 2004; Vulliet, Wiest, Baudot, & Grenier-Loustalot, 2008).

Conventional EOCs Treatment Methods

Oxidation-reduction or "redox" reactions play an important role in transformation and/or speciation of organic chemical and redox-active inorganic contaminants, such as EOCs, nitrate, perchlorate, chromate and uranium in both natural and engineered systems. Researchers have explored various treatment processes, primarily oxidation-based processes, to reduce environmental estrogenicity of estrogens in waters.

Ozonation, UV radiation, and chemical oxidation have been found effective in reducing estrogenicity in surface waters (Huber, Canonica, Park, & Von Gunten, 2003; Jiang, Huang, Chen, & Chen, 2009; Jiang, Pang, Ma, & Liu, 2012; Ohko et al., 2002; Rosenfeldt & Linden, 2004). However, current technologies are not applicable or less effective for degrading the contaminants in soil and groundwater. For instances, ozonation and UV cannot reach the contaminants in the subsurface. Other oxidants have been applied in wastewater treatment to degrade EOCs, including chlorine, ferrate(VI), and permanganate (Fan, Hu, An, & Yang, 2013; Guan, He, Ma, & Chen, 2010; Lee, Yoon, & Von Gunten, 2005; Jiang et al., 2012; Rule, Ebbett, & Vikesland, 2005; Sharma, Li, Graham, & Doong, 2008). Although free chlorine is highly effective in treating various emerging micro-pollutants including some EOCs and pharmaceuticals, the primary chlorinated products can act as precursors of disinfection byproducts (DBPs) (Deborde & Von Gunten, 2008). Ferrate(VI) can be a useful method for oxidizing some EOCs during water treatment. However, the field application of ferrate(VI) is largely limited by its instability in soil/groundwater and high cost for preparation and storage (Guan et al., 2010). Permanganate has been also widely used in EOCs removal in drinking water and groundwater. However, considering the unpleasant color of permanganate and strong oxidizing power, only very low concentrations of permanganate concentration are allowed to be delivered into the environment. On the other hand, the competitive reactions with non-target compounds (the background oxidant demand), such as natural organic matter, often requires elevated doses of the oxidant (Urynowicz, 2008). Moreover, delivering permanganate solution into the subsurface may cause spreading of the chemical and the strong oxidizing ability may adversely affect the subsurface environment. Consequently, innovative *in-situ* remediation technologies are desired to mitigate the estrogenicity of these EOCs in contaminated soil and groundwater.

Manganese Oxide Oxidation of EOCs

Manganese oxide (MnO_2) is a common component in soil and sediment and it can facilitate a variety of redox reactions in the subsurface environment. Compared to other chemical oxidants, manganese oxide is considered more environmental friendly. The birnessite group of hydrous manganese oxides is among the most important naturally existing reactants and catalysts in soil and sediment (Post, 1999). Birnessite-like minerals have been commonly synthesized and widely studied oxidation of inorganic contaminants. Conventional synthetic birnessite-like MnO_2 (δ-MnO_2) is typically a porous particulate material, which has a specific surface area from dozens to hundreds of m^2/g and a low pH of point of zero charge (PZC) of 2.25 (Murray, 1974). Birnessite (δ-MnO_2) has been widely tested for treating various EDCs such as E2, triclosan, chlorophen, tetrabromobisphenol A (TBBPA), phenol and bisphenol A (Jiang et al., 2009; Lin, Liu, & Gan, 2009a, b; Ukrainczyk & McBride, 1992; Xu, Xu, Zhao, Qiu, & Sheng, 2008; Zhang & Huang, 2003). Because of their large surface area and low PZC, manganese oxides are also good sorbents of heavy metals, radionuclides, and nutrients, behaving as natural sinks for metal contaminants. For example, birnessite can effectively remove Pb(II), Cd(II), Cu(II) through direct adsorption, and oxidize Sb(III) to Sb(V), and As(III) to As(V), thereby facilitating adsorption and separation of the contaminants (Fendorf & Zasoski, 1992; Liu et al., 2015; Manning, Fendorf, Bostick, & Suarez, 2002; Wang, et al., 2015). Recently Lee et al. (2015) reported enhanced uranium removal by surface functionalized manganese ferrite nanocrystals ($MnFe_2O_4$), where uranyl was first reduced and then adsorbed on the particle surface.

Typically, the oxidative transformation of an organic contaminant by manganese oxide involves a first rapid adsorption of the solute on the particle surface to form a surface complex followed by oxidation

Table 1. Comparison of the second order oxidation rate constants (k_2) of three selected EDCs by different oxidants.

Compound	Structure	Oxidant	k_2 (M^{-1} s^{-1})	References
Bisphenol A [1]		Mn(VII)	29[a]	(Zhang et al., 2014)
		Fe(VI)	640[a]	(Lee, Yoon, & Von Gunten, 2005)
		MnO$_2$	18[b]	(Lu, Lin, & Gan, 2011)
17β-estradiol[2]		Mn(VII)	82[a]	(Jiang, Pang, Ma, & Liu, 2012)
		Fe(VI)	720[a]	(Lee, Yoon, & Von Gunten, 2005)
		MnO$_2$	30[c]	(Han, Zhang, Zhao, & Feng, 2015)
Triclosan[3]		Mn(VII)	130[a]	(Jiang, Pang, & Ma, 2009)
		Fe(VI)	755[a]	(Yang, et al., 2010)
		MnO$_2$	5[c]	(Zhang & Huang, 2003)

[a] Conducted at pH 7. [b] Conducted at pH 5.5. [c] Conducted at pH 5.

and electron acquisition (Lin et al., 2009a, b; Stone, 1987; Zhang & Huang, 2003, Mahamallik, Saha, & Pal, 2015). Experimental studies suggest that either the formation of the surface complex or the first electron transfer may be the rate-limiting step in the overall oxidation reaction (Remucal & Ginder-Vogel, 2014). The oxidation effectiveness of manganese oxide depends on both the contaminant concentration and the material dosage. Typically, manganese oxide is applied far beyond the theoretical demand, where the initial reaction rate is considered to follow the pseudo first-order rate law. However, because the rate decreases with time after the initial stage (several minutes to several tens of minutes) due to inhibition from the products (e.g., Mn^{2+}), the reaction rate often deviates from the pseudo first-order kinetics. Consequently, the retarded first-order model or second order rate law are often invoked (Han et al. 2015). To facilitate a comparison of the reaction rates of manganese oxide and other commonly used oxidants including ferrate (Fe(VI)) and permanganate (Mn(VII)), Table 1 lists the second-order rate constants for these oxidants in oxidation of three model phenolic EOCs (bisphenol A, E2, and triclosan). It is evident that while all the three oxidants are highly effective, the reaction rate for the three oxidants follows the order of: Fe(VI) > Mn(VII) > MnO$_2$, . The slower rate of MnO$_2$ can be attributed to the heterogeneous reaction mechanism. The trade-off is that MnO$_2$ may offer longer reactive life and less toxic effects on the local microbial systems.

There has been no evidence showing any adverse effects of natural or synthetic MnO$_2$ particles on microbial activities under environmentally relevant conditions. On the contrary, MnO$_2$ may more effectively tackle some persistent organic contaminants than microbial degradation due to the stronger oxidizing potential and different reaction mechanisms, and thus, facilitate the microbial processes. For instance, rapid degradation of toluene under strictly anaerobic conditions and in the presence of MnO$_2$ was observed, and the direct contact and facilitated electron transfer between the manganese-reducing bacteria and MnO$_2$ were found responsible for the enhanced reaction rate (Langenhoff et al., 1997).

Numerous studies showed that synthesized δ-MnO$_2$ (powder or aggregates) could effectively oxidize E2 in aquatic systems. However, little has been known about the potential uses for soil and groundwater remediation. Because conventional granular or powder MnO_2 particles are too big to be delivered into soil, they are not suitable for *in-situ* degradation of the target contaminants in contaminated soil and groundwater. To facilitate *in-situ* degradation of soil-sorbed E2 in aquifers, the desired MnO_2 nanoparticles will need to meet some key criteria: 1) they must be soil deliverable, 2) they must be reactive toward the target contaminants, and 3) they must be environmental friendly. So far the key technical barrier has been the soil deliverability, i.e., synthesizing stabilized MnO_2 nanoparticles that can be delivered into contaminated soil and facilitate the desired *in-situ* oxidation of EDCs.

Conventional In-Situ Remediation Technologies and the Challenges

Various strategies have been studied for in-situ chemical adsorption and/or oxidation of organic contaminants in groundwater and soil. Permeable reactive barriers (PRBs) have been widely installed in the field for in-situ remediation of contaminated groundwater (Powell et al., 1998). For instance, Valhondo et al. (1995) employed a PRB, which is comprised of vegetable compost and minor amounts of clay and iron oxide, to remove selected pharmaceuticals (i.e., atenolol, cetirizine, gemfibrozil, and carbamazepine). Three of the four chemicals (i.e., atenolol, gemfibrozil, and cetirizine) were removed effectively, of which atenolol was lowered to below the detectable level. However, the operation of this remediation technology depends largely on the subsurface hydrogeology and distribution to assure contaminants will flow through the reactive zone. In addition, the effectiveness of PRBs is often limited by the slow desorption rate of a contaminant from soil, and thus, may take prolonged remediation time. Moreover, PRBs cannot reach deep aquifers or those underneath surface obstacles groundwater aquifer. Pump-and-treat represents another common practice. Yet, it bears with the similar drawbacks as for PRBs. Consequently, excavation and landfill is often practiced, which is not only costly but also environmentally disruptive.

Nanoparticles for In-Situ Remediation

Over the last decade or so, the application of nanoparticles for *in-situ* remediation of soil and groundwater has attracted increasing interest from both academia and industries (Bennett, He, Zhao, Aiken, & Feldman, 2010; He, Zhao, & Paul, 2010; Liu, Su, Zhang, Jiang, & Yan, 2011; Theron, Walker & Cloete, 2008; Wei et al., 2010). Compared to traditional soil remediation technologies, the *in-situ* remediation using soil-deliverable nanoparticles offers some unique advantages. First, nano scale particles have much greater specific surface area, and thus more reactive sites for binding with target compounds than conventional materials. Because many environmental reactions (e.g., sorption, redox, catalysis, and ion exchange) are surface-mediated processes, increasing the surface area will increase the reaction rate or adsorption capacity. Second, discrete nanoparticles are deliverable to soil, which offers the advantage to access to contaminant plumes deep in the ground or underneath built infrastructures or other surface barriers. Third, the transport of delivered nanoparticles in the subsurface is more controllable than dissolved solutes such as permanganate, and thus will not cause secondary contamination concern while providing long-lasting reactivity or adsorption ability. Forth, application of the nanoparticles in the subsurface may boost the long-term in-situ biological activity toward the target contaminants. Nanoparticles may serve either the electron shuttle or acceptor, and the redox reactions may prime persistent contaminants and facilitate the subsequent biodegradation process. Consequently, the *in-situ* remediation technology holds the potential to greatly cut down the remediation cost.

Most nanoparticles studied for environmental remediation have been either reducing nanoparticles or adsorptive materials. For examples, zero valent iron (ZVI) nanoparticles have been extensively studied and tested for *in-situ* reductive dechlorination of chlorinated solvents in soil and groundwater aquifers (Bennett et al., 2010; He et al., 2010; Wei et al., 2010), and magnetite nanoparticles have been used as an effective adsorption for immobilizing arsenic (Liang, Zhao, Qian, Freeland, & Feng, 2011).

Various technologies have been developed to synthesize nanoparticles. In general, two strategies are employed: the top-down approach and the bottom-up method. The top-down approach is to chisel a bulk material into the nanoscale particles through, for examples, high-energy grinding, ball milling and ultrasonic radiation. This approach usually requires high energy input, and may not be suitable for highly reactive nanoparticles. In addition, this method may not be able to generate very small nanoparticles or nanoscale particles of narrow size distribution. The bottom-up strategy is to assemble nanoscale particles from even smaller units (i.e., atoms or molecules) through controlled chemical reactions including nucleation/crystallization and particle growth. This strategy usually promises a better chance to obtain less defect, more homogeneous chemical compositions, and better ordered structure.

However, nanoparticles typically carry extremely high surface energy. As a result, nanoparticles prepared in either way tend to aggregate into larger particles, which largely compromises the unique advantages of the nanoscale particles such as high specific surface area and soil deliverability. To prevent particle aggregation, proper stabilizers are often required before the particle aggregates are formed (pre-agglomeration stabilization). A stabilizer can reduce agglomeration of nanoparticles through 1) electrostatic repulsion (i.e. introducing a charged stabilizer onto the particle surface results in an increased electrical double layer, which repels the like-charged particles), and/or 2) steric hindrance (i.e. coating a stabilizer on the nanoparticles impedes particle attractions through osmotic repulsion) (Lin & Wiesner, 2012). Some researchers applied stabilizers after aggregates are formed (Suslick, Fang, & Hyeon, 1996), which is often preceded by ultrasonic breaking down of the aggregates. Such post-aggregation stabilization method is less effective compared to the pre-aggregation approach because breaking the formed aggregates into finer nanoparticles becomes increasingly unfavorable as the particle size gets smaller.

Various stabilizers have been employed to prepare stabilized nanoparticles with controlled size distribution, such as surfactants, humic acids, and polymers (Cushing, Kolesnichenko, & O'Connor, 2004; Li, Schnablegger, & Mann, 1999; Pelley & Tufenkji, 2008; Phenrat et al., 2010). For instance, Zhao's group has synthesized various mineral nanoparticles of controllable size and much improved soil deliverability using low-cost, food-grade starch and sodium carboxymethyl cellulose (CMC) (He & Zhao, 2007; Xiong, He, Zhao, & Barnett, 2009; An & Zhao, 2012; An, Liang, & Zhao, 2011).

Previous studies showed that CMC can be used to stabilize a variety of nanoparticles such as ZVI, FeS, Fe-Mn oxides, Fe_3O_4 (He & Zhao, 2007; Xiong et al., 2009; An & Zhao, 2012; An et al., 2011). CMC is a modified polysaccharide with a pK_a value of 4.3. The CMC macromolecules sorbed on particle surface are able to stabilize the nanoparticles through concurrent steric and electrostatic repulsions. Compared to other surfactants or stabilizers, CMC is not only more effective, but also "greener" and cost-effective. It was reported that the CMC-stabilized nanoparticles are deliverable into model soil or sediment and can facilitate *in-situ* remediation of contaminated soil and groundwater (Liang & Zhao, 2014; He, Zhang, Qian, & Zhao, 2009). For instance, CMC-stabilized ZVI nanoparticles were tested at a field site that was contaminated by various chlorinated solvents (He et al., 2010), the nanoparticles were not only transportable through the aquifer, but were also able to effectively degrade chlorinated solvents. However, information on the preparation and environmental application of stabilized oxidizing nanoparticles has been limited.

Figure 1. Procedure for synthesizing CMC-stabilized MnO2 nanoparticles

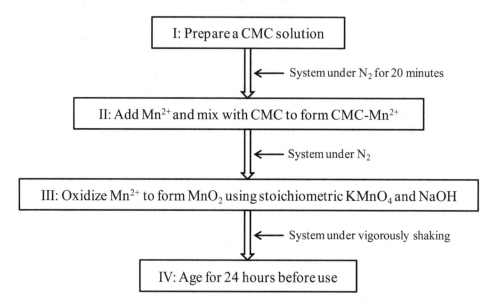

This chapter aims to introduce a new class of stabilized oxidative nanoparticles, manganese dioxide (MnO_2) nanoparticles and a new in-situ remediation technology based on the nanoparticles. Specifically, this chapter introduces the particle stabilization technique and principle of the remediation technology, and presents laboratory experimental data on the effectiveness of the nanoparticles when used for degradation of E2 as a model emerging contaminant in soil and groundwater.

2. SYNTHESIS AND CHARACTERIZATION OF CMC-STABILIZED MnO$_2$ NANOPARTICLES

Han et al. (2015) synthesized a new class of CMC-stabilized MnO_2 nanoparticles, and demonstrated that CMC can serve as an effective stabilizer to obtain highly stable and water dispersible MnO_2 nanoparticles with a narrow size distribution. Figure 1 depicts the steps for synthesizing CMC-stabilized MnO_2 nanoparticles.

First, prepare a CMC solution (55.6 mg/L) using a CMC of molecular weight = 90,000, and a $MnCl_2$ solution (4 mM) with deoxygenated water. Then, add an aliquot amount of the $MnCl_2$ solution into a predetermined volume of the CMC solution, to yield a solution of Mn^{2+}–CMC complexes. Subsequently, add dropwise stoichiometric amounts of $KMnO_4$ and NaOH (Equation 1) from the respective working solutions under vigorously shaking to obtain the desired concentration of CMC-stabilized MnO_2 suspensions. To ensure complete reaction and full growth of the nanoparticles, freshly prepared nanoparticles were allowed to age for 24 hours before use. X-ray diffraction (XRD) analysis of the stabilized nanoparticles showed no distinctive peaks, indicating that the synthetic MnO_2 is largely amorphous, which agrees with the observation by Murray (1974) and Zhang and Huang (2003), who studied the crystallinity of non-stabilized synthetic MnO_2 following a similar approach found that the synthesized MnO_2 particles were amorphous birnessite (δ-MnO_2).

Figure 2. Photographs and TEM images of: a) non-stabilized MnO_2, and b) CMC-stabilized MnO_2 particles. Pictures were taken 30 minutes after synthesis. $MnO_2 = 8\times10^{-4}$ M (0.07 g/L) for both cases, and CMC/MnO_2 molar ratio = 1.39×10^{-3} for case b.
(Han et al., 2015)

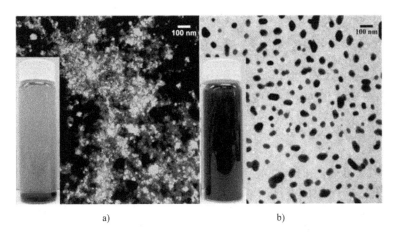

a)   b)

$$3Mn^{2+} + 2MnO_4^- + 4OH^- \rightarrow 5MnO_2 + 2H_2O \tag{1}$$

Figure 2 compares the digital photographs and transmission electron microscopy (TEM) images of non-stabilized versus CMC-stabilized MnO_2 particles. Non-stabilized MnO_2 particles appeared as irregular, inter-bridged flocs, which settled under gravity within 30 min. In contrast, CMC stabilized MnO_2 nanoparticles all appeared as fully suspended discrete nanoparticles, which remained suspended in water for more than 10 months. Based on the TEM images of more than 300 particles, the mean size of the CMC-stabilized nanoparticles was determined to be 37±10 nm (mean ± standard deviation). According to He and Zhao (2007), CMC molecules are adsorbed on metal oxide particles mainly through interactions between the carboxymethyl and hydroxyl groups of CMC molecules and the metal oxides, resulting in a surface of highly negative zeta potential (zeta potential=-55.8±0.9 mV when Fe=0.2 g/L, pH=8.1, I=15 mM) (He et al., 2009), thereby preventing the like-charged particles from aggregating. The synthesized $CMC-MnO_2$ nanoparticles are relatively poly-disperse in both size and shape, with a size spanning from 14 nm to 84 nm and geometric shapes encompassing sphere, oval, cube, cylinder, and other irregular shapes.

The hydrodynamic diameter based on dynamic light scattering (DLS) measurements serves as a novel indicator of the effective size of fine particles, as DLS measurements count for both the core MnO_2 and the CMC coating, while the TEM-based particle size measures only the electron dense metal oxide core (He & Zhao, 2008). Figure 3 shows the volume weighted hydrodynamic size distributions of the nanoparticles at various CMC/MnO_2 molar ratios. The results indicate that fully stabilized MnO_2 nanoparticles were obtained at a CMC/MnO_2 molar ratio of $\geq 8.33\times10^{-4}$, and the mean DLS size of the fully stabilized nanoparticles was less than 40 nm. Lower CMC concentrations resulted in larger particles or more aggregation. For instance, at a CMC/MnO_2 molar ratio of 2.78×10^{-4}, the particle mean DLS-based size increased to 60±2 nm, and when the CMC/MnO_2 molar ratio was lowered to 5.56×10^{-5}, the particles cannot be fully stabilized (i.e., partial precipitation appeared). The most uniform (monodisperse) nanoparticles were obtained at the CMC/MnO_2 molar ratio of 1.39×10^{-3}, where 88% of the nanoparticles were between 20–50 nm, and the mean hydrodynamic diameter was 39.5±0.8 nm. Higher CMC concentrations result in larger portions of smaller particles, but broader size distribution. In addition, the zeta potential was measured at -42 mV (MnO_2=17.4 mg/L or 2×10^{-4} M, pH= 5.1, I=0.24 μM) for CMC-stabilized MnO_2 nanoparticles, indicating that electrostatic repulsion plays an important role in the particle stabilization.

3. OXIDATIVE DEGRADATION OF E2 IN WATER AND SOIL BY STABILIZED MnO_2

Degradation of E2 in Water

Xu et al. (2008) and Jiang et al. (2009) tested the effectiveness of oxidative transformation of E2 using synthetic non-stabilized MnO_2 particles/aggregates in water, and they observed low concentrations of MnO_2 (MnO_2=0.87 mg/L or 1×10^{-5} M) particles/aggregates can effectively degrade E2. The researchers found that MnO_2 oxidizes E2 to form primary estrone (E1, $C_{18}H_{22}O_2$) and 2-hydroxyestradiol ($C_{18}H_{24}O_3$) (Equations 2 and 3), thereby reducing or eliminating the estrogenicity without generating secondary contaminations (Jiang et al., 2009; Xu et al., 2008).

$$C_{18}H_{24}O_2 + MnO_2 + 2H^+ \rightarrow C_{18}H_{22}O_2 + Mn^{2+} + 2H_2O \tag{2}$$

$$C_{18}H_{24}O_2 + MO_2 + 2H^+ \rightarrow C_{18}H_{24}O_3 + Mn^{2+} + H_2O \tag{3}$$

Han et al. (2015) demonstrated that CMC-stabilized MnO_2 nanoparticles can effectively oxidize E2 in aqueous solution. Figure 4 compares the E2 oxidation kinetics by bare or CMC-stabilized MnO_2 nanoparticles at various pH levels. Evidently, the presence of 2×10^{-5} M of CMC-stabilized MnO_2 nanoparticles were able to degrade over 95% of 4×10^{-6} M (1.09 mg/L) of E2 in the solution within 1 day at pH 5. For both bare and stabilized MnO_2, the reaction rate gradually diminished over time due to the loss of reactive sites, which is attributed to blocking of the sites with accumulation of the reaction by-products (i.e., Mn^{2+} and organic products) (Jiang et al., 2009; Lin et al., 2009a, b; Ukrainczyk & McBride, 1992; Xu et al., 2008; Zhang & Huang, 2003). Compared to the bare particles, CMC-stabilized particles showed a slower initial reaction rate but faster longer-term rate, suggesting that CMC-stabilized nanoparticles are more resistant to the reactivity fouling by the reaction byproducts. In fact, the CMC coating plays some opposing roles in the E2 oxidation process. First, the CMC coating may add a mass transfer barrier, which inhibits the accessibility of the reactive sites for E2, and therefore decreases the initial reaction rate; Second, the CMC stabilization results in smaller particles sizes of the core MnO_2, which gives higher specific surface area and thus favors E2 removal; and lastly the CMC layer may also prevent adsorption of the reaction products (e.g., Mn^{2+}) on the reaction sites on the nanoparticles, thereby mitigating the inhibitive effects of the byproducts.

Like for non-stabilized MnO_2, lower solution pH favors the oxidation of E2 by CMC-stabilized MnO_2. Similar pH effect patterns were also observed when bare MnO_2 particles were used for treating other organic compounds (Xu et al., 2008; Lin et al., 2009a, b; Zhang & Huang, 2003; Rubert & Pedersen,

Figure 3. Hydrodynamic size distribution and mean hydrodynamic diameter of MnO_2 nanoparticles prepared at various $CMC:MnO_2$ molar ratios. Note: $MnO_2 = 2 \times 10^{-4}$ M (17.4 mg/L), $CMC = 1.1 \times 10^{-8}$-2.8×10^{-6} M (1-250 mg/L); D: mean hydrodynamic diameter. Error bars indicate deviation from the mean. Data are plotted as mean of duplicates.
(Han et al., 2015)

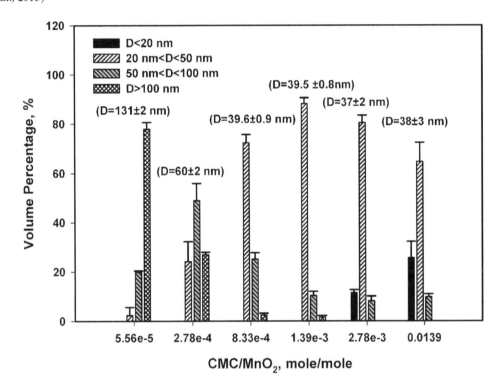

2006). At pH 6 and 7, the stabilized nanoparticles outperformed the bare particles after ~2-4 hours, and degraded 15% and 9% more E2 than non-stabilized particles in 24 hours, respectively.

The first-order retarded equation (Equation 4) was employed to interpret the E2 degradation kinetics for both non-stabilized and CMC-stabilized MnO_2 (Figure 4).

$$C_t = C_0 \left(1 + \pm t\right)^{-k_a/\alpha} \tag{4}$$

where C_0 and C_t are E2 concentrations (μM) at time 0 and t, respectively, k_a is the apparent rate constant (h^{-1}), which is analogous to the initial pseudo first order rate constant adapted to the retarded rate equation, and α is the retardation factor indicating the extent of departure from the pseudo-first-order behavior (Weber, 2001). The retarded equation is a modification of the first-order rate law by dividing the rate constant by a "sliding" factor (Lin et al., 2009b).

The first-order retarded equation provides adequate data fittings for CMC-stabilized MnO_2 at all the pH levels, with a correlation coefficient (R^2) >0.97 (Table 2).

Physically, this model agrees with the postulate that the generation and accumulation of reaction products (i.e., Mn^{2+} and organic byproducts) on the particle surface inhibits the E2 oxidation rate. For

Figure 4. Kinetics of E2 oxidation by non-stabilized or CMC-stabilized MnO_2 particles at a) pH=5, b) pH=6, c) pH=7, and d) pH=8. Initial E2 concentration = 4×10^{-6} M, CMC = 2.5 mg/L (CMC/MnO_2 molar ratio = 1.39×10^{-3}).
(Han et al., 2015)

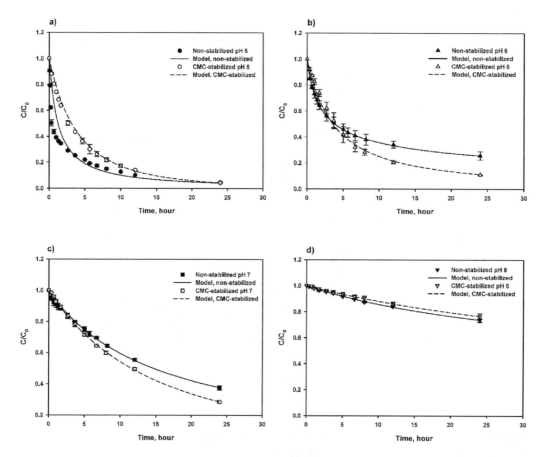

non-stabilized MnO_2, the retarded model failed to pick up the rapid initial degradation rate in the first two hours at pH 5 because the retardation effect in the initial stage was insignificant (i.e. mass transfer barrier, Mn^{2+} production and supply of H^+ (Equation 2 and 3)). However, at elevated pH (6-8), the retarded equation was able to adequately simulate the degradation rates for non-stabilized MnO_2 as well ($R^2 > 0.98$). Table 2 gives the best-fitted retardation model parameters. In all cases, the k_a values decreased with increasing pH, and CMC-stabilized nanoparticles showed a lower k_a and also a lower α value than non-stabilized MnO_2.

To test the reactive lifetime of CMC-stabilized and non-stabilized MnO_2 particles, the same particles were subjected to multiple cycles of batch degradation experiments (Han et al., 2015). Figure 5 shows the degradation kinetic profiles of E2 degradation in consecutive spikes of E2 (E2 = 4×10^{-6} M in each spike). CMC-stabilized MnO_2 remained highly reactive in three consecutive runs and degraded cumulatively 1.21×10^{-5} (±0.3%) M of E2, whereas bare MnO_2 exhausted at the end of the second run and degraded a total of 7.50×10^{-6} (±0.01%) M of E2. This observation confirms that the CMC stabilization resulted in greater surface area, more reactive surface sites, and longer reactive lifetime of the nanoparticles.

Table 2. Retarded Rate Model Parameters. (Han et al., 2015)

MnO$_2$ Particles	pH	k_a (Hour^{-1})	α (Hour^{-1})	Model Fit (R^2)
Non-stabilized	5	0.701	0.772	0.891
CMC-stabilized	5	0.252	0.096	0.997
Non-stabilized	6	0.487	1.218	0.986
CMC-stabilized	6	0.277	0.238	0.973
Non-stabilized	7	0.067	0.065	0.995
CMC-stabilized	7	0.071	0.034	0.999
Non-stabilized	8	0.019	0.046	0.994
CMC-stabilized	8	0.015	0.028	0.981

Degradation of E2 in Soil: Batch Tests

Han et al. (2015) demonstrated the effectiveness of CMC-stabilized MnO$_2$ nanoparticles for degradation of E2 pre-spiked in a model soil through a series of batch tests. The model soil (SL1) was collected from the E.V. Smith Research Center of Auburn University at Shorter, AL, USA. The soil was classified as sandy loam (pH = 6.7, cation exchange capacity = 3.20 meq/100g, and organic matter content = 0.4% (Han et al., 2015). Figure 6 compares the degradation of E2 in water and in SL1 (0.0019 mg-E2/g-soil) by CMC-stabilized MnO$_2$ nanoparticles at various particle dosages. More than 97% of aqueous E2 (1.09 mg/L) was degraded in 48 hours by 2×10^{-5} M (1.74 mg/L) CMC-stabilized MnO$_2$ particles in the absence of soil. In contrast, less than 5% of soil-sorbed E2 (0.0019 mg/g) was degraded with the same concentration of CMC-MnO$_2$. When the nanoparticle dosage was increased to 8×10^{-5} M and 2×10^{-4} M, 20% and 80% of soil-sorbed E2 were degraded, respectively.

The much lower degradation rate for soil-sorbed E2 than that for the aqueous E2 can be attributed to: 1) the desorption rate of E2 from the soil can limit the oxidation rate and extent by the nanoparticles, 2) adsorption of the nanoparticles on soil particles reduces particle mobility and also losing portion of the reactive sites, and 3) the organic compounds (e.g., tannic acid) and metal ions (e.g., Ca^{2+}, Mn^{2+}) released from soil may also competitively react with MnO$_2$ (Xu et al., 2008). Consequently, a larger dosage of CMC-stabilized MnO$_2$ is required to treat E2 sorbed in contaminated soil.

Figure 7 compares the effectiveness of non-stabilized MnO$_2$ and CMC-stabilized MnO$_2$ nanoparticles in degradation of soil-sorbed E2. After 96 hours of reaction, 83% and 70% of E2 were degraded in the heterogeneous systems by 2×10^{-4} M of CMC-stabilized and non-stabilized MnO$_2$ particles, respectively. Evidently, CMC-stabilized MnO$_2$ outperformed non-stabilized MnO$_2$ particles for degrading soil-sorbed E2 in both degradation rate and extent.

Several factors can lead to the much greater reactivity of CMC-stabilized MnO$_2$ nanoparticles for soil-sorbed E2. First, CMC-stabilized MnO$_2$ nanoparticles have lower adsorption potential to soil than non-stabilized ones. Han et al. (2015) observed that zeta potential of CMC-stabilized and non-stabilized MnO$_2$ at pH 6 were -32 mV and -27 mV, respectively. As a result, non-stabilized particles are more easily adsorbed to the negatively charged sandy loam soil (SL1) surface (pH$_{PZC}$ = 5.2). In addition, the CMC coating on the nanoparticles shield the nanoparticles from the auto-inhibitive effects of Mn^{2+} and other inhibiting components from soil (Han et al., 2015).

Figure 5. Batch kinetic data of E2 degradation in four consecutive spikes of E2 using the same MnO_2 material: a) non-stabilized MnO_2; b) CMC-stabilized MnO_2. In both cases, $MnO_2 = 2\times10^{-5}$ M, E2 in each spike = 4×10^{-6} M, pH = 5 (buffered by HAc/NaAc), and CMC = 2.5 mg/L (for stabilized MnO_2 only). C_0: Initial E2 concentration in each run, and C: E2 concentration at time t.
(Han et al., 2015)

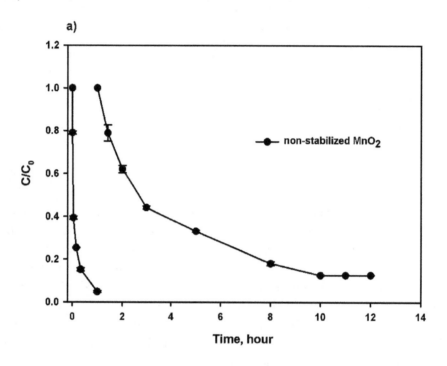

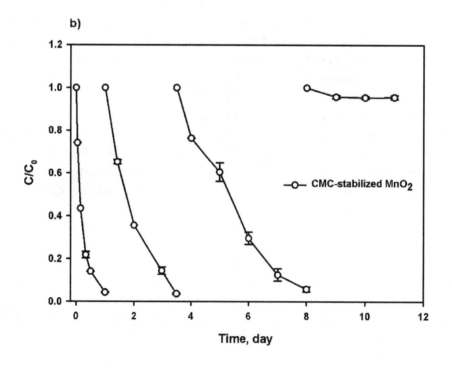

Figure 6. Degradation of aqueous and soil-sorbed estradiol at various dosages of MnO_2 (2×10^{-5} to 2×10^{-4} M). Initial estradiol in soil = 0.0019 mg/g, CMC/MnO_2 molar ratio = 1.389×10^{-3} in all cases, pH = 6±0.2 (buffered by $NaH_2PO_4/NaOH$). M_0 = Initial estradiol in the reactor system (77.6% in soil); M_t = estradiol remaining in the whole system at time t. Soil mass = 6 g, Solution/suspension volume = 60 mL. (Han et al., 2015)

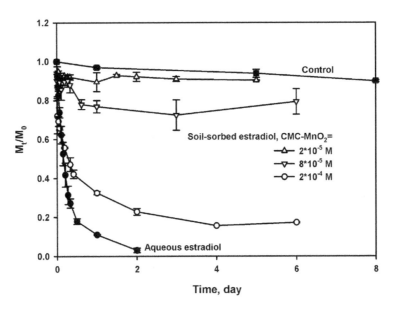

Figure 7. Degradation of soil-sorbed estradiol with CMC-stabilized and non-stabilized MnO_2 particles. Initial estradiol in SL1 soil was 0.0019 mg/g, MnO_2 concentration was 2×10^{-4} M, CMC/MnO_2 molar ratio was 1.389×10^{-3} for stabilized particles, pH = 6±0.2 (buffered by $NaH_2PO_4/NaOH$). M_0 = Initial estradiol in the system (77.6% in soil); M_t = estradiol remaining in the system at time t (including both aqueous and soil sorbed E2). Soil mass = 6 g, Solution/suspension volume = 60 mL. (Han et al., 2015)

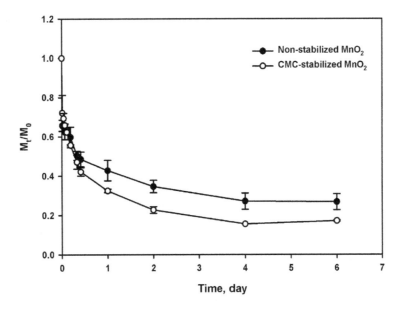

Some water leachable substances in soil may consume the reactivity of MnO_2. Due to the low pH_{PZC} value (2.25) of MnO_2, the strong electrostatic interactions between metal cations and the negatively charged particle surface can inhibit E2 oxidation. Xu et al. (2008) studied effects of typical metal cations including Zn^{2+}, Cu^{2+}, Fe^{3+}, and Mn^{2+} on E2 removal by non-stabilized MnO_2, and observed that all the cations inhibited the E2 oxidation. Xu et al. (2008) also studied the effect of some organic compounds including humic acid and tannic acid on E2 removal from water by MnO_2, and they found that the presence of low concentrations (≤ 1 mg/L) of humic acid increased the reaction rate while tannic acid decreased the reaction rate. The inhibitive effect of tannic acid was attributed to the competitive oxidation of tannic acid by MnO_2, whereas the promoting effect of humic acid was due to its binding with Mn^{2+}, which alleviates its inhibitive effects as Mn^{2+} is continuously produced during the redox reaction.

To investigate effects of water leachable soil substances, Han et al. (2015) compared the aqueous phase E2 degradation kinetics with and without soil extracts by non-stabilized and CMC-stabilized MnO_2 (Figures 8a and 8b). The soil extracts solution was prepared by mixing the sandy loam soil (SL1) with DI water at a soil to solution ratio of 1(g):10 (mL) for 3 days, and then collecting the supernatant upon centrifugation (805 g-force) followed by filtration through a 0.22 μm membrane. The soil extracts contained 6±2 mg/L dissolved organic matter measured as total organic carbon (TOC), and the major effective metal ions were 2.5±0.5 mg/L Ca^{2+} and 1.9±0.3 mg/L Mn^{2+}. The results show that the presence of the soil extracts inhibited the reaction rate in the initial stage (12–16 h), but facilitated the longer term reaction rate. Han et al. (2015) tested effects of cations such as Mn^{2+} and Ca^{2+}, and humic acid on E2 oxidation, and found that the presence of 0.01 M Mn^{2+} and Ca^{2+} remarkably inhibited the degradation rates, while addition of 5 mg/L humic acid (as TOC) enhanced the reaction rate. In the early stage, the rapid initial uptake of the soil extracts (OC, Ca^{2+}, Mn^{2+} and their complexes) diminish the reactive sites and increase the mass transfer resistance at the particle surface. Later, the organic matter acts as a scavenger by binding with produced Mn^{2+}, thereby alleviating further inhibitive effects. A comparison of Figures 8a and 8b shows that the soil extracts exerted less impact on CMC-stabilized MnO_2 than non-stabilized MnO_2. This is attributed to the protective effect of the sorbed CMC that reduces the accessibility of soil released components to MnO_2 surface sites. Han et al. (2015) reported that the CMC coating can greatly shield the nanoparticles from the auto-inhibitive effects of Mn^{2+}: while Mn^{2+} (1.92±0.31 mg/L) almost ceased the E2 oxidation for non-stabilized MnO_2, nearly 50% of E2 was degraded by CMC-stabilized MnO_2.

Degradation of E2 Using CMC-Stabilized MnO_2 Nanoparticles: Column Tests

For *in-situ* application, the delivered nanoparticles should offer high reactivity for the target compounds. Figure 9 shows the elution histories of E2 from soil SL1 when a suspension of CMC-stabilized MnO_2 nanoparticles was pumped through the soil bed (column test). For comparison, the same elution test was also carried out using a solution containing the same concentration of CMC without the nanoparticles under otherwise identical conditions.

The 22 PVs of CMC solution eluted ~72% of E2 initially sorbed in the soil, while the nanoparticle suspension eluted ~56% of E2 (Figure 9), i.e., the nanoparticle treatment lowered the overall E2 elution by 16% due to the in-situ oxidation of E2.

To determine the total mass of E2 degraded by the nanoparticles, E2 remaining in the soil bed was determined through the hot methanol extraction method, where the soil was mixed with methanol at a soil-to-methanol ratio of 3 (g):20 (ml) and in a hot water bath at 70°C for 4 hours. The E2 recovery from the soil by this method was determined to be 94±4%. Following the CMC solution elution, E2 remain-

Figure 8. Oxidation kinetics of estradiol in water with or without soil extracts: a) non-stabilized MnO_2, and b) CMC-stabilized MnO_2. Initial estradiol = 4×10^{-6} M, MnO_2 = 2×10^{-5} M, CMC = 2.5 mg/L, pH = 6 (buffered by $NaH_2PO_4/NaOH$). Data plotted as mean of duplicates and errors calculated as standard deviation.
(Han et al., 2015)

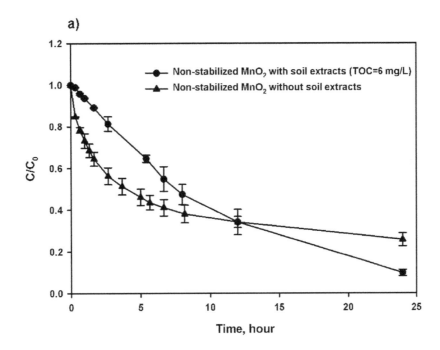

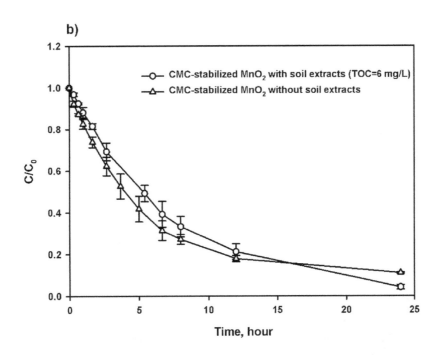

Figure 9. Elution histories of estradiol from an E2-loaded soil (SL1) column when subjected to a CMC solution without particles (control) or a CMC-stabilized MnO_2 nanoparticles suspension. Experimental conditions: E2 in soil = 0.0019 mg-estradiol/g-soil, MnO_2 = 0.17 g/L (2×10^{-3} M), CMC = 250 mg/L, soil packed in column = 12 g, soil bulk density = 1.596 g/mL, bed porosity = 0.384, hydraulic conductivity = 3.95×10^{-3} cm/min, inlet flow rate = 0.05 mL/min, pore velocity = 0.169 cm/min, empty bed contact time (EBCT) = 147.7 min.

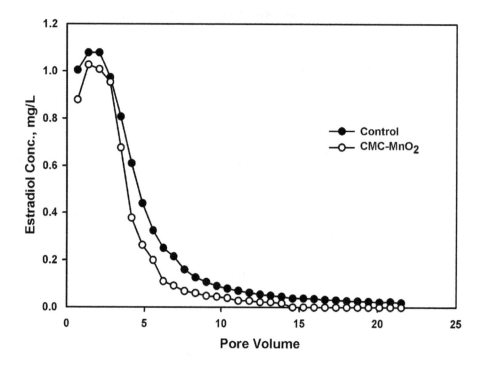

ing in soil was ~20% of the total mass loaded; in contrast, upon the nanoparticle treatment, only 17% E2 remained in the soil. The results showed that the nanoparticle treatment also lowered soil sorbed E2 by 3%. Overall, the nanoparticles degraded 19% of E2 in the system, of which 16% is water leachable E2 that is more bioavailable. The residual E2 in soil is likely much less desorbable and less available.

The sharp E2 elution peaks with an elution concentration of >1 mg/L for both cases indicate that a fraction of soil-sorbed E2 is loosely bound to SL1. In both cases, most of water leachable E2 was eluted in ~10 PVs. When the nanoparticle suspension was passed through the soil bed, MnO_2 nanoparticles are subject to the transport retardation effect, i.e., the particles will travel at a slower rate than water, and a fraction of the nanoparticles are retained in the soil bed (Han et al., 2015). Consequently, the part of E2 eluted by water traveling ahead of the nanoparticles may not be degraded due the lack of contact with the nanoparticles. Therefore, the desorption rate of E2 from soil is an important parameter controlling the degradation efficiency of water-leachable E2. Consequently, the technology is expected to be more effective for soil that releases the contaminants more slowly. Because the predominant adsorption mechanism for E2 is reported to be hydrophobic partitioning (Lee et al. 2003), higher soil organic matter content would give a higher adsorption capacity and slower desorption rate, which would allow for sufficient contact and reaction time with the nanoparticles, and thus more degradation of E2.

4. CONCLUSION AND SUGGESTIONS FOR FUTURE WORK

Steroidal hormones in soil and groundwater represent an emerging environmental problem. Because of the extent, scope, and location of the contaminants, cost-effective remediation technologies are direly needed. The *in-situ* remediation technology based on CMC-stabilized MnO_2 nanoparticles may offer a useful remediation alternative under certain geochemical conditions. The stabilized nanoparticles offer some unique properties, such as soil deliverability, high specific surface area, and good reactivity under subsurface conditions. Compare to traditional technologies such as PRBs or pump-and-treat, the *in-situ* remediation technology offers some unique advantages, including: 1) the nanoparticles may be delivered directly into the contaminant source zone and degrade the contaminants in-situ, 2) the nanoparticles can reach contaminants that are not reachable by other technologies, 3) the technology is potentially less environmentally disruptive and may cut down the remediation cost substantially, and 4) the application of CMC-stabilized MnO_2 may boost a long-term in-situ manganese reducing biological degradation of target compounds using CMC as the carbon source and manganese oxide as the electron acceptor. Preliminary bench-scale studies revealed that the nanoparticles can effectively degrade E2 in both water and soil slurries, and can effectively degrade soil sorbed E2 under simulated soil bed. At typical aquatic pH 6 and 7, CMC-stabilized nanoparticles outperform bare particles, and can degrade 15% and 9% more E2 in water, respectively. Moreover, CMC-stabilized nanoparticles display even greater reactivity than bare particles for treating soil-sorbed E2 due to their greater resistance to the inhibitive effects from soil exudates (e.g., metal ions such as Mn^{2+}, and DOM). When an E2-laden soil was treated using 22 PVs of the stabilized MnO_2 suspension ($MnO_2 = 0.17$ g/L) in a column configuration, 19% of E2 loaded in the soil was degraded. The effectiveness of the technology may be limited by the desorption rate of the contaminant from the soil, and the technology is more effective for soil that releases the contaminants more slowly than the nanoparticle transport rate.

Further researches are needed to confirm the practical viability of the technology. First, more detailed studies are needed to demonstrate the effectiveness of CMC-stabilized MnO_2 nanoparticles for various EDCs and under various soil/water chemistry conditions. Second, the technology feasibility, cost-effectiveness and process constraints will need to be demonstrated or defined through pilot-scale and field-scale experiments. Third, the effects of the delivered nanoparticles on the local hydraulic and biogeochemical conditions are yet to be determined.

REFERENCES

An, B., Liang, Q. Q., & Zhao, D. Y. (2011). Removal of arsenic(V) from spent ion exchange brine using a new class of starch-bridged magnetite nanoparticles. *Water Research, 45*(5), 1961–1972. doi:10.1016/j.watres.2011.01.004 PMID:21288549

An, B., & Zhao, D. Y. (2012). Immobilization of As(III) in soil and groundwater using a new class of polysaccharide stabilized Fe-Mn oxide nanoparticles. *Journal of Hazardous Materials, 211*, 332–341. doi:10.1016/j.jhazmat.2011.10.062 PMID:22119304

Baronti, C., Curini, R., D'Ascenzo, G., Di Corcia, A., Gentili, A., & Samperi, R. (2000). Monitoring natural and synthetic estrogens at activated sludge sewage treatment plants and in a receiving river water. *Environmental Science & Technology, 34*(24), 5059–5066. doi:10.1021/es001359q

Bennett, P., He, F., Zhao, D. Y., Aiken, B., & Feldman, L. (2010). In situ testing of metallic iron nanoparticle mobility and reactivity in a shallow granular aquifer. *Journal of Contaminant Hydrology*, *116*(1-4), 35–46. doi:10.1016/j.jconhyd.2010.05.006 PMID:20542350

Colborn, T., Saal, F. S. V., & Soto, A. M. (1993). Developmental effects of endocrine-disrupting chemicals in wildlife and humans. *Environmental Health Perspectives*, *101*(5), 378–384. doi:10.1289/ehp.93101378 PMID:8080506

Cushing, B. L., Kolesnichenko, V. L., & O'Connor, C. J. (2004). Recent advances in the liquid-phase syntheses of inorganic nanoparticles. *Chemical Reviews*, *104*(9), 3893–3946. doi:10.1021/cr030027b PMID:15352782

Deborde, M., & von Gunten, U. (2008). Reactions of chlorine with inorganic and organic compounds during water treatment-kinetics and mechanisums: A critical review. *Water Research*, *42*(1-2), 13–51. doi:10.1016/j.watres.2007.07.025 PMID:17915284

Diamanti-Kandarakis, E., Bourguignon, J. P., Giudice, L. C., Hauser, R., Prins, G. S., Soto, A. M., & Gore, A. C. (2009). Endocrine-disrupting chemicals: An endocrine society scientific statement. *Endocrine Reviews*, *30*(4), 293–342. doi:10.1210/er.2009-0002 PMID:19502515

Fan, Z. L., Hu, J. Y., An, W., & Yang, M. (2013). Detection and occurrence of chlorinated byproducts of bisphenol A, nonylphenol, and estrogens in drinking water of China: Comparison to the parent compounds. *Environmental Science & Technology*, *47*(19), 10841–10850. doi:10.1021/es401504a PMID:24011124

Fendorf, S. E., & Zasoski, R. J. (1992). Chromium(III) oxidation by delta-MnO_2. 1. Characterization. *Environmental Science & Technology*, *26*(1), 79–85. doi:10.1021/es00025a006

Finlay-Moore, O., Hartel, P. G., & Cabrera, M. L. (2000). 17 beta-estradiol and testosterone in soil and runoff from grasslands amended with broiler litter. *Journal of Environmental Quality*, *29*(5), 1604–1611. doi:10.2134/jeq2000.00472425002900050030x

Han, B. (2015). *Degradation of estradiol in water and soil using a new class of stabilized manganese dioxide nanoparticles and hydrodechlorination of triclosan using supported palladium catalysts.* (Unpublished doctoral dissertation). Auburn University, Auburn, AL.

Han, B., Zhang, M., Zhao, D. Y., & Feng, Y. C. (2015). Degradation of aqueous and soil-sorbed estradiol using a new class of stabilized manganese oxide nanoparticles. *Water Research*, *70*, 288–299. PMID:25543239

He, F., Zhang, M., Qian, T. W., & Zhao, D. Y. (2009). Transport of carboxymethyl cellulose stabilized iron nanoparticles in porous media: Column experiments and modeling. *Journal of Colloid and Interface Science*, *334*(1), 96–102. doi:10.1016/j.jcis.2009.02.058 PMID:19383562

He, F., & Zhao, D. Y. (2007). Manipulating the size and dispersibility of zerovalent iron nanoparticles by use of carboxymethyl cellulose stabilizers. *Environmental Science & Technology*, *41*(17), 6216–6221. doi:10.1021/es0705543 PMID:17937305

He, F., & Zhao, D. Y. (2008). Comment on "Manipulating the size and dispersibility of zerovalent iron nanoparticles by use of carboxymethyl cellulose stabilizers" – Response. *Environmental Science & Technology*, *42*(9), 3480–3480. doi:10.1021/es8004255

He, F., Zhao, D. Y., & Paul, C. (2010). Field assessment of carboxymethyl cellulose stabilized iron nanoparticles for in situ destruction of chlorinated solvents in source zones. *Water Research, 44*(7), 2360–2370. doi:10.1016/j.watres.2009.12.041 PMID:20106501

Herzig, J. P., Leclerc, D. M., & Legoff, P. (1970). Flow of suspensions through porous media - Application to deep filtration. *Industrial & Engineering Chemistry, 62*(5), 8–35. doi:10.1021/ie50725a003

Hohenblum, P., Gans, O., Moche, W., Scharf, S., & Lorbeer, G. (2004). Monitoring of selected estrogenic hormones and industrial chemicals in groundwaters and surface waters in Austria. *The Science of the Total Environment, 333*(1-3), 185–193. doi:10.1016/j.scitotenv.2004.05.009 PMID:15364528

Huber, M. M., Canonica, S., Park, G. Y., & Von Gunten, U. (2003). Oxidation of pharmaceuticals during ozonation and advanced oxidation processes. *Environmental Science & Technology, 37*(5), 1016–1024. doi:10.1021/es025896h PMID:12666935

Jiang, J., Pang, S. Y., Ma, J., & Liu, H. (2012).Oxidation of phenolic endocrine disrupting chemicals by potassium permanganate in synthetic and real waters.[PubMed]. *Environmental Science & Technology, 46*(3), 1774–1781. doi:10.1021/es2035587

Jiang, L. Y., Huang, C., Chen, J. M., & Chen, X. (2009).Oxidative Transformation of 17 beta-estradiol by MnO2 in aqueous solution. *Archives of Environmental Contamination and Toxicology, 57*(2), 221–229.

Kolodziej, E. P., Harter, T., & Sedlak, D. L. (2004). Dairy wastewater, aquaculture, and spawning fish as sources of steroid hormones in the aquatic environment. *Environmental Science & Technology, 38*(23), 6377–6384. doi:10.1021/es049585d PMID:15597895

Kolpin, D. W., Furlong, E. T., Meyer, M. T., Thurman, E. M., Zaugg, S. D., Barber, L. B., & Buxton, H. T. (2002). Pharmaceuticals, hormones, and other organic wastewater contaminants in US streams, 1999-2000: A national reconnaissance. *Environmental Science & Technology, 36*(6), 1202–1211. doi:10.1021/es011055j PMID:11944670

Langenhoff, A. A. M., Brouwers-Ceiler, D. L., Engelberting, J. H. L., Quist, J. J., Wolkenfelt, J. G. P. N., Zehnder, A. J. B., & Schraa, G. (1997). Microbial reduction of manganese coupled to toluene oxidation. *FEMS Microbiology Ecology, 22*(2), 119–127. doi:10.1111/j.1574-6941.1997.tb00363.x

Lee, L. S., Strock, T. J., Sarmah, A. K., & Rao, P. S. C. (2003). Sorption and dissipation of testosterone, estrogens, and their primary transformation products in soils and sediment. *Environmental Science & Technology, 37*(18), 4098–4105. doi:10.1021/es020998t PMID:14524441

Lee, S. S., Li, W. L., Kim, C., Cho, M. J., Lafferty, B. J., & Fortner, J. D. (2015). Surface functionalized manganese ferrite nanocrystals for enhanced uranium sorption and separation in water. *Journal of Materials Chemistry A, 3*(43), 21930–21939. doi:10.1039/C5TA04406E

Lee, Y., Yoon, J., & von Gunten, U. (2005). Kinetics of the oxidation of phenols and phenolic endocrine disruptors during water treatment with ferrate (Fe(VI)). *Environmental Science & Technology, 39*(22), 8978–8984. doi:10.1021/es051198w PMID:16323803

Li, M., Schnablegger, H., & Mann, S. (1999). Coupled synthesis and self-assembly of nanoparticles to give structures with controlled organization. *Nature, 402*(6760), 393–395. doi:10.1038/46509

Liang, Q. Q., & Zhao, D. Y. (2014). Immobilization of arsenate in a sandy loam soil using starch-stabilized magnetite nanoparticles. *Journal of Hazardous Materials, 271*, 16–23. doi:10.1016/j.jhazmat.2014.01.055 PMID:24584068

Liang, Q. Q., Zhao, D. Y., Qian, T. W., Freeland, K., & Feng, Y. C. (2011). Effects of stabilizers and water chemistry on arsenate sorption by polysaccharide-stabilized magnetite nanoparticles. *Industrial & Engineering Chemistry Research, 51*(5), 2407–2418. doi:10.1021/ie201801d

Lin, K., Liu, W., & Gan, J. (2009a). Oxidative removal of bisphenol A by manganese dioxide: Efficacy, products, and pathways. *Environmental Science & Technology, 43*(10), 3860–3864. doi:10.1021/es900235f PMID:19544899

Lin, K. D., Liu, W. P., & Gan, J. (2009b). Reaction of tetrabromobisphenol A (TBBPA) with manganese dioxide: Kinetics, products, and pathways. *Environmental Science & Technology, 43*(12), 4480–4486. doi:10.1021/es803622t PMID:19603665

Lin, S. H., & Wiesner, M. R. (2012). Theoretical Investigation on the Steric Interaction in Colloidal Deposition. *Langmuir, 28*(43), 15233–15245. doi:10.1021/la302201g PMID:22978750

Liu, R. P., Xu, W., He, Z., Lan, H. C., Liu, H. J., & Prasai, T. et al. (2015). Adsorption of antimony(V) onto Mn(II)-enriched surfaces of manganese-oxide and Fe-Mn binary oxide. *Chemosphere, 138*, 616–624. PMID:26218341

Liu, Y. Y., Su, G. X., Zhang, B., Jiang, G. B., & Yan, B. (2011). Nanoparticle-based strategies for detection and remediation of environmental pollutants. *Analyst (London), 136*(5), 872–877. doi:10.1039/c0an00905a PMID:21258678

Lu, Z. J., Lin, K. D., & Gan, J. (2011). Oxidation of bisphenol F (BPF) by manganese dioxide. *Environmental Pollution, 159*(10), 2546–2551. doi:10.1016/j.envpol.2011.06.016 PMID:21741139

Mahamallik, P., Saha, S., & Pal, A. (2015). Tetracycline degradation in aquatic environment by highly porous MnO2 nanosheet assembly. *Chemical Engineering Journal, 276*, 155–165. doi:10.1016/j.cej.2015.04.064

Manning, B. A., Fendorf, S. E., Bostick, B., & Suarez, D. L. (2002). Arsenic(III) oxidation and arsenic(V) adsorption reactions on synthetic birnessite. *Environmental Science & Technology, 36*(5), 976–981. doi:10.1021/es0110170 PMID:11918029

Mansell, B. L., & Drewes, J. E. (2004). Fate of steroidal hormones during soil-aquifer treatment. *Ground Water Monitoring and Remediation, 24*(2), 94–101. doi:10.1111/j.1745-6592.2004.tb00717.x

Murray, J. W. (1974). Surface chemistry of hydrous manganese-dioxide. *Journal of Colloid and Interface Science, 46*(3), 357–371. doi:10.1016/0021-9797(74)90045-9

Murray, J. W. (1975). The interaction of metal ions at the manganese dioxide-solution interface. *Geochimica et Cosmochimica Acta, 39*(4), 505–519. doi:10.1016/0016-7037(75)90103-9

Ohko, Y., Iuchi, K. I., Niwa, C., Tatsuma, T., Nakashima, T., Iguchi, T., & Fujishima, A. et al. (2002). 17 beta-estrodial degradation by TiO_2 photocatalysis as means of reducing estrogenic activity. *Environmental Science & Technology, 36*(19), 4175–4181. doi:10.1021/es011500a PMID:12380092

Pal, A., Gin, K. Y. H., Lin, A. Y. C., & Reinhard, M. (2010). Impacts of emerging organic contaminants on freshwater resources: Review of recent occurrences, sources, fate and effects. *The Science of the Total Environment*, *408*(24), 6062–6069. doi:10.1016/j.scitotenv.2010.09.026 PMID:20934204

Pelley, A. J., & Tufenkji, N. (2008).Effect of particle size and natural organic matter on the migration of nano- and microscale latex particles in saturated porous media. *Journal of Colloid and Interface Science*, *321*(1), 74–83.

Phenrat, T., Cihan, A., Kim, H. J., Mital, M., Illangasekare, T., & Lowry, G. V. (2010).Transport and deposition of polymer-modified Fe-0 nanoparticles in 2-D heterogencous porous media: Effects of particle concentration, Fe-0 content, and coatings. *Environmental Science & Technology*, *44*(23), 9086–9093.

Post, J. E. (1999). Manganese oxide minerals: Crystal structures and economic and environmental significance. *Proceedings of the National Academy of Sciences of the United States of America*, *96*(7), 3447–3454. doi:10.1073/pnas.96.7.3447 PMID:10097056

Powell, R. M., Puls, R. W., Blowes, D. W., Vogan, J. L., Gillham, R. W., Powell, P. D. ... Landis, R. (1998). Permeable reactive barrier technologies for contaminant remediation. EPA/600/R-98/125, US EPA, Washington DC.

Remucal, C. K., & Ginder-Vogel, M. (2014). A critical review of the reactivity of manganese oxides with organic contaminants. *Environmental Science Processes & Impacts*, *16*(6), 1247–1266. doi:10.1039/c3em00703k PMID:24791271

Rivera-Utrilla, J., Sanchez-Polo, M., Ferro-Garcia, M. A., Prados-Joya, G., & Ocampo-Perez, R. (2013). Pharmaceuticals as emerging contaminants and their removal from water. A review. *Chemosphere*, *93*(7), 1268–1287. doi:10.1016/j.chemosphere.2013.07.059 PMID:24025536

Rosenfeldt, E. J., & Linden, K. G. (2004). Degradation of endocrine disrupting chemicals bisphenol A, ethinyl estradiol, and estradiol during UV photolysis and advanced oxidation processes. *Environmental Science & Technology*, *38*(20), 5476–5483. doi:10.1021/es035413p PMID:15543754

Rubert, K. F., & Pedersen, J. A. (2006). Kinetics of oxytetracycline reaction with a hydrous manganese oxide. *Environmental Science & Technology*, *40*(23), 7216–7221. doi:10.1021/es060357o PMID:17180969

Rule, K. L., Ebbett, V. R., & Vikesland, P. J. (2005). Formation of chloroform and chlorinated organics by free-chlorine-mediated oxidation of triclosan. *Environmental Science & Technology*, *39*(9), 3176–3185. doi:10.1021/es048943+ PMID:15926568

Sakthivadivel, R. (1966). *Theory and mechanism of filtration of non-colloidal fines through a porous medium. Hydraulic Engineering Laboratory*. Berkeley: University of Califonia.

Sakthivadivel, R. (1969). *Clogging of a granular porous medium by sediment. Hydraulic Engineering Laboratory*. Berkeley: University of Califonia.

Sang, W., Morales, V. L., Zhang, W., Stoof, C. R., Gao, B., Schatz, A. L., & Steenhuis, T. S. et al. (2013). Quantification of colloid retention and release by straining and energy minima in variably saturated porous media. *Environmental Science & Technology*, *47*(15), 8256–8264. PMID:23805840

Sharma, V. K., Li, X. Z., Graham, N., & Doong, R. A. (2008). Ferrate(VI) oxidation of endocrine disruptors and antimicrobials in water. *Journal of Water Supply: Research & Technology - Aqua, 57*(6), 419–426. doi:10.2166/aqua.2008.077

Shore, L. S., & Shemesh, M. (2003). Naturally produced steroid hormones and their release into the environment. *Pure and Applied Chemistry, 75*(11-12), 1859–1871. doi:10.1351/pac200375111859

Stone, A. T. (1987). Reductive dissolution of manganese (III/IV) oxides by substituted phenols. *Environmental Science & Technology, 21*(10), 979–988. doi:10.1021/es50001a011 PMID:19994996

Suslick, K. S., Fang, M. M., & Hyeon, T. (1996). Sonochemical synthesis of iron colloids. *Journal of the American Chemical Society, 118*(47), 11960–11961. doi:10.1021/ja961807n

Tabata, A., Kashiwada, S., Ohnishi, Y., Ishikawa, H., Miyamoto, N., Itoh, M., & Magara, Y. (2001). Estrogenic influences of estradiol-17 beta, p-nonylphenol and bis-phenol-A on japanese medaka (Oryzias latipes) at detected environmental concentrations. *Water Science and Technology, 43*(2), 109–116. PMID:11380168

Ternes, T. A., Stumpf, M., Mueller, J., Haberer, K., Wilken, R. D., & Servos, M. (1999). Behavior and occurrence of estrogens in municipal sewage treatment plants - I. Investigations in Germany, Canada and Brazil. *The Science of the Total Environment, 225*(1-2), 81–90. doi:10.1016/S0048-9697(98)00334-9 PMID:10028705

Theron, J., Walker, J. A., & Cloete, T. E. (2008). Nanotechnology and water treatment: Applications and emerging opportunities. *Critical Reviews in Microbiology, 34*(1), 43–69. doi:10.1080/10408410701710442 PMID:18259980

Tufenkji, N., & Elimelech, M. (2004). Correlation equation for predicting single-collector efficiency in physicochemical filtration in saturated porous media. *Environmental Science & Technology, 38*(2), 529–536. doi:10.1021/es034049r PMID:14750730

Tufenkji, N., Miller, G. F., Ryan, J. N., Harvey, R. W., & Elimelech, M. (2004). Transport of cryptosporidium oocysts in porous media: Role of straining and physicochemical filtration. *Environmental Science & Technology, 38*(22), 5932–5938. doi:10.1021/es049789u PMID:15573591

Ukrainczyk, L., & McBride, M. B. (1992). Oxidation of phenol in acidic aqueous suspensions of manganese oxides. *Archive of Clays and Clay Minerals, 40*(2), 157–166. doi:10.1346/CCMN.1992.0400204

Urynowicz, M. A. (2008). In situ chemical oxidation with permanganate: Assessing the competitive interactions between target and nontarge compounds. *Soil & Sediment Contamination, 17*(1), 53–62. doi:10.1080/15320380701741412

Valhondo, C., Carrera, J., Ayora, C., Tubau, I., Martinez-Landa, L., Nodler, K., & Licha, T. (2015). Characterizing redox conditions and monitoring attenuation of selected pharmaceuticals during artificial recharge through a reactive layer. *The Science of the Total Environment, 512*, 240–250. doi:10.1016/j.scitotenv.2015.01.030 PMID:25625636

Vulliet, E., Wiest, L., Baudot, R., & Grenier-Loustalot, M.-F. (2008). Multi-residue analysis of steroids at sub-ng/L levels in surface and ground-waters using liquid chromatography coupled to tandem mass spectrometry. *Journal of Chromatography. A, 1210*(1), 84–91. doi:10.1016/j.chroma.2008.09.034 PMID:18823894

Wang, H. Y., Gao, B., Wang, S. S., Fang, J., Xue, Y. W., & Yang, K. (2015). Removal of Pb(II), Cu(II), and Cd(II) from aqueous solutions by biochar derived from $KMnO_4$ treated hickory wood. *Bioresource Technology*, *197*, 356–362. doi:10.1016/j.biortech.2015.08.132 PMID:26344243

Weber, W. J. Jr. (2001). *Environmental Systems and Processes: Principles, Modeling, and Design*. New York: Wiley-Interscience.

Wei, Y. T., Wu, S. C., Chou, C. M., Che, C. H., Tsai, S. M., & Lien, H. L. (2010). Influence of nanoscale zero-valent iron on geochemical properties of groundwater and vinyl chloride degradation: A field case study. *Water Research*, *44*(1), 131–140. doi:10.1016/j.watres.2009.09.012 PMID:19800096

Xiong, Z., He, F., Zhao, D. Y., & Barnett, M. O. (2009). Immobilization of mercury in sediment using stabilized iron sulfide nanoparticles. *Water Research*, *43*(20), 5171–5179. doi:10.1016/j.watres.2009.08.018 PMID:19748651

Xu, L., Xu, C., Zhao, M. R., Qiu, Y. P., & Sheng, G. D. (2008). Oxidative removal of aqueous steroid estrogens by manganese oxides. *Water Research*, *42*(20), 5038–5044. doi:10.1016/j.watres.2008.09.016 PMID:18929389

Yang, B., Ying, G. G., Zhao, J. L., Zhang, L. J., Fang, Y. X., & Nghiem, L. D. (2010). Oxidation of triclosan by ferrate: Reaction kinetics, products identification and toxicity evaluation. *Journal of Hazardous Materials*, *186*(1), 227–235. doi:10.1016/j.jhazmat.2010.10.106 PMID:21093982

Zhang, H. C., & Huang, C. H. (2003). Oxidative transformation of triclosan and chlorophene by manganese oxides. *Environmental Science & Technology*, *37*(11), 2421–2430. doi:10.1021/es026190q PMID:12831027

Zhang, J., Sun, B., Xiong, X. M., Gao, N. Y., Song, W. H., Du, E. D., & Zhou, G. M. et al. (2014). Removal of emerging pollutants by Ru/TiO_2-catalyzed permanganate oxidation. *Water Research*, *63*, 262–270. doi:10.1016/j.watres.2014.06.028 PMID:25016299

Zhuang, J., Qi, J., & Jin, Y. (2005). Retention and transport of amphiphilic colloids under unsaturated flow conditions: Effect of particle size and surface property. *Environmental Science & Technology*, *39*(20), 7853–7859. doi:10.1021/es050265j PMID:16295847

KEY TERMS AND DEFINITIONS

Carboxymethyl Cellulose (CMC): Carboxymethyl cellulose is a cellulose derivative with carboxymethyl groups ($-CH_2-COOH$) bound to the polysugar chain. It is an environmental friendly stabilizer for preparing stabilized nanoparticles for cleanup of contaminated soil and groundwater.

Emerging Organic Contaminants (EOCs): Emerging organic contaminants are synthetic or naturally occurring organic contaminants that have appeared only recently or have been in the environment for a while but concerns have been raised recently.

Endocrine Disrupting Chemicals (EDCs): Endocrine disrupting chemicals can interfere with hormonal and homeostatic system functions. EDCs can cause adverse developmental, reproductive, neurological, and immune effects in both humans and wildlife.

In-Situ Remediation by Stabilized Nanoparticles: The in-situ remediation technology is to degrade (for organics) or immobilize (for non-degradable chemicals such as metals and metalloids) through directly delivering reactive nanoparticles as a reactive agent or adsorbent into the contaminated soil.

Steroidal Estrogens: Steroidal estrogens are one class of EDCs, which include endogenous estrogens released by human and wildlife as well as synthetic estrogens widely used in oral contraceptive pills. When released into the environment, these hormones may cause human or wildlife abnormal development, reproductive disorders, sexual disorders as well as cancers.

Chapter 7
Light Sensitized Disinfection with Fullerene

Kyle Moor
Yale University, USA

Samuel Snow
Michigan State University, USA

Jaehong Kim
Yale University, USA

ABSTRACT

Fullerene has drawn wide interest across many fields due to its favorable electronic and optical properties, which has spurred its use in a myriad of applications. One of the hallmark properties of fullerene is its ability to act as a photosensitizer and efficiently generate $^1O_2$, a form of Reactive Oxygen Species (ROS), upon visible irradiation when dispersed in organic solvents. However, the application of fullerene in environmental systems has been somewhat limited due to fullerene's poor solubility in water, which causes individual fullerene molecules to aggregate and form large colloidal species, quenching much of fullerene's $^1O_2$ production. This is unfortunate given that $^1O_2$ provides many advantages as an oxidant compared to ROS produced from typical advanced oxidation processes, such as OH radicals, due to $^1O_2$'s greater chemical selectivity and its ability to remain unaffected by the presence of background water constituents, such as natural organic matter and carbonate. Hence, fullerene materials may hold great potential for the oxidation and disinfection of complex waters. Herein, we chronicle the advances that have been made to propel fullerene materials towards use in emerging water disinfection technologies. Two approaches to overcome fullerene aggregation and the subsequent loss of $^1O_2$ production in aqueous systems are herein outlined: 1) addition of hydrophilic functionality to fullerene's cage, creating highly photoactive colloidal fullerenes; and 2) covalent attachment of fullerene to solid supports, which physically prevents fullerene aggregation and allows efficient $^1O_2$ photo-generation. An emphasis is placed on the inactivation of MS2 bacteriophage, a model for human enteric viruses, highlighting the potential of fullerene materials for light-activated disinfection technologies.

DOI: 10.4018/978-1-5225-0585-3.ch007

1. INTRODUCTION

The development of novel materials for the inactivation of waterborne pathogens is in a critical need for much of the world today. Materials that, in response to visible light, can efficiently inactivate viruses and spore forming bacteria that survive in dry conditions would be transformative for many of the challenges in developing countries and in the context of bioterrorism defense. In particular, such materials would be highly useful for advancing solar disinfection (SODIS) and antimicrobial/biocidal coating technology. Semiconductor photocatalysts, such as the archetypical TiO_2, have been intensely pursued in hopes of realizing such disinfection technologies,(Min Cho, Chung, Choi, & Yoon, 2005a; Lonnen, Kilvington, Kehoe, Al-Touati, & McGuigan, 2005) but have been somewhat hindered by their inefficient visible light utilization. Hence, many researchers have put forth serious efforts on expanding TiO_2's visible absorption through various doping schemes,(Pelaez et al., 2012; Rehman, Ullah, Butt, & Gohar, 2009) while others have focused on pursing new small-band gap semiconductors such as WO_3,(Kim, Lee, & Choi, 2010; Zhu, Xu, Fu, Zhao, & Zhu, 2007) graphitic carbon nitride,(H. Wang et al., 2014; Xu, Wang, & Zhu, 2013) and CdS quantum dots(Bessekhouad, Robert, & Weber, 2004). Moving beyond conventional inorganic photocatalysts, photosensitizing organic dye molecules, which efficiently harvest visible light, have gained recent attention with a particular focus on buckminsterfullerenes.(Chae, Hotze, & Wiesner, 2009; Jaesang Lee et al., 2009) Buckminsterfullerenes, or simply fullerenes, and their functional derivatives have been proposed by researchers as effective antimicrobial agents, via photocatalytic production of singlet oxygen ($^1O_2$) and subsequent microbial inactivation (Liyi Huang et al., 2010; Jaesang Lee et al., 2009; Q. Li et al., 2008; George P. Tegos et al., 2005). In contrast with the ubiquitous semiconductor photocatalysts, fullerenes have the advantage of being able to be activated by visible light, especially with functionalization, and covalently anchored to a host material such as polymers. Photocatalysts used for disinfection should be recoverable/reusable, completely conserved (no escape into the environment), activated by visible light, and able to retain their catalytic properties over repeated use. Fullerene derivatives are very attractive as photocatalysts because they can potentially exhibit these essential characteristics when properly functionalized and covalently anchored onto a supporting structure.

This chapter chronicles the advances of fullerene (specifically C_{60} and C_{70}) science and technology related to their application as photocatalysts for disinfection. Fullerene photocatalysis is affected by several factors that are relevant to disinfection applications. Functionalization of the fullerene cage directly impacts the electronic structure of the chromophore, altering the energy levels and efficiencies of the intermediate excited states and transitions. Further, aggregation of fullerenes in the aqueous phase can severely diminish the photoactivity of the fullerenes. Functionalization and immobilization of fullerenes can both lessen the degree or effects of aggregation of fullerenes on their photochemistry. The utility of fullerenes applied as photocatalysts is directly proportional to the efficiency of their photosensitization of $^1O_2$. The use of fullerenes as disinfection agents, and the methods of enhancing their capabilities, are discussed here with case examples from our past publications.

2. BACKGROUND

Comprised only of carbon atoms, caged fullerene molecules represent the third allotrope of carbon besides graphite and diamond.(Kroto, Heath, O'Brien, Curl, & E, 1985) Among fullerenes (e.g., C_{70}, C_{76}, C_{78}, C_{84} and C_{90}), C_{60} has been the most prolific in fullerene research and application as C_{60} is available with high purity and in comparatively vast quantities. Yet many researchers have also pursued higher fullerenes for various applications, notably using C_{70}, the fullerene produced in the second largest quantities in the arc-discharge process (10% of total fullerenes (Scrivens, Cassell, North, & Tour, 1994)), due to the advantageous properties incurred by the slightly larger fullerene cage. As a result of such great interest, the physical and chemical properties of both C_{60} and C_{70} are well established.(Bagrii & Karaulove, 2001; Chase, Herron, & Holler, 1992; Diederich & Thilgen, 1996; Hebard, 1993; Hedberg et al., 1997; Karaulove & Bagrii, 1999; Thilgen, Herrmann, & Diederich, 1997) C_{60} consists of 60 carbon atoms structured as a perfectly symmetrical cage (I_h symmetry) with the configuration of a soccer ball, *i.e.* a network of 12 pentagonal and 20 hexagonal rings. C_{70} is similar to C_{60}, yet has 10 carbon atoms added equatorial into the C_{60} fullerene cage, resulting in an elongated ellipsoidal structure (D_{5H} symmetry) that is composed of 25 hexagonal and 12 pentagonal rings. With increasing commercial interest in fullerene's unique properties, the manufacture and the use of C_{60} and C_{70} are expected to grow very rapidly over the next decade. Projected uses of fullerene include fuel cell development, diamond manufacturing, super conductivity devices, drug delivery agents, and high temperature lubricants.(Bocquillon, Bogicevic, Fabre, & Rassat, 1993; Gupta, Bhushan, Capp, & Coe, 1994; Haddon et al., 1991; Iqbal et al., 1991; Kelty, Chen, & Lieber, 1991; Mort, Ziolo, Machonkin, Huffman, & Ferguson, 1991; Regueiro, Monceau, & Hodeau, 1992; Y. Wang, 1992; Wharton, Kini, Mortis, & Wilson, 2001)

1.1. Fullerene Photochemical Reactivity

Unique photochemical properties of fullerenes were recognized by scientific community not long after its proposed existence by Kroto *et al.* in 1985.(Kroto et al., 1985) While C_{60} does not fluoresce or phosphoresce in the ambient environment, C_{60} in organic solvent is readily excited from its ground singlet state ($^1C_{60}$) to a singlet state $^1C_{60}^*$ with quantum yield of nearly 1.0 (100%) upon UV and visible light irradiation.(J. W. Arbogast et al., 1991) Since the highest occupied molecular orbital (HOMO) - lowest unoccupied molecular orbital (LUMO) gap of pure C_{60} is 2.3 eV, comparable to that of iron oxide, the photo-excitation of C_{60} can be accomplished by photons with wavelength below 550 nm, which are abundant in the solar spectrum. The produced $^1C_{60}^*$ can either convert back to the ground state by fluorescence emission and internal conversion process or transform into triplet state ($^3C_{60}^*$) through intersystem crossing (ISC). The resulting triplet state $^3C_{60}^*$ is subject to three quenching pathways: 1) triplet quenching including energy transfer to ground-state triplet oxygen ($^3O_2$), which results in photochemical generation of singlet oxygen ($^1O_2$); 2) self-quenching through interaction between triplet and ground state C_{60}; and 3) triplet-triplet annihilation between adjacent triplet state C_{60}. In the presence of oxygen, energy transfer to oxygen (*i.e.*, the first quenching mechanism) is dominant with very high yield (*e.g.*, $\Phi(^1O_2)$ (355 nm) = 0.76 in benzene).(J. W. Arbogast et al., 1991) These processes are indicated in Figure 1. Note that C_{70} similarly exhibits the aforementioned photophysical pathways, resulting in efficient production of $^1O_2$ with large quantum yields ($\Phi = 0.81 \pm 0.15$ in benzene). (James W Arbogast & Foote, 1991)

1.2. Oxidation of Organic Pollutants and Microbial Inactivation by $^1O_2$

Due to a somewhat unusual electronic configuration, the oxygen molecule in singlet state, $^1O_2$, is higher in energy than triplet state (Kearns, 1971). This non-radical, transient species is highly energetic and exhibits strong, yet substrate-specific oxidizing power (Kearns, 1971; F. Wilkinson & Brummer, 1981). $^1O_2$ can be generated via: 1) energy transfer by photosensitizing chemicals (*i.e.* chemicals that absorb

Figure 1. Jablonski diagram for fullerene (C_{60} as an example) depicting various photophysical processes, including photochemical $^1O_2$ generation

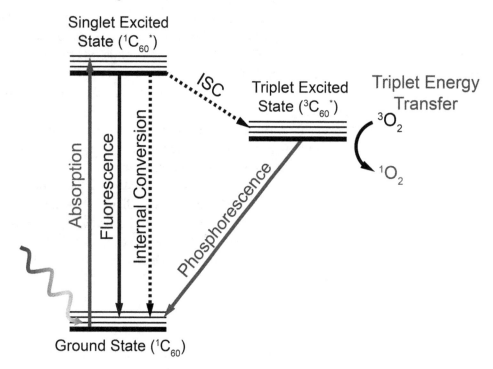

light energy and subsequently transfer to another molecules such as oxygen), 2) pulse radiolysis (in the presence of an energy mediator such as benzene), 3) microwave discharge, and 4) chemical pathways (*e.g.*, oxidation of hydrogen peroxide by hypochlorite/hypobromite/molybdate anion, hydrolysis of potassium superoxide and decomposition of aryl peroxides or ozonides) (Kearns, 1971; F. Wilkinson & Brummer, 1981). Since the role of metastable $^1O_2$ in dye-sensitized photo-oxygenation reactions was first suggested by Kautsky et al in 1939 (Kautsky, 1939), its oxidation potential has been exploited in applications including specific chemical synthesis (Leach & Houk, 2002; Stephenson, 1980) (*e.g.*, Ene and Diels-Alder reactions) and cancer and tumor therapies (Moura, Oliveira Campos, & Griffiths, 1997; Ogawa, Dy, Kobuke, Ogura, & Okura, 2007). Based on the finding that $^1O_2$ generated by photo-activation of natural organic matter (NOM) plays a critical role in degrading micropollutants in natural aquatic environment (Haag, Hoigne, Gassman, & Braun, 1984), $^1O_2$ has been considered as an alternative oxidant for water treatment. Recent studies have suggested that $^1O_2$ produced using photosensitizers including natural and synthetic dyes, polyaromatics (e.g., acridine), and metal-centered macromolecules (e.g., metallo-phthalocyanine and porphyrin) efficiently triggers oxidation of various pollutants such as substituted phenols, malodorous organic sulfides, and benzimidazole fungicides (Escalada et al., 2006; J. S. Miller, 2005; A. H. Sun, Xiong, & Xu, 2008). Concurrently, application of $^1O_2$ as a disinfectant has been also reported (Jimenez-Hernandez, Manjon, Garcia-Fresnadillo, & Orellana, 2006; Rengifo-Herrera et al., 2007). Dyes such as rose bengal and methylene blue have been employed to mediate transfer of solar energy to oxygen to produce $^1O_2$, as an agent to inactivate bacteria and virus. One recent study (Badireddy, Hotze, Chellam, Alvarez, & Wiesner, 2007) employed multihydroxylated fullerene (fullerol) to produce $^1O_2$ as a disinfection agent.

1.3. Challenges in Application of Fullerene in Aqueous Phase

Strong photocatalytic activity of fullerene could be instrumental for producing $^1O_2$, which can be used to degrade organic pollutants and inactivate pathogenic microorganisms in water. However, its intrinsic hydrophobicity and extremely low water solubility (Ruoff, Tse, Malhotra, & Lorents, 1993) ($< 10^{-9}$ mg/L) make the direct application of pristine C_{60} and C_{70} in the aqueous phase impossible. Two commonly employed approaches to make fullerenes available in water are colloidal formation and surfactant encapsulation. First, fullerene can be dispersed in water as negatively-charged colloidal forms (often referred to as nC_{60} or nC_{70}) of sizes ranging from 60 to 400 nm (Andrievsky, Kosevich, Vovk, Shelkovsky, & Vashchenko, 1995; Deguchi, Alargova, & Tsujii, 2001; J. D. Fortner et al., 2005; Ma & Bouchard, 2009). Such colloidal particles could be useful as a nanosized catalyst which provides a large surface area for reaction. However, recent studies have demonstrated that the unique photochemical properties of C_{60} to produce ROS are significantly reduced as C_{60} forms aggregates in water (Hotze, Labille, Alvarez, & Wiesner, 2008; Lee, Fortner, Hughes, & Kim, 2007; Lee, Yamakoshi, Hughes, & Kim, 2008). The same phenomenon was also reported for C_{70} in the aqueous phase (K. J. Moor, Snow, & Kim, 2015). Second, fullerene can be suspended in the aqueous phase through association with foreign molecules such as surfactants, polymers, lipids, and carbohydrates (Andersson, Nilsson, Sundahl, Westman, & Wennerstrom, 1992; Beeby, Eastoe, & Heenan, 1994; Y. N. Yamakoshi, Yagami, Fukuhara, Sueyoshi, & Miyata, 1994). The encapsulating agents function as host molecules with hydrophilic groups orienting toward water and embedding hydrophobic fullerene molecules as guests in their hydrophobic core. This allows C_{60} or C_{70} to disperse in the aqueous phase as an individual molecule or relatively small aggregates. Past studies suggested that C_{60} in this form retains its photochemical property, which is comparable to pristine-C_{60} in organic solvent (Dimitrijevic & Kamat, 1993; Y. Yamakoshi et al., 2003). However, application of fullerene associated with encapsulating agents such as surfactants is impractical, since a large quantity of surfactants (applied above critical micelle concentration) is a secondary pollutant that needs to be removed.

3. FUNCTIONALIZED FULLERENES

The potential for using C_{60} as a photocatalyst for water treatment brought much excitement to researchers, but this enthusiasm was quickly dampened by the real and difficult challenges of fullerene aggregation. Consequently, researchers turned first to hydrophilic functionalization as a potential remedy. Much effort was spent on covalently functionalizing fullerenes to enhance their water solubility for various applications (Nakamura & Isobe, 2003). Functionalization has been accomplished by many synthetic approaches, but the two most popular methods have been the Bingel-Hirch and Prato-Maggini reactions (Wudl, 2002). Interestingly, the type of bond lifted in an addition reaction ([5,6] compared to [6,6]) was found to be more important, in terms of retaining the fullerene's unique electron configuration than the number of groups added to the fullerene cage (Y. P. Sun et al., 2000). When a functional group is added to a [5,6] bond, C_{60}'s pi-conjugated system is significantly impaired, exhibiting markedly less fluorescence than other derivatives (Y. P. Sun et al., 2000). Therefore, most studies deal only with [6,6] functionalization, which, conveniently, is often the most favorable reaction pathway (Diederich & Thilgen, 1996). The nature of the functional groups used to modify C_{60} is also essential to how functionalization affects the photophysical properties of C_{60}. For example, Collini *et al.* successfully added a distyrylbenzene group to C_{60} that allowed for a two-photon absorption excitation mechanism, where the C_{60} moiety received excitation from two low energy photons from its leaf-like functional groups, resulting in $^1O_2$ production in response to IR light (Collini et al., 2010). Huang *et al.* synthesized various cationic derivatives and found that quaternary ammonium groups added to the C_{60} cage enhanced both water solubility and antimicrobial efficacy of the photosensitizer (L. Huang et al., 2010). Still others attached a short DNA sequence complimentary to a double helix and found that the fullerene-DNA conjugate would selectively bind to a DNA strand and cleave it in the presence of light (Nakamura & Isobe, 2003). There is virtually no limit to the possible innovative fullerene derivatives, and therefore it is critical to understand the underlying mechanisms of how functionalization affects C_{60}'s physicochemical properties.

Solvent interactions with functional groups, alterations of the π-conjugated electron configuration of C_{60}, and geometric changes in bond angles due to the functional groups are suggested as the primary ways functionalization affects the photophysical properties of fullerenes. Regarding solvent interactions, pristine fullerene is highly non-polar by nature and therefore not compatible with polar solvents. Functionalization of C_{60} with hydrophilic, or polar, groups allows more favorable arrangements for the polar H_2O molecules around the C_{60} derivative, decreasing the tendency of the fullerenes to aggregate. As described above, reducing aggregation enhances C_{60} photocatalysis in water by reducing quenching mechanisms. However, altering the π-conjugation of the fullerene cage can affect the C_{60}'s photochemical properties via a completely different mechanism. Upon functionalizing C_{60}, there is a disruption of the sp^2 conjugation due to the removal of a [6,6] double bond, which would blue-shift the ground state absorption spectra (Guldi & Asmus, 1997; Prat et al., 1999; Y. P. Sun et al., 2000) and thereby reducing the photocatalytic yield in the visible range. Alternatively, the functional group itself could provide additional π-conjugation to the system via sp^2 hybridized structures (*e.g.*, cyclopropane rings, phenyl rings, carbonyl groups) resulting in the enhanced visible absorption band often observed in functionalized fullerenes (Guldi & Asmus, 1997; Hamano et al., 1997; Prat et al., 1999; Y. P. Sun et al., 2000). A net red-shift in a fullerene's ground-state absorption with selected functional groups would tend to increase the quantum yield of $^1O_2$. However, it should be noted that the efficiency of energy transfer to the triplet state is also critical for $^1O_2$ production. While the first excited singlet state may be reduced in energy, hence the red shift, the triplet state tends to increase in energy with addition of functional groups due to a local relaxation in the bond geometry around the functional group (Guldi & Asmus, 1997; D. M. Guldi & M. Prato, 2000; Prat et al., 1999). This rationale comes from the fact that the ISC process is enhanced by a departure from planarity in sp^2 hybridized systems; in other words, increasing the number of addends was observed to lead to a net decrease in $^1O_2$ quantum yield (Hamano et al., 1997; Prat et al., 1999).

Of the various functionalization strategies employed to increase hydrophilicity of the fullerene cage, cationic functionalizations of the C_{60} cage have been the most promising, particularly for pharmaceutical and disinfection applications (M. Cho et al., 2010; Gilbert & Moore, 2005; L. Huang et al., 2010; Huang et al., 2012; J. Lee et al., 2009; Lu et al., 2010; Mashino et al., 2003; G. P. Tegos et al., 2005). Utilizing quaternary ammonium functional groups attached to the cage, several researchers have demonstrated that cationically-functionalized fullerenes to exhibit remarkable activity, suitable for various applications (M. Cho et al., 2010; L. Huang et al., 2010; J. Lee et al., 2009; Lu et al., 2010; Mashino et al., 2003; G. P. Tegos et al., 2005). C_{60} tris-functionalized with methyl pyrrolidinium groups, called **B3** here to be consistent with the nomenclature of our previous work (Snow, Lee, & Kim, 2012; Samuel D. Snow, KyoungEun Park, & Jae-Hong Kim, 2014), is the molecule in focus here. Viral inactivation experiments are very useful for measuring the photo-inactivation potential of fullerenes and are often performed using MS2 bacteriophage, grown with *Escherichia coli* as the virus host. In our work MS2 and fullerenes were added to a reactor containing phosphate buffered solution (PBS) at a neutral pH and placed under several different light conditions to induce $^1O_2$ sensitization and subsequent MS2 inactivation (Samuel D. Snow et al., 2014). *E. coli* inactivation experiments are similarly useful and were performed in an identical manner. The remainder of this section will detail our case example of utilizing the B3 fullerene to photoinactivate MS2 viruses (Samuel D. Snow et al., 2014).

Photo-inactivation experiments shown here were performed using a system consisting of fluorescent lamps (FLs) with a UV cutoff filter (400 nm) or Black Light Blue (BLB) lamps directly above a stirred reaction vessel. Light intensities were measured and reported using a UVX Radiometer (UVP, LLC) for the BLB lamps and a blue light sensor (PMA 2121, Solar Light Co.) for the FLs. Sunlight experiments were also performed in Atlanta, Georgia (33°46'25" N, 84°23'38" W) using a similar reaction setup under open sunlight. $^1O_2$ production was measured using a probe molecule, furfuryl alcohol (FFA), with a known reaction rate with $^1O_2$ (Haag & Hoigne, 1986; Maurette, Oliveros, Infelta, Ramsteiner, & Braun, 1983). Photoinactivation experiments using **B3** to inactivate MS2 under visible light demonstrated extremely fast kinetics (Figure 2). A 5-log (99.999%) inactivation of MS2 by 1 μM **B3** was achieved in less than 2 min. Control experiments under dark conditions showed no virus loss, indicating that the adsorption of viruses onto fullerene particles does not contribute to MS2 inactivation observed in these experiments. Concentration dependency of inactivation was clearly observed from 100 to 1,000 nM, with a 2-log inactivation achieved after 5 min with 100 nM **B3**. No significant inactivation was observed for both 10 and 50 nM during extended experiments under visible light. When UVA irradiation was used with these low concentration suspensions, efficient inactivation was observed, with an exceptional 4-log kill in 4 min by 50 nM. The inactivation for 10 nM under UVA was also increased, but only partially, achieving a 2 log removal after *ca.* 20 min. The 2-log inactivation that **B3** achieved at 10 nM, or *ca.* 9 μg/L was remarkable. As a reference, a recent report demonstrated that wastewater NOM, which is a known $^1O_2$ sensitizer, achieved a 2-log MS2 inactivation at mg/L concentrations under simulated sunlight after 12 h (Kohn & Nelson, 2007).

Given the extremely efficient viral inactivation by **B3**, its efficiency under environmentally relevant conditions were also probed. Figure 3 displays the results from sunlight experiments with and without NOM, using 250 nM **B3**. The sunlight conditions were generally comparable to the laboratory conditions with similar visible (blue) light intensity to the FL reactor and slightly lower UV light intensity than that of the UVA reactor. The inactivation of MS2 by 250 nM **B3** with NOM was slightly slower than in PBS alone, likely due to a quenching or protective effect by NOM (Kong, Mukherjee, Chan, & Zepp, 2013). It is known that NOM can act as a photosensitizer of $^1O_2$ and cause inactivation of viruses (Kohn & Nelson, 2007), but NOM's contribution to $^1O_2$-mediated MS2 inactivation in this case would be too minor to be observed in the short experimental timeframe. Even under simulated environmental conditions **B3** appears to be highly photochemically active at sub-micromolar concentrations.

4. SUPPORTED FULLERENES

Although colloidal fullerene systems have achieved some success as illustrated above, there may be key limitations in their practical use in engineered systems, where nanomaterial recovery for both catalyst reuse and to deter release into the environment is of paramount concern. An alternative option is to covalently attach fullerene to support media, where immobilization not only prevents aggregation, allowing

Figure 2. Concentration dependence of MS2 inactivation by B3 aggregates under visible or UVA irradiation
(Samuel D. Snow et al., 2014).

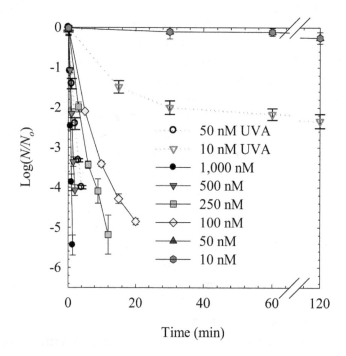

Figure 3. MS2 inactivation by B3 under sunlight, with and without NOM on the 4th and 12th of September, 2013. Experiments were conducted between 1:00 and 1:30 pm EST, the ambient temperature was measured to be 36 °C
(Samuel D. Snow et al., 2014)

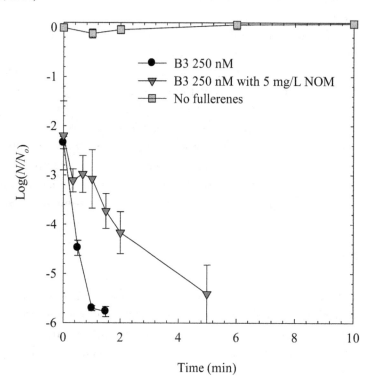

$^1O_2$ generation in water, but also permits recovery and recycling of the material while preventing environmental release. Numerous studies have focused on such heterogeneous supported-fullerene systems for chromatographic separations (Chiou & Shih, 2000), optical limiting devices (Maggini et al., 1995), and photooxidations in organic solvents (Jensen & Daniels, 2003; Latassa et al., 2002) utilizing various attachment chemistries including direct amine addition to pristine C_{60}'s cage (Hino, Anzai, & Kuramoto, 2006; Jensen & Daniels, 2003), diels-alder reaction with C_{60}'s cage (Guhr, Greaves, & Rotello, 1994), and diimide chemistry with a carboxylic-acid-functionalized fullerene derivative (Latassa et al., 2002). Of these reports, the only studies that focused on photochemical properties detailed $^1O_2$ production in the organic phase and not in water, where C_{60} exhibits significantly lower solubility (Korobov & Smith, 2000) and decreased $^1O_2$ lifetimes (Foote & Clennan, 1996). The only investigations of $^1O_2$ production in the aqueous phase utilized a positively charged, water-soluble aminofullerene derivative, which was immobilized to a support via diimide chemistry (Lee et al., 2011; Lee et al., 2010). The fullerene material successfully produced $^1O_2$ in water, degraded select micropollutants through $^1O_2$ mediated damage (Lee et al., 2011), and inactivated MS2 bacteriophage under visible light illumination (Lee et al., 2010). However, complex synthesis requirements in both preparation of the water-soluble fullerene derivative and covalent attachment to support media seriously limit the prospects of aminofullerene-based photocatalysts in real world applications.

In order to advance the potential of C_{60} as a photocatalyst in water systems, our recent study chose to pursue a much simpler method of producing supported fullerene materials (K. Moor & Kim, 2014) relying on the well-established nucelophilic reactivity of C_{60}'s cage with terminal amines (Hirsch, Li, & Wudl, 1991; G. P. Miller, 2006). This amine chemistry allows for easier synthesis methods than previous work involving fullerene functionalization, which require a complex mixture of coupling agents and chemicals, herein avoided with our chosen chemistry. In addition, this route utilizes pristine fullerene and not a fullerene derivative, thus obviating the need for further synthetic and separation procedures that are inherently associated with fullerene derivatives. Supported fullerene prepared via the direct amine addition will be detailed in this section, along with illustrations of its abilities to generate $^1O_2$ and inactivate a model virus under visible light irradiation. Accordingly, this route is expected to provide a more sustainable and cheaper approach towards using fullerene in water treatment applications.

The procedure for producing supported fullerene materials involves the nucleophilic addition of a primary amine across a [6,6] fullerene double bond, followed by proton transfer under mild condition (Hirsch et al., 1991; G. P. Miller, 2006) as depicted in Figure 4. The primary amine, located on the support media, reacts directly with fullerene's cage, causing minimal perturbation of the cage structure. We chose two commercially-available support media functionalized with similar primary amines: 1) a propyl-amine functionalized silica gel (1 mmol amine/g; 40-63 µm, as provided by manufacturer) and; 2) an amine functionalized crosslinked polystyrene resin (4 mmol amine/g; 37-74 µm, as provided by manufacturer).

The micron-sized support materials were selected due to their commercial availability and served as a benchmark to determine if using pristine fullerene-based photocatalysts is feasible in water. Additional gains in material performance may be achieved in the future by tailoring the support size to provide larger surface area or greater fullerene coverage. However, the use of a readily available support material illustrates the ease and simplicity of this preparation method. The support media were placed into solutions of C_{60} in toluene, heated to 38 °C, and stirred for 70 h. After the reaction the suspensions were filtered and rinsed with fresh toluene to remove unreacted fullerenes. The resulting polystyrene-supported C_{60} and silica-supported C_{60} will be herein referred to as C_{60}/PS and C_{60}/SiO_2, respectively.

Figure 4. (a) Depiction of reaction scheme and SEM images of (b) PS resin without immobilized C_{60}, (c) C_{60}/PS (d) SiO_2 gel without immobilized C_{60}, and (e) C_{60}/SiO_2. Samples were coated with 12 nm of chromium before microscopy
(K. Moor & Kim, 2014)

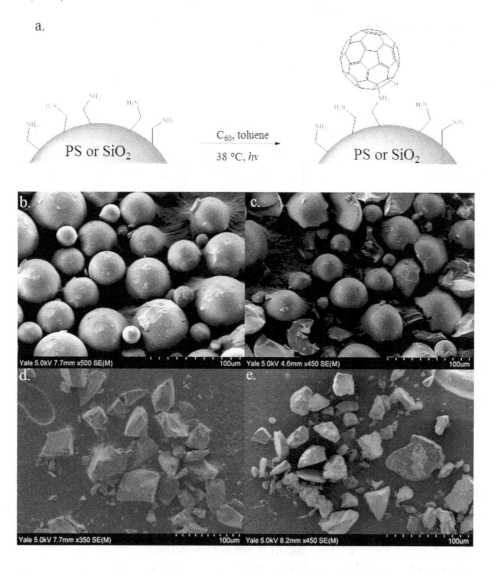

Fluorescent lamp (FL) irradiation (1.1×10^{-6} Einstein L^{-1} s^{-1} as determined with ferrioxalate actinometry) was used throughout material preparation to excite fullerene to its triplet-state ($^3C_{60}$*), producing a more efficient electron acceptor than the ground singlet-state ($^1C_{60}$)(Dirk M Guldi & Maurizio Prato, 2000) and thus increasing its reactivity towards the amine addition (J. W. Li & Liang, 2007). Consequently, higher loadings of C_{60} were achieved when solutions were irradiated compared to samples prepared in the dark with comparable heating, corresponding to loadings of 85.7 and 54.9 µmol C_{60}/g-SiO_2-support for materials prepared on the silica gel support with and without irradiation, respectively. Note that C_{60} was not photodegraded after 70 h of light exposure as evidenced by no apparent changes in the UV/Vis spectrum of C_{60} before and after the reaction.

After fullerene attachment, the support materials were significantly darker in color than the supports before fullerene immobilization, progressing from white to shades of dark brown, indicating surface-bound fullerene. Diffuse reflectance spectra for both C_{60}/SiO_2 and C_{60}/PS displayed an increasing absorption from 700 to 375 nm, while the bare support materials exhibited no significant change in absorption. This broadband increase in absorption over the visible range into the long UV is characteristic of fullerene on solid supports and has previously been reported for C_{60} on silica surfaces (Kyriakopoulos et al., 2012).

In order to assess the stability of the amine-fullerene bond, a necessary requirement to avoid fullerene leaching and safe environmental application, supported materials were refluxed in either toluene or water for 24 h. Negligible amounts of fullerene were lost from C_{60}/SiO_2 and C_{60}/PS when refluxed in water as evidenced by no observable UV/vis spectroscopy signals. When refluxed in toluene, both C_{60}/PS and C_{60}/SiO_2 lost small quantities of C_{60} into solution, corresponding to 11% and 12% of the total fullerene loading for C_{60}/PS and C_{60}/PS, respectively. Overall, these experiments suggest that the amine-fullerene bond is stable under potential engineered conditions. In order to confirm that the reaction between the terminal amine and C_{60} occurred, supports without amine functionalization (bare PS and SiO_2) were used as support media in analogous procedures as for the preparation of C_{60}/PS and C_{60}/SiO_2. After stirring in fullerene solutions, the supports appeared to have adsorbed C_{60}, which was quickly removed upon refluxing with toluene, thus restoring the original support materials. Taken together with the stability of covalently attached C_{60} as outlined above, this finding suggests that C_{60} has reacted with the terminal amine on the solid supports resulting in stable, covalent immobilization. Note that attempts at verifying covalent immobilization using FTIR spectroscopy were unsuccessful likely due to the relatively low density of surface functionalities and low sensitivity of the spectrometer.

Representative scanning electron microscopy (SEM) images of C_{60}/PS, C_{60}/SiO_2, and the support materials without fullerene are shown in Figure 4. The silica gel media did not appear to undergo significant changes in morphology or size after the reaction with C_{60}, whereas the PS resin appeared to fracture into smaller particles after reaction with C_{60}. This damage to the PS resin was likely due to the vigorous stirring employed in material preparation, but did not severely impact the photocatalytic properties of C_{60}/PS because C_{60} was distributed homogeneously throughout the resin. High-resolution transmission electron microscopy (TEM) of C_{60}/SiO_2 did not conclusively locate fullerene (molecules or aggregates) on the surface of the silica gel. This finding indicated that significant fullerene aggregation is prevented and suggested possible monolayer coverage of fullerene on the silica support, which would be difficult to detect with TEM. Other researchers have likewise experienced difficulties in using TEM to locate fullerenes on silica gel surfaces (Kyriakopoulos et al., 2012).

The photodynamic inactivation of MS2 bacteriophage by C_{60}/SiO_2 under various light sources is depicted in Figure 5. C_{60}/SiO_2 inactivated MS2 the most rapidly under black light irradiation, followed by FL and then visible (FL with UV cutoff filters) illumination. This overall trend correlates well with C_{60}'s action spectrum, absorbing strongly in the UV and more weakly in the visible range, and the amount of $^1O_2$ produced under each light source. Negative control experiments, including inactivation in the presence of L-hisitidine, a potent $^1O_2$ scavenger ($k = 1.5 \times 10^8$ $M^{-1}s^{-1}$) (Francis Wilkinson, Helman, & Ross, 1995), and in the dark, showed minimal MS2 loss.

These controls confirmed that MS2 is not lost to adsorption processes, but rather to inactivation from photochemically produced $^1O_2$ by C_{60}/SiO_2. Initially, C_{60}/SiO_2 exhibited substantial MS2 adsorption (ca. 1 log) and hence was allowed to equilibrate for 30 minutes in the dark before the start of photochemical experiments. This significant MS2 adsorption may have benefited inactivation rates by increasing the proximity of viruses to the photo-produced $^1O_2$ source, thereby effectively increasing the amount of MS2

Figure 5. Microbial inactivation of MS2 bacteriophage with C_{60}/SiO_2 with loading of 57 µmol $C_{60}/$ g support (solid lines) and controls (dotted lines) as a function of time under various illumination conditions. $[C_{60}/SiO_2]$ = 0.3 g/L; $[MS2]_o$ = 5 × 10⁴ PFUs/mL (K. Moor & Kim, 2014).

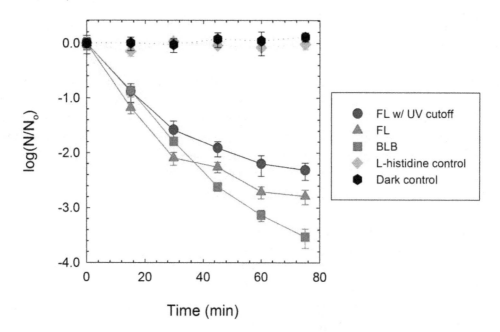

particles within the diffusion length of $^1O_2$ (~125 nm within one lifetime) (Redmond & Kochevar, 2006). At first glance, C_{60}/SiO_2 appears to exhibit similar MS2 inactivation rates under visible light compared to previously studied aminofullerene materials, which exhibited between 2 to 2.5 log inactivation for the same period of time. However, it remains difficult to accurately compare their performance due to differences in MS2 titer and photoreactor light intensities. Based on $^1O_2$ generation rates, where C_{60}/SiO_2 exhibited somewhat decreased $^1O_2$ production compared to aminofullerene materials, it is expected that C_{60}/SiO_2 will display slightly slower inactivation kinetics. However, C_{60}/SiO_2's ease of preparation and cheaper production compared to aminofullerene materials may more than compensate for this expected slight decrease in inactivation kinetics, where additional material may be applied at the same cost to achieve desired inactivation levels.

5. SUPPORTED HIGHER FULLERENE

C_{60} has long been the focal point of fullerene research, but other higher fullerenes (e.g. C_{70}, C_{76}, C_{84}, etc.) have also begun to draw considerable attention, in particular C_{70}, which has received a recent surge for use in organic photovoltaics (He, Zuo, Chen, Xiao, & Ding, 2014; Pfuetzner, Meiss, Petrich, Riede, & Leo, 2009; Wienk et al., 2003). Increasing the fullerene cage by 10 carbon atoms from C_{60} to C_{70} slightly elongates the fullerene cage, resulting in a concurrent relaxation in symmetry from I_H to D_{5H} point groups that allows electronic transitions once symmetry forbidden in C_{60} to occur for C_{70} (James W Arbogast & Foote, 1991; Hare, Kroto, & Taylor, 1991). Hence, C_{70} absorbs strongly in the visible range relative to

C_{60}, while maintaining high photoactivity and may thus hold particle advantage in visible light activated photocatalyst systems. Note that other higher fullerenes also possess considerable absorption in the visible range, but are produced in exceedingly small quantities in the arc-discharge process compared to C_{70}, leading to their incredibly high cost that somewhat limits their potential application. Based on C_{70}'s relative high availability, low cost, and strong photoactivity, we chose to further improve supported pristine C_{60} materials (C_{60}/SiO_2) by focusing on incorporation of C_{70}. Herein, we will detail the preparation of a highly visible light photoactive fullerene material using similar direct-amine addition chemistry as outlined in Section X.4. This approach allows for greatly expanding the visible light absorption of fullerene materials without relying on past methods such as functionalization of C_{60}, which not only require complex synthesis and separation methods but may also detrimentally alter fullerene's intrinsic photophysical properties. We present our recent work(K. J. Moor, Valle, Li, & Kim, 2015) using a well-known mesoporous silica support, MCM-41, which provides a high surface area substrate for fullerene attachment. MCM-41 was functionalized with an aminosilane and C_{70} was covalently immobilized using nucleophilic amine addition directly to C_{70}'s cage. This approach maintains simple preparation methods while using prisitine fullerene, yet significantly expands the visible light photoactivity of supported-fullerene materials, thus greatly advancing fullerene materials for water treatment.

The MCM-41 support was prepared via a surfactant templated sol-gel processes as previously reported (Grun, Unger, Matsumoto, & Tstutsumi, 1999). The particles were polydisperse with diameters ranging from 0.5-2 µm as determined via SEM (Figure 6), agreeing with dynamic light scattering (DLS) results that showed an average particle diameter of 2.4 ± 0.4 µm when dispersed in water. The prepared material was found to exhibit high surface areas of 857 ± 8 m^2/g, slightly lower than typical values of 900-1000 m^2/g (Piumetti, Hussain, Fino, & Russo, 2015; Yang, Gai, & Lin, 2012). The overall mesoporous structure of MCM-41 was confirmed by TEM, which displayed ordered pores with diameters of 2.6 ± 0.4 nm (Figure 6) and agreed with previous reports for MCM-41 prepared with CTAB as a template, creating pores with diameters of *ca.* 3 nm (Grun et al., 1999; Kruk, Jaroniec, Kim, & Ryoo, 1999). X-ray diffraction (XRD) analysis provides further proof that a hexagonal, long-range network of the pores was achieved, as evidenced by three diffraction signals at *ca.* 2.1, 3.7, and 4.3 2Θ, in which the intensity and peak location agreed with past work (Grun et al., 1999).

Amine functionalization of MCM-41 was achieved by refluxing 20.5 mmol 3-aminopropyl triethoxysilane (APTES) with *ca.* 1.2 g MCM-41 in toluene for 8 h. The subsequent amine functionalized MCM-41 (MCM-41-NH$_2$) was retrieved by filtration and washed with toluene to remove excess APTES. Amine decoration of MCM-41 was confirmed with Fourier transform infrared (FTIR) spectroscopy before and after reaction. Specifically, the appearance of two small peaks at 3370 and 3300 cm^{-1} and a secondary peak at *ca.* 1600 cm^{-1}, indicative of the N-H stretching and bending modes of a primary amine, respectively, suggested successful APTES reaction with MCM-41 (Bacsik, Atluri, Garcia-Bennett, & Hedin, 2010). Furthermore, an increase in signal intensity at *ca.* 2970-2840 cm^{-1} was observable, corresponding to the C-H stretching mode of CH$_2$ groups in the organic portion of immobilized propylamine. Zeta potential values provided further evidence of successful reaction; bare MCM-41 displayed a zeta potential of -29.9 ± 0.6 mV and became increasingly positive after reaction with APTES, reaching 10.2 ± 0.6 mV due to the positive charge of the surface amine groups at neutral pHs (pK_a ~ 10.5) (Hall, 1957).

Supported fullerene materials were prepared by adding MCM-41-NH$_2$ to solutions of either C_{60} (99.5%; Sigma-Aldrich) or C_{70} (98%; SES Research) in toluene (100 µmol fullerene/ g MCM-41-NH$_2$) and heating to 40 °C for up to 3 days, where the amount of fullerene and reaction time was adjusted to achieve desired fullerene loadings. The supported fullerene materials (herein termed C_{60}/MCM-41 or

Figure 6. Representative SEM images of (a) MCM-41 and (b) C_{70}/MCM-41 and TEM images of (c) MCM-41 and (d) C_{70}/MCM-41. Insets in TEM images depict a magnified view of the porous structure (Moor et al., 2015)

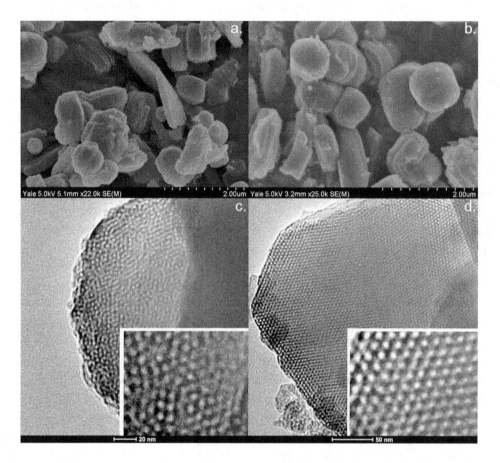

C_{70}/MCM-41) were retrieved from the reaction suspension by vacuum filtration and were thoroughly rinsed with toluene to remove non-attached fullerene. After C_{60} or C_{70} attachment, the MCM-41 support experienced a drastic color change, turning from bright white to tan-brown. Concurrently, the zeta potential decreased from a positive value for the amine-functionalized support to -16.3 ± 0.5 mV upon C_{70} immobilization. In order to confirm that fullerene was covalently attached through reaction with amines instead of only physisorption to the support, C_{70}/MCM-41 was refluxed in toluene for 8 h and the amount of C_{70} in solution was analyzed via UV-vis spectroscopy. The covalently-bound fullerene was found to be stable upon toluene reflux and the material lost an insignificant amount of C_{70} into solution. To serve as a control, bare MCM-41 (no amine decoration) was used instead of MCM-41-NH_2 in an analogous reaction as for C_{70}/MCM-41 preparation. The resulting material possessed only adsorbed C_{70}, which was rapidly removed upon toluene reflux. Given that C_{70} and C_{60} exhibit similar reactivities towards nucleophiles (Thilgen et al., 1997), together with the abundance of reports documenting the reactivity of fullerenes with primary amines (Hirsch et al., 1991; G. P. Miller, 2006; Seshadri, Govindaraj, Nagarajan, Pradeep, & Rao, 1992), the above results collectively suggested reaction of the supported amine with fullerene's cage, resulting in the stable, covalent attachment of fullerene onto MCM-41.

SEM and TEM analysis did not detect any significant change in the overall morphology and mesoporous structure of MCM-41 due to fullerene immobilization (Figure 6). In addition, DLS measurements found that MCM-41's dimensions were not significantly altered during reaction with C_{70}, which possessed diameters of 2.4 ± 0.4 μm and 1.6 ± 0.5 μm before and after reaction with C_{70}. Fullerene agglomeration was not observed on the surface of MCM-41 during TEM analysis of C_{70}/MCM-41, indicating that fullerene was possibly distributed in a monolayer on the support, similar to conclusions drawn for C_{60}/SiO_2. Given the estimated pore diameter of ca. 2.4 nm for MCM-41 and the diameter of the fullerene molecule (nucleus to nucleus diameter of ca. 0.7 nm),(Kroto et al., 1985) it is possible that C_{70} was attached within the mesoporous structure. If this did occur, photo-generated $^1O_2$ could still diffuse out of the pores into the bulk solution given the relatively large diffusion pathlength of ca. 125 nm in water,(Redmond & Kochevar, 2006) yet it is uncertain how this affected the total amount of photo-produced $^1O_2$. Diffuse reflectance-UV/vis measurements found that both C_{60}/MCM-41 and C_{70}/MCM-41 exhibited an increasing absorption from 400 to 800 nm across nearly the entire visible range, whereas bare MCM-41exhibited negligible absorption. C_{70}/MCM-41 possessed a more intense visible absorption than C_{60}/MCM-41 despite a lower fullerene loading, which is likely attributed to the larger molar extinction coefficient of C_{70} compared to C_{60} in the visible range.

The visible-light activated C_{70}/MCM-41 inactivation kinetics for MS2 bacteriophage are depicted in Figure 7. Initially, C_{70}/MCM-41 exhibited significant adsorption of MS2 (ca. 1 log), which may be due to electrostatic interactions between positively charged secondary amines used for fullerene attachment and negatively charged MS2 particles. Even though C_{70}/MCM-41 possessed an overall negative zeta potential, the positively charged amine groups may still govern local electrostatic interactions with MS2. Given the pore size of MCM-41 (2.6 ± 0.4 nm), too small for entry of MS2 particles (28 nm in diameter (Kuzmanovic, Elashvili, Wick, O'Connell, & Krueger, 2003)), it is expected that adsorption processes occurred on the exterior of MCM-41 and not inside the mesoporous structure. A one h dark equilibration period before illumination was used to differentiate between inactivation and adsorption processes. MS2 adsorption appeared to be complete after 1 h given the negligible loss of MS2 in a dark control, thus any removal after the dark equilibration was attributed to inactivation. C_{70}/MCM-41 displayed rapid MS2 inactivation kinetics, reaching greater than 3-log inactivation within 30 min of visible light irradiation. When considering both adsorption and inactivation processes, C_{70}/MCM-41 achieved greater than 4-log reduction. A control experiment with L-histidine, a strong $^1O_2$ quencher, exhibited considerably slower kinetics, suggesting that inactivation of MS2 was primarily related to $^1O_2$ mediated damage produced by photo-activated C_{70}/MCM-41. However, inactivation was not fully quenched in the presence of L-histidine, suggesting that MCM-41 may present some other form of $^1O_2$-independent antiviral properties that are currently unknown. Compared to a conventional visible light active photocatalyst, porous N-doped TiO_2 (N-TiO_2), C_{70}/MCM-41 exhibited remarkably faster inactivation kinetics. In fact, N-TiO_2 exhibited negligible removal of MS2 under visible light illumination. This poor inactivation is not surprising given that MS2 is relatively stable towards OH radical oxidation,(Min Cho, Chung, Choi, & Yoon, 2005b; Jaesang Lee et al., 2009) together with the fact that visible light sensitization of N-TiO_2 forms reduced amounts of hydroxyl radicals compared to typical UVA illumination of TiO_2 suspensions.

In general, sunlight illuminated C_{70}/MCM-41 possessed similar MS2 inactivation trends to that of visible light experiments, with C_{70}/MCM-41 again displaying significantly faster kinetics than N-TiO_2 (Figure 7). Note that sunlight illumination alone did not inactivate MS2. C_{70}/MCM-41's MS2 inactivation varied widely due to differences in sunlight intensity between experiments (65.2 ± 18.8 mW/cm^2), but broadly followed the inactivation rates achieved by visible light illumination. At first glance, C_{70}/

Figure 7. MS2 inactivation kinetics of C_{70}/MCM-41 with loading of 14 μmol C_{70}/ g support and porous N-TiO_2 under (a) visible light and (b) sunlight illumination. Various controls are also included. [C_{70}/MCM-41] = [N-TiO_2] = 0.3 g/L; [MS2]$_0$ = 3×10^8 PFUs/mL; [L -histidine]$_0$ = 250 mM (Moor et al., 2015)

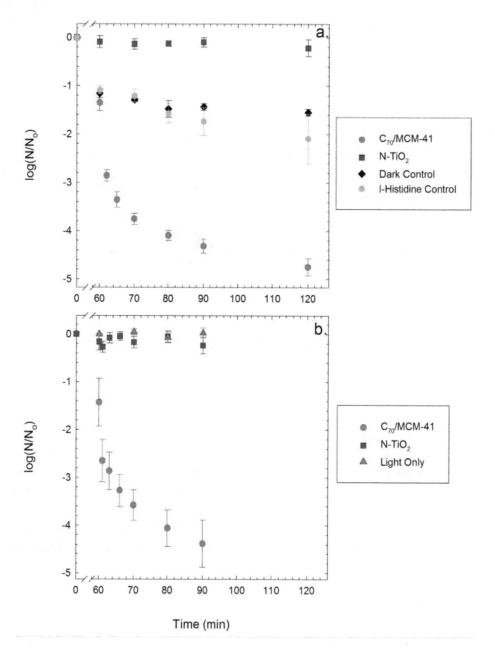

MCM-41 appeared to exhibit faster MS2 inactivation kinetics than previously studied aminofullerene materials. However, differences in light intensity (only one 4-W bulb used for aminofullerene materials), initial MS2 concentrations (2×10^5 PFUs/mL for aminofullerene materials), and fullerene solution concentrations (*ca.* 4 μM and 15 μM for C_{70}/MCM-41 and aminofullerene materials, respectively) make it difficult to make a direct quantitative comparison. Based on $^1O_2$ production rates, C_{70}/MCM-41 should produce comparable, if not better, MS2 inactivation rates compared to aminofullerene materials. In relation to C_{60}/SiO_2, which achieved 1.5-log inactivation under similar conditions, C_{70}/MCM-41 displayed enhanced inactivation kinetics despite lower fullerene loadings. C_{70}/MCM-41's increased efficacy against MS2 may be attributed to C_{70}'s better utilization of visible light compared to C_{60}.

6. CONCLUSION

Nanomaterials are at the forefront of environmental technologies, taking advantage of the unique properties of materials at the nanoscale for improved efficiency and novel performance in various applications, including photocatalysis, adsorption, filtration, and disinfection processes (Qu, Brame, Li, & Alvarez, 2012). The integration of nanomaterials with conventional engineering processes has created the opportunity for new, highly efficient next-generation systems for producing clean drinking water, treating wastewaters, and remediating various pollutants in the environment. Fullerene is an exemplar nanomaterial for use in nano-enabled technologies, due to its overall robustness, excellent photoactivity, and its ease of manipulation, making its practical engineered application achievable. Fullerene-mediated $^1O_2$ based disinfection holds particular advantages for the treatment of complex water matrixes, where background water constituents such as NOM and carbonate do not considerably affect treatment efficacy due to $^1O_2$'s selective reactivity, unlike most other photocatalysts that produce non-specific hydroxyl radicals (Brame, Long, Li, & Alvarez, 2014). For this reason, fullerene-based disinfection may have great potential in point-of-use applications, where impaired or minimally treated water (e.g. sand filtered) are commonly used as the source drinking water. Due to the relatively large size of the support media, supported-fullerene materials can be easily separated and retrieved using sedimentation and/or microfiltration processes, which is a key design constraint for common nanoscale photocatalyst systems.

As noted throughout the chapter, fullerene materials have shown exceptional promise for the photo-inactivation of viruses. Yet, they have shown somewhat lackluster performance against bacteria (Jaesang Lee et al., 2009; Samuel D Snow, KyoungEun Park, & Jae-hong Kim, 2014) due to the relative robustness of the bacteria cell membrane towards $^1O_2$ oxidation compared to the virus capsid. In order to succeed against the diverse microbial populations found in the real world, there is a great need to incorporate additional bactericidal functionality (*e.g.* silver nanoparticles, copper nanoparticles, polycationic charges) into fullerene materials, creating composites that are potently effective towards both viruses and bacteria. In addition, other functionality that interacts with microorganisms (*e.g.* cationic charges by electrostatic interactions), thereby bringing them closer to the ROS source, may drastically improve material efficacy.

Although nanomaterials hold great promise in next-generation water treatment technologies, their emerging use may present two imposing obstacles: elevated device costs and unintended environmental release as a result of large-scale manufacturing processes. To combat the first potential issue, Mitsubishi Corporation has embarked on a major effort to drastically reduce the cost of fullerene through new, efficient production methods, generating several tons of fullerene annually. (Loutfy, Lowe, Moravsky, & Katagiri, 2002) As a result, the cost of C_{60} has been driven down to roughly 20 USD per g, comparable to various noble metal (Au, Pt, Pd) catalysts, which receive widespread use in industrial applications. As fullerene based photonic technologies mature an even greater drop in price can be expected. As with any emerging material, in-depth toxicity and life-cycle assessments are necessary to clearly understand the long-term impacts of large-scale fullerene production and use. Many studies have already addressed certain aspects of the behavior of fullerene in the environment, detailing such topics as the transport of fullerene in granular media,(Yonggang Wang et al., 2008) potential biodegradation pathways,(Avanasi, Jackson, Sherwin, Mudge, & Anderson, 2014; Navarro, Kookana, McLaughlin, & Kirby, 2016) and the varying degrees of toxicity to microbial and aquatic organisms.(J D Fortner et al., 2005; Lovern, Strickler, & Klaper, 2007; Lyon, Adams, Falkner, & Alvarez, 2006) From these studies it is clear that fullerene and fullerene derivatives may possess some impact to natural systems, yet it remains unclear the magnitude of such events at environmentally relevant concentrations.

REFERENCES

Andersson, T., Nilsson, K., Sundahl, M., Westman, G., & Wennerstrom, O. (1992). C_{60} embedded in r-cyclodextrin - a water-soluble fullerene. *Journal of the Chemical Society. Chemical Communications*, (8): 604–606. doi:10.1039/C39920000604

Andrievsky, G. V., Kosevich, M. V., Vovk, O. M., Shelkovsky, V. S., & Vashchenko, L. A. (1995). On the production of an aqueous colloidal solution of fullerenes. *Journal of the Chemical Society. Chemical Communications*, (12): 1281–1282. doi:10.1039/c39950001281

Arbogast, J. W., Darmanyan, A. P., Foote, C. S., Rubin, Y., Diederich, F. N., Alvarez, M. M., & Whetten, R. L. et al. (1991). Photophysical properties of C_{60}. *Journal of Physical Chemistry*, 95(1), 11–12. doi:10.1021/j100154a006

Arbogast, J. W., & Foote, C. S. (1991). Photophysical Properties of C_{70}. *Journal of the American Chemical Society*, 113(23), 8886–8889. doi:10.1021/ja00023a041

Avanasi, R., Jackson, W. A., Sherwin, B., Mudge, J. F., & Anderson, T. A. (2014). C60 Fullerene Soil Sorption, Biodegradation, and Plant Uptake. *Environmental Science & Technology*, 48(5), 2792–2797. doi:10.1021/es405306w

Bacsik, Z., Atluri, R., Garcia-Bennett, A. E., & Hedin, N. (2010). Temperature-Induced Uptake of CO_2 and Formation of Carbamates in Mesocaged Silica Modified withn-Propylamines. *Langmuir*, 26(12), 10013–10024. doi:10.1021/la1001495

Badireddy, A. R., Hotze, E. M., Chellam, S., Alvarez, P., & Wiesner, M. R. (2007). Inactivation of Bacteriophages via photosensitization of fullerol nanoparticles. *Environmental Science & Technology*, 41(18), 6627–6632. doi:10.1021/es0708215

Bagrii, E. I., & Karaulove, E. N. (2001). New in fullerene chemistry (a review). *Petroleum Chemistry*, 41(5), 295–313.

Beeby, A., Eastoe, J., & Heenan, R. K. (1994). Solubilization of C_{60} in aqueous micellar solution. *Journal of the Chemical Society. Chemical Communications*, (2): 173–175. doi:10.1039/c39940000173

Bessekhouad, Y., Robert, D., & Weber, J. V. (2004). Bi_2S_3/TiO_2 and CdS/TiO_2 Heterojunctions as an Available Configuration for Photocatalytic Degradation of Organic Pollutant. *Journal of Photochemistry and Photobiology A Chemistry*, 163(3), 569–580. doi:10.1016/j.jphotochem.2004.02.006

Bocquillon, G., Bogicevic, C., Fabre, C., & Rassat, A. (1993). C_{60} fullerene as carbon source for diamond synthesis. *Journal of Physical Chemistry, 97*(49), 12924–12927. doi:10.1021/j100151a047

Brame, J., Long, M., Li, Q., & Alvarez, P. (2014). Trading oxidation power for efficiency: Differential inhibition of photo-generated hydroxyl radicals versus singlet oxygen. *Water Research, 60,* 259–266. doi:10.1016/j.watres.2014.05.005

Chae, S.-R., Hotze, E. M., & Wiesner, M. R. (2009). Evaluation of the Oxidation of Organic Compounds by Aqueous Suspensions of Photosensitized Hydroxylated-C_{60} Fullerene Aggregates. *Environmental Science & Technology, 43*(16), 6208–6213. doi:10.1021/es901165q

Chase, B., Herron, N., & Holler, E. (1992). Vibrational spectroscopy of fullerenes (C60 and C70). Temperature dependant studies. *Journal of Physical Chemistry, 96*(11), 4262–4266. doi:10.1021/j100190a029

Chiou, C.-S., & Shih, J.-S. (2000). Fullerene C_{60}-Cryptand Chromatographic Stationary Phase for Separations of Anions/Cations and Organic Molecules. *Analytica Chimica Acta, 416*(2), 169–175. doi:10.1016/S0003-2670(00)00906-5

Cho, M., Chung, H., Choi, W., & Yoon, J. (2005a). Different Inactivation Behaviors of MS-2 Phage and Escherichia coli in TiO_2 Photocatalytic Disinfection. *Applied and Environmental Microbiology, 71*(1), 270–275. doi:10.1128/AEM.71.1.270-275.2005

Cho, M., Chung, H., Choi, W., & Yoon, J. (2005b). Different Inactivation Behaviors of MS-2 Phage and Escherichia coli in TiO_2 Photocatalytic Disinfection. *Applied and Environmental Microbiology, 71*(1), 270–275. doi:10.1128/AEM.71.1.270-275.2005

Cho, M., Lee, J., Mackeyev, Y., Wilson, L. J., Alvarez, P. J. J., Hughes, J. B., & Kim, J. H. (2010). Visible Light Sensitized Inactivation of MS-2 Bacteriophage by a Cationic Amine-Functionalized C(60) Derivative. *Environmental Science & Technology, 44*(17), 6685–6691. doi:10.1021/es1014967

Collini, E., Fortunati, I., Scolaro, S., Signorini, R., Ferrante, C., Bozio, R., & Silvestrini, S. et al. (2010). A fullerene-distyrylbenzene photosensitizer for two-photon promoted singlet oxygen production. *Physical Chemistry Chemical Physics, 12*(18), 4656–4666. doi:10.1039/b922740g

Deguchi, S., Alargova, R. G., & Tsujii, K. (2001). Stable dispersions of fullerenes, C_{60} and C_{70}, in water. Preparation and characterization. *Langmuir, 17*(19), 6013–6017. doi:10.1021/la010651o

Diederich, F., & Thilgen, C. (1996). Covalent fullerene chemistry. *Science, 271*(5247), 317–323. doi:10.1126/science.271.5247.317

Dimitrijevic, N. M., & Kamat, P. V. (1993). Excited-state behavior and one-electron reduction of C_{60} in aqueous r-cyclodextrin solution. *Journal of Physical Chemistry, 97*(29), 7623–7626. doi:10.1021/j100131a035

Escalada, J. P., Pajares, A., Gianotti, J., Massad, W. A., Bertolotti, S., Amat-Guerri, F., & Garcia, N. A. (2006). Dye-sensitized photodegradation of the fungicide carbendazim and related benzimidazoles. *Chemosphere, 65*(2), 237–244. doi:10.1016/j.chemosphere.2006.02.057

Foote, C. S., & Clennan, E. L. (1996). Properties & Reactions of Singlet Dioxygen. In C. S. Foote, J. S. Valentine, A. Greenburg, & J. F. Liebman (Eds.), *Active Oxygen in Chemistry* (pp. 105–140). London: Blackie Academic & Professional. doi:10.1007/978-94-007-0874-7

Fortner, J. D., Lyon, D. Y., Sayes, C. M., Boyd, A. M., Falkner, J. C., Hotze, E. M., & Hughes, J. B. et al. (2005). C_{60} in Water: Nanocrystal Formation and Microbial Response. *Environmental Science & Technology*, *39*(11), 4307–4316. doi:10.1021/es048099n

Fortner, J. D., Lyon, D. Y., Sayes, C. M., Boyd, A. M., Falkner, J. C., Hotze, E. M., & Hughes, J. B. et al. (2005). C_{60} in water: Nanocrystal formation and microbial response. *Environmental Science & Technology*, *39*(11), 4307–4316. doi:10.1021/es048099n

Gilbert, P., & Moore, L. E. (2005). Cationic antiseptics: Diversity of action under a common epithet. *Journal of Applied Microbiology*, *99*(4), 703–715. doi:10.1111/j.1365-2672.2005.02664.x

Grun, M., Unger, K. K., Matsumoto, A., & Tstutsumi, K. (1999). Novel pathways for the preparation of mesoporous MCM-41 materials: Control of porosity and morphology. *Microporous and Mesoporous Materials*, *27*(2-3), 207–216. doi:10.1016/S1387-1811(98)00255-8

Guhr, K. I., Greaves, M. D., & Rotello, M. (1994). Reversible Covalent Attachment of C_{60} to a Polymer Support. *Journal of the American Chemical Society*, *116*(13), 5997–5998. doi:10.1021/ja00092a072

Guldi, D. M., & Asmus, K. D. (1997). Photophysical properties of mono- and multiply-functionalized fullerene derivatives. *The Journal of Physical Chemistry A*, *101*(8), 1472–1481. doi:10.1021/jp9633557

Guldi, D. M., & Prato, M. (2000). Excited-State Properties of C_{60} Fullerene Derivatives. *Accounts of Chemical Research*, *33*(10), 695–703. doi:10.1021/ar990144m

Guldi, D. M., & Prato, M. (2000). Excited-state properties of C(60) fullerene derivatives. *Accounts of Chemical Research*, *33*(10), 695–703. doi:10.1021/ar990144m

Gupta, B. K., Bhushan, B., Capp, C., & Coe, J. V. (1994). Materials characterization and effect of purity and ion implantation on the friction wear of sublimed fullerene films. *Journal of Materials Research*, *9*(11), 2823–2838. doi:10.1557/JMR.1994.2823

Haag, W. R., & Hoigne, J. (1986). Singlet Oxygen in Surface Waters. 3. Photochemical Formation and Steady-State Concentrations in Various Types of Waters. *Environmental Science & Technology*, *20*(4), 341–348. doi:10.1021/es00146a005

Haag, W. R., Hoigne, J., Gassman, E., & Braun, A. M. (1984). Singlet oxygen in surface waters. 2. Quantum yields of its production by some natural humic materials as a function of wavelength. *Chemosphere*, *13*(5-6), 641–650. doi:10.1016/0045-6535(84)90200-5

Haddon, R. C., Hebard, A. F., Rosseinsky, M. J., Murphy, D. W., Duclos, S. J., Lyons, K. B., & Thiel, F. A. et al. (1991). Conducting films of C_{60} and C_{70} by alkali-metal doping. *Nature*, *350*(6316), 320–322. doi:10.1038/350320a0

Hall, H. K. Jr. (1957). Correlation of the Base Strengths of Amines. *Journal of the American Chemical Society*, *79*(20), 5441–5444. doi:10.1021/ja01577a030

Hamano, T., Okuda, K., Mashino, T., Hirobe, M., Arakane, K., Ryu, A., & Nagano, T. et al. (1997). Singlet oxygen production from fullerene derivatives: Effect of sequential functionalization of the fullerene core. *Chemical Communications*, (1): 21–22. doi:10.1039/a606335g

Hare, J. P., Kroto, H. W., & Taylor, R. (1991). Preparation and UV/Visible Spectra of Fullerenes C_{60} and C_{70}. *Chemical Physics Letters, 177*(4-5), 394-398.

He, D., Zuo, C., Chen, S., Xiao, Z., & Ding, L. (2014). A highly efficient fullerene acceptor for polymer solar cells. *Physical Chemistry Chemical Physics, 16*(16), 7205–7208. doi:10.1039/c4cp00268g

Hebard, A. F. (1993). Buckminsterfullerene. *Annual Review of Materials Science, 23*(1), 159–191. doi:10.1146/annurev.ms.23.080193.001111

Hedberg, K., Hedberg, L., Buhl, M., Bethune, D. S., Brown, C. A., & Johnson, R. D. (1997). Molecular Structure of Free Molecules of the Fullerene C_{70} from Gas-Phase Electron Diffraction. *Journal of the American Chemical Society, 119*(23), 5314–5320. doi:10.1021/ja970110e

Hino, T., Anzai, T., & Kuramoto, N. (2006). Visible-Light Induced Solvent-Free Photooxygenations of Organic Substrates by Using [C_{60}] Fullerene-Linked Silica Gels as Heterogeneous Catalysts and as Solid-Phase Reaction Fields. *Tetrahedron Letters, 47*(9), 1429–1432. doi:10.1016/j.tetlet.2005.12.081

Hirsch, A., Li, Q., & Wudl, F. (1991). Globe-Trotting Hydrogens on the Surface of the Fullerene Compound $C_{60}H_6(N(CH_2CH_2)_2O)_6$. *Angewandte Chemie International Edition, 30*(10), 1309–1310. doi:10.1002/anie.199113091

Hotze, E. M., Labille, J., Alvarez, P., & Wiesner, M. R. (2008). Mechanisms of photochemistry and reactive oxygen production by fullerene suspensions in water. *Environmental Science & Technology, 42*(11), 4175–4180. doi:10.1021/es702172w

Huang, L., Terakawa, M., Zhiyentayev, T., Huang, Y.-Y., Sawayama, Y., Jahnke, A., & Hamblin, M. R. et al. (2010). Innovative cationic fullerenes as broad-spectrum light-activated antimicrobials. *Nanomedicine; Nanotechnology, Biology, and Medicine, 6*(3), 442–452. doi:10.1016/j.nano.2009.10.005

Huang, L., Terakawa, M., Zhiyentayev, T., Huang, Y. Y., Sawayama, Y., Jahnke, A., & Hamblin, M. R. et al. (2010). Innovative cationic fullerenes as broad-spectrum light-activated antimicrobials. *Nanomedicine (London), 6*(3), 442–452. doi:10.1016/j.nano.2009.10.005

Huang, L., Xuan, Y., Koide, Y., Zhiyentayev, T., Tanaka, M., & Hamblin, M. R. (2012). Type I and Type II mechanisms of antimicrobial photodynamic therapy: An in vitro study on gram-negative and gram-positive bacteria. *Lasers in Surgery and Medicine, 44*(6), 490–499. doi:10.1002/lsm.22045

Iqbal, Z., Baughman, R. H., Ramakrishna, B. L., Khare, S., Murthy, N. S., Bornemann, H. J., & Morris, D. E. (1991). Superconductivity at 45-K in Rb/Tl codoped C_{60} and C_{60}/C_{70} mixtures. *Science, 254*(5033), 826–829. doi:10.1126/science.254.5033.826

Jensen, A. W., & Daniels, C. (2003). Fullerene-Coated Beads as Reusable Catalysts. *The Journal of Organic Chemistry, 68*(2), 207–210. doi:10.1021/jo025926z

Jimenez-Hernandez, M. E., Manjon, F., Garcia-Fresnadillo, D., & Orellana, G. (2006). Solar water disinfection by singlet oxygen photogenerated with polymer-supported Ru(II) sensitizers. *Solar Energy*, *80*(10), 1382–1387. doi:10.1016/j.solener.2005.04.027

Karaulove, E. N., & Bagrii, E. I. (1999). Fullerenes: Functionalisation and prospects for the use of derivatives. *Russian Chemical Reviews*, *68*(11), 889–907. doi:10.1070/RC1999v068n11ABEH000499

Kautsky, H. (1939). Quenching of luminescence by oxygen. *Transactions of the Faraday Society, 35*(1), 216-218.

Kearns, D. R. (1971). Physical and chemical properties of singlet molecular oxygen. *Chemical Reviews*, *71*(4), 395–427. doi:10.1021/cr60272a004

Kelty, S. P., Chen, C. C., & Lieber, C. M. (1991). Superconductivity at 30-K in Cesium-doped C_{60}. *Nature*, *352*(6332), 223–225. doi:10.1038/352223a0

Kim, J., Lee, C. W., & Choi, W. (2010). Platinized WO_3 as an Environmental Photocatalyst that Generates OH Radicals under Visible Light. *Environmental Science & Technology*, *44*(17), 6849–6854. doi:10.1021/es101981r

Kohn, T., & Nelson, K. L. (2007). Sunlight-Mediated Inactivation of MS2 Coliphage via Exogenous Singlet Oxygen Produced by Sensitizers in Natural Waters. *Environmental Science & Technology*, *41*(1), 192–197. doi:10.1021/es061716i

Kong, L., Mukherjee, B., Chan, Y. F., & Zepp, R. G. (2013). Quenching and Sensitizing Fullerene Photoreactions by Natural Organic Matter. *Environmental Science & Technology*, *47*(12), 6189–6196. doi:10.1021/es304985w

Korobov, M. V., & Smith, A. L. (2000). Solubility of the Fullerenes. In K. M. Kadish & R. S. Ruoff (Eds.), *Fullerenes: Chemistry, Physics, and Technology* (pp. 54–60). New York: Wiley-Interscience.

Kroto, H. W., Heath, J. R., O'Brien, S. C., Curl, R. F., & Smalley, R. E. (1985). C_{60}: Buckminsterfullerene. *Nature*, *318*(14), 162–163. doi:10.1038/318162a0

Kruk, M., Jaroniec, M., Kim, J. M., & Ryoo, R. (1999). Characterization of Highly Ordered MCM-41 Silicas Using X-ray Diffraction and Nitrogen Adsorption. *Langmuir*, *15*(16), 5279–5284. doi:10.1021/la990179v

Kuzmanovic, D. A., Elashvili, I., Wick, C., O'Connell, C., & Krueger, S. (2003). Bacteriophage MS2: Molecular Weight and Spatial Distribution of the Protein and RNA Components by Small-Angle Neutron Scattering and Virus Counting. *Structure (London, England)*, *11*(11), 1339–1348. doi:10.1016/j.str.2003.09.021

Kyriakopoulos, J., Tzirakis, M. D., Panagiotou, G. D., Alberti, M. N., Triantafyllidis, K. S., Giannakaki, S., & Lycourghiotis, A. et al. (2012). Highly active catalysts for the photooxidation of organic compounds by deposition of [60] fullerene onto the MCM-41 surface: A green approach for the synthesis of fine chemicals. *Applied Catalysis B: Environmental*, *117-118*, 36–48. doi:10.1016/j.apcatb.2011.12.024

Latassa, D., Enger, O., Thilgen, C., Habicher, T., Offermanns, H., & Diederich, F. (2002). Polysiloxane-Supported Fullerene Derivative as a New Heterogeneous Sensitiser for the Selective Photooxidation of Sulfides to Sulfoxides by $^1O_2$. *Journal of Materials Chemistry, 12*(7), 1993–1995. doi:10.1039/b201141g

Leach, A. G., & Houk, K. N. (2002). Diels-Alder and Ene reactions of singlet oxygen, nitroso compounds and triazolinediones: Transition states and mechanisms from contemporary theory. *Chemical Communications*, (12): 1243–1255. doi:10.1039/b111251c

Lee, J., Fortner, J. D., Hughes, J. B., & Kim, J. H. (2007). Photochemical production of reactive oxygen species by C_{60} in the aqueous phase during UV irradiation. *Environmental Science & Technology, 41*(7), 2529–2535. doi:10.1021/es062066l

Lee, J., Hong, S., Mackeyev, Y., Lee, C., Chung, E., Wilson, L. J., & Alvarez, P. J. J. et al. (2011). Photosensitized Oxidation of Emerging Organic Pollutants by Tetrakis C_{60} Aminofullerene-Derivatized Silica under Visible Light Irradiation. *Environmental Science & Technology, 45*(24), 10598–10604. doi:10.1021/es2029944

Lee, J., Mackeyev, Y., Cho, M., Li, D., Kim, J.-H., Wilson, L. J., & Alvarez, P. J. J. (2009). Photochemical and Antimicrobial Properties of Novel C_{60} Derivatives in Aqueous Systems. *Environmental Science & Technology, 43*(17), 6604–6610. doi:10.1021/es901501k

Lee, J., Mackeyev, Y., Cho, M., Li, D., Kim, J. H., Wilson, L. J., & Alvarez, P. J. J. (2009). Photochemical and Antimicrobial Properties of Novel C60 Derivatives in Aqueous Systems. *Environmental Science & Technology, 43*(17), 6604–6610. doi:10.1021/es901501k

Lee, J., Mackeyev, Y., Cho, M., Wilson, L. J., Kim, J.-H., & Alvarez, P. J. J. (2010). C_{60} Aminofullerene Immobilized on Silica as a Visible-Light-Activated Photocatalyst. *Environmental Science & Technology, 44*(24), 9488–9495. doi:10.1021/es1028475

Lee, J., Yamakoshi, Y., Hughes, J. B., & Kim, J. H. (2008). Mechanism of C_{60} photoreactivity in water: Fate of triplet state and radical anion and production of reactive oxygen species. *Environmental Science & Technology, 42*(9), 3459–3464. doi:10.1021/es702905g

Li, J. W., & Liang, W. J. (2007). Loss of Characteristic Absorption Bands of C_{60} Conjugation Systems in the Addition with Aliphatic Amines. *Spectrochimica Acta Part A, 67*(5), 1346–1350. doi:10.1016/j.saa.2006.10.022

Li, Q., Mahendra, S., Lyon, D. Y., Brunet, L., Liga, M. V., Li, D., & Alvarez, P. J. J. (2008). Antimicrobial Nanomaterials for Water Disinfection and Microbial Control: Potential Applications and Implications. *Water Research, 42*(18), 4591–4602. doi:10.1016/j.watres.2008.08.015

Lonnen, J., Kilvington, S., Kehoe, S. C., Al-Touati, F., & McGuigan, K. G. (2005). Solar and Photocatalytic Disinfection of Protozoan, Fungal and Bacterial Microbes in Drinking Water. *Water Research, 39*(5), 877–883. doi:10.1016/j.watres.2004.11.023

Loutfy, R. O., Lowe, T. P., Moravsky, A. P., & Katagiri, S. (2002). Commercial Production of Fullerenes and Carbon Nanotubes. In E. Ōsawa (Ed.), *Perspectives of Fullerene Nanotechnology* (pp. 35–46). Dordrecht: Springer Netherlands. doi:10.1007/0-306-47621-5_4

Lovern, S. B., Strickler, J. R., & Klaper, R. (2007). Behavioral and Physiological Changes in Daphnia magna when Exposed to Nanoparticle Suspensions (Titanium Dioxide, Nano-C60, and C60HxC70Hx). *Environmental Science & Technology, 41*(12), 4465–4470. doi:10.1021/es062146p

Lu, Z. S., Dai, T. H., Huang, L. Y., Kurup, D. B., Tegos, G. P., Jahnke, A., & Hamblin, M. R. et al. (2010). Photodynamic therapy with a cationic functionalized fullerene rescues mice from fatal wound infections. *Nanomedicine (London), 5*(10), 1525–1533. doi:10.2217/nnm.10.98

Lyon, D. Y., Adams, L. K., Falkner, J. C., & Alvarez, P. J. J. (2006). Antibacterial Activity of Fullerene Water Suspensions: Effects of Preparation Method and Particle Size. *Environmental Science & Technology, 40*(14), 4360–4366. doi:10.1021/es0603655

Ma, X., & Bouchard, D. (2009). Formation of Aqueous Suspensions of Fullerenes. *Environmental Science & Technology, 43*(2), 330–336. doi:10.1021/es801833p

Maggini, M., Scorrano, G., Prato, M., Brusatin, G., Innocenzi, P., Guglielmi, M., & Bozio, R. et al. (1995). C_{60} derivatives embedded in sol-gel silica films. *Advanced Materials, 7*(4), 404–406. doi:10.1002/adma.19950070414

Mashino, T., Nishikawa, D., Takahashi, K., Usui, N., Yamori, T., Seki, M., & Mochizuki, M. et al. (2003). Antibacterial and antiproliferative activity of cationic fullerene derivatives. *Bioorganic & Medicinal Chemistry Letters, 13*(24), 4395–4397. doi:10.1016/j.bmcl.2003.09.040

Maurette, M. T., Oliveros, E., Infelta, P. P., Ramsteiner, K., & Braun, A. M. (1983). Singlet Oxygen and Superoxide - Experimental Differentiation and Analysis. *Helvetica Chimica Acta, 66*(2), 722–733. doi:10.1002/hlca.19830660236

Miller, G. P. (2006). Reactions Between Aliphatic Amines and [C_{60}] Fullerene: A Review. *Comptes Rendus. Chimie, 9*(7-8), 952–959. doi:10.1016/j.crci.2005.11.020

Miller, J. S. (2005). Rose bengal-sensitized photooxidation of 2-chlorophenol in water using solar simulated. *Water Research, 39*(2-3), 412–422. doi:10.1016/j.watres.2004.09.019

Moor, K., & Kim, J.-H. (2014). Simple Synthetic Method Toward Solid Supported C_{60} Visible Light Activated Photocatalysts. *Environmental Science & Technology, 48*(5), 2785–2791. doi:10.1021/es405283w

Moor, K. J., Snow, S. D., & Kim, J.-H. (2015). Differential Photoactivity of Aqueous [C60] and [C70] Fullerene Aggregates. *Environmental Science & Technology, 49*(10), 5990–5998. doi:10.1021/acs.est.5b00100

Moor, K. J., Valle, D., Li, C., & Kim, J.-H. (2015). *Improving the Visible Light Photoactivity of Supported Fullerene Photocatalysts Through the Use of* [C70]. Fullerene.

Mort, J., Ziolo, R., Machonkin, M., Huffman, D. R., & Ferguson, M. I. (1991). Electrical conductivity studies of undoped solid films of $C_{60/70}$. *Chemical Physics Letters, 186*(2-3), 284–286. doi:10.1016/S0009-2614(91)85142-J

Moura, J. C. V. P., Oliveira Campos, A. M. F., & Griffiths, J. (1997). Synthesis and evaluation of phenothiazine singlet oxygen sensitising dyes for application in cancer phototherapy. *Phosphorus, Sulfur, and Silicon and the Related Elements, 120*(1), 459–460. doi:10.1080/10426509708545597

Nakamura, E., & Isobe, H. (2003). Functionalized fullerenes in water. The first 10 years of their chemistry, biology, and nanoscience. *Accounts of Chemical Research*, *36*(11), 807–815. doi:10.1021/ar030027y

Navarro, D. A., Kookana, R. S., McLaughlin, M. J., & Kirby, J. K. (2016). Fullerol as a Potential Pathway for Mineralization of Fullerene Nanoparticles in Biosolid-Amended Soils. *Environmental Science & Technology Letters*, *3*(1), 7–12. doi:10.1021/acs.estlett.5b00292

Ogawa, K., Dy, J. T., Kobuke, Y., Ogura, S. I., & Okura, I. (2007). Singlet oxygen generation and photocytotoxicity against tumor cell by two-photon absorption. *Molecular Crystals and Liquid Crystals*, *471*(1), 61–67. doi:10.1080/15421400701545270

Pelaez, M., Nolan, N. T., Pillai, S. C., Seery, M. K., Falaras, P., Kontos, A. G., & Dionysiou, S. D. et al. (2012). A review on the visible light active titanium dioxide photocatalysts for environmental applications. *Applied Catalysis B: Environmental*, *125*, 331–349. doi:10.1016/j.apcatb.2012.05.036

Pfuetzner, S., Meiss, J., Petrich, A., Riede, M., & Leo, K. (2009). Improved bulk heterojunction organic solar cells employing C_{70} fullerenes. *Applied Physics Letters*, *94*(22), 223307. doi:10.1063/1.3148664

Piumetti, M., Hussain, M., Fino, D., & Russo, N. (2015). Mesoporous silica supported Rh catalysts for high concentration N_2O decomposition. *Applied Catalysis B: Environmental*, *165*(0), 158–168. doi:10.1016/j.apcatb.2014.10.008

Prat, F., Stackow, R., Bernstein, R., Qian, W. Y., Rubin, Y., & Foote, C. S. (1999). Triplet-state properties and singlet oxygen generation in a homologous series of functionalized fullerene derivatives. *The Journal of Physical Chemistry A*, *103*(36), 7230–7235. doi:10.1021/jp991237o

Qu, X., Brame, J., Li, Q., & Alvarez, P. J. J. (2012). Nanotechnology for a Safe and Sustainable Water Supply: Enabling Integrated Water Treatment and Reuse. *Accounts of Chemical Research*, *46*(3), 834–843. doi:10.1021/ar300029v

Redmond, R. W., & Kochevar, I. E. (2006). Spatially Resolved Cellular Responses to Singlet Oxygen. *Photochemistry and Photobiology*, *82*(5), 1178–1186. doi:10.1562/2006-04-14-IR-874

Regueiro, M. N., Monceau, P., & Hodeau, J. L. (1992). Crushing C_{60} to diamond at room temperature. *Nature*, *355*(6357), 237–239. doi:10.1038/355237a0

Rehman, S., Ullah, R., Butt, A. M., & Gohar, N. D. (2009). Strategies of Making TiO_2 and ZnO Visible Light Active. *Journal of Hazardous Materials*, *170*(2-3), 560–569. doi:10.1016/j.jhazmat.2009.05.064

Rengifo-Herrera, J. A., Sanabria, J., Machuca, F., Dierolf, C. F., Pulgarin, C., & Orellana, G. (2007). A comparison of solar photocatalytic inactivation of waterborne E-coli using tris (2,2 '-bipyridine)-ruthenium(II), rose bengal, and TiO_2. *Journal of Solar Energy Engineering-Transactions of the Asme*, *129*(1), 135–140. doi:10.1115/1.2391319

Ruoff, R. S., Tse, D. S., Malhotra, R., & Lorents, D. C. (1993). Solubility of C_{60} in a variety of solvents. *Journal of Physical Chemistry*, *97*(13), 3379–3383. doi:10.1021/j100115a049

Scrivens, W. A., Cassell, A. M., North, B. L., & Tour, J. M. (1994). Single Column Purification of Gram Quantities of C_{70}. *Journal of the American Chemical Society*, *116*(15), 6939–6940. doi:10.1021/ja00094a060

Seshadri, R., Govindaraj, A., Nagarajan, R., Pradeep, T., & Rao, C. N. R. (1992). Addition of Amines and Halogens to Fullerenes C_{60} and C_{70}. *Tetrahedron Letters*, *33*(15), 2069–2070. doi:10.1016/0040-4039(92)88144-T

Snow, S. D., Lee, J., & Kim, J. H. (2012). Photochemical and photophysical properties of sequentially functionalized fullerenes in the aqueous phase. *Environmental Science & Technology*, *46*(24), 13227–13234. doi:10.1021/es303237v

Snow, S. D., Park, K., & Kim, J.-. (2014). Cationic Fullerene Aggregates with Unprecedented Virus Photoinactivation Efficiencies in Water. *Environmental Science & Technology Letters*, *1*(6), 290–294. doi:10.1021/ez5001269

Snow, S. D., Park, K., & Kim, J.-H. (2014). Cationic Fullerene Aggregates with Unprecedented Virus Photoinactivation Efficiencies in Water. *Environmental Science & Technology Letters*, *1*(6), 290–294. doi:10.1021/ez5001269

Stephenson, L. M. (1980). Mechanism of the singlet oxygen Ene reaction. *Tetrahedron Letters*, *21*(11), 1005–1008. doi:10.1016/S0040-4039(00)78824-1

Sun, A. H., Xiong, Z. G., & Xu, Y. M. (2008). Removal of malodorous organic sulfides with molecular oxygen and visible light over metal phthalocyanine. *Journal of Hazardous Materials*, *152*(1), 191–195. doi:10.1016/j.jhazmat.2007.06.105

Sun, Y. P., Guduru, R., Lawson, G. E., Mullins, J. E., Guo, Z. X., Quinlan, J., & Gord, J. R. et al. (2000). Photophysical and electron-transfer properties of mono- and multiple-functionalized fullerene derivatives. *The Journal of Physical Chemistry B*, *104*(19), 4625–4632. doi:10.1021/jp0000329

Tegos, G. P., Demidova, T. N., Arcila-Lopez, D., Lee, H., Wharton, T., Gali, H., & Hamblin, M. R. (2005). Cationic fullerenes are effective and selective antimicrobial photosensitizers. *Chemistry & Biology*, *12*(10), 1127–1135. doi:10.1016/j.chembiol.2005.08.014

Tegos, G. P., Demidova, T. N., Arcila-Lopez, D., Lee, H., Wharton, T., Gali, H., & Hamblin, M. R. (2005). Cationic Fullerenes Are Effective and Selective Antimicrobial Photosensitizers. *Chemistry & Biology*, *12*(10), 1127–1135. doi:10.1016/j.chembiol.2005.08.014

Thilgen, C., Herrmann, A., & Diederich, F. (1997). The Covalent Chemistry of Higher Fullerenes: C_{70} and Beyond. *Angewandte Chemie International Edition*, *36*(21), 2268–2280. doi:10.1002/anie.199722681

Wang, H., Su, Y., Zhao, H., Yu, H., Chen, S., Zhang, Y., & Quan, X. (2014). Photocatalytic Oxidation of Aqueous Ammonia Using Atomic Single Layer Graphitic-C3N4. *Environmental Science & Technology*, *48*(20), 11984–11990. doi:10.1021/es503073z

Wang, Y. (1992). Photoconductivity of fullerene doped polymers. *Nature*, *356*(6370), 585–587. doi:10.1038/356585a0

Wang, Y., Li, Y., Fortner, J. D., Hughes, J. B., Abriola, L. M., & Pennell, K. D. (2008). Transport and Retention of Nanoscale C_{60} Aggregates in Water-Saturated Porous Media. *Environmental Science & Technology*, *42*(10), 3588–3594. doi:10.1021/es800128m

Wharton, T., Kini, V. U., Mortis, R. A., & Wilson, L. J. (2001). New non-ionic, highly water soluble derivatives of C_{60} designed for biological compatibility. *Tetrahedron Letters, 42*(31), 5159–5162. doi:10.1016/S0040-4039(01)00956-X

Wienk, M. M., Kroon, J. M., Verhees, W. J. H., Knol, J., Hummelen, J. C., Van Hal, P. A., & Janssen, R. A. J. (2003). Efficient Methano[70]fullerene/MDMO-PPV Bulk Heterojunction Photovoltaic Cells. *Angewandte Chemie International Edition, 42*(29), 3371–3375. doi:10.1002/anie.200351647

Wilkinson, F., & Brummer, J. G. (1981). Rate constants for the decay and reactions of the lowest electronically excited singlet-state of molecular-oxygen in solution. *Journal of Physical and Chemical Reference Data, 10*(4), 809-1000.

Wilkinson, F., Helman, W. P., & Ross, A. B. (1995). Rate Constants for the Decay and Reactions of the Lowest Electronically Excited Singlet State of Molecular Oxygen in Solution. An Expanded and Revised Compilation. *Journal of Physical and Chemical Reference Data, 24*(2), 663-677. doi:10.1063/1.555965

Wudl, F. (2002). Fullerene materials. *Journal of Materials Chemistry, 12*(7), 1959–1963. doi:10.1039/b201196d

Xu, J., Wang, Y., & Zhu, Y. (2013). Nanoporous Graphitic Carbon Nitride with Enhanced Photocatalytic Performance. *Langmuir, 29*(33), 10566–10572. doi:10.1021/la402268u

Yamakoshi, Y., Umezawa, N., Ryu, A., Arakane, K., Miyata, N., Goda, Y., & Nagano, T. et al. (2003). Active oxygen species generated from photoexcited fullerene (C_{60}) as potential medicines: O_2^- versus $^1O_2$. *Journal of the American Chemical Society, 125*(42), 12803–12809. doi:10.1021/ja0355574

Yamakoshi, Y. N., Yagami, T., Fukuhara, K., Sueyoshi, S., & Miyata, N. (1994). Solubilization of fullerenes into water with polyvinylpyrrolidone applicable to biological tests. *Journal of the Chemical Society. Chemical Communications*, (4): 517–518. doi:10.1039/c39940000517

Yang, P., Gai, S., & Lin, J. (2012). Functionalized mesoporous silica materials for controlled drug delivery. *Chemical Society Reviews, 41*(9), 3679–3698. doi:10.1039/c2cs15308d

Zhu, S., Xu, T., Fu, H., Zhao, J., & Zhu, Y. (2007). Synergetic Effect of Bi_2WO_6 Photocatalyst with C_{60} and Enhanced Photoactivity under Visible Irradiation. *Environmental Science & Technology, 41*(17), 6234–6239. doi:10.1021/es070953y

Chapter 8
Nanotechnology Applications for Sustainable Crop Production

Gaurav Mishra
Rain Forest Research Institute, India

Antara Dutta
Rain Forest Research Institute, India

Shailesh Pandey
Rain Forest Research Institute, India

Krishna Giri
Rain Forest Research Institute, India

ABSTRACT

Innovations in nanotechnology revolutionized the world in all sectors, including medicine, biotechnology, electronics, material science, energy sectors etc. In fact, the existing literature also points towards the potential application of nanotechnology in global food production and alarming situation of food scarcity. With the advancement of nanotechnology, use of nanofertilizers for yield enhancement, nanopesticides for insect pests and disease control and nanosensors to monitor soil and plant health becomes one of the most fascinating and promising lines of investigation to achieve sustainability. Furthermore, nanobiotechnology application enables gene transfer to expand the genetic base of crop varieties against different biotic and abiotic stresses. Some lacunas which may resist commercialization of nanotechnology are cost, industrial setup, and public attitude towards the health and food safety issues. Sincere attempts are required to answer these questions and develop suitable strategies to solve any problems, we might encounter in near future.

INTRODUCTION

Rise of human civilization was started with agriculture and cultivation of crops nurtured the development of human society. In developing countries, agriculture is known as the backbone of the national economy, as their livelihood depends on agriculture (Brock et al., 2011). Unfortunately, current agriculture is plagued by many problems; some of them are natural and some others are manmade. Food security with ever increasing population, limited availability of land and water, changing climate, pest and disease incidence and accumulation of pesticides and fertilizers in food stuffs are the biggest global challenges. In view of this, different technological innovations like synthetic pesticides, hybrid seeds,

DOI: 10.4018/978-1-5225-0585-3.ch008

fertilizers, high yielding varieties and transgenic were developed to boost the current agriculture system worldwide. These practices felicitates via enhancing agriculture productivity, by precisely using the same inputs and conserving soil and water resources (Prasad et al., 2012). However, extreme reliance on these innovations has many side effects as well. Problems using improper fertilization involve micronutrient imbalance, nitrate pollution and eutrophication (Savsi, 2012). Similarly, heavy pesticide application made agriculture sector paralyzed with serious problems, i.e. death of non-target organisms, pesticide resistance, bio-magnification and other human and environmental health hazards. Genetically Modified (GM) crops are one of the greatest attempts to minimize chemical treatments. Unfortunately, pests may also develop resistant against GM crops, as they have already developed resistance to many pesticides. Furthermore, environmentalists suspect novel genes may trigger unknown side effects with more severe, long-standing ecological and economic consequences. High yielding crop varieties are more prone to pest and diseases because of their narrow genetic base. Each of these modern agriculture innovations, except new strains of plants is totally reliant on the energy resources, especially petroleum. Global petroleum production is predicted to arrive at a maximum in the coming decades and to decline thereafter, a phenomenon known as peak petroleum. (Frumpkin et al 2009). The world's population is projected to reach 8 billion by 2025 and 9 billion by 2050, which could place an unprecedented pressure on the global food system (Sekhon, 2014). This situation calls the production of an additional 1 billion tones of cereals and 200 million tones of meat annually (Ghasemzadeh, 2012). Furthermore, the rising demand for meat puts huge pressure on agricultural land because farmers need to grow crops to produce animal feed (Sekhon, 2014). In view of the facts described above, there is an urgent need to conserve the natural resources along with sustainable agriculture production, so that negative effects on environment will be minimized (Densilin et al., 2011). In this context, nanotechnology has a remarkable potential to revolutionize agriculture and allied fields by target farming that involves the use of nano-sized particles with unique properties to enhance crop and livestock productivity (Batsmanova et al., 2014; Scoot and Chen, 2014). Nanotechnology is one of the most promising approaches for sustainable agriculture with high crop productivity, limited deterioration of natural resources and ability to feed the world's rapidly-growing population (Anonymous, 2009). The term "nano" is adapted from the Greek word "nanos" meaning "dwarf." A nanometer, thus, is one billionth of a meter. Particles have at least one dimension between 1 and 10 nanometers are considered as "nanoparticles" (Thakkar et al., 2010). The name nanotechnology describes diverse technologies performed on a nanometer scale with extensive applications in medicine, biotechnology, electronics, material science, agriculture, and energy sectors (Shekon, 2014; Parisi et al., 2014). It has been emerged as one of the most exciting areas of science and technology with promising applications in agriculture, including the targeted delivery of genes, fertilizers, pesticides and phytohormones (Torney et al., 2007; Melendi et al., 2008; Chen and Yada, 2011; Ghormade et al., 2011; Khot et al., 2012; Campos et al., 2014; de Oliveira et al., 2014). Applications of nanotechnology in modern agriculture have large impact on human life through the development of nano particles for better seed germination and seedling growth (Raskar and Laware, 2014), nanofertilizers for controlled release of nutrients (Naderi and Shahraki, 2013), nanocapsules for herbicide delivery (Grillo et al., 2013), nanosensors for pesticide detection (Menon et al., 2013), encapsulation of botanical insecticides (De Olveria et al., 2014) etc. Nanoparticles based agrochemical formulations enables slow release of the active ingredient and extension of its duration of action (Kah et al., 2013; Kah and Hofmann, 2014). As compared to conventional formulations, additional benefits associated with the use of nanoformulations are better protection against untimely degradation and improved uptake of the active ingredient, thus allowing reductions in pesticides dosage, application frequency and environmental contamination (Kah et al., 2013; Kah and Hofmann, 2014). These technological developments have become one of the most prominent options to tackle the situation of declining agriculture and sustainable production (Zhang, 2004). Nanotechnology promises to improve current agriculture practices through the enhancement of management and conservation of inputs in crops, animal production, and fisheries. This innovative technology has an enormous potential to augment the yield, along with the enriched quality and provides the nutritional benefits. One of the best example is of zeolites, with application of nanotechnology, they can be used to supply water and nutrients to plants in a well-organized and profitable manner. (Zhang, 2004; Harja et al., 2012). Hence, nanotechnology is likely to have a noteworthy impact on global economy within the next 10 to 15 years, growing in importance over the longer term as further scientific and technology breakthroughs are achieved. The focus of this chapter is to highlight nanotechnology developments and its role in Sustainable Agriculture

HOW NANOPARTICLES WORK

In agriculture, nanotechnology is going to make a significant impact by minimizing the loss of applied chemical products through smart release system. Relevance of nanotechnology in agriculture is already established as nanofertilizers, nanoherbicides, nanoremediation and nanobiotechnology. Furthermore, it has been also reported that nanoparticles can effectively mineralize the pesticides, and the concentration of nanoparticles enhances the rate of mineralization (Manimegalai et al.,2014). Owing to their small size, nanoparticles enter the cell membranes easily. But, little is known about how nanoparticles interact within the cell due to lack of related research and complexity of the interactions. Nanoparticles constitute a high surface area to volume ratio, have large amount of atoms with many exposed groups and unbalanced surface charges, where chemical reaction occurs. Due to these properties, nanomaterials are being able to perform variety of functions, like regulation of water and nutrient supply to plants and changing soil composition.

BASIC APPLICATIONS OF NANOTECHNOLOGY

Practical nanotechnology is essentially the increasing ability to manipulate (with precision) matter on previously impossible scales, presenting possibilities which many could never have imagined - it therefore seems unsurprising that a few areas of human technology are exempt from the benefits which nanotechnology could potentially bring. The Report of the World Summit on Sustainable Development (2002) states that nanotechnology can be targeted for the world's most critical sustainable development problems in the areas of water, energy, health and environment, agriculture, and biodiversity and ecosystem management. Nanotechnology has wide applications in medicine (diagnostics, drug delivery, tissue engineering) (Zhou et al., 2004), chemistry and environment (catalysis, filtration) (Shi et al., 2007), energy (reduction of energy consumption, increasing the efficiency of energy production, the use of more environmentally friendly energy system, recycling of batteries) (Das et al., 2007), information and communication (memory storage, novel semiconductor devices, novel optoelectronic devices, displays, quantum computers) (Hillie, 2007), heavy industry (aerospace, construction, refineries, vehicle manufacturers) (Lo et al., 2010) and consumer goods (foods, households, optics, textiles, cosmetics, agriculture). Despite nanotechnology has been exploited wide range of applicants in natural science, medicine and engineering, its role in agricultural sciences is yet to be fully explored.

NANOTECHNOLOGY APPLICATIONS IN AGRICULTURE

Recent advances in agriculture sector and green revolution contributed significantly in enhancing the global food supply, but associated environment risks alarms the need of more efficient and sustainable agriculture technology (Tillman et al., 2002). Imbalanced use of fertilizers and pesticides not only degraded soil quality, but also introduced the problem of eutrophication of aquatic ecosystem (Mukhopadhyay, 2014). Deteriorated soil and aquatic ecosystems become a severe warning to human health and culture. These degraded ecosystem services moved the modern agriculture towards conservation approach and the concept of conservation agriculture (Hobbs et al., 2008). However, by the end of last century, nanotechnology evolved as a most prominent and reliable technique in agriculture sector to cope with the problem of food scarcity along with having the potential to minimize environmental burdens (Lal, 2008). According to United States Department of Agriculture (USDA) National Planning Workshop (2003), possible areas of nanotechnology with potential applications in agriculture (Figure 1) include nanofertilizers for slow release of nutrients, encapsulated herbicides for controlled release, nanoemulsions for greater efficiency of insecticides, nanoparticles for soil conservation, delivery of nutrients and drugs for livestock and fisheries; nanobrushes and membranes for soil and water purification, cleaning of fishponds and nanosensors for soil quality and plant health monitoring besides precision agriculture. Some of the nanotechnology innovations are presented in Table 1.

Figure 1. Nanotechnology applications in agriculture

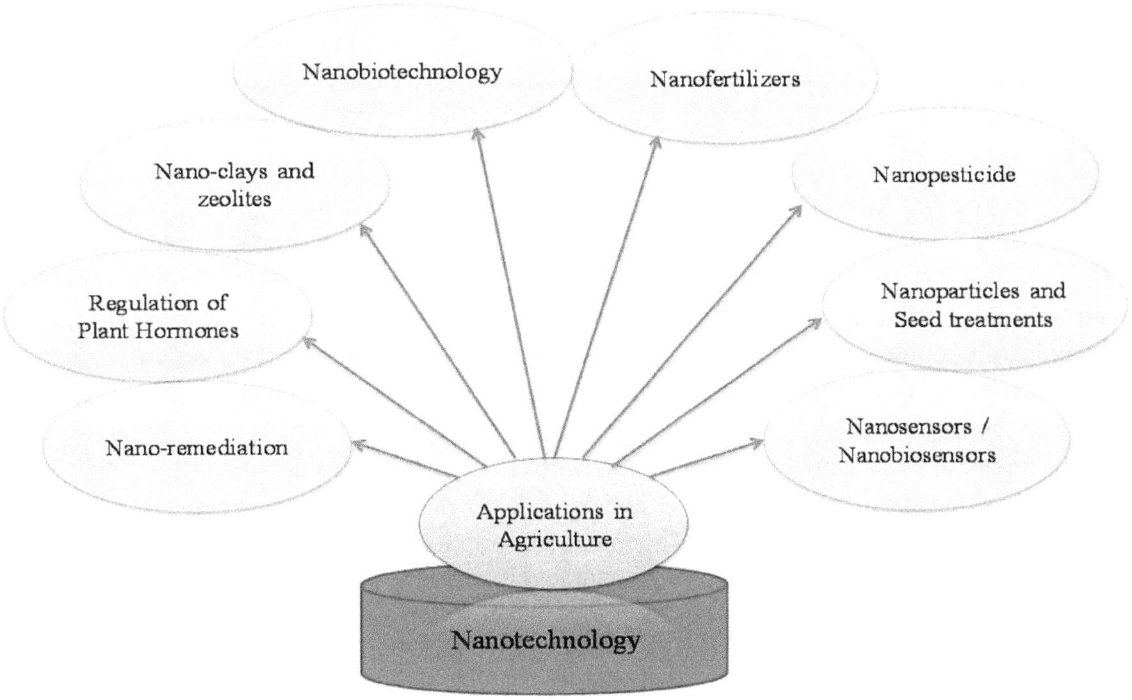

Table 1. Nanoagrochemicals and Nanomaterials Under Development

Type of Product	Product Name and Manufacturer	Nano Content	Purpose
Nano-Agrochemicals			
Super" combined fertilizer and pesticide	Pakistan-US Science and Technology Cooperative Program	Nanoclay capsule contains growth stimulants and biocontrol agents	Slow release of active ingredients, Reducing application rates
Herbicide	Tamil Nadu Agricultural University (India) and Technologico de Monterry (Mexico)	Nanoformulated	Designed to attack the seed coat of weeds, destroy soil seed banks and prevent weed germination
Pesticides, including herbicides	Australian Common wealth Scientific and Industrial Research Organization	Nanoencapsulated active ingredients	Very small size of nanocapsules increases their potency and may enable targeted release of active ingredients
Nano-Materials			
Nutritional supplement	Nanoceuticals 'mycrohydrin' powder, RBC Life sciences	Molecular cages 1-5 nm diameter made from silica mineral hydride comple	Nanosized mycrohydrin has increased potency and bioavailability. Exposure to moisture releases H- ions and acts as a powerful antioxidant.
Nutritional drink	Oat Chocolate Nutritional Drink Mix, Toddler Health	300nm particles of iron (SunActive Fe)	Nanosized iron particles have increased reactivity and bioavailability.
Food packaging	Adhesive for McDonald's burger containers, Ecosynthetix	50-150 nm starch nanospheres	These nanoparticles have 400 times the surface area of natural starch particles. When used as an adhesive they require less water and thus less time and energy to dry.
Food additive	Aquasol preservative, AquaNova	Nanoscale micelle (capsule) of lipophilic or water insoluble substances	Surrounding active ingredients within soluble nanocapsules increases absorption within the body (including individual cell)
Plant growth treatment	Primo Maxx, Syngenta	100nm particle size emulsion	Nano-sized particles increase the potency of active ingredients, potentially reducing the quantity to be applied.

Source: http://www.foeeurope.org/activities/nanotechnology/Documents/Nano_food_report.pdf

A key motivation for nanotechnology research in agriculture and food systems is based on the research evidences in the developed countries in promising areas of agriculture and food production systems (Opara, 2004; Ward and Dutta, 2005). Potential applications of nanotechnology in developing countries identified agricultural productivity enhancement as the second most critical area of application for attaining the millennium development goals while energy conversion and storage was ranked first and water treatment as the third areas needs to be focused (Buentello et al., 2005). Nanotechnology in agriculture has great potential to modernize agriculture, crop production, and better conditions of farmer and reduction of environmental pollution. A number of state-of-the art techniques are available for the improvement of precision farming practices and will allow a precise control at nanometer scale (De et al., 2014; Ngo and Van de Voorde, 2014). The detailed description of the applications of nanotechnology in sustainable agricultural crop production is given in the following section of this chapter.

NANOMATERIAL SEEDS TREATMENTS AND GERMINATION

To meet the increasing food demand, development of an efficient and ecofriendly production technology to increase seedling vigour and plant establishment through physical seed treatments is required (Siddiqui and Al-Whaibi, 2014). Literature available regarding application of nanomaterials and positive response in germination of various crops is quite interesting and indicated in Table 2. The effect of nanoparticles on plants varies from plant to plant and species to species. Nanomaterials like nanosilicon dioxide (nSiO_2), carbon nanotubes and nano-TiO_2 has a significant impact on the seed germination of various crops (Siddiqui and Al-Whaibi, 2014; Jiang et al., 2013; Castiglionc et al., 2011). These findings are useful and important as increase in germination parameters ha significant impact in increasing yield and sustainable crop production. By constrast, nano-Zn and nano-ZnO was found to inhibit seed

Table 2. Effects of Nanoparticles on Germination and Growth of Different crops

Nanoparticles	Crop	Mode of Exposure	Concentration	Comments	References
ZnO	Peanut (*Arachis hypogaea*)	Foliar and root	1000 mg kg^{-1} in soil	Improved seed germination and seedling vigor, high chlorophyll content and increased growth and yield	Prasad et al. (2012)
	Pearl Millet (*Pennisetum americanum*)	Foliar	10 mg L^{-1}	Increased shoot length, root length and area, along with chlorophyll and soluble leaf protein content, higher enzyme activities	Tarafdar et al. (2014)
	Clusterbean (*Cyamopsis tetragonoloba* L.)	Foliar	10 mg L^{-1}	Improved shoot-root growth, chlorophyll content, total soluble leaf protein content, rhizospheric microbial population, and P nutrient-mobilizing enzymes.	Raliya and Tarafdar (2013)
SiO_2	Tomato (*Lycopersicum esculentum* Mill.)	Roots	8 g L^{-1}	Improved seed germination	Siddiqui and Al-Whaibi (2014)
Carbon nanotubes	Tomato (*L. esculentum* M.)	Roots	50 ug mL^{-1}	Better seed germination and root growth, improved fresh and dry biomass and alteration in gene expression	Khodakovskaya et al. (2009)
	Tobacco (*Nicotiana tabacum*) cells	Culture	0.1–500 ug mL^{-1}	Increased cell growth and cell division, through stimulate water channel protein.	Khodakovskaya et al. (2012)
TiO_2	Spinach (*Spinacia oleracea*)	Roots	0.25–6%	Fast seed germination, growth rate and chlorophyll content, improved rubisco activity and photosynthetic rate	Zheng et al. (2005)
	Wheat (*Triticum aestivum* L.)	Foliar	0.01–0.03%	Increase in yield attributes along with higher gluten and starch content	Jaberzadeh et al. (2013)

germination of ryegrass and corn respectively (Ling and Xing, 2007). Furthermore nano-Zn or nano-ZnO terminates root elongation in radish, rape, ryegrass, lettuce, corn, and cucumber (Ling and Xing, 2007). These findings are significant in terms of use of engineered nanoparticles and their disposal. Plants also synthesize certain nanomaterials which are necessary for their growth and development (Wang et al., 2001) but the mechanism of interaction between nanomaterials and plants is still behind the curtains. Attempts are required to study the effect of nanomaterials on plant biological system, especially at the molecular level (Khodakovskaya et al., 2011).

NANOFERTILIZERS

Fertilizers play a major role to enhance the production across the spectrum of crops. The nutrient use efficiencies (NUE) of N, P, and K hardly exceed 30–35%, 18–20%, and 35–40% respectively, and remained constant for the past several decades (Subramanayam et al., 2015). Improvement of NUE is a pre-requisite for better crop production into marginal lands with low nutrient availability. So far, little success has been achieved to increase the NUE in conventional fertilizer formulations (Kottegoda et al., 2011). By constrast, the emerging nanofertilizer based approaches indicate that, due to their high surface area to volume ratio, nanofertilizers are predicted to be far more efficient than even polymer-coated conventional slow-release fertilizers (Hussein et al., 2002; Sastry et al., 2010; Hossain et al., 2008; de Rosa et al., 2010). Nanofertilizers are intended to improve the NUE while preventing the nutrient ions from either getting fixed or lost in the environment (Subramanayam et al., 2008; Naderi and Danesh, 2013). This smart delivery can only be accomplished by encapsulation, i.e., restricting the nutrients to interact with soil, water and microorganisms, and releasing nutrients only when they can be directly internalized by the plant (De Rosa et al., 2010). Release of nutrients form is regulated due to the coating and binding of nano and sub nanocomposites (Liu et al., 2006). Nanotechnology based preparations also lower bulk density, improves cation exchange capacity, water holding capacity, soil pH, nutrients release as per plant requirement and crop production (Rahale, 2010). Researchers showed that nanofertilizers containing nitrogen, phosphorus, potassium and micronutrients increase the uptake of nutrients in different crops (Guo, 2004). Nitrogen is one of the most important elements in fertilizers when judged with regard to the energy required for its synthesis, tonnage used and monetary value. However, compared with amounts of nitrogen applied to soil, the nitrogen use efficiency by crops is very low (de Rosa et al., 2010). Urea is one of the commercially available, fast-release nitrogen fertilizers, which is easy to use in agricultural fields. Reducing the release rate of urea can increase its efficiency of use and reduce nitrogen pollution. A green slow release urea-modified hydroxyapatite nanoformulation encapsulated under pressure into cavities of the soft wood of *Gliricidia sepium,* showed an early burst followed by a slow-release even on day 60, as compared to the commercial fertilizer that showed a heavily early burst and subsequent release of low and non-uniform quantities until around day 30 (Kottegoda et al., 2011). The major advantages of using slow-release nanofertilizer is improved NUE efficiency and higher crop yield. Furthermore, nanofertilizer application result in less environmental burdens from leaching of nitrogen, compared to conventional water-soluble fertilizers. (Kottegoda et al., 2011). Chitosan (CS) is a polymer of great interest in controlled release for NPK fertilizers because of its biodegradable, bioabsorbable and bactericidal properties (No et al., 2007; Corradini et al 2010; Hasaneen et al 2014). Future studies are required to understand the mechanism and to optimize the incorporation of NPK elements into chitosan based nanoparticles. Iron deficiency is a widespread problem, mainly in high pH and calcareous soils. Application of iron nanoparticles were found to be a promising approach, as highest values of spike weight 1000 grain weight, biological yield, grain yield and protein content were achieved using Fe Nano-oxide solution as a foliar spray in wheat (Bakhtiari et al., 2015). This result indicates that combining Fe compounds with nanotechnology may be a solution to combat Fe deficiency. Though, as being an infant know-how, the ethical and safety issues related to this emerging technology are limitless and must be assessed cautiously before adapting these intelligently controlled release function in agricultural fields (Solanki et al., 2015)

NANOPESTICIDES

Destruction of beneficial parasites and predators, evolution of pesticide resistance in pathogens, insect pests and weeds (Stuart, 2003), killing of birds and mammals (CWS, 2003), ground and surface water contamination (Cornell, 2003) are some major risk associated with heavy pesticide use. Nanopesticides or nano plant protection products represent a promising scientific expansion that offer a variety of benefits including increased effectiveness, durability, and a reduction in the amounts of active ingredients that is being used in protecting crops against diseases and insect pests (Kookana, 2014). A variety of formulation including nanoemulsions, nanocapsules and products containing pristine engineered nanoparticles, such as metals, metal oxides, and nanoclays has been suggested (Kah et al., 2013; Gogos et al., 2012; Khot et al., 2012). Polymer-based formulations have received the utmost interest, followed by formulations having inorganic nanoparticles (e.g., silica, titanium dioxide) and nanoemulsions (Kah and Hofmann, 2014). Porous hollow silica nanoparticles (PHSNs) have a hopeful carrier in agriculture, especially for controlled pesticide delivery (Liu et al., 2006). Nanopesticides are more efficacious and required in lower doses than the conventional formulation that results in lower environmental burdens. Nanosilver combined with silica molecules and water soluble polymer at 0.3 ppm, effectively controlled powdery mildews of pumpkin in both field and greenhouse tests (Jun et al.,2006). Similarly, chitosan based nano-formulations at 0.1% concentration showed 89.5, 63.0 and 60.1% growth inhibition of *A. alternata*, *M. phaseolina* and *R. solani*, respectively (Saharan et al.,2013). *B. cinerea* is considered as one of the major postharvest pathogens of table grapes (Latorre et al., 1994), while *P. expansum* is the main cause behind rotting of stored apples and pears (Cabanas et al., 2009). Both the pathogens are responsible for heavy postharvest fruit losses, even after the application of most superior postharvest technologies (Spadaro et al. 2004). Zinc oxide nanoparticles (ZnO NPs) at concentrations greater than 3 mmol l/1 was found to cause significant growth reductions of *B. cinerea* and *P. expansum* (He et al., 2011). As compared to commercial sulphur, signifcantly higher bioactivity of nanosulphur was reported against *Erysiphe cichoracearum* (Powdery mildew fungi) and *Tetranychus urticae* (red spider mite) (Gopal et al., 2012). A novel *Serratia* sp. (BHU-S4) was found to be a successful candidate to synthesize silver nanoparticles (AgNPs) with strong antifungal action against *Bipolaris sorokiniana* (Mishra et al., 2014), the cause of one of the most dreadful diseases of wheat (spot blotch), causing 20% yield loss in South Asia (Saari, 1998). Investigations to determine the ecological fate of nanopesticides is still scare. Pioneering study to determine the fate of a nanoformulation of atrazine indicates that the release of atrazine from the polymer nanocarriers occurred rapidly relative to the degradation kinetics and atrazine associated with the nanocarriers was subject to biotic or abiotic degradation (Kah et al., 2014). The present state of knowledge does not seem enough for a trustworthy measurement to be made of the associated benefits and hazards of nanopesticides (Kah and Hofmann, 2014). In view of this, sincere attempts are required for the development and expansion of reliable protocols to generate reliable fate properties. Furthermore, detailed investigations into the bioavailability, stability and evaluation of environmental risk assessment approaches would be of immense importance.

NANOHERBICIDES

Controlling weeds and their destruction becomes one of the biggest challenges in the agriculture sector. Using a single herbicide or permanent exposure of weeds with different herbicides, resulted in the development of herbicide resistance (Chinnamuthu and Kokiladevi, 2007). Besides herbicide resistance, toxic residues of herbicides and their longer persistence time can create germination problem for the next crop (Chinnamuthu and Boopathi, 2009). The important aspect in controlling the multiplication of any weed is the destruction of seed bank in soil, to restrict their germination. Using nanotechnology, target specific herbicides are developed which enter through the roots and get transported in all plant parts. While interacting with metabolic pathway inside the plants, they restrict glycolysis of food reserves and finally causing the death of the target due to starvation (Ali et al., 2014). Furthermore, owing to their small size, nanoherbicides are being able to easily mix in soil and eradicate the target species without leaving any contradictory effects (Prasad et al., 2014). Nanoencapsulation can also improve herbicide application to avoid phytotoxicity on the crop, providing better penetration through cuticles and tissues, and allowing slow and constant release of the active substances. (Pérez-de-Luque and Rubiales 2009). Eighty eight percent detoxification of Atrazine can be achieved by applying Carboxy Methyl Cellulose (CMC) nanoparticles (Satapanajaru et al., 2000).

NANOBIOTECHNOLOGY

Nanomaterials having potential to modify the genetic makeup of crops and enhance agriculture productivity have been developed (Kuzma, 2007; Maysinger, 2007). Nanoparticles work as magic bullets and ensure successful transfer of genes in proficient way (Pérez-de-Luque and Rubiales, 2009). Successful gene transfer has been done using mesoporus nonisillica nanoparticles in tobacco and corn (Torney et al., 2007). Similarly, carbon nanotubes or carbon nanofibers are being used in gene transfer and other molecules (Liu et al., 2006) and revolutionized the food industries. Research is now heading towards synthetic biology (dealing with genetic engineering, nanotechnology and informatics) to develop novel plant varieties (Sekhon, 2014).

NANOSENSORS

There are so many marked applications of nanosensors in different areas including agriculture. Nanosensors for water, nutrients, and chemicals have been developed to increase the agriculture productivity (Rai et al., 2012). Nanosensor has three components, biological (cell, enzyme or antibody), small transducer (supplies power) and receptor. Nanosensors can play significant role in precision farming, as they provide efficient management practices with reduction in input usages and proved to be user friendly. Some of the potential applications of nanosensors are (1) Nutrient Management- Nanosensors are known to be used in detection of soil nutrients and their management (Brock et al., 2011). They also play a key role in increasing the uptake of essential nutrients from the soil (Rameshaiah et al., 2015). They also allows the controlled use of fertilizers, reduces their consumption and pollution problems (Ingale and Chaudhari, 2013). (2) Disease Assessment- Nanosensors can also be used to identify soil born diseases. Diseases caused by viruses, bacteria, and fungi will be diagnosed via measuring the oxygen respired by microbes in the soil. (Rai et al., 2012). (3) Food products- Nanobiosensors act as a reliable tool to check the concentration of contaminants, i.e. residues of pesticides in vegetables (Amine et al., 2006). These instruments are known to quantify bacteria and viruses with accuracy and ensures customer food safety (Otles and Yalcin, 2010). (4)Detection of DNA and protein- Nanosensors like ssDNA-CNTs probes are able to detect and sequence particular kinds of DNA oligonucleotides (Cao et al., 2008 and Zhang et al., 2008). Although, biosensors based on protein nanoparticles are also developed to identify particular protein molecules (Rai et al., 2012). Innovation of these DNA and protein detecting biosensors can revolutionize the agriculture sector, as they can play pivotal role in detection of crop abnormalities.(5) Regulation of Plant Hormones- Auxin is one of the most important plant hormones responsible for root growth and seedling establishment. A team of researchers from Purdue University developed a nanosensor, which accounts for auxin concentration inside plants. This is an advance step in hormone research, as now it is possible to study the adaption of plant roots in their environment, with special reference to marginal soils (McLamore et al., 2010).

NANOREMEDIATION OF IRRIGATION WATER

With the advancement of nanotechnology, reliable purification of water can be done using nanofiltration (Hillie and Hlophe, 2007). Nanopores based nanofilters not only separate microorganisms but also able to reduce the concentration of heavy metals from irrigation water (Gao et al., 2014; Zhu et al., 2014). A powerful reductant named as Nanoscale Zero Valent Iron (NZVI) is developed for the removal of oxidized pollutants and pilot studies are under progress to check the effectiveness of NZVI (He et al., 2010). Similarly, nanoceram filter with positively charged surface removes the biological impurities like microbial endotoxins, genetic materials, viruses, other micro-sized particles etc. (Argonide, 2005; Gibbons et al., 2010).

NANO CLAYS AND ZEOLITES

Nanoclays and zeolites are naturally occurring crystalline aluminum silicates mineral, having honeycomb-like layered structure and super porous nature (Chinnamuthu and Boopathi, 2009). They act as a natural wetting agent and assists in infiltration and retention of water in soil. Their application can significantly influence water retention of sandy and porosity of clay soils (Prasad et al., 2014). There layered structure can be packed with nutrients to ensure their slow release. Application of zeolite in agriculture can control the dynamics (capture, storage and slow release) of nitrogen in soil (Leggo, 2000). Zeolities can be used for different nutrients in various ways like urea- fertilized zeolite chips (Millan et al., 2008), mixture of zeolite and phosphate rock (Allen et al., 1993) as slow release fertilizers and surfactant-modified zeolite to control nitrate release (Li, 2003).

NANO RISK

The potential of nanomaterials for better crop protection appears to improve the productivity of agriculture and food sectors but the risks coupled with their usage remain poorly understood. The possible hazards which can arise with the acceptance of technology can be categorized in three areas. Firstly, health hazards, i.e. mainly related to human health. The risk associated with the usage of nanomaterials may be diverse as these tiny particles can enter the human body through inhalation, ingestion, dermal contact, etc. and may exert adverse effects like interruption of cellular, enzymatic and organ functioning (Bhattacharyya et al., 2014). Secondly, environmental hazards, i.e. effects related to the plant health and environment. Silver nano particles, which are used to deliver the nutrients to plants, have certain menace like retarding the growth and membrane damage in grasses and weakening of photosynthesis in *Chlamydomonas reinnardtii* (Rameshaiah et al., 2015). The ecological effects of crops which have been genetically engineered using nanomaterials and their consequences on soil microorganisms in natural systems remain inadequately researched. Third is social, i.e. dealing with their availability in community, public acceptance and the interaction between society and government. So, jeopardy of the technology is heterogeneous, it may be environmentally, health related, professional or socio-economic. Besides having lot of benefits with nanotechnology, it cannot be forgotten that nanoparticles are non-biodegradable and their disposal is a serious issue (Jones, 2006; Rajathi & Sridhar, 2013). In view of the facts described above, these entire lacunas require urgent need for further research.

CONCLUSION

Nanotechnology is evolving as a promising approach and has the potential to contribute in an excellent way for sustainable crop production via better management and conservation of inputs for maintaining high crop yields and improved soil health. The scientific community can provide a plethora of benefits to the society by widening the scope of nanotechnology in agriculture and food systems (Sugunan and Dutta, 2004). Nanotechnology presents a novel platform to help global agriculture system with the problem of food security and environmental safety. During the last decade, applied research has been published regarding the target delivery of agrochemicals, i.e. nanofertilizers, nanopesticide, nanoherbicides and disease and pest detection through biosensors for better crop production. (Scrinis and Lyons, 2007; Scott, 2007). It is expected that agriculture sector would be able to solve the challenge of food security by means of nanotechnology as nanoparticles and their applications are going to be recognized (Owolade et al., 2008). Hence, role and scope of nanotechnology in revolutionizing the agriculture sector in the near future can't be underestimated. Some lacunas which resist commercialization of nanotechnology are cost, industrial setup, and public attitude towards the health and food safety issues. Public acceptance or the approval by the stakeholders and society is a prerequisite for any new technology; therefore, sincere research attempts are required to decipher the unforeseen ecological risk and challenges along with the rewards and positive potential associated with nanotechnology. There is an urgent need to study the effects of unrelieved disclosure of nanomaterials, all possible interactions and bioaccumulation effects, before transferring the technology from lab to land. Future research should aim towards the effective and sound usage of nanoagrochemicals, analysis of the risk associated, if any, to obtain a sustainable and eco-friendly production worldwide.

REFERENCES

Ali, M. A., Rehman, I., Iqbal, A., Din, S., Rao, A. Q., Latif, A., & Husnain, T. et al. (2014). Nanotechnology, a new frontier in Agriculture. *Advancements in Life Sciences, 1*(3), 129–138.

Allen, E. R., Hossner, L. R., Ming, D. W., & Henninger, D. L. (1993). Solubility and cation exchange in phosphate rock and saturated clinoptilolite mixtures. *Soil Science Society of America Journal, 57*(5), 1368–1374. doi:10.2136/sssaj1993.03615995005700050034x PMID:11537990

Amine, A., Mohammadi, H., Bourais, I., & Palleschi, G. (2006). Enzyme inhibition-based biosensors for food safety and environmental monitoring. *Biosensors & Bioelectronics, 21*(8), 1405–1423. doi:10.1016/j.bios.2005.07.012 PMID:16125923

Anonymous, N. (2009). Nanotechnology and nanoscience applicatios: revolution in India and beyond. *Strategic Appl. Integrating Nano Sciences.* Retrieved from www.sainsce.com

Argonide Corporation Nano Ceramic Sterilization Filter. (n.d.). Retrieved from http://sbir.nasa.gov/SBIR/successes/ss/9-072text.html

Batsmanova, L. M., Gonchar, L. M., Taran, N. Y., & Okanenko, A. A. (2013). Using a colloidal solution of metal nanoparticles as micronutrient fertiliser for cereals. *Proceedings of the International Conference on Nanomaterials: Applications and Properties.* Retrieved from http://nap.sumdu.edu.ua/index.php/nap/nap2013/paper/view/1097/504

Bhattacharyya, A., Chandrasekar, R., Chandra, A. K., Epidi, T. T., & Prakasham, R. S. (2014). Application of Nanoparticles in sustainable Agriculture: Its Current Status. *Short Views on Insect Biochemistry and Molecular Biology, 2*, 429–448.

Brock, D. A., Douglas, T. E., Queller, D. C., & Strassmann, J. E. (2011). Primitive agriculture in a social amoeba. *Nature, 469*(7330), 393–396. doi:10.1038/nature09668 PMID:21248849

Buentello, S., Persad, D. L., Court, E. B., Martin, D. K., Daar, A. S., & Peter, A. (2005). Nanotechnology and the Developing World. *PLoS Medicine, 2*(5), e97. doi:10.1371/journal.pmed.0020097 PMID:15807631

Cabanas, R., Abarca, M. L., Bragulat, M. R., & Cabanes, F. J. (2009). In vitro activity of imazalil against *Penicillium expansum*: Comparison of the CLSI M38-A broth microdilution method with traditional techniques. *International Journal of Food Microbiology, 129*(1), 26–29. doi:10.1016/j.ijfoodmicro.2008.10.025 PMID:19059665

Campos, E. V. R., Oliveira, J. L., & Fraceto, L. F. (2014). Applications of controlled release systems for fungicides, herbicides, acaricides, nutrients,and plant growth hormones: A review. *Adv. Sci. Eng. Med., 6*(4), 1–15. doi:10.1166/asem.2014.1538

Canadian Wildlife Service (CWS). (2002). *Pesticides and Wild Birds website.* Retrieved from http://www.cwsscf.ec.gc.ca/hww-fap/hww-fap.cfm?ID_species=90&lang=e

Cao, C., Kim, J. H., Yoon, D., Hwang, E. S., Kim, Y. J., & Baik, S. (2008). Optical Detection of DNA Hybridization Using Absorption Spectra of Single-Walled Carbon Nanotubes. *Materials Chemistry and Physics, 112*(3), 738–741. doi:10.1016/j.matchemphys.2008.07.129

Castiglione, M. R., Giorgetti, L., Geri, C., & Cremonini, R. (2011). The effects of nano-TiO$_2$ on seed germination, development and mitosis of root tip cells of *Vicia narbonensis* L. and *Zea mays* L. *Journal of Nanoparticle Research, 13*(6), 2443–2449. doi:10.1007/s11051-010-0135-8

Chen, H., & Yada, R. (2011). Nanotechnologies in agriculture: New tools for sustainable development. *Trends in Food Science & Technology, 22*(11), 585–594. doi:10.1016/j.tifs.2011.09.004

Chinnamuthu, C. R., & Boopathi, P. M. (2009). Nanotechnology and Agroecosystem. *Madras Agriculture Journal, 96*(1-6), 17–31.

Chinnamuthu, C. R., & Kokiladevi, E. (2007). Weed management through nanoherbicides. In C. R. Chinnamuthu, B. Chandrasekaran, & C. Ramasamy (Eds.), *Application of Nanotechnology in Agriculture*. Coimbatore, India: Tamil Nadu Agricultural University.

Cornell. (2003). *Common Pesticides in Groundwater*. Retrieved from http: //pmep.cce.cornell.edu/facts-slides-self/slide-set/ gwater09.html

Corradini, E., de Moura, M. R., & Mattoso, L. H. C. (2010). A preliminary study of the incorparation of NPK fertilizer into chitosan nanoparticles. *Polymer Letters, 4*(8), 509–515. doi:10.3144/expresspolymlett.2010.64

Das, S., Gates, A. J., Abd, H. A., Rose, G. S., Picconatto, C. A., & Ellenbogen, J. C. (2007). Designs for Ultra-tiny, special-Purpose Nanoeleectronic Circuits. *IEEE Transactions on Circuits and Systems, 154*(11), 2528–2540. doi:10.1109/TCSI.2007.907864

De, A., Bose, R., Kumar, A., & Mozumdar, S. (2014). Management of insect pests using nanotechnology: as modern approaches. In *Targeted delivery of pesticides using biodegradable polymeric nanoparticles* (pp. 29–33). New Delhi: Springer. doi:10.1007/978-81-322-1689-6_8

De Oliveira, J. L., Campos, E. V. R., Bakshi, M., Abhilash, P. C., & Fraceto, L. F. (2014). Application of nanotechnology for the encapsulation of botanicalinsecticides for sustainable agriculture: Prospects and promises. *Biotechnology Advances, 32*(8), 1550–1561. doi:10.1016/j.biotechadv.2014.10.010 PMID:25447424

De Rosa, M. C., Monreal, C., Schnitzer, M., Walsh, R., & Sultan, Y. (2010). Nanotechnology in fertilizers. *Nature Nanotechnology, 5*(2), 91. doi:10.1038/nnano.2010.2 PMID:20130583

Densilin, D. M., Srinivasan, S., Manju, P., & Sudha, S. (2011). Effect of Individual and Combined Application of Biofertilizers, Inorganic Fertilizer and Vermi compost on the Biochemical Constituents of Chilli (Ns - 1701). *Journal of Biofertilizers & Biopesticides, 2*(02), 106. doi:10.4172/2155-6202.1000106

DeRosa, M. C., Monreal, C., Schnitzer, M., Walsh, R., & Sultan, Y. (2010). Nanotechnology in fertilizers. *Nature Nanotechnology, 5*(2), 91. doi:10.1038/nnano.2010.2 PMID:20130583

Gao, J., Sun, S. P., Zhu, W. P., & Chung, T. S. (2014). Polyethyleneimine (PEI) cross-linked P84 nanofiltration (NF) hollow fiber membranes for Pb2+ removal. *Journal of Membrane Science, 452*, 300–310. doi:10.1016/j.memsci.2013.10.036

Ghasemzadeh, A. (2012). Global issues of food production. *Agrotechnol, 1*(2), 1–2. doi:10.4172/2168-9881.1000e102

Ghormade, V., Deshpande, M. V., & Paknikar, K. M. (2011). Perspectives fornano-biotechnology enabled protection and nutrition of plants. *Biotechnology Advances, 29*(6), 792–803. doi:10.1016/j.biotechadv.2011.06.007 PMID:21729746

Gibbons, C., Rodriguez, R., Tallon, L., & Sobsey, M. (2010). Evaluation of positively charged alumina nanofibre cartridge filters for the primary concentration of noroviruses, adenoviruses and male-specific coliphages from seawater. *Journal of Applied Microbiology, 109*(2), 635–641. PMID:20202019

Gogos, A., Knauer, K., & Bucheli, T. (2012). Nanomaterials in plant protection and fertilization: Current state, foreseen applications, and research priorities. *Journal of Agricultural and Food Chemistry, 60*(39), 9781–9792. doi:10.1021/jf302154y PMID:22963545

Gopal, M., Kumar, R., & Goswami, A. (2012). Nano-pesticides - A recent approach for pest control. *The Journal of Plant Protection Sciences, 4*(2), 1–7.

Grillo, R., Rosa, A. H., & Fraceto, L. F. (2013). Poly(ε-caprolactone) nanocapsules carrying the herbicide atrazine: Effect of chitosan-coating agent on physico-chemical stability and herbicide release profile. *International Journal of Environmental Science and Technology, 11*(6), 1691–1700. doi:10.1007/s13762-013-0358-1

Guo, J. (2004). Synchrotron radiation, soft X-ray spectroscopy and nanomaterials. *International Journal of Nanotechnology, 1*(1), 193–225. doi:10.1504/IJNT.2004.003729

Harja, M., Bucur, D., Cimpeanu, S. M., Ciocinta, R. C., & Gurita, A. A. (2012). Conversion of ash on zeolites for soil application. *Journal of Food Agriculture and Environment, 10*(2), 1056–1059.

Hasaneen, M. N. A., & Abdel-Aziz, H. M. M. (2014). Preparation of chitosan nanoparticles for loading with NPK fertilizer. *African Journal of Biotechnology, 13*(31), 3158–3164. doi:10.5897/AJB2014.13699

He, F., Zhao, D. Y., & Paul, C. (2010). Field assessment of carboxymethyl cellulose stabilized iron nanoparticles for in situ destruction of chlorinated solvents in source zones. *Water Research, 44*(7), 2360–2370. doi:10.1016/j.watres.2009.12.041 PMID:20106501

Hillie, T. (2007). *Nanocomputers and Swarm Intelligence* (p. 26). London: ISTE.

Hillie, T., & Hlophe, M. (2007). Nanotechnology and the challenge of clean water. *Nature Nanotechnology, 2*(11), 663–664. doi:10.1038/nnano.2007.350 PMID:18654395

Hobbs, P. R., Sayre, K., & Gupta, R. (2008). The role of conservation agriculture in sustainable agriculture. *Philosophical Transactions of the Royal Society of London. Series B, Biological Sciences, 363*(1491), 543–555. doi:10.1098/rstb.2007.2169 PMID:17720669

Hossain, K. Z., Monreal, C. M., & Sayari, A. (2008). Adsorption of urease on PE-MCM-41 and its catalytic effect on hydrolysis of urea. *Colloid Surf. B., 62*(1), 42–50. doi:10.1016/j.colsurfb.2007.09.016 PMID:17961995

Hussein, M. Z., Zainal, Z., Yahaya, A. H., & Foo, D. W. V. (2002). Controlled release of a plant growth regulator, 1-naphthaleneacetate from the lamella of Zn–Al-layered double hydroxide nanocomposite. *Journal of Controlled Release, 82*(2-3), 417–427. doi:10.1016/S0168-3659(02)00172-4 PMID:12175754

Ingale, A. G., & Chaudhari, A. N. (2013). Biogenic synthesis of nanoparticles and potential applications: An eco-friendly approach. *Journal of Nanomedicine & Nanotechnology, 4*, 165. doi:10.4172/2157-7439.1000165

Jaberzadeh, A., Moaveni, P., Tohidi Moghadam, H. R., & Zahedi, H. (2013). Influence of bulk and nanoparticles titanium foliar application on some agronomic traits, seed gluten and starch contents of wheat subjected to water deficit stress. *Notulae Botanicae Horti Agrobotanici, 41*(1), 201–207.

Jain, K. K. (2003). Nanodiagnostics: Application of nanotechnology in molecular diagnostics. *Expert Review of Molecular Diagnostics, 3*(2), 153–161. doi:10.1586/14737159.3.2.153 PMID:12647993

Jiang, Y., Hua, Z., Zhao, Y., Liu, Q., Wang, F., & Zhang, Q. (2013). The Effect of Carbon Nanotubes on Rice Seed Germination and Root Growth. *Proceedings of the 2012 International Conference on Applied Biotechnology* (ICAB 2012).

Jones, P. B. C. (2006). *A Nanotech Revolution in Agriculture and the Food Industry*. Retrieved from http://www.isb.vt.edu/articles/jun0605.htm

Jun, P. H., Ho, K. S., Jung, K. H., & Ho, C. S. (2006). A New Composition of Nanosized Silica-Silver for Control of Various Plant Diseases. *The Plant Pathology Journal, 22*(3), 295–302. doi:10.5423/PPJ.2006.22.3.295

Kah, M., Beulke, S., Tiede, K., & Hofmann, T. (2013). Nano-pesticides: State of knowledge, environmental fate and exposure modelling. *Critical Reviews in Environmental Science and Technology, 43*(16), 1823–1867. doi:10.1080/10643389.2012.671750

Kah, M., & Hofmann, T. (2014). Nanopesticide research: Current trends and future priorities. *Environment International, 63*, 224–235. doi:10.1016/j.envint.2013.11.015 PMID:24333990

Kah, M., Machinski, P., Koerner, P., Tiede, K., Grillo, R., Fraceto, L. F., & Hofmann, T. (2014). Analysing the fate of nanopesticides in soil and the applicability of regulatory protocols using a polymer-based nanoformulation of atrazine. *Environmental Science and Pollution Research International, 21*(20), 11699–11707. doi:10.1007/s11356-014-2523-6 PMID:24474560

Khodakovskaya, M., Dervishi, E., Mahmood, M., Xu, Y., Li, Z., Watanabe, F., & Biris, A. S. (2009). Carbon nanotubes are able to penetrate plant seed coat and dramatically affect seed germination and plant growth. *ACS Nano, 3*(10), 3221–3227. doi:10.1021/nn900887m PMID:19772305

Khodakovskaya, M. V., De Silva, K., Biris, A. S., Dervishi, E., & Villagarcia, H. (2012). Carbon nanotubes induce growth enhancement of tobacco cells. *ACS Nano, 6*(3), 2128–2135. doi:10.1021/nn204643g PMID:22360840

Khodakovskaya, M. V., De Silva, K., Nedosekin, D. A., Dervishi, E., Biris, A. S., Shashkov, E. V., & Zharov, V. P. et al. (2011). Complex genetic, photo thermal, and photo acoustic analysis of nanoparticle-plant interactions. *Proceedings of the National Academy of Sciences of the United States of America, 108*(3), 1028–1033. doi:10.1073/pnas.1008856108 PMID:21189303

Khot, L. R., Sankaran, S., Maja, J. M., Ehsani, R., & Schuster, E. (2012). Application of nanomaterials in agricultural production and crop protection: A review. *Crop Protection (Guildford, Surrey)*, *35*, 64–70. doi:10.1016/j.cropro.2012.01.007

Khot, L. R., Sankaran, S., Maja, J. M., Ehsani, R., & Schuster, E. W. (2012). Applications of nanomaterials in agricultural production and crop protection: A review. *Crop Protection (Guildford, Surrey)*, *35*, 64–70. doi:10.1016/j.cropro.2012.01.007

Kuzma, J. (2007). Moving forward responsibly: Oversight for the nanotechnology-biology interface. *Journal of Nanoparticle Research*, *9*(1), 165–182. doi:10.1007/s11051-006-9151-0

Lal, R. (2008). Promise and limitations of soils to minimize climate change. *Journal of Soil and Water Conservation*, *63*(4), 113A–118A. doi:10.2489/jswc.63.4.113A

Latorre, B., Flores, V., Sara, A. M., & Roco, A. (1994). Dicarboximide resistant strains of *Botrytis cinerea* from table grapes in Chile: Survey and characterization. *Plant Disease*, *7*(10), 990–994. doi:10.1094/PD-78-0990

Leggo, P. J. (2000). An investigation of plant growth in an organo–zeolitic substrate and its ecological significance. *Plant and Soil*, *219*(1/2), 135–146. doi:10.1023/A:1004744612234

Li, Z. (2003). Use of surfactant-modified zeolite as fertilizer carriers to control nitrate release. *Microporous and Mesoporous Materials*, *61*(1-3), 181–188. doi:10.1016/S1387-1811(03)00366-4

Lin, D., & Xing, B. (2007). Phytotoxicity of nanoparticles: Inhibition of seed germination and root growth. *Environmental Pollution*, *150*(2), 243–250. doi:10.1016/j.envpol.2007.01.016 PMID:17374428

Liu, F., Wen, L. X., Li, Z. Z., Yu, W., Sun, H. Y., & Chen, J. F. (2006). Porous hollow silica nanoparticles as controlled delivery system for water soluble pesticide. *Materials Research Bulletin*, *41*(12), 2268–2275. doi:10.1016/j.materresbull.2006.04.014

Liu, X., Feng, Z., Zhang, S., Zhang, J., Xiao, Q., & Wang, Y. (2006). Preparation and testing of cementing nano-subnano composites of slow or controlled release of fertilizers. *Scientia Agricultura Sinica*, *39*, 1598–1604.

Lo, P. K., Karam, P., & Sleiman, H. F. (2010). Loading and selective release of Cargo in DNA nano tubes with logtitudinal variation. *Nature Chemistry*, *2*(4), 319–328. doi:10.1038/nchem.575 PMID:21124515

Manimegalai, G. S., Shanthakumar, S., & Sharma, C. (2014). Silver nanoparticles: Synthesis and application in mineralization of pesticides using membrane support. *Int Nano Lett.*, *4*(105), 1–5.

Maysinger, D. (2007). Nanoparticles and cells: Good companions and doomed partnerships. *Organic & Biomolecular Chemistry*, *5*(15), 2335–2342. doi:10.1039/b704275b PMID:17637950

McLamore, E. S., Diggs, A., Calvo Marzal, P., Shi, J., Blakeslee, J. J., Peer, W. A., & Porterfield, D. M. et al. (2010). Noninvasive quantification of endogenous root auxin transport using an integrated flux microsensor technique. *The Plant Journal*, *63*(6), 1004–1016. doi:10.1111/j.1365-313X.2010.04300.x PMID:20626658

Menon, S. K., Modi, N. R., Pandya, A., & Lodha, A. (2013). Ultrasensitive and specific detection of dimethoate using a p-sulphonato-calix[4]resorcinarene functionalized silver nanoprobe in aqueous solution. *RSC Advances*, *3*(27), 10623–10627. doi:10.1039/c3ra40762d

Millán, G., Agosto, F., & Vázquez, M. (2008). Use of clinoptilolite as a carrier for nitrogen fertilizers in soils of the Pampean regions of Argentina. *Ciencia e investigación agraria. Agr*, *35*(3), 293–302.

Mishra, S., Singh, B. R., Singh, A., Keswani, C., Naqvi, A. H., & Singh, H. B. (2014). Biofabricated Silver Nanoparticles Act as a Strong Fungicide against *Bipolaris sorokiniana* Causing Spot Blotch Disease in Wheat. *PLoS ONE*, *9*(5), e97881. doi:10.1371/journal.pone.0097881 PMID:24840186

Mukhopadhyay, S. S. (2014). Nanotechnology in agriculture: Prospects and constraints. *Nanotechnology, Science and Applications*, *7*, 63–71. doi:10.2147/NSA.S39409 PMID:25187699

Naderi, M. R., & Danesh-Shahraki, A. (2013). Nanofertilizers and their roles in sustainable agriculture. *Intl J Agri Crop Sci*, *5*(19), 2229–2232.

Naderi, M. R., & Danesh-Shahraki, A. (2013). Nanofertilizers and their roles in sustainable agriculture. *International Journal of Agriculture and Crop Sciences*, *5*(19), 2229–2232.

Ngo, C., Van, D., & Voorde, M. H. (2014). Nanotechnologies in agriculture and food. In *Nanotechnology in a nutshell* (pp. 233–247). New York: Springer. doi:10.2991/978-94-6239-012-6_13

No, H. K., Meyers, S. P., Prinyawiwatkul, W., & Xu, Z. (2007). Applications of chitosan for improvement of quality and shelf life of foods: A review. *Journal of Food Science*, *72*(5), 87–100. doi:10.1111/j.1750-3841.2007.00383.x PMID:17995743

Opara, L. (2004). *Emerging Technological Innovation Triad for Smart Agriculture in the 21st Century. Part I. Prospects and Impacts of Nanotechnology in Agriculture. In Agricultural Engineering International: the CIGR Journal of Scientific Research and Development* (Vol. 6). Invited Overview Paper.

Otles, S., & Yalcin, B. (2010). Nano-biosensors as new tool for detection of food quality and safety. *LogForum*, *6*(4), 67–70.

Owolade, O. F., Ogunleti, D. O., & Adenekan, M. O. (2008). Titanium dioxide affects disease development and yield of edible cowpea. *Electronic Journal of Environmental Agricultural and Food Chemistry*, *7*(50), 2942–2947.

Perez-de-Luque, A., & Rubiales, D. (2009). Nanotechnology for parasitic plant control. *Pest Management Science*, *65*(5), 540–545. doi:10.1002/ps.1732 PMID:19255973

Prasad, R., Bagde, U. S., & Varma, A. (2012). Intellectual property rights and agricultural biotechnology: An overview. *African Journal of Biotechnology*, *11*(73), 13746–13752. doi:10.5897/AJB12.262

Prasad, R., Kumar, V., & Prasad, K. S. (2014). Nanotechnology in sustainable agriculture: Present Concerns and future aspects. *African Journal of Biotechnology*, *13*(6), 705–713. doi:10.5897/AJBX2013.13554

Prasad, T. N. V. K. V., Sudhakar, P., Sreenivasulu, Y., Latha, P., Munaswamy, V., Reddy, K. R., & Pradeep, T. et al. (2012). Effect of nanoscale zinc oxide particles on the germination, growth and yield of peanut. *Journal of Plant Nutrition*, *35*(6), 905–927. doi:10.1080/01904167.2012.663443

Rahale, S. (2010). *Nutrient release pattern of nano – fertilizer formulations.* (Thesis, Ph.D). TNAU, Tamil Nadu.

Rai, S., Kookana, R. S., Boxall, A. B. A., Reeves, P. T., Ashauer, R., Beulke, S., & Van den Brink, P. J. et al. (2014). Nanopesticides: Guiding Principles for Regulatory Evaluation of Environmental Risks. *Journal of Agricultural and Food Chemistry, 62*(19), 4227–4240. doi:10.1021/jf500232f PMID:24754346

Rai, V., Acharya, S., & Dey, N. (2012). Implications of nanobiosensors in agriculture. *Journal of Biomaterials and Nanobiotechnology, 3*(02), 315–324. doi:10.4236/jbnb.2012.322039

Rajathi, K., & Sridhar, S. (2013). Green Synthesized Silver Nanoparticles From The Medicinal Plant Wrightia Tinctoria and Its Antimicrobial Potential. *International Journal of ChemTech Research, 5*(4), 1707–1713.

Raliya, R., & Tarafdar, J. C. (2013). ZnO nanoparticle biosynthesis and its effect on phosphorous-mobilizing enzyme secretion and gum contents in Clusterbean (*Cyamopsis tetragonoloba* L.). *Agriculture Research, 2*(1), 48–57. doi:10.1007/s40003-012-0049-z

Rameshaiah, G. N., Pallavi, J., & Shabnam, S. (2015). Nanofertilizers and nanosensors – an attempt for developing smart agriculture. *International Journal of Engineering Research and General Science, 3*(1), 314–320.

Ramsden, J. J. (2005). What is nanotechnology? *Nanotechnology Perceptions., 1*(1), 3–17. doi:10.4024/N03RA05/01.01

Raskar, S. V., & Laware, S. L. (2014). Effect of zinc oxide nanoparticles on cytology and seed germination in onion. *Int.J.Curr.Microbiol.App.Sci, 3*(2), 467–473.

Saari, E. E. (1998). Leaf blight diseases and associated soilborne fungal pathogens of wheat in south and southeast Asia. In E. Duveiller, H. J. Dubin, J. Reeves, & A. McNab (Eds.), *Helminthosporium Blights of Wheat: Spot Blotch and Tan Spot* (pp. 37–51). CIMMYT.

Saharan, V., Mehrotra, A., Khatik, R., Rawal, P., Sharma, S. S., & Pal, A. (2013). Synthesis of chitosan based nanoparticles and their in vitro evaluation against phytopathogenic fungi. *International Journal of Biological Macromolecules, 62*, 677–683. doi:10.1016/j.ijbiomac.2013.10.012 PMID:24141067

Sastry, R. K., Rashmi, H. B., Rao, N. H., & Ilyas, S. M. (2010). Integrating nanotechnology into agri-food systems research in India: A conceptual framework. *Technol. Forecast. Soc., 77*(4), 639–648. doi:10.1016/j.techfore.2009.11.008

Satapanajaru, T., Anurakpongsatorn, P., Pengthamkeerati, P., & Boparai, H. (2008). Remediation of atrazine-contaminated soil and water by nano zerovalent iron. *Water, Air, and Soil Pollution, 192*(1-4), 349–359. doi:10.1007/s11270-008-9661-8

Scott, N., & Chen, H. (2002). *Nanoscale science and engineering for agriculture and food systems.* National Planning Workshop, Washington, DC. Available from: http://www.nseafs.cornell.edu/web.roadmap.pdf

Scott, N. R. (2007). Nanotechnology opportunities in agriculture and food systems. Biological and environmental engineering, Cornell university NSF nanoscale science and engineering grantees conference, Arlington, VA.

Scrinis, G., & Lyons, K. (2007). The emerging nano-corporate paradigm: Nanotechnology and the transformation of nature, food and agri-food systems. *International Journal of Sociology of Agriculture and Food, 15*, 22–44.

Scrinis, G., & Lyons, K. (2007). The emerging nano-corporate paradigm: Nanotechnology and the transformation of nature, food and agri-food systems. *International Journal of Sociology of Agriculture and Food, 15*, 22–44.

Sekhon, B. S. (2014). Nanotechnology in agri-food production: An overview. *Nanotechnology, Science and Applications, 7*, 31–53. doi:10.2147/NSA.S39406 PMID:24966671

Shi, X., Wang, S., Meshinchi, S., Van Antwerp, M. E., Bi, X., Lee, I., & Baker, J. R. (2007). Dendrimer – entrapped gold nanoparticles as a platform for cancer cell targeting and imaging. *Small, 3*(7), 1245–1252. doi:10.1002/smll.200700054 PMID:17523182

Siddiqui, M. H., & Al-Whaibi, M. H. (2014). Role of nano-SiO2 in germination of tomato (*Lycopersicum esculentum* seeds Mill.). *Saudi Journal of Biological Sciences, 21*(1), 13–17. doi:10.1016/j.sjbs.2013.04.005 PMID:24596495

Solanki, P., Bhargava, A., Chhipa, H., Jain, N., & Panwar, J. (2015). In M. Rai et al. (Eds.), *Nano-fertilizers and Their Smart Delivery System* (pp. 81–101). Nanotechnologies in Food and Agriculture, Springer International Publishing Switzerland; doi:10.1007/978-3-319-14024-7_4

Spadaro, D., Garibaldi, A., & Martines, G. F. (2004). Control of *Penicillium expansum* and *Botrytis cinerea* on apple combining a biocontrol agent with hot water dipping and acibenzolar-S-methyl, baking soda, or ethanol application. *Postharvest Biology and Technology, 33*(2), 141–151. doi:10.1016/j.postharvbio.2004.02.002

Stuart, S. (2003). *Development of Resistance in Pest Populations*. Retrieved from Http://Www.Nd.Edu/Chem191/E2.Html

Subramanian, K. S., Manikandan, A., Thirunavukkarasu, M., & Rahale, C. S. (2015). Nano-fertilizers for Balanced Crop Nutrition in Nanotechnologies in Food and Agriculture. Springer international Publishing.

Subramanian, K. S., Paulraj, C., & Natarajan, S. (2008). Nanotechnological approaches in nutrient management. In C. R. Chinnamuthu, B. Chandrasekaran, & C. Ramasamy (Eds.), *Nanotechnology applications in agriculture, TNAU technical bulletin* (pp. 37–42). Coimbatore: TNAU.

Sugunan, A., & Dutta, J. (2008). Nanotechnology: Environmental Aspects (vol. 2). Wiley-VCH.

Suman, P. R., Jain, V. K., & Varma, A. (2010). Role of nanomaterials in symbiotic fungus growth enhancement. *Current Science, 99*, 1189–1191.

Tarafdar, J. C., Raliya, R., Mahawar, H., & Rathore, I. (2014). Development of zinc nanofertilizer to enhance crop production in pearl millet (*Pennisetum americanum*). *Agriculture Research, 3*(3), 257–262. doi:10.1007/s40003-014-0113-y

Tarafdar, J. C., Sharma, S., & Raliya, R. (2013). Nanotechnology: Interdisciplinary science of applications. *African Journal of Biotechnology, 12*(3), 219–226.

Thakkar, M. N., Mhatre, S., & Parikh, R. Y. (2010). Biological synthesis of metallic nanoparticles. *Nanotecho.l Biol. Med., 6*, 257–262.

Tillman, D., Cassman, K. G., Matson, P. A., Naylor, R., & Polasky, S. (2002). Agricultural sustainability and intensive production practices. *Nature, 418*(6898), 671–677. doi:10.1038/nature01014 PMID:12167873

Torney, F., Trewyn, B. G., Lin, V. S., & Wang, K. (2007). Mesoporous silica nanoparticles deliver DNA and chemicals into plants. *Nature Nanotechnology, 2*(5), 295–300. doi:10.1038/nnano.2007.108 PMID:18654287

Torney, F., Trewyn, B. G., Lin, V. S. Y., & Wang, K. (2007). Mesoporous silica nanoparticles deliver DNA and chemicals into plants. *Nature Nanotechnology, 2*(5), 295–300. doi:10.1038/nnano.2007.108 PMID:18654287

Wang, L. J., Guo, Z. M., Li, T. J., & Li, M. (2001). The nano structure SiO_2 in the plants. *Chinese Science Bulletin, 46*, 625–631.

Ward, H. C., & Dutta, J. (2005). Nanotechnology for agriculture and food systems-A view.*Proc. Of the 2nd International Conference on Innovations in Food Processing Technology and Engineering.*

Zhang, W., Yang, T., Huang, D. M., & Jiao, K. (2008). Electro-chemical Sensing of DNA Immobilization and Hybridization Based on Carbon Nanotubes/Nano Zinc Oxide/ Chitosan Composite Film. *Chinese Chemical Letters, 19*(5), 589–591. doi:10.1016/j.cclet.2008.03.012

Zhang, Z. D. (2004). Nanocapsules Encyclopedia of Nanoscience and Nanotechnology. Academic Press.

Zheng, L., Hong, F. S., Lu, S. P., & Liu, C. (2005). Effect of nano-TiO_2 on strength of naturally and growth aged seeds of spinach. *Biological Trace Element Research, 104*(1), 83–91. doi:10.1385/BTER:104:1:083 PMID:15851835

Zhou, B., Hermans, S., & Gabor, A. (2004). *Nanotechnology in catalysis.* doi:10.1007/978-1-4419-9048-8

Zhu, W. P., Sun, S. P., Gao, J., Fu, F. J., & Chung, T. S. (2014). Dual-layer polybenzimidazole/polyethersulfone (pbi/pes) nanofiltration (nf) hollow fiber membranes for heavy metals removal from wastewater. *Journal of Membrane Science, 456*, 117–127. doi:10.1016/j.memsci.2014.01.001

ADDITIONAL READING

Ditta, A., Arshad, M., & Ibrahim, M. (2015). *Nanoparticles in Sustainable Agricultural Crop Production: Applications and Perspectives. Springer International Publishing.* doi:10.1007/978-3-319-14502-0_4

Prasad, R., Kumar, V., & Prasad, K. S. (2014). Nanotechnology in sustainable agriculture: Present Concerns and future aspects. *African Journal of Biotechnology, 13*(6), 705–713. doi:10.5897/AJBX2013.13554

KEY TERMS AND DEFINITIONS

Agrochemicals: A term used to describe all the chemical inputs used in agriculture field. It includes fertilizers, pesticides, herbicides, growth regulators etc.

Nanosensors: A customized unit having three components, biological (cell, enzyme or antibody) small transducer (supplies power) and receptor.

Nanoparticles: A collection of particles/ discrete nanometer (10^{-9} m) entities having assemblies of atoms in all three dimensions.

Nano Risk: Potential risks associated with the usage of nanotechnology, whether related to human health, environment or societal.

Nanotechnology: Nanotechnology is the science to modify the materials at molecular level, with desired properties which are different from those at a larger scale.

Nanozeolites: Crystalline aluminum silicates mineral, having honeycomb-like layered structure and super porous nature.

Pesticides: A chemical compound, used to depress, incapacitate or kill the target pests that destroy crop or spread diseases.

Sustainable Agriculture: An integrated system of package and practices having ability to enhance crop productivity with limited deterioration of natural resources on long term basis.

Synthetic Biology: A new stream of science, dealing with genetic engineering, nanotechnology, and informatics, to develop novel plant varieties.

Chapter 9
Developments in Antibacterial Disinfection Techniques:
Applications of Nanotechnology

Nicolas Augustus Rongione
University of Miami, USA

Scott Alan Floerke
University of Miami, USA

Emrah Celik
University of Miami, USA

ABSTRACT

One of the most daunting challenges facing nations today is controlling the spread of increasingly lethal bacteria. Today, a handful of bacteria can no longer be treated with traditional antibiotics and show antibacterial resistance. In this regard, nanotechnology possesses tremendous potential for the development of novel tools which help prevent and combat the spread of unwanted microorganisms. These tools can provide unique solutions for the challenges of the traditional disinfection methods, such as increased antibacterial activity, cost reduction, biocompatibility and personalized treatment. Despite its great potential, nanotechnology remains in its infancy and continued research efforts are required to achieve its full potential. In this chapter, traditional methods and their associated limitations are reviewed for their efficacy against microbial spread, and potential solutions in nanotechnology are described. A review of the state of the art disinfection techniques using nanotechnology is presented, and promising new areas in the field are discussed.

INTRODUCTION AND BACKGROUND

Colonization of surfaces by infectious microbes is a serious threat to public health (Beyth, Houri-Haddad, Domb, Khan, & Hazan, 2015; Guo, Yuan, Lu, & Li, 2013). The family of microbes consists of bacteria, fungi, and other protozoans. Individuals coming into contact with a surface hosting these microorgan-

DOI: 10.4018/978-1-5225-0585-3.ch009

isms may become infected and mobile carriers of diseases. In the modern era, such individuals can cross state lines and international borders in remarkably short time periods and therefore, microbes can travel thousands of miles alongside human passengers to foreign soil. Consequently, the task of tracking and preventing outbreaks of disease is even more formidable than in past years.

Microbes are ubiquitous organisms. They infest a variety of surfaces, from tools to hospital walls. Today, once a surface becomes contaminated, the proper response is to dispose of it. However, cost is a major concern in this procedure and it may not be a long term solution to disinfect the environment. As an example, washing hands becomes relatively ineffective if the working environment is heavily infested with microbes, such as has been the case with methicillin-resistant *Staphylococcus aureus* (MRSA) in hospitals, (Page, Wilson, & Parkin, 2009).

Another concern regarding microbial infestation is on food spoilage. Large food corporations must maintain vast distributions networks, and moreover ensure that the food arrives without spoiling by microbes. If these microbes are present during sealing, then they will proliferate continuously in the presence of nutrition and water before reaching their destination, resulting in spoilage. Once the microbes infest the source, it is a challenging task to stop further contamination.

In naval applications, the humidity provides a suitable environment for microbial growth. Due to the confined space in a vessel, microbial life has the ability to run rampant. Moreover, ships can be restrained to remain at sea for weeks or months at a time. In consequence, there is a strong need to protect food supplies from spoiling to alleviate the risk of endangering the mission.

Similar to the naval missions, the microgravity environment of space presents unique challenges for combatting microbial life. If an infection is able to spread, disposal is not an option since it would require additional cleaning and disinfection materials supplied from Earth, which is a gross time and cost investment. Life support systems, water treatment and sewage need to be carefully monitored in order to prevent the growth of dangerous microbial life, (Balagna et al., 2012).

Traditionally, antibiotics have been utilized to help mitigate complications caused by microbial borne illness. However, a handful of bacteria are no longer susceptible to these agents (Page et al., 2009). Over the past decade, a plethora of research has gone into better understanding the antibacterial properties of various materials to help combat the decline the efficacy of antibiotics. The cost of research and development for new antibiotics has become a diminishing return on investment, where the appearance of drug resistant microbes is outpacing new drug production; studies show that a new drug takes approximately 8 to 15 years of development time and a cost of $800 million dollars (Gwynn, Portnoy, Rittenhouse, & Payne, 2010). Efforts were made in the late 1990's after the first bacteria genome was fully sequenced in 1995 to find a genetic basis for targeted drug development, although the results were poor at best (Payne, Gwynn, Holmes, & Pompliano, 2007).

Besides antibiotics, using other biocidal chemicals such as bleach is not a practical solution for preventing the initial growth of a biofilm since these chemicals are only applied after a growth has taken place. Therefore, antimicrobial biocides can effectively remove the microbes on the surface but the biofilm which serves as the optimal environment for the spread underneath the contaminated surface may remain unaffected. Furthermore, such chemicals are not only toxic to microbes but also to humans. Bleach is one of the most commonly used chemical disinfectant of this type and it is used in many applications including food packaging for sterilization. It is however well documented that bleach similar to the other biocidal chemicals is harmful to human health even if trace amounts are consumed (McGlynn).

As described previously, bacteria have a strong ability to attach to solid surfaces and then form a biofilm, which serves as a reservoir for the development of pathogens, leading to health threats. Once it

Developments in Antibacterial Disinfection Techniques

develops, a biofilm is extremely difficult to remove and often requires disposal of the device (Beyth et al., 2015; Guo et al., 2013). The interest on superhydrophobic surfaces is growing rapidly since bacterial adhesion and biofilm formation are inhibited effectively on these hydrophobic surfaces. Superhydrophobicity can be achieved by applying chemical surface treatment methods or physically patterning the surface at the nanoscale similar to that of a lotus leaf (Watson et al., 2015).

Different types of antimicrobial materials have been the subject of most recent investigations including metals, metal oxides, polymers, and quaternary ammonium compounds. These various materials can be categorized into the broad, dimensional categories (0-D, 1-D, 2-D), and classified as organic, inorganic, or hybrid materials. Additionally, a slew of alternative methods have also been the subject of investigation that range from quantum dots to shrimp shells to antimicrobial peptides. The field is then further enriched by a range of synthesis techniques for each material, which are briefly discussed in this text.

Antibacterial metallic compounds such as silver, copper and gold have become widely popular due to their safety and efficiency when used against infectious microorganisms. Since copper is a cost effective substance compared to rare metals like silver and gold, copper and its alloys find a wide range of applications in typical touch surfaces like door knobs and railings; they are especially prevalent in hospitals (Iarikov et al., 2014; Page et al., 2009). Not only is copper noted for its antibacterial properties, it also has the potential to eliminate eukaryotic fungi (Cioffi et al., 2005). The mechanism of action of copper is believed to be mainly due to the contact of microbial cells and the released copper ions, where the microbial cellular membrane is damaged when exposed to these metallic ions. This damage on the cell membrane then leads to leaking of intracellular contents and cell death. Additionally, if copper is allowed inside the cell, hydrogen peroxide can be generated and cause oxidative harm to the cell.

Silver is also noted for its indiscriminant ability to eliminate bacterial populations, although its action against fungal populations is minimal. It is believed that the release of silver ions from a surface is primarily responsible for the death of bacterial cells similar to that of copper. These positively charged ions interfere with the negatively charged cell wall and cause extensive damage. In addition, these ions can enter the cell membrane and inactivate respiration, defense and growth mechanisms of these cells (Holt & Bard, 2005; Kalishwaralal, BarathManiKanth, Pandian, Deepak, & Gurunathan, 2010). In addition to release of metallic ions, researchers have showed that the bacterial contact killing mechanism also contributes to their enhanced antimicrobial effect, although killing is mainly governed by the silver release mechanism. (Agnihotri, Mukherji, & Mukherji, 2013; Fu, Ji, Fan, & Shen, 2006; Gunawan et al., 2011; Jose Ruben et al., 2005). The prohibitively high cost of silver, unfortunately, limits the extensive incorporation of silver and silver alloys in antimicrobial surfaces.

Metal oxides such as MgO, TiO_2, Al_2O_3, CuO, CeO_2, ZnO have unique advantages in their bactericidal properties, (Jones, Ray, Ranjit, & Manna, 2008). These oxides work to neutralize bacteria by producing reactive oxygen species which release free radicals that damage nucleoids and membranes, (Perelshtein et al., 2015). Compared to metals, oxides are good alternatives with high stability and significantly reduced toxicity to humans and animals.

Titanium dioxide surface coatings are particularly popular among the aforementioned metallic oxides for antimicrobial applications because of their ability to be actively controlled (Foster, Ditta, Varghese, & Steele, 2011) and their self-cleaning properties (Banerjee, Dionysiou, & Pillai, 2015; Charpentier et al., 2012; Karimi, Yazdanshenas, Khajavi, Rashidi, & Mirjalili, 2014; Navabpour et al., 2014; Wang, Wang, He, Lv, & Wang, 2010). When the titanium-dioxide is exposed to a UV light source, photocatalytic disinfection takes place which then gives rise to the complex self-cleaning process. Photocatalytic disinfection via UV activated titanium-dioxide has proven to be a powerful means of killing a broad spectrum of microorganisms. Bacteria, algae, protozoa, and viruses are all susceptible to the effects of photocatalysis. Free radicals are produced in this method during the disinfection process which are thought to destroy the cell membrane of microorganisms. The destruction of the cell membrane leads to leaking of intracellular contents and ultimately the death of the cell.

Polymers are being engineered with antimicrobial activity, superior surface adhesion and enhanced biocompatibility. As an example, these polymers are being applied to glass surfaces. Bacteria proliferation on glass surfaces has been a concern in wide range of applications including packaging, appliances, and electronics (Iarikov et al., 2014). As of today, glass monitors of laptops and cellphones are in use by millions of people for extended time intervals on a daily basis. Covalent bonding of the cationic polymer polyallylamine (PA) to glass using 3-glycidoxypropyltrimethoxy silane (GOTPS) can be a powerful means of coating glass surfaces, as demonstrated by Iarikov et al. These PA coatings exhibit a killing efficiency of about 97% relative to a control surface against S. epidermis and S. aureus.

Abrasive surface coatings which are not detectable by humans are hostile to many microbes, including viruses, bacteria, and fungi. Organosilanes are a type of such coating. Upon contact with an organosilane coated surface, microbial cellular membranes are physically ripped apart (Vishnupriya, Chaudhari, Jagannathan, & Pradeep, 2013). Colonization of the surface is thereby prevented. Moreover, organosilanes can be applied to soft and hard surfaces, such as with surface bonded organosilicon quaternary ammonium chloride, (Isquith, Abbott, & Walters, 1972). Applicable surfaces include but are not limited to clothing, carpet, and walls.

A novel method to address the increasing threat posed by microbial life to human health is the modification of a surface by bacteriophage infusion (Page et al., 2009). Bacteriophages only infect prokaryotic cells, and therefore, do not cause harm to eukaryotic human cells. After binding to the surface of a host bacterial cell, a phage injects its own genetic material into the cell. Replication occurs within the cell, and more phages are generated. After enough time has elapsed, cell lysis occurs and death results. During lysis, scores of new phages are released which are now able to infect new bacterial cells. The downside to bacteriophages, as Page et al. mention, is that they tend to target an individual species of bacteria. To combat multiple species of bacteria, a combination of bacteriophages would have to be implemented on a surface. Even if this precaution was taken, certain bacteria could still elude infection. Nevertheless, existing bacteriophages could guard against a considerably broad spectrum of harmful bacteria.

Ultimately, the difficulties in the design of antimicrobial materials stem from various requirements based on the application, and constant struggle between being bactericidal and biocompatible. Antimicrobial materials are implemented in a broad spectrum of applications from aerospace missions to packaging materials to sewage systems. Often a new material needs to be designed to optimize for the specific needs of a given application. For instance, on Earth, superhydrophobic surfaces are desirable to prevent microbial life from leaching; however, in space, there is a need for hydrophilic surfaces that are able to aggregate and control the spread of microbes. Since the desire for antimicrobial materials is prevalent throughout different facets of human life, material designers need to be aware of the interdependence of microbes, the environment, and human needs. Killing bacteria may be important for food packaging during shipment, but upon disposal, certain unintended consequences could kill useful bacteria that help breakdown waste.

Nanotechnology promises to provide unique solutions for disinfection across platforms. Nanomaterials with enhanced surface area, reactivity and stability can be used to achieve elevated levels of antimicrobial efficacy. Since a minute amount of material is required to achieve improved results, nanotechnology also offers cost effective solutions. Furthermore, nanomaterials are able to be doped and integrated into

Developments in Antibacterial Disinfection Techniques

existing surfaces and equipment through modified manufacturing techniques. Active and passive control, self-cleaning, superhydrophobicity, and biocompatibility can all be achieved using nanotechnology. Most importantly, nanomaterials offer the ability to prevent the formation of biofilms by degrading microbial attachment and spread.

This chapter focuses on new, cutting edge techniques made possible through nanotechnology. Advantages and challenges facing these techniques are compared and contrasted against traditional methods. The mechanisms of action that cause physical damage to microbial cells are presented to help facilitate and promote critical thinking for future efforts in this field. Consequently, the task of tracking and preventing outbreaks of disease is even more formidable than in past years.

CLASSIFICATION OF NANOMATERIALS

Geometrically nanomaterials are classified into three categories; zero-dimensional (0-D) nanoparticles, one dimensional (1-D) nanowires and nanotubes and two-dimensional (2-D) nanofilms. Each type of nanomaterials finds unique applications due to their physical geometry. In this section, the benefits as well as inherent limitations of 0-D, 1-D, and 2-D nanomaterials for use in combatting bacterial proliferation are presented. Each domain of nanomaterial offers considerable promise in this respect.

Nanoparticles: 0-Dimensional Nanomaterials

Nanoparticles are classified as zero-dimensional nanomaterials; they are often integrated into existing materials to enhance desired properties. Numerous recent studies confirm the effectiveness of nanoparticles in antimicrobial applications. Silver and copper nanoparticles in particular have proven to be invaluable agents in the fight against bacteria and fungi. The high surface area to volume ratio of these materials enables them to interact more effectively with the cell membranes of microbes, more so than microparticles, (Cruz-Romero, Murphy, Morris, Cummins, & Kerry, 2013; Watanabe et al., 2014). The smaller quantity needed also helps reduce cost. Nanoparticles cause physical damage to the membranes which leads to leaking of intracellular contents and cellular death. An evolutionary adaption to the physical mechanism of action of nanoparticles against microbial life is deemed highly unlikely. Consequently, there is a strong appeal for this class of nanomaterials in the field of antimicrobial sanitation.

Nanotubes and Nanowires: 1-Dimensional Nanomaterials

Within this class of nanomaterials, nanowires and nanotubes offer the greatest promise as antimicrobial agents. In a nanoparticle-infused coating, only a small portion of the nanoparticles are capable of coming into contact with the cellular membranes of pathogenic microbes. In colloidal nanoparticle systems, cleaning nanoparticles from the disinfected medium can be challenging. As an alternative to nanoparticles, an anchored array of nanowires with a high surface area of contact can be used. This may lead to enhanced antimicrobial properties. Although nanowires are arguably more difficult to fabricate compared to nanoparticles; recent advances in nanotechnology have made the fabrication of both classes of nanomaterial much more efficient.

Nanofilms: 2-Dimensional Nanomaterials

Nanofilms constitute the broadest class of nanomaterials having two dimensions in macroscale. These films are typically metallic and ionic in nature. They may be easily deposited via ion implantation, physical or chemical vapor deposition. Nanometer-thick materials with antibacterial properties are deposited onto a material of interest on top of the original surface. In addition to providing a safe environment from microbial life, these coatings can significantly enhance the lifespan of the surface. In this regard, ion implantation is particularly important and commonly used to coat surfaces such as stainless steel.

NANOMATERIALS AS ANTIBACTERIAL AGENTS

The primary factors influencing the choice of nanomaterial are the intended application, ease of implementation, financial cost, and degree of toxicity. When determining the optimum choice of nanomaterial for a given task, all of the above factors must be evaluated. Oftentimes, trade-offs will have to be made between parameters of interest. For instance, in environments where bacterial proliferation must be avoided to the utmost degree, the desire to obtain minimal financial cost may have to be cast aside in favor of optimal antimicrobial effectiveness. Ideally, an antibacterial coating must be simultaneously easy to implement, low-cost, exhibiting minimal toxicity to humans, and highly lethal to antimicrobial life. In reality, most coatings cannot simultaneously meet all of these criteria to the degree a potential user might demand; as such, innovative solutions need to be devised to enhance nanomaterials and recognize which existing nanomaterials provide the properties desired. This is commonly achieved by taking the compliment of two nanomaterials whose diverse properties come together to meet the needs of a new application, as with placing metallic nanoparticles in a composite matrix to create an antimicrobial material with sound structural integrity.

Metallic and Metal Oxide Nanomaterials

Metallic nanoparticles have been used extensively as antibacterial agents because of their proven efficiency in killing microbial life. They are nanoscale derivatives of pure metals or metal alloys. Gold, silver and copper are the major metallic materials used for this application. Engineering these metals and metallic alloys in nanoscale form brings two unique advantages, increased antibacterial activity and reduction of material cost. Increased surface area and particle dispersion are the main reasons behind increased activity as antibacterial material at the nanoscale. Decrease in material cost is also essential to develop affordable and effective technologies to fight microbial activity. Considering the cost of bulk antibacterial metallic materials such as gold and silver, nanoscale production of these materials will greatly benefit the field.

The concentration and dispersion of nanoparticles are essential parameters in determining the efficiency of antimicrobial activity, especially for silver nanoparticles which have the tendency to aggregate, (Visnapuu et al., 2013; Zhang, Niu, Yan, & Cai, 2011). Silver nanoparticles "inhibit important protein function, including biosynthesis, gene expression, energy production, and nitrification to further cause toxicity to *N. europaea*", a Gram-negative bacteria (Yuan et al., 2013). Spinel ferrites, MFe_2O_4 (M = Co, Mn, or Ni), are typically found in magnetic products, and naturally possess Gram-negative but not Gram-positive antibacterial properties, (Baraliya & Joshi, 2014). By carefully designing a magnetic

Developments in Antibacterial Disinfection Techniques

silica matrix to inlay the silver nanoparticles, the particles can maintain a better dispersion during application to increase both stability and bactericidal properties. Furthermore, the Fe_3O_4-SiO_2 silica matrix can separate back out of the magnetic composite Fe_3O_4-SiO_2-Ag in water through magnetic separation to reduce waste and prevent the adverse consequences of releasing disinfectant to the environment. In space applications, a silver nanocluster silica composite has been developed for the commercial polymeric film, Combitherm®, both to help combat microbial life by increasing particle dispersion, and also to have the special property of air impermeability (Balagna et al., 2012).

Combining CuO metal oxide with silver nanoparticles can improve bactericidal properties through the collaborative effects of each nanoparticle type. Embedding CuO into polymers also revealed a correlation between the release of ions and optimum killing. In particular, the release of ions from both Ag and Cu particles in solution causes local changes in pH and conductivity in viruses, and disrupts the membrane and enzyme functions of bacteria. Both of these effects lead to cell death. By implementing metals and metal oxides into other objects and surfaces in a hospital setting such as equipment, clothing and bedding exterior the antimicrobial activity of the environment could be better controlled, (Ren et al., 2009). Other metals oxide nanoparticles which can be used for this purpose extend to Al_2O_3, ZnO, MgO, TiO_2, and CaO.

Although metallic nanoparticles possess serious potential due to their aforementioned properties, some of these nanoparticles are considered as highly toxic and detrimental to human cells when inhaled or absorbed into the body, (Bai, Sandukas, Appleford, Ong, & Rabiei, 2012; Sondi & Salopek-Sondi, 2004). These materials also have the tendency to leach metal ions into the surrounding liquid medium, which may pose a health risk. As a result, caution must be taken as nanoparticles are increasingly incorporated into consumer products. While substantial praise has been awarded to nanoparticles on account of their antimicrobial ability, less thought has been given to the potential human toxicity of these compounds. Researchers must be aware that the differences between nanoparticles and bulk materials are not solely apparent in antimicrobial activity. In general, bulk forms of silver and titanium dioxide are considered safe. The same is not necessarily true for nanoscale silver and titanium dioxide. Researchers at MIT have reported damage to DNA brought on by exposure to nanoparticles of titanium dioxide and silver alike, (Pirela et al., 2015). Therefore, the same nanoparticles employed to eliminate microbial life and maintain our own health may degrade the integrity of our genetic material as well. If new methods are found to eliminate the adverse interaction of nanoparticles with DNA—perhaps by coating the particles with a nanothin layer of amorphous silica—then the public may be more receptive to the abundance of nanoparticles employed in consumer products.

Although stainless steel has many attractive qualities, it does not display any antimicrobial activity. Bacteria can proliferate on stainless steel, especially in nutrient rich environments. In order to provide stainless steel with exceptional antimicrobial capability, silver ion implantation has been executed. Researchers have found that, in the absence of protective coating films or protein layer barriers, silver ion implantation has the capacity to eliminate microbial populations, (R. Chen et al., 2013). The aforementioned barriers can adversely affect the quantity of silver ions available to engage in antimicrobial activity. Bacterial viability stains illustrate that stainless steel coupons implanted with silver ions alone are extremely hostile to microbial life, whereas their untreated counterparts pose no threat to potentially harmful microbes. Hence, silver ion implantation is a proven, surface treatment methodology to enhance antimicrobial efficacy.

Copper was the first recognized solid antimicrobial agent by the U.S. Environmental Protection Agency. Recent study has showed that vertically aligned copper nanotubes have significant bactericidal and bacteriostatic (inhibits reproduction) potential (Razeeb et al., 2014). In order to help compensate for the cytotoxicity of pure copper, self-organized, biocompatible and antimicrobial Cu-containing nanotubes and nano pore oxide structures have been constructed (X. G. Wang et al., 2014).

UV Light activated titanium nanofilms have been shown to further enhance antimicrobial activity (Huppmann et al., 2014). Photoactivated titanium dioxide reacts with nearby water molecules, producing hydroxide reactive oxygen species which decompose and inactivate microorganisms on the contact surface. High energy photons released act as a second wave of attack to help finish the job. In addition, nanofilm surface transforms into superhydrophilic surface by metastable hydroxyl groups after activation (Muranyi, Schraml, & Wunderlich, 2010; Watanabe et al., 1999) Superhydrophilicity allows the water molecules to wash away decomposed and detached microorganisms from the surface. This initiates a self-cleaning process. Intriguingly, photoactivated titanium dioxide surfaces are hostile to Anthrax spores. Anthrax spores are airborne pathogens with killing potential for the people who inhale them. Although antibiotics can remove the bacterial source, Bacillus anthracis, toxins built up over time in the body can still damage the afflicted person. Hence, methods for disrupting airborne Anthrax spores in air are highly coveted. The TiO_2 based air purifiers such as AiroCide has been reported to reduce nearly all airborne pathogens including anthrax ("Airing Out Anthrax,"). The machine operates continuously, and can repeatedly deliver a new dose against the threat upon recirculation through the system to reduce the chance of death by reducing the quantity of spores.

TiO_2 has been extensively studied for applications in environmental cleanup, from its use in wastewater treatment to its antimicrobial activity. It has been shown that increased photoactivity is achieved using nanoparticles over micro scale particles. Furthermore, TiO_2 is a good example of how common materials doped with nanoparticles can be used to enhance base material properties. For example blending TiO_2 nanoparticles into rubber makes a composite with controllable antimicrobial properties, (Lin, Tian, Lu, Zhang, & Zhang, 2006). Currently, porous silica-coated titanium nano-enhancers (TiO_2-SiO_2) are being embedded in whey protein isolates and corn zein protein films for enhancing antimicrobial activity in food packaging, (Kadam et al., 2014). By being sensitive to light, the antimicrobial behavior can be actively controlled.

Carbon Nanotubes and Graphene

Carbon nanotubes can be either single-walled and multi-walled variants for antimicrobial activity, where the single-walled has been shown to be more toxic, (K. L. Chen & Bothun, 2014). Single-walled carbon nanotubes (SWNT) antimicrobial properties can be enhanced by integration with lysozyme in order to achieve a more biomimetic response.

Molecular design of hybrid organic-inorganic biopolymers and SWNT's is an emerging field, (Nepal, Balasubramanian, Simonian, & Davis, 2008). Chitosan is one such organic compound. In order to improve manufacturability, biocompatibility, and biodegradability of chitosan, "nano-hybrid membranes (NHM) containing multi-walled carbon nanotubes (MWCNTs) have been modified by perfluorooctanesulfonyl fluoride (PFOSF)", (Song, Gao, Cheng, & Xie, 2015). By applying these membranes, the combination of structural benefits from carbon nanotubes with chitosan results in a material that has increased antibacterial properties and is superhydrophobic.

Developments in Antibacterial Disinfection Techniques

Carbon nanotubes (CNTs) can be cytotoxic to both human and bacteria cells, whereas graphene can be used as a biocompatible substrate for L-929, neuroendocrine P12, oligodendroglia, and osteoblast cells. Mass produced, graphene oxide nanosheets provide economical, flexible paper solutions with some antibacterial properties and minimal cytotoxicity to humans, (Hu et al., 2010). In fact, graphene oxide is already used in nanocargo transport of water-insoluble drug delivery to cells.

Quantum Dots

Nanocrystals with quantum-level attributes due to confined the spatial motion of electrons are classified as quantum dots., Similar to that of metallic nanomaterials, the mechanism for antimicrobial activity behind quantum dots is through reactive oxygen species (Lu, Li, Bao, Qiao, & Bao, 2009). It has been reported that CdTe quantum dots impair antioxidative genes and enzymes. In this process, heavy metal ions are suspected to be responsible for the antibacterial mechanism of action. Concentration dependence of quantum dots is observed in the previous studies which suggests scalability. However, biocompatibility remains the biggest issue with quantum dots for antimicrobial materials (Li, Lu, & Li, 2013). Since the field is relatively new, it is not yet recommended for clinical application until further research is done to verify the short and long term effects on human health.

Hybrid Nanomaterials

Silver containing nanofilms that are able to overcome the inherent cytotoxicity to humans have been the subject of many developments. For biomedical implants, hydroxyapatite coatings can be effectively doped with silver and enhance antibacterial property of these implants. Although these implants can effectively kill bacteria, they do not harm osteoblasts and epithelial cell growth. However, negative biocompatibility effects are observed with increased silver concentration, (Bai et al., 2012; Zhao, Chu, Zhang, & Wu, 2009). Silver nanoparticles also suffer from aggregation issues. By integrating and dispersing the silver particles into either a polymer or biopolymer hybrid coating using quaternary ammonium compounds or chitosan, respectively, bactericidal properties can be achieved that pose significantly less cytotoxicity or even promote the growth of human cells, (Guo et al., 2013).

Biodegradable polymer nanocomposites that integrate silver nanoparticles into food packaging are another alternative method with increased biocompatibility. These nanocomposites are able to control the release of the particles through the use of nano-sensors, (Echegoyen & Nerin, 2013). Similarly, depositing silver nanoparticles into a dried alginate hydrogel inhibits antimicrobial activity while allowing for growth of human cells, and can be used at wound zone without impairing wound healing process. (C. Wang et al., 2014). A ZnO nanoparticle doped aloe vera formulation enhances the antimicrobial activity, biocompatibility, and it can be produced in an eco-friendly way ensuring that no trace, toxic chemical residue remains on the surface after nanoparticle synthesis, which is quite common in chemical synthesis methods (Qian, Yao, Russel, Chen, & Wang, 2015). Hydroxyapatite whisker (bone substitute)/nano zinc oxide biomaterial has also been fabricated and shown to enhance biocompatibility over traditional alternatives, providing the benefits of increased biosecurity, durability, stability and antimicrobial activity, even in dark environment without illumination (Yu et al., 2015).

Unique alternatives to metal and metal oxides are being synthesized that have antimicrobial properties without the cytotoxicity to human cells, such as with polyetheretherketone/nano- fluorohydroxyapatite (PEEK/nano-FHA) composite biomaterial for implants, (L. X. Wang et al., 2014). In this case, the standard dental implant material PEEK is doped with nano structured FHA material to provide the antimicrobial properties. This is an example of nanotechnology being applied to existing materials to enhance their desired properties. Similar to silver and copper, iodine has also long been known as an antiseptic. Iodophors are complexes that serve as iodine carriers to control the solubility and release of iodine and thus overcome the skin staining properties of traditional iodine antiseptics. Taking advantage of this knowledge, using nanotechnology iodophor micro-and-nanoparticles of modified starches can be added to a biodegradable polymer compound without reducing the film strength to gain the antiseptic properties, (Danilovas, Navikaite, & Rutkaite, 2014).

Recent developments in green manufacturing technologies focus on combining traditional antimicrobial materials and their nano equivalents with biomaterials to increase biocompatibility and biodegradability, (Baraliya & Joshi, 2014). To this end, unique applications of these new materials are applying nanosilver into non-woven polyethylene fabric (Deng et al., 2015); either silver or modified TiO_2 particles can be woven into cotton fabrics to microbial spread and odor problems (Maryan, Montazer, & Harifi, 2015; L. M. Wang et al., 2014). Similarly, a nanosilver particle coating has been developed on leather, (Lkhagvajav, Koizhaiganova, Yasa, Celik, & Sari, 2015). Lecithin biomaterial is used to load silver nanoparticles in wool fabric resulting in an antibacterial agent that is non-toxic to humans, (Barani et al., 2014). Organic-inorganic nanohybrids (organic metal nanoaggregates) composed from biocompatible, inorganic Ag, Cu, or Au and organic dihydropyrimidones show promising results for antimicrobial drug development (Raj, Billing, Kaur, & Singh, 2015). . In space applications, silver nanoparticles are integrated onto hydrophilic zeolite to assist in the absence of gravity, whereas on Earth hydrophobicity is usually desired to help avoid surface breeding grounds for microbes, (McDonnell, Beving, Wang, Chen, & Yan, 2005). This is mentioned to emphasize the often competing requirements in the design of antimicrobial materials for diverse applications.

Biomimetic Materials with Modern Methods

As swiftly as antibiotics, chemistry and drug development dominated 20th century in medicine, modern era research is returning to nature to search for alternatives in the quest for biomimetic materials. Traditional sterilization methods are being re-explored via modern techniques such as nanotechnology to better understand the underlying mechanisms of action in these traditional methods (Kwakman & Zaat, 2012). Plant extracts such as garlic and clove have been shown to be effective at certain antibacterial applications where antibiotics have failed to succeed. Yet, much work remains to better understand the mechanism of action in these cases, as well as with many other traditional plant antiseptics (Ríos & Recio, 2005). Recent research has looked into the green manufacturing of calcium oxide nano-plates starting from shrimp shells that is effective against both Gram-positive and Gram-negative bacteria, (Gedda, Pandey, Lin, & Wu, 2015).

As stated earlier, a bacteriophage is a human-safe type of virus that only infects specific bacteria and that can be used as an antimicrobial material. This virus takes over the body of the bacteria, replicates itself, kills the bacteria, and releases its copies to go infect other bacteria. Using nano-emulsion techniques, a bacteriophage can now be stabilized and delivered for targeted treatment, (Esteban et al., 2014).

Similarly, multicellular organisms possess unique antimicrobial peptides as part of their natural immune response, (Zasloff, 2002). A peptide is a chain of amino acid monomers. These peptides are differentiated to a species, and hence it seem unlikely that bacteria will grow resistant to them. Researchers are now analyzing how they can design drugs or nanomaterials that could mimic the response of these peptides to take advantage of their superior, targeted antimicrobial properties.

Developments in Antibacterial Disinfection Techniques

Gecko skin has self-cleaning and superhydrophobic antibacterial properties. It possesses scales on which the spinules (hairs) resemble a carbon nanotube forest. The skin is lethal to Gram-negative bacteria, but not to eukaryotic human cells. Valleys within the skin help promote the self-cleaning process. The skin also appears to be selective in its bactericidal properties (Watson et al., 2015).

Research in bio-inspired design utilizing nanotechnology for antimicrobial materials development is in its infancy. However, the antibacterial properties of cellular organisms in nano and macro level are extremely effective, and thus mimicking these properties via bioinspired materials has the potential to transform the field of sterilization and self-cleaning technologies. Understanding biomimetic materials and their mechanisms of action are prospective areas of research moving forward. However the level of complexity of these fields is not yet fully comprehended, (Yan et al., 2014). In fact, it is their complexity that makes them attractive and prospective areas to pursue future research.

CHALLENGES AND RECOMMENDATIONS

From 0-D, 1-D, to 2-D, and a range of materials including metals, metal oxides, polymers and hybrids, there is no shortage of development in nanotechnology. Since material and technology development is application specific, any attempt to research antimicrobial solutions should first identify the type of antibacterial application. The next step is to understand which properties must be achieved for that application, such as efficacy, biocompatibility or biodegradability. Metals and metal oxides nanoparticles often on their own are incapable of providing desired levels of biocompatibility and antimicrobial properties. Rather, they often are doped into a specific polymer or base material such as rubber in order to achieve the properties desired.

Nanofabrication technologies have seen tremendous improvement in the recent years due to their novelty in scientific applications. Nanomaterial fabrication techniques however still requires significant labor and pricey instrumentation to achieve desired results. In addition, the techniques of material synthesis and implementation needs further standardization, although ridden with unique challenges to solve. Any new endeavor to develop a nanoscale antimicrobial device is likely to end up with a similar but differentiated product. Considering the recent developments of nanotechnology and investment on this field, these limitations can be expected to overcome in near future transforming the antimicrobial coating technology to a higher level.

Before making the decision to work with different scales of nanomaterials (0-D, 1-D and 2-D), one should consider a few following points. Nanoparticles and nanofilms are easier to synthesize, whereas the difficulties in fabricating 1-D materials remain more difficult. Enhanced microbe-material contact area and length scalability of these 1-D structures however are superior to other nanomaterials. Therefore, the material should be designed into a nanoscale geometry that will provide the desired variety of properties based on the specific application.

FUTURE RESEARCH DIRECTIONS

A broad slew of antimicrobial materials have been classified by dimension, identified by type, and studied based on the mechanism(s) of action. Application of nanotechnology offers great potential to develop cost effective, highly antimicrobial and biocompatible devices for future applications. Among current research, particular interest is given to self-cleaning technologies and biocompatibility properties which will shape the future direction of this field. Smart and active control of antimicrobial activity for food packaging, self-cleaning technologies, and biodegradable materials are becoming increasingly important to mitigate waste production. Similarly, recyclability of materials and the ability to recover material out of composite will be important for preventing unintended cross contamination into waste facilities where antimicrobial activity is not needed and can even be harmful. Addressing these 'big picture' type dilemmas will be critical to any mass production of antimicrobial products. Eco-friendly synthesis and production techniques for nanomaterials will help ensure minimal unintended by-products or unexpected side-effects from nanotechnology will be transferred to consumers. A better understanding of how nanomaterials interact with humans, as well as of how the mechanisms of action work is essential to the sustainability of antimicrobial research.

Since the targeted drug development based on the bacteria genome process showed little hope in the late 1990's, biopolymers, hybrid materials, and antimicrobial peptides today are shaping a new direction and hope for antimicrobial research that overcome the shortfalls of other materials, including the likelihood that a bacteria resistant strain would emerge, but are more complex. There is no shortage of development for the field, and addressing any one of these issues will help lead to a more sophisticated understanding and standardized approach to how to deal with microbial life in the different facets of human life.

CONCLUSION

As the role of antibiotics in the fight against bacterial infections wanes with ever-increasing antibiotic immunity, nanomaterials will assume a position of increased significance. Since a broad array of virulent bacterial infections no longer respond to antibiotic treatment, the need for rapid development and implementation of antimicrobial surface treatments must be recognized. In this regard, nanotechnology offers unique alternative solutions. Nanomaterials with enhanced antibacterial activity have proven to be effective in preventing microbial spread under various environmental conditions. It is important to note that a complete understanding of the mechanisms of cytotoxicity of nanomaterials against microbial spread is not yet fully understood, can be unique for each material, and hence this is an important area of research. There are multiple mechanisms of action for antimicrobial activity depending on the type of nanomaterial used. The primary materials considered as antibacterial agent are metals, metal oxides, quantum dots, carbon nanotubes, graphene polymers, and hybrid materials. In addition, new biomimetic nanomaterials are under investigation to achieve the high efficiency of antibacterial systems found in nature. It is expected that forthcoming research in nanotechnology will lead to discovery and innovation of novel antimicrobial materials which are highly specific to targeted microorganisms, cost effective and compatible with the human life cycle. These materials will overcome the limitations of the existing antimicrobial agents and make a significant impact to the public health and the fight against microbial spread and infection.

REFERENCES

Agnihotri, S., Mukherji, S., & Mukherji, S. (2013). Immobilized silver nanoparticles enhance contact killing and show highest efficacy: Elucidation of the mechanism of bactericidal action of silver. *Nanoscale*, *5*(16), 7328–7340. doi:10.1039/c3nr00024a PMID:23821237

Airing Out Anthrax. (n.d.). Retrieved from http://ntrs.nasa.gov/archive/nasa/casi.ntrs.nasa.gov/20020080174.pdf

Bai, X., Sandukas, S., Appleford, M., Ong, J. L., & Rabiei, A. (2012). Antibacterial effect and cytotoxicity of Ag-doped functionally graded hydroxyapatite coatings. *Journal of Biomedical Materials Research. Part B, Applied Biomaterials*, *100*(2), 553–561. doi:10.1002/jbm.b.31985 PMID:22121007

Balagna, C., Perero, S., Ferraris, S., Miola, M., Fucale, G., Manfredotti, C., & Ferraris, M. et al. (2012). Antibacterial coating on polymer for space application. *Materials Chemistry and Physics*, *135*(2-3), 714–722. doi:10.1016/j.matchemphys.2012.05.049

Baraliya, J. D., & Joshi, H. H. (2014). Surface-engineered Core-Shell Nano-Size Ferrites and their Antimicrobial activity. *Solid State Physics: Proceedings of the 58th Dae Solid State Physics Symposium 2013, Pts a & B*, *1591*, 429-431. doi:10.1063/1.4872627

Barani, H., Montazer, M., Samadi, N., Toliyat, T., Zadeh, M. K., & de Smeth, B. (2014). Application of Nano Silver/Lecithin on Wool through Various Methods: Antibacterial Properties and Cell Toxicity. *Journal of Engineered Fibers and Fabrics*, *9*(4), 126-134.

Beyth, N., Houri-Haddad, Y., Domb, A., Khan, W., & Hazan, R. (2015). Alternative Antimicrobial Approach: Nano-Antimicrobial Materials. *Evidence-Based Complementary and Alternative Medicine*. doi:10.1155/2015/246012

Charpentier, P. A., Burgess, K., Wang, L., Chowdhury, R. R., Lotus, A. F., & Moula, G. (2012). Nano-TiO 2 /polyurethane composites for antibacterial and self-cleaning coatings. *Nanotechnology*, *23*(42), 425606. doi:10.1088/0957-4484/23/42/425606 PMID:23037881

Chen, K. L., & Bothun, G. D. (2014). Nanoparticles meet cell membranes: Probing nonspecific interactions using model membranes. *Environmental Science & Technology*, *48*(2), 873–880. doi:10.1021/es403864v PMID:24341906

Chen, R., Ni, H., Zhang, H., Yue, G., Zhan, W., & Xiong, P. (2013). A preliminary study on antibacterial mechanisms of silver ions implanted stainless steel. *Vacuum*, *89*, 249–253. doi:10.1016/j.vacuum.2012.05.025

Cioffi, N., Torsi, L., Ditaranto, N., Tantillo, G., Ghibelli, L., Sabbatini, L., & Traversa, E. et al. (2005). Copper nanoparticle/polymer composites with antifungal and bacteriostatic properties. *Chemistry of Materials*, *17*(21), 5255–5262. doi:10.1021/cm0505244

Cruz-Romero, M. C., Murphy, T., Morris, M., Cummins, E., & Kerry, J. P. (2013). Antimicrobial activity of chitosan, organic acids and nano-sized solubilisates for potential use in smart antimicrobially-active packaging for potential food applications. *Food Control*, *34*(2), 393–397. doi:10.1016/j.foodcont.2013.04.042

Danilovas, P. P., Navikaite, V., & Rutkaite, R. (2014). Preparation and Characterization of Potentially Antimicrobial Polymer Films Containing Starch Nano- and Microparticles. *Materials Science-Medziagotyra, 20*(3), 283–288. doi:10.5755/j01.ms.20.3.5426

Deng, X. L., Nikiforov, A., Vujosevic, D., Vuksanovic, V., Mugosa, B., Cvelbar, U., & Leys, C. et al. (2015). Antibacterial activity of nano-silver non-woven fabric prepared by atmospheric pressure plasma deposition. *Materials Letters, 149*, 95–99. doi:10.1016/j.matlet.2015.02.112

Echegoyen, Y., & Nerin, C. (2013). Nanoparticle release from nano-silver antimicrobial food containers. *Food and Chemical Toxicology, 62*, 16–22. doi:10.1016/j.fct.2013.08.014 PMID:23954768

Esteban, P. P., Alves, D. R., Enright, M. C., Bean, J. E., Gaudion, A., Jenkins, A. T. A., & Arnot, T. C. et al. (2014). Enhancement of the Antimicrobial Properties of Bacteriophage-K via Stabilization using Oil-in-Water Nano-Emulsions. *Biotechnology Progress, 30*(4), 932–944. doi:10.1002/btpr.1898 PMID:24616404

Foster, H. A., Ditta, I. B., Varghese, S., & Steele, A. (2011). Photocatalytic disinfection using titanium dioxide: Spectrum and mechanism of antimicrobial activity. *Applied Microbiology and Biotechnology, 90*(6), 1847–1868. doi:10.1007/s00253-011-3213-7 PMID:21523480

Fu, J., Ji, J., Fan, D., & Shen, J. (2006). Construction of antibacterial multilayer films containing nanosilver via layer-by-layer assembly of heparin and chitosan-silver ions complex. *Journal of Biomedical Materials Research. Part A, 79A*(3), 665–674. doi:10.1002/jbm.a.30819 PMID:16832825

Gedda, G., Pandey, S., Lin, Y. C., & Wu, H. F. (2015). Antibacterial effect of calcium oxide nano-plates fabricated from shrimp shells. *Green Chemistry, 17*(6), 3276–3280. doi:10.1039/C5GC00615E

Gunawan, P., Guan, C., Song, X., Zhang, Q., Leong, S. S. J., Tang, C., & Xu, R. et al. (2011). Hollow Fiber Membrane Decorated with Ag/MWNTs: Toward Effective Water Disinfection and Biofouling Control. *ACS Nano, 5*(12), 10033–10040. doi:10.1021/nn2038725 PMID:22077241

Guo, L., Yuan, W., Lu, Z., & Li, C. M. (2013). Polymer/nanosilver composite coatings for antibacterial applications. *Colloids and Surfaces. A, Physicochemical and Engineering Aspects, 439*, 69–83. doi:10.1016/j.colsurfa.2012.12.029

Gwynn, M. N., Portnoy, A., Rittenhouse, S. F., & Payne, D. J. (2010). Challenges of antibacterial discovery revisited. *Annals of the New York Academy of Sciences, 1213*(1), 5–19. doi:10.1111/j.1749-6632.2010.05828.x PMID:21058956

Holt, K. B., & Bard, A. J. (2005). Interaction of Silver(I) Ions with the Respiratory Chain of Escherichia coli: An Electrochemical and Scanning Electrochemical Microscopy Study of the Antimicrobial Mechanism of Micromolar Ag+†. *Biochemistry, 44*(39), 13214–13223. doi:10.1021/bi0508542 PMID:16185089

Hu, W., Peng, C., Luo, W., Lv, M., Li, X., Li, D., & Fan, C. et al. (2010). Graphene-Based Antibacterial Paper. *ACS Nano, 4*(7), 4317–4323. doi:10.1021/nn101097v PMID:20593851

Huppmann, T., Yatsenko, S., Leonhardt, S., Krampe, E., Radovanovic, I., Bastian, M., & Wintermantel, E. (2014). Antimicrobial Polymers - the Antibacterial Effect of Photoactivated Nano Titanium Dioxide Polymer Composites. *Proceedings of Pps-29: The 29th International Conference of the Polymer - Conference Papers, 1593*, 440-443. doi:10.1063/1.4873817

Iarikov, D. D., Kargar, M., Sahari, A., Russel, L., Gause, K. T., Behkam, B., & Ducker, W. A. (2014). Antimicrobial surfaces using covalently bound polyallylamine. *Biomacromolecules, 15*(1), 169–176. doi:10.1021/bm401440h PMID:24328284

Isquith, A. J., Abbott, E. A., & Walters, P. A. (1972). Surface-Bonded Antimicrobial Activity of an Organosilicon Quaternary Ammonium Chloride. *Applied Microbiology, 24*(6), 859-863. Retrieved from http://aem.asm.org/content/24/6/859.full.pdf

Jones, N., Ray, B., Ranjit, K. T., & Manna, A. C. (2008). Antibacterial activity of ZnO nanoparticle suspensions on a broad spectrum of microorganisms. *FEMS Microbiology Letters, 279*(1), 71–76. doi:10.1111/j.1574-6968.2007.01012.x PMID:18081843

Jose Ruben, M., Jose Luis, E., Alejandra, C., Katherine, H., Juan, B. K., Jose Tapia, R., & Miguel Jose, Y. (2005). The bactericidal effect of silver nanoparticles. *Nanotechnology, 16*(10), 2346–2353. doi:10.1088/0957-4484/16/10/059 PMID:20818017

Kadam, D. M., Wang, C., Wang, S., Grewell, D., Lamsal, B., & Yu, C. (2014). Microstructure and Antimicrobial Functionality of Nano-Enhanced Protein-Based Biopolymers. *Transactions of the Asabe, 57*(4), 1141-1150.

Kalishwaralal, K., BarathManiKanth, S., Pandian, S. R. K., Deepak, V., & Gurunathan, S. (2010). Silver nanoparticles impede the biofilm formation by Pseudomonas aeruginosa and Staphylococcus epidermidis. *Colloids and Surfaces. B, Biointerfaces, 79*(2), 340–344. doi:10.1016/j.colsurfb.2010.04.014 PMID:20493674

Karimi, L., Yazdanshenas, M. E., Khajavi, R., Rashidi, A., & Mirjalili, M. (2014). Using graphene/TiO2 nanocomposite as a new route for preparation of electroconductive, self-cleaning, antibacterial and antifungal cotton fabric without toxicity. *Cellulose (London, England), 21*(5), 3813–3827. doi:10.1007/s10570-014-0385-1

Kwakman, P. H. S., & Zaat, S. A. J. (2012). Antibacterial components of honey. *IUBMB Life, 64*(1), 48–55. doi:10.1002/iub.578 PMID:22095907

Li, X. L., Lu, Z. S., & Li, Q. (2013). Multilayered films incorporating CdTe quantum dots with tunable optical properties for antibacterial application. *Thin Solid Films, 548*, 336–342. doi:10.1016/j.tsf.2013.09.088

Lin, G., Tian, M., Lu, Y.-L., Zhang, X.-J., & Zhang, L.-Q. (2006). Morphology, Antimicrobial and Mechanical Properties of Nano-TiO2/Rubber Composites Prepared by Direct Blending. *Polymer Journal, 38*(5), 498–502. doi:10.1295/polymj.38.498

Lkhagvajav, N., Koizhaiganova, M., Yasa, I., Celik, E., & Sari, O. (2015). Characterization and antimicrobial performance of nano silver coatings on leather materials. *Brazilian Journal of Microbiology, 46*(1), 41–48. doi:10.1590/S1517-838220130446 PMID:26221087

Lu, Z., Li, C. M., Bao, H., Qiao, Y., & Bao, Q. (2009). Photophysical mechanism for quantum dots-induced bacterial growth inhibition. *Journal of Nanoscience and Nanotechnology, 9*(5), 3252–3255. doi:10.1166/jnn.2009.022 PMID:19452999

Maryan, A. S., Montazer, M., & Harifi, T. (2015). Synthesis of nano silver on cellulosic denim fabric producing yellow colored garment with antibacterial properties. *Carbohydrate Polymers, 115*, 568–574. doi:10.1016/j.carbpol.2014.08.100 PMID:25439933

McDonnell, A. M. P., Beving, D., Wang, A. J., Chen, W., & Yan, Y. S. (2005). Hydrophilic and antimicrobial zeolite coatings for gravity-independent water separation. *Advanced Functional Materials, 15*(2), 336–340. doi:10.1002/adfm.200400183

McGlynn, W. (n.d.). *Guidelines for the Use of Chlorine Bleach as a Sanitizer in Food Processing Operations*. Retrieved from http://ucfoodsafety.ucdavis.edu/files/26437.pdf

Muranyi, P., Schraml, C., & Wunderlich, J. (2010). Antimicrobial efficiency of titanium dioxide-coated surfaces. *Journal of Applied Microbiology, 108*(6), 1966–1973. doi:10.1111/j.1365-2672.2009.04594.x PMID:19886892

Navabpour, P., Ostovarpour, S., Hampshire, J., Kelly, P., Verran, J., & Cooke, K. (2014). The effect of process parameters on the structure, photocatalytic and self-cleaning properties of TiO2 and Ag-TiO2 coatings deposited using reactive magnetron sputtering. *Thin Solid Films, 571*(Part 1), 75–83. doi:10.1016/j.tsf.2014.10.040

Nepal, D., Balasubramanian, S., Simonian, A. L., & Davis, V. A. (2008). Strong antimicrobial coatings: Single-walled carbon nanotubes armored with biopolymers. *Nano Letters, 8*(7), 1896–1901. doi:10.1021/nl080522t PMID:18507479

Page, K., Wilson, M., & Parkin, I. P. (2009). Antimicrobial surfaces and their potential in reducing the role of the inanimate environment in the incidence of hospital-acquired infections. *Journal of Materials Chemistry, 19*(23), 3819. doi:10.1039/b818698g

Payne, D. J., Gwynn, M. N., Holmes, D. J., & Pompliano, D. L. (2007). Drugs for bad bugs: confronting the challenges of antibacterial discovery. *Nat Rev Drug Discov, 6*(1), 29-40. Retrieved from http://www.nature.com/nrd/journal/v6/n1/pdf/nrd2201.pdf10.1038/nrd2201

Perelshtein, I., Lipovsky, A., Perkas, N., Gedanken, A., Moschini, E., & Mantecca, P. (2015). The influence of the crystalline nature of nano-metal oxides on their antibacterial and toxicity properties. *Nano Research, 8*(2), 695–707. doi:10.1007/s12274-014-0553-5

Pirela, S. V., Miousse, I. R., Lu, X., Castranova, V., Thomas, T., Qian, Y., & Demokritou, P. et al. (2015). Effects of Laser Printer-Emitted Engineered Nanoparticles on Cytotoxicity, Chemokine Expression, Reactive Oxygen Species, DNA Methylation, and DNA Damage: A Comprehensive Analysis in Human Small Airway Epithelial Cells, Macrophages, and Lymphoblasts. *Environmental Health Perspectives*. doi:10.1289/ehp.1409582 PMID:26080392

Qian, Y. G., Yao, J., Russel, M., Chen, K., & Wang, X. Y. (2015). Characterization of green synthesized nano-formulation (ZnO-A. vera) and their antibacterial activity against pathogens. *Environmental Toxicology and Pharmacology, 39*(2), 736–746. doi:10.1016/j.etap.2015.01.015 PMID:25723342

Raj, T., Billing, B. K., Kaur, N., & Singh, N. (2015). Design, synthesis and antimicrobial evaluation of dihydropyrimidone based organic-inorganic nano-hybrids. *RSC Advances*, *5*(58), 46654–46661. doi:10.1039/C5RA08765A

Razeeb, K. M., Podporska-Carroll, J., Jamal, M., Hasan, M., Nolan, M., McCormack, D. E., & Pillai, S. C. et al. (2014). Antimicrobial properties of vertically aligned nano-tubular copper. *Materials Letters*, *128*, 60–63. doi:10.1016/j.matlet.2014.04.130

Ren, G., Hu, D., Cheng, E. W., Vargas-Reus, M. A., Reip, P., & Allaker, R. P. (2009). Characterisation of copper oxide nanoparticles for antimicrobial applications. *International Journal of Antimicrobial Agents*, *33*(6), 587–590. doi:10.1016/j.ijantimicag.2008.12.004 PMID:19195845

Ríos, J. L., & Recio, M. C. (2005). Medicinal plants and antimicrobial activity. *Journal of Ethnopharmacology*, *100*(1-2), 80–84. doi:10.1016/j.jep.2005.04.025 PMID:15964727

Sondi, I., & Salopek-Sondi, B. (2004). Silver nanoparticles as antimicrobial agent: A case study on E. coli as a model for Gram-negative bacteria. *Journal of Colloid and Interface Science*, *275*(1), 177–182. doi:10.1016/j.jcis.2004.02.012 PMID:15158396

Song, K. L., Gao, A. Q., Cheng, X., & Xie, K. L. (2015). Preparation of the superhydrophobic nano-hybrid membrane containing carbon nanotube based on chitosan and its antibacterial activity. *Carbohydrate Polymers*, *130*, 381–387. doi:10.1016/j.carbpol.2015.05.023 PMID:26076639

Vishnupriya, S., Chaudhari, K., Jagannathan, R., & Pradeep, T. (2013). Single-Cell Investigations of Silver Nanoparticle-Bacteria Interactions. *Particle & Particle Systems Characterization*, *30*(12), 1056–1062. doi:10.1002/ppsc.201300165

Visnapuu, M., Joost, U., Juganson, K., Kunnis-Beres, K., Kahru, A., Kisand, V., & Ivask, A. (2013). Dissolution of Silver Nanowires and Nanospheres Dictates Their Toxicity to Escherichia coli. *Biomed Res Int*. doi:10.1155/2013/819252

Wang, C., Huang, X. B., Deng, W. L., Chang, C. L., Hang, R. Q., & Tang, B. (2014). A nano-silver composite based on the ion-exchange response for the intelligent antibacterial applications. *Materials Science & Engineering C-Materials for Biological Applications*, *41*, 134–141. doi:10.1016/j.msec.2014.04.044 PMID:24907746

Wang, L. M., Ding, Y., Shen, Y., Cai, Z. S., Zhang, H. F., & Xu, L. H. (2014). Study on properties of modified nano-TiO_2 and its application on antibacterial finishing of textiles. *Journal of Industrial Textiles*, *44*(3), 351–372. doi:10.1177/1528083713487758

Wang, L. X., He, S., Wu, X. M., Liang, S. S., Mu, Z. L., Wei, J., & Wei, S. C. et al. (2014). Polyetheretherketone/nano-fluorohydroxyapatite composite with antimicrobial activity and osseointegration properties. *Biomaterials*, *35*(25), 6758–6775. doi:10.1016/j.biomaterials.2014.04.085 PMID:24835045

Wang, R.-M., Wang, B.-Y., He, Y.-F., Lv, W.-H., & Wang, J.-F. (2010). Preparation of composited Nano-TiO_2 and its application on antimicrobial and self-cleaning coatings. *Polymers for Advanced Technologies*, *21*(5), 331–336. doi:10.1002/pat.1432

Wang, X. G., Qiao, J. W., Yuan, F. X., Hang, R. Q., Huang, X. B., & Tang, B. (2014). In situ growth of self-organized Cu-containing nano-tubes and nano-pores on Ti-90 (-) (x)Cu(10)Alx (x=0,45) alloys by one-pot anodization and evaluation of their antimicrobial activity and cytotoxicity. *Surface and Coatings Technology, 240*, 167–178. doi:10.1016/j.surfcoat.2013.12.036

Watanabe, T., Fujimoto, R., Sawai, J., Kikuchi, M., Yahata, S., & Satoh, S. (2014). Antibacterial Characteristics of Heated Scallop-Shell Nano-Particles. *Biocontrol Science, 19*(2), 93-97. Retrieved from https://www.jstage.jst.go.jp/article/bio/19/2/19_93/_pdf

Watanabe, T., Nakajima, A., Wang, R., Minabe, M., Koizumi, S., Fujishima, A., & Hashimoto, K. (1999). Photocatalytic activity and photoinduced hydrophilicity of titanium dioxide coated glass. *Thin Solid Films, 351*(1–2), 260–263. doi:10.1016/S0040-6090(99)00205-9

Watson, G. S., Green, D. W., Schwarzkopf, L., Li, X., Cribb, B. W., Myhra, S., & Watson, J. A. (2015). A gecko skin micro/nano structure - A low adhesion, superhydrophobic, anti-wetting, self-cleaning, biocompatible, antibacterial surface. *Acta Biomaterialia, 21*, 109–122. doi:10.1016/j.actbio.2015.03.007 PMID:25772496

Yan, X. F., Jie, Z. Q., Zhao, L. H., Yang, H., Yang, S. P., & Liang, J. (2014). High-efficacy antibacterial polymeric micro/nano particles with N-halamine functional groups. *Chemical Engineering Journal, 254*, 30–38. doi:10.1016/j.cej.2014.05.114

Yu, J., Zhang, W. Y., Li, Y., Wang, G., Yang, L. D., Jin, J. F., & Huang, M. H. et al. (2015). Synthesis, characterization, antimicrobial activity and mechanism of a novel hydroxyapatite whisker/nano zinc oxide biomaterial. *Biomedical Materials, 10*(1). doi. Artn, 015001. doi:10.1088/1748-6041/10/1/015001

Yuan, Z., Li, J., Cui, L., Xu, B., Zhang, H., & Yu, C. P. (2013). Interaction of silver nanoparticles with pure nitrifying bacteria. *Chemosphere, 90*(4), 1404–1411. doi:10.1016/j.chemosphere.2012.08.032 PMID:22985593

Zasloff, M. (2002). Antimicrobial peptides of multicellular organisms. *Nature, 415*(6870), 389-395. Retrieved from http://www.nature.com/nature/journal/v415/n6870/pdf/415389a.pdf

Zhang, X., Niu, H., Yan, J., & Cai, Y. (2011). Immobilizing silver nanoparticles onto the surface of magnetic silica composite to prepare magnetic disinfectant with enhanced stability and antibacterial activity. *Colloids and Surfaces. A, Physicochemical and Engineering Aspects, 375*(1-3), 186–192. doi:10.1016/j.colsurfa.2010.12.009

Zhao, L., Chu, P. K., Zhang, Y., & Wu, Z. (2009). Antibacterial coatings on titanium implants. *Journal of Biomedical Materials Research. Part B, Applied Biomaterials, 91B*(1), 470–480. doi:10.1002/jbm.b.31463 PMID:19637369

KEY TERMS AND DEFINITIONS

Antimicrobial: Effective at reducing or stymieing bacteria, fungi, and other protozoan.
Biofilm: Microbes that fester together on a surface.
Biomaterial: Materials designed to integrate with organisms.

Biopolymer: Materials constructed from organic material from living organisms.
Cytotoxic: Toxicity to cells.
Mechanism of Action: Method by which an antimicrobial material effectively causes cellular death.
Toxicity: Severity of harm induced on exposed organism.

Chapter 10
Assessment of Advanced Biological Solid Waste Treatment Technologies for Sustainability

Duygu Yasar
University of Miami, USA

Nurcin Celik
University of Miami, USA

ABSTRACT

53.8% of annually generated US Municipal Solid Waste was discarded in landfills by 2012. However, landfills fail to provide a sustainable solution to manage the waste. The State of Florida has responded to the need of establishing sustainable SWM systems by setting an ambitious 75% recycling goal to be achieved by 2020. To this end, Advanced Biological Treatment (ABT) and Thermal Treatment (ATT) of municipal solid waste premise a sustainable solution to manage the waste as it drastically reduces the volume of waste discarded in landfills and produces biogas that can be used to generate energy. In this chapter, ABT and ATT technologies are analyzed; and their advantages and disadvantages are examined from a sustainability perspective. A comprehensive top-to-bottom assessment of ABT technologies is provided for Florida using Analytic Hierarchy Process based on the collected subject matter expert rankings.

INTRODUCTION

According to World Energy Council, global energy use will double by 2050 with the growing world population (World Energy Council, 2013). Currently, 80% of the world's energy need is met by burning fossil fuels which is an unsustainable way of obtaining energy due to its resultant greenhouse gas emissions, the most important reason of global warming. Global warming is one of the most serious problems that people encounter in recent years. In order to prevent global warming, there need to be either substantial changes in people's energy consumption patterns or the necessary energy need to be

DOI: 10.4018/978-1-5225-0585-3.ch010

Figure 1. Renewable energy from waste

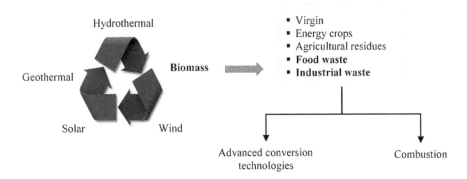

obtained from other renewable sources. Wind power, solar power, geothermal energy, biomass energy, hydroelectric power, and hydrokinetic energy are different types of such renewable energy sources. Biomass energy is obtained from the organic portion of a variety of materials using combustion or advanced conversion technologies (Figure 1).

Renewable energy production from waste brings different benefits on earth other than generating energy. It reduces the amount of waste discarded in landfills and the pollution caused by landfills. Furthermore, waste is a renewable source which is produced everyday by households and industries. Thus, the waste should be managed in a way that people benefit from, instead of being discarded in landfills which brings several environmental problems.

Municipal solid waste (MSW) is defined as the garbage such as food waste, papers, packaging, furniture which is discarded by households and industries after daily use (EPA, 2015). MSW generation continues to increase every year due to growing world population and consumption. Amount of MSW generated globally is approximately 1.3 billion tons per year (World Bank Group, 2014) whereas approximately 254 million tons of the total MSW was generated only by United States during 2013 (EPA, 2015). 52.8% of this waste was discarded in landfills. However, discarding waste in landfills brings a number of environmental problems such as air pollution caused by emitted greenhouse gas emissions and ground water contamination from leachates. Furthermore, it diminishes the available space for humans, animals, and other living species since waste requires large amount of lands to be disposed. For approximately 30 years after closure of landfills, they still remain as a threat as emissions continue to be released. Until that time, they require post maintenance. Such and similar issues raised by the landfill disposal of waste has led to exploration and development of new and sustainable methods for MSW management. Advanced biological treatment (ABT) of MSW is one of these emerging methods for integrated and sustainable waste management. While they have been widely used for treatment of sewage sludge, their application for MSW treatment is relatively new. Another method for sustainable waste treatment is advanced thermal treatment (ATT) technologies which have recently been used for non-organic portion of MSW. Finally, nanotechnology has been recently studied to convert waste into energy and to form nanomaterials from waste. Nanotechnology is defined as the ability to work at the atomic and molecular levels to create structures with larger surfaces and new properties (Roco, 2005). The development of nanomaterials with new physical and chemical characteristics, and the ways to identify and assemble them have brought a breakthrough approach to energy conversion (Tegart, 2009). Nanomaterials can be found in many forms such as nanoparticles, nanorods, nanowires, nanofibers,

Figure 2. Total MSW generation by material in the U.S.
(EPA, 2015)

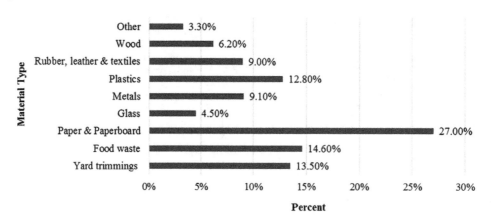

nanoporous, and nanotubes in the size ranging between 1 and 100 nanometer (nm) (Chen et al., 2012). Their major advantage for energy production is the process efficiency increase due to their very large surface area per unit volume which results in a higher surface activity than in the bulk material (Tegart, 2009). From the waste management point of view, nanotechnology have been making its way in waste management area including organic waste management, fly ash management, nuclear waste management, etc. Hydrothermal carbonization, pyrolysis, and high energy ball milling are the examples of processes that have been studied to manage waste by nanotechnology.

MSW is an inhomogeneous waste, which may be divided into three main sub fractions of digestible organic fraction such as kitchen wastes and grass cuttings, slow digestible or indigestible combustible fraction such as wood, paper, cardboard, plastics and other synthetics, and inert fraction such as stones, sand, glass, metals, and bones (Braber, 1995). Figure 2 shows MSW generation by material type in United States during 2013. The largest portion of the MSW continue to be organic waste (food waste and yard trimmings) with a rate of approximately 28.1% (Braber, 1995). Treatment of organic portion of MSW can significantly reduce greenhouse gas emissions and landfill utilization.

Advanced thermal treatment (ATT) technologies can process non-organic portion of the waste. Process steps of ATT technologies are similar, however, temperature and amount of oxygen in the environment differ amongst them. Gasification, plasma arc gasification, and pyrolysis are three main ATT technologies that are discussed in this study. Advanced biological treatment (ABT) technologies can only process biodegradable organic materials such as food and green waste. During ABT of MSW, biodegradable waste decomposed by living microbes. Two types of environment where microbes are able to live are aerobic and anaerobic conditions. While both of them can process the biodegradable materials, some materials are more suitable for only one of the treatment methods. For example, composting process is better suited to green waste with much bulky wood material than the anaerobic process.

Furthermore, each technology has different advantages and disadvantages. For example, anaerobic processes require less energy input than aerobic composting and create much lower amounts of biologically produced heat (DEFRA, 2013). Compost from aerobic composting and biogas from anaerobic digestion are some of the useful outputs obtained from ABT processes. Biogas can be converted into fuel for vehicles and boilers to produce heat (hot water and steam). The selection of the most appropriate technology largely depends on the preferences and strategies of given community for their integrated waste management system. Hence, a comprehensive evaluation and comparison of alternatives based on various criteria is required before their implementation. In the case study presented here, an evaluation of available ABT technologies for waste management in the State of Florida is completed in order to present an assessment that considers environmental, social, and economic aspects. The set of criteria was defined considering the sustainable waste management requirements. Evaluation was performed using Analytical Hierarchy Process (AHP) which is a multi-criteria decision-making tool. Solid waste management divisions of Floridian counties presented their rankings on the defined criteria in terms of their importance. The case study presented in this study brings a state-of-the-art approach to current literature of ABT technologies by presenting not only the most appropriate system for the State of Florida but also ranking the alternative ABT technologies from the best suitable to the least suitable one.

BACKGROUND AND LITERATURE REVIEW

In recent years, managing solid waste has become more arduous as a result of unsatisfactory performance of the conventional waste management methods namely landfilling and combustion. These methods have failed to meet the sustainability requirements due to the various problems that they cause. A sustainable waste management method should be environmentally friendly, economically feasible, and socially acceptable. However, discharging waste in landfills and combustion of waste release greenhouse gas (GHG) emissions to the environment, cause odors and smell while at the same time raising public concern and discomfort.

Disposal of MSW in landfills and combustion are the least favorable waste management methods whereas the most favorable solution is the prevention of the generation of such waste at the first place. However, it is difficult to implement this solution and some amount of waste remains to be managed. Recycling is the second favorable option, but not all materials in MSW can be recycled. Utilizing advanced conversion technologies can be noted as the third favorable option in the sustainable waste management hierarchy after prevention and recycling. It is an emerging method for municipalities which have started to look for sustainable ways to manage the growing waste and reduce environmental problems caused by it. Additionally, this method premises generation of renewable energy generation sources as well as marketable products such as chemicals and compost from waste.

Various research has been conducted to review, evaluate, and compare these technologies from different aspects such as economical, technical, social, and sustainable. Murphy and McKeogh (2004) conducted a study to evaluate and compare incineration, gasification, and anaerobic digestion in terms of their capital costs, gate fees, conversion efficiency, and greenhouse gas emissions. A decision support software program was written for the study to model their considered systems. They concluded that GHG emissions reduce as system's total energy production increases. Incineration has the highest gate fee whereas anaerobic digestion has the least one. They also stated that anaerobic digestion and thermal technologies are not alternatives for each other due to the fact that they can process different portions of the waste.

Yasar et al. (2016) conducted a study to evaluate gasification, pyrolysis, and plasma arc gasification using the analytical hierarchy process (AHP) which is a multi-criteria decision making process. Revenue, tipping fees, capital cost, operation costs, development period, flexibility of process, land requirement of facility, net conversion efficiency, ease of permitting, marketability, environmental impact, public acceptability, and number of facilities in US were taken into consideration as criteria in the evaluation.

The evaluation was conducted for 67 Floridian counties by dividing them into 8 groups. It was concluded that gasification had the highest global priority while pyrolysis had the lowest global priority for all groups of counties.

Williams et al. (2003) carried out a project to investigate waste conversion technologies for California Integrated Waste Management. Gasification, pyrolysis, anaerobic digestion, and aerobic composting were among the investigated technologies. In their work, a comprehensive database was created with the information obtained from identified technology vendors and literature.

A brief for review of ABT options was prepared by DEFRA in 2005 (DEFRA, 2013). In this brief, ABT was divided into two major groups as aerobic and anaerobic processes. Technical issues, markets for their outputs, obtaining necessary permits, and their costs were discussed and case studies were provided in the brief.

Another evaluation of alternative solid waste processing technologies was performed by URS Corporation for City of Los Angeles (URS Corporation, 2005). They divided alternative technologies into three major groups, namely thermal, physical, and biological and chemical processing technologies. Anaerobic and aerobic digestion processes were evaluated under the biological processing technology group. Electricity production, net efficiency, diversion rate, air emissions, solid wastes, regulatory issues, capital cost, revenues, and tipping fees were taken into consideration for ranking of technology suppliers. Pyrolysis and gasification were chosen as the best suitable technologies for processing black bin source separated MSW which is the landfilled portion of MSW. In this work, the ABT technologies were evaluated for all MSW to be landfilled even though in practice they can only process the organic fraction of MSW.

Wilson et al. conducted a study to assess MSW conversion technologies namely incineration, gasification, plasma arc gasification, pyrolysis, anaerobic digestion, and aerobic digestion in terms of their design, operation, waste treatment capability, conversion efficiency, economic performance, and environmental impact (Wilson, Williams, Liss, & Wilson, 2013). They suggested that anaerobic and aerobic digestion can be implemented with thermal conversion technologies for integrated waste management, since they can process only the biodegradable portion of MSW.

California Integrated Waste Management Board conducted a survey with 69 technology providers to gather information about waste conversion technologies namely pyrolysis, gasification, plasma arc gasification, and biochemical processes (Hackett et al., 2004). They elaborated biochemical processes under 3 main categories which are anaerobic digestion, composting, and fermentation. They concluded that biochemical technologies can be used jointly with thermochemical technologies for integrated SWM although they have a limited application due to the fact that they can only be applied to biodegradable fraction of the waste.

In a white paper prepared by ISWA Working Group on Energy Recovery, pyrolysis, plasma arc gasification, and gasification were analyzed in terms of technical aspects, experiences on technology, available information in the literature, and overview of potential risks (ISWA, 2013). They stated that the lack of available information on advanced conversion technologies makes it challenging to compare them with the conventional methods.

This work describes the mechanical treatment, ATT, and ABT technologies for MSW management in detail and presents a case-study that evaluates the ABT technologies for the State of Florida in terms of their capital cost, annual cost, land requirement, efficiency, environmental impact, operation time, and public acceptance. A review of the studies conducted about using nanotechnology for waste treatment is presented in the "Nanotechnology for Waste Treatment" Section.

NANOTECHNOLOGY FOR WASTE TREATMENT

Nanotechnology is one of the promising methods to generate sustainable energy. Its application to waste conversion has become more attractive and been studied in recent years. In this section, applying nanotechnology for waste management is reviewed along with the studies that were conducted. The examples of high energy ball milling, catalytic conversion of waste using pyrolysis, hydrothermal carbonization, and upcycling plastic waste into carbon nanotubes are explained with the help of related studies. Finally, some examples from the nanomaterials that are used for wastewater treatment are presented.

Using nanotechnology to reduce the particle size of fly ash was studied by Paul et al. (2007). Fly ash which is the largest globally produced industrial waste consists of inorganic and incombustible part of the combusted coal (Paul et al., 2007). The size of fly ash can be modified from micro to nano levels using nanotechnology to increase the surface area. Increased surface area changes the fly ash from an inert state to a more reactive one. A high-energy planetary ball mill was utilized in the Paul's study to reduce the particle sizes from as large as 60 micrometer to as small as 148 nanometer (nm). During the high energy ball milling process, a powder mixture is placed in the ball mill and exposed to high energy from the balls (Cao, 2016). It was concluded that the total surface free energy of fly ash has been improved by 300% following a ball milling process for 60 hours whereas the surface area has been enlarged by 102 times (Paul et al., 2007). The study suggested that the nanostructured fly ash obtained from the process can be used to strengthen filler in polymer matrices.

Alves et al. (2011) analyzed the catalytic conversion of wastes discarded by bioethanol industry into carbon nanomaterials. Biomass residues were processed using pyrolysis which involves a molecular breakdown of large molecules. Formed gases were used to grow carbon nanomaterials. Carbon nanomaterials have exceptional mechanical, thermal, and electrical properties (Alves et al., 2011). Furthermore, they are used in many areas including energy management, electronics, structural materials and chemical processing. The major advantage of using waste from bioethanol industry to produce carbon nanomaterials is that it reduces the raw material cost.

Another advanced method to convert waste streams into carbon nanomaterials is hydrothermal carbonization (HTC). Berge et al. (2011) defined HTC as a wet, relatively low temperature (180-350 °C) process to transform carbohydrates into a carbonaceous residue named as hydrochar. Lu et al. (2012) conducted experiments in order to examine the hydrothermal carbonization (HTC) process of solid waste streams. In Lu et al. (2012)'s study shredded office paper, crushed rabbit food, and mixed MSW (shredded office paper, shredded plastic bottles, crushed glass bottles, crushed rabbit food, and shredded aluminum cans) were mixed to represent typical MSW. Results of this study supported the fact that carbonizing paper, food, and mixed MSW results in reduced CO_2 emissions than if the materials were landfilled. The reason is that a huge part of the carbon (45-75%) initially presented to process remains attached to hydrochar (Lu et al., 2012).

Production of carbon based nanomaterials are resource and energy intensive processes and use of plastic waste to produce them can bring a cost effective and environmentally friendly solution to this problem. Zhuo and Levendis (2014) reviewed the prior work on upcycling plastic waste (the major constituent is carbon) into carbon nanotubes (CNTs), fullerenic materials. The variable nature of MSW was stated as one of the major challenges in converting plastic waste into CNTs since the variability in feedstock reduces the quality of CNTs. Another major challenge explained in the study is the complexity of the process. The process should be investigated in detail to keep the quality of CNTs under control.

Using nanotechnology for waste management is not only limited to converting waste into carbon based nanomaterials. Using nanomaterials for wastewater treatment has also garnered much attention recently. One of them is iron oxide nanomaterials with low toxicity, chemical inertness, and biocompatibility (Xu et al., 2012). Iron oxide nanomaterials can break down the contaminants in wastewater to less toxic constituents and also be used for heavy metal contaminant adsorption (Xu et al., 2012). Antimicrobial nanomaterials such as chitosan can also be used for wastewater treatment (Li et al., 2008).

To this end, nanotechnology brings an avant-garde perspective to waste management. The property, size, and shape of waste are modified to form more valuable materials that can be used as raw materials in the industry. Furthermore, materials with higher energy and larger and more reactive surfaces are obtained. The complexity of the processes and the lack of information related to this field pose some challenges to use nanotechnology for waste management. However, it is also a unique opportunity to deal with the waste problem in an advanced and innovative way.

SUSTAINABLE WASTE MANAGEMENT

Overview

As mentioned earlier, significant amount of solid waste generated every year should be managed properly in order to prevent pollution caused by greenhouse gas emissions, protect human health, and reduce landfill utilization. In recent years, integrated solid waste management has gained importance to manage MSW collection, transportation, treatment, recovery, and disposal activities in a sustainable way and in a way that meets the local communities' needs. Integrated waste management includes encouraging reduction in consumption, reuse of discarded products, providing an effective collection and transportation of waste from different sources, and treatment and disposal of waste based on regulations, laws, needs of stakeholders, and existing technology.

The most favorable and sustainable option is to prevent the waste generation which can be accomplished by public awareness and support of manufacturers. Preventing waste generation remove the need for the rest of the chain for waste management. Waste collection, transportation, treatment, and disposal activities can be eliminated by preventing the waste generation (London Borough of Lambeth, 2010). Some of the waste prevention methods can be listed as: using electronic mails for communication, reusing supermarket bags, and using food waste for animal feeding.

Second option in the waste management hierarchy is reusing the materials discarded without any processing. Product life can be extended by reusing electronic parts. Recycling is the most environmentally friendly option for waste conversion. However, only certain material types such as paper, glass, and plastics can be recycled. The rest of the materials should be treated in a sustainable way to reduce landfill disposal. For organic portion of waste, composting is the most commonly used method. Backyard composting is one of the recent applications encouraged to increase the waste diversion. This application also reduces collection and transportation costs since waste is processed at the point it is created.

There are other emerging technologies for waste treatment. Figure 3 shows the categorization of MSW treatment methods. In order to increase landfill diversion of MSW, advanced thermal, biological, and mechanical conversion technologies have been utilized recently.

Advanced conversion technologies generate energy and marketable outputs out of waste. Waste should be sorted and its size should be reduced prior to actual process. They reduce greenhouse gas emissions

Figure 3. MSW management methods

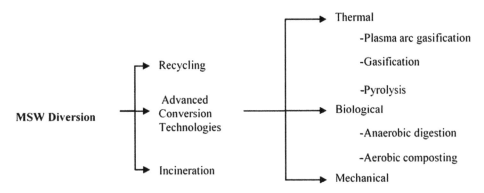

by reducing the amount of waste discarded in landfills. Advanced thermal treatment (ATT) technologies are more suitable for plastic waste and utilize heat to process the waste. ABT technologies are more suitable for organic waste such as food and uses microorganisms for conversion of waste. Obtaining permit for ATT technologies is more challenging than obtaining for ABT technologies.

In order to recycle or convert waste into energy and products by ATT and ABT, they need to be sorted at the point they are thrown away or after they are collected. In single stream recycling (SSR), all recyclables are collected in a single bin and they are separated in material recovery facilities. SSR makes waste collection easier and less costly. It increases the amount of recyclables collected due to its convenience for customers. However, operating costs are higher and recyclables are contaminated by other types of waste. For example paper and plastics are contaminated by broken glass. Another option for recyclable collection is dual stream recycling (DSR). In DSR, paper is collected in a separate bin while other recyclables are collected in a single bin. This system increases the collection costs. However, it prevents paper contamination. Communities need to decide which collection, transportation, and treatment method they will use based on their strategies.

Waste Management Worldwide

Many governments around the world have established ambitious recycling goals and MSW management strategies in order to increase their recycling rates and manage their MSW in a sustainable way. Regulations have been increased throughout the world to develop sustainable waste management practices.

MSW generation in the United States has steadily increased over the past 15 years because of a larger population, urbanization, industry, and technological impacts (Bastani and Celik, 2015). United States Environmental Protection Agency (EPA) has established several goals in order to exceed federal waste diversion requirements which are as follows (EPA, 2012):

- Achieve a 50% waste diversion rate by fiscal year (FY) 2015.
- Achieve a 50% construction and demolition waste diversion rate by FY 2015.
- Reduce paper use.
- Increase diversion of organic material, such as food scraps and yard trimmings.

Reducing waste generation is their top priority. It is followed by managing to divert 55% of the non-hazardous waste by Fiscal Year 2015, diverting at least 50% of the construction and demolition materials and debris, reducing printing paper use, and increasing the percentage of compostable and organic materials diversion.

California which is the most populous state of the U.S. has established a 75% recycling rate goal to be achieved by 2020. They defined main focus areas which are important to achieve 75% recycling goal: improving recycling structure, reducing organic material disposal in landfills, increasing commercial recycling, improving producing responsibility, reforming beverage container program which currently requires glass container manufacturers to use a certain percentage of post filled glass, increasing the demand for post-consumer recycled materials, exploring new models for funding waste management projects in the State, focusing source reduction and encouraging backyard and vermicomposting applications (Cal-Recycle, 2012).

European Commission requires its member states to recycle at least 50% of its MSW and 70% of construction waste by 2020 (European Commission, 2010). European Environment Agency prepared a report to investigate and analyze recycling rates of member countries from 2001 to 2010 (EEA, 2013). Austria, Belgium, Germany, the Netherlands, and Switzerland are among the countries which exceed the 50% recycling rate. Ireland, Italy, Luxemburg, Slovenia, Sweden, and United Kingdom can achieve the targeted recycling rate by 2020 if they can keep their recycling rate increase during 2001-2009 period. However, the rest of the member countries need to accelerate their recycling rate increases (EEA, 2013).

The first country in the world which take up the challenge of Zero Waste in landfills as national policy is New Zealand (Envision New Zealand, 2003). The first council that adopted the policy was Opotiki District Council. The Council managed an 85% reduction in the amount of waste discarded in landfills in 5 years (Envision New Zealand, 2003). There are communities which adopt Zero Waste policy in the following countries: England, Canada, Philippines, and Australia. The first and most important step for achieving the goal is to provide public involvement whereas the last step is to treat the waste in a sustainable way.

On the other hand, there are countries which have insufficient solid waste management practices and high landfill disposal rates. Russia is one of these countries. 95% of the generated MSW is disposed in landfills. Furthermore, 30% of the landfills have practices which do not comply with regulations (IFC, 2012).

Governments encourage waste generation prevention and reuse in the first place. This can be done by public outreach and obtaining manufacturer's support. However, significant amounts of waste still remain to be managed in most communities. To this end, ATT and ABT technologies can be utilized for the waste which cannot be recycled or prevented from generation.

Mechanical Treatment Technologies

MSW has a heterogeneous structure. Before ATT and ABT technologies are utilized, they need to be sorted by material type and shredded. Mechanical treatment technologies are used for this purpose. Mechanically sorted and treated waste stream needs less operation time afterwards and has less damage on ABT and ATT equipment. Their processing becomes simpler after mechanical treatment, since their surface area is increased by mechanical treatment. Pretreatment increases the efficiency of post treatment processes. Firstly, bulky materials such as carpets are removed from the waste stream. Then they are sorted as recyclables, non-recyclables, inert, and biodegradables. Air classification technology is used for separation of light and heavy materials such as paper and plastic from others. Magnetic separation is used for separation of ferrous metals. After different materials are separated, size of waste stream is reduced to have more homogenous input for ATT and ABT.

One of the processes used for waste size reduction is hammer mills. Their shaft design can be vertical or horizontal. Brittle materials are more suitable for this process. Stringy materials can wrap around the equipment (Fitzgerald, 2009). Shredders, rotating drums, ball mills, wet rotating drums, and bag splitters are other methods that can be used for waste size reduction and separation (DEFRA, 2013).

Advanced Thermal Treatment Technologies

Due to several problems that conventional thermal waste treatment methods cause, advanced thermal treatment of waste started to be implemented in recent years. Plasma arc gasification, gasification, and pyrolysis are three main types of thermal treatment technologies which have been commercially used around the world. The main difference between conventional thermal (combustion) and ATT technologies is that the process parameters can be controlled precisely in ATT technologies. Another difference is the scale of the facility. ATT technologies can process at smaller scales than the conventional ones which makes them suitable to be utilized for small municipalities. They have lower carbon footprint than combustion. They produces syngas at high temperatures. Syngas can be used for energy generation after further treatment. The waste should also be treated before the process in order to reduce their sizes and sort the unsuitable materials out for ATT. Pretreatment and post treatment requirements increase the operating costs. Advantages and disadvantages of ATT technologies are summarized in Table 1.

Brief information about the facilities is provided as the following:

- A plasma gasification facility is operated by Eco-valley in Japan. The facility is operated since 2003. Approximately 220 tons per day of MSW is processed in the facility. If MSW is mixed with auto shredder residue, 165 tons per day of the mixture can be processed (Willis et al., 2010).

Table 1. Advantages and disadvantages of ATT technologies

	Advantages	Disadvantages
Plasma Arc Gasification	• Different types of waste can be processed together. • Hazardous waste can be processed. • Electricity can be produced from syngas.	• It requires high electricity input. • It requires high capital cost. • Comparison with conventional methods is hard due to the lack of available information.
Gasification	• Their modular design allows variation in the amount of input. • It can be used in small scales. • Lower oxygen environment prevents waste oxidation.	• It requires high capital cost. • Tars, heavy metals, halogens and alkaline compounds are released which damage the equipment (Zafar, 2009).
Pyrolysis	• The system has a simple structure. • It can be used spacecraft applications (Serio et al., 2001). • It has a flexible design which can adapt to changes in feedstock composition (Serio et al., 2000).	• Commercial scale use of technology is limited. • Product gas needs further treatment due to the high CO_2 concentrations (Serio et al., 2001).

I Plasma Arc Gasification

Plasma arc gasification process takes place at high temperatures (higher than 3000°F). High temperatures reveal a significant amount of energy from waste. Different kind of wastes can be processed together. Plasma arc torches are used for heating the waste. Heated waste molecules are broken down into syngas and slag. Contaminants in syngas should be removed before it is used for power generation.

II Gasification

Solid waste is partially degraded in the presence of oxygen at temperatures above 1300°F. Oxygen concentrations are far lower than the necessary oxygen amount for combustion. Ash and syngas are the outputs of the process. Syngas can be used for a wide variety of applications such as power generation or methanol production. Before syngas is used for power generation, it needs to be cleaned for emission purification. Tars, heavy metals, alkaline are released during the process and can damage the equipment which increases the maintenance costs. Two main types of reactors used in gasification process are fixed beds and fluidized beds (Klein, 2002). Fixed bed design is more suitable for small and medium scale projects. Fluidized bed design is more suitable for processing MSW.

One of the most important parameters of gasification process is the reactor temperature. The temperature affects the gas composition and gas yield as well as the conversion efficiency. He et al. (2009) conducted a study to observe the effect of the reactor temperature on the gas yield, gas composition, steam decomposition, low heating value (LHV), cold gas efficiency and carbon conversion efficiency at 700°C, 750°C, 800°C, 850°C, and 900°C. The experiment results for product distribution, carbon conversion efficiency, syngas content, and product gas yield are shown in Table 2.

In the material balance calculations the total amount exceed 100% as can be seen in Table 2. The reason for that is the steam introduced to gasification process (He et al., 2009). As the reactor temperature increases from 700 to 900°C, the char and the tar distributions decreased, while gas increased remarkably from 94.52% to as high as 145.23%. Furthermore, higher temperature increased the conversion of waste into syngas.

Table 2. Effect of reactor temperature on gasification process

Temperature (°C)	700	750	800	850	900
Product distribution (wt %)					
Gas	94.52	108.25	118.05	132.40	145.23
Tar	12.94	8.51	5.11	3.57	2.62
Char	19.15	15.14	13.41	12.08	12.65
Syngas	37.25	42.84	51.77	60.83	64.35
Carbon conversion efficiency (wt %)	67.87	69.09	70.22	80.68	86.54
Product gas yield (N m^3/kg)	1.22	1.31	1.47	1.81	2.04

(He et al., 2009)

III Pyrolysis

Solid waste is processed in the absence of oxygen. It is an endothermic process which is heated by an energy source from outside. Process does not require any chemical substance for reactions. However, moisture content of waste should be lower than 10% which makes pyrolysis infeasible for large scale conversion. A solid residue and syngas are obtained from the conversion of waste.

Pyrolysis temperature, particle size of waste, feed rate, and flow rate are some of the important parameters that affects the pyrolysis process. Heo et al. (2010) conducted a study to observe the effect of changing a set of process parameters on fast pyrolysis of waste furniture sawdust in a fluidized-bed reactor to produce bio-oil. The moisture content of sawdust was lower than 1% in the study. The results of the study (Heo et al., 2010) showed that:

- The temperature was increased from 400°C to 550°C. As the temperature was increased, bio-oil yields were maximized to 58.1% at 450°C. Higher temperatures than 450°C decreased the yields.
- Extremely small or large feed size decreased the bio-oil yields. Increasing the feed size to 1.3 millimeter (mm) at 450 °C decreased the yield by approximately 5%. Furthermore, decreasing the feed size to 0.3 mm reduced the yield by 3%. The optimal particle size was found to be 0.7 mm.
- Higher flow and feeding rates were more effective for the production of bio-oil due to reduce residence time of vapor in the reactor; however, they did not create a significant variation in the bio-solid yields.

Advanced Biological Treatment Technologies

Biodegradable materials such as food and green waste can be used as feedstock of ABT processes. ABT offers the opportunity to convert the wet fraction of MSW which is unsuitable for thermal treatment. Organic rich fraction of municipal solid waste (MSW) should be separated from mixed waste before they are used as feedstock for ABT processes. ABT technologies are divided into two major groups as aerobic and anaerobic processes. Windrow composting, aerated static pile composting, and in-vessel composting are evaluated under the category of aerobic processes in this study. Before MSW is processed using ABT technologies, source separation of MSW should be performed to sort out the biodegradable materials. It is followed by particle size reduction stage which provides obtaining a better quality output. The quality of output can also be improved by increasing curing time. Pretreatment requirement is a disadvantage of ABT technologies since it increases operating costs. Another disadvantage is that there is not adequate available information about the implementation of ABT technologies since they are not established technologies. However, they offer a sustainable solution for management of waste. Further advantages and disadvantages of each ABT technology are summarized in Table 3.

IV Aerobic Composting Process

Aerobic process of MSW is performed by aerobic microbes in an oxygen rich environment. Bacteria, actinomycetes and fungi are three microorganisms that are mainly active in different stages of the process. Firstly, bacteria and fungi increases the process temperature to approximately 70°C. At this temperature bacteria and fungi becomes inactive whereas actinomycetes becomes active. Due to the fact that composting process is mainly carried out by heat sensitive microorganisms, temperature should be

Table 3. Advantages and disadvantages of ABT technologies

	Advantages	Disadvantages
Windrow Composting	• Uniform compost is produced. • It requires lower capital cost than other ABT technologies. • It requires low energy input than other ABT processes. • It can process large quantities of waste and produce large quantities of compost.	• Large land area is required. • Process parameters can be affected easily by weather changes. • Process emits odor, dust, and other airborne particles. • Their sizes and capacities are limited by the size of turner.
Aerated Static Pile Composting	• The system forestall the creation of anaerobic conditions. • It does not require any pile rotation which reduce the need for manpower.	• Large land area is required. • It takes a long time to produce compost.
In-vessel Composting	• Process parameters and environment can be monitored and controlled precisely. • Weather changes do not affect the process parameters. • Less odors are released during the process than windrow and aerated static pile composting. • Less land area is required for the facility than windrow and aerated pile composting. • High quality compost is produced • System is highly automated and does not require any manual turning.	• It requires high capital cost and high operating cost. • System requires qualified workers such as engineers. • System is more complex than windrow and static pile composting. • It is not flexible to handle different types of bio-solids. • Their capacities are limited by the size of vessel.
Anaerobic Digestion	• Renewable energy can be generated by capturing the methane released. • System can be installed for any scale. • Less greenhouse gas emissions are released because the released methane is captured.	• It requires high capital costs due to expensive equipment • It requires high operating costs. • Big vessels are required for large scales • Different equipment and system need to be established for reuse of biogas produced.

kept under stringent control during the process. Microbes uses biodegradable material as their inputs and convert this material into carbon dioxide, water, and heat. The residual material is called as compost and contains non-biodegradable materials and microbes. The produced compost can be used in a wide variety of applications. It can be used as solid amendment, fertilizer, and mulch.

Carbon and nitrogen content, pH level, process temperature, moisture content, feedstock type and size, aeration, and process time are the most important factors for the quality of obtained compost. Particle size of the material is minimized to obtain a large surface area and the structure of the biodegradable material should be porous so that oxygen can penetrate into it. High solid content materials are more suitable for aerobic process (moisture content around 60%) (DEFRA, 2007).

Hamoda et al. (1998) conducted a series of experiments to evaluate the effects of temperature, moisture content, waste particle size, and the ratio of carbon amount to nitrogen (C/N) in composting process. In order to determine the optimum temperature and moisture contents, 20°C, 40°C, and 60°C process temperatures and 45%, 60%, and 75% moisture contents were tested. Furthermore, five different particle sizes of waste (5 mm, 10 mm, 20 mm, 30 mm, and 40 mm) were evaluated to determine the most effective particle size. Finally, C/N ratio (15, 20, and 30) was set as the variable parameter to determine the best initial C/N ratio. The results of the experiments demonstrated that (Hamoda et al., 1998):

- Optimum starting temperature is 40°C.

- Optimum moisture content is 60%. Higher moisture levels than suggested might reduce the oxygen dispersion over the compost whereas the lower levels of moisture might not be enough for organic matter to be solubilized.
- The study suggested 40 mm as the best particle size which resulted in a higher rate of organic matter decomposition than other sizes. It was stated that the higher space in larger particle sizes allows the oxygen transformation in the organic matter which yields in a faster degradation.
- The most favorable C/N of raw waste is 30.

Three main types of aerobic composting processes, namely, windrow composting, static pile composting, and in vessel composting, are discussed in this section in detail in terms of the aforementioned parameters as well as their process descriptions. Even though each composting facility is designed unique for the communities' demands and needs, given characteristics are common for all of them. Figure 4 shows the main inputs and outputs of composting system.

Brief information about the facilities is provided as the following:

An in-vessel composting facility is operated by Z-Best Company in City of Gilroy, California. The facility receives food waste up to 600 tons per day since 2001 and diversion rate of in-vessel composting is 78%. Non compostable materials are removed prior to composting in an enclosed building. Compostable items are turned and cured during a four months retention time. The produced compost pass through a screening process before it is sold.

A turned windrow composting facility is operated by Nature's Green-Releaf in Franklinton, North California. The facility uses the turned windrow composting process. The materials that arrive to facility first are sorted by type followed by a particle size reduction process. Carbon to Nitrogen of 30:1, moisture content of 50%, and an appropriate pile structure that allows 5% oxygen content are the necessary process environment for composting process. Facility sells the output compost to horticultural market.

IV.A Windrow Composting

One of the most prevalently used methods for large scale composting is windrow composting. The main component of windrow composting is rows of long turning piles named windrows. Organic waste is fed into piles as they turn periodically. Turning the windrows more frequently reduces the odor problems and creates aeration to maintain the oxygen rate inside the windrows. Turning process is followed by a screening process of end material to take away the impurities. The width and height of the windrows are

Figure 4. Composting system inputs and outputs

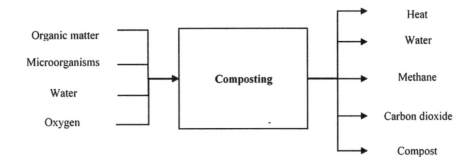

determined based on the aeration requirements, available land, structure and the size of feedstock. There should be enough space between the windrows to let the equipment rotation. Due to this requirement and the nature of the equipment, the process requires large land area to be established. The height of the pile should be suitable for producing enough heat and maintaining the required oxygen flow. Even though windrow composting systems can be either open or closed systems, the size of the equipment allows mostly the use of open systems. Use of food as feedstock can cause environmental problems such as unpleasant odors. Periodic rotation of windrows results in a high operating cost and fast wear of the equipment. Further disadvantages and advantages of the system can be seen in Table 3.

IV.B Aerated Static Pile Composting

Aerated static pile system forestall the creation of anaerobic conditions by an air pipe which provides oxygen. They are similar to windrow systems however, they do not require to be turned. Hence, their size is not limited by the design capacity of turner and can be larger than the windrows. Organic waste is fed into a large pile in aerated static process whereas in windrow process it is fed into rows of piles. Blowers which are attached to pipes remove odors from air. This process is suitable for large volumes of waste. Covering the pile with finished compost can reduce the hazardous odors and keep high temperatures all over the pile. Because there is no turning, temperature should be monitored carefully to maintain uniform distribution of heat over the pile.

IV.C In Vessel Aerobic Composting

In vessel composting systems are closely monitored aeration systems (British Columbia Ministry of Agriculture, 1996) which produces compost in an enclosed container such as reactor. Temperature, moisture, and other process parameters can be closely monitored during the process. The most costly composting systems are in vessel systems but they demand less amount of land and process time. Their cost makes them infeasible for large scale. In vessel composting systems require no manual turning which results in a less labor cost. Odor control is easier and emission is lower than open systems since the emitting surface is decreased to the lowest.

V Anaerobic Digestion Process

Anaerobic digestion (AD) of MSW involves conversion of biodegradable waste into water and biogas by microbes in the absence of oxygen. The process should be carried out in an indoor vessel to keep the process environment out of oxygen. Four types of microbes perform the process: hydrolytic, fermentative, acetogenic, and methanogenic.

Anaerobic digestion is performed in three main steps: hydrolysis, acetogenesis, and methanogenesis. Fermentative bacteria hydrolyzes the insoluble portion of organic matter to soluble molecules such as sugar, alcohol, amino acids in the first step. Second step is carried out by acetogenic bacteria which converts the outputs of first step into carbon dioxide, hydrogen and simple organic acids (Verma, 2002). Produced biogas which consists of methane and carbon dioxide can be converted into heat and electricity, or its quality can be improved to be distributed to grid system. In the last step, methane formers produce methane. AD process scheme can be seen in Figure 5.

C/N is an important process parameter in anaerobic digestion. The most favorable C/N ratio for process is between 20 and 30. A high ratio shows that nitrogen is rapidly consumed by microbes, which

Figure 5. Anaerobic digestion process scheme

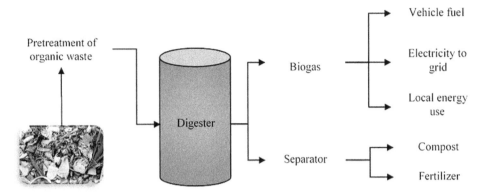

Table 4. Effect of temperature and HRT on anaerobic digestion of food waste (Kim et al., 2006)

Digestion Condition		Gas Production (l/d)		Methane Content in Biogas (%)	Methane Gas Yield (l CH_4/kg)
Temperature (°C)	HRT (days)	Biogas	Methane		
40	10	7.3	4.5	61.6	145
	12	6.1	4.0	65.6	154
45	10	8.7	5.5	63.2	177
	12	7.4	4.9	66.2	187
50	10	10.4	6.7	64.4	216
	12	8.6	5.8	67.4	223
55	10	6.8	3.7	54.4	119
	12	5.6	3.3	58.9	129

results in a low gas generation. Then again, a low C/N ratio increases the pH value which is unfavorable for methanogenic bacteria.

The retention time and the temperature of the process are other important factors that affect biogas production as well as the methane yield. Kim et al. (2006) conducted an experimental study to observe the effect of aforementioned parameters on the anaerobic digestion of the food waste. The temperature of the reactor was increased from 30°C to 55°C at intervals of 5°C. Gas production, methane content in biogas, and methane gas yield were monitored at each given temperature for 10 and 12 days of hydraulic retention times (HRT). The highest biogas production was achieved at 50°C when the HRT is 10 days. A process temperature of 50°C and a HRT of 12 days achieved the maximum methane yield. The study concluded that a thermophilic anaerobic digestion of food waste with a 10 days HRT is favorable. The results of the study are shown in Table 4.

There are three ways to classify anaerobic processes: based on the number of process stages, based on operation temperature, and based on the solid content. Table 5 presents an overview for categorization of anaerobic digestion processes.

The process can take place at mesophilic (20-40°C) or thermophilic (50-65°C) operating temperatures (Baere, 2006). The optimal conditions can be obtained by a constant temperature around 30-35°C

Table 5. Categorization of AD process (Griffin, 2012)

Moisture Content	Operation Temperature	Process Stages
Wet/Dry	Mesophilic	Single
		Multiple
	Thermophilic	Single
		Multiple

and a neutral pH. Second way of classifying anaerobic digestion is based on the number of reactors used. Anaerobic digestion systems might have single or multiple reactors. Hydrolysis, acetogenesis, and methanogenesis phases take place in a single vessel in single stage systems, whereas hydrolysis occurs in a separate vessel in multistage systems. This feature of single stage systems reduces the cost and complexity of process, however, limits the loading rate. On the other hand, multistage systems offers the opportunity to optimize the process conditions for each stage but comes with the expense of high capital costs. Another way to classify the anaerobic digestion process is performed based on the total solid content. Solid content of low solids systems is less than 10%, medium solid systems ranges between 15%-20%, and high solid systems ranges from 22% to 40% (Verma, 2002).

Another categorization method for anaerobic digestion processes is made based on the way of loading the feedstock. If the feedstock is constantly loaded to digesters, the process is named as continuous, if it is loaded only at the beginning of the process, it is named as a batch system.

Captured biogas from anaerobic digestion can be utilized for vehicle fuel or electricity generation after post-processing. However, anaerobic digestion cannot be utilized itself only for obtaining revenue from energy generation.

Brief information about the facilities is provided as the following:

An anaerobic digestion facility is operated by Zero Waste Energy Development Company in San Jose, California. It is the first large-scale commercial anaerobic digestion facility in United States. It can process 90,000 tons of organic waste per year and converts it into biogas. The facility has 16 high solids dry fermentation digesters and four in-vessel composting tunnels.

Case Study

ABT technologies that are examined in this study may help the State of Florida achieve its 75% recycling goal to be achieved by 2020 while managing solid waste in a most sustainable way against the ever increasing amount of waste discarded by households and industries. In this section, the ABT technologies discussed in the earlier section are evaluated for the State of Florida and the most suitable technology is selected based on the collected criteria. A commonly used multi-criteria decision making method, namely Analytical Hierarchy Process (AHP), is used for evaluation. The explained process can be used by other states or countries for evaluation of ABT and ATT technologies or conventional methods for waste treatment.

Assessment of Advanced Biological Solid Waste Treatment Technologies for Sustainability

Methodology and Data Collection

AHP is a multi-criteria decision making method which was first introduced by Saaty (Saaty, 1980). It consists of a hierarchical structure of goal, a set of criteria, and alternatives. The method depends on making pairwise comparisons between the elements at each level. Pairwise comparisons can be performed using verbal or numerical judgements. Verbal judgments namely *equal, moderately more, strongly more, very strongly more, extremely more* are converted into values in a scale from 1 to 9. Alternatives are compared at the lowest level with respect to each criterion and this provides the local priorities of alternatives. It is followed by pairwise comparison of criteria which provides the criteria weights. Criteria weights are derived from subject matter expert (SME) opinions. SMEs rank the criteria set based on their importance and the rankings are converted into pairwise comparison matrices using the same 1-9 scale. The best alternative is determined by the combination of criteria weights and the local priorities of alternatives.

In this case study, rankings of the set of criteria are obtained from 37 out of 67 Floridian counties via surveys. Subject matter experts from solid waste management divisions of each county ranked the criteria based on their importance. The set of criteria consists of capital cost, annual cost, land requirement, operation time, public acceptability, waste reduction, and emitted odors. Here, the *capital cost* includes the costs occurred by planning, designing, and establishing the facility (i.e., purchase of equipment or land for facility, construction of facility, etc.). *Annual operating cost* is the totality of yearly costs for operating the facility, including the utility expenses, employee salaries etc. *Land requirement* is the size of the land where the facility is built on. Larger land requirement pose further disadvantages with associated increase in cost and effort to locate in targeted areas. *Operation time* is the time required to produce useful outputs such as compost, and may vary based on different variables. For instance, compost production with static pile composting takes longer than that with windrow composting. *Public acceptability* evaluates the people's approach for the new establishments of ABT facilities for waste conversion. *Waste reduction efficiency* indicates the percentage of input waste that can be recycled by the ABT technology. High conversion rates reduce the amount of waste discarded in landfills. *Emitted odors* evaluate the released odors during the processes and environmental issues related to using ABT process for conversion of waste.

AHP model structure is depicted in Figure 6. The proposed model consists of four alternative ABT technologies which are evaluated with respect to seven criteria. Data for the aforementioned parameters of each ABT technology are collected from publicly available sources. All collected data are entered to Expert Choice Decision Support System after they are converted into pairwise comparison matrices using 1-9 scale.

During the evaluation process, 37 rankings are obtained from SMEs and converted into 37 different pairwise comparison matrices. Aggregation of individual pairwise comparison matrices into a single matrix is done by calculating the geometric mean of 37 matrices. The matrix obtained from the geometric mean operation is called the consolidated matrix (see Table 6). Consolidated matrix is entered to Expert Choice Decision Support Software and the criteria weights are obtained from the software.

The data collected for alternative technologies are used for pairwise comparison of alternatives with respect to each criterion. Data consist of the technology data for 7 different criteria in different measurement units. Additionally, for some of the criteria, there is only qualitative data available. For instance, capital cost of anaerobic digestion technology is higher than capital cost of static pile composting technology. AHP has an advantage to put different qualitative and quantitative data on the same scale and calculate the highest scored option. The collected data for alternatives are converted into values from

Figure 6. Analytical hierarchy process for evaluating ABT technologies

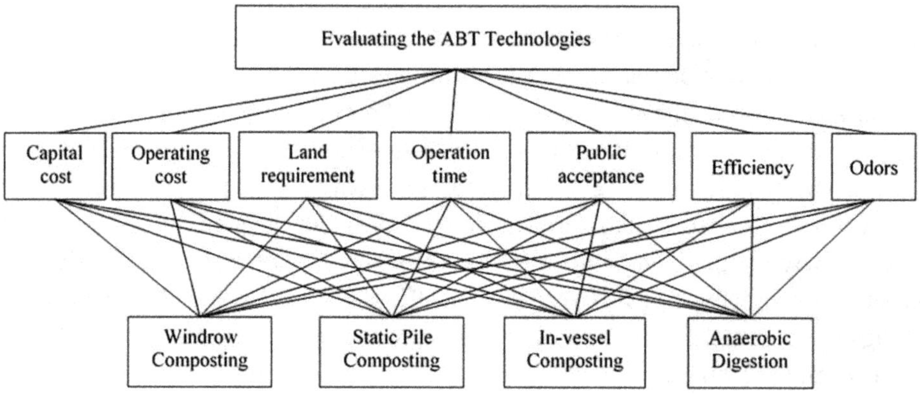

Table 6. Pairwise comparison of the set of criteria

With Respect to Goal	Capital Cost	Annual Cost	Operation Time	Land Requirement	Efficiency	Odors	Public Acceptance
Capital cost	1	1.39	2.53	3.04	2.78	1.33	1.39
Operating cost		1	2.23	2.74	2.43	1.18	1.28
Operation time			1	1.31	1.08	0.57	0.63
Land requirement				1	0.8	0.44	0.45
Efficiency					1	0.48	0.67
Odors						1	1.21
Public acceptance							1

Table 7. Pairwise comparison matrices for the alternatives with respect to each criterion C_1: Windrow composting, C_2: Static pile composting, C_3: In-vessel aerobic composting, C_4: Anaerobic digestion

Alternative Pairwise	Capital Cost				Operating Cost				Land Requirement				Operation Time			
	A_1	A_2	A_3	A_4	A_1	A_2	A_3	A_4	A_1	A_2	A_3	A_4	A_1	A_2	A_3	A_4
A_1	1	1/2	2	4	1	1/2	2	1/4	1	1/2	1/4	1/6	1	2	1/4	1/2
A_2		1	4	6		1	4	1/2		1	1/2	1/4		1	1/6	1/4
A_3			1	2			1	1/6			1	1/2			1	2
A_4				1				1				1				1
Alternative Pairwise	Public Acceptance				Efficiency				Odors							
	A_1	A_2	A_3	A_4	A_1	A_2	A_3	A_4	A_1	A_2	A_3	A_4				
A_1	1	1/2	1/4	2	1	2	1/4	1/2	1	1/3	1/5	1/7				
A_2		1	1/2	4		1	1/6	1/4		1	1/2	1/4				
A_3			1	6			1	2			1	1/2				
A_4				1				1				1				

Assessment of Advanced Biological Solid Waste Treatment Technologies for Sustainability

1 to 9 based on their relative contribution on the goal where 1 means two criteria contributes equally on the goal whereas a 9 means the first one has extremely more contribution on the goal than the other.

Pairwise comparison matrices for alternatives are provided in Table 7. In total, seven 4×4 pairwise comparison matrices for alternative comparison and a single 7×7 pairwise comparison matrix for criteria comparison are created. Local priorities of alternatives are calculated using given pairwise comparison matrices. Values that are less than 1 shows that the performance of the alternative in the row is lower than the alternative in the column. After local priorities of alternatives are obtained from these matrices, the last step is to combine the priorities and the criteria weights to find the most favorable alternative in terms of their contribution to the objective. In given matrices in Table 5, if the value for (A_1, A_2) is equal to x, then the value for (A_2, A_1) is equal to $1/x$ which is the reciprocal of x. The diagonal values of each matrix is equal to 1 since the performance of A_i compared to A_j with respect to any criterion is same.

Results

Criteria weights, local priorities, and global priorities of alternatives are obtained from the AHP model. Criteria weights are shown in Figure 7. The most and least important criterion for selecting the best ABT technology for the State of Florida is found to be the capital cost and land requirement with a weight of 0.233 and 0.072, respectively. These weights might change per SMEs opinions. However, in this work, the inconsistency ratio (IR) is calculated as 0.00159 ensuring that model does not require any further revision or updating in terms of SMEs' judgements (the smaller the better for inconsistency ratios). In other cases, where bias is noticed in human judgements, SMEs may be asked to reevaluate their criteria rankings to repeat the pairwise comparison.

Another advantage of AHP is that it allows the decision maker to give different weights to criteria based on their importance. That means each criterion has a different weight on selecting the most appropriate alternative. Based on the obtained results, capital cost has more impact than other criteria to select an ABT technology for the State of Florida. It should be noted here that the most important criterion and the ranking of the set of criteria can change if the judgments of SMEs or the selected SME group are changed.

The relative performances of alternatives with respect the each criterion can be seen in Figure 8. For example, static pile composting performs better than other alternatives with respect to capital cost, however, its performance is lower than others with respect to operation time and efficiency.

Global priorities of alternatives are calculated and are shown in Figure 9. For each alternative, it is calculated by multiplication of local priority of the alternative with the related criterion weight. Preliminary results indicate that the most appropriate alternative for the State of Florida is anaerobic digestion with a weight of 0.303 whereas the least favorable alternative is windrow composting with a weight of 0.152 (see Figure 9). It should be noted that AHP results may vary based on the considered states as well as the selected set of criteria.

RECOMMENDATIONS

In this study, the ABT technologies are assessed for the State of Florida with an aim to shed light to future research and decision making. The results obtained from the AHP model suggest anaerobic digestion as

Figure 7. Criteria weights obtained from the AHP model

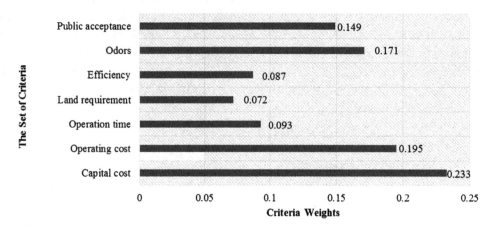

Figure 8. Global priorities of alternatives with respect to each criterion

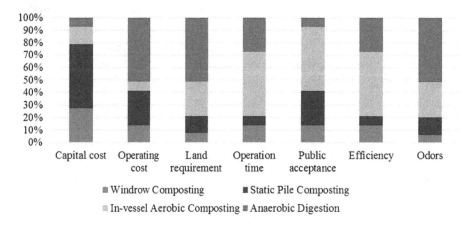

Figure 9. Global priorities of alternatives

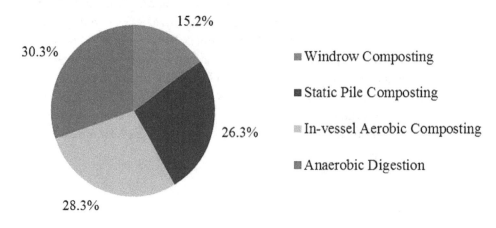

the most appropriate technology based on the defined criteria set while being subject to state needs, and budgetary limitations. The technologies those are ranked first and second, respectively, can be evaluated from these aspects in detail as there is a small difference between their weights. Anaerobic digestion has a weight of 0.303 whereas in-vessel aerobic composting has a weight of 0.283. Giving a decision in the direction of anaerobic digestion will bring a second investment requirement in order to generate energy from biogas. On the other hand, it will create a renewable energy source for the households and industry around the anaerobic digestion facility. If the land requirement of the considered facility is not large, anaerobic digestion may prove as a more doable option for the state. On the other hand, if there is a significant budget limitation, in-vessel composting may prove more beneficial with its lower capital cost even though it only produce compost as a useful output. It also does not require a second system for energy production.

FUTURE RESEARCH DIRECTIONS

Managing solid waste with the utilization of ATT, ABT technologies as well as nanotechnology provides two main benefits: preventing waste disposal in landfills and generating renewable energy and marketable products. Preventing waste disposal in landfills reduce greenhouse gas emissions while leave a clean environment for the living creatures. Many state governments which are aware of this fact have established aggressive recycling goals to be achieved in a short periods of times. Emerging methods presented here to treat MSW may help reach this goal. Yet, only limited research has been conducted to investigate and evaluate them so far. A case study to evaluate ABT technologies for the State of Florida has been presented here. States and countries which have established ambitious recycling rate goals can conduct similar evaluation studies and can customize the criteria set based on the strategies and current management systems of the given countries.

Furthermore, AHP can be utilized for comparison of conventional methods and advanced conversion technologies. For example, combustion, landfill disposal, ATT, and ABT technologies can be evaluated with AHP. This comparison might display the performance differences on advanced conversion technologies versus the conventional ones based on a set of criteria. Costs of landfill disposal and combustion are lower than advanced treatment technologies whereas environmental issues that they cause are a lot more than advanced treatment technologies. Another improvement might be done is to define a more detailed criteria set by creating a sub-criteria set. For example, capital cost, operating cost, and revenue can be evaluated under the title of financial issues, greenhouse gas emissions, odors, and wastewater production can be evaluated under environmental issues. Defining sub-criteria set makes the evaluation more detailed at the expense of a more detailed and careful evaluation required from SMEs.

Consequently, utilizing advanced technologies is an emerging way of managing MSW. Limited information is available for their evaluation. Most of the studies which have been conducted so far present data collection from vendors. There is a great need for detailed evaluation and assessment studies. The process that we present in this work serves as a basis for this line of research.

CONCLUSION

MSW generation has increased dramatically in recent years. Sustainable waste management strategies started to be implemented because of various environmental problems caused by the currently utilized conventional MSW management methods. While preventing and reducing waste generation, reusing waste materials, and recycling may be primal to ensured sustainability, there is always an amount of waste to be treated after all efforts. Advanced conversion technologies have been utilized recently for waste treatment due to the aforementioned reasons. They reduce greenhouse gas emissions, provide outputs to generate renewable energy and marketable products such as compost from waste. On the other hand, they require high capital costs and operating costs. They can be categorized mainly as ATT and ABT technologies. Furthermore, nanotechnology has garnered attention in recent years. The waste that cannot be treated by other waste treatment technologies can be reduced in size to nano levels to form valuable nanomaterials that can be used as raw material in industry and for energy generation. In this work, ATT, ABT technologies, and nanotechnology alternatives are discussed for treatment of MSW. ATT and ABT technologies may treat different portions of the waste and their main characteristics, advantages, and disadvantages are provided in this study. In the case study, ABT technologies are evaluated for the State of Florida in terms of their capital cost, operating cost, land requirement, operation time, public acceptance, efficiency, and odor issues. AHP is used as a decision making tool. SME judgements on criteria rankings are collected from 37 Floridian counties via surveys. The rankings are converted into pairwise comparison matrices using 1-9 scale. Capital cost is ranked as the most important criterion and anaerobic digestion is ranked the highest in AHP. Preliminary recommendations are discussed for the State of Florida based on the results obtained from AHP, where they are subject to change per different waste management strategies and state policies.

REFERENCES

Alves, J. O., Zhuo, C., Levendis, Y. A., & Tenório, J. A. (2011). Catalytic conversion of wastes from the bioethanol production into carbon nanomaterials. *Applied Catalysis B: Environmental, 106*(3), 433–444. doi:10.1016/j.apcatb.2011.06.001

Baere, L. (2006). Will anaerobic digestion of solid waste survive in the future? *Water Science and Technology, 53*(8), 187–194. doi:10.2166/wst.2006.249 PMID:16784203

Bastani, M., & Celik, N. (2015). Assessment of occupational safety risks in Floridian solid waste systems using Bayesian analysis. *Waste Management & Research*.

Berge, N. D., Ro, K. S., Mao, J., Flora, J. R., Chappell, M. A., & Bae, S. (2011). Hydrothermal carbonization of municipal waste streams. *Environmental Science & Technology, 45*(13), 5696–5703. doi:10.1021/es2004528 PMID:21671644

Braber, K. (1995). Anaerobic digestion of municipal solid waste: A modern waste disposal option on the verge of breakthrough. *Biomass and Bioenergy, 9*(1), 365–376. doi:10.1016/0961-9534(95)00103-4

British Columbia Ministry of Agriculture. Food and Fisheries. (1996). *Composting methods factsheet*. Retrieved August 2015, from http://www.al.gov.bc.ca/resmgmt/publist/300Series/382500-5.pdf

CalRecycle. (2012). *California's new goal: 75% recycling*. Retrieved June 2015, from http://www.calrecycle.ca.gov/75percent/Plan.pdf

Cao, W. *Synthesis of nanomaterials by high energy ball milling*. Retrieved January 2016, from http://www.understandingnano.com/nanomaterial-synthesis-ball-milling.html

Chen, X., Li, C., Grätzel, M., Kostecki, R., & Mao, S. S. (2012). Nanomaterials for renewable energy production and storage. *Chemical Society Reviews, 41*(23), 7909–7937. doi:10.1039/c2cs35230c PMID:22990530

Department for Environment Food & Rural Affairs (DEFRA). (2007). *Advanced biological treatment of municipal solid waste*. Retrieved April 2015, from http://www.recycleforgloucestershire.com/CHttpHandler.ashx?id=59162&p=0

Department for Environment Food & Rural Affairs (DEFRA). (2013). *Advanced biological treatment of municipal solid waste*. Retrieved May 2015, from https://www.gov.uk/government/uploads/system/uploads/attachment_data/file/221037/pb13887-advanced-biological-treatment-waste.pdf

Department for Environment Food & Rural Affairs (DEFRA). (2013). *Mechanical biological treatment of municipal solid waste*. Retrieved May 2015, from https://www.gov.uk/government/uploads/system/uploads/attachment_data/file/221039/pb13890-treatment-solid-waste.pdf

Envision New Zealand. (2003). *The road to zero waste*. Retrieved August 2015, from https://www.zerowaste.co.nz/assets/Reports/roadtozerowaste150dpi.pdf

European Commission. (2010). *Being wise with waste: The EU's approach to waste management*. Retrieved August 2015, from http://ec.europa.eu/environment/waste/pdf/WASTE%20BROCHURE.pdf

European Environment Agency (EEA). (2013). *Managing municipal solid waste – A review of achievements in 32 European counties*. ISSN 1725-9177. Retrieved June 2015, from file:///C:/Users/Duygu/Downloads/Managing%20municipal%20solid%20waste.pdf

Fitzgerald, G. C., & Themelis, N. J. (2009). *Technical and economic analysis of pre-shredding municipal solid wastes prior to disposal*. (Unpublished MS thesis). Columbia University, New York, NY.

Griffin, L. P. (2012). *Anaerobic digestion of organic wastes: The impact of operating conditions on hydrolysis efficiency and microbial community composition*. (Doctoral dissertation). Colorado State University, Fort Collins, CO.

Hackett, C., Durbin, T. D., Welch, W., & Pence, J. (2004). *Evaluation of conversion technology processes and products*. Retrieved April 2015, from http://energy.ucdavis.edu/files/05-06-2013-2004-evaluation-of-conversion-technology-processes-products.pdf

Hamoda, M. F., Qdais, H. A., & Newham, J. (1998). Evaluation of municipal solid waste composting kinetics. *Resources, Conservation and Recycling, 23*(4), 209–223. doi:10.1016/S0921-3449(98)00021-4

He, M., Xiao, B., Hu, Z., Liu, S., Guo, X., & Luo, S. (2009). Syngas production from catalytic gasification of waste polyethylene: Influence of temperature on gas yield and composition. *International Journal of Hydrogen Energy, 34*(3), 1342–1348. doi:10.1016/j.ijhydene.2008.12.023

Heo, H. S., Park, H. J., Park, Y. K., Ryu, C., Suh, D. J., Suh, Y. W., & Kim, S. S. et al. (2010). Bio-oil production from fast pyrolysis of waste furniture sawdust in a fluidized bed. *Bioresource Technology*, *101*(1), S91–S96. doi:10.1016/j.biortech.2009.06.003 PMID:19560915

IFC Advisory Services in Eastern Europe and Central Asia. (2012). *Municipal solid waste management: Opportunities for Russia, summary of key findings*. Retrieved July 2015, from http://www.ifc.org/wps/wcm/connect/a00336804bbed60f8a5fef1be6561834/PublicationRussiaRREP-SolidWasteMngmt-2012-en.pdf?MOD=AJPERES

International Solid Waste Association (ISWA). (2013). *Alternative waste conversion technologies*. Retrieved April 2015, from file:///C:/Users/Duygu/Downloads/ISWA_WGER_WP_on__ATT_FINAL_2013_04_15%20(1).pdf

Kim, J. K., Oh, B. R., Chun, Y. N., & Kim, S. W. (2006). Effects of temperature and hydraulic retention time on anaerobic digestion of food waste. *Journal of Bioscience and Bioengineering*, *102*(4), 328–332. doi:10.1263/jbb.102.328 PMID:17116580

Klein, A. (2002). *Gasification: An alternative process for energy recovery and disposal of municipal solid wastes*. (Doctoral dissertation). Columbia University, New York, NY.

Li, Q., Mahendra, S., Lyon, D. Y., Brunet, L., Liga, M. V., Li, D., & Alvarez, P. J. (2008). Antimicrobial nanomaterials for water disinfection and microbial control: Potential applications and implications. *Water Research*, *42*(18), 4591–4602. PMID:18804836

London Borough of Lambeth. (2010). *Municipal solid waste management strategy 2011-2031*. Retrieved August 2015, from https://www.lambeth.gov.uk/sites/default/files/rr-lambeth-waste-strategy-waste-prevention-plan.pdf

Lu, X., Jordan, B., & Berge, N. D. (2012). Thermal conversion of municipal solid waste via hydrothermal carbonization: Comparison of carbonization products to products from current waste management techniques. *Waste Management (New York, N.Y.)*, *32*(7), 1353–1365. doi:10.1016/j.wasman.2012.02.012 PMID:22516099

Murphy, J. D., & McKeogh, E. (2004). Technical, economic and environmental analysis of energy production from municipal solid waste. *Renewable Energy*, *29*(7), 1043–1057. doi:10.1016/j.renene.2003.12.002

Paul, K. T., Satpathy, S. K., Manna, I., Chakraborty, K. K., & Nando, G. B. (2007). Preparation and characterization of nano structured materials from fly ash: A waste from thermal power stations, by high energy ball milling. *Nanoscale Research Letters*, *2*(8), 397–404. doi:10.1007/s11671-007-9074-4

Roco, M. C. (2005). International perspective on government nanotechnology funding in 2005. *Journal of Nanoparticle Research*, *7*(6), 707–712. doi:10.1007/s11051-005-3141-5

Saaty, T. L. (1980). *The analytic hierarchy process*. New York, NY: McGraw-Hill.

Serio, M. A., Chen, Y., Wójtowicz, M. A., & Suuberg, E. M. (2000). Pyrolysis processing of mixed solid waste streams. *ACS Div. of Fuel Chem. Prepr*, *45*(3), 466–474.

Serio, M. A., Kroo, E., Bassilakis, R., Wójtowicz, M. A., & Suuberg, E. M. (2001). *A prototype pyrolyzer for solid waste resource recovery in space* (No. 2001-01-2349). SAE Technical Paper.

Tegart, G. (2009). Energy and nanotechnologies: Priority areas for Australia's future. *Technological Forecasting and Social Change, 76*(9), 1240–1246. doi:10.1016/j.techfore.2009.06.010

United States Environmental Protection Agency (EPA). (2012). *Greening EPA: Federal waste diversion and recycling requirements.* Retrieved June 2015, from http://www.epa.gov/greeningepa/waste/requirements.htm

United States Environmental Protection Agency (EPA). (2015). *Advancing sustainable materials management: Assessing trends in material generation, recycling and disposal in the United States.* Retrieved June 2015, from http://www.epa.gov/wastes/nonhaz/municipal/pubs/2013_advncng_smm_fs.pdf

United States Environmental Protection Agency (EPA). (2015). *Municipal solid waste.* Retrieved July 2015, from http://www.epa.gov/epawaste/nonhaz/municipal/

URS Corporation. (2005). *Evaluation of alternative solid waste processing technologies.* Retrieved June 2015, from http://lacitysan.org/solid_resources/strategic_programs/alternative_tech/PDF/final_report.pdf

Verma, S. (2002). *Anaerobic digestion of biodegradable organics in municipal solid wastes.* (Master's thesis). Columbia University, New York, NY.

Williams, R. B., Jenkins, B. M., & Nguyen, D. (2003). Solid waste conversion: A review and database of current and emerging technologies. Report prepared for California Integrated Waste Management Board, Davis, CA.

Willis, K. P., Osada, S., & Willerton, K. L. (2010). Plasma gasification: lessons learned at Eco-Valley WTE facility. In *18th Annual North American Waste-to-Energy Conference* (pp. 133-140). New York, NY: American Society of Mechanical Engineers. doi:10.1115/NAWTEC18-3515

Wilson, B., Williams, N., Liss, B., & Wilson, B. (2013). *A comparative assessment of commercial technologies for conversion of solid waste to energy.* Retrieved June 2015, from http://www.itigroup.co/uploads/files/59_Comparative_WTE-Technologies-Mar-_2014.pdf

World Bank Group. (2014). Results-based financing for municipal solid waste (91861 v2). Washington, DC: Author.

World Energy Council. (2013). *World Energy Council issues official statement ahead of 22nd World Energy Congress.* Retrieved August 2015, from https://www.worldenergy.org/news-and-media/news/world-energy-council-issues-official-statement-ahead-of-22nd-world-energy-congress/

Xu, P., Zeng, G. M., Huang, D. L., Feng, C. L., Hu, S., Zhao, M. H., & Liu, Z. F. et al. (2012). Use of iron oxide nanomaterials in wastewater treatment: A review. *The Science of the Total Environment, 424,* 1–10. doi:10.1016/j.scitotenv.2012.02.023 PMID:22391097

Yasar, D., Celik, N., & Sharit, J. (2016). Evaluation of advanced thermal solid waste management technologies for sustainability in Florida. *International Journal of Performability Engineering, 12*(01), 63–78.

Zafar, S. (2009). *Gasification of municipal solid waste.* Retrieved May 2015, from http://www.altenergymag.com/content.php?issue_number=09.06.01&article=zafar

Zhuo, C., & Levendis, Y. A. (2014). Upcycling waste plastics into carbon nanomaterials: A review. *Journal of Applied Polymer Science, 131*(4), n/a. doi:10.1002/app.39931

KEY TERMS AND DEFINITIONS

Ash: A substance formed after solid waste is processed by heat.
Biogas: A gas that is produced by the degradation of orgonic portion of waste.
Combustion: A process where the waste is burned.
Landfill: A land area where the wastes are disposed in.
Methane: A greenhouse gas that is released by degradation of organic portion of waste.
Plasma: The fourth state of the matter.
Recycling: A process which treat the waste to be used again.
Syngas: A gas output which is generated by thermal degradation of waste at high temperatures and mainly consists of carbon monoxide, carbon dioxide, and hydrogen.
Windrow: Long turning rows which is used for processing waste and producing compost.

Chapter 11
Hybrid Nanostructures:
Synthesis and Physicochemical Characterizations of Plasmonic Nanocomposites

Ahmed Nabile Emam
National Research Centre, Egypt

Ahmed Sadek Mansour
Cairo University, Egypt

Emad Girgis
National Research Centre, Egypt

Mona Bakr Mohamed
Cairo University, Egypt

ABSTRACT

The recent extensive interest of nanostructure materials associated with their unique properties is motivated to develop new hybrid nanocomposites that couple two nano-components together in the form of Core/Shell, nanoalloys, and doped nanostructures. Hybrid nanostructure provides another opportunity for tuning the physical and chemical properties at the nanoscale. This opens the door for the discovery of new properties and potential for more applications. This chapter is devoted to present, and discuss the recent advances and progress relevance for Plasmonic hybrid nanocomposites. In addition, literature reviewed on different attempts to obtain high quality plasmonic nanocomposites via chemical routes, and their physico-chemical aspects for this class of novel nanomaterials. The authors presented their recent published work regarding Plasmonic hybrid nanostructure regarding plasmonic-semiconductor, plasmonic magnetic and plasmonic graphene nanocomposites.

1. INTRODUCTION

Hybrid nanostructures are composed of two or more nanoscale materials to form nano-composites. In such materials, their individual components combined their functionalities, and advantages to be integrated into one nano-object (Kickelbick 2007, and Merhari 2009). Hybrid nanocomposite offer the possibility and flexibility to obtain new properties and functions, which different from that of their isolated components through varying the composition of the materials and related parameters such as morphology and interface for wide-range of applications.(Bogue 2011, Kickelbick 2007, Berthelot 1999, and Pandya

DOI: 10.4018/978-1-5225-0585-3.ch011

2015) Importantly, their morphologies are versatile, such as zero dimensional (0D) nanoparticles (i.e. sphere), 1D dimension (i.e. rods, or wires).(Huang et al. 2014)

Most of the nanocomposites that have been developed and demonstrated technological importance can be classified according to the type of filler, as following (Figure 1): (Huang et al. 2014)

1. Plasmonic based nanocomposites.
2. Magnetic based nanocomposites.
3. Polymer based nanocomposites.
4. Semiconductor based nanocomposites.
5. Carbon based nanocomposites.

This chapter is devoted to explore the synthesis and characterizations of plasmonic hybrid nanocomposites including core-shell, nanoalloys, doped plasmonic nanomaterials and plasmonic-organic hybrid nanocomposite, and their applications as photo-catalysts in photo-degradation of organic wastes and waste water treatment.

2. PHYSICS OF PLASMONIC NANOSTRUCTURES

The optical properties of metal nanoparticles are dominated by what is called Plasmon (i.e. the collective oscillation of conduction electrons resulting from their interaction with electromagnetic radiation as shown in Figure (2). (Mody et al. (2010), Faraday (1857), Murray& Barnes (2007), Bruda et al. (2005) and Kreibig & Vollmer (1995)) Plasmon are resonant modes that involve the interaction between free charges (i.e. –ve electrons) and light.(Bruda et al. (2005)) When the size of metallic particles reduced within nanoscale, in which the particle size is smaller than the electron mean free path, the electrons motion will consequently confined within nanoscale (1-100 nm). In this case, unusual phenomena called plasmonic effects (i.e. surface plasmon band) are observed.(Creighton (1991), and Kreibig & Vollmer (1995)) The electric field of the incoming radiation induces the formation of a dipole in the nanoparticle. A restoring force in the nanoparticle tries to recover for this, resulting in a unique resonance wavelength,

Figure 1. The classification of Hybrid Nanocomposites

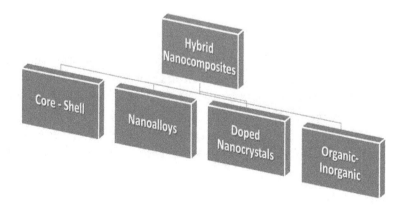

Hybrid Nanostructures

(Bruda et al. (2005)) which depend on a number of factors such as, particle size and shape, as well as the nature of the surrounding medium.(Murray & Barnes (2007), Bruda et al. (2005), Creighton (1991), Kreibig & Vollmer (1995), Link & El-Sayed (2000 & 2003), Kerker (1969), Bohren & Huffman(1983), and Lee et al. (2005))

The oscillation frequency of the surface Plasmon band (SP) is determined by four factors: the density of electrons, the effective electron mass, the shape of the charge distribution, and the size of the charge distribution. The frequency and width of SP depend on the size and shape of the metal nanoparticles as well as on the dielectric constant of the metal itself and the surrounding medium (Figure 3).(Kreibig & Vollmer (1995), Link & El-Sayed (2003), and Papavassiliou (1979))

For example the SPR absorption in spherical Au NPs occurs at about 520 nm. This absorption is absent for clusters (i.e. << 2 nm), as well as bulk Au. The rod-shaped Au NPs have two absorption bands. The first one, appears at ~ 520 nm, corresponds to the oscillation of the electrons perpendicular to the long rod axis and is called transverse plasmon absorption (Figure 4a). This band is insensitive to the nanorod length, and coincides with the surface plasmon absorption band of the gold nanospheres. The second absorption band appears at a lower energy and is caused by the oscillation of the free electrons along the long rod axis and is known as the longitudinal surface plasmon absorption (Figure 4b). This longitudinal plasmon band is very sensitive to the aspect ratio (length/width) of the rods where a red shift occurs as the aspect ratio increases.(Murphy (2002)) Furthermore, the relative intensity ratio of the longitudinal to the transverse mode increases with increasing aspect ratio.(Murphy (2002))

Figure 2. Schematic diagram illustrate the surface plasmon resonance (SPR) absorption in plasmonic nanomaterials.
Adapted with permission from (Burda, C., Chen, X., Narayanan, R., & El-Sayed, M. A. (2005). Chemical reviews, 105(4), 1025-1102). Copyright (2005) American Chemical Society

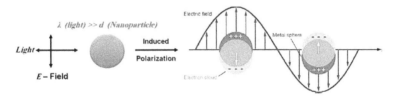

Figure 3. The SPR absorption band dependence on Size (a), and Shape (b)

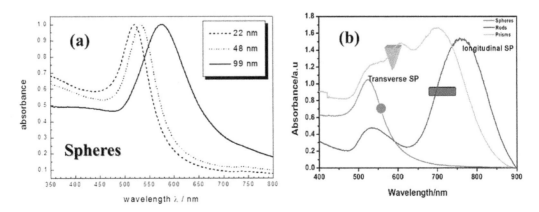

Figure 4. The optical response of rod-like nanoparticles to an electric field E. Two oscillating modes can be possible: (a) transverse oscillation and (b) the longitudinal oscillation

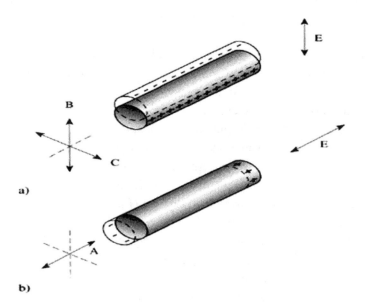

Moreover, the plasmon resonance band of triangular shaped plasmonic nanoparticles splits into show three bands, longitudinal mode, transverse mode (i.e. in-plane dipole resonance), and quadrupole modes (i.e. in-plane quadrupole and out-of-plane quadrupole resonances). The in-plane dipole resonance is very large field enhancements at their sharp tips as shown in Figure (5).(Jin et al. (2001 & 2003), and Millstone et al. (2005)) In addition, the quadrupole band, and the longitudinal band depend on the length of the three axes of the prism. As the size (length and thickness) of these particles increases, these bands shift to lower energy. The higher energy bands are assigned to oscillation along the C3 axis perpendicular to the triangular faces and the broad low energy band is assigned to the in-plane oscillation on the triangular faces.(Jin et al. (2001), Callegari et al. (2003), and Sherry et al. (2006)) Kelly and co-worker (Kelly et al. (2003)), and Rang and co-worker (Rang et al. (2008)) demonstrated the localized filed enhancement in prism shape using discrete dipole approximation (DDA). They showed that the maximum enhancement for the dipole resonance is at the tips, while for the quadrupole resonance the significant regions of enhancement occur at the sides. Also, it's noticed that at the particle tips, is that the field decays away from the surface faster for the quadrupole than for the dipole as shown in Figure (6).(Kelly et al. (2003))

It is well known that the dielectric constant of the surrounding material affect the oscillation frequency of the Plasmon resonance, which attributed to the alteration in the ability of the surface to accommodate the electron density from the nanoparticles.(Eustis & El-Sayed (2006) and Jain et al. (2007) Although dielectric constant is dependent on the surrounding solvent, but the capping material is the most important in determining the shift of the plasmon resonance due to their effect on the surface of the nanoparticle. Therefore, any chemically bonded molecules can be detected by any change induced from the variation in the electron density on the surface, which results in change of the position surface plasmon absorption band. (Eustis & El-Sayed (2006))

In metal nanoshell, the core material can be dielectric or semiconducting, whereas the shell material is from metallic in nanoscale range. These hybrid nanostructures give a good evidence for a fact, in

Figure 5. The optical extinction spectrum for Ag nanoprism based on dimension of prism snips.
Reproduced with permission from (Kelly, K. L., Coronado, E., Zhao, L. L., & Schatz, G. C. (2003). The optical properties of metal nanoparticles: the influence of size, shape, and dielectric environment. The Journal of Physical Chemistry B, 107(3), 668-677). Copyright (2003) American Chemical Society

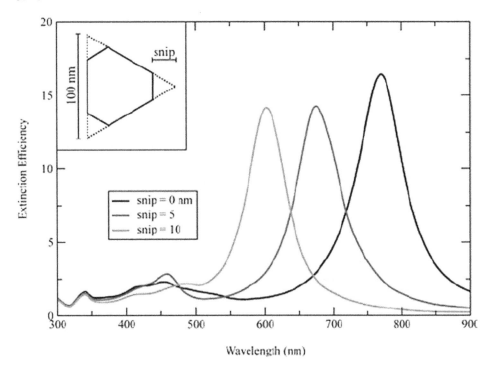

Figure 6. The E-Field enhancement contours external to Ag nanoprism
Reproduced with permission from (Kelly, K. L., Coronado, E., Zhao, L. L., & Schatz, G. C. (2003). The optical properties of metal nanoparticles: the influence of size, shape, and dielectric environment. The Journal of Physical Chemistry B, 107(3), 668-677). Copyright (2003) American Chemical Society

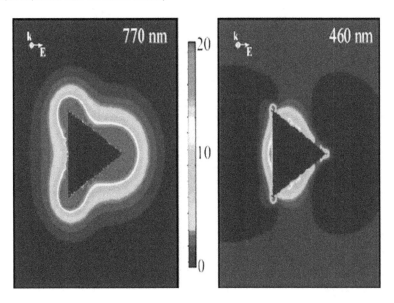

which the surface plasmon resonance is strongly dependent on the relative thickness of the nanoparticle core and its metallic shell. Therefore, the position of the plasmon band can be tuned anywhere across the visible or infrared regions of the optical spectrum, by varying the core and shell thicknesses as shown in Figure (7). (Oldenburg et al. (1998)) Oldenburg and co-worker who the first group which realized the nanoshell structure by formation of metal nanoshell from gold on the surface of silica dielectric core. (Oldenburg et al. (1998))

Starting from a solid spherical particle, as the shell thickness decreases, The SPR band red-shifted from the visible to the NIR at a thin shell thickness. (Oldenburg et al. (1998), and Jain & El-Sayed (2007)) (Figure 7). This is due to increased coupling between the inner and outer shell surface plasmons. (Prodan et al. (2003) and Jain & El-Sayed (2007)). In 2007, Jain and co-worker demonstrated thatany decrease shell thickness/core radius ratio, will results in exponentially decrease of SPR frequency. (Jain & El-Sayed (2007)). This means that the position of SPR bandis independent on each the nanoshell size, core material, shell metal, or surrounding medium, thus making it easy to design nanoshells with a desired optical resonance.(Jain & El-Sayed (2007), Jain et al. (2008), and Ghosh Chaudhuri & Paria (2011)).

Figure 7. (a) theoretical calculation of SPR of metal nanoshell
Reproduced with permission from (Oldenburg, S. J., Averitt, R. D., Westcott, S. L., & Halas, N. J. (1998).Nanoengineering of optical resonances. Chemical Physics Letters, 288(2), 243-247.).Copyright (1998) El-Sevier

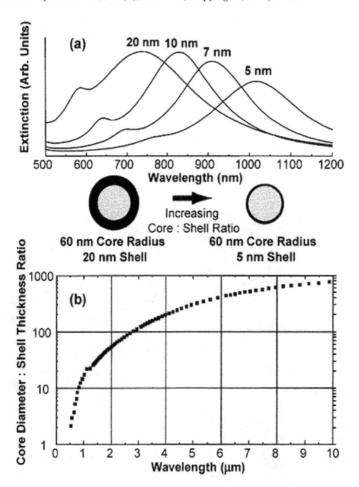

Hybrid Nanostructures

In nanoalloys, a linear dependence of the plasmon absorption maximum on the composition of the nanoparticles was found.(Link et al. (1999)) In case of spherical plasmonic nanoparticles aremixedwith magnetic or semiconducting within a nano-object, they exhibit a strong red shift SPR.(Shi et al. (2006), Ghosh Chaudhuri & Paria (2011), Lee & El-Sayed (2006), Barcaro et al. (2015), and Ferrando et al. (2008)) Girgis et al. show that illustrated that the red shift in SPR of Au−Co bimetallic nanoalloys than Au NPs due to the homogeneous mixture of metal−metal bond between gold and cobalt leads to the formation of an intermetallic or alloyed structure, where Co^{2+} ions diffuse into the Au NP host crystal. (Girgis et al. (2012)) as shown in Figure (8a). Moreover, alteration of pure Au surface via electronic charging caused a plasmon band shift.(Templeton et al. (2008)) But in case of rod-like shape, Emam et al. showed that both of transverse and longitudinal SPR in Au-Co bimetallic nanoalloys is blue shifted than in case of Pure Au nanorods.(Emam et al. (2015)) as shown in Figure (8b). This is due to loss of continuous density of states in gold matrix, as a results of variation in both of electron mean free path, and surface charge density upon the presence of another metals in-situ gold matrix.(Xu et al. (2007), Lica et al. (2004), and Boyer (2010))

3. SYNTHETIC ROUTS AND STRATEGIE OF PLASMONIC NANOSTRUCTURES

A wide variety of different techniques are used to produce nanoparticles: wet chemical or solution method, mechanical size reduction, gas phase synthesis, etc. All of these are being used commercially and each has its own merits and drawbacks.

3.1. Chemical Reduction Method

It is the most widely and oldest method used to prepare metallic nanoparticles. This method of preparation generally involves the reduction of metal salts using mild reducing agents (Figure 9) such as sodium citrate,(Turkevich & Kim (1970)) sodium borohydride, hydrogen, carbon monoxide, acetylene, or phosphorus.(Turkevich & Kim (1970)) The reaction proceeds then followed by nucleation and growth in the presence of stabilizing agents.(Turkevich & Kim (1970)) The additions of stabilizing agents as protective capping materials are necessary to prevent aggregation and further growth of the particles. (Turkevich & Kim (1970), Hirai et al. (1978 & 1979), Franke et al. (1996), Vidoni et al. (1999), Tanori & Pileni (1997), Reetz & Lohmer (1996), Kiwi & Grätzel (1979), Sinzig et al. (1998), and Turkevich et al. (1954)) These capping agent may be metal salts,(Turkevich & Kim (1970), Hirai et al. (1978)) polymers,(Hirai et al. (1979) and Franke et al. (1996)) organic solvents, (Vidoni et al. (1999), Tanori & Pileni (1997)) long chain alcohol,(Reetz & Lohmer (1996)) surfactants,(Kiwi & Grätzel (1979), Sinzig et al. (1998)) and organometallics.(Turkevich et al. (1954))

One of the most pioneer work has been achieved by Turkavich and co-workers. (Turkevich & Kim (1970), Turkevich (1985), and Tano & Meguro (1989)) In their study, they developed a method to prepare gold nanoparticles of different sizes and shapes depending on the reducing agent as shown in Figures (10-12). The reduction by hot sodium citrate yields spherical particles of 20 nm diameter of very narrow size distribution, while boiling with citric acid produces large numbers of flat triangles and hexagonal plates, spheres and bipyramidal shapes. (Turkevich & Kim (1970)) On the other hand, reduction with carbon monoxide gives elongated cylinders whereas acetylene produces irregular flat plates. (Turkevich & Kim (1970))

Figure 8. (a) Red shift SPR band in case of spherical Au-Co nanoalloys, and (b) Blue shift SPR band in case of Au-Co alloyed nanorods
Reproduced with permission from (Emam, A. N., Mohamed, M. B., Girgis, E., & Rao, K. V. (2015). Hybrid magnetic–plasmonic nanocomposite: embedding cobalt clusters in gold nanorods. RSC Advances, 5(44), 34696-34703.).Copyright (2015) Royal Society of Chemistry

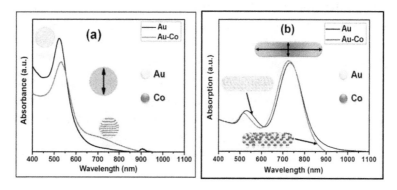

Figure 9. Schematic diagram for Chemical Reduction method

Figure 10. Synthesis of Gold Nanoparticles according
(Turkevich& Kim (1970))

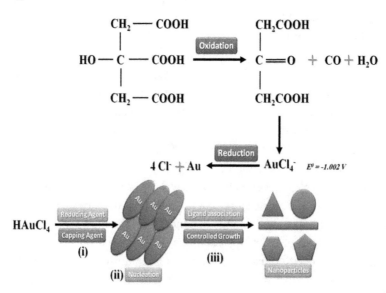

Figure 11. TEM micrograph of Gold (a), and Silver (b)Nanoparticles synthesized in via chemical reduction method using tri-sodium citrate as reducing agent

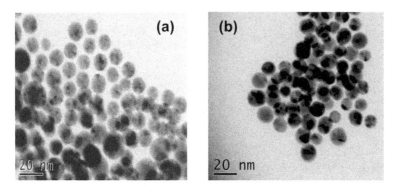

Figure 12. TEM micrograph of Gold (a), and Silver (b) Nanoparticles synthesized via chemical reduction method using ethyl alcohol as reducing agent

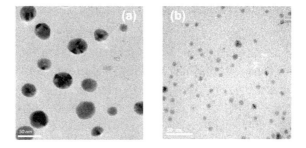

The same method could be used to prepare metallic nanoalloys and core shell nanoparticles. In case of bimetallic nanoalloys; the reduction of the mixture of two or more metal salts with desired molar ratio occurs using same mild reducing agents as shown in Figure (13). (Link et al. (1999), and Girgis et al. (2012)) While, the core-shell nanoparticles have been prepared, by the successive reduction of one metal over the nuclei of the other as shown in Figure (14). (Schimid (1994))

For example, Girgis *et al.* reported that gold–cobalt alloyed nanoparticles (Au–Co NPs), have been prepared by substituting predetermined moles of cobalt atoms in the form of cobalt acetate (i.e. mixed particles are with cobalt mole fractions xCo), using chemical reduction for both of Au^{3+} and Co^{2+} ions precursors within the same time, using $NaBH_4$ as areducing agent. (Girgis et al. (2012)) as shown in

Figure 13. Schematic diagram for Chemical Reduction method of nanoalloys

Figure 14. Schematic diagram for Chemical Reduction method of core-shell nanoparticles

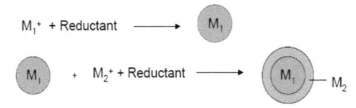

Figure (15). Whereas, gold-silver (Au-Ag) alloy particles are prepared in by substituting a predetermined number of moles of gold atoms by the equivalent number of moles of silver atoms in the form of silver nitrate $AgNO_3$, and mixed with gold mole fractions xAu in presence of Tri-sodium Citrate as mild reducing agent as shown in Figure (16).(Link et al. (1999))

3.2. Organometallic Pyrolysis Method

Metallic nanoparticles can also be prepared by pyrolysis. In this method, chemical precursors decompose under suitable thermal treatment into one solid compound and unwanted waste evaporates away. (Fleming & Williams (2004)) Generally; each of the atomic species that will form the nanocrystal (NC) is introduced into a reactor in the form of a *precursor*. A *precursor* is an organometallic compound such as acetylacetonate (Okasha et al. (2010), AbouZeid et al. (2011), Son et al. (2004), and Marquardt et al. (2011)), carbonyl, (Bera et al. (2010)) and dithiocarbazate. (Murray et al. (1993)) Once the precursors are introduced into the reaction flask, they react or decompose, generating the reactive monomers that will cause the nucleation and growth of the nanocrystals (NCs). The energy required to decompose the precursors and to crystallize the nanoparticles, is provided by the liquid of the reactor, either by thermal collisions or a chemical reaction between the liquid medium and the precursors, or by a combination of these two mechanisms. (Qu et al. (2001), Peng & Peng (2001), and Puntes et al. (2001))

The key parameter in the size and shape controlled growth of colloidal nanocrystals (NCs) is the presence of one or more organic molecules in the reactor, here broadly termed as "*surfactants*". Surfactants are amphiphilic compounds, i.e. molecules composed by one hydrophilic part and one hydrophobic part

Figure 15. The absorption spectra for both of Au, and Au-Co NPs (a), TEM (b), and HR-TEM (c) images of Au-Co NPs.
Reproduced with permission from (Girgis, E., Khalil, W. K. B., Emam, A. N., Mohamed, M. B., & Rao, K. V. (2012). Nanotoxicity of gold and gold–cobalt nanoalloy. Chemical research in toxicology, 25(5), 1086-1098.).Copyright (2012) American Chemical Society

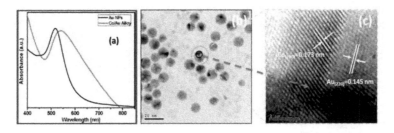

Hybrid Nanostructures

Figure 16. TEM micrographs of Au-Ag NPs
Reproduced with permission from (Link, S., Wang, Z. L., & El-Sayed, M. A. (1999). Alloy formation of gold-silver nanoparticles and the dependence of the plasmon absorption on their composition. The Journal of Physical Chemistry B, 103(18), 3529-3533). Copyright (1999) American Chemical Society

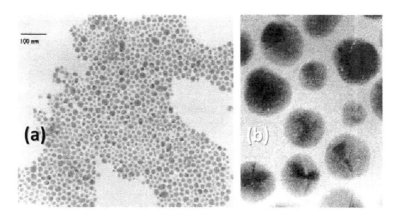

(i.e. hydrocarbon chains). Growth of nanocrystals (NCs) can be properly controlled in many cases by the use of special molecules that behave as "terminating" agents. Such molecules direct the growth of nanostructures by dynamically coordinating their surface under the reaction conditions. Some examples of suitable surfactants include molecules carrying functional groups with electron donor atoms, such as carboxylic, alkyl thiols, phosphines, phosphine oxides, phosphates, phosphonates, amides or amines, carboxylic acids, and nitrogen-containing aromatics.(Fleming & Williams (2004), Okasha et al. (2010), AbouZeid et al. (2011), Son et al. (2004), and Marquardt et al. (2011), and Bera et al. (2010)) The choice of surfactants varies from case to case: a molecule that binds too strongly to the surface of the quantum dot is not useful, as it would not allow the nanocrystal to grow.(Qu et al. (2001), and Peng & Peng (2001))

Puntes et al. achieved various particle shapes using the idea first described by Lohmer (Reetz & Lohmer (1996)) that in sol processes monodisperse nanostructures require a single, short nucleation event followed by the slower growth on the existing nuclei. (Peng & Peng (2001)) This may be achieved by rapid addition of reagents into a reaction vessel containing a hot coordinating solvent (Figures 17-18)). (Puntes et al. (2001 & 2001) and Zhu et al. (2000))

Figure 17. TEM micrographs for as-prepared gold (a), and silver (b) nanoparticles via thermal decomposition noble metal precursor (i.e. Ag, and Au) in presence of oleic acid and oleylamine
Reprinted with permission from (AbouZeid, K. M., Mohamed, M. B., & El-Shall, M. S. (2011). Small, 7(23), 3299-3307). Copyright (2011) John Wiley and Sons

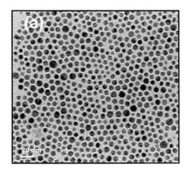 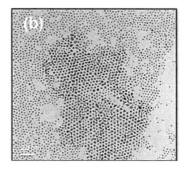

Figure 18. Absorption spectra (a) and TEM micrograph (b) of silver nanoparticles via thermal decomposition Ag precursor in presence of TOPO and HDA
Reprinted with permission from (Okasha, A. M., Mohamed, M. B., Abdallah, T., Basily, A. B., Negm, S., & Talaat, H. (2010, March). Characterization of core/shell (Ag/CdSe) nanostructure using photoacoustic spectroscopy. In Journal of Physics: Conference Series (Vol. 214, No. 1, p. 012131). Copyright (2010) IOP Publishing[1]

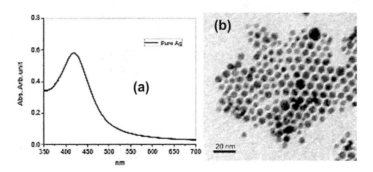

3.3. Microwave Irradiation (MWI) Method

Microwave irradiation (MWI) is considered one of the most facile synthetic routs of nanomaterials with controllable size, and shape.(Liang et al.(2002), Gerbec et al. (2005), Panda et al. (2007 & 2008), Mohamed et al. (2010), and Chen et al. (2005)) The main advantages of MWI over other conventional methods can be summarized as following; (i) High temperature and pressure not required, (ii) Time saving includes rapid heating, and reaction rate, (iii) No hot-wall effects and hence homogenous almost mono-dispersed crystal growth.(Chen et al. (2005)) By using metal precursors that have large microwave absorption cross sections relative to the solvent, very high effective reaction temperatures can be achieved. This allows the rapid decomposition of the precursors, thus creating highly supersaturated solutions where nucleation and growth can take place to produce the desired nanomaterials. (Chen et al. (2005))

Several parameters can play an important role in tailoring the particle size, and shape via using MWI method such as; varying the MWI reaction time, and the relative concentrations of different organic surfactants with variable binding strengths to the initial precursors and to the nanocrystals as shown in Figure (19-21). Moreover, MW power can be considered. (Chen et al. (2005))

Few reports for controlling the shape of the plasmonic nanocrystals via MWI method.(Chen et al. (2005), Abdelsayed et al. (2009), Pastoriza-Santos& Liz-Marzán (2002), Kundu et al. (2008), Okitsu (2011)) Mohamed et al. demonstrated a facile one-pot MWI method to fabrication of different shapes of plasmonic nanoparticles capped with a mixture of oleylamine (O.Am) and oleic acid (O.Ac). The size, shape, and morphology of the nanocrystals could be tailored by varying the ratio of oleylamine to oleic acid, the microwave time, and the concentration of the metal ions. (Chen et al. (2005)) El-Shall and co-workers present into several reports the synthesis of different nanomaterials such as; plasmonic, semiconductor nanoparticles and nanoalloys with different size and shapes using MWI.(Panda et al. (2007), Mohamed et al. (2010), Pastoriza-Santos& Liz-Marzán (2002))

Kundu *et al.* reported a MWI method to synthesize multi-shapes gold nanoparticles in ammonium bromide (CTAB) as a capping material.(Kudu et al. (2008))

Hybrid Nanostructures

Figure 19. (a) UV-Vis spectra for Au NPs via MWI using different ratios of O.Ac: O.Am at 1 min MWI time. TEM Micrograph of Au NPs using 0:1 (b), and 1:1 ratio of O.Ac: O.Am (c)
Reprinted with permission from (Mohamed, M. B., AbouZeid, K. M., Abdelsayed, V., Aljarash, A. A., & El-Shall, M. S. (2010). Growth mechanism of anisotropic gold nanocrystals via microwave synthesis: formation of dioleamide by gold nanocatalysis. ACS nano, 4(5), 2766-2772).Copyright (2010) American Chemical Society

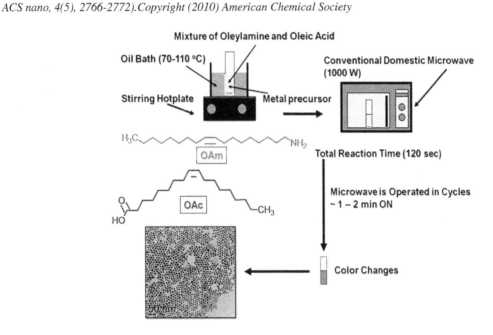

Figure 20. (a) UV-Vis spectra for different anisotropic shapes of Au NPs via MWI using different ratios with different molar ratios of O.Ac: O.Am according to (a) 50, (b) 60, (c) 70, (d) 80, (e) 90, and (f) 100%. TEM micrograph for Au NPs at different molar ratios of O.Ac: O.Am according to (i) 50, (ii) 60, (iii) 70, and (iv) 80
Reproduced with permission from (Mohamed, M. B., AbouZeid, K. M., Abdelsayed, V., Aljarash, A. A., & El-Shall, M. S. (2010). Growth mechanism of anisotropic gold nanocrystals via microwave synthesis: formation of dioleamide by gold nanocatalysis. ACS nano, 4(5), 2766-2772).Copyright (2010) American Chemical Society

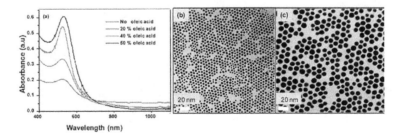

3.4. Sonochemical Method

This method is based on the sonochemical reduction of the corresponding metal ions in aqueous solutions. The main advantages of such methods are cost effective, and its simplicity as shown in Figure (22).(Mason (2009), Bang & Suslick (2010), Gutierrez et al. (1987), Okitsu et al.(2009))The reduction mechanism is suggested to be due to the reactions with reducing species formed from the sonolysis of

Figure 21. TEM images of bimetallic nanoalloys prepared by the MWI method, and Digital photographs of metallic and bimetallic nanocrystals colloidal solutions
Reproduced (Abdelsayed, V., Aljarash, A., El-Shall, M. S., Al Othman, Z. A., & Alghamdi, A. H. (2009). Microwave synthesis of bimetallic nanoalloys and CO oxidation on ceria-supported nanoalloys. Chemistry of Materials, 21(13), 2825-2834). Copyright (2009) American Chemical Society

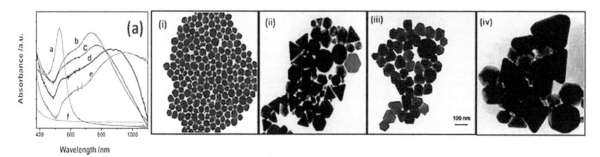

organic additives (i.e. Polymeric capping agent) and water. The rate of reduction of metal ions can be changed by changing the types and concentration of organic additives. In addition, various parameters such as ultrasound intensity, ultrasound frequency, dissolved gas, position of reaction.(Mason (2009))

It is important to understand the reduction mechanism of metal ions, because the reduction processes can be applied to the synthesis of various metal nanoparticles and nanostructured materials. The reduction and growth mechanism was summarized in following steps as reported by Gutierrez et al.:(Gutierrez et al. (1987))

Figure 22. Schematic diagram of sonochemical device to prepare various nanostructured inorganic materials
Reproduced with permission from (Bang, J. H., & Suslick, K. S. (2010).Applications of ultrasound to the synthesis of nanostructured materials. Advanced materials, 22(10), 1039-1059).Copyright (2010) American Chemical Society

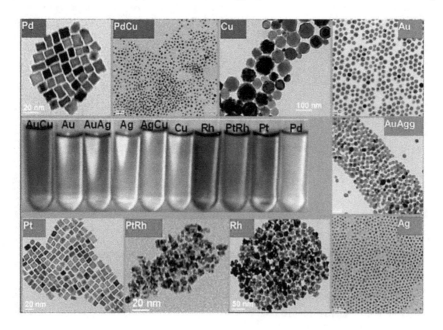

Step 1: $H_2O \rightarrow OH + H$

Step 2: $RH + OH(H) \rightarrow Reducing\, species + H_2O(H_2)$

Step 3: $RH + H_2O \rightarrow Pyrolysis\, radicals\, and\, unstale\, products$

Step 4: $M^n + Reductants \rightarrow M^0$

Step 5: $nM^0 \rightarrow (M^0)_n$

Step 6: $M^0 + (M^0)_n \rightarrow (M^0)_{n+1}$

Where M^{n+} corresponds to a metal ion and RH corresponds to an organic additive. Various metallic colloids have been prepared via the sonochemical route by many research groups.(Nemamcha et al. (2006), Zhang et al. (2006), Su et al. (2003), Caruso et al. (2002), Mizukoshi et al. (1999), and Okitsu et al.(2005)) In their studies, Grieser and coworkers carried out a systematic study on sonochemical reduction to unveil the complex reduction mechanism and to understand the effect of each parameter (e.g., time, concentration, ultrasonic frequency, and different organic additives) on particle size and shape.(Mizukoshi et al. (1999), Okitsu et al.(2005), Vinodgopal et al. (2006), Brotchie et al. (2008), and Anandan et al. (2008))

3.5. Seed Mediated Method

The seed-mediated growth method is one of the most reliable and versatile methods to control the shapes of noble metal nanocrystals (i.e. Spheres, rods, wires, flowers and prisms etc…). This strategy has been used to synthesize large Au nanoparticles with excellent monodispersity as reported by Brown and co-worker in 2000.(Brown et al. (2000)) This method don't limited to prepare spherical particles with large size, but its employed to obtain different anisotropic structures of noble metallic nanoparticles such as gold nanorods.(Jana et al. (2001), Nikoobakht & El-Sayed (2003), and Sau & Murphy (2004)) During the last two decades several reports has been demonstrated the development of seed-mediated method into a reliable and versatile approach to control the shapes of Au and Ag NCs.(Jana et al. (2001), Nikoobakht & El-Sayed (2003), Sau & Murphy (2004), Perez-Juste et al. (2004), Attia et al. (2014))

This method involves two steps as shown in Figure (23). *The First step*; is the synthesis of seed nanoparticles by a chemical reduction of the metal salt is reduced by reducing agents in the presence of stabilizers. Sodium borohydride is the commonly used reducing agent. *The Second step;* the growth solution contains a surfactant or a shaping agent and a mild reducing agent. In this process, metal salts are reduced on the surface of the seed nanoparticles. The surfactant molecules form suitable templates that facilitate the growth process to yield nanoparticles of desired morphology. The size of the nanoparticles can be tuned by changing the amount of seed nanoparticles added.(Nikoobakht & El-Sayed (2003))

Several parameters control the shape of plasmonic nanoparticles (i.e. Au, and Ag). These parameters are listed as following; (i) surfactant concentration, (iii) Silver ions contents depending on the molar concentration of $AgNO_3$, which has an important role to enhance the yield formation of nanorods than spherical particles. Finally; the concentration of seed solution injected into the growth solution.(Nikoobakht & El-Sayed (2003), and Attia et al. (2014))

Figure 23. Experimental Setup for preparing of different anisotropic structure of plasmonic nanoparticles via seed – mediated method

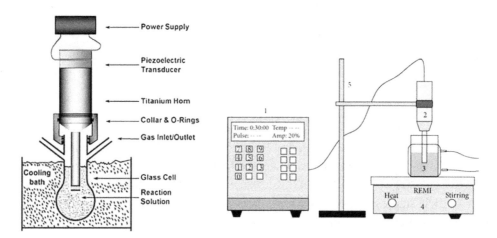

3.6. Biosynthesis Method

Although chemical method is one of the most facile approaches to obtain nanoparticles, with high yield and in good quality, using capping agents is necessary for size stabilization of the nanoparticles. These chemicals reagents might be toxic and lead to byproducts that are not environment benign. A variety of natural sources including plants, fungi, yeast, action-mycetes, bacteria have been used for nanoparticle synthesis. Additionally, the unicellular and multicultural organisms can produce intracellular and extra cellular inorganic nanoparticles.(Vithiya & Sen (2011)) This approach based on the natural origin of biological candidates such as extra-cellular enzymes into bacteria,(Klaus et al. (1999), Nair & Pradeep (2002), Sweeney et al. (2004), and Lengke et al. (2006)) and Fungi,(Ahmad et al. (2003), Mukherjee et al. (2001), Bhainsa & D'souza (2006), Kowshik et al. (2003), Bharde et al. (2006), and Kathiresan et al. (2009)) which act as a mild reducing agent. In addition, the natural capping agents, which supplied by plants (Table 1),(Shankar et al. (2004), Armendariz et al. (2004), Chandran et al. (2006), Mubarak Ali et al. (2011), Sedki et al. (2015)) and yeast.(Dameron et al. (1989))

4. SYNTHESIS, CHARACTERIZATION AND POTENTIAL APPLICATIONS OF PLASMONIC HYBRID NANOSTRUCTURES

4.1. Magneto-Plasmonic Nanocomposite

In the recent years a new technological term has been emerged as an active topic of research, this term is known as "Magneto-plasmonics".(Armelles et al. (2013)) In such system, both of plasmonic and magnetic functionalities are combined within a unified nano-system, where the magneto-optical (MO) features of the resulting composite become coexist and interrelated.(Armelles et al. (2013), and Bonanni et al. (2011))In magneto-plasmonic nanocomposites two significant effects can be observed; (i) localized enhancement of an electromagnetic field associated with various plasmon resonance polaritions

Table 1. Summarize sources of Biosynthesis of Nanoparticles

Source	Name	Nanoparticles	Size (nm)	Ref.
Plant	Azadirachta indica (Neem)	Au, Ag	10 -100	Shankar et al. 2004
	Geranium leaves plant extract	Ag	16-40	Shankar et al. 2004
	Avena sativa (Oat)	Au	5-85	Armendariz et al. 2004
	Aloe vera	Au	50-530	Chandran et al. 2006
	Mentha piperita (Lamiaceae)	Au, Ag	150	MubarakAli et al. 2011
	Potamogeton pectinatus	Ag	11-20	Sedki et al. 2015
Fungi	Fusarium oxysporum	Au	20-40	Ahmad et al. 2003
	Verticillium sp	Ag	12-25	Mukherjee et al. 2001
	Aspergillus fumigatus	Ag	5-25	Bhainsa et al. 2006
	Schizosaccharomycepombe	CdS	200	Kowshik et al. 2003
	Fusarium oxysporum and Verticillium sp.	Fe_3O_4	20-50	Bharde et al. 2006
	Penicillium fellutanum	Ag	50	Kathiresan et al. 2009
Bacteria	Pseudomonas stutzeri	Ag	200	Klaus et al. 1999
	Lactobacillus strains	Au, Ag	N/A	Nair et al. 2002
	Escherichia coli	CdS	2-5	Sweeney et al. 2004
	Klebsiella pneumonia	Au	5-32	Klaus et al. 1999
	Cyanobacteria	Au	< 10	Lendke et al. 2006
Yeast	Yeast strain MKY3	Ag	2-5	Bhainsa et al. 2006
	Candida glabrata	CdS	200	Dameron et al. 1989
	Schizosaccharomyce pombe	CdS	200	Dameron et al. 1989

(SPPs) can significantly alter the magneto-optical activity.; ii) Concurrently, the plasmonic properties can be controlled by an external magnetic field depending with the off-diagonal elements of the dielectric tensor.(Armelles et al. (2013), Bonanni et al. (2011), and Zhou et al. (2014)) Whereas, these effects could offer the opportunity for further development of active nano-plasmonic devices such as in magnetic field sensors, and data storage.(Armelles et al. (2008)) Moreover, open the door for wide range of applications relevance in biotechnological, and optoelectronics.(Zhou et al. (2014)) This new direction has brought forward a numerous of theoretical studies of MO phenomena, upon the light-matter coupling in magneto-plasmonic nanocomposites.(Bonanni et al. (2011), Melle et al. (2003), and Liu et al. (2009)) These studies showed that the SPR in ferromagnetic materials typically suffer from severe damping than in noble metals. Therefore, the common strategy is to overcome this strong damping by embedding noble metals, and ferromagnetic materials within the same object, where the noble metal increases the plasmonic response.(Chen et al. (2011), Gonzalez-Diaz et al. (2010), Banthi et al. (2012), and Temnov et al. (2010))

Several approaches for fabrication of binary and ternary phase segregated hybrid magneto-plasmonic nanocomposites with different anisotropic structures (i.e. spherical core–shell, core-satellite, heterodimer, nanoalloys, and non-spherical core–shell architectures) have been reported in the literature.(Cho et al. (2005), Lyon et al. (2004), Mikhaylova et al. (2004), Lim et al. (2008)) These synthetic routes includes laser ablation,(Zhang et al. (2006)) layer-by-layer electrostatic deposition,(Spasova et al. (2005)) redox trans-metalation,(Lee et al. (2007)) the reverse-micelle method micro-emulsion,(Mikhaylova et al. (2004)) sonochemical reaction,(Wu et al. (2007)) and chemical reduction in aqueous and organic phases, among others.(Bao et al. (2009), and Umut et al. (2012)) Au@Fe_xO_y as an example of magneto-plasmonic nanocomposites has attracted broad attention due to; (i) the low reactivity, (ii) high chemical stability, (iii) biocompatibility, and (iv) good affinity of the outer Au layer for so many terminal groups.(Zhou et al. (2014)) Zhou et al. reported an efficient route to obtain spherical core-shell from Fe_3O_4@Au NPs, with a particle size ~ 20 nm using sodium citrate reducing agent.(Zhou et al. (2012)) Xu et al. demonstrated a comparison study for both binary, and ternary core-shell systems from Au and Ag- coated magnetite NPs. They reported that the plasmonic properties of these core–shell NPs can be fine-tuned by varying the coating thickness and coating material.(Xu et al. (2007))

Another hetero-structure of Au@Fe_3O_4 composed of two nanoparticles from gold and magnetic nanoparticles, attached to each other inter-facially via sequential growth of magnetic nanoparticles onto the surface of gold nanocrystals forming a structure denoted as *"heterodimers"*. Wang et al. illustrate in their study a facile recipe to conjugates gold and magnetite nanoparticles heterogeneously using polyethylene glycol (PEG) to be used as MRI contrast agent.(Wang et al. (2012)) Another approach has been reported by Lee and co-worker (Lee et al. (2010)) They developed a unique procedures to grow Fe_3O_4NPs onto the surface of Au NPs (Figure 24) in a single system via etching using metal halide in acidic medium for catalysis applications. (Lee et al. (2010))

Although; the hybrid nanocomposites including metal-metal interface with a rod-like (1D one dimensional) shape are powerful in many applications such as high-density magnetic recording devices, or a contrast agents for different biomedical imaging techniques, but there is no more work reported on wide range.(Wetz et al. (2007)) Wang *et al.* who the first group has been synthesized CoPt rod-like alloyed nanostructures using ionic liquids method.(Wang et al. (2005)) Also, CuPt nanorods with controlled

Figure 24. TEM micrographs of as-prepared Au@Fe_3O_4 heterodimers (a & b), and Scheme for growth mechanism of Au-Fe_3O_4 heterodimers NPs (c)
Reprinted with permission from (Lee, Y., Garcia, M. A., Frey Huls, N. A., & Sun, S. (2010).Synthetic tuning of the catalytic properties of Au-Fe_3O_4 nanoparticles. Angewandte Chemie,122(7), 1293-1296.). Copyright (2010) John Wiley and Sons

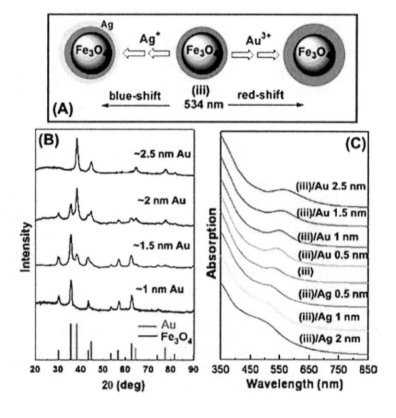

Hybrid Nanostructures

aspect ratio has been prepared using a mixture from oleic acid and oleylamine.(Wang et al. (2009)) Teng *et al.* reported the synthesis of Pt/Au hybrid nanowires via galvanic replacement reaction between Pt nanowires and $AuCl_3$.(Teng et al. (2008 & 2009)) While Wu *et al.* prepared FePt–Au hetero-structured nanostructures by hetero-epitaxial growth of Au nanoparticles onto FePt nanorods.(Wu et al. (2011)) Recently, Brullot *et al.* studied the magneto-optical properties of $Au-Co_xFe_{3-x}O_4$ core–shell nanowires embedded in porous alumina membranes prepared via atomic layer deposition method (ALD).(Brullot et al. (2014)) Zhu *et al.* developed a facile fabrication method 1D Pd-based alloy nanowires including Pd-Pt and Pd-Au using Te nanowires as a sacrificial template and reducing agent.(Zhu et al. (2012))

In case of 1D structure magneto-plasmonic nanoalloys from Au-Co, several attempts have been carried out to obtain this kind of structures. Hall *et al.* focused on a template fabrication of different lengths from Co/Au core-shell nanorods partially embedded in polycarbonate track-etch (PCTE) and anodic aluminum oxide (AAO) nano-porous membranes.(Hall et al. (2005)) Wetz et al. studied the parameters affecting the nucleation of Au nanoparticles (Au NPs) on the Co nanorods. They determined the optimum conditions to grow gold on the surface of Co nanorods by galvanic displacement or by heterogeneous growth.(Wetz et al. (2007)) Another structure of 1D magneto-plasmonic nanoalloy has been fabricated through formation of thin film made of Au/Co/Au trilayers via several techniques such as lithography,(González-Díaz et al. (2008)) ion beam etching,(Armelles et al. (2008)) magnetron sputtering,(Temnov et al. (2010), Armelles et al. (2008), and Torrado et al. (2010)) and molecular beam epitaxy (MBE).(Schouteden and Haesendonck (2010))

Recently, Emam et al. developed a convenient method to fabricate a rod-like Au–Co hybrid nanocomposite of controlled composition and aspect ratio via chemical routes based on modified seed mediated method.(Emam et al. (2015)) The only modification is to use cobalt ions instead of Au^{3+} in the preparation of the seed solution to obtain bimetallic gold-cobalt nanorods as shown in Figure (25). They reported that the length of the nanorods can be tailored by adjusting the cobalt ion concentration within the seed solution as shown in Figure (30). (Emam et al. (2015))

Moreover, the blue shift in the optical absorption spectrum of rod shaped gold-cobalt hybrid nanostructure than that of the pure gold nanorods. This shift is attributed the strong damping of continuous electron density of state within the gold matrix. This damping is originated from the embedding of cobalt clusters in-situ gold matrix, resulting in change of the surface charge density, and the mean free path of electron.(Xu et al. (2007), Boyer et al . (2010), Lica et al. (2004))

In addition, the proposed growth mechanism in their study was based on the galvanic replacement reaction.(Emam et al. (2015)) In such mechanism, growth of the nanorods is strongly dependent on the redox potentials of nanomaterials (i.e. Co clusters), where the redox potential of Co^{2+} (~-0.28Volt)

Figure 25. Scheme for the synthesis of anisotropic structure of Au-Co bimetallic nanoalloys, based on seed mediated method

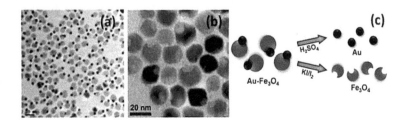

is lower than Au³⁺(~+1.50Volt) and Ag¹⁺(~+0.80Volt). Thus, the cobalt seed particles (i.e. Clusters) will be adsorbed faster on the surface of the gold particle.(Ozkar & Finke (2002), and Reetz & Maase (1999)) The role of these clusters act is to enhance the growth of the gold nanorods forming bimetallic Au-Co nanorods (Figure 26). This proposed mechanism is agreed with the mechanism reported by Liu and Guyot-Sinnnest.(Liu &Guyot-Sinnnest (2005))

4.2. Plasmonic-Semiconductor Nanocomposite

II-VI group semiconductor nanocrystals are very attractive because of their unique optoelectronic properties, which attributed to electron-hole recombination, and quantum size effect.(Efors & Rosen (2000)) They are an excellent candidate for a wide applications such as, light-emitting devices,(Colvin et al. (1994)) nonlinear optics,(Shaviv & Banin (2010)) biological labels,(Derfus et al. (2004)) and photo-electron transfer devices.(Yang et al. (2010), Talapin et al. (2009), and Beecroft & Ober (1997)) Additionally, plasmonic nanoparticles (i.e. Au, and Ag NPs) show unique optical properties, which attributed to localized surface plasmon resonance (LSPR),(Mody et al. (2010), Link & El-Sayed (2003), and Thomas & Kamat (2003)) and surface enhanced Raman scattering (SERS).(Cao et al. (2002)) Therefore, they are considered as potential components in highly compacted optoelectronic and nonlinear optical devices as well as sensors.(McPhillips et al. (2010))

Hybrid metal-semiconductor nanocomposites have brought numerous studies, due to their promising applications in development of electronic and optoelectronic nano-devices. Therefore, the capability to tailoring each of size, shape, and chemical composition for these composite constituents, as well as to exploit the combination of properties for both of semiconductor and plasmonic nanostructures, not to provide many viable devices, such as ultra-small transistors, memory elements,(Kong et al. (2004)) light-emitting elements,(Lin et al. (2006), and Achermann (2010)) and sensors.(Lee & El-Sayed (2006), and Lee et al. (2007)) But also generate new phenomena based on the intra-particles interaction between the metal and the semiconductor nanocrystals at their interface.(Shaviv & Banin (2010), Costi et al. (2009), Yuhas et al. (2009), de Paiva & Di Felice (2008)) Despite of this huge potential of these nano-hybrid, except most current studies limited so far to isotropic shapes, and little investigated.(de Paiva & Di Felice

Figure 26. Shows the proposed mechanism of rod shaped Au-Co hybrid nanostructure
Reprinted with permission from (Emam, A. N., Mohamed, M. B., Girgis, E., & Rao, K. V. (2015). Hybrid magnetic–plasmonic nanocomposite: embedding cobalt clusters in gold nanorods. RSC Advances, 5(44), 34696-34703.). Copyright (2010) Royal Society of Chemistry

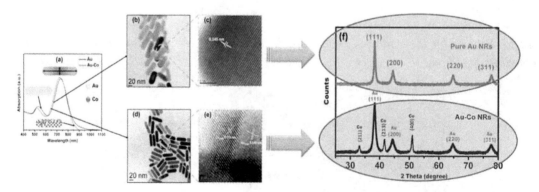

(2008)) These limitations were especially due to difficulties in their fabrication techniques. Moreover, understanding the metal-semiconductor interactions and how it depends on; (i) size, (ii) shape, and (iii) the spacing between the metallic core and the semiconductor shell within the hybrid nanocomposites. (AbouZeid et al. (2011), Achermann (2010), and Mokari et al. (2004)) The latter is crucial for many potential applications of such as biological labeling,(Li et al. (2009)) photocatalysis,(Zeng et al. (2008), Dawson & Kamat (2001), and Halder et al. (2012)) photoelectronics,(Yu & Wang (2010)) and photo-voltaic.(Ichikawa et al. (2008), and Har-Lavan et al. (2008))

The presence of metal-semiconductor interface promotes effective charge separation carrier transfers,(Costi et al. (2009)) which subsequently enhances the photo-catalytic activity. In addition, the metal core can provide an anchor point for electrical or chemical connections to the functional semiconductor part.(Mokari et al. (2004 & 2005)) There are two main postulates according to the distance between plasmonic, and semiconductor nanocrystals. The first one, where plasmonic is part separated from semiconductor part by a dielectric spacer. In this case, light–mater interactions will be enhanced near metal surface which based on the enhancement of local fields associated with the surface plasmon (SP) of the metallic part. This effect could enhance the fluorescence of the semiconductor part significantly. (Lin et al. (2006), and Achermann (2010)) The second, is the non-radiative dumping due to either energy transfer between the semiconductor and the metal or the electron transfer from semiconductor to the metal, when the particles in close proximity contact.(Achermann (2010), and Shevchenko et al. (2008))

Although the physico-chemical properties of metal-semiconductor hybrid nanostructures are strongly dependent on their size, shape, composition, and spatial distribution of each component. Therefore, the find out of new synthetic routes for these hybrid nanostructures has become increasingly important. Several attempts have been devoted in order to obtain different architectures from metal-semiconductor nanocomposites. Sanuders et al. reported the growth mechanism of gold nanocrystals onto preformed CdS nanorods to form hybrid metal-semiconductor nanorod colloids. They showed that by manipulating the growth conditions, it is possible to obtain nanostructures exhibiting Au nanocrystal growth at only one nanorod tip, at both tips, or at multiple locations (Figure 27) along the nanorod surface.(Saunders et al. (2006)).

Ma et al. found the resonant energy transfer and increase of the two-photon absorption coefficient in Au-CdS composites.(Ma et al. (2004)) Yang *et al.* reported the enhancement of third-order optical nonlinearity in densely packed Au-CdS composite films.(Yang et al. (2005)) Tang *et al.* assembled the CdSe-CdS core-shell quantum dots on a chemically modified Au surfaces for photoreception uses. (Tang et al. (2003)) Granot *et al.* prepared CdS semiconductor nanoparticle monolayer on Au surfaces for photoelectron-chemical applications.(Granot et al. (2004)) Therefore, it should be very interesting to assemble noble metal nanoparticles on II-VI group semiconductor nanorods and study the property of such 1D superstructure.

Liang et al. represents a facile method for synthesis of Au-AgCdSe hybrid nanorods (NRs) through the deposition of silver (Ag) tips at the ends of Au NRs as a seed solution, followed by selenization of the silver (Ag) tips, and overgrowth of CdSe on these sites.(Liang et al . (2012)) This is method has been carried out by controlling the pH value.(Liang et al . (2012))Das et al.(Das et al. (2013)) demonstrate a method to fabricate Au decorated CdSe nanowires (NWs) based on the wet chemical method (Mokari et al. (2005), Kuno (2008), and Puthussery et al. (2009)) to be employed as surface enhanced Raman spectroscopy (SERS) substrates.(Das et al. (2013)) In this synthetic route, the Au NPs prefer to nucleate on lattice defects at the lateral facets of the CdSe NWs, which have a great effect to obtain a homogeneous

Figure 27. Comparison of gold tipped CdSe nanorods under argon (a) and under air (b) after 3 h
Reprinted with permission from (Saunders, A. E., Popov, I., & Banin, U. (2006). Synthesis of hybrid CdS-Au colloidal nanostructures. The Journal of Physical Chemistry B, 110(50), 25421-25429). Copyright (2006) American Chemical Society

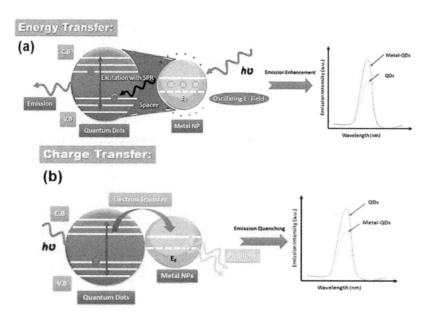

distribution of Au NPs on the nanowire.(Das et al. (2013)) Bala et al. (Bala et al. (2013)) demonstrated an effective procedure based on phase transfer between the aqueous and organic media in order to form Au tips on cadmium chalcogenide NPs and NRs. This phase transfer process can be achieved in presence of organic ligand which act as phase transfer and reducing agent for Au^{3+} ions.(Bala et al. (2013)) In another attempt to design Au-CdSe nanorods heterostructure, Li et al. represent an approach for the selective deposition of metals (i.e. Pd, and Fe) on Au-tipped CdSe-seeded CdS nanorods that upon light-induced deposition.(Li et al. (2010)) Pena et al. developed an facile route based on electrochemical replication of porous aluminum oxide (AAO) and polycarbonate track etch membranes with pore diameters of 350 and 70 nm, respectively.(Pena et al. (2002))

Lu et al. (Lu et al. (2005)) demonstrated a facile route for small Au-CdSe core/shell spherical particles based on Au-Cd nano-alloy as a precursor, which reacted with the TOP-Se complex at high temperature leading to formation of Au/CdSe core/shell structure (Figure 28). (Lu et al. (2005)) Pullabhotla *et al.* reported synthesis of cysteine capped Au–CdSe hybrid nanoparticles (Figure 29), using a simple solution based route. The hybrid material iswater soluble allowing for the conjugation to biomolecules in potential biomedical applications.(Salant et al. (2006))

Another heterostructures from metal-semiconductor hybrid nanocomposites (i.e. tetrapods, dumbles, hypber-branched,flowers, Pentapods) have been achieved. In 2004, Banin and co-workers reported a straightforward procedure enabling selective growth of gold tips onto the apexes of CdSe nanorods and tetrapods, resulting in the formation of dumbbell-shaped metallic-semiconductor hybrids.(Mokari et al. (2004)) They found an obvious emission quenching and the increase of the conductivity for CdSe nanorods. The gold deposition method was later extended to other CdSe structures, enabling deposition of single gold tips onto quantum dots.(Mokari et al. (2005))

Figure 28. High-resolution images of Au/CdSe core/shell particles (a-e). High-resolution image of a CdSe nanoparticle without Au core(f). Optical absorption spectra for Au particles of 4.0 and 2.2 nm, Au/CdSe particles of 6.0 nm, CdSe byproduct (fromtop to bottom) (g)
Reprinted with permission from (Lu, W., Wang, B., Zeng, J., Wang, X., Zhang, S., & Hou, J. G. (2005). Synthesis of core/shell nanoparticles of Au/CdSe via Au-Cd bialloy precursor. Langmuir, 21(8), 3684-3687.). Copyright (2005) American Chemical Society

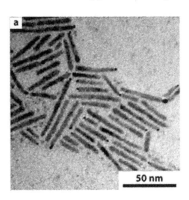
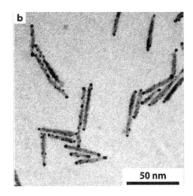

Figure 29. TEM images (taken prior to separation) of (a) dimers of NDBs in different conjugation angles, (b) trimers, and (c) flowers (50 nm scale bars)
Reprinted with permission from (Salant, A., Amitay-Sadovsky, E., & Banin, U. (2006). Directed self-assembly of gold-tipped CdSe nanorods. Journal of the American Chemical Society,128(31), 10006-10007). Copyright (2006) American Chemical Society

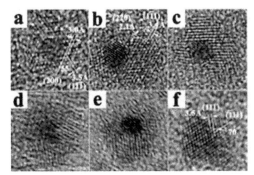
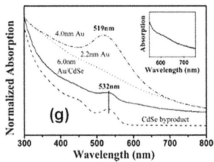

Salant et al. modified a hybrid metal-tipped semiconductor nanorods system known as nano-dumbbells, and used the gold tips as an anchor points for self-assembly using simple thiol molecules (Figure 29). (Salant et al. (2006))

Khalavka and Sonnichsen reported on the synthesis of a novel metal–semiconductor hybrid based on hyperbranched CdTe particles (Figure 30). (Khalavka & Sonnivhsen (2008)) Gold particles have been grown selectively onto the tips of preformed semiconductor substrates at ambient conditions. Hyperbranched CdTe particles have been synthesized in tri-octyl-phosphine oxide (TOPO) by the reaction of tri-octyl-phosphine telluride (TOP=Te) with Cd-phosphonic acid (PA) complexes under an argon atmosphere. The metal tips have been added to the hyperbranched particles in a second step at ambient temperature by mixing with a gold growth solution containing $HAuCl_4$, dodecylamine (DDA), and didodecyl diamine ammonium bromide (DDAB) in toluene; the tips of the hyperbranched particles act as nucleation centers for the growth of gold. (Khalavka & Sonnivhsen (2008))

Figure 30. TEM images showing growth of gold onto the tips of hyperbranched particles. a) As-prepared hyperbranched CdTe nanoparticles. b–d) Hybrid (gold on CdTe) structures grown on the particles in (a) with molar ratios of CdTe:gold of b) 1:0.5, c) 1:1, and d) 1:2. Changes from some tips covered by gold (b) to non-selective growth (d) are clearly apparent. The scale bars represent 100 nm. (e) HRTEM image of the gold-coated tip of a CdTe branch
Reprinted with permission from (Khalavka, Y., & Sönnichsen, C. (2008). Growth of gold tips onto hyperbranched CdTe nanostructures. Advanced Materials, 20(3), 588-591). Copyright (2008) John Wiley and Sons

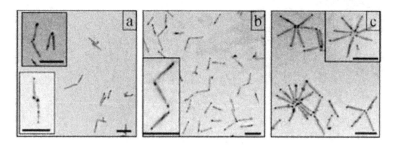

In (2011), AbouZeid et al. reported a facile method in order to control the shape (i.e. Nano-flowers) of the metal-semiconductor hybrid nanocomposites. This method based on organometallic pyrolysis, leading to heterogeneous growth and nucleation of CdSe nanostructures on the growth facets of metallic core (i.e. Au, and Ag NPs) as shown in Figures (31 and 32.(AboZeid et al. (2011))

Recently; we developed in our research group a facile synthetic route based on to fabricate a new heterostructures (i.e. Tetrapods-like shape) from the Au/CdSe hybrid nanocomposites. The growth mechanism of such heterostructures was based on the "*Direct heterogeneous deposition*" mechanism (Figure 33), in which mechanism the heterogeneous nucleation growth of CdSe arms (nanorods) structure at the nucleating sites on the surface defects of gold nano-clusters.(Mansour et al. (2016)) In our study, we monitored a remarkable fluorescence enhancement upon their loading on graphene oxide as spacer.(Mansour et al. (2016))

Other common semiconductor–metal nanocomposites include Au-ZnS, Ag–ZnO, Au-ZnO, Pt–TiO_2, and Au–TiO_2 that are important to be used in so many applications. Herring and co-workers develop a convenient "one-pot" route for the synthesis of hybrid Au-ZnO hexagonal nano-pyramids, which based on sequential heterogeneous nucleation steps involving both Au and Zn ions using MWI technique as shown in Figure (34).(Herring et al. (2011))

4.3. Plasmonic-Graphene Nanocomposite

The most popular strategies used for the synthesis of graphene–Au nanocomposites is the direct chemical reduction of Au precursor (e.g. $HAuCl_4$) in the presence of graphene oxide (GO) or reduced graphene oxide (RGO) sheets using reducing agent such as amines, $NaBH_4$, and ascorbic acid.(Huang et al. (2010 & 2012), and Lu et al. (2009)) Besides spherical Au NPs, (Vinodgopal et al. (2010)) other unique anisotropic Au nanostructures have also been synthesized on graphene-based materials. (Huang et al. (2010 & 2012), Lu et al. (2009), and Jasuja & Berry (2009)) Jasuja and Berry developed a facile method to obtain snow-flake-shaped Au nanostructures on a GO surface based seeded mediated method as shown in Figure (35).(Lu et al. (2009)) In such method, Au NPs formed upon reduction of $HAuCl_4$ using hydroxyl-amine

Hybrid Nanostructures

Figure 31. (a) Absorption spectra as a function of reaction time of the hybrid Au–CdSe nanocrystals prepared at 150 ° C. TEM images of samples obtained at different reaction times: (b) 0 min (Au nanocrystals only), (c–e) 5 min, (f) 25 min, and (g) 40 min
Reprinted with permission from (AbouZeid, K. M., Mohamed, M. B., & El-Shall, M. S. (2011). Small, 7(23), 3299-3307).
Copyright (2011) John Wiley and Sons

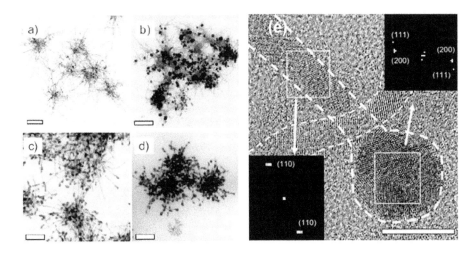

Figure 32. (a) Absorption spectra as a function of reaction time of the hybrid Ag–CdSe nanocrystals prepared at 150 ° C. TEM images of samples obtained at different reaction times: (b) 0 min (Ag nanocrystals only); (c) 35 min; (d) 85 min; and (e) 140 min
Reprinted with permission from (AbouZeid, K. M., Mohamed, M. B., & El-Shall, M. S. (2011). Small, 7(23), 3299-3307).
Copyright (2011) John Wiley and Sons

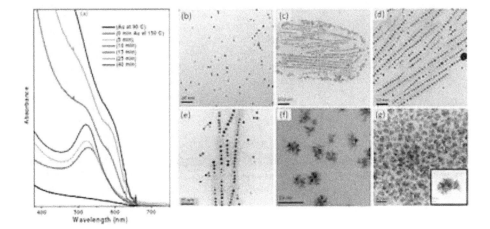

as the reduction agent. (Lu et al. (2009)) Au nanorods were obtained on RGO films or seed modified RGO films by a simple electro-deposition method at a low temperature.(Arif et al. (2011))

Huang and co-workers succeeded in their reports to synthesize several anisotropic structures of Au NPs (i.e. dots, wires, and sheets) on GO and RGO as a template for Au NPs, via photochemical reduction method of $HAuCl_4.3H_2O$ (Figure 36).((Huang et al. (2010 & 2012))

Figure 33. Scheme for the proposed growth mechanism of Au/CdSe tetrapod-like shape Heterostructure

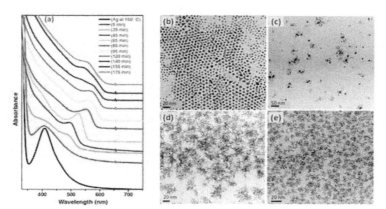

Figure 34. TEM images of as-prepared Au-ZnO hybrid nano-pyramids
Reprinted with permission from (Herring, N. P., AbouZeid, K., Mohamed, M. B., Pinsk, J., & El-Shall, M. S. (2011). Formation mechanisms of gold–zinc oxide hexagonal nanopyramids by heterogeneous nucleation using microwave synthesis. Langmuir, 27(24), 15146-15154). Copyright (2011) American Chemical Society

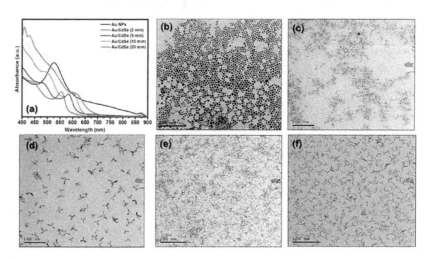

Figure 35. Formation mechanism of snow-flake-shaped gold nanostructures on graphene oxide (GO)
Reprinted with permission from (Jasuja, K., & Berry, V. (2009). Implantation and growth of dendritic gold nanostructures on graphene derivatives: electrical property tailoring and Raman enhancement. ACS Nano, 3(8), 2358-2366). Copyright (2009) American Chemical Society

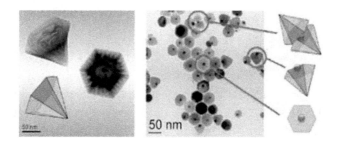

Figure 36. (a) TEM image of ordered Au nanodots photochemically synthesized on rGO
Reprinted with permission from (Huang, X., Zhou, X., Wu, S., Wei, Y., Qi, X., Zhang, J., Boey, F., Zhang, H. (2010). Reduced Graphene OxideTemplated Photochemical Synthesis and in situ Assembly of Au Nanodots to Orderly Patterned Au Nanodot Chains. Small,6(4), 513-516). Copyright (2010) John Wiley & Sons, Inc.(b) TEM image of Au nanowires grown on GO. Reprinted with permission from (Huang, X., Li, S., Wu, S., Huang, Y., Boey, F., Gan, C. L., & Zhang, H. (2012). Graphene OxideTemplated Synthesis of Ultrathin or TadpoleShaped Au Nanowires with Alternating hcp and fcc Domains. Advanced Materials, 24(7), 979-983). Copyright (2012) John Wiley & Sons, Inc

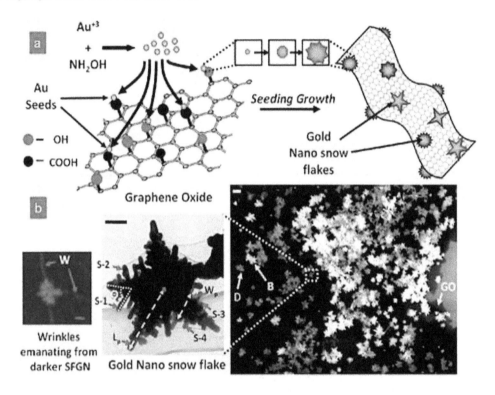

In our research group, we developed a one-pot process using sunlight, UV light or halogen lamp to synthesize shape controlled gold nanoparticles loaded into graphene (i.e. spherical, rods, wires, and flake) in the presence of a mixture of oleylamine and oleic acid as shown in Figure (37). We found that, the final particle shape and the growth rate depend on the ratio of the two surfactants (OAc:OAm), as well as the concentration of gold ions and the Graphene oxide.(Rady et al. (2016))

Besides Au nanostructures which have been successfully incorporated with graphene has been mentioned in so many reports.(Zhou et al. (2009), Lu et al. (2011), Lu et al. (2009), Pasricha et al. (2009), Wen et al. (2010), Li & Liu (2010), and Ren et al. (2011), Zhu et al. (2011)) In the synthesis of Ag-Graphene nanocomposites, silver nitrate ($AgNO_3$) has been used as metal precursor, in presence of mild reducing agent including amines, borohydride, and ascorbic acid (Figure (38)).(Zhou et al. (2009), Lu et al. (2011), Lu et al. (2009), Pasricha et al. (2009), Wen et al. (2010), Li & Liu (2010), and Ren et al. (2011), Zhu et al. (2011)) In such case, Ag NPs are physically adsorbed on the surface graphene sheets via electrostatic force.(Lu et al. (2009)). Various graphene–Pd, and Pt composites have been prepared by similar synthetic strategies to those used for the preparation of graphene based Au, Ag, and Pt nanostructures.(Bong et al. (2010), Yang et al. (2012), Yin et al. (2012), Wu et al. (2012), and Qin & Li (2010))

Figure 37. TEM image of ordered Au (a) Rods, (b) wires, (c) spheres, and (d) flakes photo-chemically synthesized on GO using sun-light

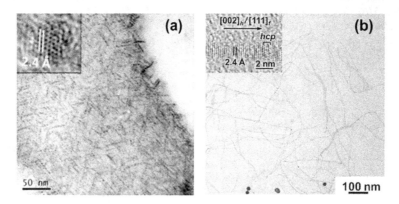

Figure 38. (a) Scanning electron microscopy (SEM) image of Ag NPs grown on GO
Reprinted with permission from (Zhou, X., Huang, X., Qi, X., Wu, S., Xue, C., Boey, F. Y., Yan, Q., Chen P., Zhang, H. (2009). In situ synthesis of metal nanoparticles on single-layer graphene oxide and reduced graphene oxide surfaces. The Journal of Physical Chemistry C,113(25), 10842-10846). Copyright (2009) American Chemical Society.(b) Typical SEM images of the synthesized Ag/AgBr/GO. Reprinted with permission from (Zhu, M., Chen, P., & Liu, M. (2011). Graphene oxide enwrapped Ag/AgX (X= Br, Cl) nanocomposite as a highly efficient visible-light plasmonic photocatalyst. ACS nano, 5(6), 4529-4536). Copyright (2011) American Chemical Society

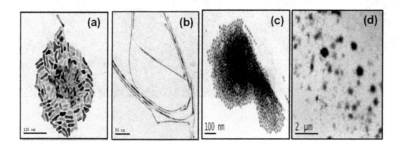

Graphene-supported Ru NPs were synthesized through microwave irradiation in ionic liquids,(Choi et al. (2011)) or via a wet-chemical method combined with thermal exfoliation of graphite.(Sutter et al. (2012)) Recently, Rh nanoclusters assembled on graphene were reported.(Choi et al. (2011)) In addition, the highly water dispersible graphene–Rh nanocomposites were successfully synthesized by the simple reduction of Rhon (PEO/PPO/PEO) tri block copolymer or pluronic-stabilized GO nano sheets with $NaBH_4$.(Chandra et al. (2011))

5. CONCLUSION

In conclusion, we introduced complete survey about the optical properties of plasmonic nanomaterials and their fabrication methods. The dependence of the surface Plasmon resonance on size, shape and chemical compositions has explored in details. Hybrid plasmonic nanostructures consisting of metallic nanoparticles coupled with magnetic, semiconductor or graphene were discussed in details. The coupling between the localized plasmon of the metal component and the excitons in semiconductor nanomaterials and the magnetic field of the magnetic nanomaterials opens new doors for muti-functional materials.

REFERENCES

Abdelsayed, V., Aljarash, A., El-Shall, M. S., Al Othman, Z. A., & Alghamdi, A. H. (2009). Microwave synthesis of bimetallic nanoalloys and CO oxidation on ceria-supported nanoalloys. *Chemistry of Materials*, *21*(13), 2825–2834. doi:10.1021/cm9004486

AbouZeid, K. M., Mohamed, M. B., & El-Shall, M. S. (2011). Hybrid Au–CdSe and Ag–CdSe Nanoflowers and Core–Shell Nanocrystals via One-Pot Heterogeneous Nucleation and Growth. *Small*, *7*(23), 3299–3307. doi:10.1002/smll.201100688 PMID:21994186

Achermann, M. (2010). Exciton−Plasmon Interactions in Metal− Semiconductor Nanostructures. *The Journal of Physical Chemistry Letters*, *1*(19), 2837–2843. doi:10.1021/jz101102e

Ahmad, A., Mukherjee, P., Senapati, S., Mandal, D., Khan, M. I., Kumar, R., & Sastry, M. (2003). Extracellular biosynthesis of silver nanoparticles using the fungus Fusarium oxysporum. *Colloids and Surfaces. B, Biointerfaces*, *28*(4), 313–318. doi:10.1016/S0927-7765(02)00174-1

Anandan, S., Grieser, F., & Ashokkumar, M. (2008). Sonochemical synthesis of Au− Ag core− shell bimetallic nanoparticles. *The Journal of Physical Chemistry C*, *112*(39), 15102–15105. doi:10.1021/jp806960r

Arif, M., Heo, K., Lee, B. Y., Lee, J., Seo, D. H., Seo, S., & Hong, S. et al. (2011). Metallic nanowire–graphene hybrid nanostructures for highly flexible field emission devices. *Nanotechnology*, *22*(35), 355709. doi:10.1088/0957-4484/22/35/355709 PMID:21828894

Armelles, G., Cebollada, A., García-Martín, A., & González, M. U. (2013). Magnetoplasmonics: Combining magnetic and plasmonic functionalities. *Advanced Optical Materials*, *1*(1), 10–35. doi:10.1002/adom.201200011

Armelles, G., González-Díaz, J. B., García-Martín, A., García-Martín, J. M., Cebollada, A., González, M. U., & Badenes, G. et al. (2008). Localized surface plasmon resonance effects on the magneto-optical activity of continuous Au/Co/Au trilayers. *Optics Express*, *16*(20), 16104–16112. doi:10.1364/OE.16.016104 PMID:18825249

Armendariz, V., Herrera, I., Jose-yacaman, M., Troiani, H., Santiago, P., & Gardea-Torresdey, J. L. (2004). Size controlled gold nanoparticle formation by Avena sativa biomass: Use of plants in nanobiotechnology. *Journal of Nanoparticle Research*, *6*(4), 377–382. doi:10.1007/s11051-004-0741-4

Attia, Y. A., Buceta, D., Blanco-Varela, C., Mohamed, M. B., Barone, G., & López-Quintela, M. A. (2014). Structure-Directing and High-Efficiency Photocatalytic Hydrogen Production by Ag Clusters. *Journal of the American Chemical Society*, *136*(4), 1182–1185. doi:10.1021/ja410451m PMID:24410146

Bala, T., Singh, A., Sanyal, A., O'Sullivan, C., Laffir, F., Coughlan, C., & Ryan, K. M. (2013). Fabrication of noble metal-semiconductor hybrid nanostructures using phase transfer. *Nano Research*, *6*(2), 121–130. doi:10.1007/s12274-013-0287-9

Bang, J. H., & Suslick, K. S. (2010). Applications of ultrasound to the synthesis of nanostructured materials. *Advanced Materials*, *22*(10), 1039–1059. doi:10.1002/adma.200904093 PMID:20401929

Banthí, J. C., Meneses-Rodríguez, D., García, F., González, M. U., García-Martín, A., Cebollada, A., & Armelles, G. (2012). High Magneto-Optical Activity and Low Optical Losses in Metal-Dielectric Au/Co/Au–SiO2 Magnetoplasmonic Nanodisks. *Advanced Materials*, *24*(10), OP36–OP41. PMID:22213149

Bao, F., Yao, J. L., & Gu, R. A. (2009). Synthesis of magnetic Fe2O3/Au core/shell nanoparticles for bioseparation and immunoassay based on surface-enhanced Raman spectroscopy. *Langmuir*, *25*(18), 10782–10787. doi:10.1021/la901337r PMID:19552373

Barcaro, G., Sementa, L., Fortunelli, A., & Stener, M. (2015). Optical properties of nanoalloys. *Physical Chemistry Chemical Physics*, *17*(42), 27952–27967. doi:10.1039/C5CP00498E PMID:25875393

Beecroft, L. L., & Ober, C. K. (1997). Nanocomposite materials for optical applications. *Chemistry of Materials*, *9*(6), 1302–1317. doi:10.1021/cm960441a

Bera, P., Kim, C. H., & Seok, S. I. (2010). High-yield synthesis of quantum-confined CdS nanorods using a new dimeric cadmium (II) complex of S-benzyldithiocarbazate as single-source molecular precursor. *Solid State Sciences*, *12*(4), 532–535. doi:10.1016/j.solidstatesciences.2009.12.020

Berthelot, J. M. (1999). *Composite materials: mechanical behavior and structural analysis*. Springer-Verlag Berlin Heidelberg New York. doi:10.1007/978-1-4612-0527-2

Bhainsa, K. C., & D'souza, S. F. (2006). Extracellular biosynthesis of silver nanoparticles using the fungus Aspergillus fumigatus. *Colloids and Surfaces. B, Biointerfaces*, *47*(2), 160–164. doi:10.1016/j.colsurfb.2005.11.026 PMID:16420977

Bharde, A., Rautaray, D., Bansal, V., Ahmad, A., Sarkar, I., Yusuf, S. M., & Sastry, M. (2006). Extracellular biosynthesis of magnetite using fungi. *Small*, *2*(1), 135–141. doi:10.1002/smll.200500180 PMID:17193569

Bogue, R. (2011). Nanocomposites: A review of technology and applications. *Assembly Automation*, *31*(2), 106–112. doi:10.1108/01445151111117683

Bohren, C. F., & Huffman, D. R. (1983). *Absorption and Scattering of Light by Small Particles*. New York: Wiley.

Bonanni, V., Bonetti, S., Pakizeh, T., Pirzadeh, Z., Chen, J., Nogués, J., & Dmitriev, A. et al. (2011). Designer magnetoplasmonics with nickel nanoferromagnets. *Nano Letters*, *11*(12), 5333–5338. doi:10.1021/nl2028443 PMID:22029387

Bong, S., Uhm, S., Kim, Y. R., Lee, J., & Kim, H. (2010). Graphene supported Pd electrocatalysts for formic acid oxidation. *Electrocatalysis*, *1*(2-3), 139–143. doi:10.1007/s12678-010-0021-2

Boyer, P., Ménard, D., & Meunier, M. (2010). Nanoclustered Co– Au Particles Fabricated by Femtosecond Laser Fragmentation in Liquids. *The Journal of Physical Chemistry C, 114*(32), 13497–13500. doi:10.1021/jp1037552

Brotchie, A., Grieser, F., & Ashokkumar, M. (2008). Sonochemistry and sonoluminescence under dual-frequency ultrasound irradiation in the presence of water-soluble solutes. *The Journal of Physical Chemistry C, 112*(27), 10247–10250. doi:10.1021/jp801763v

Brown, K. R., Walter, D. G., & Natan, M. J. (2000). Seeding of colloidal Au nanoparticle solutions. 2. Improved control of particle size and shape. *Chemistry of Materials, 12*(2), 306–313. doi:10.1021/cm980065p

Brullot, W., Strobbe, R., Bynens, M., Bloemen, M., Demeyer, P. J., Vanderlinden, W., & Verbiest, T. et al. (2014). Layer-by-Layer synthesis and tunable optical properties of hybrid magnetic–plasmonic nanocomposites using short bifunctional molecular linkers. *Materials Letters, 118*, 99–102. doi:10.1016/j.matlet.2013.12.057

Burda, C., Chen, X., Narayanan, R., & El-Sayed, M. A. (2005). Chemistry and properties of nanocrystals of different shapes. *Chemical Reviews, 105*(4), 1025–1102. doi:10.1021/cr030063a PMID:15826010

Callegari, A., Tonti, D., & Chergui, M. (2003). Photochemically grown silver nanoparticles with wavelength-controlled size and shape. *Nano Letters, 3*(11), 1565–1568. doi:10.1021/nl034757a

Cao, Y. C., Jin, R., & Mirkin, C. A. (2002). Nanoparticles with Raman spectroscopic fingerprints for DNA and RNA detection. *Science, 297*(5586), 1536–1540. doi:10.1126/science.297.5586.1536 PMID:12202825

Caruso, R. A., Ashokkumar, M., & Grieser, F. (2002). Sonochemical formation of gold sols. *Langmuir, 18*(21), 7831–7836. doi:10.1021/la020276f

Chandra, S., Bag, S., Bhar, R., & Pramanik, P. (2011). Sonochemical synthesis and application of rhodium–graphene nanocomposite. *Journal of Nanoparticle Research, 13*(7), 2769–2777. doi:10.1007/s11051-010-0164-3

Chandran, S. P., Chaudhary, M., Pasricha, R., Ahmad, A., & Sastry, M. (2006). Synthesis of gold nanotriangles and silver nanoparticles using Aloevera plant extract. *Biotechnology Progress, 22*(2), 577–583. doi:10.1021/bp0501423 PMID:16599579

Chen, J., Albella, P., Pirzadeh, Z., Alonso-González, P., Huth, F., Bonetti, S., & Hillenbrand, R. (2011). Plasmonic nickel nanoantennas. *Small, 7*(16), 2341–2347. doi:10.1002/smll.201100640 PMID:21678553

Chen, W., Zhao, J., Lee, J. Y., & Liu, Z. (2005). Microwave heated polyol synthesis of carbon nanotubes supported Pt nanoparticles for methanol electrooxidation. *Materials Chemistry and Physics, 91*(1), 124–129. doi:10.1016/j.matchemphys.2004.11.003

Cho, S. J., Idrobo, J. C., Olamit, J., Liu, K., Browning, N. D., & Kauzlarich, S. M. (2005). Growth mechanisms and oxidation resistance of gold-coated iron nanoparticles. *Chemistry of Materials, 17*(12), 3181–3186. doi:10.1021/cm0500713

Choi, S. M., Seo, M. H., Kim, H. J., & Kim, W. B. (2011). Synthesis and characterization of graphene-supported metal nanoparticles by impregnation method with heat treatment in H 2 atmosphere. *Synthetic Metals*, *161*(21), 2405–2411. doi:10.1016/j.synthmet.2011.09.008

Colvin, V. L., Schlamp, M. C., & Alivisatos, A. P. (1994). Light-emitting diodes made from cadmium selenide nanocrystals and a semiconducting polymer. *Nature*, *370*(6488), 354–357. doi:10.1038/370354a0

Costi, R., Cohen, G., Salant, A., Rabani, E., & Banin, U. (2009). Electrostatic force microscopy study of single Au−CdSe hybrid nanodumbbells: Evidence for light-induced charge separation. *Nano Letters*, *9*(5), 2031–2039. doi:10.1021/nl900301v PMID:19435381

Creighton, J. A., & Eadon, D. G. (1991). Ultraviolet–visible absorption spectra of the colloidal metallic elements. *Journal of the Chemical Society, Faraday Transactions*, *87*(24), 3881–3891. doi:10.1039/FT9918703881

Dameron, C. T., Reese, R. N., Mehra, R. K., Kortan, A. R., Carroll, P. J., Steigerwald, M. L., & Winge, D. R. et al. (1989). Biosynthesis of cadmium sulphide quantum semiconductor crystallites. *Nature*, *338*(6216), 596–597. doi:10.1038/338596a0

Das, G., Chakraborty, R., Gopalakrishnan, A., Baranov, D., Di Fabrizio, E., & Krahne, R. (2013). A new route to produce efficient surface-enhanced Raman spectroscopy substrates: Gold-decorated CdSe nanowires. *Journal of Nanoparticle Research*, *15*(5), 1–9. doi:10.1007/s11051-013-1596-3

Dawson, A., & Kamat, P. V. (2001). Semiconductor-metal nanocomposites. Photoinduced fusion and photocatalysis of gold-capped TiO2 (TiO2/gold) nanoparticles. *The Journal of Physical Chemistry B*, *105*(5), 960–966. doi:10.1021/jp0033263

de Paiva, R., & Di Felice, R. (2008). Atomic and Electronic Structure at Au/CdSe Interfaces. *ACS Nano*, *2*(11), 2225–2236. doi:10.1021/nn8004608 PMID:19206387

Derfus, A. M., Chan, W. C., & Bhatia, S. N. (2004). Intracellular delivery of quantum dots for live cell labeling and organelle tracking. *Advanced Materials*, *16*(12), 961–966. doi:10.1002/adma.200306111

Efros, A. L., & Rosen, M. (2000). The Electronic Structure of Semiconductor Nanocrystals. *Annual Review of Materials Science*, *30*(1), 475–521. doi:10.1146/annurev.matsci.30.1.475

Emam, A. N., Mohamed, M. B., Girgis, E., & Rao, K. V. (2015). Hybrid magnetic–plasmonic nanocomposite: Embedding cobalt clusters in gold nanorods. *RSC Advances*, *5*(44), 34696–34703. doi:10.1039/C5RA01918D

Eustis, S., & El-Sayed, M. A. (2006). Why gold nanoparticles are more precious than pretty gold: Noble metal surface plasmon resonance and its enhancement of the radiative and nonradiative properties of nanocrystals of different shapes. *Chemical Society Reviews*, *35*(3), 209–217. doi:10.1039/B514191E PMID:16505915

Faraday, M. (1857). The Bakerian lecture: Experimental relations of gold (and other metals) to light. *Philosophical Transactions of the Royal Society of London*, *147*(0), 145–181. doi:10.1098/rstl.1857.0011

Ferrando, R., Jellinek, J., & Johnston, R. L. (2008). Nanoalloys: From theory to applications of alloy clusters and nanoparticles. *Chemical Reviews*, *108*(3), 845–910. doi:10.1021/cr040090g PMID:18335972

Fleming, D. A., & Williams, M. E. (2004). Size-controlled synthesis of gold nanoparticles via high-temperature reduction. *Langmuir, 20*(8), 3021–3023. doi:10.1021/la0362829 PMID:15875823

Franke, R., Rothe, J., Pollmann, J., Hormes, J., Bönnemann, H., Brijoux, W., & Hindenburg, T. (1996). A Study of the Electronic and Geometric Structure of Colloidal Ti0⊙ 0.5 THF. *Journal of the American Chemical Society, 118*(48), 12090–12097. doi:10.1021/ja953525d

Gerbec, J. A., Magana, D., Washington, A., & Strouse, G. F. (2005). Microwave-enhanced reaction rates for nanoparticle synthesis. *Journal of the American Chemical Society, 127*(45), 15791–15800. doi:10.1021/ja052463g PMID:16277522

Ghosh Chaudhuri, R., & Paria, S. (2011). Core/shell nanoparticles: Classes, properties, synthesis mechanisms, characterization, and applications. *Chemical Reviews, 112*(4), 2373–2433. doi:10.1021/cr100449n PMID:22204603

Girgis, E., Khalil, W. K. B., Emam, A. N., Mohamed, M. B., & Rao, K. V. (2012). Nanotoxicity of gold and gold–cobalt nanoalloy. *Chemical Research in Toxicology, 25*(5), 1086–1098. doi:10.1021/tx300053h PMID:22486372

González-Díaz, J. B., Sepúlveda, B., García-Martín, A., & Armelles, G. (2010). Cobalt dependence of the magneto-optical response in magnetoplasmonic nanodisks. *Applied Physics Letters, 97*(4), 043114. doi:10.1063/1.3474617

Granot, E., Patolsky, F., & Willner, I. (2004). Electrochemical assembly of a CdS semiconductor nanoparticle monolayer on surfaces: Structural properties and photoelectrochemical applications. *The Journal of Physical Chemistry B, 108*(19), 5875–5881. doi:10.1021/jp038004o

Gutierrez, M., Henglein, A., & Dohrmann, J. K. (1987). Hydrogen atom reactions in the sonolysis of aqueous solutions. *Journal of Physical Chemistry, 91*(27), 6687–6690. doi:10.1021/j100311a026

Haldar, K. K., Sinha, G., Lahtinen, J., & Patra, A. (2012). Hybrid colloidal Au-CdSe pentapod heterostructures synthesis and their photocatalytic properties. *ACS Applied Materials & Interfaces, 4*(11), 6266–6272. doi:10.1021/am301859b PMID:23113704

Hall, J., Dravid, V. P., & Aslam, M. (2005). Templated Fabrication of Co@Au Core-Shell Nanorods Structure. *Nanoscape, 2*(1), 67–71.

Har-Lavan, R., Ron, I., Thieblemont, F., & Cahen, D. (2008, April). Hybrid photovoltaic junctions: metal/molecular organic insulator/semiconductor MOIS solar cells. In *Photonics Europe* (pp. 70020M–70020M). International Society for Optics and Photonics. doi:10.1117/12.768652

Herring, N. P., AbouZeid, K., Mohamed, M. B., Pinsk, J., & El-Shall, M. S. (2011). Formation mechanisms of gold–zinc oxide hexagonal nanopyramids by heterogeneous nucleation using microwave synthesis. *Langmuir, 27*(24), 15146–15154. doi:10.1021/la201698k PMID:21819068

Hirai, H., Nakao, Y., & Toshima, N. (1978). Colloidal rhodium in poly (vinylpyrrolidone) as hydrogenation catalyst for internal olefins. *Chemistry Letters*, (5): 545–548. doi:10.1246/cl.1978.545

Hirai, H., Nakao, Y., & Toshima, N. (1979). Preparation of colloidal transition metals in polymers by reduction with alcohols or ethers. *Journal of Macromolecular Science. Chemistry*, *13*(6), 727–750. doi:10.1080/00222337908056685

Huang, X., Li, S., Wu, S., Huang, Y., Boey, F., Gan, C. L., & Zhang, H. (2012). Graphene Oxide-Templated Synthesis of Ultrathin or Tadpole-Shaped Au Nanowires with Alternating hcp and fcc Domains. *Advanced Materials*, *24*(7), 979–983. doi:10.1002/adma.201104153 PMID:22252895

Huang, X., Tan, C., Yin, Z., & Zhang, H. (2014). 25th Anniversary Article: Hybrid Nanostructures Based on Two-Dimensional Nanomaterials. *Advanced Materials*, *26*(14), 2185–2204. doi:10.1002/adma.201304964 PMID:24615947

Huang, X., Zhou, X., Wu, S., Wei, Y., Qi, X., Zhang, J., & Zhang, H. et al. (2010). Reduced Graphene Oxide-Templated Photochemical Synthesis and in situ Assembly of Au Nanodots to Orderly Patterned Au Nanodot Chains. *Small*, *6*(4), 513–516. doi:10.1002/smll.200902001 PMID:20077425

Ichikawa, Y., Kobayashi, M., Sasase, M., & Suemasu, T. (2008). Molecular beam epitaxy of semiconductor (BaSi2)/metal (CoSi2) hybrid structures on Si (111) substrates for photovoltaic application. *Applied Surface Science*, *254*(23), 7963–7967. doi:10.1016/j.apsusc.2008.04.011

Jain, P. K., & El-Sayed, M. A. (2007). Surface plasmon resonance sensitivity of metal nanostructures: Physical basis and universal scaling in metal nanoshells. *The Journal of Physical Chemistry C*, *111*(47), 17451–17454. doi:10.1021/jp0773177

Jain, P. K., & El-Sayed, M. A. (2007). Universal scaling of plasmon coupling in metal nanostructures: Extension from particle pairs to nanoshells. *Nano Letters*, *7*(9), 2854–2858. doi:10.1021/nl071496m PMID:17676810

Jain, P. K., Huang, X., El-Sayed, I. H., & El-Sayed, M. A. (2007). Review of some interesting surface plasmon resonance-enhanced properties of noble metal nanoparticles and their applications to biosystems. *Plasmonics*, *2*(3), 107–118. doi:10.1007/s11468-007-9031-1

Jain, P. K., Huang, X., El-Sayed, I. H., & El-Sayed, M. A. (2008). Noble metals on the nanoscale: Optical and photothermal properties and some applications in imaging, sensing, biology, and medicine. *Accounts of Chemical Research*, *41*(12), 1578–1586. doi:10.1021/ar7002804 PMID:18447366

Jana, N. R., Gearheart, L., & Murphy, C. J. (2001). Wet chemical synthesis of high aspect ratio cylindrical gold nanorods. *The Journal of Physical Chemistry B*, *105*(19), 4065–4067. doi:10.1021/jp0107964

Jasuja, K., & Berry, V. (2009). Implantation and growth of dendritic gold nanostructures on graphene derivatives: Electrical property tailoring and Raman enhancement. *ACS Nano*, *3*(8), 2358–2366. doi:10.1021/nn900504v PMID:19702325

Jin, R., Cao, Y., Mirkin, C. A., Kelly, K. L., Schatz, G. C., & Zheng, J. G. (2001). Photoinduced conversion of silver nanospheres to nanoprisms. *Science*, *294*(5548), 1901–1903. doi:10.1126/science.1066541 PMID:11729310

Jin, R., Cao, Y. C., Hao, E., Métraux, G. S., Schatz, G. C., & Mirkin, C. A. (2003). Controlling anisotropic nanoparticle growth through plasmon excitation. *Nature*, *425*(6957), 487–490. doi:10.1038/nature02020 PMID:14523440

Kathiresan, K., Manivannan, S., Nabeel, M. A., & Dhivya, B. (2009). Studies on silver nanoparticles synthesized by a marine fungus, Penicillium fellutanum isolated from coastal mangrove sediment. *Colloids and Surfaces. B, Biointerfaces*, *71*(1), 133–137. doi:10.1016/j.colsurfb.2009.01.016 PMID:19269142

Kelly, K. L., Coronado, E., Zhao, L. L., & Schatz, G. C. (2003). The optical properties of metal nanoparticles: The influence of size, shape, and dielectric environment. *The Journal of Physical Chemistry B*, *107*(3), 668–677. doi:10.1021/jp026731y

Kerker, M. (1969). *The Scattering of Light and Other Electromagnetic Radiation*. New York: Academic.

Khalavka, Y., & Sönnichsen, C. (2008). Growth of gold tips onto hyperbranched CdTe nanostructures. *Advanced Materials*, *20*(3), 588–591. doi:10.1002/adma.200701518

Kickelbick, G. (Ed.). (2007). *Hybrid materials: synthesis, characterization, and applications*. Germany: John Wiley & Sons.

Kiwi, J., & Grätzel, M. (1979). Projection, size factors, and reaction dynamics of colloidal redox catalysts mediating light induced hydrogen evolution from water. *Journal of the American Chemical Society*, *101*(24), 7214–7217. doi:10.1021/ja00518a015

Klaus, T., Joerger, R., Olsson, E., & Granqvist, C. G. (1999). Silver-based crystalline nanoparticles, microbially fabricated. *Proceedings of the National Academy of Sciences of the United States of America*, *96*(24), 13611–13614. doi:10.1073/pnas.96.24.13611 PMID:10570120

Kong, X. Y., Ding, Y., & Wang, Z. L. (2004). Metal-semiconductor Zn-ZnO core-shell nanobelts and nanotubes. *The Journal of Physical Chemistry B*, *108*(2), 570–574. doi:10.1021/jp036993f

Kowshik, M., Ashtaputre, S., Kharrazi, S., Vogel, W., Urban, J., Kulkarni, S. K., & Paknikar, K. M. (2003). Extracellular synthesis of silver nanoparticles by a silver-tolerant yeast strain MKY3. *Nanotechnology*, *14*(1), 95–100. doi:10.1088/0957-4484/14/1/321

Kreibig, U., & Vollmer, M. (1995). *Optical properties of metal clusters*. Berlin, Germany: Springer. doi:10.1007/978-3-662-09109-8

Kundu, S., Peng, L., & Liang, H. (2008). A new route to obtain high-yield multiple-shaped gold nanoparticles in aqueous solution using microwave irradiation. *Inorganic Chemistry*, *47*(14), 6344–6352. doi:10.1021/ic8004135 PMID:18563880

Kuno, M. (2008). An overview of solution-based semiconductor nanowires: Synthesis and optical studies. *Physical Chemistry Chemical Physics*, *10*(5), 620–639. doi:10.1039/B708296G PMID:19791445

Lee, B. I., Qi, L., & Copeland, T. (2005). Nanoparticles for materials design: Present & future. *Journal of Ceramic Processing & Research*, *6*(1), 31–40.

Lee, J., Hernandez, P., Lee, J., Govorov, A. O., & Kotov, N. A. (2007). Exciton–plasmon interactions in molecular spring assemblies of nanowires and wavelength-based protein detection. *Nature Materials*, *6*(4), 291–295. doi:10.1038/nmat1869 PMID:17384635

Lee, K. S., & El-Sayed, M. A. (2006). Gold and silver nanoparticles in sensing and imaging: Sensitivity of plasmon response to size, shape, and metal composition. *The Journal of Physical Chemistry B*, *110*(39), 19220–19225. doi:10.1021/jp062536y PMID:17004772

Lee, Y., Garcia, M. A., Frey Huls, N. A., & Sun, S. (2010). Synthetic tuning of the catalytic properties of Au-Fe3O4 nanoparticles. *Angewandte Chemie*, *122*(7), 1293–1296. doi:10.1002/ange.200906130 PMID:20077449

Lengke, M. F., Fleet, M. E., & Southam, G. (2006). Morphology of gold nanoparticles synthesized by filamentous cyanobacteria from gold (I)-thiosulfate and gold (III)-chloride complexes. *Langmuir*, *22*(6), 2780–2787. doi:10.1021/la052652c PMID:16519482

Li, J., & Liu, C. Y. (2010). Ag/graphene heterostructures: Synthesis, characterization and optical properties. *European Journal of Inorganic Chemistry*, *2010*(8), 1244–1248. doi:10.1002/ejic.200901048

Li, L., Daou, T. J., Texier, I., Kim Chi, T. T., Liem, N. Q., & Reiss, P. (2009). Highly luminescent CuInS2/ZnS core/shell nanocrystals: Cadmium-free quantum dots for in vivo imaging. *Chemistry of Materials*, *21*(12), 2422–2429. doi:10.1021/cm900103b

Li, X., Lian, J., Lin, M., & Chan, Y. (2010). Light-induced selective deposition of metals on gold-tipped CdSe-seeded CdS nanorods. *Journal of the American Chemical Society*, *133*(4), 672–675. doi:10.1021/ja1076603 PMID:21174430

Liang, J., Deng, Z., Jiang, X., Li, F., & Li, Y. (2002). Photoluminescence of tetragonal ZrO2 nanoparticles synthesized by microwave irradiation. *Inorganic Chemistry*, *41*(14), 3602–3604. doi:10.1021/ic025532q PMID:12099861

Liang, S., Liu, X. L., Yang, Y. Z., Wang, Y. L., Wang, J. H., Yang, Z. J., & Zhang, Z. et al. (2012). Symmetric and Asymmetric Au–AgCdSe Hybrid Nanorods. *Nano Letters*, *12*(10), 5281–5286. doi:10.1021/nl3025505 PMID:22947073

Lica, G. C., Zelakiewicz, B. S., Constantinescu, M., & Tong, Y. (2004). Charge dependence of surface plasma resonance on 2 nm octanethiol-protected Au nanoparticles: Evidence of a free-electron system. *The Journal of Physical Chemistry B*, *108*(52), 19896–19900. doi:10.1021/jp045302s

Lim, J., Eggeman, A., Lanni, F., Tilton, R. D., & Majetich, S. A. (2008). Synthesis and Single-Particle Optical Detection of Low-Polydispersity Plasmonic-Superparamagnetic Nanoparticles. *Advanced Materials*, *20*(9), 1721–1726. doi:10.1002/adma.200702196

Lin, J. M., Lin, H. Y., Cheng, C. L., & Chen, Y. F. (2006). Giant enhancement of bandgap emission of ZnO nanorods by platinum nanoparticles. *Nanotechnology*, *17*(17), 4391–4394. doi:10.1088/0957-4484/17/17/017

Link, S., & El-Sayed, M. A. (2000). Shape and size dependence of radiative, non-radiative and photothermal properties of gold nanocrystals. *International Reviews in Physical Chemistry, 19*(3), 409–453. doi:10.1080/01442350050034180

Link, S., & El-Sayed, M. A. (2003). Optical properties and ultrafast dynamics of metallic nanocrystals. *Annual Review of Physical Chemistry, 54*(1), 331–366. doi:10.1146/annurev.physchem.54.011002.103759 PMID:12626731

Link, S., Wang, Z. L., & El-Sayed, M. A. (1999). Alloy formation of gold-silver nanoparticles and the dependence of the plasmon absorption on their composition. *The Journal of Physical Chemistry B, 103*(18), 3529–3533. doi:10.1021/jp990387w

Liu, M., & Guyot-Sionnest, P. (2005). Mechanism of silver (I)-assisted growth of gold nanorods and bipyramids. *The Journal of Physical Chemistry B, 109*(47), 22192–22200. doi:10.1021/jp054808n PMID:16853888

Liu, Z., Shi, L., Shi, Z., Liu, X. H., Zi, J., Zhou, S. M., & Xia, Y. J. et al. (2009). Magneto-optical Kerr effect in perpendicularly magnetized Co/Pt films on two-dimensional colloidal crystals. *Applied Physics Letters, 95*(3), 032502. doi:10.1063/1.3182689

Lu, G., Li, H., Liusman, C., Yin, Z., Wu, S., & Zhang, H. (2011). Surface enhanced Raman scattering of Ag or Au nanoparticle-decorated reduced graphene oxide for detection of aromatic molecules. *Chemical Science, 2*(9), 1817–1821. doi:10.1039/c1sc00254f

Lu, G., Mao, S., Park, S., Ruoff, R. S., & Chen, J. (2009). Facile, noncovalent decoration of graphene oxide sheets with nanocrystals. *Nano Research, 2*(3), 192–200. doi:10.1007/s12274-009-9017-8

Lu, W., Wang, B., Zeng, J., Wang, X., Zhang, S., & Hou, J. G. (2005). Synthesis of core/shell nanoparticles of Au/CdSe via Au-Cd bialloy precursor. *Langmuir, 21*(8), 3684–3687. doi:10.1021/la0469250 PMID:15807621

Lu, Y. H., Zhou, M., Zhang, C., & Feng, Y. P. (2009). Metal-embedded graphene: A possible catalyst with high activity. *The Journal of Physical Chemistry C, 113*(47), 20156–20160. doi:10.1021/jp908829m

Lyon, J. L., Fleming, D. A., Stone, M. B., Schiffer, P., & Williams, M. E. (2004). Synthesis of Fe oxide core/Au shell nanoparticles by iterative hydroxylamine seeding. *Nano Letters, 4*(4), 719–723. doi:10.1021/nl035253f

Ma, G. H., He, J., Rajiv, K., Tang, S. H., Yang, Y., & Nogami, M. (2004). Observation of resonant energy transfer in Au: CdS nanocomposite. *Applied Physics Letters, 84*(23), 4684–4686. doi:10.1063/1.1760220

Mansour, A.S., Emam, A.N., Mohamed, M.B. (2016). *Remarkable Enhancement of Optical and Photocatalytic Properties of Au-CdSe Tetrapods upon Loading into Graphene Oxide.* (Unpublished to be submitted in Preparation).

Marquardt, D., Vollmer, C., Thomann, R., Steurer, P., Mülhaupt, R., Redel, E., & Janiak, C. (2011). The use of microwave irradiation for the easy synthesis of graphene-supported transition metal nanoparticles in ionic liquids. *Carbon, 49*(4), 1326–1332. doi:10.1016/j.carbon.2010.09.066

Mason, T. J. (2009). Sonoelectrochemical synthesis of nanoparticles. *Molecules (Basel, Switzerland)*, *14*(10), 4284–4299. doi:10.3390/molecules14104284 PMID:19924064

McPhillips, J., Murphy, A., Jonsson, M. P., Hendren, W. R., Atkinson, R., Hook, F., & Pollard, R. J. et al. (2010). High-performance biosensing using arrays of plasmonic nanotubes. *ACS Nano*, *4*(4), 2210–2216. doi:10.1021/nn9015828 PMID:20218668

Melle, S., Menéndez, J. L., Armelles, G., Navas, D., Vázquez, M., Nielsch, K., & Gösele, U. et al. (2003). Magneto-optical properties of nickel nanowire arrays. *Applied Physics Letters*, *83*(22), 4547–4549. doi:10.1063/1.1630840

Merhari, L. (Ed.). (2009). *Hybrid nanocomposites for nanotechnology*. Springer, US. doi:10.1007/978-0-387-30428-1

Mikhaylova, M., Kim, D. K., Bobrysheva, N., Osmolowsky, M., Semenov, V., Tsakalakos, T., & Muhammed, M. (2004). Superparamagnetism of magnetite nanoparticles: Dependence on surface modification. *Langmuir*, *20*(6), 2472–2477. doi:10.1021/la035648e PMID:15835712

Millstone, J. E., Park, S., Shuford, K. L., Qin, L., Schatz, G. C., & Mirkin, C. A. (2005). Observation of a quadrupole plasmon mode for a colloidal solution of gold nanoprisms. *Journal of the American Chemical Society*, *127*(15), 5312–5313. doi:10.1021/ja043245a PMID:15826156

Mizukoshi, Y., Oshima, R., Maeda, Y., & Nagata, Y. (1999). Preparation of platinum nanoparticles by sonochemical reduction of the Pt (II) ion. *Langmuir*, *15*(8), 2733–2737. doi:10.1021/la9812121

Mody, V. V., Siwale, R., Singh, A., & Mody, H. R. (2010). Introduction to metallic nanoparticles. *Journal of Pharmacy and Bioallied Sciences*, *2*(4), 282. doi:10.4103/0975-7406.72127 PMID:21180459

Mohamed, M. B., AbouZeid, K. M., Abdelsayed, V., Aljarash, A. A., & El-Shall, M. S. (2010). Growth mechanism of anisotropic gold nanocrystals via microwave synthesis: Formation of dioleamide by gold nanocatalysis. *ACS Nano*, *4*(5), 2766–2772. doi:10.1021/nn9016179 PMID:20392051

Mokari, T., Rothenberg, E., Popov, I., Costi, R., & Banin, U. (2004). Selective growth of metal tips onto semiconductor quantum rods and tetrapods. *Science*, *304*(5678), 1787–1790. doi:10.1126/science.1097830 PMID:15205530

Mokari, T., Sztrum, C. G., Salant, A., Rabani, E., & Banin, U. (2005). Formation of asymmetric one-sided metal-tipped semiconductor nanocrystal dots and rods. *Nature Materials*, *4*(11), 855–863. doi:10.1038/nmat1505

Mubarak Ali, D., Thajuddin, N., Jeganathan, K., & Gunasekaran, M. (2011). Plant extract mediated synthesis of silver and gold nanoparticles and its antibacterial activity against clinically isolated pathogens. *Colloids and Surfaces. B, Biointerfaces*, *85*(2), 360–365. doi:10.1016/j.colsurfb.2011.03.009 PMID:21466948

Mukherjee, P., Ahmad, A., Mandal, D., Senapati, S., Sainkar, S. R., Khan, M. I., & Sastry, M. (2001). Fungus-mediated synthesis of silver nanoparticles and their immobilization in the mycelial matrix: A novel biological approach to nanoparticle synthesis. *Nano Letters*, *1*(10), 515–519. doi:10.1021/nl0155274

Murphy, C. J. (2002). Nanocubes and nanoboxes. *Science, 298*(5601), 2139–2141. doi:10.1126/science.1080007 PMID:12481122

Murray, C., Norris, D. J., & Bawendi, M. G. (1993). Synthesis and characterization of nearly monodisperse CdE (E= sulfur, selenium, tellurium) semiconductor nanocrystallites. *Journal of the American Chemical Society, 115*(19), 8706–8715. doi:10.1021/ja00072a025

Murray, W. A., & Barnes, W. L. (2007). Plasmonic materials. *Advanced Materials, 19*(22), 3771–3782. doi:10.1002/adma.200700678

Nair, B., & Pradeep, T. (2002). Coalescence of nanoclusters and formation of submicron crystallites assisted by Lactobacillus strains. *Crystal Growth & Design, 2*(4), 293–298. doi:10.1021/cg0255164

Nemamcha, A., Rehspringer, J. L., & Khatmi, D. (2006). Synthesis of palladium nanoparticles by sonochemical reduction of palladium (II) nitrate in aqueous solution. *The Journal of Physical Chemistry B, 110*(1), 383–387. doi:10.1021/jp0535801 PMID:16471546

Nikoobakht, B., & El-Sayed, M. A. (2003). Preparation and growth mechanism of gold nanorods (NRs) using seed-mediated growth method. *Chemistry of Materials, 15*(10), 1957–1962. doi:10.1021/cm020732l

Okasha, A. M., Mohamed, M. B., Abdallah, T., Basily, A. B., Negm, S., & Talaat, H. (2010, March). Characterization of core/shell (Ag/CdSe) nanostructure using photoacoustic spectroscopy.[). IOP Publishing.]. *Journal of Physics: Conference Series, 214*(1), 012131. doi:10.1088/1742-6596/214/1/012131

Okitsu, K. (2011). Sonochemical synthesis of metal nanoparticles. In *Theoretical and Experimental Sonochemistry Involving Inorganic Systems* (pp. 131–150). Springer Netherlands.

Okitsu, K., Ashokkumar, M., & Grieser, F. (2005). Sonochemical synthesis of gold nanoparticles: Effects of ultrasound frequency. *The Journal of Physical Chemistry B, 109*(44), 20673–20675. doi:10.1021/jp0549374 PMID:16853678

Okitsu, K., Sharyo, K., & Nishimura, R. (2009). One-pot synthesis of gold nanorods by ultrasonic irradiation: The effect of pH on the shape of the gold nanorods and nanoparticles. *Langmuir, 25*(14), 7786–7790. doi:10.1021/la9017739 PMID:19545140

Oldenburg, S. J., Averitt, R. D., Westcott, S. L., & Halas, N. J. (1998). Nanoengineering of optical resonances. *Chemical Physics Letters, 288*(2), 243–247. doi:10.1016/S0009-2614(98)00277-2

Özkar, S., & Finke, R. G. (2002). Nanocluster formation and stabilization fundamental studies. 2. Proton sponge as an effective H+ scavenger and expansion of the anion stabilization ability series. *Langmuir, 18*(20), 7653–7662. doi:10.1021/la020225i

Panda, A. B., Glaspell, G., & El-Shall, M. S. (2007). Microwave synthesis and optical properties of uniform nanorods and nanoplates of rare earth oxides. *The Journal of Physical Chemistry C, 111*(5), 1861–1864. doi:10.1021/jp0670283

Panda, A. B., Glaspell, G., & El-Shall, M. S. (2008). Microwave Synthesis of Highly Aligned Ultra Narrow Semiconductor Rods and Wires[J. Am. Chem. Soc. 2006, 128, 2790-2791]. *Journal of the American Chemical Society, 130*(12), 4203–4203. doi:10.1021/ja8003717 PMID:16506744

Pandya, S. (n.d.). *Nanocomposites and its Applications – Review*. Retrieved July 20, 2015 from https://www.academia.edu/3038972/Nanocomposites_and_its_Applications-Review

Papavassiliou, G. C. (1979). Optical properties of small inorganic and organic metal particles. *Progress in Solid State Chemistry*, *12*(3), 185–271. doi:10.1016/0079-6786(79)90001-3

Pasricha, R., Gupta, S., & Srivastava, A. K. (2009). A Facile and Novel Synthesis of Ag–Graphene-Based Nanocomposites. *Small*, *5*(20), 2253–2259. doi:10.1002/smll.200900726 PMID:19582730

Pastoriza-Santos, I., & Liz-Marzán, L. M. (2002). Formation of PVP-protected metal nanoparticles in DMF. *Langmuir*, *18*(7), 2888–2894. doi:10.1021/la015578g

Pena, D. J., Mbindyo, J. K., Carado, A. J., Mallouk, T. E., Keating, C. D., Razavi, B., & Mayer, T. S. (2002). Template growth of photoconductive metal-CdSe-metal nanowires. *The Journal of Physical Chemistry B*, *106*(30), 7458–7462. doi:10.1021/jp0256591

Peng, Z. A., & Peng, X. (2001). Formation of high-quality CdTe, CdSe, and CdS nanocrystals using CdO as precursor. *Journal of the American Chemical Society*, *123*(1), 183–184. doi:10.1021/ja003633m PMID:11273619

Perez-Juste, J., Liz-Marzan, L. M., Carnie, S., Chan, D. Y., & Mulvaney, P. (2004). Electric-field-directed growth of gold nanorods in aqueous surfactant solutions. *Advanced Functional Materials*, *14*(6), 571–579. doi:10.1002/adfm.200305068

Prodan, E., Radloff, C., Halas, N. J., & Nordlander, P. (2003). A hybridization model for the plasmon response of complex nanostructures. *Science*, *302*(5644), 419–422. doi:10.1126/science.1089171 PMID:14564001

Puntes, V. F., Krishnan, K., & Alivisatos, A. P. (2002). Synthesis of colloidal cobalt nanoparticles with controlled size and shapes. *Topics in Catalysis*, *19*(2), 145–148. doi:10.1023/A:1015252904412

Puntes, V. F., Krishnan, K. M., & Alivisatos, A. P. (2001). Colloidal nanocrystal shape and size control: The case of cobalt. *Science*, *291*(5511), 2115–2117. doi:10.1126/science.1057553 PMID:11251109

Puntes, V. F., Krishnan, K. M., & Alivisatos, P. (2001). Synthesis, self-assembly, and magnetic behavior of a two-dimensional superlattice of single-crystal ε-Co nanoparticles. *Applied Physics Letters*, *78*(15), 2187–2189. doi:10.1063/1.1362333

Puntes, V. F., Zanchet, D., Erdonmez, C. K., & Alivisatos, A. P. (2002). Synthesis of hcp-Co nanodisks. *Journal of the American Chemical Society*, *124*(43), 12874–12880. doi:10.1021/ja027262g PMID:12392435

Puthussery, J., Kosel, T. H., & Kuno, M. (2009). Facile Synthesis and Size Control of II–VI Nanowires Using Bismuth Salts. *Small*, *5*(10), 1112–1116. doi:10.1002/smll.200801838 PMID:19334010

Qin, W., & Li, X. (2010). A theoretical study on the catalytic synergetic effects of Pt/graphene nanocomposites. *The Journal of Physical Chemistry C*, *114*(44), 19009–19015. doi:10.1021/jp1072523

Qu, L., Peng, Z. A., & Peng, X. (2001). Alternative routes toward high quality CdSe nanocrystals. *Nano Letters*, *1*(6), 333–337. doi:10.1021/nl0155532

Rady, H. S., Emam, A. N., Mohamed, M. B., & El-shall, M. S. (2016). (Manuscript submitted for publication). Graphene Interface Enhances the Photochemical Synthesis, Stability and Photothermal Effect of Plasmonic Nanostructures. *Carbon*.

Rang, M., Jones, A. C., Zhou, F., Li, Z. Y., Wiley, B. J., Xia, Y., & Raschke, M. B. (2008). Optical near-field mapping of plasmonic nanoprisms. *Nano Letters*, *8*(10), 3357–3363. doi:10.1021/nl801808b PMID:18788789

Reetz, M. T., & Lohmer, G. (1996). Propylene carbonate stabilized nanostructured palladium clusters as catalysts in Heck reactions. *Chemical Communications*, (16): 1921–1922. doi:10.1039/cc9960001921

Reetz, M. T., & Maase, M. (1999). Redox-Controlled Size-Selective Fabrication of Nanostructured Transition Metal Colloids. *Advanced Materials*, *11*(9), 773–777. doi:10.1002/(SICI)1521-4095(199906)11:9<773::AID-ADMA773>3.0.CO;2-1

Ren, W., Fang, Y., & Wang, E. (2011). A binary functional substrate for enrichment and ultrasensitive SERS spectroscopic detection of folic acid using graphene oxide/Ag nanoparticle hybrids. *ACS Nano*, *5*(8), 6425–6433. doi:10.1021/nn201606r PMID:21721545

Salant, A., Amitay-Sadovsky, E., & Banin, U. (2006). Directed self-assembly of gold-tipped CdSe nanorods. *Journal of the American Chemical Society*, *128*(31), 10006–10007. doi:10.1021/ja063192s PMID:16881617

Sau, T. K., & Murphy, C. J. (2004). Room temperature, high-yield synthesis of multiple shapes of gold nanoparticles in aqueous solution. *Journal of the American Chemical Society*, *126*(28), 8648–8649. doi:10.1021/ja047846d PMID:15250706

Saunders, A. E., Popov, I., & Banin, U. (2006). Synthesis of hybrid CdS-Au colloidal nanostructures. *The Journal of Physical Chemistry B*, *110*(50), 25421–25429. doi:10.1021/jp065594s PMID:17165989

Schimid, G. (1994). *Clusters and colloids: from theory to application*. New York: VCH. doi:10.1002/9783527616077

Schouteden, K., & Van Haesendonck, C. (2010). Narrow Au (111) terraces decorated by self-organized Co nanowires: A low-temperature STM/STS investigation. *Journal of Physics Condensed Matter*, *22*(25), 255504. doi:10.1088/0953-8984/22/25/255504 PMID:21393803

Sedki, M., Mohamed, M. B., Fawzy, M., Abdelrehim, D. A., & Abdel-Mottaleb, M. M. (2015). Phytosynthesis of silver–reduced graphene oxide (Ag–RGO) nanocomposite with an enhanced antibacterial effect using Potamogeton pectinatus extract. *RSC Advances*, *5*(22), 17358–17365.

Shankar, S. S., Rai, A., Ahmad, A., & Sastry, M. (2004). Rapid synthesis of Au, Ag, and bimetallic Au core–Ag shell nanoparticles using Neem (Azadirachta indica) leaf broth. *Journal of Colloid and Interface Science*, *275*(2), 496–502. doi:10.1016/j.jcis.2004.03.003 PMID:15178278

Shankar, S. S., Rai, A., Ahmad, A., & Sastry, M. (2004). Biosynthesis of silver and gold nanoparticles from extracts of different parts of the geranium plant. *Applications in Nanotechnology*, *1*, 69–77.

Shaviv, E., & Banin, U. (2010). Synergistic Effects on Second Harmonic Generation of Hybrid CdSe–Au Nanoparticles. *ACS Nano*, *4*(3), 1529–1538. doi:10.1021/nn901778k PMID:20192238

Sherry, L. J., Jin, R., Mirkin, C. A., Schatz, G. C., & Van Duyne, R. P. (2006). Localized surface plasmon resonance spectroscopy of single silver triangular nanoprisms. *Nano Letters*, *6*(9), 2060–2065. doi:10.1021/nl061286u PMID:16968025

Shevchenko, E. V., Ringler, M., Schwemer, A., Talapin, D. V., Klar, T. A., Rogach, A. L., & Alivisatos, A. P. et al. (2008). Self-assembled binary superlattices of CdSe and Au nanocrystals and their fluorescence properties. *Journal of the American Chemical Society*, *130*(11), 3274–3275. doi:10.1021/ja710619s PMID:18293987

Shi, W., Zeng, H., Sahoo, Y., Ohulchanskyy, T. Y., Ding, Y., Wang, Z. L., & Prasad, P. N. et al. (2006). A general approach to binary and ternary hybrid nanocrystals. *Nano Letters*, *6*(4), 875–881. doi:10.1021/nl0600833 PMID:16608302

Sinzig, J., de Jongh, L. J., Bönnemann, H., Brijoux, W., & Köppler, R. (1998). Antiferromagnetism of colloidal [MnO·0.3THF] x. *Applied Organometallic Chemistry*, *12*(5), 387–391. doi:10.1002/(SICI)1099-0739(199805)12:5<387::AID-AOC743>3.0.CO;2-S

Son, S. U., Jang, Y., Yoon, K. Y., Kang, E., & Hyeon, T. (2004). Facile synthesis of various phosphine-stabilized monodisperse palladium nanoparticles through the understanding of coordination chemistry of the nanoparticles. *Nano Letters*, *4*(6), 1147–1151. doi:10.1021/nl049519+

Spasova, M., Salgueiriño-Maceira, V., Schlachter, A., Hilgendorff, M., Giersig, M., Liz-Marzán, L. M., & Farle, M. (2005). Magnetic and optical tunable microspheres with a magnetite/gold nanoparticle shell. *Journal of Materials Chemistry*, *15*(21), 2095–2098. doi:10.1039/b502065d

Su, C. H., Wu, P. L., & Yeh, C. S. (2003). Sonochemical synthesis of well-dispersed gold nanoparticles at the ice temperature. *The Journal of Physical Chemistry B*, *107*(51), 14240–14243. doi:10.1021/jp035451v

Sutter, E., Wang, B., Albrecht, P., Lahiri, J., Bocquet, M. L., & Sutter, P. (2012). Templating of arrays of Ru nanoclusters by monolayer graphene/Ru Moirés with different periodicities. *Journal of Physics Condensed Matter*, *24*(31), 314201. doi:10.1088/0953-8984/24/31/314201 PMID:22820349

Sweeney, R. Y., Mao, C., Gao, X., Burt, J. L., Belcher, A. M., Georgiou, G., & Iverson, B. L. (2004). Bacterial biosynthesis of cadmium sulfide nanocrystals. *Chemistry & Biology*, *11*(11), 1553–1559. doi:10.1016/j.chembiol.2004.08.022 PMID:15556006

Talapin, D. V., Lee, J. S., Kovalenko, M. V., & Shevchenko, E. V. (2009). Prospects of colloidal nanocrystals for electronic and optoelectronic applications. *Chemical Reviews*, *110*(1), 389–458. doi:10.1021/cr900137k PMID:19958036

Tang, J., Birkedal, H., McFarland, E. W., & Stucky, G. D. (2003). Self-assembly of CdSe/CdS quantum dots by hydrogen bonding on Au surfaces for photoreception. *Chemical Communications (Cambridge)*, (18): 2278–2279. doi:10.1039/b306888a PMID:14518873

Tano, T., Esumi, K., & Meguro, K. (1989). Preparation of organopalladium sols by thermal decomposition of palladium acetate. *Journal of Colloid and Interface Science*, *133*(2), 530–533. doi:10.1016/S0021-9797(89)80069-4

Tanori, J., & Pileni, M. P. (1997). Control of the shape of copper metallic particles by using a colloidal system as template. *Langmuir, 13*(4), 639–646. doi:10.1021/la9606097

Temnov, V. V., Armelles, G., Woggon, U., Guzatov, D., Cebollada, A., Garcia-Martin, A., & Bratschitsch, R. et al. (2010). Active magneto-plasmonics in hybrid metal–ferromagnet structures. *Nature Photonics, 4*(2), 107–111. doi:10.1038/nphoton.2009.265

Templeton, A. C., Pietron, J. J., Murray, R. W., & Mulvaney, P. (2000). Solvent refractive index and core charge influences on the surface plasmon absorbance of alkanethiolate monolayer-protected gold clusters. *The Journal of Physical Chemistry B, 104*(3), 564–570. doi:10.1021/jp991889c

Teng, X., Feygenson, M., Wang, Q., He, J., Du, W., Frenkel, A. I., & Aronson, M. et al. (2009). Electronic and magnetic properties of ultrathin Au/Pt nanowires. *Nano Letters, 9*(9), 3177–3184. doi:10.1021/nl9013716 PMID:19645434

Teng, X., Han, W., Wang, Q., Li, L., Frenkel, A. I., & Yang, J. C. (2008). Hybrid Pt/Au nanowires: Synthesis and electronic structure. *The Journal of Physical Chemistry C, 112*(38), 14696–14701. doi:10.1021/jp8054685

Thomas, K. G., & Kamat, P. V. (2003). Chromophore-functionalized gold nanoparticles. *Accounts of Chemical Research, 36*(12), 888–898. doi:10.1021/ar030030h PMID:14674780

Torrado, J. F., González-Díaz, J. B., González, M. U., García-Martín, A., & Armelles, G. (2010). Magneto-optical effects in interacting localized and propagating surface plasmon modes. *Optics Express, 18*(15), 15635–15642. doi:10.1364/OE.18.015635 PMID:20720945

Turkevich, J. (1985). Colloidal gold. Part II. *Gold Bulletin, 18*(4), 125–131. doi:10.1007/BF03214694

Turkevich, J., Garton, G., & Stevenson, P. C. (1954). The color of colloidal gold. *Journal of Colloid Science, 9*, 26–35. doi:10.1016/0095-8522(54)90070-7

Turkevich, J., & Kim, G. (1970). Palladium: Preparation and catalytic properties of particles of uniform size. *Science, 169*(3948), 873–879. doi:10.1126/science.169.3948.873 PMID:17750062

Umut, E., Pineider, F., Arosio, P., Sangregorio, C., Corti, M., Tabak, F., & Ghigna, P. et al. (2012). Magnetic, optical and relaxometric properties of organically coated gold–magnetite (Au–Fe_3O_4) hybrid nanoparticles for potential use in biomedical applications. *Journal of Magnetism and Magnetic Materials, 324*(15), 2373–2379. doi:10.1016/j.jmmm.2012.03.005

Vidoni, O., Philippot, K., Amiens, C., Chaudret, B., Balmes, O., Malm, J. O., & Casanove, M. J. et al. (1999). Novel, spongelike ruthenium particles of controllable size stabilized only by organic solvents. *Angewandte Chemie International Edition, 38*(24), 3736–3738. doi:10.1002/(SICI)1521-3773(19991216)38:24<3736::AID-ANIE3736>3.0.CO;2-E PMID:10649342

Vinodgopal, K., He, Y., Ashokkumar, M., & Grieser, F. (2006). Sonochemically prepared platinum-ruthenium bimetallic nanoparticles. *The Journal of Physical Chemistry B, 110*(9), 3849–3852. doi:10.1021/jp060203v PMID:16509663

Vinodgopal, K., Neppolian, B., Lightcap, I. V., Grieser, F., Ashokkumar, M., & Kamat, P. V. (2010). Sonolytic design of graphene–Au nanocomposites. simultaneous and sequential reduction of graphene oxide and Au (III). *The Journal of Physical Chemistry Letters*, *1*(13), 1987–1993. doi:10.1021/jz1006093

Vithiya, K., & Sen, S. (2011). Biosynthesis of nanoparticles. *Int J Pharm Sci Res*, *2*, 2781–2785.

Wang, C., Yin, H., Chan, R., Peng, S., Dai, S., & Sun, S. (2009). One-pot synthesis of oleylamine coated AuAg alloy NPs and their catalysis for CO oxidation. *Chemistry of Materials*, *21*(3), 433–435. doi:10.1021/cm802753j

Wang, G., Van Hove, M. A., Ross, P. N., & Baskes, M. I. (2005). Monte Carlo simulations of segregation in Pt-Ni catalyst nanoparticles. *The Journal of Chemical Physics*, *122*(2), 024706. doi:10.1063/1.1828033 PMID:15638613

Wang, M., Wang, C., Young, K. L., Hao, L., Medved, M., Rajh, T., & Stamenkovic, V. R. et al. (2012). Cross-linked heterogeneous nanoparticles as bifunctional probe. *Chemistry of Materials*, *24*(13), 2423–2425. doi:10.1021/cm300381f

Wen, Y., Xing, F., He, S., Song, S., Wang, L., Long, Y., & Fan, C. et al. (2010). A graphene-based fluorescent nanoprobe for silver (I) ions detection by using graphene oxide and a silver-specific oligonucleotide. *Chemical Communications*, *46*(15), 2596–2598. doi:10.1039/b924832c PMID:20449319

Wetz, F., Soulantica, K., Falqui, A., Respaud, M., Snoeck, E., & Chaudret, B. (2007). Hybrid Co–Au nanorods: Controlling Au nucleation and location. *Angewandte Chemie*, *119*(37), 7209–7211. doi:10.1002/ange.200702017 PMID:17685370

Wu, J., Hou, Y., & Gao, S. (2011). Controlled synthesis and multifunctional properties of FePt-Au heterostructures. *Nano Research*, *4*(9), 836–848. doi:10.1007/s12274-011-0140-y

Wu, S., He, Q., Zhou, C., Qi, X., Huang, X., Yin, Z., & Zhang, H. et al. (2012). Synthesis of Fe_3O_4 and Pt nanoparticles on reduced graphene oxide and their use as a recyclable catalyst. *Nanoscale*, *4*(7), 2478–2483. doi:10.1039/c2nr11992g PMID:22388949

Wu, W., He, Q., Chen, H., Tang, J., & Nie, L. (2007). Sonochemical synthesis, structure and magnetic properties of air-stable Fe_3O_4/Au nanoparticles. *Nanotechnology*, *18*(14), 145609. doi:10.1088/0957-4484/18/14/145609

Xu, Z., Hou, Y., & Sun, S. (2007). Magnetic core/shell Fe_3O_4/Au and Fe_3O_4/Au/Ag nanoparticles with tunable plasmonic properties. *Journal of the American Chemical Society*, *129*(28), 8698–8699. doi:10.1021/ja073057v PMID:17590000

Yang, J., Tian, C., Wang, L., Tan, T., Yin, J., Wang, B., & Fu, H. (2012). In situ reduction, oxygen etching, and reduction using formic acid: An effective strategy for controllable growth of monodisperse palladium nanoparticles on graphene. *ChemPlusChem*, *77*(4), 301–307. doi:10.1002/cplu.201100058

Yang, Y., Nogami, M., Shi, J., Chen, H., Ma, G., & Tang, S. (2005). Enhancement of third-order optical nonlinearities in 3-dimensional films of dielectric shell capped Au composite nanoparticles. *The Journal of Physical Chemistry B*, *109*(11), 4865–4871. doi:10.1021/jp045854a PMID:16863140

Yang, Z., Chu, H., Jin, Z., Zhou, W., & Li, Y. (2010). Preparation and properties of CdS/Au composite nanorods and hollow Au tubes. *Chinese Science Bulletin*, *55*(10), 921–926. doi:10.1007/s11434-010-0061-2

Yin, Z., He, Q., Huang, X., Zhang, J., Wu, S., Chen, P., & Zhang, H. et al. (2012). Real-time DNA detection using Pt nanoparticle-decorated reduced graphene oxide field-effect transistors. *Nanoscale*, *4*(1), 293–297. doi:10.1039/C1NR11149C PMID:22089471

Yu, C., & Wang, H. (2010). Light-Induced Bipolar-Resistance Effect Based on Metal–Oxide–Semiconductor Structures of Ti/SiO2/Si. *Advanced Materials*, *22*(9), 966–970. doi:10.1002/adma.200903070 PMID:20217821

Yuhas, B. D., Habas, S. E., Fakra, S. C., & Mokari, T. (2009). Probing compositional variation within hybrid nanostructures. *ACS Nano*, *3*(11), 3369–3376. doi:10.1021/nn901107p PMID:19813744

Zeng, H., Cai, W., Liu, P., Xu, X., Zhou, H., Klingshirn, C., & Kalt, H. (2008). ZnO-based hollow nanoparticles by selective etching: Elimination and reconstruction of metal– semiconductor interface, improvement of blue emission and photocatalysis. *ACS Nano*, *2*(8), 1661–1670. doi:10.1021/nn800353q PMID:19206370

Zhang, J., Du, J., Han, B., Liu, Z., Jiang, T., & Zhang, Z. (2006). Sonochemical formation of single-crystalline gold nanobelts. *Angewandte Chemie*, *118*(7), 1134–1137. doi:10.1002/ange.200503762 PMID:16389606

Zhang, J., Post, M., Veres, T., Jakubek, Z. J., Guan, J., Wang, D., & Simard, B. et al. (2006). Laser-assisted synthesis of superparamagnetic Fe@Au core-shell nanoparticles. *The Journal of Physical Chemistry B*, *110*(14), 7122–7128. doi:10.1021/jp0560967 PMID:16599475

Zhou, H., Lee, J., Park, T. J., Lee, S. J., Park, J. Y., & Lee, J. (2012). Ultrasensitive DNA monitoring by Au–Fe$_3$O$_4$ nanocomplex. *Sensors and Actuators. B, Chemical*, *163*(1), 224–232. doi:10.1016/j.snb.2012.01.040

Zhou, H., Zou, F., Koh, K., & Lee, J. (2014). Multifunctional magnetoplasmonic nanomaterials and their biomedical applications. *Journal of Biomedical Nanotechnology, 10*(10), 2921-2949.

Zhou, X., Huang, X., Qi, X., Wu, S., Xue, C., Boey, F. Y., & Zhang, H. et al. (2009). In situ synthesis of metal nanoparticles on single-layer graphene oxide and reduced graphene oxide surfaces. *The Journal of Physical Chemistry C*, *113*(25), 10842–10846. doi:10.1021/jp903821n

Zhu, C., Guo, S., & Dong, S. (2012). PdM (M= Pt, Au) Bimetallic Alloy Nanowires with Enhanced Electrocatalytic Activity for Electro-oxidation of Small Molecules. *Advanced Materials*, *24*(17), 2326–2331. doi:10.1002/adma.201104951 PMID:22473584

Zhu, J., Palchik, O., Chen, S., & Gedanken, A. (2000). Microwave assisted preparation of CdSe, PbSe, and Cu$_{2-x}$Se nanoparticles. *The Journal of Physical Chemistry B*, *104*(31), 7344–7347. doi:10.1021/jp001488t

Zhu, M., Chen, P., & Liu, M. (2011). Graphene oxide enwrapped Ag/AgX (X= Br, Cl) nanocomposite as a highly efficient visible-light plasmonic photocatalyst. *ACS Nano*, *5*(6), 4529–4536. doi:10.1021/nn200088x PMID:21524132

Chapter 12
Hybrid Plasmonic Nanostructures:
Environmental Impact and Applications

Ahmed Nabile Emam
National Research Centre, Egypt

Emad Girgis
National Research Centre, Egypt

Ahmed Sadek Mansour
Cairo University, Egypt

Mona Bakr Mohamed
Cairo University, Egypt

ABSTRACT

Plasmonic hybrid nanostructure including Semiconductor-metallic nanoparticles, and graphene-plasmonic nanocomposites have great potential to be used as photocatalyst for hydrogen production and for photodegradation of organic waste. Also, they are potential candidate as active materials in photovoltaic devices. Plasmonic-magnetic nanocomposites could be used in photothermal therapy and biomedical imaging. This chapter will focus on the environmental impact of these materials and their in-vitro and in-vivo toxicity. In addition, the applications of these hybrid nanostructures in energy and environment will be discussed in details.

1. INTRODUCTION

Recently, population growth and fast urbanization and industrialization have been increased dramatically. Therefore, several serious environmental issues such as; climate changes, global warming and pollution of water resources were still a main challenge for the whole world. These issues may become even worse if no efficient/sustainable solutions has been developed to reduce their reasons.(Li et al. (2015)) Due to these tremendous advances, the search for the suitable platforms, and materials to be used in wide-variety of environmental applications became the main target for several research groups.

Plasmonic materials and their hybrid nanostructures have unique and intriguing physico-chemical and biological properties compared to their bulk counterparts. This is attributed to their high *surface-to-volume* ratio, shape and size dependent optical properties and their localized surface Plasmon.(Li et al. (2001), Sharma et al. (2008), Iglesias-Silva et al. (2007), Huang and Yang (2008) and El-Nour et al. (2010)) Additionally, the ease of surface modification functionalization to be utilized in different ap-

DOI: 10.4018/978-1-5225-0585-3.ch012

plications.(Qu et al. (2013), Cuenya (2010), and Zhang et al. (2010)) Therefore, plasmonic nanomaterials and their hybrid nanocomposites represent new class of materials which could be good candidates to provide promising solutions for the most environmental issues.(Guo and Wang (2011), Cao et al. (2013), and Corma and Serna (2006)) These possible applications can varies either from their catalytic and photocatalytic applications, light harvesting, sensing, and antimicrobial applications (Figure 1). (Li et al. (2015))

2. ENVIROMENTAL APPLICATIONS OF PLASMONIC NANOPARTICLES

a. Silver Nanoparticles

Silver nanoparticles exhibit unique optical, catalytic, sensing and antimicrobial characteristics which made them have huge industrial demands, pegged at 16,000 to 24,000 tons per year. As an antimicrobial agent, silver nanoparticles are expected to have annual demand of 2800 tons/year in the fields of 'food, hygiene and water purification, in addition to 3125 tons/year for medicine. Only 36% of the silver consumed in these industrial processes is being recycled and the rest is let out to the environment. The widespread application of Ag NPs in our daily life will unavoidably increase the human and ecosystem exposure. Furthermore, during their production, transport, erosion, washing or disposal of Ag NP containing prod-

Figure 1. Schematic illustration of the possible applications od plasmonic nanoparticles in the environmental field. Reproduced from[Li, J., Zhao, T., Chen, T., Liu, Y., Ong, C.N. and Xie, J., 2015. Engineering noble metal nanomaterials for environmental applications. Nanoscale, 7(17), pp.7502-7519] Published by The Royal Society of Chemistry

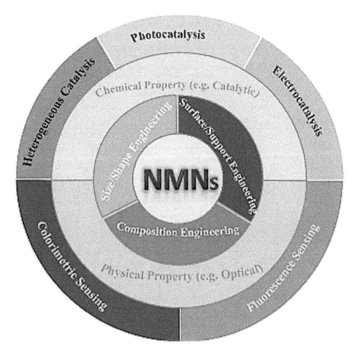

ucts, results in their release to the environment. Though the long historical use of silver has not shown obvious adverse effects, there is concern about the potential risks of Ag NPs in the environment.(Lima et al. (2012) and Yu et al. (2013))

i. Antimicrobial Applications

Ag NPs could be used as antimicrobial agents in the health industry, food storage, textile coatings, cosmetics and a large number of environmental applications. Silver nanoparticles were applied in a wide range of applications from disinfecting medical devices and home appliances to water treatment (Bosetti et al. (2002), Cho et al. (2005), Jain and Pradeep (2005), and Li et al. (2008)) Textile industry use Ag NPs in different textile fabrics. In this direction, silver nanocomposite fibers were prepared containing silver nanoparticles incorporated inside the fabric.(Montazer et al. (2012), Xue et al. (2012), Khalil-Abad et al. (2010)), In addition, Ag NPs might be used as surface disinfection agent.(Kumar et al. (2008), Roe et al. (2008), Jain et al. (2009), Gottesman et al. (2011), and Freeman et al. (2012)) This property can be achieved through different ways, such as embedding into bactericidal coating,(Kumar et al. (2008)), surface functionalization of plastic catheters and other medical devices,(Roe et al. (2008)), gel formulation for topical use,(Jain et al. (2009)) coating of packing papers to be used in food preservation,(Gottesman et al. (2011)), or by impregnating fabrics for clinical clothing.(Freeman et al. (2012)).

The mechanism of silver nanoparticles as antimicrobial has been studies against many types of bacteria and biofilms. Table 1 summarizes the mechanism of action and Figure (2) describes how silver nanoparticles can be toxic to bacteria.

Table 1. Details of Ag NPs and their mechanisms of action against bacteria and biofilms

Bacterial Strain	Mechanism of Action
Acinetobacterbaumannii	Alteration of cell wall and cytoplasm
Escherichia coli	Alteration of membrane permeability and respiration
Enterococcus faecalis	Alteration of cell wall and cytoplasm
Klebsiellapneumonia	Alteration of membrane
Listeria monocytogenes	Morphological changes, separation of the cytoplasmic membrane from the cell wall, plasmolysis
Micrococcus luteus	Alteration of membrane
Nitrifying bacteria	Inhibits respiratory activity
Pseudomonas aeruginosa	Irreversible damage on bacterial cells; Alteration of membrane permeability and respiration
Proteus mirabilis	Alteration of cell wall and cytoplasm
Staphylococcus aureus	Irreversible damage on bacterial cells
Staphylococcus epidermidis	Inhibition of bacterial DNA replication, bacterial cytoplasm membranes damage, modification of intracellular ATP levels
Salmonella typhi	Inhibition of bacterial DNA replication, bacterial cytoplasm membranes damage, modification of intracellular ATP levels
Vibrio cholera	Alteration of membrane permeability and respiration

Reprinted from [Franci, G., Falanga, A., Galdiero, S., Palomba, L., Rai, M., Morelli, G. and Galdiero, M., 2015. Silver nanoparticles as potential antibacterial agents.Molecules, 20(5), pp.8856-8874.] Published by MDPI AG, Basel, Switzerland

Figure 2. Mechanisms of Ag NPs' toxic action
Reprinted from [Franci, G., Falanga, A., Galdiero, S., Palomba, L., Rai, M., Morelli, G. and Galdiero, M., 2015. Silver nanoparticles as potential antibacterial agents. Molecules, 20(5), pp.8856-8874].Published by MDPI AG, Basel, Switzerland

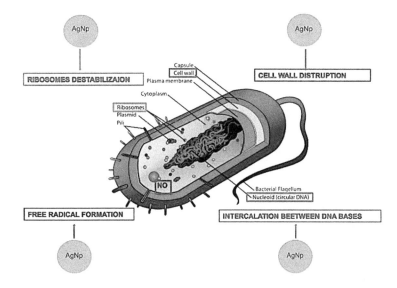

Several products include containing Ag NPs have been approved by a range of accredited authorities such as, FDA, EPA, and SIAA of Japan.(Azonano,El-Nour et al. (2010), Jia et al.(2008) and Bhattacharya and Mukherjee(2008))

For these purpose, the toxicity of silver nanoparticles has been studied in different mammalian cell systems, including rat and human cells. (i.e. rat liver cells, (Hussain et al. (2005)) human keratinocytes and fibroblasts cultures,(Burd et al. (2007)) and human spermatogonial stem cells.(Braydich-Stolle et al. (2005)) *In vitro* studies shows that an elevated dose of nano silver induces oxidative stress (liberation of reactive oxygen species) as a mechanism of cytotoxicity.(Arora et al. (2008)) At the safe concentration range, silver nanoparticles have been described to cause an exert anti-inflammatory effects as: acceleration of wound healing,(Tian et al. (2007)), inhibition of allergic contact dermatitis in mice, suppression of the expression of TNF-α and IL-12, and induction of apoptosis of inflammatory cells.(Bhol and Schechter (2005)), modulation of cytokine production and induction of peripheral blood mononuclear cells proliferation (Figure 2),(Shin et al. (2007)).

ii. Water Purification and Disinfictents

Water is one of the most important substances on Earth and is essential to all living things. Although 70% of the Earth is covered with water, only 0.6% is suitable for human uses. According to the WHO, at least 1 billion people do not have access to safe drinking water. Contamination of drinking water are the leading cause of death in many developing nations.(Pradeep (2009), Tran et al. (2013) and Gusseme et al. (2010))

Complexity of contaminants, population growth, presence of resistive pathogens and their impact on the environment create a demand for advanced technologies for clean and safe drinking water. Several chlorine and halo-compounds such as tri-halo methanes, halo-acetic acids, halo-acetonitrile, haloketones and other DBPs were found to cause colon, rectal and bladder cancers and adverse reproductive disorders. Ag NPs are very ideal to be used in water disinfection. It could be easily incorporated to polymeric membranes to disinfect the water contaminated from bacteria and viruses.(Gusseme et al. (2010 and 2011), Lv et al. (2009), Yakub and Soboyejo (2012), Jain and Pradeep (2005), Gangadharan et al. (2010), Zhang et al. (2012), and Sheng and Liu (2011)) Moreover, the addition of silver-based NPs could prevent bacterial/viral attachment and biofilm formation in filtration medium.(Sintubin et al. (2011) and Fernández et al. (2015)). The application of silver-based NPs is of utmost importance to prevent outbreaks of waterborne diseases related to poor treatment of drinking water.

b. Gold Nanoparticles

Gold nanoparticles have unique optical and electronic properties due to their surface Plasmon resonance. Thus, many researchers developed the use of gold nanoparticles and their composites in many areas such as biological tagging, chemical and biological sensing, optoelectronics, photo thermal therapy, biomedical imaging, DNA labeling, microscopy and photo-acoustic imaging, drug delivery, surface-enhanced Raman spectroscopy, tracking and drug delivery, catalysis and cancer therapy and imaging.(Kim and Jon (2012), Khlebtsov and Dykman (2010), and Tedsco et al. (2010))

Environmental applications of gold nanoparticles has been limited to fabrication of gold nanoparticles based sensors to detect various metal ions by working on the principle of colour change due to the aggregation of gold nanoparticles. Such types of sensors have been widely used for the detection of copper, mercury lead and arsenic in water (Nalawade et al. (2012)). Another very promising application is using them in catalysis to convert harmful gasses into less toxic or environmental friendly gasses. In particular, using gold nanoparticles, or nanoalloy of gold and another metal loaded into metal oxides are catalytically very active for a wide range of reactions, (Kim et al (2004)) including CO oxidation,(Yoon et al. (2005) and Gluhoi and Nieuwenhuys (2004)) water gas shift reaction,(Liu et al. (2005)) NO_x reduction, CO and CO_2 hydrogenation, NO oxidation, alkene epoxidation and selective oxidation of hydrocarbons. (Hughes et al. (2005))

c. Enviromental Appliations of Plasmonic Hybrid Nanocomposites

Magnetic nanoparticles, especially iron (Fe^0), magnetite (Fe_3O_4) and maghemite ($\gamma\text{-}Fe_2O_3$) nanoparticles, have sparked an enormous interest in applications for treatment of polluted water or subsurface environments.(Yantasee et al. (2007), Shen et al. (2009 and 2011), Tang and Lo (2013)). Magneto-plasmonic nanostructures open an opportunity for some many environmental applications, such as sunlight photocatalytic inactivation of Bacteria (Zhai et al. (2011)). Zhai and co-workers illustrated a convenient method to fabricate noble metals based super-paramagnetic nano hybrid hetero-structures (i.e. Ag-Fe_3O_4 and Au-AgCl@Fe_3O_4). They reported in their study that Au-AgCl@Fe_3O_4 nanotubes was an excellent candidate for photocatalytic inactivation of bacteria upon irradiation with sunlight (Figure 3). (Zhai et al. (2011))

Sun and co-worker (Sun et al. (2015)) developed another system based on dumbbell-like structure $\alpha\text{-}Fe_2O_3$/Ag/AgX heterostructures via *in-situ* oxidation reaction. In such system showed a higher photocatalytic activity than both of $\alpha\text{-}Fe_2O_3$ and $\alpha\text{-}Fe_2O_3$/Ag systems as shown Figure (4).

Metal oxide semiconductors nanoparticles hold great promise for wide range applications, such as energy conversion and storage, environmental remediation, and etc... However, critical factors including the high charge – carrier recombination rate and narrow light absorption have restricted more practical

Hybrid Plasmonic Nanostructures

Figure 3. Schematic depiction of the synthesis of petal-like Ag/Fe$_3$O$_4$, Ag nanocube/Fe$_3$O$_4$, Ag nanowire/Fe$_3$O$_4$, and Fe$_3$O$_4$/Au-AgCl nanotube composites. And Survival rate of E. coli under direct sunlight (700 W/m^2) without hybrids and with 0.016 mg/mL Fe$_3$O$_4$/Au/AgCl nanotubes

Reprinted with permission from (Zhai, Y., Han, L., Wang, P., Li, G., Ren, W., Liu, L., Wang, E. and Dong, S., 2011. Superparamagnetic plasmonic nanohybrids: shape-controlled synthesis, TEM-induced structure evolution, and efficient sunlight-driven inactivation of bacteria. ACS nano, 5(11), 8562-8570.Copyright (2011) American Chemical Society

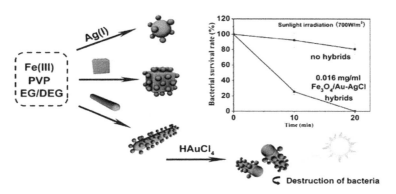

Figure 4. (a) UV–vis spectral evolution of RhB as a function of light irradiation time over α-Fe$_2$O$_3$@Ag/AgCl; (b) Photocatalytic degradation of RhB solution over as-prepared samples as a function of irradiation time under simulated sunlight irradiation; (c) kinetic linear simulation curves of the α-Fe$_2$O$_3$/Ag/AgX (X = Cl, Br, I)photo-catalysts for the photo-degradation of RhB under simulated sunlight irradiation. (d) Schematic illustration of the photocatalytic process for dumbbell-like α-Fe$_2$O$_3$/Ag/AgX (X = Cl, Br, I)

Adopted with permission from (Sun, L., Wu, W., Tian, Q., Lei, M., Liu, J., Xiao, X., Zheng, X., Ren, F. and Jiang, C., (2015). In situ Oxidation and Self-Assembly Synthesis of Dumbbell-Like α-Fe2O3/Ag/AgX (X= Cl, Br, I) Heterostructures with Enhanced Photocatalytic Properties. ACS Sustainable Chemistry & Engineering..Copyright (2015) American Chemical Society

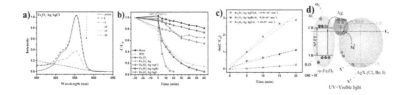

and viable applications. ((Pelaez et al. (2012)) Particularly, TiO$_2$ is one of the most candidates which got a lot of interests from several research groups, due to its low cost and high stability. (Fang et al. and Pelaez et al. (2012), Chehadi et al. (2016)). However, TiO$_2$ has large energy band-gap of 3.2 eV and only absorbs solar radiations in the ultraviolet (UV) region, while the UV lights only contribute to less than 10% of total solar radiations; the visible lights. On the other hand, contribute to 50% of the solar radiations.(Pelaez et al. (2012)) Furthermore, the photo-catalytic activity of TiO$_2$ decreases substantially with prolonging the service time; and the deactivation of TiO$_2$ – based photo-catalysts is a crucial drawback for the application.(Wang et al. (2012)) There have been many research efforts devoted to improve the photocatalytic activity by incorporating metals or ions (i.e. Ag, and Au NPs),(Arunachalam et al. (2012) and Suvith and Philip (2014), Bhakya et al. (2015)) to broaden the absorption of solar radiations and to re-activate and regenerate TiO$_2$-based photocatalysts.(Sangpour et al. (2010), Wu et al. (2010), Linic et al. (2011), Kochuveedu et al. (2013))

Figure 5. (a) Absorption Spectra of Au-TiO₂ Hybrid Nanocomposites. Copyright © D M Fouad and M B Mohamed, "Comparative Study of the Photocatalytic Activity of Semiconductor Nanostructures and Their Hybrid Metal Nanocomposites on the Photodegradation of Malathion," Journal of Nanomaterials, vol. 2012, Article ID 524123, 8 pages, 2012. (b) Fermi level equilibrium in semiconductor-metal nanocomposite system, (c) Equilibrium of semiconductor –metal nanocomposites with redox couple after irradiation

Adopted with permission from (Subramanian, V., Wolf, E.E. and Kamat, P.V., 2004.Catalysis with TiO2/gold nanocomposites. Effect of metal particle size on the Fermi level equilibration. Journal of the American Chemical Society, 126(15), pp.4943-4950 Copyright (2004) American Chemical Society

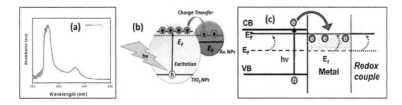

Over the past decade, the idea of fabricating hybrid nano-objects that composed of plasmonic metals and semiconductors was emerged, due to their localized surface Plasmon resonance and the formation of effective Schottky junctions.(Subramanian et al. (2003 and 2004), Hirakawa and Kamat (2005)). In such nano-hybrid system; plasmonic nanoparticles are able to absorb visible light intensely due to the surface plasmon resonance (SPR) effect. Through the SPR effect, the electromagnetic field of incident light couples with the oscillations of conduction electrons in the plasmonic nanoparticles, resulting in surface plasmon absorption and an enhancement of the local electromagnetic fields close to the surface of plasmonic nanoparticles. Therefore, plasmonic nanostructures of noble metals in combination with semiconductors offer a promising future for the next generation of energy and environmental needs (Figure 5).(Subramanian et al. (2003 and 2004), Sangpour et al. (2010), Wu et al. (2010), Linic et al. (2011), Kochuveedu et al. (2013))

Contamination of waste water with organic compounds (i.e. dyes, pesticides, drugs, phenolic compounds and polymers) has huge hazardous effect in the environment and spontaneously affects human health. Conventional treatment method of organic pollutant is based upon removing large amounts of organic compounds from waste water streams by adsorption. The disadvantage of this method is that it is non-destructive for the organic pollutants, which simply transfers the pollutants from one phase/substance to another. Photocatalytic degradation of organic pollutants using UV or sun light in presence of nano-catalyst has been a great interest because it has the capability of converting these organic hazards into non-toxic compounds. In the recent years, numerous metal oxides, plasmonic and plasmonic graphene hybrid nanocomposites have attracted growing attentions for photo-degradation of organic pollutants. (Wu et al. (2010), Linic et al. (2011), Pelaez et al. (2012), Wang et al. (2012), Kochuveedu et al. (2013))

Phenolic based – organic hazardous compounds are considered one of the most and highly spread highly toxic wastes. This is due to the difficulty of their biological degradation, leading to a setting up of restricted level of phenol in the environment.(Hu et al. (2009)) Phenolic based water pollutant compounds, include 2-chlorophenol (2-CP), 2,4-dichlorophenol (2,4- DCP), and trichlorophenol (TCP) have been used in photo-catalytic activity assessment, upon irradiation via sunlight.(Hu et al. (2009)) Several reports has been devoted to use a hybrid nano-objects based on combination between plasmonic nano-structures and metal oxide as a catalyst supports.(Hu et al. (2009) and Chehadi et al. (2016)) Chehadi and co-worker demonstrated the influence of Au NPs as an example of the plasmonic nanostructure on the photocatalytic efficacy of various metal oxide (i.e. TiO_2, ZnO, and Al_2O_3) as a catalyst supports on the photo-degradation of Bisphenol A (BSA) upon laser irradiation. In addition, investigate the effect of different parameter such as, irradiation time, laser power, and catalyst type. They reported that Au/TiO_2 showed a fastest degradation rate than other composite systems. (Chehadi et al. (2016)) Another study has been illustrated by Hu et al. (Hu et al. (2009)). They developed metal-semiconductor nanocomposites base on coupling between plasmonic nanostructures Ag-AgI with mesoporous alumina using deposition-precipitation, and photo-reduction techniques.(Hu et al. (2009)) They reported that their configuration showed higher photo- sensitivity, stability and catalytic activity for the degradation and mineralization of toxic persistent organic pollutants especially Phenolic based organic pollutants upon irradiation using simulated sun light.(Hu et al. (2009)) in addition the capability to be used not only as a photo-catalyst, but also photovoltaic fuel cells (Figure 6).(Hu et al. (2009))

Different shapes of hybrid CdSe/Au penta-pods and nano-dumbblls shows photocatalytic activity for the photo-degradation of organic dyes by irradiation using UV light. Presence of plasmonic at the interface of CdSe Semiconductor nanoparticles enhances the photocatalytic activity three times more (Figures 7 and 8). This is due to the electron transfer from CdSe to Au Fermi level. (Costi et al. (2008), Gao et al. (2011), Tongying et al. (2012) and Haldar et al. (2012))

Figure 6. Temporal course of photodegradation of 2-chlorophenol (10 mg L; 60 mL) in aqueous dispersions (containing catalyst: 1.6 g L-1): (a) Ag-AgI/Al2O3 in dark, (b) no catalyst; (d) AgI/Al2O3, (e) Ag/Al2O3 (f) TiO2-xNx, (g) Ag-AgI/Al2O3 with λ > 420 nm; and (c) AgI/Al2O3, (h) Ag-AgI/ Al2O3 with λ > 450 nm

Adopted with permission from (Hu, C., Peng, T., Hu, X., Nie, Y., Zhou, X., Qu, J. and He, H., (2009). Plasmon-induced photodegradation of toxic pollutants with Ag− AgI/Al2O3 under visible-light irradiation. Journal of the American Chemical Society,132(2), 857-862.Copyright (2009) American Chemical Society

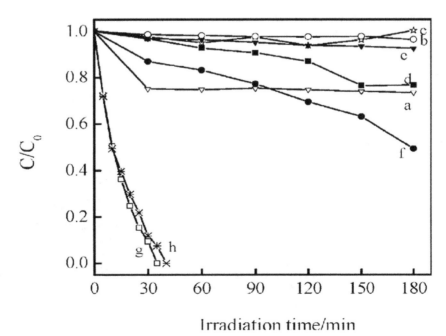

Figure 7. TEM (upper panels) and HRTEM (lower panels) images of (a) green (QD1), (b) yellow (QD2), and (c) red (QD3) emitting QD and QD/Au. (d) Schematic representation of the entire decay process for CdSe and CdSe/Au

Reprinted with copyright permission from [Tongying, P., Plashnitsa, V.V., Petchsang, N., Vietmeyer, F., Ferraudi, G.J., Krylova, G. and Kuno, M., 2012. Photocatalytic hydrogen generation efficiencies in one-dimensional CdSe heterostructures. The journal of physical chemistry letters, 3(21), pp.3234-3240] American Chemical Society (2012)

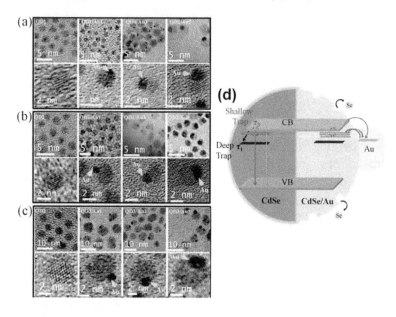

Figure 8. Set of absorption spectra for photocatalytic reduction MB by using (a) QD1, (b) QD1/Au3, (c) QD1/Au5, (e) QD2, (f) QD3/Au3, (g) QD3/Au5, (i) QD3, (j) QD3/Au3 and (k) QD3/Au5. Change in normalized absorption with respect to time for (d) QD1, QD1/Au3, QD1/Au5, (h) QD2, QD2/Au3, QD2/Au5, and (l) QD3, QD3/Au3, QD3/Au5

Reprinted with copyright permission from [Tongying, P., Plashnitsa, V.V., Petchsang, N., Vietmeyer, F., Ferraudi, G.J., Krylova, G. and Kuno, M., 2012. Photocatalytic hydrogen generation efficiencies in one-dimensional CdSe heterostructures. The journal of physical chemistry letters, 3(21), pp.3234-3240] American Chemical Society (2012)

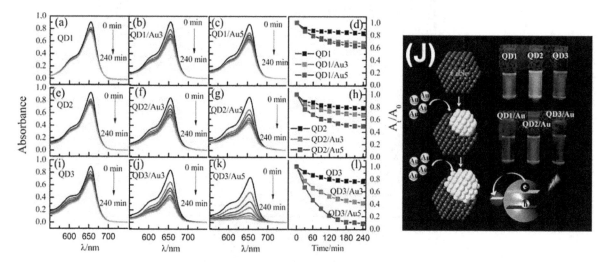

Recently, Zhou et al. (Zhou et al. (2015)) developed a ternary hybrid nanocomposites of semiconductor-plasmonic hetero-nanostructures loaded into graphene has been developed and display extraordinary photocatalytic efficiency induced by the plasmonic energy that operates in the Ag@CdSe-rGO hybrid ternary composites. The obtained plasmonic photo-catalysts in nanoscale were fabricated by using a one-step hydrothermal method, during which the in situ nucleation of Ag@CdSe core–shell nanoparticles and the reduction of GO to rGO occurred simultaneously. The enhanced photocatalytic and optical properties toward photo-degrading of organic pollutant under visible light irradiation can be ascribed to plasmonic enhancement of Ag by charge and energy transfer and to the carrier rGO further enhancing the separation of photo-generated e^-/h^+ pairs. Three major roles of Ag core, electronic acceptor by the Schottky barrier formed in Ag@CdSe hetero-structure and induced by the local Plasmon resonance of Ag, have facilitated the separation of photo-generated e^-/h^+ in CdSe QDs, and the plasmonic energy transfer from Ag to CdSe has strengthened the excitation and charge separation of CdSe QDs. The introduction of rGO carrier in the Ag@CdSe-rGO system has enhanced the photocatalytic activity for its high charge carrier mobility and ambipolar field effect Figure (9).

3. CONCLUSION

Plasmonic nanoparticles such as Ag NPs and Au NPs have unique optical, catalytic and electronic properties which made their uses in industrial applications grow tremendously in the last few years. The super-antiseptic and antimicrobial effect of silver nanoparticles lead to new generation of disinfectant, infection control and antibiotics based nano-silver products. The catalytic and photothermal properties of gold nanoparticles developed several based applications in medicine and environment such as drug delivery, cancer therapy and imaging, sensors for detection of heavy metals and bacteria in environment Plasmonic-hybrid nanocomposites including magnetic-plasmonic, metal oxide-plasmonic, or quantum

Figure 9. Illustrations of three different roles of Ag core in enhancing the separation of the photogenerated e^-/h^+ transfer process in the Ag@CdSerGO system: (I) electronic acceptor; (II) DET and PIRET effect induced by the LPR effect of Ag core, (III) The decrease (C/C_0) of 20 mg/L TC·HCl versus irradiation time for the TC·HCl without catalysts, CdSe, Ag@CdSe, and Ag@CdSe-rGO

Adopted with permission from (Zhou, M., Li, J., Ye, Z., Ma, C., Wang, H., Huo, P., Shi, W. and Yan, Y. (2015). Transferring charge and energy of Ag@CdSe QDs-rGO core-shell plasmonic photocatalyst for enhanced visible light photocatalytic activity. ACS Applied Materials & Interfaces, 7 (51), 28231–28243. Copyright (2015) American Chemical Society

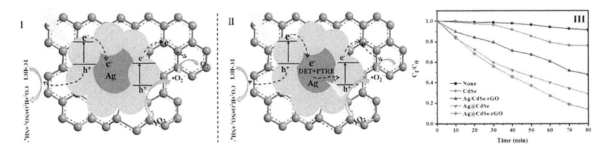

dots/plasmonic opens a new class of photo-catalysts used for the photo-degradation of organic pollutant and waste water purifications, light harvesting and hydrogen productions. Extraordinary photocatalytic activity had been achieved via loading these plasmonic hybrid nanostructures on graphene sheets.

REFERENCES

Arora, S., Jain, J., Rajwade, J. M., & Paknikar, K. M. (2008). Cellular responses induced by silver nanoparticles: In vitro studies. *Toxicology Letters*, *179*(2), 93–100. doi:10.1016/j.toxlet.2008.04.009 PMID:18508209

Arunachalam, R., Dhanasingh, S., Kalimuthu, B., Uthirappan, M., Rose, C., & Mandal, A. B. (2012). Phytosynthesis of silver nanoparticles using Coccinia grandis leaf extract and its application in the photocatalytic degradation. *Colloids and Surfaces. B, Biointerfaces*, *94*, 226–230. doi:10.1016/j.colsurfb.2012.01.040 PMID:22348986

Bhakya, S., Muthukrishnan, S., Sukumaran, M., Muthukumar, M., Senthil Kumar, T., & Rao, M. V. (2015). Catalytic Degradation of Organic Dyes using Synthesized Silver Nanoparticles: A Green Approach. *Journal of Bioremediation & Biodegradation*, *6*(5), 1–9.

Bhattacharya, R., & Mukherjee, P. (2008). Biological properties of "naked" metal nanoparticles. *Advanced Drug Delivery Reviews*, *60*(11), 1289–1306. doi:10.1016/j.addr.2008.03.013 PMID:18501989

Bhol, K. C., & Schechter, P. J. (2005). Topical nanocrystalline silver cream suppresses inflammatory cytokines and induces apoptosis of inflammatory cells in a murine model of allergic contact dermatitis. *The British Journal of Dermatology*, *152*(6), 1235–1242. doi:10.1111/j.1365-2133.2005.06575.x PMID:15948987

Bosetti, M., Masse, A., Tobin, E., & Cannas, M. (2002). Silver coated materials for external fixation devices: In vitro biocompatibility and genotoxicity. *Biomaterials*, *23*(3), 887–892. doi:10.1016/S0142-9612(01)00198-3 PMID:11771707

Braydich-Stolle, L., Hussain, S., Schlager, J. J., & Hofmann, M. C. (2005). In vitro cytotoxicity of nanoparticles in mammalian germline stem cells. *Toxicological Sciences*, *88*(2), 412–419. doi:10.1093/toxsci/kfi256 PMID:16014736

Burd, A., Kwok, C. H., Hung, S. C., Chan, H. S., Gu, H., Lam, W. K., & Huang, L. (2007). A comparative study of the cytotoxicity of silver-based dressings in monolayer cell, tissue explant, and animal models. *Wound Repair and Regeneration*, *15*(1), 94–104. doi:10.1111/j.1524-475X.2006.00190.x PMID:17244325

Cao, Y., Li, D., Jiang, F., Yang, Y., & Huang, Z. (2013). Engineering metal nanostructure for SERS application. *Journal of Nanomaterials*.

Chehadi, Z., Alkees, N., Bruyant, A., Toufaily, J., Girardon, J. S., Capron, M., & Jradi, S. et al. (2016). Plasmonic enhanced photocatalytic activity of semiconductors for the degradation of organic pollutants under visible light. *Materials Science in Semiconductor Processing*, *42*, 81–84. doi:10.1016/j.mssp.2015.08.044

Cho, M., Chung, H., Choi, W., & Yoon, J. (2005). Different inactivation behaviors of MS-2 phage and Escherichia coli in TiO2 photocatalytic disinfection. *Applied and Environmental Microbiology, 71*(1), 270–275. doi:10.1128/AEM.71.1.270-275.2005 PMID:15640197

Corma, A., & Serna, P. (2006). Chemoselective hydrogenation of nitro compounds with supported gold catalysts. *Science, 313*(5785), 332–334. doi:10.1126/science.1128383 PMID:16857934

Costi, R., Saunders, A. E., Elmalem, E., Salant, A., & Banin, U. (2008). Visible light-induced charge retention and photocatalysis with hybrid CdSe-Au nanodumbbells. *Nano Letters, 8*(2), 637–641. doi:10.1021/nl0730514 PMID:18197720

Cuenya, B. R. (2010). Synthesis and catalytic properties of metal nanoparticles: Size, shape, support, composition, and oxidation state effects. *Thin Solid Films, 518*(12), 3127–3150. doi:10.1016/j.tsf.2010.01.018

De Gusseme, B., Hennebel, T., Christiaens, E., Saveyn, H., Verbeken, K., Fitts, J.P., Boon, N. & Verstraete, W., (2011). Virus disinfection in water by biogenic silver immobilized in polyvinylidene fluoride membranes. *Water Research, 45*(4), 1856-1864.

De Gusseme, B., Sintubin, L., Baert, L., Thibo, E., Hennebel, T., Vermeulen, G., & Boon, N. et al. (2010). Biogenic silver for disinfection of water contaminated with viruses. *Applied and Environmental Microbiology, 76*(4), 1082–1087. doi:10.1128/AEM.02433-09 PMID:20038697

De Gusseme, B., Sintubin, L., Hennebel, T., Boon, N., Verstraete, W., Baert, L., & Uyttendaele, M. (2010). Inactivation of viruses in water by biogenic silver: innovative and environmentally friendly disinfection technique. In *Bioinformatics and Biomedical Engineering (iCBBE), 2010 4th International Conference on* (pp. 1-5). IEEE. doi:10.1109/ICBBE.2010.5515631

El-Nour, K. M. A., Eftaiha, A. A., Al-Warthan, A., & Ammar, R. A. (2010). Synthesis and applications of silver nanoparticles. *Arabian Journal of Chemistry, 3*(3), 135-140.

Fang, J., Cao, S. W., Wang, Z., Shahjamali, M. M., Loo, S. C. J., Barber, J., & Xue, C. (2012). Mesoporous plasmonic Au–TiO2 nanocomposites for efficient visible-light-driven photocatalytic water reduction. *International Journal of Hydrogen Energy, 37*(23), 17853–17861. doi:10.1016/j.ijhydene.2012.09.023

Fernández, J. G., Almeida, C. A., Fernández-Baldo, M. A., Felici, E., Raba, J., & Sanz, M. I. (2016). Development of nitrocellulose membrane filters impregnated with different biosynthesized silver nanoparticles applied to water purification. *Talanta, 146*, 237–243. doi:10.1016/j.talanta.2015.08.060 PMID:26695258

Franci, G., Falanga, A., Galdiero, S., Palomba, L., Rai, M., Morelli, G., & Galdiero, M. (2015). Silver nanoparticles as potential antibacterial agents. *Molecules (Basel, Switzerland), 20*(5), 8856–8874. doi:10.3390/molecules20058856 PMID:25993417

Freeman, A. I., Halladay, L. J., & Cripps, P. (2012). The effect of silver impregnation of surgical scrub suits on surface bacterial contamination. *Veterinary Journal (London, England), 192*(3), 489–493. doi:10.1016/j.tvjl.2011.06.039 PMID:22015140

Gangadharan, D., Harshvardan, K., Gnanasekar, G., Dixit, D., Popat, K. M., & Anand, P. S. (2010). Polymeric microspheres containing silver nanoparticles as a bactericidal agent for water disinfection. *Water Research*, *44*(18), 5481–5487. doi:10.1016/j.watres.2010.06.057 PMID:20673945

Gao, B., Lin, Y., Wei, S., Zeng, J., Liao, Y., Chen, L., & Hou, J. et al. (2012). Charge transfer and retention in directly coupled Au-CdSe nanohybrids. *Nano Research*, *5*(2), 88–98. doi:10.1007/s12274-011-0188-8

Geiser, M., Rothen-Rutishauser, B., Kapp, N., Schürch, S., Kreyling, W., Schulz, H., & Gehr, P. et al. (2005). Ultrafine particles cross cellular membranes by nonphagocytic mechanisms in lungs and in cultured cells. *Environmental Health Perspectives*, *113*(11), 1555–1560. doi:10.1289/ehp.8006 PMID:16263511

Gluhoi, A. C., Lin, S. D., & Nieuwenhuys, B. E. (2004). The beneficial effect of the addition of base metal oxides to gold catalysts on reactions relevant to air pollution abatement. *Catalysis Today*, *90*(3), 175–181. doi:10.1016/j.cattod.2004.04.025

Gottesman, R., Shukla, S., Perkas, N., Solovyov, L. A., Nitzan, Y., & Gedanken, A. (2010). Sonochemical coating of paper by microbiocidal silver nanoparticles. *Langmuir*, *27*(2), 720–726. doi:10.1021/la103401z PMID:21155556

Guo, J. Z., Cui, H., Zhou, W., & Wang, W. (2008). Ag nanoparticle-catalyzed chemiluminescent reaction between luminol and hydrogen peroxide. *Journal of Photochemistry and Photobiology A Chemistry*, *193*(2), 89–96. doi:10.1016/j.jphotochem.2007.04.034

Guo, S., & Wang, E. (2011). Noble metal nanomaterials: Controllable synthesis and application in fuel cells and analytical sensors. *Nano Today*, *6*(3), 240–264. doi:10.1016/j.nantod.2011.04.007

Hahm, J. I., & Lieber, C. M. (2004). Direct ultrasensitive electrical detection of DNA and DNA sequence variations using nanowire nanosensors. *Nano Letters*, *4*(1), 51–54. doi:10.1021/nl034853b

Haldar, K. K., Sinha, G., Lahtinen, J., & Patra, A. (2012). Hybrid colloidal Au-CdSe pentapod heterostructures synthesis and their photocatalytic properties. *ACS Applied Materials & Interfaces*, *4*(11), 6266–6272. doi:10.1021/am301859b PMID:23113704

Hirakawa, T., & Kamat, P. V. (2005). Charge separation and catalytic activity of Ag@ TiO2 core-shell composite clusters under UV-irradiation. *Journal of the American Chemical Society*, *127*(11), 3928–3934. doi:10.1021/ja042925a PMID:15771529

Holtz, R. D., Lima, B. A., Souza Filho, A. G., Brocchi, M., & Alves, O. L. (2012). Nanostructured silver vanadate as a promising antibacterial additive to water-based paints. *Nanomedicine; Nanotechnology, Biology, and Medicine*, *8*(6), 935–940. doi:10.1016/j.nano.2011.11.012 PMID:22197722

Hu, C., Peng, T., Hu, X., Nie, Y., Zhou, X., Qu, J., & He, H. (2009). Plasmon-induced photodegradation of toxic pollutants with Ag− AgI/Al2O3 under visible-light irradiation. *Journal of the American Chemical Society*, *132*(2), 857–862. doi:10.1021/ja907792d PMID:20028089

Huang, H., & Yang, Y. (2008). Preparation of silver nanoparticles in inorganic clay suspensions. *Composites Science and Technology*, *68*(14), 2948–2953. doi:10.1016/j.compscitech.2007.10.003

Hughes, M. D., Xu, Y. J., Jenkins, P., McMorn, P., Landon, P., Enache, D. I., & Stitt, E. H. et al. (2005). Tunable gold catalysts for selective hydrocarbon oxidation under mild conditions. *Nature*, *437*(7062), 1132–1135. doi:10.1038/nature04190 PMID:16237439

Hussain, S. M., Hess, K. L., Gearhart, J. M., Geiss, K. T., & Schlager, J. J. (2005). In vitro toxicity of nanoparticles in BRL 3A rat liver cells. *Toxicology In Vitro*, *19*(7), 975–983. doi:10.1016/j.tiv.2005.06.034 PMID:16125895

Iglesias-Silva, E., Rivas, J., Isidro, L. L., & Lopez-Quintela, M. A. (2007). Synthesis of silver-coated magnetite nanoparticles. *Journal of Non-Crystalline Solids*, *353*(8), 829–831. doi:10.1016/j.jnoncrysol.2006.12.050

Jain, J., Arora, S., Rajwade, J. M., Omray, P., Khandelwal, S., & Paknikar, K. M. (2009). Silver nanoparticles in therapeutics: Development of an antimicrobial gel formulation for topical use. *Molecular Pharmaceutics*, *6*(5), 1388–1401. doi:10.1021/mp900056g PMID:19473014

Jain, P., & Pradeep, T. (2005). Potential of silver nanoparticle-coated polyurethane foam as an antibacterial water filter. *Biotechnology and Bioengineering*, *90*(1), 59–63. doi:10.1002/bit.20368 PMID:15723325

Jia, X., Ma, X., Wei, D., Dong, J., & Qian, W. (2008). Direct formation of silver nanoparticles in cuttlebone-derived organic matrix for catalytic applications. *Colloids and Surfaces. A, Physicochemical and Engineering Aspects*, *330*(2), 234–240. doi:10.1016/j.colsurfa.2008.08.016

Khalil-Abad, M. S., & Yazdanshenas, M. E. (2010). Superhydrophobic antibacterial cotton textiles. *Journal of Colloid and Interface Science*, *351*(1), 293–298. doi:10.1016/j.jcis.2010.07.049 PMID:20709327

Khlebtsov, N. G., & Dykman, L. A. (2010). Optical properties and biomedical applications of plasmonic nanoparticles. *Journal of Quantitative Spectroscopy & Radiative Transfer*, *111*(1), 1–35. doi:10.1016/j.jqsrt.2009.07.012

Kim, D., & Jon, S. (2012). Gold nanoparticles in image-guided cancer therapy. *Inorganica Chimica Acta*, *393*, 154–164. doi:10.1016/j.ica.2012.07.001

Kim, W. B., Voitl, T., Rodriguez-Rivera, G. J., & Dumesic, J. A. (2004). Powering fuel cells with CO via aqueous polyoxometalates and gold catalysts. *Science*, *305*(5688), 1280–1283. doi:10.1126/science.1100860 PMID:15333837

Kochuveedu, S. T., Jang, Y. H., & Kim, D. H. (2013). A study on the mechanism for the interaction of light with noble metal-metal oxide semiconductor nanostructures for various photophysical applications. *Chemical Society Reviews*, *42*(21), 8467–8493. doi:10.1039/c3cs60043b PMID:23925494

Köhler, J. M., Abahmane, L., Wagner, J., Albert, J., & Mayer, G. (2008). Preparation of metal nanoparticles with varied composition for catalytical applications in microreactors. *Chemical Engineering Science*, *63*(20), 5048–5055. doi:10.1016/j.ces.2007.11.038

Kumar, A., Vemula, P. K., Ajayan, P. M., & John, G. (2008). Silver-nanoparticle-embedded antimicrobial paints based on vegetable oil. *Nature Materials*, *7*(3), 236–241. doi:10.1038/nmat2099 PMID:18204453

Li, J., Zhao, T., Chen, T., Liu, Y., Ong, C. N., & Xie, J. (2015). Engineering noble metal nanomaterials for environmental applications. *Nanoscale*, *7*(17), 7502–7519. doi:10.1039/C5NR00857C PMID:25866322

Li, L. S., Hu, J., Yang, W., & Alivisatos, A. P. (2001). Band gap variation of size-and shape-controlled colloidal CdSe quantum rods. *Nano Letters*, *1*(7), 349–351. doi:10.1021/nl015559r

Li, N., Sioutas, C., Cho, A., Schmitz, D., Misra, C., Sempf, J., & Nel, A. et al. (2003). Ultrafine particulate pollutants induce oxidative stress and mitochondrial damage. *Environmental Health Perspectives*, *111*(4), 455–460. doi:10.1289/ehp.6000 PMID:12676598

Li, Q., Mahendra, S., Lyon, D. Y., Brunet, L., Liga, M. V., Li, D., & Alvarez, P. J. (2008). Antimicrobial nanomaterials for water disinfection and microbial control: Potential applications and implications. *Water Research*, *42*(18), 4591–4602. doi:10.1016/j.watres.2008.08.015 PMID:18804836

Lima, R., Seabra, A. B., & Durán, N. (2012). Silver nanoparticles: A brief review of cytotoxicity and genotoxicity of chemically and biogenically synthesized nanoparticles. *Journal of Applied Toxicology*, *32*(11), 867–879. doi:10.1002/jat.2780 PMID:22696476

Linic, S., Christopher, P., & Ingram, D. B. (2011). Plasmonic-metal nanostructures for efficient conversion of solar to chemical energy. *Nature Materials*, *10*(12), 911–921. doi:10.1038/nmat3151 PMID:22109608

Liu, P., & Zhao, M. (2009). Silver nanoparticle supported on halloysite nanotubes catalyzed reduction of 4-nitrophenol (4-NP). *Applied Surface Science*, *255*(7), 3989–3993. doi:10.1016/j.apsusc.2008.10.094

Liu, Z. P., Jenkins, S. J., & King, D. A. (2005). Origin and activity of oxidized gold in water-gas-shift catalysis. *Physical Review Letters*, *94*(19), 196102. doi:10.1103/PhysRevLett.94.196102 PMID:16090190

Lv, Y., Liu, H., Wang, Z., Liu, S., Hao, L., Sang, Y., & Boughton, R. I. et al. (2009). Silver nanoparticle-decorated porous ceramic composite for water treatment. *Journal of Membrane Science*, *331*(1), 50–56. doi:10.1016/j.memsci.2009.01.007

Manno, D., Filippo, E., Di Giulio, M., & Serra, A. (2008). Synthesis and characterization of starch-stabilized Ag nanostructures for sensors applications. *Journal of Non-Crystalline Solids*, *354*(52), 5515–5520. doi:10.1016/j.jnoncrysol.2008.04.059

Montazer, M., Alimohammadi, F., Shamei, A., & Rahimi, M. K. (2012). Durable antibacterial and cross-linking cotton with colloidal silver nanoparticles and butane tetracarboxylic acid without yellowing. *Colloids and Surfaces. B, Biointerfaces*, *89*, 196–202. doi:10.1016/j.colsurfb.2011.09.015 PMID:21978552

Montazer, M., Shamei, A., & Alimohammadi, F. (2012). Stabilized nanosilver loaded nylon knitted fabric using BTCA without yellowing. *Progress in Organic Coatings*, *74*(1), 270–276. doi:10.1016/j.porgcoat.2012.01.003

Mpenyana-Monyatsi, L., Mthombeni, N. H., Onyango, M. S., & Momba, M. N. (2012). Cost-effective filter materials coated with silver nanoparticles for the removal of pathogenic bacteria in groundwater. *International Journal of Environmental Research and Public Health*, *9*(1), 244–271. doi:10.3390/ijerph9010244 PMID:22470290

Nalawade, P., Mukherjee, T., & Kapoor, S. (2012). High-yield synthesis of multispiked gold nanoparticles: Characterization and catalytic reactions. *Colloids and Surfaces. A, Physicochemical and Engineering Aspects*, *396*, 336–340. doi:10.1016/j.colsurfa.2012.01.018

Nel, A., Xia, T., Mädler, L., & Li, N. (2006). Toxic potential of materials at the nanolevel. *Science*, *311*(5761), 622–627. doi:10.1126/science.1114397 PMID:16456071

Oberdörster, G., Maynard, A., Donaldson, K., Castranova, V., Fitzpatrick, J., Ausman, K., & Olin, S. et al. (2005). Principles for characterizing the potential human health effects from exposure to nanomaterials: Elements of a screening strategy. *Particle and Fibre Toxicology*, *2*(1), 1. doi:10.1186/1743-8977-2-8 PMID:16209704

Pelaez, M., Nolan, N. T., Pillai, S. C., Seery, M. K., Falaras, P., Kontos, A. G., & Entezari, M. H. et al. (2012). A review on the visible light active titanium dioxide photocatalysts for environmental applications. *Applied Catalysis B: Environmental*, *125*, 331–349. doi:10.1016/j.apcatb.2012.05.036

Porter, A. E., Gass, M., Muller, K., Skepper, J. N., Midgley, P., & Welland, M. (2007). Visualizing the uptake of C60 to the cytoplasm and nucleus of human monocyte-derived macrophage cells using energy-filtered transmission electron microscopy and electron tomography. *Environmental Science & Technology*, *41*(8), 3012–3017. doi:10.1021/es062541f PMID:17533872

Pradeep, T., & Anshup, . (2009). Noble metal nanoparticles for water purification: A critical review. *Thin Solid Films*, *517*(24), 6441–6478. doi:10.1016/j.tsf.2009.03.195

Qu, F., Li, N. B., & Luo, H. Q. (2013). Transition from nanoparticles to nanoclusters: Microscopic and spectroscopic investigation of size-dependent physicochemical properties of polyamine-functionalized silver nanoclusters. *The Journal of Physical Chemistry C*, *117*(7), 3548–3555. doi:10.1021/jp3091792

Rai, M., Yadav, A., & Gade, A. (2009). Silver nanoparticles as a new generation of antimicrobials. *Biotechnology Advances*, *27*(1), 76–83. doi:10.1016/j.biotechadv.2008.09.002 PMID:18854209

Roe, D., Karandikar, B., Bonn-Savage, N., Gibbins, B., & Roullet, J. B. (2008). Antimicrobial surface functionalization of plastic catheters by silver nanoparticles. *The Journal of Antimicrobial Chemotherapy*, *61*(4), 869–876. doi:10.1093/jac/dkn034 PMID:18305203

Sangpour, P., Hashemi, F., & Moshfegh, A. Z. (2010). Photoenhanced degradation of methylene blue on cosputtered M: TiO2 (M= Au, Ag, Cu) nanocomposite systems: a comparative study. *The Journal of Physical Chemistry C*, *114*(33), 13955–13961. doi:10.1021/jp910454r

Savić, R., Luo, L., Eisenberg, A., & Maysinger, D. (2003). Micellar nanocontainers distribute to defined cytoplasmic organelles. *Science*, *300*(5619), 615–618. doi:10.1126/science.1078192 PMID:12714738

Sergeev, G. B. (2003). Cryochemistry of metal nanoparticles. *Journal of Nanoparticle Research*, *5*(5-6), 529–537. doi:10.1023/B:NANO.0000006153.65107.42

Sergeev, G. B., & Shabatina, T. I. (2008). Cryochemistry of nanometals. *Colloids and Surfaces. A, Physicochemical and Engineering Aspects*, *313*, 18–22. doi:10.1016/j.colsurfa.2007.04.064

Sharma, V. K., Yngard, R. A., & Lin, Y. (2009). Silver nanoparticles: Green synthesis and their antimicrobial activities. *Advances in Colloid and Interface Science*, *145*(1), 83–96. doi:10.1016/j.cis.2008.09.002 PMID:18945421

Shen, C., Shen, Y., Wen, Y., Wang, H., & Liu, W. (2011). Fast and highly efficient removal of dyes under alkaline conditions using magnetic chitosan-Fe(III) hydrogel. *Water Research, 45*, 5200-5210

Shen, Y.F., Tang, J., Nie, Z.H., Wang, Y.D., Ren, Y., & Zuo, L. (2009). Tailoring size and structural distortion of Fe3O4 nanoparticles for the purification of contaminated water. *Bioresource Technology, 100*, 4139-4146.

Sheng, Z., & Liu, Y. (2011). Effects of silver nanoparticles on wastewater biofilms. *Water Research, 45*(18), 6039–6050. doi:10.1016/j.watres.2011.08.065 PMID:21940033

Singh, M., Singh, S., Prasad, S., & Gambhir, I. S. (2008). Nanotechnology in medicine and antibacterial effect of silver nanoparticles. *Digest Journal of Nanomaterials and Biostructures, 3*(3), 115–122.

Sintubin, L., Awoke, A. A., Wang, Y., Van der Ha, D., & Verstraete, W. (2012). Enhanced disinfection efficiencies of solar irradiation by biogenic silver. *Annals of Microbiology, 62*(1), 187–191. doi:10.1007/s13213-011-0245-2

Sondi, I., & Salopek-Sondi, B. (2004). Silver nanoparticles as antimicrobial agent: A case study on E. coli as a model for Gram-negative bacteria. *Journal of Colloid and Interface Science, 275*(1), 177–182. doi:10.1016/j.jcis.2004.02.012 PMID:15158396

Subramanian, V., Wolf, E. E., & Kamat, P. V. (2003). Influence of metal/metal ion concentration on the photocatalytic activity of TiO2-Au composite nanoparticles. *Langmuir, 19*(2), 469–474. doi:10.1021/la026478t

Subramanian, V., Wolf, E. E., & Kamat, P. V. (2004). Catalysis with TiO2/gold nanocomposites.Effect of metal particle size on the Fermi level equilibration. *Journal of the American Chemical Society, 126*(15), 4943–4950. doi:10.1021/ja0315199 PMID:15080700

Sun, L., Wu, W., Tian, Q., Lei, M., Liu, J., Xiao, X., & Jiang, C. et al. (2015). *In situ Oxidation and Self-Assembly Synthesis of Dumbbell-Like α-Fe2O3/Ag/AgX (X= Cl, Br, I)*. Heterostructures with Enhanced Photocatalytic Properties. ACS Sustainable Chemistry & Engineering.

Suvith, V. S., & Philip, D. (2014). Catalytic degradation of methylene blue using biosynthesized gold and silver nanoparticles. *Spectrochimica Acta. Part A: Molecular and Biomolecular Spectroscopy, 118*, 526–532. doi:10.1016/j.saa.2013.09.016 PMID:24091344

Tang, S. C., & Lo, I. M. (2013). Magnetic nanoparticles: Essential factors for sustainable environmental applications. *Water Research, 47*(8), 2613–2632. doi:10.1016/j.watres.2013.02.039 PMID:23515106

Tedesco, S., Doyle, H., Blasco, J., Redmond, G., & Sheehan, D. (2010). Oxidative stress and toxicity of gold nanoparticles in Mytilus edulis. *Aquatic Toxicology (Amsterdam, Netherlands), 100*(2), 178–186. doi:10.1016/j.aquatox.2010.03.001 PMID:20382436

Tian, D. S., Dong, Q., Pan, D. J., He, Y., Yu, Z. Y., Xie, M. J., & Wang, W. (2007). Attenuation of astrogliosis by suppressing of microglial proliferation with the cell cycle inhibitor olomoucine in rat spinal cord injury model. *Brain Research, 1154*, 206–214. doi:10.1016/j.brainres.2007.04.005 PMID:17482149

Tongying, P., Plashnitsa, V. V., Petchsang, N., Vietmeyer, F., Ferraudi, G. J., Krylova, G., & Kuno, M. (2012). Photocatalytic hydrogen generation efficiencies in one-dimensional CdSe heterostructures. *The Journal of Physical Chemistry Letters, 3*(21), 3234–3240. doi:10.1021/jz301628b PMID:26296035

Tran, Q. H., & Le, A. T. (2013). Silver nanoparticles: Synthesis, properties, toxicology, applications and perspectives. *Advances in Natural Sciences: Nanoscience and Nanotechnology*, *4*(3), 033001.

Wang, P., Huang, B., Dai, Y., & Whangbo, M. H. (2012). Plasmonic photocatalysts: Harvesting visible light with noble metal nanoparticles. *Physical Chemistry Chemical Physics*, *14*(28), 9813–9825. doi:10.1039/c2cp40823f PMID:22710311

Williams, D. (2008). The relationship between biomaterials and nanotechnology. *Biomaterials*, *29*(12), 1737–1738. doi:10.1016/j.biomaterials.2008.01.003 PMID:18249439

Wu, Z. C., Zhang, Y., Tao, T. X., Zhang, L., & Fong, H. (2010). Silver nanoparticles on amidoxime fibers for photo-catalytic degradation of organic dyes in waste water. *Applied Surface Science*, *257*(3), 1092–1097. doi:10.1016/j.apsusc.2010.08.022

Xue, C. H., Chen, J., Yin, W., Jia, S. T., & Ma, J. Z. (2012). Superhydrophobic conductive textiles with antibacterial property by coating fibers with silver nanoparticles. *Applied Surface Science*, *258*(7), 2468–2472. doi:10.1016/j.apsusc.2011.10.074

Yakub, I., & Soboyejo, W. O. (2012). Adhesion of E. coli to silver-or copper-coated porous clay ceramic surfaces. *Journal of Applied Physics*, *111*(12), 124324. doi:10.1063/1.4722326

Yantasee, W., Warner, C.L., Sangvanich, T., Addleman, R.S., Carter, T.G., Wiacek, R.J., Fryxell, G.E., Timchalk, C., & Warner, M.G. (2007). Removal of heavy metals from aqueous systems with thiol functionalized superparamagnetic nanoparticles. *Environmental Science and Technology, 41*, 5114-5119.

Yoon, B., Häkkinen, H., Landman, U., Wörz, A. S., Antonietti, J. M., Abbet, S., & Heiz, U. et al. (2005). Charging effects on bonding and catalyzed oxidation of CO on Au8 clusters on MgO. *Science*, *307*(5708), 403–407. doi:10.1126/science.1104168 PMID:15662008

Yu, S. J., Yin, Y. G., & Liu, J. F. (2013). Silver nanoparticles in the environment. *Environmental Science: Processes & Impacts*, *15*(1), 78–92. PMID:24592429

Zhai, Y., Han, L., Wang, P., Li, G., Ren, W., Liu, L., & Dong, S. et al. (2011). Superparamagnetic plasmonic nanohybrids: Shape-controlled synthesis, TEM-induced structure evolution, and efficient sunlight-driven inactivation of bacteria. *ACS Nano*, *5*(11), 8562–8570. doi:10.1021/nn201875k PMID:21951020

Zhang, J., Zhang, Y., Chen, Y., Du, L., Zhang, B., Zhang, H., & Wang, K. et al. (2012). Preparation and characterization of novel polyethersulfone hybrid ultrafiltration membranes bending with modified halloysite nanotubes loaded with silver nanoparticles. *Industrial & Engineering Chemistry Research*, *51*(7), 3081–3090. doi:10.1021/ie202473u

Zhang, Q., Xie, J., Yu, Y., Yang, J., & Lee, J. Y. (2010). Tuning the crystallinity of Au nanoparticles. *Small*, *6*(4), 523–527. doi:10.1002/smll.200902033 PMID:20108243

Zhou, M., Li, J., Ye, Z., Ma, C., Wang, H., Huo, P., & Yan, Y. et al. (2015). Transferring charge and energy of Ag@CdSe QDs-rGO core-shell plasmonic photocatalyst for enhanced visible light photocatalytic activity. *ACS Applied Materials & Interfaces*, *7*(51), 28231–28243. doi:10.1021/acsami.5b06997 PMID:26669327

Chapter 13
Ecotoxicity and Toxicity of Nanomaterials with Potential for Wastewater Treatment Applications

Verónica Inês Jesus Oliveira Nogueira
University of Porto, Portugal

Ana Gavina
University of Porto, Portugal

Sirine Bouguerra
Engineering School of Sfax, Tunisia

Tatiana Andreani
University of Porto, Portugal & CITAB-University of Trás-os-Montes and Alto Douro, Portugal

Isabel Lopes
University of Aveiro, Portugal

Teresa Rocha-Santos
University of Aveiro, Portugal

Ruth Pereira
University of Porto, Portugal

ABSTRACT

Nanotechnology holds the promise of develop new processes for wastewater treatment. However, it is important to understand what the possible impacts on the environment of NMs. This study joins all the information available about the toxicity and ecotoxicity of NMs to human cell lines and to terrestrial and aquatic biota. Terrestrial species seems more protected, since effects are being recorded for concentrations higher than those that could be expected in the environment. The soil matrix is apparently trapping and filtering NMs. Further studies should focus more on indirect effects in biological communities rather than only on effects at the individual level. Aquatic biota, mainly from freshwater ecosystems, seemed to be at higher risk, since dose effect concentrations recorded were remarkable lower, at least for some NMs. The toxic effects recorded on different culture lines, also give rise to serious concerns regarding the potential effects on human health. However, few data exists about environmental concentrations to support the calculation of risks to ecosystems and humans.

DOI: 10.4018/978-1-5225-0585-3.ch013

INTRODUCTION

Surface and ground water resources are continuously facing profound changes and quality deterioration, caused by anthropogenic activities, such as, mining operations, manufacturing and agro-industries. With the industrial development, the generation and accumulation of waste products has tremendously increased and one of the major challenges is the proper management and safe disposal of the vast amount and array of such solid and liquid wastes. Industrial wastewaters are one of the major sources of direct and often continue input of pollutants into aquatic ecosystems (Kanu & Achi, 2011). Due to the lack of effluent treatment facilities, proper treatment methodologies and disposal systems, huge amounts of industrial wastewater, containing high loads of organic and inorganic chemicals with high toxicity and recalcitrant properties, are being discharged into aquatic environments. Depending on the type of industry, the wastewater produced can contain different pollutants such as dyes, phenolic compounds, surfactants, pharmaceuticals, pesticides, organic solvents, chlorinated by-products, metals and microorganisms, which can cause the increase in biological oxygen demand (BOD), chemical oxygen demand (COD) and total dissolved solids (TDS) in the receiving water systems promoting their deterioration.

In the last few years, nanomaterials (NMs) with their unique proprieties have been extensively studied for water and wastewater treatment. Nanotechnology holds the promise of enhancing the performance of existing treatment technologies but also offers the potential for developing new treatment solutions (Qu, Alvarez, & Li, 2013). In 2011, the European Commission (EC) has adopted the following definition for NMs: "a natural, incidental or manufactured material containing particles, in an unbound state or as an aggregate or as an agglomerate and where, for 50% or more of the particles in the number size distribution, one or more external dimensions is in the size range 1–100 nm"(EC, 2011). Within this definition, several NMs have received considerable attention for wastewater treatment, namely metal (silver (Ag), gold (Au)) and metal or metalloid oxides of titanium (TiO_2), silica (SiO_2), iron (Fe_2O_3), cerium (CeO_2) (García et al., 2011; Jarvie et al., 2009; Kaegi et al., 2011). Due to their reduced size, NMs display unique physical, chemical and biological proprieties compared to their bulk counterparts (Stone et al., 2010; Zhang & Fang, 2010). At the nanoscale, the specific surface area and the surface/volume ratio increases, leading to an increase in the number of surface atoms and therefore contributing for enhanced optical, electrical and magnetic properties (Mohmood et al., 2013). The properties that make NMs suitable for wastewater treatment, in particular, include high surface areas with more active sites available for adsorption; high reactivity and catalytic potential for use in photocatalysis process, antimicrobial activity, high mobility in solution, as well as, super magnetism proprieties for particle separation (Hariharan, 2006; Qu, Brame, Li, & Alvarez, 2013; Sánchez et al., 2011). Therefore, these unique characteristics can be useful for efficient removal of metals or to degrade persistent organic compounds.

Although there are great advances with the use of nanotechnology for wastewater treatment, health effects and environmental impacts associated to NMs are attracting considerable concern in the scientific field and regulatory agencies. The unique proprieties of NMs, which make them so appealing, can also be responsible for ecotoxicological effects. Hund-Rink & Simon (2006) for example, were one of the first authors reporting the effects caused by the potential formation of reactive oxygen species (ROS) during UV-irradiation of TiO_2 NMs used as photocatalysts on *Daphnia magna* and *Desmodesmus subspicatus* (Hund-Rinke & Simon, 2006). Also, it was recognized that the nanosize of these materials can favour the cross through cell membranes and their interaction with cellular components (Colvin, 2003). In the last two decades, some reviews were published joining data about the toxicity of NMs to several organisms (Baun, Hartmann, Grieger, & Kusk, 2008; Menard, Drobne, & Jemec, 2011; Navarro et al.,

2008; Peralta-Videa et al., 2011). However, there still exists a considerable gap between the available ecotoxicological data and the large amount of NMs that are being produced worldwide. Further, it is important to take into account that few is known about the fate of NMs in the environment, and their toxic effects will not only depend on morphologic proprieties, composition, size or synthesis method, but also on the physicochemical characteristics of the receptor medium (Xiaoke Hu, Cook, Wang, & Hwang, 2009; Lowry, Gregory, Apte, & Lead, 2012). Therefore, it is imperative to access the eco/cytotoxicity of NMs before their use for water and wastewater treatments purposes, as well as to perceive their stability and fate in the environment.

In this context, the aim of the present chapter is to highlight the potential health and environmental impacts of NMs with potential for wastewater treatment, in order to join information that will contribute for safe uses in the future.

NANOMATERIALS AND THEIR APPLICATION FOR WASTEWATER TREATMENT

The ability of NMs to trap metals is well known. Recillas et al. (2011) explored the ability of CeO_2, Fe_3O_4 and TiO_2 NMs for the removal of Pb(II) through adsorption, reporting adsorption capacities of 189 mg Pb g^{-1}, 83 mg Pb g^{-1} and 159 mg Pb g^{-1}, respectively (Recillas et al., 2011). Pena et al. (2005) reported another example, in which nanocrystalline TiO_2 were used both as adsorbent to remove arsenate [As(V)] and arsenite [As(III)] as well as to complete convert As(III) to As(V), through photocatalytic oxidation (Pena, Korfiatis, Patel, Lippincott, & Meng, 2005). In a recent review, Hua et al. (2012) reported several studies demonstrating the efficiency of nanosized metal oxides for metal's removal, as well as other strategies involving the functionalization of NMs, such as surface modification with amino groups, supporting with zeolites or coating with poly(3,4-ethylenedioxythiophene) to enhance the efficiency of metal removal (Hua et al., 2012). Furthermore, research on the removal of organic compounds has also being performed. Iron oxide NMs found application on the efficient removal of organic pollutants (Iram, Guo, Guan, Ishfaq, & Liu, 2010; Parham, Zargar, & Rezazadeh, 2012; Hui Wang & Huang, 2011). Special attention has being focused on the use of NMs for photocatalysis to degrade a variety of organic compounds such as dyes (M. Faisal, Abu Tariq, & Muneer, 2007; Giwa, Nkeonye, Bello, & Kolawole, 2012; Shu, Chang, & Chang, 2009), phenols (Chiou, Wu, & Juang, 2008; Morales-Flores, Pal, & Sánchez Mora, 2011), pesticides (Mahmoodi, Arami, Limaee, & Gharanjig, 2007), drugs (El-Kemary, El-Shamy, & El-Mehasseb, 2010), as well chlorinated aromatic compounds (Lu et al., 2011; Selli, Bianchi, Pirola, Cappelletti, & Ragaini, 2008). The major drawbacks in available studies are typically the use of synthetic water that obviously does not represent real wastewaters. Further, several of these studies focused on individual compounds forgetting the interaction in the complex mixtures where they are included like industrial wastewaters.

Ecotoxicity Assessment of NMs for Aquatic Compartments

The increasingly use of NMs for wastewater and water treatment purposes as well as for environmental remediation will lead to the release of substantial amounts of these materials in the aquatic environment. Although the environmental concentrations of NMs are still largely unknown there are evidences of the presence of NMs in wastewater streams and waste leachates which in turn will be discharged into rivers and lakes endangering aquatic organisms (Brar, Verma, Tyagi, & Surampalli, 2010; Hennebert,

Table 1. Toxicity of metal nanomaterials in different aquatic organisms

NMs	Size (nm)	Species	Endpoint	Effect	Reference
TiO$_2$	21	*Daphnia magna* Japanese medaka	Acute phototoxicity under simulated solar radiation (SSR)	48 h LC$_{50}$ = 29.8 mg L^{-1} 96 h LC$_{50}$ = 2.2 mg L^{-1}	(Hongbo Ma, Brennan, & Diamond, 2012b)
	10 30 300	*Raphidocelis subcapitata*	72 h growth inhibition	EC$_{50}$ = 241 mg L^{-1} EC$_{50}$ = 71.1 mg L^{-1} EC$_{50}$ = 145 mg L^{-1}	(Hartmann et al., 2010)
	15–30	*Thalassiosira pseudonana* *Skeletonema costatum* *Dunaliella tertiolecta* *Isochrysis galbana*	96 h growth inhibition under UV light (Photocatalytic activity)	NOEC = 1 mg L^{-1} -------- NOEC = 1 mg L^{-1} NOEC < 1 mg L^{-1}	(Miller, Bennett, Keller, Pease, & Lenihan, 2012)
	5-10	*Lemna minor*	7 days growth inhibition	Inhibited plant growth at high concentrations (> 200 mg L^{-1})	(G. Song et al., 2012)
	----	*Danio renio*	Reproduction	29.5% reduction in the cumulative number of zebrafish eggs after 13 weeks	(J. Wang, et al., 2011)
NiO	20	*Chlorella vulgaris*	72 h growth inhibition	EC$_{50}$ = 32.28 mg L^{-1}	(Gong et al., 2011)
	10-20 100	*Vibrio fischeri* *Raphidocelis subcapitata* *Lemna minor* *Daphnia magna* *Brachionus plicatilis* *Artemia salina*	Bioluminescence inhibition 72 h growth inhibition 7 days growth inhibition Immobilization and reproduction Immobilization Immobilization	-------- EC$_{50}$ (10-20) = 15.2 mg L^{-1} EC$_{50}$ (100) = 8.25 mg L^{-1} EC$_{50}$ (100) = 4.39 mg L^{-1} NOEC (10-20) = 6.55 mg L^{-1} LC$_{50}$ (100) = 9.74 mg L^{-1} LC$_{50}$ (10-20) = 9.76 mg L^{-1} NOEC (100) = 0.04 mg L^{-1} NOEC (10-20) = 0.11 mg L^{-1} -------- --------	(Nogueira et al., 2015)
CuO	≈ 30	*Vibrio fischeri* *Daphnia magna* *Thamnocephalus platyurus*	Growth inhibition 48 h Mortality 24 h Mortality	EC$_{50}$ = 79 mg L^{-1} LC$_{50}$ = 3.2 mg L^{-1} LC$_{50}$ = 2.1 mg L^{-1}	(Heinlaan, Ivask, Blinova, Dubourguier, & Kahru, 2008)
	≈ 30	*Raphidocelis subcapitata*	72 h growth inhibition	EC$_{50}$ = 0.71 mg L^{-1}	(Aruoja, et al., 2009)
	40–500 Average 197	*Danio rerio*	Hatching and malformation	Affects hatching and increase malformations at concentrations ≥ 1 mg L^{-1}	(Vicario-Parés et al., 2014)
ZnO	50–70	*Vibrio fischeri* *Daphnia magna* *Thamnocephalus platyurus*	Growth inhibition 48 h Mortality 24 h Mortality	EC$_{50}$ = 1.9 mg L^{-1} LC$_{50}$ = 3.2 mg L^{-1} LC$_{50}$ = 0.18 mg L^{-1}	(Heinlaan, et al., 2008)
	100	Marine algae *Dunaliella tertiolecta*	96h growth inhibition	EC$_{50}$ = 2.42 mg L^{-1}	(Manzo, Miglietta, Rametta, Buono, & Di Francia, 2013)
	40–100	*Xenopus laevis*	Malformations	EC$_{50}$ = 10.3 mg L^{-1}	(Nations et al., 2011)
	150–1000 (DLS) <100 (TEM)	*Danio rerio*	Hatching and malformation	Affects hatching at ≥ 5 mg L^{-1}	(Vicario-Parés, et al., 2014)

continued on following page

Table 1. Continued

NMs	Size (nm)	Species	Endpoint	Effect	Reference
Fe_2O_3	50-150	*Chironomus tentans* larvae	Survival and growth (10 days)	NOEC = 5 mg kg^{-1}	(Oberholster et al., 2011)
	30	*Danio rerio*	Embryo/larva survival Embryo hatching rate	LC_{50} = 53.35 mg L^{-1} EC_{50} = 36.06 mg L^{-1}	(X. Zhu, Tian, & Cai, 2012)
Fe_3O_4	6	*Vibrio fischeri* *Daphnia magna*	Bioluminescence inhibition Immobilization	EC_{50} = 0.24 mg L^{-1} LC_{50} = 0.00023 mg L^{-1}	(García, et al., 2011)
	20–40	*Ceriodaphnia dubia*	Immobilization Bioaccumulation	*C. dubia* significantly accumulated nano-Fe_2O_3 in the gut, with the maximum accumulation achieved after 6 h of exposure.	(J. Hu, Wang, Wang, & Wang, 2012)
Ag	50	*Chlorella vulgaris* *Dunaliella tertiolecta*	Cellular viability	Decrease in chlorophyll content, viable algal cells, increased ROS formation and lipids peroxidation	(Oukarroum, Bras, Perreault, & Popovic, 2012)
	50	*Lemna paucicostata*	Growth inhibition	EC_{50} = 13.8 mg L^{-1}	(E. Kim et al., 2011)
	5-25 ≈16.6 20	*Daphnia magna*	Immobilization	EC_{50} = 0.004 mg L^{-1} EC_{50} = 0.002 mg L^{-1} EC_{50} = 0.187 mg L^{-1}	(Asghari et al., 2012)
	<100	*Chironomus riparius*	Gene expression	Induction of genes related to oxidative stress and detoxificatio	(Nair, Park, & Choi, 2013)

TEM: Transmission electron microscope

Avellan, Yan, & Aguerre-Chariol, 2013; Westerhoff, Song, Hristovski, & Kiser, 2011). In particular this applies to NMs that are already being massively produced, such as metal-based and metal oxides, and applied in water/wastewater treatment (Baker, Tyler, & Galloway, 2014). Thus, since water resources are particularly exposed to the contamination with NMs it is imperative to address their potential toxicity to aquatic organisms, before the damage is unavoidable. In Table 1, we summarize some toxicity studies of NMs applied in wastewater treatment to several aquatic species.

Nano-TiO_2 is probably one of the most extensively studied NMs, whether if it is for water and wastewater treatment applications or for evaluating its ecotoxicological effects on aquatic organism. TiO_2 was efficiently applied to remove different metals but also several organic compounds (Khattab et al., 2012; Mahdavi, Jalali, & Afkhami, 2012; Mu et al., 2010; Recillas, et al., 2011). However, several studies also described the toxicity of nano-TiO_2. Arouja et al. (2009) reported an effective concentration (EC_{50}) for *Pseudokirchneriella subcapitata* of 5.83 mg L^{-1}, six times lower than the bulk-TiO_2 (Aruoja, Dubourguier, Kasemets, & Kahru, 2009). In another study, Zhu et al. (2010) found that the toxicity of nano-TiO_2 to *Daphnia magna* increased when the exposure time of the acute test was extended from 48 h to 72 h (EC_{50} > 100 mg L^{-1} and EC_{50}=1.62 mg L^{-1}, respectively) (Xiaoshan Zhu, Chang, & Chen, 2010). These authors also reported sub-lethal effects on reproduction at 0.1 mg L^{-1} and mortality at 2 mg L^{-1} after 21 days of exposure (Xiaoshan Zhu, et al., 2010). Wang et al. (2011) also demonstrated that a chronic exposure (13 week) to 0.1 mg L^{-1} nano-TiO_2 can negatively affect the reproduction of zebrafish (*Danio rerio*) (J. Wang et al., 2011).

Ecotoxicity and Toxicity of Nanomaterials with Potential for Wastewater Treatment Applications

There are several studies focusing the toxicity of other metal-based NMs to aquatic biota, besides nano-TiO$_2$, that are also used in wastewater treatment (Bour et al., 2015; García et al., 2012; Heinlaan, et al., 2008; S.-W. Lee, Kim, & Choi, 2009; Nations, et al., 2011; Nogueira, et al., 2015; H. Zhu, Han, Xiao, & Jin, 2008). For example, Nogueira et al. (2015) tested the toxicity of NiO (with two different particle sizes), TiO$_2$ and Fe$_2$O$_3$ NMs in several aquatic organisms (Nogueira, et al., 2015). Nano-sized iron oxides (including nano-Fe$_2$O$_3$) are promising for industrial wastewater treatment due to their strong adsorption ability and photocatalytic activity (Xu et al., 2012). The same occurs for nickel oxide NMs that have been efficiently used in the photocatalytic degradation of phenol, adsorption of dyes and removal of metals from wastewaters (Hayat, Gondal, Khaled, & Ahmed, 2011; Hristovski, Baumgardner, & Westerhoff, 2007). Nogueira et al. (2015) found that apparently Fe$_2$O$_3$ NMs seemed to be the one with less risk for the aquatic environment when applied to wastewater treatment. On the contrary, the freshwater species (*Raphidocelis subcapitata*, *Lemna minor* and *D. magna*) were acutely and chronically affected by NiO NMs (Nogueira, et al., 2015). In opposition, to freshwater species, these authors did not record any toxicity, for the range of concentration tested, in the assays with marine species (*Vibrio fischeri*, *Artemia salina* and *Brachionus plicatilis*) (Nogueira, et al., 2015).

In another study, the acute effects of Fe$_2$O$_3$, TiO$_2$, ZnO and CuO NMs on *Xenopus laevis* embryos was evaluated (Nations, et al., 2011). The exposure to these metal oxides caused no mortality, however, CuO and ZnO NMs induced the development of abnormalities at very low concentrations. Zinc oxide NMs is a semiconductor, very similar to nano-TiO$_2$, however, ZnO presents a greatest advantage since it adsorbs a large fraction of the solar spectrum (Sakthivel et al., 2003). Studies already proved the efficiency of ZnO as photocatalyst in the degradation of some organic compounds (Jang, Simer, & Ohm, 2006; Q. I. Rahman, Ahmad, Misra, & Lohani, 2013; Huihu Wang et al., 2007). However, when in aquatic environment, ZnO NMs have the tendency to dissolve and this soluble form of ionic zinc (Zn^{2+}) has been recognized as the main factor of toxicity to several aquatic organism (Hongbo Ma, Williams, & Diamond, 2013). Arouja et al. (2009) showed that the toxicity of nano-ZnO to the algae *P. subcapitata* was attributed to soluble Zn^{2+} ions (Aruoja, et al., 2009). Besides, the toxicity has been attributed to the dissolution of ZnO, the generation of ROS is also the mode of action for this NM, specially associated with its photocatalytic activity (Hongbo Ma et al., 2014). Metal-oxide NMs have been the subject of numerous toxicological and ecotoxicological tests, however only few studies address the photoreactivity of some of these NMs, which is the property that makes them particularly attractive for wastewater treatment applications. Miller et al. (2012) reported that even low levels of UV lamps can induce phototoxicity of nano-TiO$_2$ to marine phytoplankton (*Thalassiosira pseudonanan*, *Skeletonema marinoi*, *Isochrysis galbabana*, and *Dunaliella tertiolecta*), however, when the UV light was blocked, no toxic effects were recorded (Miller, et al., 2012). In a more realistic study, conducted under simulated solar radiation, (Hongbo Ma, et al., 2012b) assessed the acute toxicity of nano-TiO$_2$ to both aquatic species *D. magna* and *Oryzias latipes*. These authors reported a 48 h lethal concentration (LC$_{50}$) of 29.8 mg L^{-1} for *D. magna* and a 96 h LC$_{50}$ of 2.2 mg L^{-1} for *Oryzias latipes*. The phototoxicity of nano-TiO$_2$, under simulated sunlight, increased by two to four orders of magnitude when compared to the toxicity under laboratory light. Being nano-TiO$_2$ a photocatalyst (as well other semiconductor metal oxides such ZnO) with a band gap energy of 3.2 eV (for anatase TiO$_2$) it can be photo-activated in the presence of UV light enhancing the generation of ROS, the main factor responsible for breaking down contaminants in wastewater treatments (Robichaud, Uyar, Darby, Zucker, & Wiesner, 2009). Unfortunately, the generation of ROS has been proposed as one of the main mechanisms responsible for the toxicity of NMs (Fu, Dionysiou, & Liu, 2014; Kahru, Dubourguier, Blinova, Ivask, & Kasemets, 2008) to several aquatic organisms, such as marine phytoplankton (Miller, et al., 2012), freshwater planktonic organisms (K. T. Kim, Klaine, Cho, Kim, & Kim, 2010; Hongbo Ma, Brennan, & Diamond, 2012a; Hongbo Ma, et al., 2012b) and benthic organisms (S. Li, Wallis, Diamond, Ma, & Hoff, 2014; S. Li, Wallis, Ma, & Diamond, 2014; Wallis et al., 2014).

Besides metal oxide NMs, several other metal-based materials are being applied to water treatment. Nano-Zero valent iron (nZVI) has been used effectively for the treatment of toxic contaminants present in soil, groundwater and wastewaters (Fu, et al., 2014). The environmental applications of nZVI has been widely accepted mainly due to the low costs associated and also because the low toxicity of iron (Crane & Scott, 2012). However, the environmental risks associated with their use are still poorly understood. For example, Keller et al. (2012) tested the effect of three commercial forms (uncoated, organic coating, and iron oxide coating) of nZVI to three species of marine phytoplankton (*Isochrysis galbana*, *Dunaliella tertiolecta* and *Thalassiosira pseudonana*), one species of freshwater phytoplankton (*P. subcapitata*), and a freshwater zooplankton species (*D. magna*) (Keller, Garner, Miller, & Lenihan, 2012). The results showed that nZVI can be toxic to aquatic organisms living either in freshwater streams or marine environments (Keller, et al., 2012). Another metal-based NMs that has received considerable attention for applications in water treatment, due to their antibacterial properties, are silver NMs (Q. Li et al., 2008). However, the biological activity responsible by their antibacterial properties could also endanger other organisms once discharged into the environment. The ecotoxicity of silver NMs has been reported to various aquatic organisms, including algae (Miao et al., 2010; Miao et al., 2009; Oukarroum, et al., 2012), aquatic plants (Gubbins, Batty, & Lead, 2011; S. Kim et al., 2009), aquatic invertebrates (Asghari, et al., 2012; Nair, et al., 2013) and fish (Chae et al., 2009; George et al., 2014). The toxicity of nano-Ag is a subject of great debate. While some studies suggested that the release of silver ions caused the toxicity, other studies related the toxicity not only with silver ions but also with the nano-size of these particles (Behra et al., 2013). For example, Griffitt et al. (2008) exposed several aquatic organisms (*Danio rerio*, *Daphnia pulex*, *Ceriodaphnia dubia* and *Pseudokirchneriella subcapitata*) to silver NMs and $AgNO_3$ (Griffitt, Luo, Gao, Bonzongo, & Barber, 2008). These authors found that during the exposure the dissolution of nano-Ag was relatively low, so the recorded mortality cannot be attributed solely to the particle solubilization but also to the nano-Ag particles (Griffitt, et al., 2008). However, Miao et al. (2009) found that the toxicity of nano-Ag to the marine microalgae (*Thalassiosira weisflogii*) was entirely mediated by the dissolution of Ag ions from the NMs (Miao, et al., 2009).

It is recognized that the development and application of nanotechnology can play an important role in solving or ameliorating issues related to water and wastewater treatment. However we cannot deny the toxic effects of these NMs and caution should be taken when using NMs for water treatment.

Ecotoxicity of Nanomaterials for the Soil Biota

Once in the soil, NMs can be degraded by several processes (biotic and abiotic), and be transported to the aquatic compartment through runoff and leaching (Boxall, Tiede, & Chaudhry, 2007). Physical and chemical characteristics, such as the size of NMs, their chemical composition or their functionalization, will determine what their fate in the environment (Brar, et al., 2010). Some NMs can be carried over larger distances before becoming trapped in the soil matrix or they can be immediately adsorb to soil particles, becoming more immobile (Soni, Naoghare, Saravanadevi, & Pandey, 2015). Additionally, soil microorganisms may also adsorb or degrade NMs (Wiesner, Lowry, Alvarez, Dionysiou, & Biswas, 2006).

Table 2. Toxicity effect of NMs towards plants

NMs	Size Supports by Manufacturer (nm)	Plant	Concentration Tested	Observations	Reference
ZnO	20±5	*Lolium perenne* (ryegrass)	10 - 1000 mg L^{-1}	Biomass decrease, root tips shrank, root cells vacuolated or collapsed.	(D. Lin & Xing, 2008)
	50	*Allium cepa* (onion bulbs)	5 - 20 mg L^{-1}	Inhibition of root elongation. Accumulation in both the cellular and the chromosomal modules.	(Ghodake, Seo, & Lee, 2011)
	< 100	*Triticum aestivum* (wheat)	≈ 25 mg L^{-1}	Biomass decrease.	(Du et al., 2011)
	< 100	*Brassica juncea* (cabbage)	200 – 1500 mg L^{-1}	Decrease in plant biomass. Increase in proline content and lipid peroxidation.	(Rao & Shekhawat, 2014)
TiO$_2$	15, 25, 32	*Linum usitatissimum* (flax seeds)	0.01 – 100 mg L^{-1}	Inhibition of germination and root biomass.	(Clément, Hurel, & Marmier, 2013)
	Anastase 14, 25, 140 Rutile 22, 36, 655	*Triticum aestivum* (wheat)	10, 50 and 100 mg L^{-1}	Seed germination, vegetative development, photosynthesis and redox balance was not affected. Wheat roots accumulate NPs with a primary diameter lower than 140 nm and the NPs can translocate to the leaves if the primary size is smaller than 36 nm.	(C. Larue et al., 2012)
	27±4	*Cucumis sativus* (cucumber)	0 to 750 mg kg^{-1}	The NPs were translocated without biotransformation or to the edible part of cucumber plants.	(Servin et al., 2013)
nZVI	----	*Linum usitatissimum L.* (flax) *Lolium perenne L.* (ryegrass) *Hordeum vulgare L.* (two-rowed barley)	0–5000 mg L^{-1}	Inhibitory effects in aqueous suspensions at 250 mg L^{-1} and complete inhibition of germination observed at 1000-2000 mg L^{-1}. Inhibitory effects observed at 300 mg L^{-1} in soil. Complete inhibition observed at 750 and 1500 mg L^{-1} in sandy soil for flax and ryegrass, respectively.	(Yehia S. El-Temsah & Erik J. Joner, 2012)
Fe/ Fe$_3$O$_4$	50 - 60	*Lactuca sativa* (lettuce)	10 and 20 mg L^{-1}	Did not affect lettuce growth and chlorophyll content	(Trujillo-Reyes, Majumdar, Botez, Peralta-Videa, & Gardea-Torresdey, 2014)
Cu/ CuO	20 - 30	*Lactuca sativa* (lettuce)	10 and 20 mg L^{-1}	Affect plant growth, water content, dry biomass production, and concentration of several nutrients.	
Ag	----	*Arabidopsis thaliana* (thale cress)	0.2 – 3.0 mg L^{-1}	Inhibition of growth and root elongation. Decrease photosynthetic pigment content. Disruption of thylakoid membrane structure.	(Qian et al., 2013)
	5 to 25 (average 10)	*Phaseolus radiatus* (mung bean) *Sorghum bicolor* (sorghum)	0 - 40 mg L^{-1}	The exposure medium influences the phytotoxicity. Agar medium: *P. radiates* EC$_{50}$ = 13 mg L^{-1} *S. bicolor* EC$_{50}$ = 26 mg L^{-1} Soil medium EC$_{50}$ > 2000 mg kg^{-1} for both species	(W.-M. Lee, Kwak, & An, 2012)

NMs used for water and wastewater treatments can enter the terrestrial ecosystem during manufacture/transport/application or via storage/dump of sludge in landfills or through the use of contaminated sewage sludge or biosolids for agricultural land fertilization (Brar, et al., 2010; Kiser, Ryu, Jang, Hristovski, & Westerhoff, 2010). Biosolids are being widely recommended as a cheaper alternative for improving the fertility of agricultural soils (Antolin, Muro, & Sanchez-Diaz, 2010; Suppan, 2013), while the volumes to be deposited in landfills are being reduced. In the U.S., more than 60% of biosolids produced each year are being applied to agricultural lands (Gardea-Torresdey, Rico, & White, 2014). While in Europe the risks associated with this procedure are being deeply analysed, since NMs as well as other emergent contaminants can interact and have several impacts on soil organisms (microorganisms, plants and invertebrates). Considering, that soils are essential to the sustainability of ecosystems and for the survival of the human species (O'Halloran, 2006), it is important to understand how NMs will affect the soil biota and subsequently soil quality, functions, and services.

Ecotoxicity of Nanomaterials for Plants

Plants are an essential component of terrestrial ecosystems, maintaining the carbon and nitrogen cycling and are an important food source for humans and other organisms. Further, they are responsible by many other ecosystem services as soil formation and protection, climate regulation etc. Several recent data on plants responses to NMs exposure have been gathered and reviewed (Gardea-Torresdey, et al., 2014; Klaine et al., 2008; K.-E. Li, Chang, Shen, Yaon, & M.H. Siddiqui et al. (eds.), 2015; Y. Ma et al., 2010; Miralles, Church, & Harris, 2012; Navarro, et al., 2008; Schwab et al., 2015). Table 2 shows the main toxicity studies of NMs on plants. However, a limited range of plant species has been tested in nanophytotoxicity studies, and data acquisition is also limited to some NMs.

Seeds germination, biomass production and/or root and shoots elongation have been the main endpoints used for assessing the effects of NMs on plants (Elgrabli et al., 2008; Stampoulis, Sinha, & White, 2009; Yang & Watts, 2005) frequently under short exposure periods (Gardea-Torresdey, et al., 2014). Several studies and reviews also focused on the genotoxic effects, uptake, translocation, bioaccumulation and biotransformation in food crops (Chen et al., 2015; Kumari, Mukherjee, & Chandrasekaran, 2009; Y. Ma, et al., 2010; Miralles, et al., 2012; Remédios, Rosário, & Bastos, 2012; Rico, Majumdar, Duarte-Gardea, Peralta-Videa, & Gardea-Torresdey, 2011; H. Zhu, et al., 2008). However, in the great majority of studies the plants were exposed to NMs through hydroponic conditions, filter papers or through soils directly contaminated under laboratory conditions which do not represent the real route of exposure of plants, to NMs used for wastewater treatments, which will occur through soils amendments with sludge. Thus, there is still a vast gap of knowledge that limits our understanding of the long-term risks to plants (Gardea-Torresdey, et al., 2014; S. Lin et al., 2009).

Nowadays, TiO_2 NMs is well known for its great potential for a vast array of applications and also as a photocatalyst to remove organic pollutants from treated effluents and for several other purification technologies (Adesina, 2004; Shahid, McDonagh, Kim, & Ho Shon, 2015). However, the smallest particles of TiO_2 NMs are difficult to extract from water (Shahid, et al., 2015) while on the other hand the bigger and the heavier aggregates that can be formed during treatments, can easily settle, becoming part of sediments and sludge. Several studies showed the accumulation of TiO_2 NMs in root and leaves of seedlings of wheat and rapeseed, and also an accumulation of TiO_2 NMs in nodules of garden peas and in trichomes of cucumber (Fan, Huang, Grusak, Huang, & Sherrier, 2014; Camille Larue, Julien Laurette, et al., 2012; Camille Larue, Giulia Veronesi, et al., 2012; Servin et al., 2012). Even more, concerning TiO$_2$ NMs was found accumulated in cucumber fruits (Servin, et al., 2013), does showing a great potential for trophic transfer. In opposition, several other studies demonstrated that rather than being bioaccumulated TiO$_2$ NMs have absorbed on the surface of seeds and roots of lettuce, radish and cucumber (Wu et al., 2012). The main processes contributing for such adsorption were probably the physical attachment of particles on the rough seed surfaces, electrostatic attractions and hydrophobic interactions between seeds and NMs aggregates. These authors postulated that seed wax composition may be a determinant factor in this interaction of NMs with seed coats (X. Hu, Daun, & Scarth, 1994; H. N. Zhu, Lu, & Abdollahi, 2005).

As far as iron and iron oxide NMs are considered, a study evidenced that nano-Fe$_3$O$_4$ can be uptaken, translocated, and accumulated within various tissues of pumpkin plants (H. Zhu, et al., 2008).

Phytotoxic effects, as inhibition of shoot growth of ryegrass and barley were also observed at 1000 mg nZVI kg^{-1} soil (Y.S. El-Temsah & E.J. Joner, 2012). Moreover this iron NMs displayed strong toxic effect at concentrations above 200 mg L^{-1} reducing the transpiration and growth of hybrid poplar seedlings (X. Ma, Gurung, & Deng, 2013). The authors related the phytotoxicity of this NMs with the irregular aggregates of nanoscale zero-valent iron coating plant roots surface as revealed by transmission electron microscope (TEM) analysis and, with the internalization of nZVI by poplar root cells as revealed by scanning transmission electron microscope (STEM) analysis (X. Ma, et al., 2013).

With a more ecological relevant exposure, Chen et al. (2015) evaluated the toxicogenomic responses of *Medicago truncatula* growing in soils amended with biosolids containing a mixture of different NMs, such as Ag, ZnO, and TiO$_2$, and aged for six months (Chen, et al., 2015). The data gathered demonstrated differential expression of several genes involved in oxidative stress. Further the inhibition in the growth and nodulation of *M. truncatula* was observed and it was mainly related with enhanced bioavailability of Zn ions in the biosolids amended soils (Chen, et al., 2015), thus suggesting that the suspected hazard of this NMs, persists even when integrated in soil as part of a complex organic matter matrix.

In another study, the effects of NMs (zinc and zinc oxide) on seed germination and root growth of six higher plant species (radish, rape, ryegrass, lettuce, corn, and cucumber) was investigated and the results showed that seed germination was affected by nano-Zn on ryegrass and nano-ZnO on corn at 2000 mg L^{-1}. Further, these NMs inhibited the root elongation of tested species (Daohui Lin & Xing, 2007). Similar measurements, in a more recent study showed that the effect of NMs (Ag, Cu, Si, and ZnO) on *Cucurbita pepo* (zucchini) on roots elongation was more relevant than the impacts on germination (Stampoulis, et al., 2009). Also, Wu et al. (2012) investigated the toxicity of various metal oxide NMs on lettuce, radish and cucumber seeds (Wu, et al., 2012). In this study, only CuO and NiO showed harmful impacts in germination of all species with calculated EC$_{50s}$ of lettuce: NiO: 28 mg L^{-1}; CuO: 13 mg L^{-1}; radish: NiO: 401 mg L^{-1}; CuO: 398 mg L^{-1}; cucumber: NiO: 175 mg L^{-1}; CuO: 228 mg L^{-1}. Further, a short term exposures of tomato seeds to nano-NiO resulted in a significant reduction in root growth and caused oxidative stress (Mohammad Faisal et al., 2013)

In the same way, silver NMs are considered as effective antimicrobial materials for coliform and pathogen bacteria found in wastewater (Furno et al., 2004; Morones et al., 2005; Tiwari, Behari, & Sen, 2008). These NMs showed a 71% and 57% reduction in zucchini biomass and transpiration, respectively, after an exposure at 1000 and 500 mg L^{-1} (Stampoulis, et al., 2009). Ag NMs have also shown effects on cucumber, lettuce, ryegrass and barley seeds germination (Barrena et al., 2009; Y.S. El-Temsah & E.J. Joner, 2012). In contrast, these NMs did not show any effect in growth of *Phaseolus radiates,* and only a slight inhibition in growth was observed for *Sorghum bicolor* at a range of concentrations up to 2000 mg kg^{-1} in soil (W.-M. Lee, et al., 2012). The same authors proved that the bioavailability of silver ions dissolved from nano-Ag was more reduced in soil than in other tested media. Furthermore, tomato exposure to nano-Ag resulted in a significantly decreased biomass at concentrations up to 500 mg kg^{-1} nano-Ag (U. Song et al., 2013). Kumari et al. (2009) focused on the genotoxic effects of silver nanoparticles on *Allium cepa* cells from root and revealed several cytological consequences, namely chromatin bridges and stickiness (50 ppm concentration of nano-Ag suspended in deionized water), troubled metaphase (70 ppm) and chromosomal breaks (100 ppm) (Kumari, et al., 2009).

Besides the effects of metal-based NMs on plants, several studies also concentrated their attention on impact of carbon nanotubes. Begum, Ikhtiari, & Fugetsu (2014) using 1000 mg L^{-1} and 2000 mg L^{-1} of multi-walled carbon nanotubes showed significant inhibition in shoot elongation, cell death and electrolyte leakage in red spinach, lettuce, and cucumber, after 15 days of exposure (Begum, Ikhtiari, & Fugetsu, 2014). Also an effect on root elongation of *Cucurbita pepo* (zucchini) was recorded (Stampoulis, et al., 2009). However, toxicity effects of MWNTs were different between plants species. For example, red spinach and lettuce were more sensitive than rice and cucumber, while chili, lady's finger, and soybean showed slight or no toxic effects after being exposed to these NMs (Begum, et al., 2014).

Several studies showed that plants, and in particular food crops, are sensitive and may be affected by exposure to NMs. However, the evidences about the entry, transformation and accumulation into the food chain are limited and the implications on human health/nutrition on terrestrial communities are poorly understood or even unknown. Preferably, long-term exposures and full life-cycle studies are needed. Moreover, the future studies need to focus and take into consideration the different factors that may contribute or ameliorate NMs phytotoxicity like the effect of aging of MNs in soil, the role of different soil physico-chemical properties in the bioavailability of NMs, the addition of NMs to soils via sludge amendments or irrigation with treated wastewaters, as well as the interaction of NMs with several other contaminants, already present in natural soils (e.g. metals, pesticides, PCBs, PAHs, etc.).

Ecotoxicity of NMs for Terrestrial Invertebrates

Most of the studies with terrestrial invertebrates have been made with metal oxide NMs. These studies have mainly focused in assessing the toxicity of NMs in the avoidance and reproduction of earthworms as shown in Table 3. Earthworms are widely used in ecotoxicological assessments and were the former soil organism for which standard protocols were developed. One characteristic that makes interesting to understand the effect of NMs in these organism is their permanent and direct contact with the soil, in parallel with their important roles in soil functions (Blouina et al., 2013). Heckmann et al. (2011) screened the effect of some NMs (Ag-NP, SiO$_2$-NP, TiO$_2$-NP) at a concentration of 1000 mg kg$^{-1}_{DW}$ and their corresponding metal salts (AgNO$_3$) and bulk metal oxides in reproduction of the earthworm *Eisenia fetida* (Heckmann et al., 2011). Reproduction was affected by some of the NMs tested: nano-Ag completely inhibited the earthworm reproduction, and nano-TiO$_2$ also affected reproduction.

Hu et al. (2010) evaluated the toxicity of TiO$_2$ and ZnO NMs to the earthworm *Eisenia fetida* by measuring the activity of superoxide dismutase (SOD), catalase (CAT), and content of malondialdehyde (MDA) and DNA damages (C. W. Hu, et al., 2010). TiO$_2$ NMs promoted an increase of the activity of CAT and the content of MDA at concentrations of 1 g kg^{-1} and 5 g kg^{-1}, respectively. The same was not observed for ZnO NMs. The authors observed an increase in the activity of CAT and in the content of MDA at lowest concentrations, and then a decreasing tendency was observed for the highest concentrations. DNA damages were observed in organisms exposed to artificial OECD soil spiked with ZnO and TiO$_2$ NMs (1 and 5 g kg^{-1}) (C. W. Hu, et al., 2010). Cañas et al. (2011) also observed effects

Table 3. Characterization and toxicity effect of NMs on terrestrial invertebrate

NMs	Size Supported by the Manufacturer (nm)	Particle Size by DLS (nm)	Concentration	Species	Endpoint	Effect	Reference
Ag	3-8	-	1000 mg kg^{-1}	P. pruinosus	Avoidance and feeding activity	EC$_{50}$ values of 16.0 mg kg^{-1})	(Tourinho, van Gestel, Jurkschat, Soares, & Loureiro, 2015)
Ag	30-50	235±3.73	1000 mg kg^{-1}	E. fetida	Effects on reproduction	Complete inhibited	(Heckmann, et al., 2011)
Ag	<100	14-20	0.1 and 0.5 mg L^{-1}	C. elegans	Survival, growth, reproduction, and gene expression	Considerably toxic	(Roh et al., 2009)
Cu	80	419 ± 1.46	1000 mg kg^{-1}	E. fetida	Effects on reproduction	Affected	(Heckmann, et al., 2011)
TiO$_2$	10-20	-	1 g kg^{-1}	E. fetida	Activity of CAT	Increase	(C. W. Hu et al., 2010)
TiO$_2$	10-20	-	5 g kg^{-1}	E. fetida	Content of MDA	Increase	(C. W. Hu, et al., 2010)
TiO$_2$	10-20	-	5 g kg^{-1}	E. fetida	DNA damage	Observed	(C. W. Hu, et al., 2010)
TiO$_2$	15	350-500 (sonicated dispersion) 780-970 (non-sonicated dispersion)	3000 µg g^{-1}	P. scaber	Mortality, weight, and feeding behaviour	No effect	(Jemec, Drobne, Remskar, Sepcic, & Tisler, 2008)
TiO$_2$	15	350-500 (sonicated dispersion) 780-970 (non-sonicated dispersion)	0.1, 1000 and 3000 µg g^{-1}	P. scaber	Antioxidant enzymes activities (CAT and GST)	Sub-lethal effects	(Jemec, et al., 2008)
TiO$_2$	32	-	1000 mg kg^{-1}	E. fetida	Effects on reproduction	Cocoon production decreased	(Cañas et al., 2011)
TiO$_2$	50	338-917	24-239.6 mg L^{-1}	C. elegans	Growth and reproduction	Inhibition LC$_{50}$ = 80 mg L^{-1}	(Huanhua Wang, Wick, & Xing, 2009)
ZnO	10-20	-	>0.5 mg kg^{-1}	E. fetida	Activity of SOD, CAT, and content of MDA	Decreasing tendency	(C. W. Hu, et al., 2010)
ZnO	10-20	-	5 g kg^{-1}	E. fetida	DNA damage	Observed	(C. W. Hu, et al., 2010)
ZnO	20	478-980	0.4 and 8.1 mg L^{-1}	C. elegans	Growth and reproduction	Highly toxic LC$_{50}$ = 2.2 mg L^{-1}	(Huanhua Wang, et al., 2009)
ZnO	40-100	-	1000 mg kg^{-1}	E. fetida	Effects on reproduction	Inhibited cocoon production	(Cañas, et al., 2011)
ZnO	<100	-	750 mg kg^{-1}	E. veneta	Effects on reproduction	Reduced	(Hooper et al., 2011)

continued on following page

Table 3. Continued

NMs	Size Supported by the Manufacturer (nm)	Particle Size by DLS (nm)	Concentration	Species	Endpoint	Effect	Reference
ZnO	<200	-	6400 mg kg^{-1}	*F. candida*	Survival	No effect	(Kool, Ortiz, & van Gestel, 2011)
ZnO	<200	-	1800 mg kg^{-1}	*F. candida*	Reproduction	Reduction	(Kool, et al., 2011)
nZVI	-	178 - 424	>750 mg kg^{-1}	*L. rubellus*	Avoidance	OECD soil: $EC_{50} = 582$ mg kg^{-1} $LC_{50} = 866$ mg kg^{-1} Natural soil: $EC_{50} = 532$ mg kg^{-1} $LC_{50} = 447$ mg kg^{-1}	(Yehia S. El-Temsah & Erik J. Joner, 2012)
nZVI	-	178 - 424	>750 mg kg^{-1}	*E. fetida*	Avoidance	OECD soil: $EC_{50} = 511$ mg kg^{-1} Natural soil: $EC_{50} = 563$ mg kg^{-1} $LC_{50} = 399$ mg kg^{-1}	(Yehia S. El-Temsah & Erik J. Joner, 2012)
nZVI	-	178 - 424	1000 mg kg^{-1}	*L. rubellus* and *E. fetida*	Reproduction	Completely inhibited	(Yehia S. El-Temsah & Erik J. Joner, 2012)
DWCNT	-	-	495 mg kg^{-1}	*E. veneta*	Reproduction	The production of cocoons was reduced by 60%. $EC_{10} = 37\pm73$ mg DWNT kg^{-1} $EC_{50} = 176\pm150$ mg DWNT kg^{-1}	(Scott-Fordsmand, Krogh, Schaefer, & Johansen, 2008)

on reproduction of *E. fetida* exposed to an artificial OECD soil and sand-manure spiked with ZnO and TiO$_2$ NMs (range of concentrations from 0.1 and 1000 mg kg^{-1}) (Cañas, et al., 2011). ZnO NM at the highest concentration tested (1000 mg kg^{-1}) significantly inhibited cocoon production in OECD soil. In sand-manure, cocoon production also decreased with the increase of ZnO NM concentrations. The authors observe a significant inhibition of cocoon production at 100 and 1000 mg kg^{-1}. ZnO NMs in clay loam amended with 3% of organic matter reduced in 50% the reproduction of the earthworm *E. veneta* at concentrations of 750 mg Zn kg^{-1} (Hooper, et al., 2011).

Nano zero-valent iron (nZVI) has been considered a great, inexpensive, and environmental friendly reducing agent, that has been used for remediation of numerous contaminants. However, few works have reported the effect of nZVI in soils (El-Temsah & Joner, 2013). Same authors tested the effects of nZVI in the avoidance and reproduction of *Eisenia fetida* and *Lumbricus rubellus* in natural and OECD artificial soils (Yehia S. El-Temsah & Erik J. Joner, 2012). Both species, in different soils, have a tendency

Table 4. Effect of metal and metal oxide nanoparticles on cells in vitro

NMs	Size Supported by the Manufacturer	Physicochemical Characterization by Dynamic Light Scattering (DLS)*	Cellular Model	Results	Reference
Ag	15 nm	n/a	C18-4	Cell morphology changes at 10 µg mL^{-1}. Decrease of cell viability (EC = 8.75 µg mL^{-1}). Increase of membrane leakage (EC = 2.5 µg mL^{-1})	(Braydich-Stolle, Hussain, Schlager, & Hofmann, 2005)
Ag	7-10 nm	n/a	HepG2	High toxicity and changes in cell morphology at higher doses (1.0 µg mL^{-1}).	(Kawata K., Osawa, & Okabe, 2009)
Ag stabilized with PVP	30-50 nm	121 nm; -21 mV	A549	Induction of oxidative stress leading to cyto/genotoxicity at concentrations ranging 2.5 µg mL^{-1} from 10 µg mL^{-1}	(Foldbjerg, Dang, & Autrup, 2011)
Ag stabilized by citrate	10, 20, 30, 40 and 60 nm	6, 11, 16, 58 and 68 nm, respectively; -25, -25, -24, -15 and -16 mV, respectively	Balb/3T3	Smallest NMs (10 nm) lead to highest bioavailability of Ag ions (EC=0.27 µg mL^{-1})	(Ivask et al., 2014)
TiO$_2$	≤25 nm	n/a	U-87 and HFF-I	Induction of cell death on U-87 and HFF-I cell lines with EC of 30-40 µg mL^{-1} and 40 µg mL^{-1}, respectively	(Lai et al., 2008)
TiO$_2$	n/a	93.9 nm	L929	Reduction of cell viability (≥ 60 µg mL^{-1}) after 48 h and induction of oxidative stress at higher concentration (600 µg mL^{-1})	(Jin, Zhu, Wang, & Lu, 2008)
TiO$_2$	n/a	124.9 nm; -17.6 mV	A431	Decrease of cell viability, production of ROS and oxidative stress causing DNA injury and micronucleus formation at concentrations of 8 and 80 µg mL^{-1}	(Shukla et al., 2011)
TiO$_2$	101	- 31.74 mV	HAECs and HUVECs	High expression of adhesion molecules (VCAM-1 and E-selectin) in HAECs with endothelium dysfunction at 20 µg mL^{-1}	(Alinovi et al., 2015)
ZnO	n/a	165 nm;- 26 mV	A431	Genotoxicity effects induced by oxidative stress with depletion of CAT, glutathione and SOD at concentration above 0.008 µg mL^{-1}	(Sharma et al., 2009)
ZnO	n/a	n/a	HL-60 and PBMC	Higher toxicity on HL60 (EC= 52.80 µg mL^{-1}) than PBMC cells (741.82 µg mL^{-1})	(Premanathan, Karthikeyan, Jeyasubramanian, & Manivannan, 2011)
ZnO	n/a	456.5 nm; + 14.4 mV	RAW 264.7 and BEAS-2B	Cytotoxicity was more pronounced in RAW 264.7 than in BEAS-2B at concentrations of 10, 15 and 20 µg mL^{-1}. DNA damage at doses above 10 µg mL^{-1}	(Ng et al., 2011)
ZnO rod shape and ZnO with aminopropil sylane coating	n/a	245 and 145 nm, respectively; + 23.5 and + 35 mV, respectively	RAW 264	NMs dissolution is the main cause of ZnO toxicity. The toxicity can be attributed to the metabolic perturbation leading to cell death.	(Triboulet et al., 2014)

continued on following page

Table 4. Continued

NMs	Size Supported by the Manufacturer	Physicochemical Characterization by Dynamic Light Scattering (DLS)*	Cellular Model	Results	Reference
SPIONs and SPIONs coated with PVA	n/a	158, 195, 41.2, 44.9, 199 and 262 nm.	L929	Increase of cell viability by increasing the concentration of iron salts. Coated NMs showed less toxicity events	(Mahmoudi, Simchi, Milani, & Stroeve, 2009)
Fe_3O_4	n/a	174 nm	A549	Decrease of cell viability, changes in cell morphology, gluathione depletion, induction of ROS and lipid peroxidation at cell exposed to 10, 25 and 50 µg mL^1	(Dwivedi et al., 2014)

*measured in distilled water. C18-4: Mouse spermatogonial stem cells; HepG2: human hepatocarcinoma cells; PVP: Poly vinylalcohol; A549: Human lung adenocarcinoma epithelial cells; Balb/3T3: Mouse embryonic fibroblast cells; U-87: Human glioblastoma cells; HFF-I: Human Foreskin Fibroblasts; L929: Mouse fibroblasts; A431: human epithelial carcinoma cells;; HAECs: Human aortic endothelial cells; HUVECs: Human umbilical vein endothelial cells; VCAM-1: Vascular cell adhesion molecule 1; SOD: Superoxide dismutase; HL-60: Human myeloblastic leukemia cells; PBMC: Peripheral blood mononuclear cells; RAW 264.7: Mouse macrophage cells; BEAS-2B: Human bronchial epithelial cells; SPIOs: Superparamagnetic iron oxide nanoparticles; PVA: Polyvinyl alcohol;

n/a: not available;

Figure 1. Schematic representation of NMs effects at cellular level caused by oxidative stress generation

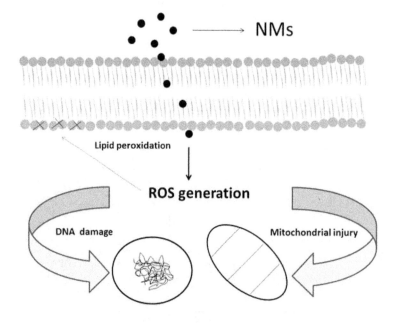

to prefer soils with the lowest concentrations of nZVI (≤ 500 mg kg^{-1}). The EC$_{50}$ values determined for avoidance were 582 mg kg^{-1} and 532 mg kg^{-1} for *L. rubellus* in OECD and natural soil, respectively. For *E. fetida* they determined an EC$_{50}$ value of 511 mg kg^{-1} in OECD soil, and of 563 mg kg^{-1} for natural soil. The same authors reported LC$_{50}$ values for *L. rubellus* (866 mg kg^{-1} in OECD soil and 447 mg kg^{-1} in natural soil). For *E. fetida* they only determined a LC$_{50}$ of 399 mg kg^{-1} in natural soil, after 14 days of exposure. The reproduction of both species was completely inhibited in all the concentrations tested, up to 1000 mg kg^{-1}. This study showed that nZVI can have serious harmful effects for the behaviour and reproduction of earthworms at high concentrations, which could be attained if this NM is directly applied to soils, for remediation purposes.

Considering non-metallic NMs, Scott-Fordsmand et al. (2008) observed the toxic effects of double-walled carbon nanotubes (DWCNT) in the earthworm *Eisenia veneta*. The production of cocoons was reduced by 60% at 495 mg DWNT kg^{-1} (Scott-Fordsmand, et al., 2008). These authors reported an EC$_{10}$ of 37±73 mg DWNT kg^{-1} and EC$_{50}$ of 176±150 mg DWNT kg^{-1} for the reproduction of this species.

Regarding other soil invertebrates, nano-TiO$_2$ at concentrations of 3000 µg g^{-1} did not caused effects on mortality, weight or feeding behaviour of the isopod *Parcellio scaber* (Jemec, et al., 2008). However, sub-lethal effects on antioxidant enzymes activities (CAT and glutathione-S-transferase – (GST)) were recorded for exposure concentrations of 0.1, 1000 and 3000 µg TiO$_2$ g^{-1} (Jemec, et al., 2008).

Roh et al. (2009) assessed the toxicity of silver NMs for *Caenorhabditis elegans* by assessing survival, growth, reproduction and gene expression endpoints (Roh, et al., 2009). They concluded that Ag NMs (0.1 and 0.5 mg L^{-1}) were considerably toxic to *C. elegans*, mainly affecting reproduction potential. These authors also concluded that the sod-3 and daf-12 genes may have been associated with the reproductive failure induced by Ag NMs in these organisms and that oxidative stress had an important role in the mechanism of toxicity of Ag NMs. Tourinho et al. (2015) also observed that Ag NMs affected the avoidance behaviour (EC$_{50}$ values of 16.0 mg kg^{-1}) and the feeding activity of the isopod *P. pruinosus* (Tourinho, et al., 2015).

The survival of collembolans *F. candida* was not affected by ZnO NMs at concentrations up to 6400 mg Zn kg^{-1}$_{DW}$ (Kool, et al., 2011). The authors also concluded that nano-ZnO contributed to a reduction of springtails reproduction for concentrations below 1800 mg Zn kg^{-1} in natural LUFA soil. Ma et al. (2009) also reported the low toxicity (for behaviour, lethality, and reproduction endpoints) of nano-ZnO to *C. elegans* (H. Ma, Bertsch, Glenn, Kabengi, & Williams, 2009). In opposition, Wang et al. (2009), testing particles with a bigger size, observed that uncoated ZnO (a range of concentrations from 0.4 and 8.1 mg L^{-1}) was highly toxic to *C. elegans* larvae (LC$_{50}$ of 2.2 mg L^{-1}) (Huanhua Wang, et al., 2009). These authors also showed the toxicity of TiO$_2$ (24 – 239.6 mg L^{-1}) that inhibited the growth and especially the reproductive capability of *C. elegans* (LC$_{50}$ of 80 mg L^{-1}).

Although few studies available suggest that NMs will affect soil invertebrates, but such effects will occur only at concentrations that could be attained with cumulative applications of sewage sludge on soils. Thus in the future studies, will be crucial to obtain more data about the toxicity of other NMs, for a wide range of soil invertebrates; to perceive the indirect effects of NMs in soil communities and their population interactions, as well as gain new insights in trans-generational effects in invertebrate populations, submitted to cumulative exposures to low levels of NMs.

In Vitro Toxicity Studies of NMs Using Cell Culture Lines

The impact of NMs at the cellular level has been widely investigated. Almost all the toxicity tests with cell lines were conducted to perceive the mechanisms of toxicity of the contaminants. Table 4 lists the *in vitro* toxicity studies with different cell lines, evaluating NMs with potential for wastewater treatment. According to the Table 4, the effect of NMS on biological systems can be related not only to their physicochemical features (size, surface charge, morphology, solubility) (Donaldson, Stone, Tran, Kreyling, & Borm, 2004) but also with the type of cell line used *in vitro* toxicity assays, which is selected depending on the expected entry pathways of the NMs (lung, skin, intestine or liver) in the body (Oberdörster et al., 2002).

The cell viability studies can be performed by a battery of standard toxicological assays, including AlamarBlue (AB), MTT (3-(4,5-dimethyl-2-thiazolyl)-2,5-diphenyl-2H-tetrazolium bromide), neutral red (to evaluate the dye uptake) and lactate dehydrogenase (LDH) (to evaluate the damage of cell membrane due to the release of LDH) methods already reported in several studies and reviews (Ahamed, Akhtar, Alhadlaq, Khan, & Alrokayan, 2015; Andreani et al., 2014; Asare et al., 2012; Doktorovova, Souto, & Silva, 2014). Many other more extensive NMs cytotoxicity assays can be conducted to evaluate DNA damages (comet assay), DNA damage and apoptosis (flow cytometry), detection of oxidative stress (ROS production and alteration in antioxidant enzymes level), lipid peroxidation (thiobarbituric acidic method-TBA) and inflammation (enzyme-linked immunosorbant assay- ELISA) (Lewinski, Colvin, & Drezek, 2008).

Similarly to ecotoxicological test organisms and according to data reported the toxicity of most tested NMs to cell lines can be related with the induction of oxidative stress (Figure 1). ROS including the superoxide anion, hydrogen peroxide and hydroxyl radicals, can accumulate in the cells leading to severe cellular injuries, including membrane lipid peroxidation and protein and DNA damages. The cellular defense mechanisms against ROS can include non-enzymatic antioxidants (e.g. α-tocopherols, carotens, ubiquinone), as well as enzymatic scavengers, such as, SOD, CAT, glutathione peroxidase (GPx) and GST enzymes (Franco, Sánchez-Olea, Reyes-Reyes, & Panayiotidis, 2009).

Several studies claim that the use of ultrafine nano-TiO_2, one of the most used NMs for wastewater treatment applications, can lead to severe cytotoxicity when exposed to UV irradiation. However, they are considered safe materials in the absence of photoactivation (Peters, Unger, Kirkpatrick, Gatti, & Monari, 2004). In contrast, several other studies showed that nano-TiO_2 (10 and 20 nm) at 10 µg mL^{-1} can lead to lipid peroxidation, DNA damage and increase of nitric oxide and hydrogen peroxide levels in human bronchial epithelial cells (BEAS-2B) (Gurr, Wang, Chen, & Jan, 2005). This NMs has also induced significant apoptotic events in Syrian hamster embryo fibroblasts cells at 1.0 µg mL^{-1} (Q. Rahman et al., 2002), decreased the intracellular level of reduced glutathione (GSH) and activated caspase-3 and chromatin condensation at concentrations above 5 µg mL^{-1} leading to the apoptosis in BEAS-2B cells (E.-J. Park et al., 2008). According to García et al. (2014), nano-TiO_2 was also able to induce oxidative stress effects on glial cells associated with an increase of GPx activity. CAT and SOD activities, which were not able to prevent lipid peroxidation and mitochondrial damage at 20 µg mL^{-1} (Huerta-García et al., 2014).

Induction of cell toxicity by Ag NMs was also shown in different cell lines (AshaRani, Low Kah Mun, Hande, & Valiyaveettil, 2009; Hackenberg et al., 2011; Y.-H. Lee et al., 2014; E.-J. Park, Yi, Kim, Choi, & Park, 2010; M. V. D. Z. Park et al., 2011). Evaluation of the cytotoxicity of nano-Ag was conducted on dermal noncancerous (HaCat) and cervical cancer (HeLa) cell lines demonstrated the induction of high levels of oxidative stress, glutathione depletion and cell membrane damage with subsequent apoptosis. However, HeLa cells demonstrated to be more sensitive upon nano-Ag exposure (lethal dose (LD) of 6.9 µg mL^{-1}) in comparison to HaCat cells (29 µg mL^{-1}) after 79 h with AB assay. This difference between the two cell lines was attributed to different capacities of antioxidant defense mechanisms (Mukherjee, O'Claonadh, Casey, & Chambers, 2012). Some studies have also demonstrated that the oxidative damage caused by nanoparticulate systems can be driven by the release of the metal ions. For instance, Beer et al. (2012), observed that a synergic effect between silver ions and nano-Ag contributed for toxicity events at lower concentrations of Ag$^+$ ions ($\leq$ 2.6%) in A549 cells (Beer, Foldbjerg, Hayashi, Sutherland, & Autrup, 2012). However, Kim et al. (2009) studied the toxicity of nano-Ag and Ag$^+$ (AgNO$_3$ source) ions using HepG2 cell line. The authors using a cation exchange treatment showed that nano-Ag produces low concentration of Ag ions and thus, the cytotoxicity and genotoxicity effects observed were induced by nano-Ag *per se* and not by silver ions (S. Kim, et al., 2009).

The cytotoxicity of nano-ZnO has also been investigated in several studies using different cells as shown in Table 1. In some studies, the toxicity of nano-ZnO was ascribed to the release of Zn^{+2} ions from the NMs (George et al., 2010). According to Xia et al. (2008), the toxicity of nano-ZnO on RAW 264 and BEAS-2B cells was also likely dependent on the release of dissolved Zn ions (Xia et al., 2008). However, Sharma et al. (2012), demonstrated that nano-ZnO *per se* induced mitochondrial apoptosis mediated by ROS generation in HepG2 cells at 14 µg mL^{-1} (Sharma, Anderson, & Dhawan, 2012). The authors evaluated the supernatant from the incubation of nano-ZnO in cell culture medium and observed no cytotoxic events. Also, they evaluated the effect of a free Zn ions source (ZnCl$_2$) and again no cell toxicity was detectable, suggesting that the Zn ions release did not contribute to the toxicity effects on HepG2 cell line.

The oxidative stress induced by nano-ZnO even a low concentration (2.5 µg mL^{-1}), was also showed by Guo et al. (2013) in rat retinal ganglion cells (RGC-5 cells), since ROS overproduction in cells leaded to apoptotic events (Guo et al., 2013). Another interesting study, conducted by Kumar et al. (2015), showed that the surface modification of nano-ZnO using reduced glutathione (GSH) and curcumin decreased the oxidative stress in human embryonic kidney cells (HEK-293) (Kumar et al., 2015).

The application of iron NMs for wastewater treatment has been recently reviewed by Xu et al. (2012). The cellular mechanism underlying ROS production and oxidative stress induced by iron NMs in cell culture lines were exhaustively discussed (Enrico Burello & Wortha, 2011; Keenan, Goth-Goldstein, Lucas, & Sedlak, 2009). Dwivedi et al. (2014) using A 549 cells, showed that nano-Fe$_3$O$_4$ induced cell morphological changes and production of intracellular ROS (23, 46 and 82% of control for 10, 25 and 50 µg mL^{-1}, respectively) leading to lipid peroxidation (15, 32 and 91% of untreated cells for 10, 25 and 50 µg mL^{-1}, respectively) and depletion of glutathione at 10 µg mL^{-1} (Dwivedi, et al., 2014).

A systematic study conducted by Naqvi et al. (2010) indicated that highest concentrations of super-paramagnetic iron oxide nanoparticles (SPIONs) (above 100 µg mL^{-1}) in murine macrophage cells (J774) resulted in elevated ROS production and cell death (Naqvi et al., 2010). Also, the cytotoxicity effects following iron oxide NMs exposures were associated with ROS generation and oxidative stress which resulted in the formation of actin stress fiber in Porcine aortic endothelial cells (PAEC) at concentration of 0.5 mg mL^{-1} (Buyukhatipoglu & Clyne, 2011).

Similarly to what was previously described for nano-ZnO above, the surface modification of Fe$_3$O$_4$ NMs showed that the coating of NMs with silica shell reduced the generation of ROS on A549 and HeLa cells due to the reduction of the free iron ions release (Malvindi et al., 2014). In summary, available data indicates that NMs with potential for wastewater treatment may also have risks on human health, as they can be cytotoxic to different human cell lines, representing different exposure pathways. It was also demonstrated that surface modifications of some of these NMs, or the combination with others may mitigate their effects on cells. Thus, in the future, more research is needed to perceive if such modifications also benefit the capability of NMs to remove organic and inorganic contaminants from water.

CONCLUSION

In recent years, the design and application of NMs on the industry are widely increasing due to their great potential for the society and the economy. Water treatment is an attractive area for the application of advanced nanotechnology based solutions in order to develop new and functional materials to capture different contaminants. Although there are great advances with the use of nanotechnology, environmental and health effects associated to MNs are attracting considerable concern in the scientific community and in the society

Studies to predict the mechanism toxicity of NMs on environmental species and in cell culture lines *in vitro* are important to guarantee their safety for ecosystems and human health. For this purpose, the collaboration between industry and academy is crucial to provide robust basis and information about the mechanisms of action of NMs, in order to support new and safe developments in NMs that are designed for being intentionally applied in the environment.

ACKNOWLEDGMENT

The authors wish to acknowledge FCT (Fundação para Ciência e Tecnologia) under the reference SFRH/BD/94902/2013 (Doctoral grant for Gavina, A.C.)

REFERENCES

Adesina, A. A. (2004). Industrial exploitation of photocatalysis progress, perspectives and prospects. *Catalysis Surveys from Asia, 8*(4), 265–273. doi:10.1007/s10563-004-9117-0

Ahamed, M., Akhtar, M. J., Alhadlaq, H. A., Khan, M. A. M., & Alrokayan, S. A. (2015). Comparative cytotoxic response of nickel ferrite nanoparticles in human liver HepG2 and breast MFC-7 cancer cells. *Chemosphere, 135,* 278–288. doi:10.1016/j.chemosphere.2015.03.079 PMID:25966046

Alinovi, R., Goldoni, M., Pinelli, S., Campanini, M., Aliatis, I., Bersani, D., & Mutti, A. et al. (2015). Oxidative and pro-inflammatory effects of cobalt and titanium oxide nanoparticles on aortic and venous endothelial cells. *Toxicology In Vitro, 29*(3), 426–437. doi:10.1016/j.tiv.2014.12.007 PMID:25526690

Andreani, T., Kiill, C. P., Souza, A. L. R., Fangueiro, J. F., Fernandes, L., Doktorovová, S., & Silva, A. M. et al. (2014). Surface engineering of silica nanoparticles for oral insulin delivery: Characterization and cell toxicity studies. *Colloid Surf B, 123,* 916–923. doi:10.1016/j.colsurfb.2014.10.047 PMID:25466464

Antolin, M. C., Muro, I., & Sanchez-Diaz, M. (2010). Application of sewage sludge improves growth, photosynthesis and antioxidant activities of nodulated alfalfa plants under drought conditions. 68. *Environmental and Experimental Botany*, *68*(1), 75–82. doi:10.1016/j.envexpbot.2009.11.001

Aruoja, V., Dubourguier, H.-C., Kasemets, K., & Kahru, A. (2009). Toxicity of nanoparticles of CuO, ZnO and TiO_2 to microalgae Pseudokirchneriella subcapitata. *The Science of the Total Environment*, *407*(4), 1461–1468. doi:10.1016/j.scitotenv.2008.10.053 PMID:19038417

Asare, N., Instanes, C., Sandberg, W. J., Refsnes, M., Schwarze, P., Kruszewski, M., & Brunborg, G. (2012). Cytotoxic and genotoxic effects of silver nanoparticles in testicular cells. *Toxicology*, *291*(1–3), 65–72. doi:10.1016/j.tox.2011.10.022 PMID:22085606

Asghari, S., Johari, S. A., Lee, J. H., Kim, Y. S., Jeon, Y. B., & Choi, H. J. et al.. (2012). Toxicity of various silver nanoparticles compared to silver ions in Daphnia magna. *Journal of Nanobiotechnology*, *10*(14). doi:10.1186/1477-3155-10-14 PMID:22472056

AshaRani, P. V., Low Kah Mun, G., Hande, M. P., & Valiyaveettil, S.AshaRani. (2009). Cytotoxicity and genotoxicity of silver nanoparticles in human cells. *ACS Nano*, *3*(2), 279–290. doi:10.1021/nn800596w PMID:19236062

Baker, T. J., Tyler, C. R., & Galloway, T. S. (2014). Impacts of metal and metal oxide nanoparticles on marine organisms. *Environmental Pollution*, *186*, 257–271. doi:10.1016/j.envpol.2013.11.014 PMID:24359692

Barrena, R., Casals, E., Colón, J., Font, X., Sánchez, A., & Puntes, V. (2009). Evaluation of the ecotoxicity of model nanoparticles. *Chemosphere*, *75*(7), 850–857. doi:10.1016/j.chemosphere.2009.01.078 PMID:19264345

Baun, A., Hartmann, N. B., Grieger, K., & Kusk, K. O. (2008). Ecotoxicity of engineered nanoparticles to aquatic invertebrates: A brief review and recommendations for future toxicity testing. *Ecotoxicology (London, England)*, *17*(5), 387–395. doi:10.1007/s10646-008-0208-y PMID:18425578

Beer, C., Foldbjerg, R., Hayashi, Y., Sutherland, D. S., & Autrup, H. (2012). Toxicity of silver nanoparticles—Nanoparticle or silver ion? *Toxicology Letters*, *208*(3), 286–292. doi:10.1016/j.toxlet.2011.11.002 PMID:22101214

Begum, P., Ikhtiari, R., & Fugetsu, B. (2014). Potential impact of multi-walled carbon nanotubes exposure to the seedling stage of selected plant species. *Nanomaterials*, *4*(2), 203–221. doi:10.3390/nano4020203

Behra, R., Sigg, L., Clift, M. J. D., Herzog, F., Minghetti, M., Johnston, B., & Rothen-Rutishauser, B. et al. (2013). Bioavailability of silver nanoparticles and ions: From a chemical and biochemical perspective. *Journal of the Royal Society, Interface*, *10*(87), 20130396. doi:10.1098/rsif.2013.0396 PMID:23883950

Blouina, M., Hodsonb, M. E., Delgadoc, E. A., Bakerd, G., Brussaarde, L., & Buttf, K. R. et al.. (2013). A review of earthworm impact on soil function and ecosystem services. *European Journal of Soil Science*, *64*(2), 161–182. doi:10.1111/ejss.12025

Bour, A., Mouchet, F., Verneuil, L., Evariste, L., Silvestre, J., Pinelli, E., & Gauthier, L. (2015). Toxicity of CeO$_2$ nanoparticles at different trophic levels – Effects on diatoms, chironomids and amphibians. *Chemosphere, 120*, 230–236. doi:10.1016/j.chemosphere.2014.07.012 PMID:25086917

Boxall, A. B. A., Tiede, K., & Chaudhry, Q. (2007). Engineered nanomaterials in soils and water: How do they behave and could they pose a risk to human health? *Nanomedicine (London), 2*(6), 919–927. doi:10.2217/17435889.2.6.919 PMID:18095854

Brar, S. K., Verma, M., Tyagi, R. D., & Surampalli, R. Y. (2010). Engineered nanoparticles in wastewater and wastewater sludge evidence and impacts. *Waste Management (New York, N.Y.), 30*(3), 504–520. doi:10.1016/j.wasman.2009.10.012 PMID:19926463

Braydich-Stolle, L., Hussain, S., Schlager, J. J., & Hofmann, M.-C. (2005). In vitro cytotoxicity of nanoparticles in mammalian germline stem cells. *Toxicological Sciences, 88*(2), 412–419. doi:10.1093/toxsci/kfi256 PMID:16014736

Buyukhatipoglu, K., & Clyne, A. M. (2011). Superparamagnetic iron oxide nanoparticles change endothelial cell morphology and mechanics via reactive oxygen species formation. *Journal of Biomedical Materials Research. Part A, 96A*(1), 186–195. doi:10.1002/jbm.a.32972 PMID:21105167

Cañas, J. E., Qi, B., Li, S., Maul, J. D., Cox, S. B., Das, S., et al. (2011). Acute and reproductive toxicity of nano-sized metal oxides (ZnO and TiO$_2$) to earthworms (Eisenia fetida). *J Environ Monit, 13*(12), 3351-3357. doi: 10.1039/c1em10497g

Chae, Y. J., Pham, C. H., Lee, J., Bae, E., Yi, J., & Gu, M. B. (2009). Evaluation of the toxic impact of silver nanoparticles on Japanese medaka (Oryzias latipes). *Aquatic Toxicology (Amsterdam, Netherlands), 94*(4), 320–327. doi:10.1016/j.aquatox.2009.07.019 PMID:19699002

Chen, C., Unrine, J. M., Judy, J. D., Lewis, W. R., Guo, J., Mcnear, D. H. J. Jr, & Tsyusko, O. V. (2015). Toxicogenomic responses of the model legume Medicago truncatula to aged biosolids containing a mixture of nanomaterials (TiO2, Ag and ZnO) from a pilot wastewater treatment plant. *Environmental Science & Technology, 49*(14), 8759–8768. doi:10.1021/acs.est.5b01211 PMID:26065335

Chiou, C.-H., Wu, C.-Y., & Juang, R.-S. (2008). Influence of operating parameters on photocatalytic degradation of phenol in UV/TiO2 process. *Chemical Engineering Journal, 139*(2), 322–329. doi:10.1016/j.cej.2007.08.002

Clément, L., Hurel, C., & Marmier, N. (2013). Toxicity of TiO$_2$ nanoparticles to cladocerans, algae, rotifers and plants - Effects of size and crystalline structure. *Chemosphere, 90*(3), 1083–1090. doi:10.1016/j.chemosphere.2012.09.013 PMID:23062945

Colvin, V. L. (2003). The potential environmental impact of engineered nanomaterials. *Nat Biotech, 21*(10), 1166-1170.

Crane, R. A., & Scott, T. B. (2012). Nanoscale zero-valent iron: Future prospects for an emerging water treatment technology. *Journal of Hazardous Materials, 211–212*, 112–125. doi:10.1016/j.jhazmat.2011.11.073 PMID:22305041

Doktorovova, S., Souto, E. B., & Silva, A. M. (2014). Nanotoxicology applied to solid lipid nanoparticles and nanostructured lipid carriers – A systematic review of in vitro data. *European Journal of Pharmaceutics and Biopharmaceutics, 87*(1), 1–18. doi:10.1016/j.ejpb.2014.02.005 PMID:24530885

Donaldson, K., Stone, V., Tran, C. L., Kreyling, W., & Borm, P. J. A. (2004). Nanotoxicology. *Occupational and Environmental Medicine, 61*(9), 727–728. doi:10.1136/oem.2004.013243 PMID:15317911

Du, W., Sun, Y., Ji, R., Zhu, J., Wu, J., & Guo, H. (2011). TiO_2 and ZnO nanoparticles negatively affect wheat growth and soil enzyme activities in agricultural soil. *Journal of Environmental Monitoring, 13*(4), 822–828. doi:10.1039/c0em00611d PMID:21267473

Dwivedi, S., Siddiqui, M. A., Farshori, N. N., Ahamed, M., Musarrat, J., & Al-Khedhairy, A. A. (2014). Synthesis, characterization and toxicological evaluation of iron oxide nanoparticles in human lung alveolar epithelial cells. *Colloid Surf B, 122*, 209–215. doi:10.1016/j.colsurfb.2014.06.064 PMID:25048357

EC. (2011). Recomendation on the definition of nanomaterial. *Off J Eur Communities, L274*, 1-40.

El-Kemary, M., El-Shamy, H., & El-Mehasseb, I. (2010). Photocatalytic degradation of ciprofloxacin drug in water using ZnO nanoparticles. *Journal of Luminescence, 130*(12), 2327–2331. doi:10.1016/j.jlumin.2010.07.013

El-Temsah, Y. S., & Joner, E. J. (2012). Ecotoxicological effects on earthworms of fresh and aged nano-sized zero-valent iron (nZVI) in soil. *Chemosphere, 89*(1), 76–82. doi:10.1016/j.chemosphere.2012.04.020 PMID:22595530

El-Temsah, Y. S., & Joner, E. J. (2012). Impact of Fe and Ag nanoparticles on seed germination and differences in bioavailability during exposure in aqueous suspension and soil. *Environmental Toxicology, 27*(1), 42–49. doi:10.1002/tox.20610 PMID:20549639

El-Temsah, Y. S., & Joner, E. J. (2013). Effects of nano-sized zero-valent iron (nZVI) on DDT degradation in soil and its toxicity to collembola and ostracods. *Chemosphere, 92*(1), 131–137. doi:10.1016/j.chemosphere.2013.02.039 PMID:23522781

Elgrabli, D., Floriani, M., Abella-Gallart, S., Meunier, L., Gamez, C., Delalain, P., & Lacroix, G. et al. (2008). Biodistribution and clearance of instilled carbon nanotubes in rat lung. *Particle and Fibre Toxicology, 5*(1), 20. doi:10.1186/1743-8977-5-20 PMID:19068117

Enrico Burello, E., & Wortha, A. P. (2011). A theoretical framework for predicting the oxidative stress potential of oxide nanoparticles. *Nanotoxicology, 5*(2), 228–235. doi:10.3109/17435390.2010.502980 PMID:21609138

Faisal, M., Abu Tariq, M., & Muneer, M. (2007). Photocatalysed degradation of two selected dyes in UV-irradiated aqueous suspensions of titania. *Dyes and Pigments, 72*(2), 233–239. doi:10.1016/j.dyepig.2005.08.020

Faisal, M., Saquib, Q., Alatar, A. A., Al-Khedhairy, A. A., Hegazy, A. K., & Musarrat, J. (2013). Phytotoxic hazards of NiO-nanoparticles in tomato: A study on mechanism of cell death. *Journal of Hazardous Materials, 250–251*, 318–332. doi:10.1016/j.jhazmat.2013.01.063 PMID:23474406

Fan, R., Huang, Y. C., Grusak, M. A., Huang, C. P., & Sherrier, D. J. (2014). Effects of nano-TiO2 on the agronomically-relevant Rhizobium–legume symbiosis. *The Science of the Total Environment, 466–467*, 503–512. doi:10.1016/j.scitotenv.2013.07.032 PMID:23933452

Foldbjerg, R., Dang, D., & Autrup, H. (2011). Cytotoxicity and genotoxicity of silver nanoparticles in the human lung cancer cell line, A549. *Archives of Toxicology, 85*(7), 743–750. doi:10.1007/s00204-010-0545-5 PMID:20428844

Franco, R., Sánchez-Olea, R., Reyes-Reyes, E. M., & Panayiotidis, M. I. (2009). Environmental toxicity, oxidative stress and apoptosis: Ménage à Trois. *Mutation Research/Genetic Toxicology and Environmental Mutagenesis, 674*(1–2), 3–22. doi:10.1016/j.mrgentox.2008.11.012 PMID:19114126

Fu, F., Dionysiou, D. D., & Liu, H. (2014). The use of zero-valent iron for groundwater remediation and wastewater treatment: A review. *Journal of Hazardous Materials, 267*, 194–205. doi:10.1016/j.jhazmat.2013.12.062 PMID:24457611

Furno, F., Morley, K. S., Wong, B., Sharp, B. L., Arnold, P. L., & Howdle, S. M. et al.. (2004). Silver nanoparticles and polymeric medical devices: A new approach to prevention of infection? *The Journal of Antimicrobial Chemotherapy, 54*(6), 1019–1024. doi:10.1093/jac/dkh478 PMID:15537697

García, A., Delgado, L., Torà, J. A., Casals, E., González, E., Puntes, V., & Sánchez, A. et al. (2012). Effect of cerium dioxide, titanium dioxide, silver, and gold nanoparticles on the activity of microbial communities intended in wastewater treatment. *Journal of Hazardous Materials, 199–200*, 64–72. doi:10.1016/j.jhazmat.2011.10.057 PMID:22088500

García, A., Espinosa, R., Delgado, L., Casals, E., González, E., Puntes, V., & Sánchez, A. et al. (2011). Acute toxicity of cerium oxide, titanium oxide and iron oxide nanoparticles using standardized tests. *Desalination, 269*(1–3), 136–141. doi:10.1016/j.desal.2010.10.052

Gardea-Torresdey, J. L., Rico, C. M., & White, J. C. (2014). Trophic transfer, transformation, and impact of engineered nanomaterials in terrestrial environments. *Environmental Science & Technology, 48*(5), 2526–2540. doi:10.1021/es4050665 PMID:24499408

George, S., Gardner, H., Seng, E. K., Chang, H., Wang, C., Yu Fang, C. H., & Chan, W. K. et al. (2014). Differential effect of solar light in increasing the toxicity of silver and titanium dioxide nanoparticles to a ish cell line and Zebrafish embryos. *Environmental Science & Technology, 48*(11), 6374–6382. doi:10.1021/es405768n PMID:24811346

George, S., Pokhrel, S., Xia, T., Gilbert, B., Ji, Z., Schowalter, M., & Nel, A. E. et al. (2010). Use of a rapid cytotoxicity screening approach to engineer a safer zinc oxide nanoparticle through iron doping. *ACS Nano, 4*(1), 15–29. doi:10.1021/nn901503q PMID:20043640

Ghodake, G., Seo, Y. D., & Lee, D. S. (2011). Hazardous phytotoxic nature of cobalt and zinc oxide nanoparticles assessed using Allium cepa. *Journal of Hazardous Materials, 186*(1), 952–955. doi:10.1016/j.jhazmat.2010.11.018 PMID:21122986

Giwa, A., Nkeonye, P. O., Bello, K. A., & Kolawole, K. A. (2012). Photocatalytic decolourization and degradation of C. I. basic blue 41 using TiO_2 nanoparticles. *J Environ Protect, 3*(9), 1063–1069. doi:10.4236/jep.2012.39124

Gong, N., Shao, K., Feng, W., Lin, Z., Liang, C., & Sun, Y. (2011). Biotoxicity of nickel oxide nanoparticles and bio-remediation by microalgae Chlorella vulgaris. *Chemosphere, 83*(4), 510–516. doi:10.1016/j.chemosphere.2010.12.059 PMID:21216429

Griffitt, R. J., Luo, J., Gao, J., Bonzongo, J.-C., & Barber, D. S. (2008). Effects of particle composition and species on toxicity of metallic nanomaterials in aquatic organisms. *Environmental Toxicology and Chemistry, 27*(9), 1972–1978. doi:10.1897/08-002.1 PMID:18690762

Gubbins, E. J., Batty, L. C., & Lead, J. R. (2011). Phytotoxicity of silver nanoparticles to Lemna minor L. *Environmental Pollution, 159*(6), 1551–1559. doi:10.1016/j.envpol.2011.03.002 PMID:21450381

Guo, D., Bi, H., Liu, B., Wu, Q., Wang, D., & Cui, Y. (2013). Reactive oxygen species-induced cytotoxic effects of zinc oxide nanoparticles in rat retinal ganglion cells. *Toxicology In Vitro, 27*(2), 731–738. doi:10.1016/j.tiv.2012.12.001 PMID:23232460

Gurr, J.-R., Wang, A. S. S., Chen, C.-H., & Jan, K.-Y. (2005). Ultrafine titanium dioxide particles in the absence of photoactivation can induce oxidative damage to human bronchial epithelial cells. *Toxicology, 213*(1-2), 66–73. doi:10.1016/j.tox.2005.05.007 PMID:15970370

Hackenberg, S., Scherzed, A., Kessler, M., Hummel, S., Technau, A., & Froelich, K. et al.. (2011). Silver nanoparticles: Evaluation of DNA damage, toxicity and functional impairment in human mesenchymal stem cells. *Toxicology Letters, 201*(1), 27–33. doi:10.1016/j.toxlet.2010.12.001 PMID:21145381

Hariharan, C. (2006). Photocatalytic degradation of organic contaminants in water by ZnO nanoparticles: Revisited. *Appl Catalysis A, 304*, 55–61. doi:10.1016/j.apcata.2006.02.020

Hartmann, N. B., Von der Kammer, F., Hofmann, T., Baalousha, M., Ottofuelling, S., & Baun, A. (2010). Algal testing of titanium dioxide nanoparticles—Testing considerations, inhibitory effects and modification of cadmium bioavailability. *Toxicology, 269*(2-3), 190–197. doi:10.1016/j.tox.2009.08.008 PMID:19686796

Hayat, K., Gondal, M. A., Khaled, M. M., & Ahmed, S. (2011). Effect of operational key parameters on photocatalytic degradation of phenol using nano nickel oxide synthesized by sol–gel method. *J Mol Catal A, 336*(1-2), 64–71. doi:10.1016/j.molcata.2010.12.011

Heckmann, L.-H., Hovgaard, M., Sutherland, D., Autrup, H., Besenbacher, F., & Scott-Fordsmand, J. (2011). Limit-test toxicity screening of selected inorganic nanoparticles to the earthworm Eisenia fetida. *Ecotoxicology (London, England), 20*(1), 226–233. doi:10.1007/s10646-010-0574-0 PMID:21120603

Heinlaan, M., Ivask, A., Blinova, I., Dubourguier, H.-C., & Kahru, A. (2008). Toxicity of nanosized and bulk ZnO, CuO and TiO_2 to bacteria Vibrio fischeri and crustaceans Daphnia magna and Thamnocephalus platyurus. *Chemosphere, 71*(7), 1308–1316. doi:10.1016/j.chemosphere.2007.11.047 PMID:18194809

Hennebert, P., Avellan, A., Yan, J., & Aguerre-Chariol, O. (2013). Experimental evidence of colloids and nanoparticles presence from 25 waste leachates. *Waste Management (New York, N.Y.), 33*(9), 1870–1881. doi:10.1016/j.wasman.2013.04.014 PMID:23746986

Hooper, H. L., Jurkschat, K., Morgan, A. J., Bailey, J., Lawlor, A. J., Spurgeon, D. J., & Svendsen, C. (2011). Comparative chronic toxicity of nanoparticulate and ionic zinc to the earthworm Eisenia veneta in a soil matrix. *Environment International, 37*(6), 1111–1117. doi:10.1016/j.envint.2011.02.019 PMID:21440301

Hristovski, K., Baumgardner, A., & Westerhoff, P. (2007). Selecting metal oxide nanomaterials for arsenic removal in fixed bed columns: From nanopowders to aggregated nanoparticle media. *Journal of Hazardous Materials, 147*(1–2), 265–274. doi:10.1016/j.jhazmat.2007.01.017 PMID:17254707

Hu, C. W., Li, M., Cui, Y. B., Li, D. S., Chen, J., & Yang, L. Y. (2010). Toxicological effects of TiO2 and ZnO nanoparticles in soil on earthworm Eisenia fetida. *Soil Biology & Biochemistry, 42*(4), 586–591. doi:10.1016/j.soilbio.2009.12.007

Hu, J., Wang, D., Wang, J., & Wang, J. (2012). Bioaccumulation of Fe_2O_3 (magnetic) nanoparticles in Ceriodaphnia dubia. *Environmental Pollution, 162*, 216–222. doi:10.1016/j.envpol.2011.11.016 PMID:22243867

Hu, X., Cook, S., Wang, P., & Hwang, H.-. (2009). In vitro evaluation of cytotoxicity of engineered metal oxide nanoparticles. *The Science of the Total Environment, 407*(8), 3070–3072. doi:10.1016/j.scitotenv.2009.01.033 PMID:19215968

Hu, X., Daun, J. K., & Scarth, R. (1994). Proportions of C18:1n−7 and C18:1n−9 fatty acids in canola seedcoat surface and internal lipids. *Journal of the American Oil Chemists' Society, 71*(2), 221–222. doi:10.1007/BF02541560

Hua, M., Zhang, S., Pan, B., Zhang, W., Lv, L., & Zhang, Q. (2012). Heavy metal removal from water/wastewater by nanosized metal oxides: A review. *Journal of Hazardous Materials, 211–212*, 317–331. doi:10.1016/j.jhazmat.2011.10.016 PMID:22018872

Huerta-García, E., Pérez-Arizti, J. A., Márquez-Ramírez, S. G., Delgado-Buenrostro, N. L., Chirino, Y. I., Iglesias, G. G., & López-Marure, R. (2014). Titanium dioxide nanoparticles induce strong oxidative stress and mitochondrial damage in glial cells. *Free Radical Biology & Medicine, 73*, 84–94. doi:10.1016/j.freeradbiomed.2014.04.026 PMID:24824983

Hund-Rinke, K., & Simon, M. (2006). Ecotoxic effect of photocatalytic active nanoparticles (TiO_2) on algae and daphnids (8 pp). *Environ Sci Pollut Res, 13*(4), 225–232. doi:10.1065/espr2006.06.311 PMID:16910119

Iram, M., Guo, C., Guan, Y., Ishfaq, A., & Liu, H. (2010). Adsorption and magnetic removal of neutral red dye from aqueous solution using Fe_3O_4 hollow nanospheres. *Journal of Hazardous Materials, 181*(1–3), 1039–1050. doi:10.1016/j.jhazmat.2010.05.119 PMID:20566240

Ivask, A., Kurvet, I., Kasemets, K., Blinova, I., Aruoja, V., Suppi, S., & Kahru, A. et al. (2014). Size-dependent toxicity of silver nanoparticles to bacteria, yeast, algae, crustaceans and mammalian cells. *PLoS ONE, 9*(7), e102108. doi:10.1371/journal.pone.0102108 PMID:25048192

Jang, Y. J., Simer, C., & Ohm, T. (2006). Comparison of zinc oxide nanoparticles and its nano-crystalline particles on the photocatalytic degradation of methylene blue. *Materials Research Bulletin, 41*(1), 67–77. doi:10.1016/j.materresbull.2005.07.038

Jarvie, H. P., Al-Obaidi, H., King, S. M., Bowes, M. J., Lawrence, M. J., Drake, A. F., & Dobson, P. J. et al. (2009). Fate of Silica Nanoparticles in Simulated Primary Wastewater Treatment. *Environmental Science & Technology, 43*(22), 8622–8628. doi:10.1021/es901399q PMID:20028062

Jemec, A., Drobne, D., Remskar, M., Sepcic, K., & Tisler, T. (2008). Effects of ingested nano-sized titanium dioxide on terrestrial isopods (Porcello scaber). *Environmental Toxicology and Chemistry, 27*(9), 1904–1914. doi:10.1897/08-036.1 PMID:19086208

Jin, C.-Y., Zhu, B.-S., Wang, X.-F., & Lu, Q.-H. (2008). Cytotoxicity of titanium dioxide nanoparticles in mouse fibroblast cells. *Chemical Research in Toxicology, 21*(9), 1871–1877. doi:10.1021/tx800179f PMID:18680314

Kaegi, R., Voegelin, A., Sinnet, B., Zuleeg, S., Hagendorfer, H., Burkhardt, M., & Siegrist, H. (2011). Behavior of metallic silver nanoparticles in a pilot wastewater treatment plant. *Environmental Science & Technology, 45*(9), 3902–3908. doi:10.1021/es1041892 PMID:21466186

Kahru, A., Dubourguier, H. C., Blinova, I., Ivask, A., & Kasemets, K. (2008). Biotests and biosensors for ecotoxicology of metal oxide nanoparticles: A Minireview. *Sensors (Basel, Switzerland), 8*(8), 5153–5170. doi:10.3390/s8085153

Kanu, I., & Achi, O. K. (2011). Industrial effluents and their impact on water quality of receiving rivers in Nigeria. *J Appl Technol Environ Sanitat, 1*(1), 75–86.

Kawata, K., Osawa, M., & Okabe, S. (2009). In vitro toxicity of silver nanoparticles at noncytotoxic doses to HepG2 human hepatoma cells. *Environmental Science & Technology, 43*(15), 6046–6051. doi:10.1021/es900754q PMID:19731716

Keenan, C. R., Goth-Goldstein, R., Lucas, D., & Sedlak, D. L. (2009). Oxidative stress induced by zero-valent iron nanoparticles and Fe(II) in human bronchial epithelial cells. *Environmental Science & Technology, 43*(12), 4555–4560. doi:10.1021/es9006383 PMID:19603676

Keller, A. A., Garner, K., Miller, R. J., & Lenihan, H. S. (2012). Toxicity of nano-zero valent iron to freshwater and marine organisms. *PLoS ONE, 7*(8), e43983. doi:10.1371/journal.pone.0043983 PMID:22952836

Khattab, I. A., Ghaly, M. Y., Österlund, L., Ali, M. E. M., Farah, J. Y., Zaher, F. M., & Badawy, M. I. (2012). Photocatalytic degradation of azo dye Reactive Red 15 over synthesized titanium and zinc oxides photocatalysts: A comparative study. *Desalination and Water Treatment, 48*(1-3), 120–129. doi:10.1080/19443994.2012.698803

Kim, E., Kim, S.-H., Kim, H.-C., Lee, S., Lee, S., & Jeong, S. (2011). Growth inhibition of aquatic plant caused by silver and titanium oxide nanoparticles. *Toxicol Environ Health Sci, 3*(1), 1–6. doi:10.1007/s13530-011-0071-8

Kim, K. T., Klaine, S. J., Cho, J., Kim, S.-H., & Kim, S. D. (2010). Oxidative stress responses of Daphnia magna exposed to TiO_2 nanoparticles according to size fraction. *The Science of the Total Environment, 408*(10), 2268–2272. doi:10.1016/j.scitotenv.2010.01.041 PMID:20153877

Kim, S., Choi, J. E., Choi, J., Chung, K.-H., Park, K., Yi, J., & Ryu, D.-Y. (2009). Oxidative stress-dependent toxicity of silver nanoparticles in human hepatoma cells. *Toxicology In Vitro*, *23*(6), 1076–1084. doi:10.1016/j.tiv.2009.06.001 PMID:19508889

Kiser, M. A., Ryu, H., Jang, H., Hristovski, K., & Westerhoff, P. (2010). Biosorption of nanoparticles to heterotrophic wastewater biomass. *Water Research*, *44*(14), 4105–4114. doi:10.1016/j.watres.2010.05.036 PMID:20547403

Klaine, S. J., Alvarez, P. J. J., Batley, G. E., Fernandes, T. F., Handy, R. D., Lyon, D. Y., & Lead, J. R. et al. (2008). Nanomaterials in the environment: Behavior, fate, bioavailability, and effects. *Environmental Toxicology and Chemistry*, *27*(9), 1825–1851. doi:10.1897/08-090.1 PMID:19086204

Kool, P. L., Ortiz, M. D., & van Gestel, C. A. M. (2011). Chronic toxicity of ZnO nanoparticles, non-nano ZnO and $ZnCl_2$ to Folsomia candida (Collembola) in relation to bioavailability in soil. *Environmental Pollution*, *159*(10), 2713–2719. doi:10.1016/j.envpol.2011.05.021 PMID:21724309

Kumar, A., Zafaryab, M., Umar, A., Rizvi, M. M. A., Ansari, H. Z. A. F., & Ansari, S. G. (2015). Relief of oxidative stress using curcumin and glutathione functionalized ZnO nanoparticles in HEK-293 cell line. *J Biomed Nanotechnol*, *11*(11), 1913–1926. doi:10.1166/jbn.2015.2166 PMID:26554152

Kumari, M., Mukherjee, A., & Chandrasekaran, N. (2009). Genotoxicity of silver nanoparticles in Allium cepa. *The Science of the Total Environment*, *407*(19), 5243–5246. doi:10.1016/j.scitotenv.2009.06.024 PMID:19616276

Lai, J. C. K., Lai, M. B., Jandhyam, S., Dukhande, V. V., Bhushan, A., & Daniels, C. K. et al.. (2008). Exposure to titanium dioxide and other metallic oxide nanoparticles induces cytotoxicity on human neural cells and fibroblasts. *International Journal of Nanomedicine*, *3*(4), 533–545. PMID:19337421

Larue, C., Laurette, J., Herlin-Boime, N., Khodja, H., Fayard, B., Flank, A.-M., & Carriere, M. et al. (2012). Accumulation, translocation and impact of TiO_2 nanoparticles in wheat (Triticum aestivum spp.): Influence of diameter and crystal phase. *The Science of the Total Environment*, *431*, 197–208. doi:10.1016/j.scitotenv.2012.04.073 PMID:22684121

Larue, C., Laurette, J., Herlin-Boime, N., Khodja, H., Fayard, B., Flank, A.-M., & Carriere, M. et al. (2012). Accumulation, translocation and impact of TiO_2 nanoparticles in wheat (Triticum aestivum spp.): Influence of diameter and crystal phase. *The Science of the Total Environment*, *431*, 197–208. doi:10.1016/j.scitotenv.2012.04.073 PMID:22684121

Larue, C., Veronesi, G., Flank, A.-M., Surble, S., Herlin-Boime, N., & Carrière, M. (2012). Comparative uptake and impact of TiO_2 nanoparticles in wheat and rapeseed. *Journal of Toxicology and Environmental Health. Part A.*, *75*(13-15), 722–734. doi:10.1080/15287394.2012.689800 PMID:22788360

Lee, S.-W., Kim, S.-M., & Choi, J. (2009). Genotoxicity and ecotoxicity assays using the freshwater crustacean Daphnia magna and the larva of the aquatic midge Chironomus riparius to screen the ecological risks of nanoparticle exposure. *Environmental Toxicology and Pharmacology*, *28*(1), 86–91. doi:10.1016/j.etap.2009.03.001 PMID:21783986

Lee, W.-M., Kwak, J. I., & An, Y.-J. (2012). Effect of silver nanoparticles in crop plants Phaseolus radiatus and Sorghum bicolor: Media effect on phytotoxicity. *Chemosphere*, *86*(5), 491–499. doi:10.1016/j.chemosphere.2011.10.013 PMID:22075051

Lee, Y.-H., Cheng, F.-Y., Chiu, H.-W., Tsai, J.-C., Fang, C.-Y., Chen, C.-W., & Wang, Y.-J. (2014). Cytotoxicity, oxidative stress, apoptosis and the autophagic effects of silver nanoparticles in mouse embryonic fibroblasts. *Biomaterials*, *35*(16), 4706–4715. doi:10.1016/j.biomaterials.2014.02.021 PMID:24630838

Lewinski, N., Colvin, V., & Drezek, R. (2008). Cytotoxicity of nanoparticles. *Small*, *4*(1), 26–49. doi:10.1002/smll.200700595 PMID:18165959

Li, K.-E., Chang, Z.-Y., Shen, C.-X., Yaon, N., Siddiqui, M. H., . . . (Eds.). (2015). Toxicity of Nanomaterials to Plants. In M. H. Siddiqui, M. H. Al-Whaibi & F. Mohammad (Eds.), Nanotechnology and Plant Sciences (pp. 101-123). Springer International Publishing. doi:10.1007/978-3-319-14502-0_6

Li, Q., Mahendra, S., Lyon, D. Y., Brunet, L., Liga, M. V., Li, D., & Alvarez, P. J. J. (2008). Antimicrobial nanomaterials for water disinfection and microbial control: Potential applications and implications. *Water Research*, *42*(18), 4591–4602. doi:10.1016/j.watres.2008.08.015 PMID:18804836

Li, S., Wallis, L. K., Diamond, S. A., Ma, H., & Hoff, D. J. (2014). Species sensitivity and dependence on exposure conditions impacting the phototoxicity of TiO_2 nanoparticles to benthic organisms. *Environmental Toxicology and Chemistry*, *33*(7), 1563–1569. doi:10.1002/etc.2583 PMID:24846372

Li, S., Wallis, L. K., Ma, H., & Diamond, S. A. (2014). Phototoxicity of TiO_2 nanoparticles to a freshwater benthic amphipod: Are benthic systems at risk? *The Science of the Total Environment*, *466–467*, 800–808. doi:10.1016/j.scitotenv.2013.07.059 PMID:23973546

Lin, D., & Xing, B. (2007). Phytotoxicity of nanoparticles: Inhibition of seed germination and root growth. *Environmental Pollution*, *150*(2), 243–250. doi:10.1016/j.envpol.2007.01.016 PMID:17374428

Lin, D., & Xing, B. (2008). Root uptake and phytotoxicity of ZnO nanoparticles. *Environmental Science & Technology*, *42*(15), 5580–5585. doi:10.1021/es800422x PMID:18754479

Lin, S., Reppert, J., Hu, Q., Hudson, J. S., Reid, M. L., & Ratnikova, T. A. et al.. (2009). Uptake, translocation, and transmission of carbon nanomaterials in rice plants. *Small*, *5*(10), 1128–1132. doi:10.1002/smll.200801556 PMID:19235197

Lowry, G. V., Gregory, K. B., Apte, S. C., & Lead, J. R. (2012). Transformations of nanomaterials in the environment. *Environmental Science & Technology*, *46*(13), 6893–6899. doi:10.1021/es300839e PMID:22582927

Lu, S.-y., Wu, D., Wang, Q.-, Yan, J., Buekens, A. G., & Cen, K.-. (2011). Photocatalytic decomposition on nano-TiO2: Destruction of chloroaromatic compounds. *Chemosphere*, *82*(9), 1215–1224. doi:10.1016/j.chemosphere.2010.12.034 PMID:21220149

Ma, H., Bertsch, P. M., Glenn, T. C., Kabengi, N. J., & Williams, P. L. (2009). Toxicity of manufactured zinc oxide nanoparticles in the nematode Caenorhabditis elegans. *Environmental Toxicology and Chemistry*, *28*(6), 1324–1330. doi:10.1897/08-262.1 PMID:19192952

Ma, H., Brennan, A., & Diamond, S. A. (2012a). Photocatalytic reactive oxygen species production and phototoxicity of titanium dioxide nanoparticles are dependent on the solar ultraviolet radiation spectrum. *Environmental Toxicology and Chemistry, 31*(9), 2099–2107. doi:10.1002/etc.1916 PMID:22707245

Ma, H., Brennan, A., & Diamond, S. A. (2012b). Phototoxicity of TiO_2 nanoparticles under solar radiation to two aquatic species: Daphnia magna and Japanese medaka. *Environmental Toxicology and Chemistry, 31*(7), 1621–1629. doi:10.1002/etc.1858 PMID:22544710

Ma, H., Wallis, L. K., Diamond, S., Li, S., Canas-Carrell, J., & Parra, A. (2014). Impact of solar UV radiation on toxicity of ZnO nanoparticles through photocatalytic reactive oxygen species (ROS) generation and photo-induced dissolution. *Environmental Pollution, 193*, 165–172. doi:10.1016/j.envpol.2014.06.027 PMID:25033018

Ma, H., Williams, P. L., & Diamond, S. A. (2013). Ecotoxicity of manufactured ZnO nanoparticles – A review. *Environmental Pollution, 172*, 76–85. doi:10.1016/j.envpol.2012.08.011 PMID:22995930

Ma, X., Gurung, A., & Deng, Y. (2013). Phytotoxicity and uptake of nanoscale zero-valent iron (nZVI) by two plant species. *The Science of the Total Environment, 443*, 844–849. doi:10.1016/j.scitotenv.2012.11.073 PMID:23247287

Ma, Y., Kuang, L., He, X., Bai, W., Ding, Y., Zhang, Z., & Chai, Z. et al. (2010). Effects of rare earth oxide nanoparticles on root elongation of plants. *Chemosphere, 78*(3), 273–279. doi:10.1016/j.chemosphere.2009.10.050 PMID:19897228

Mahdavi, S., Jalali, M., & Afkhami, A. (2012). Heavy metals removal from aqueous solutions using TiO2, MgO, and Al_2O_3 nanoparticles. *Chemical Engineering Communications, 200*(3), 448–470. doi:10.1080/00986445.2012.686939

Mahmoodi, N. M., Arami, M., Limaee, N. Y., & Gharanjig, K. (2007). Photocatalytic degradation of agricultural N-heterocyclic organic pollutants using immobilized nanoparticles of titania. *Journal of Hazardous Materials, 145*(1–2), 65–71. doi:10.1016/j.jhazmat.2006.10.089 PMID:17145132

Mahmoudi, M., Simchi, A., Milani, A. S., & Stroeve, P. (2009). Cell toxicity of superparamagnetic iron oxide nanoparticles. *Journal of Colloid and Interface Science, 336*(2), 510–518. doi:10.1016/j.jcis.2009.04.046 PMID:19476952

Malvindi, M. A., De Matteis, V., Galeone, A., Brunetti, V., Anyfantis, G. C., Athanassiou, A., & Pompa, P. P. et al. (2014). Toxicity assessment of silica coated iron oxide nanoparticles and biocompatibility improvement by surface engineering. *PLoS ONE, 9*(1), e85835. doi:10.1371/journal.pone.0085835 PMID:24465736

Manzo, S., Miglietta, M. L., Rametta, G., Buono, S., & Di Francia, G. (2013). Toxic effects of ZnO nanoparticles towards marine algae Dunaliella tertiolecta. *The Science of the Total Environment, 445*, 371–376. doi:10.1016/j.scitotenv.2012.12.051 PMID:23361041

Menard, A., Drobne, D., & Jemec, A. (2011). Ecotoxicity of nanosized TiO_2. Review of in vivo data. *Environmental Pollution, 159*(3), 677–684. doi:10.1016/j.envpol.2010.11.027 PMID:21186069

Miao, A.-J., Luo, Z., Chen, C.-S., Chin, W.-C., Santschi, P. H., & Quigg, A. (2010). Intracellular uptake: A possible mechanism for silver engineered nanoparticle toxicity to a freshwater alga Ochromonas danica. *PLoS ONE, 5*(12), e15196. doi:10.1371/journal.pone.0015196 PMID:21203552

Miao, A.-J., Schwehr, K. A., Xu, C., Zhang, S.-J., Luo, Z., Quigg, A., & Santschi, P. H. (2009). The algal toxicity of silver engineered nanoparticles and detoxification by exopolymeric substances. *Environmental Pollution, 157*(11), 3034–3041. doi:10.1016/j.envpol.2009.05.047 PMID:19560243

Miller, R. J., Bennett, S., Keller, A. A., Pease, S., & Lenihan, H. S. (2012). TiO_2 nanoparticles are phototoxic to marine phytoplankton. *PLoS ONE, 7*(1), e30321. doi:10.1371/journal.pone.0030321 PMID:22276179

Miralles, P., Church, T. L., & Harris, A. T. (2012). Toxicity, uptake, and translocation of engineered nanomaterials in vascular plants. *Environmental Science & Technology, 46*(17), 9224–9239. doi:10.1021/es202995d PMID:22892035

Mohmood, I., Lopes, C., Lopes, I., Ahmad, I., Duarte, A., & Pereira, E. (2013). Nanoscale materials and their use in water contaminants removal—a review. *Environ Sci Pollut Res, 20*(3), 1239–1260. doi:10.1007/s11356-012-1415-x PMID:23292223

Morales-Flores, N., Pal, U., & Sánchez Mora, E. (2011). Photocatalytic behavior of ZnO and Pt-incorporated ZnO nanoparticles in phenol degradation. *Appl Catalysis A, 394*(1–2), 269–275. doi:10.1016/j.apcata.2011.01.011

Morones, J. R., Elechiguerra, J. L., Camacho, A., Holt, K., Kouri, J. B., Ramírez, J. T., & Yacaman, M. J. (2005). The bactericidal effect of silver nanoparticles. *Nanotechnology, 16*(10), 2346–2353. doi:10.1088/0957-4484/16/10/059 PMID:20818017

Mu, R., Xu, Z., Li, L., Shao, Y., Wan, H., & Zheng, S. (2010). On the photocatalytic properties of elongated TiO_2 nanoparticles for phenol degradation and Cr(VI) reduction. *Journal of Hazardous Materials, 176*(1–3), 495–502. doi:10.1016/j.jhazmat.2009.11.057 PMID:19969418

Mukherjee, S. G., O'Claonadh, N., Casey, A., & Chambers, G. (2012). Comparative in vitro cytotoxicity study of silver nanoparticle on two mammalian cell lines. *Toxicology In Vitro, 26*(2), 238–251. doi:10.1016/j.tiv.2011.12.004 PMID:22198051

Nair, P. M. G., Park, S. Y., & Choi, J. (2013). Evaluation of the effect of silver nanoparticles and silver ions using stress responsive gene expression in Chironomus riparius. *Chemosphere, 92*(5), 592–599. doi:10.1016/j.chemosphere.2013.03.060 PMID:23664472

Naqvi, S., Samim, M., Abdin, M. Z., Ahmed, F. J., Maitra, A. N., & Prashant, C. K. et al.. (2010). Concentration-dependent toxicity of iron oxide nanoparticles mediated by increased oxidative stress. *International Journal of Nanomedicine, 5*, 983–989. doi:10.2147/IJN.S13244 PMID:21187917

Nations, S., Wages, M., Cañas, J. E., Maul, J., Theodorakis, C., & Cobb, G. P. (2011). Acute effects of Fe_2O_3, TiO_2, ZnO and CuO nanomaterials on Xenopus laevis. *Chemosphere, 83*(8), 1053–1061. doi:10.1016/j.chemosphere.2011.01.061 PMID:21345480

Navarro, E., Baun, A., Behra, R., Hartmann, N., Filser, J., Miao, A.-J., & Sigg, L. et al. (2008). Environmental behavior and ecotoxicity of engineered nanoparticles to algae, plants, and fungi. *Ecotoxicology (London, England)*, *17*(5), 372–386. doi:10.1007/s10646-008-0214-0 PMID:18461442

Ng, K. W., Khoo, S. P. K., Heng, B. C., Setyawati, M. I., Tan, E. C., Zhao, X., & Loo, J. S. C. et al. (2011). The role of the tumor suppressor p53 pathway in the cellular DNA damage response to zinc oxide nanoparticles. *Biomaterials*, *32*(32), 8218–8225. doi:10.1016/j.biomaterials.2011.07.036 PMID:21807406

Nogueira, V., Lopes, I., Rocha-Santos, T. A. P., Rasteiro, M. G., Abrantes, N., Gonçalves, F., & Pereira, R. et al. (2015). Assessing the ecotoxicity of metal nano-oxides with potential for wastewater treatment. *Environ Sci Pollut Res*, *22*(17), 13212–13224. doi:10.1007/s11356-015-4581-9 PMID:25940480

O'Halloran, K. (2006). Toxicological considerations of contaminants in the terrestrial environment for ecological risk assessment. *Human and Ecological Risk Assessment*, *12*(1), 74–83. doi:10.1080/10807030500428603

Oberdörster, G., Sharp, Z., Atudorei, V., Elder, A., Gelein, R., Lunts, A., & Cox, C. et al. (2002). Extrapulmonary translocation of ultrafine carbon particles following whole-body inhalation exposure of rats. *Journal of Toxicology and Environmental Health. Part A.*, *65*(20), 1531–1543. doi:10.1080/00984100290071658 PMID:12396867

Oberholster, P. J., Musee, N., Botha, M., Chelule, P. K., Focke, W. W., & Ashton, P. J. (2011). Assessment of the effect of nanomaterials on sediment-dwelling invertebrate Chironomus tentans larvae. *Ecotoxicology and Environmental Safety*, *74*(3), 416–423. doi:10.1016/j.ecoenv.2010.12.012 PMID:21216008

Oukarroum, A., Bras, S., Perreault, F., & Popovic, R. (2012). Inhibitory effects of silver nanoparticles in two green algae, Chlorella vulgaris and Dunaliella tertiolecta. *Ecotoxicology and Environmental Safety*, *78*, 80–85. doi:10.1016/j.ecoenv.2011.11.012 PMID:22138148

Parham, H., Zargar, B., & Rezazadeh, M. (2012). Removal, preconcentration and spectrophotometric determination of picric acid in water samples using modified magnetic iron oxide nanoparticles as an efficient adsorbent. *Materials Science and Engineering C*, *32*(7), 2109–2114. doi:10.1016/j.msec.2012.05.044

Park, E.-J., Yi, J., Chung, K.-H., Ryu, D.-Y., Choi, J., & Park, K. (2008). Oxidative stress and apoptosis induced by titanium dioxide nanoparticles in cultured BEAS-2B cells. *Toxicology Letters*, *180*(3), 222–229. doi:10.1016/j.toxlet.2008.06.869 PMID:18662754

Park, E.-J., Yi, J., Kim, Y., Choi, K., & Park, K. (2010). Silver nanoparticles induce cytotoxicity by a Trojan-horse type mechanism. *Toxicology In Vitro*, *24*(3), 872–878. doi:10.1016/j.tiv.2009.12.001 PMID:19969064

Park, M. V. D. Z., Neigh, A. M., Vermeulen, J. P., de la Fonteyne, L. J. J., Verharen, H. W., Briedé, J. J., & de Jong, W. H. et al. (2011). The effect of particle size on the cytotoxicity, inflammation, developmental toxicity and genotoxicity of silver nanoparticles. *Biomaterials*, *32*(36), 9810–9817. doi:10.1016/j.biomaterials.2011.08.085 PMID:21944826

Pena, M. E., Korfiatis, G. P., Patel, M., Lippincott, L., & Meng, X. (2005). Adsorption of As(V) and As(III) by nanocrystalline titanium dioxide. *Water Research*, *39*(11), 2327–2337. doi:10.1016/j.watres.2005.04.006 PMID:15896821

Peralta-Videa, J. R., Zhao, L., Lopez-Moreno, M. L., de la Rosa, G., Hong, J., & Gardea-Torresdey, J. L. (2011). Nanomaterials and the environment: A review for the biennium 2008-2010. *Journal of Hazardous Materials, 186*(1), 1–15. doi:10.1016/j.jhazmat.2010.11.020 PMID:21134718

Peters, K., Unger, R., Kirkpatrick, C. J., Gatti, A., & Monari, E. (2004). Effects of nano-scaled particles on endothelial cell function in vitro: Studies on viability, proliferation and inflammation. *Journal of Materials Science. Materials in Medicine, 15*(4), 321–325. doi:10.1023/B:JMSM.0000021095.36878.1b PMID:15332593

Premanathan, M., Karthikeyan, K., Jeyasubramanian, K., & Manivannan, G. (2011). Selective toxicity of ZnO nanoparticles toward Gram-positive bacteria and cancer cells by apoptosis through lipid peroxidation. *Nanomedicine (London), 7*(2), 184–192. doi:10.1016/j.nano.2010.10.001 PMID:21034861

Qian, H., Peng, X., Han, X., Ren, J., Sun, L., & Fu, Z. (2013). Comparison of the toxicity of silver nanoparticles and silver ions on the growth of terrestrial plant model Arabidopsis thaliana. *Journal of Environmental Sciences (China), 25*(9), 1947–1955. doi:10.1016/S1001-0742(12)60301-5 PMID:24520739

Qu, X., Alvarez, P. J. J., & Li, Q. (2013). Applications of nanotechnology in water and wastewater treatment. *Water Research, 47*(12), 3931–3946. doi:10.1016/j.watres.2012.09.058 PMID:23571110

Qu, X., Brame, J., Li, Q., & Alvarez, P. J. J. (2013). Nanotechnology for a safe and sustainable water supply: Enabling integrated water treatment and reuse. *Accounts of Chemical Research, 46*(3), 834–843. doi:10.1021/ar300029v PMID:22738389

Rahman, Q., Lohani, M., Dopp, E., Pemsel, H., Jonas, L., Weiss, D. G., & Schiffmann, D. (2002). Evidence that ultrafine titanium dioxide induces micronuclei and apoptosis in syrian hamster embryo fibroblasts. *Environmental Health Perspectives, 110*(8), 797–800. doi:10.1289/ehp.02110797 PMID:12153761

Rahman, Q. I., Ahmad, M., Misra, S., & Lohani, M. (2013). Effective photocatalytic degradation of rhodamine B dye by ZnO nanoparticles. *Materials Letters, 91*, 170–174. doi:10.1016/j.matlet.2012.09.044

Rao, S., & Shekhawat, G. S. (2014). Toxicity of ZnO engineered nanoparticles and evaluation of their effect on growth, metabolism and tissue specific accumulation in Brassica juncea. *J Environ Chem Eng, 2*(1), 105–114. doi:10.1016/j.jece.2013.11.029

Recillas, S., García, A., González, E., Casals, E., Puntes, V., Sánchez, A., & Font, X. (2011). Use of CeO_2, TiO_2 and Fe_3O_4 nanoparticles for the removal of lead from water: Toxicity of nanoparticles and derived compounds. *Desalination, 277*(1-3), 213–220. doi:10.1016/j.desal.2011.04.036

Remédios, C., Rosário, F., & Bastos, V. (2012). Environmental nanoparticles interactions with plants: Morphological, physiological, and genotoxic aspects. *Le Journal de Botanique, 2012*. doi:10.1155/2012/751686

Rico, C. M., Majumdar, S., Duarte-Gardea, M., Peralta-Videa, J. R., & Gardea-Torresdey, J. L. (2011). Interaction of nanoparticles with edible plants and their possible implications in the food chain. *Journal of Agricultural and Food Chemistry, 59*(8), 3485–3498. doi:10.1021/jf104517j PMID:21405020

Robichaud, C. O., Uyar, A. E., Darby, M. R., Zucker, L. G., & Wiesner, M. R. (2009). Estimates of upper bounds and trends in nano-TiO_2 production as a basis for exposure assessment. *Environmental Science & Technology, 43*(12), 4227–4233. doi:10.1021/es8032549 PMID:19603627

Roh, J.-y., Sim, S. J., Yi, J., Park, K., Chung, K. H., Ryu, D.-y., & Choi, J. (2009). Ecotoxicity of silver nanoparticles on the soil nematode Caenorhabditis elegans using functional ecotoxicogenomics. *Environmental Science & Technology*, *43*(10), 3933–3940. doi:10.1021/es803477u PMID:19544910

Sakthivel, S., Neppolian, B., Shankar, M. V., Arabindoo, B., Palanichamy, M., & Murugesan, V. (2003). Solar photocatalytic degradation of azo dye: Comparison of photocatalytic efficiency of ZnO and TiO2. *Solar Energy Materials and Solar Cells*, *77*(1), 65–82. doi:10.1016/S0927-0248(02)00255-6

Sánchez, A., Recillas, S., Font, X., Casals, E., González, E., & Puntes, V. (2011). Ecotoxicity of, and remediation with, engineered inorganic nanoparticles in the environment. *TrAC Trends Analytical Chem*, *30*(3), 507–516. doi:10.1016/j.trac.2010.11.011

Schwab, F., Zhai, G., Kern, M., Turner, A., Schnoor, J. L., & Wiesner, M. R. (2015). Barriers, pathways and processes for uptake, translocation and accumulation of nanomaterials in plants – Critical review. *Nanotoxicology*, 1–22. doi:10.3109/17435390.2015.1048326 PMID:26067571

Scott-Fordsmand, J. J., Krogh, P. H., Schaefer, M., & Johansen, A. (2008). The toxicity testing of double-walled nanotubes-contaminated food to Eisenia veneta earthworms. *Ecotoxicology and Environmental Safety*, *71*(3), 616–619. doi:10.1016/j.ecoenv.2008.04.011 PMID:18514310

Selli, E., Bianchi, C. L., Pirola, C., Cappelletti, G., & Ragaini, V. (2008). Efficiency of 1,4-dichlorobenzene degradation in water under photolysis, photocatalysis on TiO_2 and sonolysis. *Journal of Hazardous Materials*, *153*(3), 1136–1141. doi:10.1016/j.jhazmat.2007.09.071 PMID:17976904

Servin, A. D., Castillo-Michel, H., Hernandez-Viezcas, J. A., Diaz, B. C., Peralta-Videa, J. R., & Gardea-Torresdey, J. L. (2012). Synchrotron micro-XRF and micro-XANES confirmation of the uptake and translocation of TiO_2 nanoparticles in cucumber (Cucumis sativus) plants. *Environmental Science & Technology*, *46*(14), 7637–7643. doi:10.1021/es300955b PMID:22715806

Servin, A. D., Morales, M. I., Castillo-Michel, H., Hernandez-Viezcas, J. A., Munoz, B., Zhao, L., & Gardea-Torresdey, J. L. et al. (2013). Synchrotron Verification of TiO_2 accumulation in cucumber fruit: A possible pathway of TiO_2 nanoparticle transfer from soil into the food chain. *Environmental Science & Technology*, *47*(20), 11592–11598. doi:10.1021/es403368j PMID:24040965

Shahid, M., McDonagh, A., Kim, H., & Shon, H. K. (2015). Magnetised titanium dioxide (TiO_2) for water purification: Preparation, characterisation and application. *Desalination and Water Treatment*, *54*(4-5), 979–1002. doi:10.1080/19443994.2014.911119

Sharma, V., Anderson, D., & Dhawan, A. (2012). Zinc oxide nanoparticles induce oxidative DNA damage and ROS-triggered mitochondria mediated apoptosis in human liver cells (HepG2). *Apoptosis*, *17*(8), 852–870. doi:10.1007/s10495-012-0705-6 PMID:22395444

Sharma, V., Shukla, R. K., Saxena, N., Parmar, D., Das, M., & Dhawan, A. (2009). DNA damaging potential of zinc oxide nanoparticles in human epidermal cells. *Toxicology Letters*, *185*(3), 211–218. doi:10.1016/j.toxlet.2009.01.008 PMID:19382294

Shu, H.-Y., Chang, M.-C., & Chang, C.-C. (2009). Integration of nanosized zero-valent iron particles addition with UV/H_2O_2 process for purification of azo dye Acid Black 24 solution. *Journal of Hazardous Materials*, *167*(1–3), 1178–1184. doi:10.1016/j.jhazmat.2009.01.106 PMID:19250743

Shukla, R. K., Sharma, V., Pandey, A. K., Singh, S., Sultana, S., & Dhawan, A. (2011). ROS-mediated genotoxicity induced by titanium dioxide nanoparticles in human epidermal cells. *Toxicology In Vitro*, *25*(1), 231–241. doi:10.1016/j.tiv.2010.11.008 PMID:21092754

Song, G., Gao, Y., Wu, H., Hou, W., Zhang, C., & Ma, H. (2012). Physiological effect of anatase TiO_2 nanoparticles on Lemna minor. *Environmental Toxicology and Chemistry*, *31*(9), 2147–2152. doi:10.1002/etc.1933 PMID:22760594

Song, U., Jun, H., Waldman, B., Roh, J., Kim, Y., Yi, J., & Lee, E. J. (2013). Functional analyses of nanoparticle toxicity: A comparative study of the effects of TiO_2 and Ag on tomatoes (Lycopersicon esculentum). *Ecotoxicology and Environmental Safety*, *93*, 60–67. doi:10.1016/j.ecoenv.2013.03.033 PMID:23651654

Soni, D., Naoghare, P. K., Saravanadevi, S., & Pandey, R. A. (2015). Release, transport and toxicity of engineered nanoparticles. In D. M. Whitacre (Ed.), *Reviews of environmental contamination and toxicology* (Vol. 234). Switzerland: Springer. doi:10.1007/978-3-319-10638-0_1

Stampoulis, D., Sinha, S. K., & White, J. C. (2009). Assay-dependent phytotoxicity of nanoparticles to plants. *Environmental Science & Technology*, *43*(24), 9473–9479. doi:10.1021/es901695c PMID:19924897

Stone, V., Nowack, B., Baun, A., van den Brink, N., von der Kammer, F., Dusinska, M., & Fernandes, T. F. et al. (2010). Nanomaterials for environmental studies: Classification, reference material issues, and strategies for physico-chemical characterisation. *The Science of the Total Environment*, *408*(7), 1745–1754. doi:10.1016/j.scitotenv.2009.10.035 PMID:19903569

Suppan, S. (2013). *Nanomaterials in soil: our future food chain?*. Academic Press.

Tiwari, D. K., Behari, J., & Sen, P. (2008). Application of nanoparticles in waste water treatment. *World Appl Sci J*, *3*(3), 417–433.

Tourinho, P. S., van Gestel, C. A. M., Jurkschat, K., Soares, A. M. V. M., & Loureiro, S. (2015). Effects of soil and dietary exposures to Ag nanoparticles and $AgNO_3$ in the terrestrial isopod Porcellionides pruinosus. *Environmental Pollution*, *205*, 170–177. doi:10.1016/j.envpol.2015.05.044 PMID:26071943

Triboulet, S., Aude-Garcia, C., Armand, L., Gerdil, A., Diemer, H., Proamer, F., & Rabilloud, T. et al. (2014). Analysis of cellular responses of macrophages to zinc ions and zinc oxide nanoparticles: A combined targeted and proteomic approach. *Nanoscale*, *6*(11), 6102–6114. doi:10.1039/c4nr00319e PMID:24788578

Trujillo-Reyes, J., Majumdar, S., Botez, C. E., Peralta-Videa, J. R., & Gardea-Torresdey, J. L. (2014). Exposure studies of core-shell Fe/Fe_3O_4 and Cu/CuO NPs to lettuce (Lactuca sativa) plants: Are they a potential physiological and nutritional hazard? *Journal of Hazardous Materials*, *267*, 255–263. doi:10.1016/j.jhazmat.2013.11.067 PMID:24462971

Vicario-Parés, U., Castañaga, L., Lacave, J. M., Oron, M., Reip, P., Berhanu, D., & Orbea, A. et al. (2014). Comparative toxicity of metal oxide nanoparticles (CuO, ZnO and TiO_2) to developing zebrafish embryos. *Journal of Nanoparticle Research*, *16*(8), 2550. doi:10.1007/s11051-014-2550-8

Wallis, L. K., Diamond, S. A., Ma, H., Hoff, D. J., Al-Abed, S. R., & Li, S. (2014). Chronic TiO_2 nanoparticle exposure to a benthic organism, Hyalella azteca: Impact of solar UV radiation and material surface coatings on toxicity. *The Science of the Total Environment, 499*, 356–362. doi:10.1016/j.scitotenv.2014.08.068 PMID:25203828

Wang, H., & Huang, Y. (2011). Prussian-blue-modified iron oxide magnetic nanoparticles as effective peroxidase-like catalysts to degrade methylene blue with H_2O_2. *Journal of Hazardous Materials, 191*(1–3), 163–169. doi:10.1016/j.jhazmat.2011.04.057 PMID:21570769

Wang, H., Wick, R. L., & Xing, B. (2009). Toxicity of nanoparticulate and bulk ZnO, Al_2O_3 and TiO_2 to the nematode Caenorhabditis elegans. *Environmental Pollution, 157*(4), 1171–1177. doi:10.1016/j.envpol.2008.11.004 PMID:19081167

Wang, H., Xie, C., Zhang, W., Cai, S., Yang, Z., & Gui, Y. (2007). Comparison of dye degradation efficiency using ZnO powders with various size scales. *Journal of Hazardous Materials, 141*(3), 645–652. doi:10.1016/j.jhazmat.2006.07.021 PMID:16930825

Wang, J., Zhu, X., Zhang, X., Zhao, Z., Liu, H., George, R., & Chen, Y. et al. (2011). Disruption of zebrafish (Danio rerio) reproduction upon chronic exposure to TiO_2 nanoparticles. *Chemosphere, 83*(4), 461–467. doi:10.1016/j.chemosphere.2010.12.069 PMID:21239038

Westerhoff, P., Song, G., Hristovski, K., & Kiser, M. (2011). Occurrence and removal of titanium at full scale wastewater treatment plants: Implications for TiO_2 nanomaterials. *Journal of Environmental Monitoring, 13*(5), 1195–1203. doi:10.1039/c1em10017c PMID:21494702

Wiesner, M. R., Lowry, G. V., Alvarez, P., Dionysiou, D., & Biswas, P. (2006). Assessing the risks of manufactured nanomaterials. *Environmental Science & Technology, 40*(14), 4336–4345. doi:10.1021/es062726m PMID:16903268

Wu, S. G., Huang, L., Head, J., Chen, D.-R., Kong, I.-C., & Tang, Y. J. (2012). Phytotoxicity of metal oxide nanoparticles is related to both dissolved metals ions and adsorption of particles on seed surfaces. *J Petrol Environ Biotechnol, 3*(4).

Xia, T., Kovochich, M., Liong, M., Mädler, L., Gilbert, B., Shi, H., & Nel, A. E. et al. (2008). Comparison of the mechanism of toxicity of zinc oxide and cerium oxide nanoparticles based on dissolution and oxidative stress properties. *ACS Nano, 2*(10), 2121–2134. doi:10.1021/nn800511k PMID:19206459

Xu, P., Zeng, G. M., Huang, D. L., Feng, C. L., Hu, S., Zhao, M. H., & Liu, Z. F. et al. (2012). Use of iron oxide nanomaterials in wastewater treatment: A review. *The Science of the Total Environment, 424*, 1–10. doi:10.1016/j.scitotenv.2012.02.023 PMID:22391097

Yang, L., & Watts, D. J. (2005). Particle surface characteristics may play an important role in phytotoxicity of alumina nanoparticles. *Toxicology Letters, 158*(2), 122–132. doi:10.1016/j.toxlet.2005.03.003 PMID:16039401

Zhang, L., & Fang, M. (2010). Nanomaterials in pollution trace detection and environmental improvement. *Nano Today, 5*(2), 128–142. doi:10.1016/j.nantod.2010.03.002

Zhu, H., Han, J., Xiao, J. Q., & Jin, Y. (2008). Uptake, translocation, and accumulation of manufactured iron oxide nanoparticles by pumpkin plants. *J Environ Monit, 10*(6), 713-717. doi: 10.1039/b805998e

Zhu, H. N., Lu, Q. X., & Abdollahi, K. (2005). Seed coat structure of Pinus koraiensis. *Microscopy and Microanalysis*, *11*(S02), 1158–1159. doi:10.1017/S1431927605503635

Zhu, X., Chang, Y., & Chen, Y. (2010). Toxicity and bioaccumulation of TiO_2 nanoparticle aggregates in Daphnia magna. *Chemosphere*, *78*(3), 209–215. doi:10.1016/j.chemosphere.2009.11.013 PMID:19963236

Zhu, X., Tian, S., & Cai, Z. (2012). Toxicity assessment of iron oxide nanoparticles in Zebrafish (Danio rerio) early life stages. *PLoS ONE*, *7*(9), e46286. doi:10.1371/journal.pone.0046286 PMID:23029464

Chapter 14
Ecotoxicity Effects of Nanomaterials on Aquatic Organisms:
Nanotoxicology of Materials on Aquatic Organisms

César A Barbero
Universidad Nacional de Río Cuarto, Argentina

Edith Inés Yslas
Universidad Nacional de Río Cuarto, Argentina

ABSTRACT

The increasing production and use of engineered nanomaterials raise concerns about inadvertent exposure and the potential for adverse effects on the aquatic environment. The aim of this chapter is focused on studies of nanotoxicity in different models of aquatic organisms and their impact. Moreover, the chapter provides an overview of nanoparticles, their applications, and the potential nanoparticle-induced toxicity in aquatic organisms. The topics discussed in this chapter are the physicochemical characteristic of nanomaterials (size, aggregation, morphology, surface charge, reactivity, dissolution, etc.) and their influence on toxicity. Further, the text discusses the direct effect of nanomaterials on development stage (embryonic and adult) in aquatic organisms, the mechanism of action as well as the toxicity data of nanomaterials in different species.f action as well as the toxicity data of nanomaterials in different species.

INTRODUCTION

Nanotechnology is an emerging multidisciplinary science that involves synthesis, characterization, and applications of nanomaterials (NMs). The nanomaterials have been defined as new materials and structures with at least one dimension between 1 and 100 nm. However, related materials of somewhat larger size also show new properties (Dowling et al., 2004). They have attracted a great interest during

DOI: 10.4018/978-1-5225-0585-3.ch014

Ecotoxicity Effects of Nanomaterials on Aquatic Organisms

recent years, due to their many technologically interesting properties. Manufactured nanomaterials have numerous industrial applications including electronics, optics, and textiles, as well as applications in medical devices, drug delivery systems, chemical sensors, biosensors, and in environmental remediation. The rapid growth of nanotechnology applications in the past years has increased the release of nanoparticles (NPs) to the environment (Gottschalk & Nowack, 2011). Safe manufacturing and application of nanomaterials are an emerging issue with the bloom of nanotechnology. Understanding the interaction of nanomaterials with the biological system is essential for the realization of their safe applications. The nanoscale size of the particles can confer novel and significantly improved physical, chemical and biological properties (Ogden, 2013).

Experts agree with the fact that the environmental effects of nanoparticles cannot be predicted from the known ecotoxicity of the macroscopic material. Because of the production and use of engineered nanomaterials (ENMs) have increased exponentially during the last years, should be taken into account that these products may have impacts on environmental health. Moreover, in many circumstances, the exposure assessment or potential cannot be quantified, due either to limitations of measurement or because of technological limitations in measuring nanoparticle exposures (Ong et al., 2014). Hence, a new subdiscipline of nanotechnology called nanotoxicology has emerged (Oberdörster, Oberdörster, & Oberdörster, 2005). Nanotoxicology involves the study of the interactions of nanomaterials with biological systems with an emphasis on elucidating the relationship between the physical and chemical properties of nanomaterials with induction of toxic biological responses (Fischer & Chan, 2007). It is noteworthy that the absence of toxicity of the substance in the elemental or molecular state does not mean that the nanomaterial is harmless. The introduction of nanoparticles in aquatic ecosystems will bring about and new potentially toxic interactions in exposed organisms. Although the presence of nanoparticles in the aquatic environment is still largely undocumented, their release could certainly occur in the future and what is worrying is that little is known about of the effect on aquatic organisms or on their possible impacts (Chapman, 2006).

Aquatic ecosystems are one of the main final destinations of the released nanomaterials into the environment although the toxicology of nanotechnology products is virtually unknown. These nanomaterials may have harmful effects on the aquatic organisms, so the study of these effects is of great importance. Aquatic nanotoxicology is the assessment of toxic effects of nanomaterials on aquatic organisms. The NMs will enter, through numerous direct and indirect routes, into the water environment. For this reason, there is an urgent need to address several critical nanomaterials-associated ecotoxicological and environmental risks assessment issues, such as toxicology of NMs to aquatic organisms, the bioavailability, and bioaccumulation of NMs in water environment. Since the amount and diversity of nanomaterial applications increases, concerns about their release into the environment and their impact on natural ecosystems are growing. Evaluating the potential biological impact of nanomaterials has become increasingly important in recent years. This is particularly relevant because the rapid of nanotechnology development has not been accompanied by an investigation of its safety in the environment. The aim of this chapter is focused on studies of nanotoxicity in different models of aquatic organisms and their impact on aquatic ecosystems. The NMs are applied for their specific characteristics resulting from an altered surface area to volume ratio compared to their bulk counterparts.

Preliminary studies report that the NMs can play a role in damaging numerous important biological processes on living organisms. According to a recent report, (Henry et al., 2007) carbon nanomaterials have the highest relative frequency of occurrence in consumer products already on the market and these materials may contaminate the environment in the future. The effects of NMs in the environment will depend on the amount of released nanoparticles and their physicochemical characteristics. Size is a well-known important factor determining nanotoxicity to aquatic organisms (Angel, Batley, Jarolimek, & Rogers, 2013). Theoretical considerations suggest that smaller particles are likely to be more toxic on the account of their larger specific surface area and greater bioavailability. As surface area increases, reactivity increases due to surface atoms control of the particles properties (Nel et al., 2009). On the hand, the stability of chemical compositions of nanomaterials itself should be considered as an important factor affecting their potential environmental impacts and biological effects. The intrinsic characteristics of the nanomaterials imply a multi-variable complexity, which affects their toxicological potential. Detailed knowledge of the behavior of NPs in the environment is crucial to determining a cause and effect relationship between exposure and effects. NPs have the potential to impact organisms due to their intrinsic properties. The assessment of environmental effects requires an understanding of their mobility, reactivity, ecotoxicity, and persistence. The toxicity is a critical factor to consider when evaluating the potential use of nanocomposite materials in industries for aquatic ecosystem.

PHYSICOCHEMICAL INTERACTIONS BETWEEN NANOMATERIALS AND BIOLOGICAL ORGANISMS

Nanomaterials possess different properties compared to the same material in its coarser or bulk form. It is well known that physical-chemical properties of materials can alter dramatically at nanoscale, and the growing use of nanotechnologies requires careful assessment of unexpected toxicities and biological interactions. The toxicological community has focused on identifying key physicochemical properties of the nanomaterial or the material morphology which can influence the fate of nanoparticles in aquatic systems (Peng, Palma, Fisher, & Wong, 2011; Albanese, Tang, & Chan, 2012; Ivask et al., 2014; Yang, Liu, Yang, Zhang, & Xi, 2009). These properties are depicted in Figure 1, and the bulk form of the same material does not share them. However, these are essential for a better understanding of nanoparticle toxicity for aquatic organisms and they may cause additional damage or may induce new adverse effects (Jiang, Oberdörster, & Biswas, 2008; Griffitt, Luo, Gao, Bonzongo, & Barber, 2008; Albanese et al., 2012; Podila & Brown, 2013).

The particle size of nanoscale materials is a critical parameter in the assessment of environmental, health and safety aspects when compared to those caused by the bulk material. Particle size and surface area are important material characteristics that may adversely affect organisms. As the size of a particle decreases, its surface area increases and also allows a greater proportion of its atoms or molecules to be displayed on the surface rather than the interior of the material that could determine the uptake and toxic effects on aquatic organisms (Nel, Xia, Mädler, & Li, 2006; Lee et al., 2012). Therefore, it is known that when the size of a particle decreases to the nanoscale its optical, electrical and magnetic properties differ substantially compared with the parent material in bulk form (Klaine et al., 2008). Asharani et al., (2008) reported the first time investigation of size dependent toxic effect of silver nanoparticles (AgNPs) in zebrafish models. This study showed the health and environmental impact of AgNPs using as capping agents starch and bovine serum albumin (BSA).

In this work Ag-NP induces a dose-dependent toxicity in embryos treated embryos demonstrating a concentration-dependent increase in mortality and hatching delay was observed. Another work that confirmed of size-dependent toxic effects of AgNPs to bacteria, yeast, protozoa, algae, crustaceans and mammalian cells in vitro was reported by Ivask et al. (2014). The study revealed some differences

Figure 1. Schematic overview of important characteristics of nanomaterials that can influence their toxicity

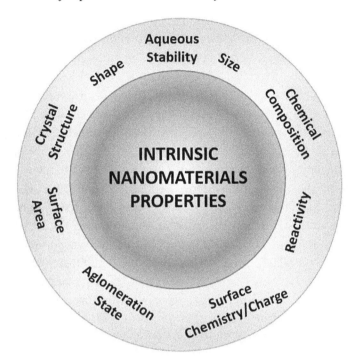

between model species. This work confirmed that the toxicity of AgNPs increased with decreasing particle size. In addition, when the toxic effect was analyzed indicate that the most sensitive organism to AgNPs was crustacean, followed by algae, bacteria, yeast, and finally fibroblasts. Fujiwara et al., (2008) reported that SiO_2 nanoparticles exhibit size-dependent toxicity toward the alga Chlorella kessleri. The cells were exposed to 5% 5 nm silica nanoparticles, 10% 26 nm silica nanoparticles and 10% 78 nm silica nanoparticles for 96 h. This report showed the 50% Inhibitory Concentrations (IC50) value for 5 nm = 0.8 +/- 0.6, 26 nm = 7.1 +/- 2.8, and 78 nm = 9.1 +/- 4.7 weight/volume % of SiO_2 nanoparticles.

A similar result was reported by Polonini et al., (2015) for barium titanate micro-nanoparticles. Li et al., (2008) reported that the Lethal Concentration 50 (LC50) of selenium nanoparticles (Nano-Se) is fivefold lower than that of selenite. The result significantly affected overall toxicity observed for some organisms (e,g. Medaka). Another aspect of NPs are the surface chemistries, it plays a pivotal role in NPs toxicity and have the potential to increase or decrease negative biological impacts, although their impacts are poorly understood (Bozich et al., 2014). The surfaces of the NPs are typically modified with functional surface groups in order to improve stability. The presence of exposed surface groups has repercussions for their fate and behavior in the environment (Tejamaya, Römer, Merrifield, & Lead, 2012). Surface groups can make the nanomaterials hydrophilic or hydrophobic, lipophilic or lipophobic, positively or negatively-charged. These properties can influence the interaction of NPs with an organism and has therefore been suggested to impact on its toxicity (Walters, Pool, & Somerset, 2014; Dominguez et al., 2015). Bozich et al., (2014) demonstrated that surface chemistry plays a significant role in NP toxicity.

The charge of the NP surface has been implicated as a major factor in toxicity while did not affect the uptake of Au NPs in Daphnia. Au NPs functionalized with a cationic surface group were more toxic than those having anionic or neutral groups charged surface. This toxicity effect caused by the positive charge on the NP surface was observed in aquatic organisms such as algae, bivalves and fish (Couling, Bernot, Docherty, Dixon, & Maginn, 2006). Other properties of nanomaterials such as shape, aggregation and solubility may also affect the addressed specific physicochemical and transport properties for example in an aqueous medium, nanoparticle aggregation of the particle. The aggregation states of nanoparticles also influence their toxicities and depend on size, surface charge, and composition. It is important to mention that high concentrations of nanoparticles favor the formation of aggregates and lead to unrealistic scenarios for toxicity assessment. For this reason, the assessment of in-situ aggregation should be an essential component of any valid ecotoxicological study (Römer et al., 2011; Bian, Mudunkotuwa, Rupasinghe, & Grassian, 2011). Other factors that affect the aggregation and may play a role in the nanoparticle toxicity are the solution pH, the presence of natural organic matter, and the ionic strength (Baalousha, Manciulea, Cumberland, Kendall, & Lead, 2008; Baalousha, 2009; Zhang, Shao, Bekaroglu, & Karanfil, 2009; Badawy et al., 2010; Bian et al., 2011). Other factors that increase the aggregate formation are the increases in cation concentration in the medium (Chen, Mylon, & Elimelech, 2007; Shipway, Lahav, Gabai, & Willner, 2000) and the concentration of dissolved organic carbon (Louie, Tilton, & Lowry, 2013), such as humic acid (Mohd Omar, Abdul Aziz, & Stoll, 2014; Baalousha et al., 2008), and fulvic acid (Domingos, Tufenkji, & Wilkinson, 2009).

Several authors revealed that the morphology of the nanoparticle also affects the toxicity, (McLaren, Valdes-Solis, Li, & Tsang, 2009; Peng et al., 2011). Bonfanti et al., (2015) demonstrated that smaller, round ZnO-NPs were more effective than bigger, rod-shaped ZnO-NPs. This result suggests the possibility of tuning the shape during the synthesis of nanoparticle to modulate the toxic effects and to produce safer NMs. It has been suggested that the toxicity of particulate is largely due to the dissolved fraction. All the above exposed is important to analyze the particle physical-chemical characteristics not only in dry form but also when suspended in cell media to ensure toxicity data are properly attributed to specific nanoparticle risks.

Effect of Nanomaterials on Development of Biological Organisms

The production and use of ENMs have increased exponentially recently with the increase industrialization and the use in technology during the last years. Therefore, new and unknown materials are introduced into the environment, generating new ecological relationships among living and non-living systems, with unpredictable scenarios for the long-term effects on aquatic environmental. Because of their use, ENMs can reach freshwater, estuarine and coastal ecosystems where they can interact with aquatic organisms eventually leading to the development of adverse effects. This is because nanoparticles have the potential to interact with biota in a fundamentally different way due to their small size and high surface reactivity (Nel et al., 2009). The aquatic ecotoxicology of ENMs, aquatic nanoecotoxicology, is a relatively new and evolving field. This necessitates a careful examination of the physicochemical characteristics of the nanomaterials and the use of models of aquatic species. As models organisms for examining the mechanisms involved including aquatic bacteria, unicellular and multicellular algae, zooplanktons, mollusks, crustaceans, amphibians, fish, etc. They have the potential of being bio-indicators of environment damage and are of great importance for investigation of the pollution aquatic ecosystems. As it was discussed above, the fate and effects of ENMs in the environment will depend on the amount of released nanoparticles and their physicochemical characteristics such as size (surface area and size distribution), surface structure (surface reactivity, surface groups, inorganic or organic coatings, etc.), shape, chemical composition (purity, crystallinity, electronic properties, etc.), solubility, and aggregation.

These physicochemical characteristics can be altered by environmental parameters (Fabrega, Luoma, Tyler, Galloway, & Lead, 2011). Although impressive from a technological viewpoint, the novel properties of ENM raise concerns about adverse effects on biological systems. Arming nanotoxicologists with this understanding, combined with knowledge of the physics, chemistry and biology of these materials is essential for maintaining the relevance of future toxicological assessments. Interaction mechanisms between nanoparticles and living systems are not yet fully understood. Due to this complexity, the particles are able to bind and interact with biological matter and change their surface characteristics, depending on the environment. For this reason, is necessary first a complete characterization of the ENMs. Then, it is required to determine the fate and behavior of ENMs in the environment using different technical of characterization and different of aquatic organism. This clearly reflects the need for interdisciplinary research involving the collaboration of physicists, chemists, biologists and materials scientists. Thus, in order to understand the interactions of nanoparticles with the aquatic organisms and their mechanism of toxic action more studies are needed (Figure 2.).

While toxicity mechanisms have not yet been completely elucidated for most NMs, possible mechanisms include damage to membrane integrity, lipid peroxidation, oxidation of proteins, DNA damage, genotoxicity, the formation of reactive oxygen species (ROS), and release of toxic soluble constituents (Figure 3.).

ROS plays an important role in eliciting the adverse effects of NPs, leading to a series of biochemical reactions that can ultimately result in cell damage and even cell death if the mechanism of defense and repair are inadequate (Clemente et al., 2013). ROS production has been found in several classes of nanomaterials (Zhu, Wang, Zhang, Chang, & Chen, 2009; Usenko, Harper, & Tanguay, 2008; F. Li et al., 2015; Nogueira, Nakabayashi, & Zucolotto, 2015; Karlsson, Cronholm, Gustafsson, & Möller, 2008; Jagadeesh, Khan, Chandran, & Khan, 2015; Aruoja et al., 2015). Formation of ROS is one of the primary mechanisms of nanoparticle toxicity; it may result in oxidative stress and consequent damage to proteins,

Figure 2. Schematic illustrating pathways of the toxicity evaluation. The process can include physical-chemical characteristic and toxicity testing

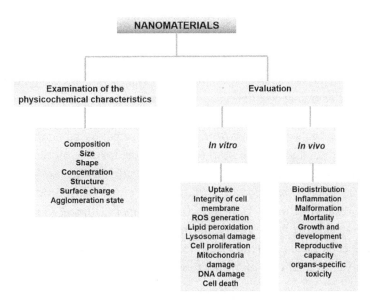

Figure 3. Possible mechanisms of nanoparticle toxicity

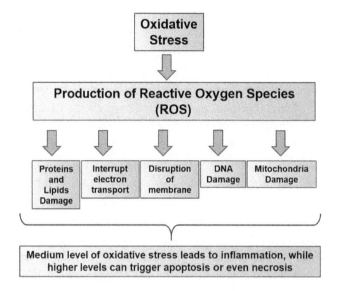

membranes, and DNA. Organisms living in environments containing NPs will incorporate them within their bodies through different routes of uptake. The routes for the uptake of NPs by aquatic organisms include direct ingestion (Duan, Yu, Li, et al., 2013), entry through gills, olfactory organs and through the skin (Bury, Walker, & Glover, 2003; L Canesi et al., 2010; Volland et al., 2015; Florence, 2005; Lovern, Owen, & Klaper, 2008; Laura Canesi et al., 2012; Gaiser et al., 2012; Griffitt et al., 2007; Bonfanti et al., 2015). The uptake of some ENMs such as metallic oxides nanoparticles was documented for algae (Aruoja, Dubourguier, Kasemets, & Kahru, 2009; Ji, Long, & Lin, 2011), amphibians (S. L. Nations, 2009; S. Nations et al., 2011; Bacchetta et al., 2015), bacteria (Heinlaan, Ivask, Blinova, Dubourguier, & Kahru, 2008; Heinlaan et al., 2008; Dasari, Pathakoti, & Hwang, 2013), crustacean (Heinlaan et al., 2008), fish (Vicario-Parés et al., 2014) and protozoa (Mortimer, Kasemets, & Kahru, 2010).

Concerns about the potential use and toxicity of metal oxide nanomaterials arise from their assumed persistence as small particles in aquatic systems (Rogers, Franklin, Apte, & Batley, 2007). Adverse effects after zinc oxide nanoparticle exposure were reported also in aquatic organisms throughout the trophic chain. Several authors have shown a significant toxicity of the metal oxide nanoparticles such as ZnO, CuO, CeO_2 and TiO_2 to aquatic organisms. However, there is little data in the literature on the chronic toxicity of metal oxide nanoparticles to aquatic species (Adam et al, 2015, Schmitt, et al., 2015; Adam, Vakurov, Knapen, & Blust, 2015; Wiench et al., 2009; Wang et al., 2015). As a result, no current risk of these nanoparticles to the aquatic environment is expected, based on the information available.

On the other hand, recent studies have been focused on the toxicity of metallic nanomaterials gold (Au), silver (Ag), copper (Cu) in aquatic environments, especially in the case Ag nanoparticles. AgNPs, have been, and continue to be, recognized worldwide as either cure or as a preventive for bacterial, fungal, and viral diseases (Suresh, Pelletier, & Doktycz, 2013; Guzman, Dille, & Godet, 2012; Rai, Yadav, & Gade, 2009). It is known that the AgNPs cause harm to both the pathogenic and the beneficial microorganisms, which cause ecological problems (Agnihotri, Mukherji, & Mukherji, 2014; Dimkpa, Calder, & Gajjar, 2011). AgNPs, in aquatic systems, were less toxic than ionic silver, but much more toxic than micron-sized silver. Toxicity was related to the release of dissolved silver, which was released more slowly from micron-sized silver due to the low surface area (Angel et al., 2013). Most of the studies have been made on zebra fish (Asharani et al., 2008; Yeo and Kang, 2008; Bar-Ilan et al., 2009; Choi et al., 2009; Griffitt et al., 2009; Powers et al., 2010). There is only one in vivo study in regard to the chronic toxicity of nanosilver in rainbow trout (Scown et al., 2010). In actuality in vivo biological models for studying the interactions of nanomaterials in biological systems have been reported (Table 1.) (Nelson et al., 2010; Usenko, Harper, & Tanguay, 2007; Hendrickson, Zherdev, Gmoshinskii, & Dzantiev, 2014).

In most studies, the researchers have been focused in the study of nanomaterials toxicity in aquatic environments on zebrafish (Table 2.). This model has recently emerged as a model for toxicological study due has exhibited great potential for studies of toxicological mechanisms (Lin, Zhao, Nel, & Lin, 2013). One new model has emerged in the year last is the use of amphibians for Frog Embryo Teratogenesis Assay-Xenopus (FETAX) protocol. This assay is a well-established tool for evaluating embryo toxicity in amphibians and it has been used also for the evaluation of nanomaterials toxicity (Bacchetta et al., 2011). Also, was employed embryonic and larval stages of *Rhinella arenarum* (amphibian) for in vivo toxicological models (Yslas et al., 2012; Ibarra et al., 2015). Although there are already some studies on potential hazard of manufactured NPs, their release into the aquatic environment and their harmful effects remain largely unknown (Moore, 2006). It was a great early challenge to adapt techniques designed for toxicants to the nascent field of nanotoxicology.

Table 1. Nanomaterial toxicity using aquatic organisms

Nanomaterials	*In Vivo* Models	Effect	Reference
Single-walled carbon nanotubes (SWCNTs)	Rare minnow (Gobiocypris rarus) (0–320 mg/L)	The SWCNTs induces apoptosis and malformation of embryos and larvae.	(Zhu, Liu, Ling, Song, & Wang, 2015)
Zinc oxide (ZnO), Titanium dioxide (TiO$_2$), Fe$_2$O$_3$, and CuO nanomaterials	Embryos Xenopus	The TiO2 or Fe$_2$O$_3$ inhibited the growth of tadpoles (1000 mg/L) but the CuO or ZnO inhibited the growth of at concentrations of 10 mg/L or greater. On the other hand, CuO and ZnO exposure also produced malformations in tadpoles.	(Nations, 2009)
Cerium dioxide nanoparticles (CeO$_2$ NPs)	The amphibian larvae Xenopus laevis and Pleurodeles waltl	This study show mortality, growth inhibition and genotoxic effects on amphibian larvae of NP1 and NP2 (0.1 to 10 mg/L) and on Pleurodeles exposed to 10 mg /L of NP2.	(Bour et al., 2015)
Single walled carbon nanotubes (SWCNTs)	Algae, crustacean, and fish	The present research show that SWCNTs could induce acute toxicity in freshwater algae (Raphidocelis subcapitata and Chlorella vulgaris), yet not in a freshwater microcrustacean (Daphnia magna) and freshwater fish (Oryzias latipes).Therefore, WCNTs would seem to have different short-term effects on different species of aquatic Organisms.	(Sohn et al., 2015)
Double-walled carbon nanotubes (DWCNTs)	Algae (Pseudokirchneriella subcapitata), Invertebrate (Daphnia pulex), and fish (Poecilia reticulata)	The toxicity of DWCNTs to the three organisms is significantly different, with D. pulex being the most sensitive followed by P. subcapitata and P. reticulata being the least sensitive.	(Lukhele, Mamba, Musee, & Wepener, 2015)
Copper oxide (CuO) nanoparticles	Xenopus laevis 0.3–2.5 mg/ L	CuONP induces mortality and affects amphibians by decreasing survivability as well as inducing physiological stresses and lower growth rates.	(Nations et al., 2015)

Table 2. Nanomaterial toxicity using zebrafish embryos

Nanomaterials	*In Vivo* Models	Effect	Reference
Multi-Walled Carbon Nanotubes Graphene Oxide, and Reduced Graphene Oxide	Zebrafish embryos (1, 5, 10, 50, 100 mg/L)	None of the 3 nanomaterials induced mortality but had some sublethal effects on the heart rate, hatching rate, and the length of larvae.	(Liu et al., 2014)
Titanium dioxide nanoparticles	Zebrafish embryos	In aqueous exposure at a low dose did not induce embryonic developmental malformations and little disturbance on neurogenesis and neuronal differentiation.	(Y.-J. Wang et al., 2014)
Chitosan nanoparticles.	zebrafish embryos (5-40 mg/L)	This nanoparticle triggers death and malformation of zebrafish embryos occurred with increasing chitosan nanoparticle concentrations. Addition embryos exposed to nanoparticle showed an increased rate of cell death, high expression of reactive oxygen species, as well as overexpression of heat shock protein 70.	(Hu, Qi, Han, Shao, & Gao, 2011)
Quantum dots	Zebrafish embryos (0, 2.5, 5, 10, 20 nM)	The QDs caused developmental embryonic toxicity. The QDs caused embryonic malformations, including head malformation, pericardial edema, yolk sac edema, bent spine, and yolk not depleted.	(Duan, Yu, Li, et al., 2013)
Zinc oxide nanoparticle	Zebrafish embryos (1 and 10 mg/l)	These aggregates caused dose-dependent toxicity, causing pericardial edema and oxidative stress may be associated with the developmental toxicity.	(Zhu et al., 2009)
Zinc oxide nanoparticle	Zebrafish embryos (0.0161-250mg/l)	This research shows the toxicity of 17 diverse ZnO NPs varying in both size and surface chemistry to developing zebrafish and illustrates how the intrinsic properties of NPs are useful to separate and identify ZnO NP toxicity to zebrafish.	(Zhou, Son, Harper, Zhou, & Harper, 2015)
Nickel and copper oxide nanoparticles	Zebrafish embryos (20-100ppm)	Both Ni and CuO NPs induces changes in the physiological serotonin concentration. This result provides evidence of the effect of NPs on the physiology of the intestine.	(Ozel, Wallace, & Andreescu, 2014)
Silver nanoparticles	Zebrafish embryos (0-0.7 nM)	The Ag NPs cause stage- and dose-dependent toxic effect on embryonic development.	(Lee, Browning, Nallathamby, Osgood, & Xu, 2013)
Silver nanoparticle	Zebrafish (5 -100 mg/L)	The nano-Ag-treated embryos exhibited severe phenotypic changes, altered physiological functions, and degradation of body parts. Also demonstrated that the effect might be more serious when the concentration is increased up to 10 mg/L or higher	(Xia et al., 2015)
Silica nanoparticles	Zebrafish (25, 50, 100, 200 mg/mL)	The embryos treated with SiNps induces mortality, hatching rate, malformation and whole-embryo cellular death. The present study demonstrates that SiNPs cause developmental embryonic toxicity resulted in persistent effects on larval behavior.	(Duan, Yu, Shi, et al., 2013)
Silicon dioxide nanoparticles	Zebrafish (0, 50, 100, 150, 200, 250, 300 mg/·L)	This study shows that the SiO2 NPs can significantly affect the development of zebrafish embryos could be resulted by oxidative damage and eventually caused death and advanced development of embryos which were not fully developed.	(Ye, Yu, Yang, Yuan, & Yang, 2013)
Graphene quantum dots	Zebrafish (12.5-200µg/mL)	The embryos exposure (200 µg/mL) result in various embryonic malformations including pericardial edema, vitelline cyst, bent spine, and bent tail. The low concentrations are non-toxic but at concentrations exceeding 50µg/mL disrupt the progression of embryonic development.	(Wang et al., 2015)
Gold nanoparticles	Zebrafish	This study clearly demonstrated that both nonfunctionalized and functionalised gold nanoparticles have no adverse effects on zebrafish development at concentrations as high as 500µM Au	(García-Cambero et al., 2012)
Cobalt ferrite nanoparticles	Zebrafish 10-500µM	$CoFe_2O_4$ NPs induce acute developmental toxicity in dose and time dependent way by arresting cell cycle, apoptosis, decreased metabolism, hatching delay, unstable heart beat, cardiac/ yolk sac edema, tail and spinal cord flexure.	(Ahmad, Liu, Zhou, & Yao, 2015)

MECHANISM OF ACTION OF NANOSILVER IN VITRO AND IN VIVO USING ZEBRAFISH

The use of AgNPs is increasing worldwide and is known for their potential adverse effects to organisms in the environment. AgNPs are used as antibacterial agents in different products like cosmetics products, food storage containers, clothes, children's toys and household appliances (Chaloupka, Malam, & Seifalian, 2010; J. S. Kim et al., 2007). The release of NPs into the aquatic environment is crucial due to they present potential ecological and environmental health risks (Maurer-Jones, Gunsolus, Murphy, & Haynes, 2013). The metal ions released from certain nanomaterials could also act as a key factor in causing cell damage and the Ag ions (Ag^+) released by AgNPs are hypothesized to be the main trigger for the adverse effects of AgNPs on organisms (Osborne et al., 2012; Newton, Puppala, Kitchens, Colvin, & Klaine, 2013). Poynton et al., (2012) propose that the toxicity triggers by AgNPs can be classified into three possible hypotheses. The first hypothesis suggests that in aqueous environments, the Ag^+ released results in toxicity. Secondly, they may cause toxicity through the metal ion independent mechanism. Thirdly, they remain as nanoparticles outside cells and release Ag^+ as they dissolve inside cells. Ag nanoparticles have been shown to be a source of ionic silver due their dissolution in a biological environment. Consequently, the dissolution and release of silver ions have been attributed as the primary mechanism of Ag nanoparticle toxicity in aquatic organisms. Lee et al., (2007) characterized transport of single AgNPs into an in vivo model system (zebrafish embryos) and investigated their effects on early embryonic development. They found that individual Ag nanoparticles can passively diffuse via chorion pore canals, create specific effects on embryonic development dose-dependent manner. Other properties can lead to toxic effects, is the size was found that single AgNPs (30-72 nm diameters) passively diffused into the embryos through in vivo chorionic pores of embryos (Lee, Browning, et al., 2012). This study confirmed the size-dependent nanotoxicity. The chorion possesses canals, the pore size of which is approximately 0.6-0.7 μm, larger than the size of the NPs.

On the other hand, K.-T. Kim and Tanguay (2014) investigated how the size-dependent toxicity of AgNPs is influenced by the presence and absence of the chorion in an embryonic zebrafish assay. However, although the chorionic pore size is known to be larger than the NP size, the toxicity may be complicated when NPs are agglomerated. This event occurs when a few Ag NPs that trapped inside chorionic pores with other Ag NPs that diffused through the same paths and they are trapped in the same pores, generating clogging the pores. Accordingly some NPs were shown to aggregate together and stay trapped inside the chorionic pores. This aggregation in the chorionic pores influences nutrient transportation, which might have adverse effects on embryonic development. This work demonstrated that embryonic toxicity in the absence of the chorion was higher than in the presence of the chorion. Therefore, the smaller 20 nm AgNPs were more toxic than the larger 110 nm AgNPs. This indicates that the chorion definitely serves as a barrier to AgNPs exposure and contact with the embryos. Embryos that were counted as malformed in the presence of the chorion showed only one malformation, for example, mainly yolk sac edema. While that the malformations observed in the absence of the chorion were jaw and eye abnormalities, pericardial and yolk sac edema, and snout and circulation defects. In AgNP-exposed zebrafish embryos have been shown to enter via the chorion pore canals and were reported to be present in the brain, heart, yolk and blood of developing (Lee et al. 2007; Asharani et al. 2008). However, the dissolution and release of Ag ions may not be the only characteristic that is of importance in Ag nanoparticle toxicity because other alternatively could be the shaped metallic nanoparticles with increased surface reactivity. George et al., (2012) investigated and compared nanosize Ag spheres, plates, and wires in a fish gill epithelial cell line (RT-W1) and in zebrafish embryos to understand the mechanism of toxicity.

In this paper, was demonstrate that plate-shaped Ag nanoparticles exhibit toxicological effects that differ from those of Ag nanospheres and Ag nanowires. Although the bioavailability and uptake of Ag are significantly lower in Ag-nanoplate-treated cells compared to cells exposed to Ag nanospheres and Ag nanowires. For this reason, they suggest that the mechanism is not mediated by a mechanism the uptake into the cells or dissolution but is direct Ag-nanoplates contact with the cell membrane, leading to the surface membrane damage results from particle-mediated surface reactivity. Therefore, to validate the cellular observations of a higher rate of Ag-nanoplate toxicity were used zebrafish embryos. In agreement with the cellular observations, the mortality rate in zebrafish embryos was the highest for Ag nanoplates and were observed also prominent sublethal effects, such as inhibition of embryo hatching and morphological defects. This research shows that should be considered as another important mechanism for Ag nanoparticle toxicity the increased surface reactivity of Ag-nanoplates that is due to the expression of crystal defects. Oxidative stress induced by metal nanoparticles is one of the important mechanisms of their toxicity. Oxidative stress is induced when the generation of ROS exceeds the cell's antioxidant capacity. AgNPs is capable of inducing oxidative stress as the mechanism of biological injury. These NPs causing oxidative stress through reactive oxygen species (ROS) formation, endoplasmatic reticulum (ER) stress response. This induction of ER stress can have several consequences including the activation of apoptotic and inflammatory pathways (Christen, Capelle, & Fent, 2013). The difference between the strong ER stress induction in vitro and the low and partial ER stress induction in zebrafish embryos may be related to the different AgNP concentrations and different exposure conditions in the in vitro and in vivo study.

Also, the exposure to AgNP affected biological processes such as induction of oxidative stress, alterations to the regulation of enzymes responsible for free radical scavenging, disrupted regulation of the cellular machinery involved in storing, detoxification and metabolism of metals and altered regulation of gene expression pathways involved in apoptosis (Choi et al., 2010; van Aerle et al., 2013). Zebrafish also shows that the AgNP can attach to cell membranes, disturbing permeability and respiration (Morones et al., 2005), and generate reactive oxygen species (ROS) (Jones, Garg, He, Pham, & Waite, 2011), which damage lipids, proteins, and DNA (Figure 4.). AgNPs also altered the gill filament morphology and global gene expression in zebrafish (Danio rerio) (Griffitt, Hyndman, Denslow, & Barber, 2009). Xin et al., (2015) suggests that AgNPs could affect the neural development of zebrafish embryos, and the toxicity may be partially attributed to the higher uptake in the head area. In other studies, have been reported the cleavage-stage embryos to the AgNPs and found that they died or developed to deformed zebrafish in a dose, size, surface charge, and chemical dependent manners.(Lee et al., 2007; Lee, Nallathamby, et al., 2012; Lee, Browning, Nallathamby, & Xu, 2013)

CONCLUSION

In conclusion, in this chapter we have focused on the important issue of nanomaterials toxicity to biological organisms, both when the nanomaterials are incorporated by purpose (e.g. when used as drug delivery systems) and are uptake from the environment as contaminants. Careful characterization studies, including size and shape, are critical prerequisites for studying the effect on environment impacts of

Figure 4. Schematic overview summarizing the toxic effect of AgNPs

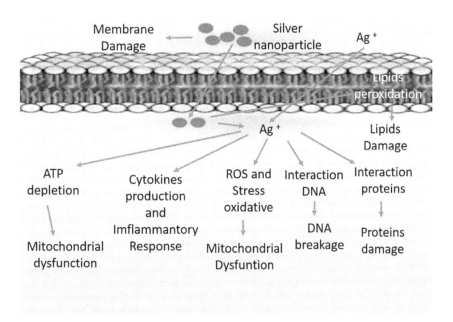

nanomaterials. A comparative review of the reported lethal and toxicological effects of different nanomaterials, using different models of aquatic organisms, is discussed. Specifically, this report was focused on the mechanism of AgNPs. Due to the AgNPs represent one of the most prevalent nanomaterials in commercial products and medical applications. This lead to the release of AgNPs and subsequent entry into the aquatic environment. For this reason is important to understand the impact of AgNPs on the early developmental stages in aquatic organisms. AgNPs have been shown to enter zebrafish embryos via the chorion pore canals and thus produce specific effects on embryonic development. The experiments showed that induce deformities mainly including spinal curvature, yolk sac edema, and malformations of the eyes and head. Also have been reported to accumulate in gill, liver and brain tissues. Therefore, have been demonstrated that exposure to AgNPs inducing respiratory toxicity, oxidative stress, and apoptosis. However, the actual mechanisms of toxicity for AgNPs are not yet clearly understood. In the next decade, it will be important to elucidate how the physicochemical properties of nanomaterials interact with aquatic organisms. This reality is reflected in the few reports that have been conducted that can be harmful to the environment and, therefore, little is known about concentrations that are harmless to the aquatic environment. The chapter concludes with a perspective on future research needs for the development of appropriate risk assessment strategies for NPs to protect the aquatic environment and more harmonized approaches for effectively advancing our understanding on the ecotoxicity of NPs.

REFERENCES

Adam, N., Schmitt, C., Galceran, J., Companys, E., Vakurov, A., & Wallace, R. (2015). *The chronic toxicity of ZnO nanoparticles and ZnCl2 to Daphnia magna and the use of different methods to assess nanoparticle aggregation and dissolution. In Nanotoxicology.* Taylor & Francis.

Adam, N., Vakurov, A., Knapen, D., & Blust, R. (2015). The chronic toxicity of CuO nanoparticles and copper salt to Daphnia magna. *Journal of Hazardous Materials, 283*, 416–422. doi:10.1016/j.jhazmat.2014.09.037 PMID:25464278

Agnihotri, S., Mukherji, S., & Mukherji, S. (2014). Size-controlled silver nanoparticles synthesized over the range 5–100 nm using the same protocol and their antibacterial efficacy. *RSC Advances, 4*(8), 3974–3983. doi:10.1039/C3RA44507K

Ahmad, F., Liu, X., Zhou, Y., & Yao, H. (2015). An in vivo evaluation of acute toxicity of cobalt ferrite (CoFe2O4) nanoparticles in larval-embryo Zebrafish (Danio rerio). *Aquatic Toxicology (Amsterdam, Netherlands), 166*, 21–28. doi:10.1016/j.aquatox.2015.07.003 PMID:26197244

Albanese, A., Tang, P. S., & Chan, W. C. W. (2012). The Effect of Nanoparticle Size, Shape, and Surface Chemistry on Biological Systems. *Annual Review of Biomedical Engineering, 14*(1), 1–16. doi:10.1146/annurev-bioeng-071811-150124 PMID:22524388

Angel, B. M., Batley, G. E., Jarolimek, C. V., & Rogers, N. J. (2013). The impact of size on the fate and toxicity of nanoparticulate silver in aquatic systems. *Chemosphere, 93*(2), 359–365. doi:10.1016/j.chemosphere.2013.04.096 PMID:23732009

Aruoja, V., Dubourguier, H.-C., Kasemets, K., & Kahru, A. (2009). Toxicity of nanoparticles of CuO, ZnO and TiO2 to microalgae Pseudokirchneriella subcapitata. *The Science of the Total Environment, 407*(4), 1461–1468. doi:10.1016/j.scitotenv.2008.10.053 PMID:19038417

Aruoja, V., Pokhrel, S., Sihtmäe, M., Mortimer, M., Mädler, L., & Kahru, A. (2015). *Toxicity of 12 metal-based nanoparticles to algae, bacteria and protozoa. Environ. Sci.: Nano.* The Royal Society of Chemistry.

Asharani, P. V., Lian Wu, Y., Gong, Z., & Valiyaveettil, S. (2008). Toxicity of silver nanoparticles in zebrafish models. *Nanotechnology, 19*(25), 255102. doi:10.1088/0957-4484/19/25/255102 PMID:21828644

Baalousha, M. (2009). Aggregation and disaggregation of iron oxide nanoparticles: Influence of particle concentration, pH and natural organic matter. *The Science of the Total Environment, 407*(6), 2093–2101. doi:10.1016/j.scitotenv.2008.11.022 PMID:19059631

Baalousha, M., Manciulea, A., Cumberland, S., Kendall, K., & Lead, J. R. (2008). Aggregation and surface properties of iron oxide nanoparticles: influence of pH and natural organic matter. *Environmental Toxicology and Chemistry / SETAC, 27*(9), 1875–1882.

Bacchetta, R., Santo, N., Fascio, U., Moschini, E., Freddi, S., & Chirico, G. (2011). *Nano-sized CuO, TiO2 and ZnO affect Xenopus laevis development. In Nanotoxicology.* Taylor & Francis.

Bacchetta, R., Santo, N., Fascio, U., Moschini, E., Freddi, S., & Chirico, G. (2015). *Nano-sized CuO, TiO2 and ZnO affect Xenopus laevis development. In Nanotoxicology.* Taylor & Francis.

Bian, S.-W., Mudunkotuwa, I. A., Rupasinghe, T., & Grassian, V. H. (2011). Aggregation and dissolution of 4 nm ZnO nanoparticles in aqueous environments: Influence of pH, ionic strength, size, and adsorption of humic acid. *Langmuir, 27*(10), 6059–6068. doi:10.1021/la200570n PMID:21500814

Bonfanti, P., Moschini, E., Saibene, M., Bacchetta, R., Rettighieri, L., Calabri, L., & Mantecca, P. et al. (2015). Do Nanoparticle Physico-Chemical Properties and Developmental Exposure Window Influence Nano ZnO Embryotoxicity in Xenopus laevis? *International Journal of Environmental Research and Public Health, 12*(8), 8828–8848. doi:10.3390/ijerph120808828 PMID:26225989

Bour, A., Mouchet, F., Verneuil, L., Evariste, L., Silvestre, J., Pinelli, E., & Gauthier, L. (2015). Toxicity of CeO2 nanoparticles at different trophic levels--effects on diatoms, chironomids and amphibians. *Chemosphere, 120*, 230–236. doi:10.1016/j.chemosphere.2014.07.012 PMID:25086917

Bozich, J. S., Lohse, S. E., Torelli, M. D., Murphy, C. J., Hamers, R. J., & Klaper, R. D. (2014). Surface chemistry, charge and ligand type impact the toxicity of gold nanoparticles to Daphnia magna. *Environ. Sci.: Nano, 1*(3), 260–270.

Bury, N. R., Walker, P. A., & Glover, C. N. (2003). Nutritive metal uptake in teleost fish. *The Journal of Experimental Biology, 206*(Pt 1), 11–23. doi:10.1242/jeb.00068 PMID:12456693

Canesi, L., Ciacci, C., Fabbri, R., Marcomini, A., Pojana, G., & Gallo, G. (2012). Bivalve molluscs as a unique target group for nanoparticle toxicity. *Marine Environmental Research, 76*, 16–21. doi:10.1016/j.marenvres.2011.06.005 PMID:21767873

Canesi, L., Fabbri, R., Gallo, G., Vallotto, D., Marcomini, A., & Pojana, G. (2010). Biomarkers in Mytilus galloprovincialis exposed to suspensions of selected nanoparticles (Nano carbon black, C60 fullerene, Nano-TiO2, Nano-SiO2). *Aquatic Toxicology (Amsterdam, Netherlands), 100*(2), 168–177. doi:10.1016/j.aquatox.2010.04.009 PMID:20444507

Chaloupka, K., Malam, Y., & Seifalian, A. M. (2010). Nanosilver as a new generation of nanoproduct in biomedical applications. *Trends in Biotechnology, 28*(11), 580–588. doi:10.1016/j.tibtech.2010.07.006 PMID:20724010

Chapman, P. M. (2006). Emerging Substances—Emerging Problems? *Environmental Toxicology and Chemistry, 25*(6), 1445. doi:10.1897/06-025.1 PMID:16764460

Chen, K. L., Mylon, S. E., & Elimelech, M. (2007). Enhanced aggregation of alginate-coated iron oxide (Hematite) nanoparticles in the presence of calcium, strontium, and barium cations. *Langmuir, 23*(11), 5920–5928. doi:10.1021/la063744k PMID:17469860

Choi, J. E., Kim, S., Ahn, J. H., Youn, P., Kang, J. S., Park, K., & Ryu, D.-Y. et al. (2010). Induction of oxidative stress and apoptosis by silver nanoparticles in the liver of adult zebrafish. *Aquatic Toxicology (Amsterdam, Netherlands), 100*(2), 151–159. doi:10.1016/j.aquatox.2009.12.012 PMID:20060176

Christen, V., Capelle, M., & Fent, K. (2013). Silver nanoparticles induce endoplasmatic reticulum stress response in zebrafish. *Toxicology and Applied Pharmacology, 272*(2), 519–528. doi:10.1016/j.taap.2013.06.011 PMID:23800688

Clemente, Z., Castro, V. L., Feitosa, L. O., Lima, R., Jonsson, C. M., Maia, A. H. N., & Fraceto, L. F. (2013). Fish exposure to nano-TiO2 under different experimental conditions: Methodological aspects for nanoecotoxicology investigations. *The Science of the Total Environment, 463-464*, 647–656. doi:10.1016/j.scitotenv.2013.06.022 PMID:23845857

Couling, D. J., Bernot, R. J., Docherty, K. M., Dixon, J. K., & Maginn, E. J. (2006). Assessing the factors responsible for ionic liquid toxicity to aquatic organisms via quantitative structure–property relationship modeling. *Green Chemistry*, *8*(1), 82–90. doi:10.1039/B511333D

Dasari, T. P., Pathakoti, K., & Hwang, H.-M. (2013). Determination of the mechanism of photoinduced toxicity of selected metal oxide nanoparticles (ZnO, CuO, Co3O4 and TiO2) to E. coli bacteria. *Journal of Environmental Sciences (China)*, *25*(5), 882–888. doi:10.1016/S1001-0742(12)60152-1 PMID:24218817

Dimkpa, C., Calder, A., & Gajjar, P. (2011). Interaction of silver nanoparticles with an environmentally beneficial bacterium, Pseudomonas chlororaphis. *Journal of Hazardous*.

Domingos, R. F., Tufenkji, N., & Wilkinson, K. J. (2009). Aggregation of titanium dioxide nanoparticles: Role of a fulvic acid. *Environmental Science & Technology*, *43*(5), 1282–1286. doi:10.1021/es8023594 PMID:19350891

Dominguez, G. A., Lohse, S. E., Torelli, M. D., Murphy, C. J., Hamers, R. J., Orr, G., & Klaper, R. D. (2015). Effects of charge and surface ligand properties of nanoparticles on oxidative stress and gene expression within the gut of Daphnia magna. *Aquatic Toxicology (Amsterdam, Netherlands)*, *162*, 1–9. doi:10.1016/j.aquatox.2015.02.015 PMID:25734859

Dowling, A., Clift, R., Grobert, N., Hutton, D., Oliver, R., & O'neill, O. et al.. (2004). Nanoscience and nanotechnologies : Opportunities and uncertainties. *London The Royal Society The Royal Academy of Engineering Report*, *46*(July), 618–618.

Duan, J., Yu, Y., Li, Y., Yu, Y., Li, Y., Huang, P., & Sun, Z. et al. (2013). Developmental toxicity of CdTe QDs in zebrafish embryos and larvae. *Journal of Nanoparticle Research*, *15*(7), 1700. doi:10.1007/s11051-013-1700-8

Duan, J., Yu, Y., Shi, H., Tian, L., Guo, C., Huang, P., & Sun, Z. et al. (2013). Toxic effects of silica nanoparticles on zebrafish embryos and larvae. *PLoS ONE*, *8*(9), e74606. doi:10.1371/journal.pone.0074606 PMID:24058598

El Badawy, A. M., Luxton, T. P., Silva, R. G., Scheckel, K. G., Suidan, M. T., & Tolaymat, T. M. (2010). Impact of environmental conditions (pH, ionic strength, and electrolyte type) on the surface charge and aggregation of silver nanoparticles suspensions. *Environmental Science & Technology*, *44*(4), 1260–1266. doi:10.1021/es902240k PMID:20099802

Fabrega, J., Luoma, S. N., Tyler, C. R., Galloway, T. S., & Lead, J. R. (2011). Silver nanoparticles: Behaviour and effects in the aquatic environment. *Environment International*, *37*(2), 517–531. doi:10.1016/j.envint.2010.10.012 PMID:21159383

Fischer, H. C., & Chan, W. C. W. (2007). Nanotoxicity: The growing need for in vivo study. *Current Opinion in Biotechnology*, *18*(6), 565–571. doi:10.1016/j.copbio.2007.11.008 PMID:18160274

Florence, A. T. (2005). Nanoparticle uptake by the oral route: Fulfilling its potential? *Drug Discovery Today. Technologies*, *2*(1), 75–81. doi:10.1016/j.ddtec.2005.05.019 PMID:24981758

Fujiwara, K., Suematsu, H., Kiyomiya, E., Aoki, M., Sato, M., & Moritoki, N. (2008). Size-dependent toxicity of silica nano-particles to Chlorella kessleri. *Journal of Environmental Science and Health. Part A, Toxic/Hazardous Substances & Environmental Engineering, 43*(10), 1167–1173. doi:10.1080/10934520802171675 PMID:18584432

Gaiser, B. K., Fernandes, T. F., Jepson, M. A., Lead, J. R., Tyler, C. R., Baalousha, M., Biswas, A., et al. (2012). Interspecies comparisons on the uptake and toxicity of silver and cerium dioxide nanoparticles. *Environmental Toxicology and Chemistry / SETAC, 31*(1), 144–54.

García-Cambero, J. P., Herranz, A. L., Díaz, G. L., Cuadal, J. S., Castelltort, M. E. R., & Calvo, A. C. (2012). *Efectos letales y subletales de nanopartículas y material soluble de oro en el desarrollo de embriones de pez cebra*. Revista de Toxicología.

George, S., Lin, S., Ji, Z., Thomas, C. R., Li, L., Mecklenburg, M., & Nel, A. E. et al. (2012). Surface defects on plate-shaped silver nanoparticles contribute to its hazard potential in a fish gill cell line and zebrafish embryos. *ACS Nano, 6*(5), 3745–3759. doi:10.1021/nn204671v PMID:22482460

Gottschalk, F., & Nowack, B. (2011). The release of engineered nanomaterials to the environment. *Journal of environmental monitoring. JEM, 13*(5), 1145–1155. PMID:21387066

Griffitt, R. J., Hyndman, K., Denslow, N. D., & Barber, D. S. (2009). Comparison of molecular and histological changes in zebrafish gills exposed to metallic nanoparticles. *Toxicological Sciences : An Official Journal of the Society of Toxicology, 107*(2), 404–15.

Griffitt, R. J., Luo, J., Gao, J., Bonzongo, J.-C., & Barber, D. S. (2008). Effects of particle composition and species on toxicity of metallic nanomaterials in aquatic organisms. *Environmental Toxicology and Chemistry / SETAC, 27*(9), 1972–8.

Griffitt, R. J., Weil, R., Hyndman, K. A., Denslow, N. D., Powers, K., Taylor, D., & Barber, D. S. (2007). Exposure to Copper Nanoparticles Causes Gill Injury and Acute Lethality in Zebrafish (Danio rerio). *Environmental Science & Technology, 41*(23), 8178–8186. doi:10.1021/es071235e PMID:18186356

Guzman, M., Dille, J., & Godet, S. (2012). Synthesis and antibacterial activity of silver nanoparticles against gram-positive and gram-negative bacteria. *Nanomedicine; Nanotechnology, Biology, and Medicine, 8*(1), 37–45. doi:10.1016/j.nano.2011.05.007 PMID:21703988

Heinlaan, M., Ivask, A., Blinova, I., Dubourguier, H.-C., & Kahru, A. (2008). Toxicity of nanosized and bulk ZnO, CuO and TiO2 to bacteria Vibrio fischeri and crustaceans Daphnia magna and Thamnocephalus platyurus. *Chemosphere, 71*(7), 1308–1316. doi:10.1016/j.chemosphere.2007.11.047 PMID:18194809

Hendrickson, O. D., Zherdev, A. V., Gmoshinskii, I. V., & Dzantiev, B. B. (2014). Fullerenes: In vivo studies of biodistribution, toxicity, and biological action. *Nanotechnologies in Russia, 9*(11-12), 601–617. doi:10.1134/S199507801406010X

Henry, T. B., Menn, F.-M., Fleming, J. T., Wilgus, J., Compton, R. N., & Sayler, G. S. (2007). Attributing effects of aqueous C60 nano-aggregates to tetrahydrofuran decomposition products in larval zebrafish by assessment of gene expression. *Environmental Health Perspectives, 115*(7), 1059–1065. doi:10.1289/ehp.9757 PMID:17637923

Hu, Y. L., Qi, W., Han, F., Shao, J. Z., & Gao, J. Q. (2011). Toxicity evaluation of biodegradable chitosan nanoparticles using a zebrafish embryo model. *International Journal of Nanomedicine, 6*, 3351–3359. PMID:22267920

Ibarra, L. E., Tarres, L., Bongiovanni, S., Barbero, C. A., Kogan, M. J., Rivarola, V. A., & Yslas, E. I. et al. (2015). Assessment of polyaniline nanoparticles toxicity and teratogenicity in aquatic environment using Rhinella arenarum model. *Ecotoxicology and Environmental Safety, 114*, 84–92. doi:10.1016/j.ecoenv.2015.01.013 PMID:25617831

Ivask, A., Kurvet, I., Kasemets, K., Blinova, I., Aruoja, V., Suppi, S., & Kahru, A. et al. (2014). Size-dependent toxicity of silver nanoparticles to bacteria, yeast, algae, crustaceans and mammalian cells in vitro.[Public Library of Science.]. *PLoS ONE, 9*(7), e102108. doi:10.1371/journal.pone.0102108 PMID:25048192

Jagadeesh, E., Khan, B., Chandran, P., & Khan, S. S. (2015). Toxic potential of iron oxide, CdS/Ag_2S composite, CdS and Ag_2S NPs on a fresh water alga Mougeotia sp. *Colloids and Surfaces. B, Biointerfaces, 125*, 284–290. doi:10.1016/j.colsurfb.2014.11.008 PMID:25465759

Ji, J., Long, Z., & Lin, D. (2011). Toxicity of oxide nanoparticles to the green algae Chlorella sp. *Chemical Engineering Journal, 170*(2-3), 525–530. doi:10.1016/j.cej.2010.11.026

Jiang, J., Oberdörster, G., & Biswas, P. (2008). Characterization of size, surface charge, and agglomeration state of nanoparticle dispersions for toxicological studies. *Journal of Nanoparticle Research, 11*(1), 77–89. doi:10.1007/s11051-008-9446-4

Jones, A. M., Garg, S., He, D., Pham, N., & Waite, T. D. (2011). Superoxide-mediated formation and charging of silver nanoparticles. *Environmental Science & Technology, 45*(4), 1428–1434. doi:10.1021/es103757c PMID:21265570

Karlsson, H. L., Cronholm, P., Gustafsson, J., & Möller, L. (2008). Copper oxide nanoparticles are highly toxic: A comparison between metal oxide nanoparticles and carbon nanotubes. *Chemical Research in Toxicology, 21*(9), 1726–1732. doi:10.1021/tx800064j PMID:18710264

Kim, J. S., Kuk, E., Yu, K. N., Kim, J.-H., Park, S. J., Lee, H. J., & Cho, M.-H. et al. (2007). Antimicrobial effects of silver nanoparticles. *Nanomedicine; Nanotechnology, Biology, and Medicine, 3*(1), 95–101. doi:10.1016/j.nano.2006.12.001 PMID:17379174

Kim, K.-T., & Tanguay, R. L. (2014). The role of chorion on toxicity of silver nanoparticles in the embryonic zebrafish assay. *Environmental Health and Toxicology, 29*, e2014021.

Klaine, S. J., Alvarez, P. J. J., Batley, G. E., Fernandes, T. F., Handy, R. D., Lyon, D. Y., & Lead, J. R. et al. (2008). Nanomaterials in the environment: Behavior, fate, bioavailability, and effects. *Environmental Toxicology and Chemistry, 27*(9), 1825. doi:10.1897/08-090.1 PMID:19086204

Lee, K. J., Browning, L. M., Nallathamby, P. D., Desai, T., Cherukuri, P. K., & Xu, X.-H. N. (2012). In vivo quantitative study of sized-dependent transport and toxicity of single silver nanoparticles using zebrafish embryos. *Chemical Research in Toxicology, 25*(5), 1029–1046. doi:10.1021/tx300021u PMID:22486336

Lee, K. J., Browning, L. M., Nallathamby, P. D., Osgood, C. J., & Xu, X.-H. N. (2013). Silver nanoparticles induce developmental stage-specific embryonic phenotypes in zebrafish. *Nanoscale*, *5*(23), 11625–11636. doi:10.1039/c3nr03210h PMID:24056877

Lee, K. J., Browning, L. M., Nallathamby, P. D., & Xu, X.-H. N. (2013). Study of charge-dependent transport and toxicity of peptide-functionalized silver nanoparticles using zebrafish embryos and single nanoparticle plasmonic spectroscopy. *Chemical Research in Toxicology*, *26*(6), 904–917. doi:10.1021/tx400087d PMID:23621491

Lee, K. J., Nallathamby, P. D., Browning, L. M., Desai, T., Cherukuri, P. K., & Xu, X.-H. N. (2012). Single nanoparticle spectroscopy for real-time in vivo quantitative analysis of transport and toxicity of single nanoparticles in single embryos. *Analyst (London)*, *137*(13), 2973–2986. doi:10.1039/c2an35293a PMID:22563577

Lee, K. J., Nallathamby, P. D., Browning, L. M., Osgood, C. J., & Xu, X.-H. N. (2007). In vivo imaging of transport and biocompatibility of single silver nanoparticles in early development of zebrafish embryos. *ACS Nano*, *1*(2), 133–143. doi:10.1021/nn700048y PMID:19122772

Li, F., Liang, Z., Zheng, X., Zhao, W., Wu, M., & Wang, Z. (2015). Toxicity of nano-TiO_2 on algae and the site of reactive oxygen species production. *Aquatic Toxicology (Amsterdam, Netherlands)*, *158*, 1–13. doi:10.1016/j.aquatox.2014.10.014 PMID:25461740

Li, H., Zhang, J., Wang, T., Luo, W., Zhou, Q., & Jiang, G. (2008). Elemental selenium particles at nano-size (Nano-Se) are more toxic to Medaka (Oryzias latipes) as a consequence of hyper-accumulation of selenium: A comparison with sodium selenite. *Aquatic Toxicology (Amsterdam, Netherlands)*, *89*(4), 251–256. doi:10.1016/j.aquatox.2008.07.008 PMID:18768225

Lin, S., Zhao, Y., Nel, A. E., & Lin, S. (2013). Zebrafish: an in vivo model for nano EHS studies. Small (Weinheim an der Bergstrasse, Germany), 9(9-10), 1608–18.

Liu, X. T., Mu, X. Y., Wu, X. L., Meng, L. X., Guan, W. B., & Ma, Y. Q. et al.. (2014). Toxicity of multi-walled carbon nanotubes, graphene oxide, and reduced graphene oxide to zebrafish embryos. *Biomedical and environmental sciences. BES*, *27*(9), 676–683. PMID:25256857

Louie, S. M., Tilton, R. D., & Lowry, G. V. (2013). Effects of molecular weight distribution and chemical properties of natural organic matter on gold nanoparticle aggregation. *Environmental Science & Technology*. Retrieved from http://www.ncbi.nlm.nih.gov/pubmed/23550560

Lovern, S. B., Owen, H. A., & Klaper, R. (2008). Electron microscopy of gold nanoparticle intake in the gut of Daphnia magna. *Nanotoxicology*, *2*(1), 43–48. doi:10.1080/17435390801935960

Lukhele, L. P., Mamba, B. B., Musee, N., & Wepener, V. (2015). *Acute Toxicity of Double-Walled Carbon Nanotubes to Three Aquatic Organisms*. Academic Press.

Maurer-Jones, M., Gunsolus, I. L., Murphy, C. J., & Haynes, C. L. (2013). Toxicity of engineered nanoparticles in the environment. *Analytical Chemistry*, *85*(6), 3036–3049. doi:10.1021/ac303636s PMID:23427995

McLaren, A., Valdes-Solis, T., Li, G., & Tsang, S. C. (2009). Shape and size effects of ZnO nanocrystals on photocatalytic activity. *Journal of the American Chemical Society, 131*(35), 12540–12541. doi:10.1021/ja9052703 PMID:19685892

Mohd Omar, F., Abdul Aziz, H., & Stoll, S. (2014). Aggregation and disaggregation of ZnO nanoparticles: Influence of pH and adsorption of Suwannee River humic acid. *The Science of the Total Environment, 468-469*, 195–201. PMID:24029691

Moore, M. N. (2006). Do nanoparticles present ecotoxicological risks for the health of the aquatic environment? *Environment International, 32*(8), 967–976. doi:10.1016/j.envint.2006.06.014 PMID:16859745

Morones, J. R., Elechiguerra, J. L., Camacho, A., Holt, K., Kouri, J. B., Ramírez, J. T., & Yacaman, M. J. (2005). The bactericidal effect of silver nanoparticles. *Nanotechnology, 16*(10), 2346–2353. doi:10.1088/0957-4484/16/10/059 PMID:20818017

Mortimer, M., Kasemets, K., & Kahru, A. (2010). Toxicity of ZnO and CuO nanoparticles to ciliated protozoa Tetrahymena thermophila. *Toxicology, 269*(2-3), 182–189. doi:10.1016/j.tox.2009.07.007 PMID:19622384

Nations, S., Long, M., Wages, M., Maul, J. D., Theodorakis, C. W., & Cobb, G. P. (2015). Subchronic and chronic developmental effects of copper oxide (CuO) nanoparticles on Xenopus laevis. *Chemosphere, 135*, 166–174. doi:10.1016/j.chemosphere.2015.03.078 PMID:25950410

Nations, S., Wages, M., Cañas, J. E., Maul, J., Theodorakis, C., & Cobb, G. P. (2011). Acute effects of Fe2O3, TiO2, ZnO and CuO nanomaterials on Xenopus laevis. *Chemosphere*.

Nations, S. L. (2009, May 1). *Acute and developmental toxicity of metal oxide nanoparticles (ZnO, TiO2, Fe2O3, and CuO) in Xenopus laevis*. Retrieved August 22, 2015, from https://repositories.tdl.org/ttu-ir/handle/2346/16511

Nel, A., Xia, T., Mädler, L., & Li, N. (2006). Toxic potential of materials at the nanolevel. *Science, 311*(5761), 622–627. doi:10.1126/science.1114397 PMID:16456071

Nel, A. E., Mädler, L., Velegol, D., Xia, T., Hoek, E. M. V., Somasundaran, P., & Thompson, M. et al. (2009). Understanding biophysicochemical interactions at the nano-bio interface. *Nature Materials, 8*(7), 543–557. doi:10.1038/nmat2442 PMID:19525947

Nelson, S. M., Mahmoud, T., Beaux, M. II, Shapiro, P., McIlroy, D. N., & Stenkamp, D. L. (2010). Toxic and teratogenic silica nanowires in developing vertebrate embryos. *Nanomedicine; Nanotechnology, Biology, and Medicine, 6*(1), 93–102. doi:10.1016/j.nano.2009.05.003 PMID:19447201

Newton, K. M., Puppala, H. L., Kitchens, C. L., Colvin, V. L., & Klaine, S. J. (2013). Silver nanoparticle toxicity to Daphnia magna is a function of dissolved silver concentration. *Environmental Toxicology and Chemistry / SETAC, 32*(10), 2356–64.

Nogueira, P. F. M., Nakabayashi, D., & Zucolotto, V. (2015). The effects of graphene oxide on green algae Raphidocelis subcapitata. *Aquatic Toxicology (Amsterdam, Netherlands), 166*, 29–35. doi:10.1016/j.aquatox.2015.07.001 PMID:26204245

Oberdörster, G., Oberdörster, E., & Oberdörster, J. (2005). Nanotoxicology: An emerging discipline evolving from studies of ultrafine particles. *Environmental Health Perspectives, 113*(7), 823–839. doi:10.1289/ehp.7339 PMID:16002369

Ogden, L. E. (2013). Nanoparticles in the Environment: Tiny Size, Large Consequences?. *Bioscience, 63*(3), 236–236. doi:10.1525/bio.2013.63.3.17

Ong, K. J., MacCormack, T. J., Clark, R. J., Ede, J. D., Ortega, V. A., Felix, L. C., & Goss, G. G. et al. (2014). Widespread nanoparticle-assay interference: Implications for nanotoxicity testing. *PLoS ONE, 9*(3), e90650. doi:10.1371/journal.pone.0090650 PMID:24618833

Osborne, O. J., Johnston, B. D., Moger, J., Balousha, M., Lead, J. R., Kudoh, T., & Tyler, C. R. (2012). Effects of particle size and coating on nanoscale Ag and TiO2 exposure in zebrafish (Danio rerio) embryos. In *Nanotoxicology*. Taylor & Francis. Retrieved January 22, 2016, from http://www.tandfonline.com/doi/abs/10.3109/17435390.2012.737484?journalCode=inan20

Ozel, R. E., Wallace, K. N., & Andreescu, S. (2014). Alterations of intestinal serotonin following nanoparticle exposure in embryonic zebrafish. *Environmental Science. Nano, 2014*(1), 27–36.

Peng, X., Palma, S., Fisher, N. S., & Wong, S. S. (2011). Effect of morphology of ZnO nanostructures on their toxicity to marine algae. *Aquatic Toxicology (Amsterdam, Netherlands), 102*(3-4), 186–196. doi:10.1016/j.aquatox.2011.01.014 PMID:21356181

Podila, R., & Brown, J. M. (2013). Toxicity of engineered nanomaterials: A physicochemical perspective. *Journal of Biochemical and Molecular Toxicology, 27*(1), 50–55. doi:10.1002/jbt.21442 PMID:23129019

Polonini, H. C., Brandão, H. M., Raposo, N. R. B., Brandão, M. A. F., Mouton, L., Couté, A., & Brayner, R. et al. (2015). Size-dependent ecotoxicity of barium titanate particles: The case of Chlorella vulgaris green algae. *Ecotoxicology (London, England), 24*(4), 938–948. doi:10.1007/s10646-015-1436-6 PMID:25763523

Poynton, H. C., Lazorchak, J. M., Impellitteri, C. A., Blalock, B. J., Rogers, K., Allen, H. J., & Govindasmawy, S. et al. (2012). Toxicogenomic responses of nanotoxicity in Daphnia magna exposed to silver nitrate and coated silver nanoparticles. *Environmental Science & Technology, 46*(11), 6288–6296. doi:10.1021/es3001618 PMID:22545559

Rai, M., Yadav, A., & Gade, A. (2009). Silver nanoparticles as a new generation of antimicrobials. *Biotechnology Advances, 27*(1), 76–83. doi:10.1016/j.biotechadv.2008.09.002 PMID:18854209

Rogers, N. J., Franklin, N. M., Apte, S. C., & Batley, G. E. (2007). The importance of physical and chemical characterization in nanoparticle toxicity studies. *Integrated Environmental Assessment and Management, 3*(2), 303–304. doi:10.1002/ieam.5630030219 PMID:17477301

Römer, I., White, T. A., Baalousha, M., Chipman, K., Viant, M. R., & Lead, J. R. (2011). Aggregation and dispersion of silver nanoparticles in exposure media for aquatic toxicity tests. *Journal of Chromatography. A, 1218*(27), 4226–4233. doi:10.1016/j.chroma.2011.03.034 PMID:21529813

Shipway, A. N., Lahav, M., Gabai, R., & Willner, I. (2000). Investigations into the electrostatically induced aggregation of Au nanoparticles. *Langmuir, 16*(23), 8789–8795. doi:10.1021/la000316k

Sohn, E. K., Chung, Y. S., Johari, S. A., Kim, T. G., Kim, J. K., Lee, J. H., Lee, Y. H., et al. (2015). *Acute Toxicity Comparison of Single-Walled Carbon Nanotubes in Various Freshwater Organisms.* Academic Press.

Suresh, A. K., Pelletier, D. A., & Doktycz, M. J. (2013). Relating nanomaterial properties and microbial toxicity. *Nanoscale, 5*(2), 463–474. doi:10.1039/C2NR32447D PMID:23203029

Tejamaya, M., Römer, I., Merrifield, R. C., & Lead, J. R. (2012). Stability of citrate, PVP, and PEG coated silver nanoparticles in ecotoxicology media. *Environmental Science & Technology, 46*(13), 7011–7017. doi:10.1021/es2038596 PMID:22432856

Usenko, C. Y., Harper, S. L., & Tanguay, R. L. (2007). In vivo evaluation of carbon fullerene toxicity using embryonic zebrafish. *Carbon, 45*(9), 1891–1898. doi:10.1016/j.carbon.2007.04.021 PMID:18670586

Usenko, C. Y., Harper, S. L., & Tanguay, R. L. (2008). Fullerene C60 exposure elicits an oxidative stress response in embryonic zebrafish. *Toxicology and Applied Pharmacology, 229*(1), 44–55. doi:10.1016/j.taap.2007.12.030 PMID:18299140

Van Aerle, R., Lange, A., Moorhouse, A., Paszkiewicz, K., Ball, K., Johnston, B. D., & Santos, E. M. et al. (2013). Molecular mechanisms of toxicity of silver nanoparticles in zebrafish embryos. *Environmental Science & Technology, 47*(14), 8005–8014. doi:10.1021/es401758d PMID:23758687

Vicario-Parés, U., Castañaga, L., Lacave, J. M., Oron, M., Reip, P., Berhanu, D., & Orbea, A. et al. (2014). Comparative toxicity of metal oxide nanoparticles (CuO, ZnO and TiO2) to developing zebrafish embryos. *Journal of Nanoparticle Research, 16*(8), 2550. doi:10.1007/s11051-014-2550-8

Volland, M., Hampel, M., Martos-Sitcha, J. A., Trombini, C., Martínez-Rodríguez, G., & Blasco, J. (2015). Citrate gold nanoparticle exposure in the marine bivalve Ruditapes philippinarum: Uptake, elimination and oxidative stress response. *Environmental Science and Pollution Research International, 22*(22), 17414–17424. doi:10.1007/s11356-015-4718-x PMID:25994271

Walters, C. R., Pool, E. J., & Somerset, V. S. (2014). Ecotoxicity of silver nanomaterials in the aquatic environment: A review of literature and gaps in nano-toxicological research. *Journal of Environmental Science and Health. Part A, Toxic/Hazardous Substances & Environmental Engineering, 49*(13), 1588–1601. doi:10.1080/10934529.2014.938536 PMID:25137546

Wang, H., Fan, W., Xue, F., Wang, X., Li, X., & Guo, L. (2015). Chronic effects of six micro/nano-Cu_2O crystals with different structures and shapes on Daphnia magna. *Environmental Pollution (Barking, Essex : 1987), 203*, 60–8. R

Wang, Y.-J., He, Z.-Z., Fang, Y.-W., Xu, Y., Chen, Y.-N., Wang, G.-Q., Yang, Y.-Q., et al. (2014). Effect of titanium dioxide nanoparticles on zebrafish embryos and developing retina. *International Journal of Ophthalmology, 7*(6), 917–23.

Wang, Z. G., Zhou, R., Jiang, D., Song, J. E., Xu, Q., & Si, J. (2015). Toxicity of Graphene Quantum Dots in Zebrafish Embryo. *Biomedical and Environmental Sciences, 28*(5), 341–351. PMID:26055561

Wiench, K., Wohlleben, W., Hisgen, V., Radke, K., Salinas, E., Zok, S., & Landsiedel, R. (2009). Acute and chronic effects of nano- and non-nano-scale TiO(2) and ZnO particles on mobility and reproduction of the freshwater invertebrate Daphnia magna. *Chemosphere, 76*(10), 1356–1365. doi:10.1016/j.chemosphere.2009.06.025 PMID:19580988

Xia, G., Liu, T., Wang, Z., Hou, Y., Dong, L., Zhu, J., & Qi, J. (2015). The effect of silver nanoparticles on zebrafish embryonic development and toxicology. *Artificial Cells, Nanomedicine, and Biotechnology*, 1–6.

Xin, Q., Rotchell, J. M., Cheng, J., Yi, J., & Zhang, Q. (2015). Silver nanoparticles affect the neural development of zebrafish embryos. *Journal of applied Toxicology*. Retrieved August 17, 2015, from http://www.ncbi.nlm.nih.gov/pubmed/25976698

Yang, H., Liu, C., Yang, D., Zhang, H., & Xi, Z. (2009). Comparative study of cytotoxicity, oxidative stress and genotoxicity induced by four typical nanomaterials: The role of particle size, shape and composition. *Journal of applied toxicology. JAT, 29*(1), 69–78. PMID:18756589

Ye, R., Yu, X., Yang, S., Yuan, J., & Yang, X. (2013). Effects of Silica Dioxide Nanoparticles on the Embryonic Development of Zebrafish. *Integrated Ferroelectrics, 147*(1), 166–174. doi:10.1080/10584587.2013.792625

Yslas, E. I., Ibarra, L. E., Peralta, D. O., Barbero, C. A., Rivarola, V. A., & Bertuzzi, M. L. (2012). Polyaniline nanofibers: Acute toxicity and teratogenic effect on Rhinella arenarum embryos. *Chemosphere, 87*(11), 1374–1380. doi:10.1016/j.chemosphere.2012.02.033 PMID:22386461

Zhang, S., Shao, T., Bekaroglu, S. S. K., & Karanfil, T. (2009). The impacts of aggregation and surface chemistry of carbon nanotubes on the adsorption of synthetic organic compounds. *Environmental Science & Technology, 43*(15), 5719–5725. doi:10.1021/es900453e PMID:19731668

Zhou, Z., Son, J., Harper, B., Zhou, Z., & Harper, S. (2015). Influence of surface chemical properties on the toxicity of engineered zinc oxide nanoparticles to embryonic zebrafish. *Beilstein Journal of Nanotechnology, 6*(1), 1568–1579. doi:10.3762/bjnano.6.160 PMID:26425408

Zhu, B., Liu, G.-L., Ling, F., Song, L.-S., & Wang, G.-X. (2015). Development toxicity of functionalized single-walled carbon nanotubes on rare minnow embryos and larvae. *Nanotoxicology, 9*(5), 579–590. doi:10.3109/17435390.2014.957253 PMID:25211547

Zhu, X., Wang, J., Zhang, X., Chang, Y., & Chen, Y. (2009). The impact of ZnO nanoparticle aggregates on the embryonic development of zebrafish (Danio rerio). *Nanotechnology, 20*(19), 195103. doi:10.1088/0957-4484/20/19/195103 PMID:19420631

Chapter 15
Copper and Copper Nanoparticles Induced Hematological Changes in a Freshwater Fish *Labeo rohita* – A Comparative Study:
Copper and Copper Nanoparticle Toxicity to Fish

Kaliappan Krishnapriya
Bharathiar University, India

Mathan Ramesh
Bharathiar University, India

ABSTRACT

In the present study, fish Labeo rohita were exposed to 20, 50 and 100 µg/L of both Cu NPs and copper sulphate ($CuSO_4$, bulk copper) for 24 h and hematological profiles were estimated. A significant (P< 0.01) change in the hemoglobin (Hb), hematocrit (Hct), white blood cells (WBC) and Mean Corpuscular Volume (MCV) levels were observed in all the three concentrations of both bulk and Cu NPs treated fish when compared to control groups. However a non significant change in red blood cells (RBC) (20 and 50 µg/L Cu NPs) and mean corpuscular hemoglobin (MCH) (20 and 50 µg/L bulk Cu) were observed. The alteration in Mean Corpuscular Hemoglobin Concentration (MCHC) value was found to be non significant both in bulk and Cu NPs treated fish. The alterations of these parameters can be used as a potential indicator to examine the health of fish in aquatic ecosystem contaminated with metal and metal based nanoparticles.

DOI: 10.4018/978-1-5225-0585-3.ch015

INTRODUCTION

Nanotechnology is a new branch of science, deals with synthesis of nano-sized particles that enhance the physical, chemical and biological properties of the metals. It is the fast growing and one of the prominent technologies in the 21st century (Chen *et al.*, 2012; Lee *et al.*, 2014; Abdel-Khalek *et al.*, 2015). Nanoparticles (NPs) ranges between 1 and 100 nm, dimensions from quantum dots to one, two or three dimensional nanoparticles (Handy *et al.*, 2008a; Lee *et al.*, 2014) which in turn increase its surface to volume ratio of the NPs and provide large surface area for binding of biomolecules. Moreover, nanotechnology has wide applications in industries, biomedical sciences, electronics, cosmetics, pharmaceuticals and research fields (Zhao *et al.*, 2011; Jovanovic and Palic, 2012). NPs have not only reached the markets, but also it is used for various domestic purposes and end up in the environment. In the environment these particles may pose a risk to the organisms.

In this juncture, the aquatic ecosystem is more susceptible to many kinds of pollutants including NPs (Scown *et al.*, 2010; Jovanovi´c and Pali´c, 2012). NPs may enter the aquatic ecosystem from its manufacturing waste, nanoproducts and its byproducts (Moore, 2006; Navarro *et al.*, 2008). Fate of nanoparticles in aquatic ecosystem is mainly governed by its solubility, dispersibility, and their interaction between biotic and abiotic factors (Brar *et al.*, 2010). NPs are toxic to aquatic organisms when exposed to higher doses as these particles can cross biological cell membranes (Griffitt *et al.*, 2007; Brar *et al.*, 2010; Siddiqui *et al.*, 2015). Recently the wide production of engineered nanoparticles (ENPs) due to their applications in many industrial processes finds their way in to the aquatic environment and cause adverse effects in aquatic organisms (Zhu *et al.*, 2008; Binelli *et al.*, 2009; Lu *et al.*, 2011; Yokel and MacPhai, 2011; Sanchez *et al.*, 2012; Qiuli *et al.*, 2013; Baker *et al.*, 2014). However, due to their unique physical and chemical properties, their fate in the aquatic organism is not clearly understood (Zhu *et al.*, 2009; Scown *et al.*, 2010; Remya *et al.*, 2015).

Copper nanoparticles (Cu NPs) have distinctive characters and commonly used as a substitute for noble metal catalysts (Cava, 1990; Tranquada *et al.*, 1995; Xu *et al.*, 1999; Zhou *et al.*, 2006; Chang *et al.*, 2012) and cheaper than the other metal oxide NPs (Machado *et al.*, 2008). Cu NPs find its application in textiles, skin products, ceramics, wood preservation, lubrication, nanofluids, bioactive coatings; electronic devices such as inkjet printing or integrated circuits; biocidal and antimicrobial activities (Yoon *et al.*, 2007; Gomes *et al.*, 2011; Santo *et al.*, 2012; Wang *et al.*, 2014; Nations *et al.*, 2015; Siddiqui *et al.*, 2015). Cu NPs are also used in medicine and as antifouling agents in paints used in boats (Kiaune and Singhasemanon, 2011; Perreault *et al.*, 2012).

Due to their low production cost and easy availability, and other specific properties such as antibacterial potency, catalytic activity, optical and magnetic properties Cu NPs has attracted the scientific community and huge quantity of Cu NPs has been synthesized (Khanna *et al.*, 2007; Kathad and Gajera, 2014). The extensive production and use of Cu NPs has led to entry of these particles in to aquatic environment and cause adverse effects in aquatic organisms (Griffitt *et al.*, 2008, 2009; Nations *et al.*, 2011; Chang *et al.*, 2012; Al-Bairuty *et al.*, 2013; Wang *et al.*, 2014; Song *et al.*, 2015; Hedayati *et al.*, 2016). While comparing with the other metal nanoparticles (MNPs) and nanotubes, Cu NPs showed greater toxicity in *in vitro* studies (Chang *et al.*, 2012). Griffitt *et al.* (2007) reported that Cu NPs produce acute toxicity in Zebra fish mainly in the gill. The LC50 value of Cu-NPs was 1.5mg/L in *Danio rerio* (Griffitt *et al.*, 2007). In addition, accumulation of these nanoparticles in aquatic organisms may transferred to higher trophic levels and poses a health hazard to animals and humans (Zhao *et al.*, 2011; Shaw *et al.*, 2012; Wang *et al.*, 2015).

For example, CuO NPs is more toxic than the bulk CuO towards many organisms (Heinlaan *et al.*, 2008; Mortimer *et al.*, 2009). In this line, Abdel-Khalek *et al.* (2015) reported that CuO (NPs) showed more toxicity than CuO (BPs) in vital organs like liver and gill tissues in the fish Nile tilapia *Oreochromis niloticus*. Ramskov *et al.* (2015) reported that even though both aqueous and particulate forms of Cu were equally bioavailable to the freshwater gastropod *P. antipodarum* only CuO NP spheres and platelets results adverse effects in growth of snail indicating possible NP-specific toxicity. Similarly, Al-Bairuty *et al.* (2013) reported that both $CuSO_4$ and Cu-NPs produce similar pathology effects in rainbow trout (*Oncorhynchus mykiss*); however the severity of injuries caused by Cu NPs was more in the organs studied. Ruiz *et al.* (2015) reported that the copper content in the mussels *Mytilus galloprovincialis* exposed to bulk CuO was lower than for those exposed to nano CuO.

The toxicity of NPs on the health condition of the aquatic organisms can be evaluated by monitoring the biological responses (Zhu *et al.*, 2008; Klaper *et al.*, 2009; Lu *et al.*, 2011). The study on the hematological profile of a fish reflects the effect of changes occurring in the aquatic environment (Gabriel, 2007; Ghaffar, 2014; Ullah, 2015). Fish blood is widely used as a bioindicators in assessing the toxic stress of pollutants (Romani *et al.*, 2003; Barcellos *et al.*, 2004, Petri *et al.*, 2006; Ramesh *et al.*, 2009; Suvetha *et al.*, 2010; Saravanan *et al.*, 2011a; Abhijith *et al.*, 2012; Al-Asgah *et al.*, 2015). Haematological parameters such as hemoglobin (Hb), hematocrit (Hct), red blood cell (RBC) count, white blood cell (WBC) count, and hematological indices like mean cellular volume (MCV), mean cellular hemoglobin (MCH) and mean cellular hemoglobin concentration (MCHC), are widely used as an indicator of toxic stress induced by environmental contaminants including nanoparticles (Llacuna, 1996; Gregory, 2001; Davis, 2008; Kavitha *et al.*, 2010; Saravanan *et al.*, 2012; Al-Asgah *et al.*, 2015; Maceda-Veiga *et al.*, 2015).

To our knowledge the comparative account on the impact of nanomaterials and bulk particles on freshwater organisms are scanty. Hence in the present study, copper sulphate (Cu SO_4) as bulk copper ($CuSO_4$ BPs) is used to compare the toxic effect of copper nano particles (Cu NPs) in a freshwater fish *Labeo rohita*. The fish *L. rohita* is a widespread species and used in carp polyculture systems in India and have a higher market demand.

MATERIALS AND METHOD

Procurement of Fish and Acclimatization

Experimental animal, *Labeo rohita* fingerlings of average length 6 – 7 cm and weight 8 g were procured from Tamil Nadu Fisheries Development Corporation Limited, Aliyar Fish Farm, Tamil Nadu, India and acclimatized to the laboratory condition for the period of 20 days (Ramesh *et al.*, 2014). During acclimatization, fish were fed (*ad libitum*) rice bran and ground nut oil cake daily one hour prior to replacement of water. Dechlorinated tap water was used for the present study. Water (three-forth) was changed regularly in order to remove excess feed and metabolic waste and following water quality was maintained during acclimatization and experimental period; temperature: 27.0 ± 2.0 C; pH: 7.0 ± 1, dissolved oxygen: 6.9 ± 0.01 mgL^{-1}; total alkalinity: 20.0 ± 5 mg L^{-1}; salinity: 0.42 ± 0.1 ppt; total hardness: 18.2 ± 0.2 mgL^{-1}; calcium: 3.4 ± 0.4 mg L^{-1} and magnesium: 2.05 ± 0.2 mg L^{-1}.

Preparation of Stock Solution

Cu NPs were purchased from Sigma Aldrich (manufacturer's information: 99.9% purity, mean particle size of < 50 nm, according to TEM). A stock solution of Cu NPs was prepared by dispersing 1.0 g Cu L^{-1} in deionized water (Millipore, ion free and unbuffered) with constant stirring for 1 h and sonication for about 30 mins in a bath-type sonicator (100 W, 40 kHz). Dosing of Cu NPs to the fish was carried out in the 12 hours interval. This stock was prepared freshly at 7 am prior to dosing and used again at 7 pm (the stock was stirred for 10 min and sonicated for about 20 min). $CuSO_4$ stock solution was prepared by dissolving 1.0 g of Cu^{2+} (3.929 g $CuSO_4 \cdot 5H_2O$) in 1 L of ultrapure water). In the present study, stock solution was prepared by following the procedure of Wang *et al.* (2014).

Acute Toxicity Study

Healthy *Labeo rohita* fingerlings were selected from the stock and exposed to three different concentrations of both Cu NPs and bulk Cu (20, 50 and 100 µg/L) based on the environmental level of Cu NPs (0.06 mg/L) as reported by Chio *et al.* (2012). A toxicant free control was maintained separately. Four replicates were maintained for each concentration and control group. Acute toxicity study was carried out for 24 hours. Feeding was withheld before 24 h and no feed was given during the experiment in order to minimize the risk of toxicant absorbing to food or metabolic waste. Death fish were removed immediately from the glass aquaria by observing the cessation of opercular movement.

Collection of Blood and Hematological Assay

Blood from control, Cu NPs and bulk Cu treated groups was collected by cardiac puncture. The collected blood sample was immediately transferred into plastic vials, which is previously rinsed with heparin. The whole blood was used for the estimation of Hb, RBC and WBC counts. Erythrocytes and leukocytes were counted by using the method of Rusia and Sood (1992) using haemocytometer. Hemoglobin content of the blood was estimated by Cyanmethemoglobin method (Drabkin, 1946). Hematocrit was estimated by microhematocrit (capillary) method (Nelson and Morris, 1989). Erythrocyte indices like MCV, MCH and MCHC were also calculated according to standard formulas.

MCV (fl) = Hct (%)/ RBC counts in millions/ mm^3 x 10
MCH (picograms) = Hb (g/dl)/RBC count in millions/ mm^3 x 10
MCHC (g/dl) = Hb (g/dl)/ Hct (%) x 100

Statistical Analysis

All values were expressed as means and analyzed by Analysis of Variance (ANOVA), followed by Duncan's Multiple Range Test (DMRT) to determine the significant differences ($P < 0.01$ and $P < 0.05$) on each parameter.

RESULTS

In the present study, no mortality was observed in all the concentrations of Cu NPs treated fish *L. rohita* whereas 80% mortality was observed in 100 µg/L of bulk copper treated fish. In 50 µg/L of bulk copper treated fish only 10% mortality was observed. Fish exposed to 100 µg/L of bulk copper showed behavioral changes like less opercular movement, loss of schooling behavior and sometimes remained at the corner of the glass aquaria.

In this study, Hb level was found to be increased significantly in 20 and 100 µg/L of Cu NPs treated fish when compared with the control group (Figure 1a). However, a significant decrease in Hb level was noticed in 50 µg/L Cu NPs treated fish. Likewise the Hb level was found to be increased in 50 and 100 µg/L of $CuSO_4$ BPs treated fish when compared with the control group. However in 20 µg/L of $CuSO_4$ BPs treated fish a reduction in Hb content was noted at the end of 24 h treatment (Figure 1). The Hct value was found to be increased in 20 and 100 µg/L of Cu NPs treated fish as compared with the control group (Fig 1b). In contrast, Hct value was found to be decreased in 50 µg/L of Cu NPs treated fish. Likewise Hct level was found to be increased in 50 and 100 µg/L of bulk Cu treated groups. However in 20 µg/L of bulk Cu treated fish Hct value was significantly decreased at the end of 24 h exposure period (Figure 2).

Figure 3. represents the data on RBC count of *L. rohita* exposed to Cu NPs and bulk copper treated groups. RBC count was found to be decreased in all the concentrations of bulk copper treated fish as compared with the control and Cu NPs treated fish. An increase of 3.720 millions/ cu.mm RBC cells

Figure 1. Hemoglobin level of a freshwater fish L. rohita treated with different concentrations of Cu NPs and bulk copper (20, 50 and 100 µg /L) for the period of 24 h

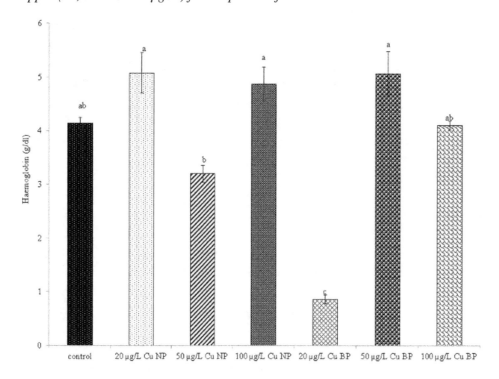

Figure 2. Hematocrit value of a freshwater fish L. rohita treated with different concentrations of Cu NPs and bulk copper (20, 50 and 100 µg /L) for the period of 24 h

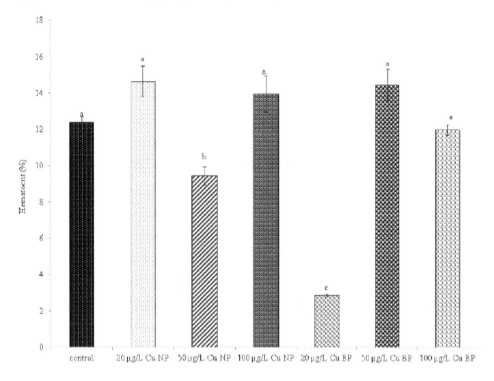

Figure 3. RBC count of a freshwater fish L. rohita treated with different concentrations of Cu NPs and bulk copper (20, 50 and 100 µg /L) for the period of 24 h

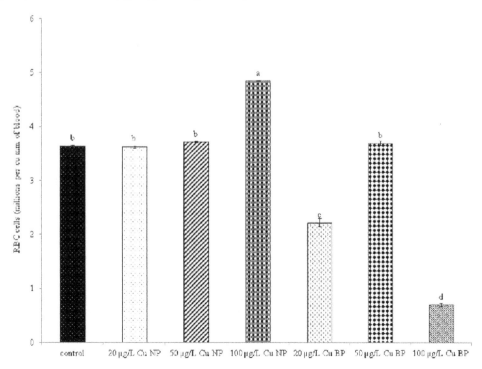

Figure 4. WBC count of a freshwater fish L. rohita treated with different concentrations of Cu NPs and bulk copper (20, 50 and 100 µg /L) for the period of 24 h

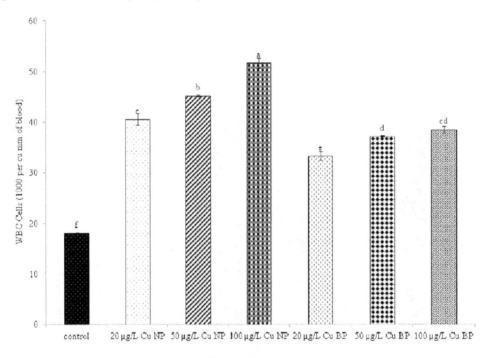

Figure 5. MCV value of a freshwater fish L. rohita treated with different concentrations of Cu NPs and bulk copper (20, 50 and 100 µg /L) for the period of 24 h

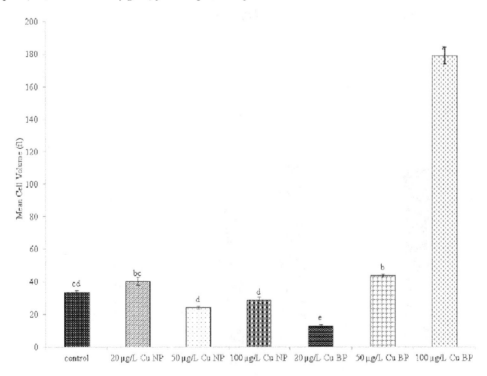

was observed in 50 µg/L Cu NPs treated fish whereas in 100 µg/L bulk copper treated fish the number of RBC cells was found to be decreased showing 0.770 millions/ cu.mm at the end of 24 h period. In Cu NPs treated fish the number of RBC cells is more or less equal to control group. WBC count was found to be increased significantly in all the concentrations of Cu NPs and bulk copper treated fish when compared with the control group. A maximum increase of 51.6 (1000/cu. mm) WBC cells was observed in 100 µg/L Cu NPs treated fish (Figure 4).

In the present investigation, MCV and MCH value was found to be decreased in 50 and 100 µg/L Cu NPs treated fish whereas MCV and MCH value was found to be increased in 20 µg/L Cu NPs treated fish when compared with the control group (Figures 5 and 6). A maximum decrease of 12.918 fl MCV and 3.908 pg of MCH value was noted in 20 µg/L bulk copper treated fish. In contrast to the above findings, MCV and MCH value was found to be increased in 100 µg/L bulk copper treated fish. MCHC level was found to be increased in all the concentrations of Cu NPs treated fish as compared to control. In bulk copper treated fish MCHC value was found to be decreased in 20 and 100 µg/L treated fish at the end of 24 h (Figure 7). However, in 50 µg/L bulk copper treated fish the MCHC value was slightly increased. A minimum decrease of 30.140 (g/dl) and a maximum increase 34.980 (g/dl) of MCHC value was noted in 20 µg/L Cu NPs and bulk copper treated fish respectively.

Figure 6. MCH value of a freshwater fish L. rohita treated with different concentrations of Cu NPs and bulk copper (20, 50 and 100 µg /L) for the period of 24 h

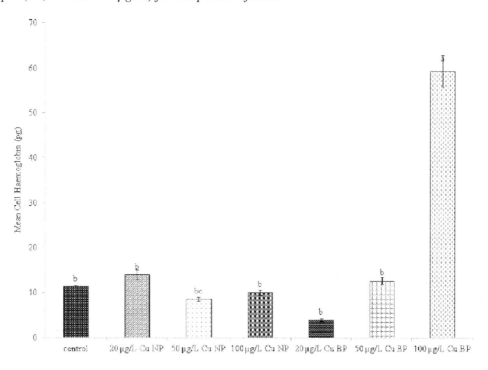

Figure 7. MCHC value of a freshwater fish L. rohita treated with different concentrations of Cu NPs and bulk copper (20, 50 and 100 μg /L) for the period of 24 h

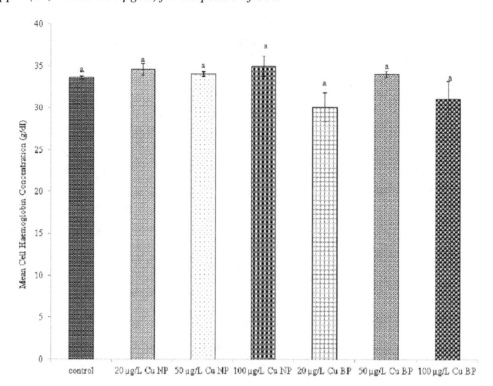

DISCUSSION

Toxicity studies on the nanoparticles are paramount important as it possess unique characteristics from its size to chemical composition, agglomeration, high persistence and biocompatibility (Zhao *et al.*, 2013). There is an urgent need to evaluate the risk of anthropogenic nanomaterials since the progress in the nanotechnology field might have an adverse effect on environment and human health (Buzea *et al.*, 2007; Arora *et al.*, 2012). For the safe use of nanoproducts in the environment, nanotoxicology is an emerging domain which is scrutinizing the level of nanoparticles and predicts the adverse effect of these particles in the environment and in model organisms in advance. Since fish are the one of the most common aquatic organisms and also occupy top most layers in the aquatic food chain, they are widely used as a model organism in toxicity studies (Farkas *et al.*, 2002; Ramesh *et al.*, 2009; Saravanan *et al.*, 2012).

The aquatic environment may act as a sink for the entry of ENPs (Hardman, 2006; Scown *et al.*, 2010; Gaiser *et al.*, 2012) and may cause harmful effects to aquatic organisms particularly fish (Chen *et al.*, 2011; Chen *et al.*, 2013; Abdel-Khalek *et al.*, 2015; Remya *et al.*, 2015). Copper is highly toxic to aquatic organisms at higher concentrations (Baldwin *et al.*, 2003). Copper being a transitional metal usually participates in Fenton and HabereWeiss reactions and help generation of ROS which results in oxidative stress (Regoli and Principato, 1995). Likewise Cu NPs may also cause oxidative stress (Fu *et al.*, 2014) and other adverse effects in aquatic organisms (Pang *et al.*, 2012; Shaw *et al.*, 2012; Wang *et al.*, 2014; Abdel-Khalek *et al.*, 2015; Song *et al.*, 2015; Adam *et al.*, 2015).

Generally, NPs can enter inside the cell through diffusion, ion channels, transporter protein molecules, endocytosis etc., (Colvin *et al.*, 2003; Moore, 2006; Verma *et al.*, 2008; Nel *et al.*, 2009; Ho *et al.*, 2010; Chang *et al.*, 2012). Moreover, NPs can easily cross the biological barriers like blood brain barriers (Lee *et al.*, 2014). Handy *et al.* (2008) reported that NPs in the aquatic environment can easily penetrate the fish mucous layer by peri-kinetic forces and bind with mucoproteins and finally get entrapped. Eventually, these NPs may undergo biodegradation or bioaccumulation in the aquatic organism followed by biomagnifications (Peralta-Videa *et al.*, 2011; Gomes *et al.*, 2012; Jun *et al.*, 2013). Soluble Cu^{2+} ions may be released from Cu NPs and non target aquatic organisms are exposed to dissolved Cu and nano size Cu^{2+} ions as well (Aruoja *et al.*, 2009; Shaw and Handy, 2011; Wang *et al.*, 2014; Ramskov *et al.*, 2015).

Recently, biomarkers are widely used to test the toxicity of NPs on the physiology of the organism and to monitor the environmental quality (Klaper *et al.*, 2009; Kavitha *et al.*, 2010). Fish blood is widely used in toxicological research as a feasible biomarker of pathophysiological changes (Adhikari *et al.*, 2004; Petri *et al.*, 2006; Velisek *et al.*, 2010). There is an intimate link among the external environment and fish circulatory system, analysis of blood assays helps to identify the physiochemical changes and health status of fish exposed to toxicant (Sampaio *et al.*, 2012; Lavanya *et al.*, 2011). Hematological variables may be used as a possible biomarker in clinical diagnosis of fish health and also helps to assess the toxic effects of xenobiotic substances (Wendelaar Bonga, 1997; Tellez-Banuelos *et al.*, 2009; Saravanan *et al.*, 2011b; Krishna Priya *et al.*, 2015). Hematological parameters are widely used to detect physiological changes under stress conditions in a number of fish species (Adhikari *et al.*, 2004; Adriana *et al.*, 2007; Alwan *et al.*, 2009; Suvetha *et al.*, 2010). Moreover, these parameters will give an insight of fish's health status (Sancho *et al.*, 2000; Harikrishnan *et al.*, 2011).

In the present investigation a significant alteration has been noted in hematological profiles of fish exposed to three different concentrations of Cu NPs and bulk copper. Hb is a protein responsible for carrying oxygen in fish body, and its concentration is closely related to red blood cell counts (Clark *et al.*, 1989). Hct is the ratio of blood volume that is occupied by red blood cells, expressed as a percentage of total blood volume. The decrease in the hemoglobin and hematocrit value of Cu nanoparticles treated fish may be due to lysing of erythrocytes due to toxicant stress whereas increased Hb and Hct values of bulk copper treated fish may be due to the starvation of fish due to toxicant exposure. This is in accordance with the work of Lavanya *et al.* (2011) and Steffens (1989). Further, the observed increase in Hb and Hct value of 20 µg/L and 100 µg/L Cu NPs and bulk copper treated fish is not statistically significant as compared to the control. This result is in agreement with the Federici *et al.* (2007), Ramsden *et al.* (2009), Shaw *et al.* (2012) and Boyle *et al.* (2013).

Erythrocytes are the most common cells seen in the fish blood (Bastami *et al.*, 2009). In the present investigation, RBC count was increased significantly in 100 µg/L Cu NPs exposed fish when compared with the rest of the concentrations of both Cu NPs and bulk copper and control fish which may be due to the reduction in the oxygen level in the blood resulted from the histological alterations in the gill lamellae due to accumulation of nanoparticles. Fish gills are susceptible to engineered NPs and these particles may enter the fish body either by absorption or penetration through the gill surface (Mazon and Fernandes, 1999; Linhua *et al.*, 2009). Moreover, gill is the primary organ for the Cu NPs toxicity and accumulation (Griffitt et al., 2007; Wang *et al.*, 2015). Damage to gill filaments and gill pavement cells has been observed in *Oncorhynchus mykiss*, *Pimephales promelas* and *Danio rerio* upon exposure to Cu NPs (Song *et al.*, 2015). Likewise, epidermal lesions were noted in the epithelial layer of fish Siberian sturgeon (*Acipenser baerii*) after exposure to Cu NPs and their frequency and severity increased as the nanoparticle concentrations increased (Ostaszewsk *et al.*, 2016). Similar to this study, a significant increase in RBC count was noted in *Labeo rohita* exposed to Fe_2O_3 NPs (Remya *et al.*, 2015). Khabbazi *et al.* (2015) reported that changes in the RBC cell count of fish *Oncorhynchus mykiss* exposed to Cu NPs might have resulted from respiratory disease due to Cu NPs toxicity. Furthermore, the decrease in RBC, Hct and Hb concentration in *Oreochromis niloticus* exposed to zinc oxide nanoparticles may be due to a decrease of the life span of RBCs (Alkaladi *et al.*, 2015).

In the present study, fish exposed to 20 and 100 µg/L bulk copper showed a significant reduction in the RBC count when compared to the control fish which may be due to the destruction of RBC cells due to metal toxicity. A fall in RBC count in *Oreochromis mossambicus* (Nussey *et al.*, 1995), *Heteropneustes fossilis* (Singh and Reddy, 1990) and in *Dicentrarchus labrax* (Gwoździński *et al.*, 1992) when exposed to copper may be due to hemolysis caused by copper toxicity. In general hemolysis causes a fall in RBC count (Hedayati *et al.*, 2016). Similar results has been found in fresh water fish treated with other metals such as arsenic (Oladimeji *et al.*, 1984; Cockell *et al.*, 1992; Lavanya *et al.*, 2011), arsenate (Kavitha *et al.*, 2010) and cadmium (Remyla *et al.*, 2008).

WBC cells are responsible for immune function of an organism (Jurd, 1985). Davis (2008) reported that variations in leucocytes counts are useful in the field of conservation physiology as they can be directly altered by stress and their role in the regulation of immunological functions. In the present investigation WBC count was significantly increased in all the three groups (20, 50 and 100 µg/L) of Cu NPs treated fish when compared to the bulk copper and control fish which may be due to the direct effect of the Cu NPs in the fish. This result is contrast to the previous literature stating that there was significant reduction in WBC cells exposed to Fe_2O_3 NPs (Remya *et al.*, 2015), TiO2 NPs (Ramsden *et al.*, 2009; Velasco-Santamaría *et al.*, 2011) and no changes in WBC count was noted in fish exposed to SWCNT NPs (Smith *et al.*, 2007). However, Alkaladi *et al.* (2015) reported that the increase in WBC count in *Oreochromis niloticus* exposed to zinc oxide nanoparticles indicate the presence of stress.

MCV and MCH are often determined as an index of health status especially in aquatic organisms (Oshode *et al.*, 2008). MCHC measure was used to assess the amount of red cell swelling (decreased MCHC) or shrinkage (increased MCHC) presents (Wepener *et al.*, 1992). In the present study, there was a significant increase in MCV and MCH value in 100 µg/L bulk copper treated fish which was not observed among the rest of the treated concentrations of both Cu NPs and bulk copper treated fish. Similar results were observed by Wepener *et al.* (1992) and Kavitha *et al.* (2010) in fish exposed to metals which may be due the swelling of RBC cells or increase in the circulating immature RBC cells due to the high concentration (toxicity) of the metal. The non significant effect in MCV, MCH and MCHC values during the study period indicate that fish were able to maintain salt and water balance in the blood (Smith *et al.*, 2007).

Hua *et al.* (2014) reported that the toxicity of different-sized copper nano- and submicron particles to zebrafish may result from the particle form of Cu particles rather than from dissolved Cu from the Cu particles. The net surface area may determine the release of ion from Cu NPs in which smaller NPs may have larger net surface area (Song *et al.*, 2013). The toxicity of Cu NPs is more than the ionic Cu due to their larger uptake of NPs (Noureen and Jabeen, 2015). The formation of copper ions during dissolution of the nanoparticles in the exposure medium may cause the adverse effects in *D. magna* (Adam *et al.*, 2015). Tavares *et al.* (2014) reported that $CuSO_4$ is more toxic than the CuONPs due to greater bioavailability of Cu ions in the test media. In addition aggregation of the NPs can decrease the release of Cu ions (Perreault *et al.*, 2014). However, Shaw *et al.* (2012) reported that Cu-NPs have similar types of toxic effects on hematology and biochemistry of rainbow trout (*Oncorhynchus mykiss*) as compared to $CuSO_4$ which can occur at lower tissue Cu concentrations than expected for the dissolved metal. Likewise, Wang *et al.* (2014) reported that the dissolved Cu was more toxic than Cu-NPs to juvenile *E. coioides*. The adverse effect of ionic copper is more in the organs like gill and liver of *E. coioides*, whereas the effect of nano copper is more in the gut (Wang *et al.*, 2014; 2015). In the present investigation also the alterations in the hematological parameters of fish was found to be more in bulk copper treated fish when compared to Cu-NPs.

However some authors have reported that Cu NPs were more toxic than the bulk Cu (Karlsson *et al.*, 2009; Zhao *et al.*, 2011). Poole and Owens (2003) reported that the unique physical properties of NPs are mostly attributed to their high surface to volume ratio, with a large proportion of the atoms being exposed on the surface compared to the bulk material. Furthermore, soluble copper is the foremost cause for the acute toxicity of the copper nanoparticles as reported by Song *et al.* (2015). Moreover, the toxicity and fate of copper nanoparticles may depend upon the physico chemical parameters of the water. For example the fate of Cu NPs in the aqueous media was more in high temperature when compared with the low temperature indicating that enhanced particle aggregation and higher rate of dissolution in higher temperature (Song *et al.*, 2015). In general the toxicity of Cu NPs depends on the size and the concentration of the particle.

In addition, ecotoxicity of NPs are influenced by many factors such as particle size/size distribution, solubility and agglomeration, shape and crystal structure, surface area, mass, environmental condition etc., (Tiede *et al.*, 2008; Ates *et al.*, 2014). For example a decrease of particle to nano-scale, the reactivity of NPs increases significantly compared to metals in bulk form because surface to volume ratio increases with decreasing particle size (Nel *et al.*, 2006). Likewise, particle shape also influences the bioaccumulation and thus toxicity (Dai *et al.*, 2015).

CONCLUSION

Cu NPs and bulk copper could cause adverse effect on the non-targeting organism *Labeo rohita* and the alterations of hematological parameters can be used as a potential indicator in analyzing the health of organism and aquatic ecosystem. However, to confirm the toxicity of nanoparticles over the bulk metal, chronic toxicity studies are needed mainly to determine the release of nanosize free metal ions from the nanoparticles and their fate in the environment. Further research is needed in the aspects of relating the physiochemical characteristics of nanoparticles and its level of toxicity in the aquatic organisms.

ACKNOWLEDGMENT

The investigators gratefully acknowledge the University Grants Commission -New Delhi, for sanctioning this major research project (MRP-MAJOR-ZOOL-2013-25987) (OBC) and providing financial assistance.

REFERENCES

Abdel-Khalek, A. A., Kadry, M. A. M., Badran, S. R., & Marie, M. A. S. (2015). Comparative toxicity of copper oxide bulk and nanoparticles in Nile Tilapia; *Oreochromis niloticus*: Biochemical and oxidative stress. *Journal of Basic & Applied Zoology*, 72, 43–57. doi:10.1016/j.jobaz.2015.04.001

Abhijith, B. D., Ramesh, M., & Poopal, R. K. (2012). Sublethal toxicological evaluation of methyl parathion on some haematological and biochemical parameters in an Indian major carp *Catla catla*. *Comparative Clinical Pathology*, *21*(1), 55–61. doi:10.1007/s00580-010-1064-8

Adam, N., Vakurov, A., Knapen, D., & Blust, R. (2015). The chronic toxicity of CuO nanoparticles and copper salt to *Daphnia magna*. *Journal of Hazardous Materials*, *283*, 416–422. doi:10.1016/j.jhazmat.2014.09.037 PMID:25464278

Adhikari, S., Sarkar, B., Chatterjee, A., Mahapatra, C. T., & Ayyappan, S. (2004). Effects of cypermethrin and carbofuran haematological parameters and prediction of their recovery in a freshwater teleost, *Labeo rohita* (Hamilton). *Ecotoxicology and Environmental Safety*, *58*(2), 220–226. doi:10.1016/j.ecoenv.2003.12.003 PMID:15157576

Adriana, B., & Almodóva, A. N. M. (2007). Antimicrobial efficacy of *Curcuma zedoaria* extract as assessed by linear regression compared with commercial mouthrinses. *Brazilian Journal of Microbiology*, *38*(3), 440–445. doi:10.1590/S1517-83822007000300011

Al-Asgah, N. A., Abdel-Warith, A. W. A., Younis, E. S. M., & Allam, H. Y. (2015). Haematological and biochemical parameters and tissue accumulations of cadmium in *Oreochromis niloticus* exposed to various concentrations of cadmium chloride. *Saudi Journal of Biological Sciences*, *22*(5), 543–550. doi:10.1016/j.sjbs.2015.01.002 PMID:26288556

Al-Bairuty, G. A., Shaw, B. J., Handy, R. D., & Henry, T. B. (2013). Histopathological effects of waterborne copper nanoparticles and copper sulphate on the organs of rainbow trout (*Oncorhynchus mykiss*). *Aquatic Toxicology (Amsterdam, Netherlands)*, *126*, 104–115. doi:10.1016/j.aquatox.2012.10.005 PMID:23174144

Alkaladi, A., Nasr El-Deen, N. A. M., Afifi, M., & Abu Zinadah, O. A. (2015). Hematological and biochemical investigations on the effect of vitamin E and C on *Oreochromis niloticus* exposed to zinc oxide nanoparticles. *Saudi Journal of Biological Sciences*, *22*(5), 556–563. doi:10.1016/j.sjbs.2015.02.012 PMID:26288558

Alwan, S. F., Hadi, A. A., & Shokr, A. E. (2009). Alterations in haematological parameter of fresh water fish. *Tilapia zilli* exposed to Aluminium. *Journal of Science and Its Applications*, *3*(1), 12–19.

Arora, S., Rajwade, J. M., & Paknikar, K. M. (2012). Nanotoxicology and in vitro studies: The need of the hour. *Toxicology and Applied Pharmacology*, *258*(2), 151–165. doi:10.1016/j.taap.2011.11.010 PMID:22178382

Aruoja, V., Dubourguier, H., Kasemets, K., & Kahru, A. (2009). Toxicity of nanoparticles of CuO, ZnO and TiO_2 to microalgae *Pseudokirchneriella subcapitata*. *The Science of the Total Environment*, *407*(4), 1461–1468. doi:10.1016/j.scitotenv.2008.10.053 PMID:19038417

Ates, M., Dugo, M. A., Demir, V., Arslan, Z., & Tchounwou, P. B. (2014). Effect of copper oxide nanoparticles to sheepshead minnow (*Cyprinodon variegatus*) at different salinities. *Digest Journal of Nanomaterials and Biostructures*, *9*(1), 369–377. PMID:25411584

Baker, T. J., Tyler, C. R., & Galloway, T. S. (2014). Impacts of metal and metal oxide nanoparticles on marine organisms. *Environmental Pollution*, *186*, 257–271. doi:10.1016/j.envpol.2013.11.014 PMID:24359692

Baldwin, D. H., Sandahl, J. F., Labenia, J. S., & Scholz, N. L. (2003). Sublethal effects of copper on coho salmon: Impacts on nonoverlapping receptor pathways in the peripheral olfactory nervous system. *Environmental Toxicology and Chemistry*, *22*(10), 2266–2274. doi:10.1897/02-428 PMID:14551988

Barcellos, L. J. G., Kreutz, L. C., Quevedo, R. M., Fioreze, I., Soso, A. B., Cericato, L., & Ritter, F. et al. (2004). Nursery rearing of jundiá, *Rhamdia quelen* (Quoy and Gaimard) in cages: Cage type, stocking density and stress response to confinement. *Aquaculture (Amsterdam, Netherlands)*, *232*(1-4), 383–394. doi:10.1016/S0044-8486(03)00545-3

Bastami, D. K., Moradlou, H. A., Zaragabadi, M. A., Mir, S. S. V., & Shakiba, M. M. (2009). Measurement of some haematological characteristics of the wild carp. *Comparative Clinical Pathology*, *18*(3), 321–323. doi:10.1007/s00580-008-0802-7

Binelli, A., Parolini, M., Cogni, D., Pedriali, A., & Provini, A. (2009). A multi-biomarker assessment of the impact of the antibacterial trimethoprim on the non-target organism zebra mussel (*Dreissena polymorpha*). *Comparative Biochemistry and Physiology*, (Part C): 150, 329–336. PMID:19481616

Boyle, D., Al-Bairuty, G. A., Ramsden, C. S., Sloman, K. A., Henry, T. B., & Handy, R. D. (2013). Subtle alterations in swimming speed distributions of rainbow trout exposed to titanium dioxide nanoparticles are associated with gill rather than brain injury. *Aquatic Toxicology (Amsterdam, Netherlands)*, *126*, 116–127. doi:10.1016/j.aquatox.2012.10.006 PMID:23178178

Brar, S. K., Verma, M., Tyagi, R. D., & Surampalli, R. Y. (2010). Engineered nanoparticles in wastewater and wastewater sludge – Evidence and impacts. *Waste Management (New York, N.Y.)*, *30*(3), 504–520. doi:10.1016/j.wasman.2009.10.012 PMID:19926463

Buzea, C., Pacheco, I. I., & Robbie, K. (2007). Nanomaterials and nanoparticles: Sources and toxicity. *Biointerphases*, *2*(4), 17–71. doi:10.1116/1.2815690 PMID:20419892

Cava, R. J. (1990). Structural chemistry and the local charge picture of copper oxide superconductors. *Science*, *247*(4943), 656–662. doi:10.1126/science.247.4943.656 PMID:17771881

Chang, Y. N., Zhang, M., Xia, L., Zhang, J., & Xing, G. (2012). The toxic effects and mechanisms of CuO and ZnO nanoparticles. *Materials (Basel)*, *5*(12), 2850–2871. doi:10.3390/ma5122850

Chen, L. Q., Kang, B., & Ling, J. (2013). Cytotoxicity of cuprous oxide nanoparticles to fish blood cells: Hemolysis and internalization. *Journal of Nanoparticle Research*, *15*(3), 1507–1513. doi:10.1007/s11051-013-1507-7

Chen, P. J., Tan, S. W., & Wu, W. L. (2012). Stabilization or oxidation of nanoscale zero valent iron at environmentally relevant exposure changes bioavailability and toxicity in medaka fish. *Environmental Science & Technology*, *46*(15), 8431–8439. doi:10.1021/es3006783 PMID:22747062

Chen, T. H., Lin, C. Y., & Tseng, M. C. (2011). Behavioral effects of titanium dioxide nanoparticles on larval zebrafish (*Danio rerio*). *Marine Pollution Bulletin*, *63*(5-12), 303–308. doi:10.1016/j.marpolbul.2011.04.017 PMID:21565364

Chio, C. P., Chen, W. Y., Chou, W. C., Hsieh, N. H., Ling, M. P., & Liao, C. M. (2012). Assessing the potential risks to zebra fish posed by environmentally relevant copper and silver nanoparticles. *The Science of the Total Environment*, *420*, 111–118. doi:10.1016/j.scitotenv.2012.01.023 PMID:22326136

Clark, I. A., Chaudhri, G., & Cowden, W. B. (1989). Some roles of free radicals in malaria. *Free Radical Biology & Medicine*, *6*(3), 315–321. doi:10.1016/0891-5849(89)90058-0 PMID:2663664

Cockell, K. A., Hilton, J. W., & Bettger, W. J. (1992). Hepatobiliary and hematological effects of dietary di sodium arsenate heptahydrate in juvenile rainbow trout (*Oncorhynchus mykiss*). *Comparative Biochemistry and Physiology*, (Part C): 103, 453–458.

Colvin, R. A., Fontaine, C. P., Laskowski, M., & Thomas, D. (2003). Zn^{2+} transporters and Zn^{2+} homeostasis in neurons. *European Journal of Pharmacology*, *479*(1-3), 171–185. doi:10.1016/j.ejphar.2003.08.067 PMID:14612148

Dai, L., Banta, G. T., Selck, H., & Forbes, V. E. (2015). Influence of copper oxide nanoparticle form and shape on toxicity and bioaccumulation in the deposit feeder, *Capitella teleta*. *Marine Environmental Research*, *111*, 99–106. doi:10.1016/j.marenvres.2015.06.010 PMID:26138270

Davis, A. K. (2008). Ontogenetic changes in erythrocyte morphology in larval mole salamanders, *Ambystoma talpoideum*, measured with image analysis. *Comparative Clinical Pathology*, *17*(1), 23–28. doi:10.1007/s00580-007-0702-2

Drabkin, D. L. (1946). Spectrometric studies, XIV: The crystallographic and optimal properties of the hemoglobin of man in comparison with those of other species. *The Journal of Biological Chemistry*, *164*, 703–723. PMID:21001166

Farkas, A., Salanki, J., & Specziar, A. (2002). Relation between growth and the heavy metal concentration in organs of bream; *Abramis brama* L. populating Lake Balaton. *Archives of Environmental Contamination and Toxicology*, *43*(2), 236–243. doi:10.1007/s00244-002-1123-5 PMID:12115050

Federici, G., Shaw, B. J., & Handy, R. D. (2007). Toxicity of titanium dioxide nanoparticles to rainbow trout (*Oncorhynchus mykiss*): Gill injury, oxidative stress, and other physiological effects. *Aquatic Toxicology (Amsterdam, Netherlands)*, *84*(4), 415–430. doi:10.1016/j.aquatox.2007.07.009 PMID:17727975

Fu, P. P., Xia, Q., Hwang, H. M., Ray, P. C., & Yu, H. (2014). Mechanisms of nanotoxicity: Generation of reactive oxygen species. *Journal of Food and Drug Analysis*, *22*(1), 64–75. doi:10.1016/j.jfda.2014.01.005 PMID:24673904

Gabriel, U. U., Amakiri, E. U., & Ezeri, G. N. O. (2007). Haematology and gill pathology of *Clarias gariepinus* exposed to refined petroleum oil under laboratory conditions. *Journal of Animal and Veterinary Advances*, *6*, 461–465.

Gaiser, B. K., Fernandes, T. F., Jepson, M. A., Lead, J. R., Tyler, C. R., Baalousha, M., & Stone, V. et al. (2012). Interspecies comparisons on the uptake and toxicity of silver and cerium dioxide nanoparticles. *Environmental Toxicology and Chemistry, 31*(1), 144–154. doi:10.1002/etc.703 PMID:22002553

Ghaffar, A., Ashraf, S., Hussain, R., Hussain, T., Shafique, M., Noreen, S., & Aslam, S. (2014). Clinico hematological disparities induced by triazophos (organophosphate) in Japanese quail. *Pakistan Veterinary Journal, 34*, 257–259.

Gomes, T., Pereira, C. G., Cardoso, C., Pinheiro, J. P., Cancio, I., & Bebianno, M. J. (2012). Accumulation and toxicity of copper oxide nanoparticles in the digestive gland of *Mytilus galloprovincialis*. *Aquatic Toxicology, 118 – 119*, 72– 79.

Gomes, T., Pinheiro, J. P., Cancio, I., Pereira, C. G., Cardoso, C., & Bebianno, M. J. (2011). Effects of copper nanoparticles exposure in the mussel *Mytilus galloprovincialis*. *Environmental Science & Technology, 45*(21), 9356–9362. doi:10.1021/es200955s PMID:21950553

Gregory, T. R. (2001). The bigger the C-value, the larger the cell: Genome size and red blood cell size in vertebrates. *Blood Cells, Molecules & Diseases, 27*(5), 830–843. doi:10.1006/bcmd.2001.0457 PMID:11783946

Griffitt, R. J., Hyndman, K., Denslow, N. D., & Barber, D. S. (2009). Comparison of molecular and histological changes in zebrafish gills exposed to metallic nanoparticles. *Toxicological Sciences, 107*(2), 404–415. doi:10.1093/toxsci/kfn256 PMID:19073994

Griffitt, R. J., Luo, J., Gao, J., Bonzango, J.-C., & Barber, D. S. (2008). Effects of particle composition and species on toxicity of metallic nanoparticles in aquatic organisms. *Environmental Toxicology and Chemistry, 27*(9), 1972–1978. doi:10.1897/08-002.1 PMID:18690762

Griffitt, R. J., Weil, R., Hyndman, K. A., Denslow, N. D., Powers, K., Taylor, D., & Barber, D. S. (2007). Exposure to copper nanoparticles causes gill injury and acute lethality in zebrafish (*Danio rerio*). *Environmental Science & Technology, 41*(23), 8178–8186. doi:10.1021/es071235e PMID:18186356

Gwoździński, K., Roche, H., & Pérès, G. (1992). The comparison of the effects of heavy metal ions on antioxidant enzyme activities in human and fish *Dicentrarchus labrax* erythrocytes. *Comparative Biochemistry and Physiology, 102*, 57–60. PMID:1358529

Handy, R. D., Kammer, F. V. D., Lead, J. R., Hassellóv, M., Owen, R., & Crane, M. (2008a). The ecotoxicology and chemistry of manufactured nanoparticles. *Ecotoxicology (London, England), 17*(4), 287–314. doi:10.1007/s10646-008-0199-8 PMID:18351458

Handy, R. D., Owen, R., & Valsami-Jones, E. (2008). The ecotoxicology of nanoparticles and nanomaterials: Current status, knowledge gaps, challenges, and future needs. *Ecotoxicology (London, England), 17*(5), 315–325. doi:10.1007/s10646-008-0206-0 PMID:18408994

Hardman, R. (2006). A toxicologic review of quantum dots: Toxicity depends on physic chemical and environmental factors. *Environmental Health Perspectives, 114*(2), 165–172. doi:10.1289/ehp.8284 PMID:16451849

Harikrishnan, R., Balasundaram, C., & Heo, M. S. (2011). Impact of plant products on innate and adaptive immune system of cultured finfish and shellfish. *Aquaculture (Amsterdam, Netherlands), 317*(1-4), 1–15. doi:10.1016/j.aquaculture.2011.03.039

Hedayati, A., Hoseini, S. M., & Hoseinifar, S. H. (2016). Response of plasma copper, ceruloplasmin, iron and ions in carp, *Cyprinus carpio* to waterborne copper ion and nanoparticle exposure. *Comparative Biochemistry and Physiology. Part C, Pharmacology, Toxicology & Endocrinology, 179*, 87–93. doi:10.1016/j.cbpc.2015.09.007 PMID:26408942

Heinlaan, M., Ivask, A., Blinova, I., Dubourguier, H. C., & Kahru, A. (2008). Toxicity of nanosized and bulk ZnO, CuO and TiO2 to bacteria *Vibrio fischeri* and crustaceans *Daphnia magna* and *Thamnocephalus platyurus*. *Chemosphere, 71*(7), 1308–1316. doi:10.1016/j.chemosphere.2007.11.047 PMID:18194809

Hernández Battez, A., Viesca, J. L., González, R., Blanco, D., Asedegbega, E., & Osorio, A. (2010). Friction reduction properties of a CuO nano lubricant used as lubricant for a NiCrBSi coating. *Wear, 268*(1-2), 325–328. doi:10.1016/j.wear.2009.08.018

Ho, C. L., Teo, S. S., Rahim, R. A., & Phang, S. M. (2010). Transcripts of *Gracilaria changii* that improve copper tolerance of *Escherichia coli*. *Asia-Pacific Journal of Molecular Biology and Biotechnology, 18*, 315–319.

Hua, J., Vijver, M. G., Ahmad, F., Richardson, M. K., & Peijnenburgy, W. J. G. M. (2014). Toxicity of different-sized copper nano- and submicron particles and their shed copper ions to zebrafish embryos. *Environmental Toxicology and Chemistry, 33*(8), 1774–1782. doi:10.1002/etc.2615 PMID:24839162

Jovanoviˊc, B., & Paliˊc, D. (2012). Immunotoxicology of non-functionalized engineered nanoparticles in aquatic organisms with special emphasis on fish—Review of current knowledge, gap identification, and call for further research. *Aquatic Toxicology, 118– 119*, 141– 151.

Jun, X., Zhou, Z. H., & Hua, L. U. G. (2013). Effects of selected metal oxide nanoparticles on multiple biomarkers in *Carassius auratus*. *Biomedical and Environmental Sciences, 26*(9), 742–749. PMID:24099608

Jurd, R. D. (1985). Specialization in teleost and anuran immune response: a comparative critique. In M. J. Manning & M. F. Tatner (Eds.), *Fish immunology* (pp. 9–28). London: Academic Press.

Karlsson, U., Ringsberg, J. W., Johnson, E., Hoseini, M., & Ulfvarson, A. (2009). Experimental and numerical investigation of bulb impact with a ship side-shell structure. *Marine Technology, 46*(1), 16–26.

Kathad, U., & Gajera, H. P. (2014). Synthesis of copper nanoparticles by two different methods and size comparison. *International Journal of Pharma and Bio Sciences, 5*(3), 533–540.

Kavitha, C., Malarvizhi, A., Senthil Kumaran, S., & Ramesh, M. (2010). Toxicological effects of arsenate exposure on hematological, biochemical and liver transaminases activity in an Indian major carp, *Catla catla*. *Food and Chemical Toxicology, 48*(10), 2848–2854. doi:10.1016/j.fct.2010.07.017 PMID:20654677

Khabbazi, M., Harsij, M., Hedayati, S. A. A., Gholipoor, H., Gerami, M. H., & Farsani, H. G. (2015). Effect of CuO nanoparticles on some hematological indices of rainbow trout *Oncorhynchus mykiss* and their potential toxicity. *Nanomedicine (London), 2*(1), 67–73.

Khanna, P. K., Gaikwad, S., Adhyapak, P. V., Singh, N., & Marimuthu, R. (2007, October). Synthesis and characterization of copper nanoparticles. *Materials Letters*, *61*(25), 4711–4714. doi:10.1016/j.matlet.2007.03.014

Kiaune, L., & Singhasemanon, N. (2011). Pesticidal copper (I) oxide: Environmental fate and aquatic toxicity. *Reviews of Environmental Contamination and Toxicology*, *213*, 1–26. doi:10.1007/978-1-4419-9860-6_1 PMID:21541846

Klaper, R., Crago, J., Barr, J., Arndt, D., Setyowati, K., & Chen, J. (2009). Toxicity biomarker expression in daphinids exposed to manufactured nanoparticles: Changes in toxicity with functionalization. *Environmental Pollution*, *157*(4), 1152–1156. doi:10.1016/j.envpol.2008.11.010 PMID:19095335

Krishna Priya, K., Ramesh, M., Saravanan, M., & Ponpandian, N. (2015). Ecological risk assessment of silicon dioxide nanoparticles in a fresh-water fish *Labeo rohita*: Hematology, ionoregulation and gill Na^+/K^+ ATPase activity. *Ecotoxicology and Environmental Safety*, *120*, 295–302. doi:10.1016/j.ecoenv.2015.05.032 PMID:26094035

Lavanya, S., Ramesh, M., Kavitha, C., & Malarvizhi, A. (2011). Hematological, biochemical and ionoregulatory responses of Indian major carp *Catla catla* during chronic sublethal exposure to inorganic arsenic. *Chemosphere*, *82*(7), 977–985. doi:10.1016/j.chemosphere.2010.10.071 PMID:21094981

Lee, J., Kim, J., Shin, Y., Ryu, J., Eom, I., Lee, J. S., & Lee, B. et al. (2014). Serum and ultra structure responses of common carp (*Cyprinus carpio* L.) during long-term exposure to zinc oxide nanoparticles. *Ecotoxicology and Environmental Safety*, *104*, 9–17. doi:10.1016/j.ecoenv.2014.01.040 PMID:24632117

Linhua, H., Zhenyu, W., & Baoshan, X. (2009). Effect of sub acute exposure to TiO_2 nanoparticles on oxidative stress and histopathological changes in Juvenile Carp (*Cyprinus carpio*). *Journal of Environmental Sciences (China)*, *21*(10), 1459–1466. doi:10.1016/S1001-0742(08)62440-7 PMID:20000003

Llacuna, S., Gorriz, A., Riera, M., & Nadal, J. (1996). Effects of air pollution on hematological parameters in passerine birds. *Archives of Environmental Contamination and Toxicology*, *31*(1), 148–152. doi:10.1007/BF00203919 PMID:8688002

Lu, G. H., Chen, W., & Li, Y. (2011). Effects of PAHs on biotransformation enzymatic activities in fish. *Chemical Research in Chinese Universities*, *27*, 413–416.

Maceda-Veiga, A., Figuerola, J., Martínez-Silvestre, A., Viscor, G., Ferrari, N., & Pacheco, M. (2015). Inside the Redbox: Applications of haematology in wildlife monitoring and ecosystem health assessment. *The Science of the Total Environment*, *514*, 322–332. PMID:25668285

Machado, A. B. M., Drummond, G. M., & Paglia, A. P. (2008). *Livro vermelho da fauna brasileira ameaçada de extinção* (Vol. II). Brasília, Belo Horizonte: MMA / Fundação Biodiversitas.

Mazon, A. F., & Fernandes, M. N. (1999). Toxicity and differential tissue accumulation of copper in the tropical freshwater fish, *Prochilodus scrofa* (Prochilodontidae). *Bulletin of Environmental Contamination and Toxicology*, *63*(6), 797–804. doi:10.1007/s001289901049 PMID:10594155

Moore, M. N. (2006). Do nanoparticles present ecotoxicological risks for the health of the aquatic environment? *Environment International*, *32*(8), 967–976. doi:10.1016/j.envint.2006.06.014 PMID:16859745

Mortimer, J. C., Coxon, K. M., Laohavisit, A., & Davies, J. M. (2009). Hemi independent soluble and membrane associated peroxidase activity of a Zea mays annexin preparations. *Plant Signaling & Behavior*, *4*(5), 428–430. doi:10.4161/psb.4.5.8297 PMID:19816107

Nations, S., Long, M., Wages, M., Maul, J. D., Theodorakis, C. W., & Cobb, G. P. (2015). Sub chronic and chronic developmental effects of copper oxide (CuO) nanoparticles on *Xenopus laevis*. *Chemosphere*, *135*, 166–174. PMID:25950410

Nations, S., Wages, M., Cañas, J. E., Maul, J., Theodorakis, C., & Cobb, G. P. (2011). Acute effects of Fe_2O_3, TiO_2, ZnO and CuO nanomaterials on *Xenopus laevis*. *Chemosphere*, *83*(8), 1053–1061. doi:10.1016/j.chemosphere.2011.01.061 PMID:21345480

Navarro, E., Baun, A., Behra, R., Hartmann, N. B., Filser, J., Miao, A. J., & Sigg, L. et al. (2008). Environmental behavior and ecotoxicity of engineered nanoparticles to algae, plants, and fungi. *Ecotoxicology (London, England)*, *17*(5), 372–386. doi:10.1007/s10646-008-0214-0 PMID:18461442

Nel, A., Xia, T., Madler, L., & Li, N. (2006). Toxic potential of materials at the nanolevel. *Science*, *311*(5761), 622–627. doi:10.1126/science.1114397 PMID:16456071

Nel, A. E., Madler, L., Velegol, D., Xia, T., Hoek, E. M. V., Somasundaran, P., & Thompson, M. et al. (2009). Understanding biophysicochemical interactions at the nano - bio interface. *Nature Materials*, *8*(7), 543–557. doi:10.1038/nmat2442 PMID:19525947

Nelson, D. A., & Morris, M. W. (1989). Basic methodology. Hematology and coagulation. In D. A. Nelson, & J. B. Henry (Eds.), Clinical Diagnosis and Management by Laboratory Methods (pp. 578 –625). Saunder Company.

Noureen, A., & Jabeen, F. (2015). The toxicity, ways of exposure and effects of Cu nanoparticles and Cu bulk salts on different organisms. *International Journal of Biosciences*, *6*(2), 147–156. doi:10.12692/ijb/6.2.147-156

Nussey, G., Van Vuren, J., & du Preez, H. H. (1995). Effect of copper on the haematology and osmoregulation of the Mozambique tilapia, *Oreochromis mossambicus* (Cichlidae). *Comparative Biochemistry and Physiology. Part C, Pharmacology, Toxicology & Endocrinology*, *111*(3), 369–380. doi:10.1016/0742-8413(95)00063-1

Oladimeji, A. A., Quadri, S. U., & DeFreitas, A. S. W. (1984). Long-term effects of arsenic accumulation in rainbow trout, *Salmo gairdneri*. *Bulletin of Environmental Contamination and Toxicology*, *32*(1), 732–741. doi:10.1007/BF01607564 PMID:6743863

Oshode, O. A., Bakare, A. A., Adeogun, A. O., Efuntoye, M. O., & Sowunmi, A. A. (2008). Ecotoxicological assessment using *Clarias gariepius* and microbial characterization of leachate from municipal solid waste landfill. *International Journal of Environmental of Research*, *2*(4), 391–400.

Ostaszewska, T., Chojnacki, M., Kamaszewski, M., & Chwalibóg, E. S. (2016). Histopathological effects of silver and copper nanoparticles on the epidermis, gills, and liver of Siberian sturgeon. *Environmental Science and Pollution Research International*, *23*(2), 1621–1633. doi:10.1007/s11356-015-5391-9 PMID:26381783

Pang, C., Selck, H., Misra, S. K., Berhanu, D., Dybowska, A., Valsami-Jones, E., & Forbes, V. E. (2012). Effects of sediment-associated copper to the deposit-feeding snail, *Potamopyrgus antipodarum*: A comparison of Cu added in aqueous form or as nano- and micro-CuO particles. *Aquatic Toxicology, 106 - 107*, 114 - 122.

Peralta-Videa, J. R., Zhao, L., Lopez-Moreno, M. L., de la Rosa, G., Hong, J., & Gardea-Torresdey, J. L. (2011). Nanomaterials and the environment: A review for the biennium 2008 - 2010. *Journal of Hazardous Materials, 186*(1), 1–15. doi:10.1016/j.jhazmat.2010.11.020 PMID:21134718

Perreault, F., Oukarroum, A., Melegari, S. P., Matias, W. G., & Popovic, R. (2012). Polymer coating of copper oxide nanoparticles increases nanoparticles uptake and toxicity in the green alga *Chlamydomonas reinhardtii*. *Chemosphere, 87*(11), 1388–1394. doi:10.1016/j.chemosphere.2012.02.046 PMID:22445953

Perreault, F., Popovic, R., & Dewez, D. (2014). Different toxicity mechanisms between bare and polymer-coated copper oxide nanoparticles in *Lemna gibba*. *Environmental Pollution, 185*, 219–227. doi:10.1016/j.envpol.2013.10.027 PMID:24286697

Petri, D., Glover, C. N., Ylving, S., Kolas, K., Fremmersvik, G., Waagbo, R., & Berntssen, M. H. G. (2006). Sensitivity of Atlantic salmon (*Salmo salar*) to dietary endosulfan as assessed by haematology, blood biochemistry, and growth parameters. *Aquatic Toxicology (Amsterdam, Netherlands), 80*(3), 207–216. doi:10.1016/j.aquatox.2006.07.019 PMID:17081631

Poole, C. P., & Owens, F. J. (2003). *Introduction to Nanotechnology*. New York: Wiley-Interscience.

Qiuli, W., Abdelli, N., Yiping, L., Min, Z., & Wei, W. (2013). Comparison of toxicities from three metal oxide nano-particles at environmental relevant concentrations in nematode *Caenorhabditis elegans*. *Chemosphere, 90*(3), 1123–1131. doi:10.1016/j.chemosphere.2012.09.019 PMID:23062833

Ramesh, M., Sankaran, M., Veera-Gowtham, V., & Poopal, R. K. (2014). Hematological, biochemical and enzymological responses in an Indian major carp *Labeo rohita* induced by sublethal concentration of waterborne selenite exposure. *Chemico-Biological Interactions, 207*, 67–73. doi:10.1016/j.cbi.2013.10.018 PMID:24183823

Ramesh, M., Srinivasan, R., & Saravanan, M. (2009). Effect of atrazine (Herbicide) on blood parameters of common carp *Cyprinus carpio* (Actinopterygii: Cypriniformes). *African Journal of Environmental Science and Technology, 3*(12), 453–458.

Ramsden, C. S., Smith, T. J., Shaw, B. J., & Handy, R. D. (2009). Dietary exposure to titanium dioxide nanoparticles in rainbow trout (*Oncorhynchus mykiss*): No effect on growth, but subtle biochemical disturbances in the brain. *Ecotoxicology (London, England), 18*(7), 939–951. doi:10.1007/s10646-009-0357-7 PMID:19590957

Ramskov, T., Croteau, M. N., Forbes, V. E., & Selck, H. (2015). Biokinetics of different-shaped copper oxide nanoparticles in the freshwater gastropod, *Potamopyrgus antipodarum*. *Aquatic Toxicology (Amsterdam, Netherlands), 163*, 71–80. doi:10.1016/j.aquatox.2015.03.020 PMID:25863028

Regoli, F., & Principato, G. (1995). Glutathione, glutathione-dependent and antioxidant enzymes in mussel, *Mytilus galloprovincialis*, exposed to metals under field and laboratory conditions: Implications for the use of biochemical biomarkers. *Aquatic Toxicology (Amsterdam, Netherlands)*, *31*(2), 143–164. doi:10.1016/0166-445X(94)00064-W

Remya, A. S., Ramesh, M., Saravanan, M., Poopal, R. K., Bharathi, S., & Nataraj, D. (2015). Iron oxide nanoparticles to an Indian major carp, *Labeo rohita*: Impacts on hematology, ionoregulation and gill Na^+/K^+ATPase activity. *Journal of King Saud University – Science*, *27*, 151 – 160.

Remyla, S. R., Ramesh, M., Sajwan, K. S., & Senthil Kumar, K. (2008). Influence of zinc on cadmium induced haematological and biochemical responses in a freshwater teleost fish *Catla catla*. *Fish Physiology and Biochemistry*, *34*(2), 169–174. doi:10.1007/s10695-007-9157-2 PMID:18649034

Romani, R., Antognelli, C., Baldracchini, F., Santis, A., Isani, G., Giovannini, E., & Rosi, G. (2003). Increased acetylcholinesterase activities in specimens of *Sparus auratus* exposed to sublethal concentrations. *Chemico-Biological Interactions*, *145*(3), 321–329. doi:10.1016/S0009-2797(03)00058-9 PMID:12732458

Ruiz, P., Katsumiti, A., Nieto, J. A., Bori, J., Romero, A. J., Reip, P., & Cajaraville, M. P. et al. (2015). Short-term effects on antioxidant enzymes and long-term genotoxic and carcinogenic potential of CuO nanoparticles compared to bulk CuO and ionic copper in mussels *Mytilus galloprovincialis*. *Marine Environmental Research*, *111*, 107–120. doi:10.1016/j.marenvres.2015.07.018 PMID:26297043

Rusia, V., & Sood, S. K. (1992). Routine hematological studies. In L. Kanai (Ed.), *Medical Laboratory Technology* (pp. 252–258). Tata McGraw Hill Publishing.

Sampaio, F. G., Boijink, C. L., Bichara dos Santos, L. R., Tie Oba, E., Kalinin, A. L., Barreto Luiz, A. J., & Rantin, F. T. (2012). Antioxidant defenses and biochemical changes in the neotropical fish pacu, *Piaractus mesopotamicus:* Responses to single and combined copper and hypercarbia exposure. *Comparative Biochemistry and Physiology*, (Part C): 156, 178–186. PMID:22796211

Sanchez, D. A., Consiglio, A., & Richaud, Y. (2012). Efficient generation of A9 midbrain dopaminergic neurons by lentiviral delivery of LMX1A in human embryonic stem cells and induced pluripotent stem cells. *Human Gene Therapy*, *23*(1), 56–69. doi:10.1089/hum.2011.054 PMID:21877920

Sancho, E., Cerón, J. J., & Ferrando, M. D. (2000). Cholinesterase activity and hematological parameters as biomarkers of sublethal molinate exposure in *Anguilla anguilla*. *Ecotoxicology and Environmental Safety*, *46*(1), 81–86. doi:10.1006/eesa.1999.1888 PMID:10805997

Santo, C. E., Quaranta, D., & Grass, G. (2012). Antimicrobial metallic copper surfaces kill *Staphylococcus haemolyticus* via membrane damage. *Microbiology Open*, *1*(1), 46–52.

Saravanan, M., Karthika, S., Malarvizhi, A., & Ramesh, M. (2011b). Ecotoxicological impacts of clofibric acid and diclofenac in common carp (*Cyprinus carpio*) fingerlings: Hematological, biochemical, ionoregulatory and enzymological responses. *Journal of Hazardous Materials*, *195*, 188–194. doi:10.1016/j.jhazmat.2011.08.029 PMID:21885190

Saravanan, M., Prabhu Kumar, K., & Ramesh, M. (2011a). Haematological and biochemical responses of freshwater teleost fish *Cyprinus carpio* (Actinopterygii: Crypriniformes) during acute and chronic sublethal exposure to lindane. *Pesticide Biochemistry and Physiology*, *100*(3), 206–211. doi:10.1016/j.pestbp.2011.04.002

Saravanan, M., Usha Devi, K., Malarvizhi, A., & Ramesh, M. (2012). Effects of Ibuprofen on hematological, biochemical and enzymological parameters of blood in an Indian major carp, *Cirrhinus mrigala*. *Environmental Toxicology and Pharmacology*, *34*(1), 14–22. doi:10.1016/j.etap.2012.02.005 PMID:22418069

Scown, T. M., van Aerle, R., & Tyler, C. R. (2010). Do engineered nanoparticles pose a significant threat to the aquatic environment? *Critical Reviews in Toxicology*, *40*(7), 653–670. doi:10.3109/10408 444.2010.494174 PMID:20662713

Shaw, B. J., Al-Bairuty, G. A., & Handy, R. D., 2012. Effects of waterborne copper nano-particles and copper sulphate on rainbow trout (*Oncorhynchus mykiss*): physiology and accumulation. *Aquatic Toxicology*, *116 – 117*, 90 – 101.

Shaw, B. J., & Handy, R. D. (2011). Physiological effects of nanoparticles on fish: A comparison of nano-metals versus metal ions. *Environment International*, *37*(6), 1083–1097. doi:10.1016/j.envint.2011.03.009 PMID:21474182

Siddiqui, S., Goddard, R. H., & Bielmyer-Fraser, G. K. (2015). Comparative effects of dissolved copper and copper oxide nanoparticle exposure to the sea anemone, *Exaiptasia pallid*. *Aquatic Toxicology (Amsterdam, Netherlands)*, *160*, 205–213. doi:10.1016/j.aquatox.2015.01.007 PMID:25661886

Singh, H., & Reddy, T. (1990). Effect of copper sulfate on hematology, blood chemistry, and hepatosomatic index of an Indian catfish, *Heteropneustes fossilis* (Bloch), and its recovery. *Ecotoxicology and Environmental Safety*, *20*(1), 30–35. doi:10.1016/0147-6513(90)90043-5 PMID:2226241

Smith, C. J., Shaw, B. J., & Handy, R. D. (2007). Toxicity of single walled carbon nanotubes on rainbow trout, (*Onchorhynchus mykiss*): Respiratory toxicity, organ pathologies and other physiological effects. *Aquatic Toxicology (Amsterdam, Netherlands)*, *82*(2), 94–109. doi:10.1016/j.aquatox.2007.02.003 PMID:17343929

Song, L., Connolly, M., Fernández-Cruz, M. L., Vijver, M. G., Fernández, M., Conde, E., & Navas, J. M. et al. (2013). Species-specific toxicity of copper nanoparticles among mammalian and piscine cell lines. *Nanotoxicology*, 1–11. PMID:23600739

Song, L., Vijver, M. G., Peijnenburg, W. J. G. M., Galloway, T. S., & Tyle, C. R. (2015). A comparative analysis on the in vivo toxicity of copper nanoparticles in three species of freshwater fish. *Chemosphere*, *139*, 181–189. doi:10.1016/j.chemosphere.2015.06.021 PMID:26121603

Steffens, W. (1989). *Principles of Fish Nutrition* (p. 384). Chichester, UK: Ellis Harwood.

Suvetha, L., Ramesh, M., & Saravanan, M. (2010). Influence of cypermethrin toxicity on ionic regulation and gill Na^+/K^+-ATPase activity of a freshwater teleost fish *Cyprinus carpio*. *Environmental Toxicology and Pharmacology*, *29*(1), 44–49. doi:10.1016/j.etap.2009.09.005 PMID:21787581

Tavares, K. P., Caloto-Oliveira, A., Vicentini, D. S., Melegari, S. P., Matias, W. G., Barbosa, S., & Kummrow, F. (2014). Acute toxicity of copper and chromium oxide nanoparticles to *Daphnia similis*. *Ecotoxicology and Environmental Contamination, 9*(1), 43–50. doi:10.5132/eec.2014.01.006

Tellez-Banuelos, M. C., Santerre, A., Casas-Solis, J., Bravo-Cuellar, A., & Zaitseva, G. (2009). Oxidative stress in macrophages from spleen of Nile tilapia (*Oreochromis niloticus*) exposed to sublethal concentration of endosulfan. *Fish & Shellfish Immunology, 27*(2), 105–111. doi:10.1016/j.fsi.2008.11.002 PMID:19049881

Tiede, K., Boxall, A. B., Tear, S. P., Lewis, J., David, H., & Hassellov, M. (2008). Detection and characterization of engineered nanoparticles in food and the environment. *Food Additives & Contaminants Part A Chemistry, Analysis, Control. Exposure & Risk Assessment., 25*(7), 795–821.

Tranquada, J. M., Sternlieb, B. J., Axe, J. D., Nakamura, Y., & Uchida, S. (1995). Evidence for stripe correlations of spins and holes in copper oxide superconductors. *Nature, 375*(6532), 561–563. doi:10.1038/375561a0

Ullah, R., Zuberi, A., Naeem, M., & Ullah, S. (2015). Toxicity to hematology of liver, brain and gills during acute exposure of Mahseer (*Tor putitora*) to cypermethrin. *Pakistan Journal of Agriculture & Biology,* 12 - 18.

Velasco-Santamaría, Y. M., Handy, R. D., & Sloman, K. A. (2011). Endosulfan affects health variables in adult zebrafish (*Danio rerio*) and induces alterations in larvae development. *Comparative Biochemistry and Physiology Part C Toxicology & Pharmacology, 153*(4), 372–380. doi:10.1016/j.cbpc.2011.01.001 PMID:21262389

Velisek, J., Svobodova, Z., & Piackova, V. (2009). Effects of acute exposure to bifenthrin on some haematological, biochemical and histopathological parameters of rainbow trout (*Oncorhynchus mykiss*). *Veterinarni Medicina, 54*(3), 131–137.

Verma, A., Uzun, O., Hu, Y., Hu, Y., Han, H. S., Watson, N., & Stellacci, F. et al. (2008). Surface structure regulated cell membrane penetration by monolayer protected nanoparticles. *Nature Materials, 7*(7), 588–595. doi:10.1038/nmat2202 PMID:18500347

Wang, T., Long, X., Cheng, Y., Liu, Z., & Yan, S. (2014). The potential toxicity of copper nanoparticles and copper sulphate on juvenile *Epinephelus coioides*. *Aquatic Toxicology (Amsterdam, Netherlands), 152*, 96–104. doi:10.1016/j.aquatox.2014.03.023 PMID:24742820

Wang, T., Long, X., Liu, Z., Cheng, Y., & Yan, S. (2015). Effect of copper nanoparticles and copper sulphate on oxidation stress, cell apoptosis and immune responses in the intestines of juvenile *Epinephelus coioides*. *Fish & Shellfish Immunology, 44*, 674–682. PMID:25839971

Wang, W., Mai, K., Zhang, W., Ai, Q., Yao, C., Li, H., & Liufu, Z. (2009). Effects of dietary copper on survival, growth and immune response of juvenile abalone, *Haliotis discus hannai* Ino. *Aquaculture (Amsterdam, Netherlands), 297*(1–4), 122–127. doi:10.1016/j.aquaculture.2009.09.006

Wendelaar Bonga, S. E. (1997). The stress response in fish. *Physiological Reviews, 77*, 591–625. PMID:9234959

Wepener, V., van-Vuren, J. H. J., & Du-Preez, H. H. (1992). The effect of iron and manganese at an acidic pH on the hematology of the banded Tilapia (*Tilapia sparrmanii*, Smith). *Bulletin of Environmental Contamination and Toxicology, 49*(4), 613–619. doi:10.1007/BF00196307 PMID:1421857

Xu, J. F., Ji, W., Shen, Z. X., Tang, S. H., Ye, X. R., Jia, D. Z., & Xin, X. Q. (1999). Preparation and characterization of CuO nanocrystals. *Journal of Solid State Chemistry, 147*(2), 516–519. doi:10.1006/jssc.1999.8409

Yokel, R. A., & MacPhail, R. C. (2011). Engineered nanomaterials: Exposures, hazards, and risk prevention. *Journal of Occupational Medicine and Toxicology (London, England), 6*(1), 7. doi:10.1186/1745-6673-6-7 PMID:21418643

Yoon, K. Y., Byeon, J. H., Park, J. H., & Hwang, J. (2007). Susceptibility constants of *Escherichia coli* and *Bacillus subtilis* to silver and copper nanoparticles. *The Science of the Total Environment, 373*(2-3), 572–575. doi:10.1016/j.scitotenv.2006.11.007 PMID:17173953

Zhao, J., Wang, Z., Liu, X., Xie, X., Zhang, K., & Xing, B. (2011). Existing research on nanotoxicity has concentrated on empirical evaluation of the toxicity of various nanoparticles. *Journal of Hazardous Materials, 197*, 304–310. doi:10.1016/j.jhazmat.2011.09.094 PMID:22014442

Zhao, X., Wang, S., Wu, Y., You, H., & Lv, L. (2013). Acute ZnO nanoparticles exposure induces developmental toxicity, oxidative stress and DNA damage in embryo-larval zebrafish. *Aquatic Toxicology, 136 – 137*, 49 – 59.

Zhou, K., Wang, R., Xu, B., & Li, Y. (2006). Synthesis, characterization and catalytic properties of CuO nanocrystals with various shapes. *Nanotechnology, 17*(15), 3939–3943. doi:10.1088/0957-4484/17/15/055

Zhu, M. T., Feng, W. Y., Wang, B., Wang, T. C., Gu, Y. Q., Wang, M., & Chai, Z. F. et al. (2008). Comparative study of pulmonary responses to nano and submicron-sized ferric oxide in rats. *Toxicology, 247*(2-3), 102–111. doi:10.1016/j.tox.2008.02.011 PMID:18394769

Zhu, X., Wang, J., Zhang, X., Chang, Y., & Chen, Y. (2009). The impact of ZnO nanoparticle aggregates on the embryonic development of zebrafish (*Danio rerio*). *Nanotechnology, 20*(19), 5103. doi:10.1088/0957-4484/20/19/195103 PMID:19420631

Chapter 16
Control of Perishable Goods in Cold Logistic Chains by Bionanosensors

David Bogataj
Universidad Politécnica de Cartagena, Spain

Damjana Drobne
University of Ljubljana, Slovenia

ABSTRACT

Nanotechnology can contribute to food security in supply chains of agri production-consumption systems. The unique properties of nanoparticles have stimulated the increasing interest in their application as biosensing. Biosensing devices are designed for the biological recognition of events and signal transduction. Many types of nanoparticles can be used as biosensors, but gold nanoparticles have sparked most interest. In the work presented here, we will address the problem of fruit and vegetable decay and rotting during transportation and storage, which could be easily generalized also onto post-harvest loss prevention in general. During the process of rotting, different compounds, including different gasses, are released into the environment. The application of sensitive bionanosensors in the storage/transport containers can detect any changes due to fruit and vegetable decay and transduce the signal. The goal of this is to reduce the logistics cost for this items. Therefore, our approach requires a multidisciplinary and an interdisciplinary approach in science and technology. The cold supply chain is namely a science, a technology and a process which combines applied bio-nanotechnology, innovations in the industrial engineering of cooling processes including sensors for temperature and humidity measurements, transportation, and applied mathematics. It is a science, since it requires the understanding of chemical and biological processes linked to perishability and the systems theory which enables the developing of a theoretical framework for the control of systems with perturbed time-lags. Secondly, it is a technology developed in engineering which relies on the physical means to assure appropriate temperature condi-

DOI: 10.4018/978-1-5225-0585-3.ch016

tions along the CSC and, thirdly, it is also a process, since a series of tasks must be performed to prepare, store, and transport the cargo as well as monitor the temperature and humidity of sensitive cargo and give proper feedback control, as it will be outlined in this chapter. Therefore, we shall discuss how to break the silos of separated knowledge to build an interdisciplinary and multidisciplinary science of post-harvest loss prevention. Considering the sensors as floating activity cells, modelled as floating nodes, in a graph of such a system, an extended Material Requirement Planning (MRP) theory will be described which will make it possible to determine the optimal feedback control in post-harvest loss prevention, based on bionanosensors. Therefore, we present also a model how to use nanotechnology from the packaging facility to the final retail. Any changes in time, distance, humidity or temperature in the chain could cause the Net Present Value (NPV) of the activities and their added value in the supply chain to be perturbed, as presented in the subchapter. In this chapter we give the answers to the questions, how to measure the effects of some perturbations in a supply chain on the stability of perishable agricultural goods in such systems and how nanotechnology can contribute with the appropriate packaging and control which preserves the required level of quality and quantity of the product at the final delivery. The presented model will not include multicriteria optimization but will stay at the NPV approach. But the annuity stream achieved by improved sensing and feedback control could be easily combined with environmental and medical/health criteria. An interdisciplinary perspective of industrial engineering and management demonstrates how the development of creative ideas born in separate research fields can be liaised into an innovative design of smart control devices and their installation in trucks and warehouses. These innovative technologies could contribute to an increase in the NPV of activities in the supply chains of perishable goods in general.

INTRODUCTION

Nanotechnology and Nanomaterials in Agri-Food Sector

Nanotechnology is the creation and use of materials or devices at 1 to 100 nanometer (nm) https://ec.europa.eu/jrc/en/research-topic/nanotechnology). At these dimensions, materials exhibit different physical properties and behaviors not observed at the microscopic level. The new properties of materials at the nanoscale could develop and transform different technologies and create an added value.[1] The properties of materials can be different from that at larger scale. For example, materials at the nanoscale are more reactive because they have a relatively larger surface area when compared to the same mass of material produced in a larger form. In addition, at nanoscale the quantum effects can begin to dominate the behavior of matter affecting the optical, electrical and magnetic behavior of materials

The nanotechnologies have been in the phase of global growth for two decades as investigated by the bibliometric analysis, the topic analysis, the citation network analysis, and by different methods and authors (Islam & Miyazaki, 2010; Chen et al., 2013; Cozzens, Cortes, Soumonni, & Woodson, 2013).

In 2011 the European Commission (EC) developed a Recommendation (2011/696/EU) on the definition of nanomaterials. Following this document, a nanomaterial is defined as a natural, incidental or manufactured material containing particles, in an unbound state or as an aggregate or as an agglomerate and where 50% or more of the particles in the number size distribution and one or more external dimensions are in the size range of 1 nm-100 nm.

According to EC (2013) and European Food Safety Authority (EFSA, 2014), *agriculture* is the backbone of most developing countries, with more than 60% of the population relying on it for their livelihood (EC, 2013; European Food Society Association, 2014). The World Bank's (2013) projection is that the *agribusiness* will represent a $ 2.9 trillion industry in global investment by 2030 (World Bank 2013). It was recognized that the emergence of nanotechnology could benefit agriculture and *food industry* (Gruère, 2012; Poudelet, 2013; Handford et al., 2014; Sabourin and Ayande, 2015). Sabourin and Ayande (2015) review commercial opportunities and the market demand for nanotechnologies in the agribusiness sector and foresee that nanotechnology could have a dramatic impact on all sectors of the agribusiness industry in the next 10 years. Nanotechnologies could make the agri-food sector more efficient, increase yields and product quality, increase the agricultural productivity, enhance flavor to food products, extend product shelf life, and increase the quality and safety of food. Many of these benefits will enhance the range, quality and quantity of food products, enable new international market opportunities, and improve profit margins. They will also offer a great potential for improvements for food and water safety and nutrition in developing countries and thus influence the economic growth of industries (Dasgupta et al., 2015). In addition, the benefits of nanotehnologies could improve nutrition, food and water safety in developing countries.

The global number of scientific publications related to nanotechnology in general and related to the application of nanotechnology in agricultural production has shown an exponentially growing trend until today (Handford et al., 2014). However, the JRC Scientific and Policy Report on "Nanotechnology for the agricultural sector: from research to the field" (Parisi, Vigani & Rodríguez-Cerezo, 2014), reported that despite many potential advantages, the agricultural sector is still comparably marginal and has not yet made it to the market to any larger extent in comparison with other sectors of nanotechnology application. Some authors did not share the view on the revolutionary role of nanotechnology (Busch, 2008). The reasons for this claim are the low funding allocation to nanotechnologies in the agri-food sector, the lack of interdisciplinary research due to disciplinary barriers and the profit potentials being far greater elsewhere. Further, Busch (2008) explains that after being badly burned by biotechnology, the funding agencies have decided that nanotechnologies are simply not worth the effort. It is also true that nanotechnologies pose enormous regulatory hurdles and some serious knowledge gaps. According to the review of Dasgupta et al. (2015) even if the obstacles listed by Busch (2008) may be true, it turned out that nanotechnology in the agro-food sector is one of the fastest growing fields in nano-research. This is evidenced by an increase in the number of publications, patents and intellectual property rights in the field of nano-agri-food and recent research trends in food processing, packaging, nutraceutical delivery, quality control and functional food. Government organizations, scientists, inventors as well as industries are coming up with new techniques, protocols and products that boast a direct application of nanotechnology in agriculture and food products (Dasgupta et al., 2015). According to some reports, *nanosensors* are among more promising nanostructured applications in the agri-food sector (Ravichandran, 2010). In the chapter presented here, we address additional application of nanomaterials /nano sensors in agri-food sector which is reducing the logistics costs.

The nanosensors use different mechanisms to be applicable in agri-food sector. The nanosensors are used either in packaging to detect food deterioration or in devices for monitoring storage conditions and detection of contaminants (Bouwmeester et al., 2009). Within the agro-food chain, metal or metal-oxide nanoparticles are largely applied (e.g., nano-Ag, nano-ZnO, nano-Cu, nano-TiO2) for purposes like detection of organic molecules, gases, moisture, micro organisms etc. A key component of the biosensing is the transduction mechanisms which are responsible for converting the responses of bioanalyte interactions in an identifiable and reproducible manner. The sensing is based on conversion of specific biochemical reaction energy into an electrical (optical) form through the use of transduction mechanisms. In the chapter presented here, we suggest the use of reaction between an enzyme and analyte, where the analyte is affected by the environmental factor being recorded.

Post-Harvest Losses (PHL)

One of the main global challenges is represented by the question on how to ensure food security and sustainable development for a growing world population. According to the Food and Agricultural Organization of the United Nations (FAO), food production will need to increase by 70% to feed the entire world population, which will reach 9 billion by 2050. One of the main ways of strengthening food security is by reducing post-harvest losses (Kiaya, 2014).

By definition, post-harvest food loss (PHL) is any loss in quantity (such as physical weight losses) and quality (loss in edibility, nutritional quality, caloric value, consumer acceptability) that occurs between the time of harvest and the time it reaches the consumer (Buzby & Jeffrey, 2011). Food loss is a subset of PHL and represents the part of the edible share of food that is available for consumption at either the retail or consumer levels but not consumed for any reason, while food waste is the subset of food loss. Parfitt *et al.* (2010) distinguishes between food losses and food wastes, arguing that the losses relate to early stages of the food supply chain (FSC) and refers to a system, which needs investment in the infrastructure to mitigate the risk and reduce loss. In the present chapter chapter we intend to provide how nanotechnology could influence food losses in a supply chain.

Post-harvest losses vary greatly among commodities and production areas and seasons. As a product moves in the post-harvest chain, PHLs may occur due to a number of causes, such as improper handling or bio-deterioration by microorganisms, insects, rodents or birds (Aulakh & Regmi, 2013). Findings of novel scientific researches such as biotechnology and nanotechnology could decrease post-harvest losses and provide sufficient and healthy food. Nanotechnology can help in many different aspects, among them with bionanosensors used for labelling products as well as by assuring the automated control of cargo and inventory in warehouses which could be of great importance.

The *development of nanotechnology* creates an excellent opportunity to address complex technical issues of *agricultural supply chains*. Although there is evidence that nanotechnology could enhance agricultural supply chains, further research is necessary for the implementation of these new technologies (Lu & Bowles, 2013), In the study presented here, we will describe the potential of using nanotechnologies in improving the supply chain management and feedback control.

Jedermann et al (2014) present current development of technology for reducing food losses by intelligent food logistics where also potential economic and social impact of such technologies is presented.

NANOSENSORS

Nanotechnology could provide nano devices incorporating multiple sensing elements for various signals. The obvious challenges are reducing the cost of materials and devices, improving reliability and packing the devices into different products (Li et al. 2010).

In the agri-food sector they are used for identifying and tracking pathogens, contaminants, nutrients, environmental characteristics (light/dark, hot/cold, wet/dry), heavy metals, particulates, and allergens (National Institute of Food and Agriculture, 2015).

The vital parameters of *nanosensors* are sensitivity and low detection ranges. Many sensors operate through the variation of a surface parameter, by surface conductivity, with analyte concentration. Hence, the effective surface area of the sensor (area actually interacting with the analyte), determines the sensitivity. In nanosensors, the surface area is increased due to the application of nanomaterials (Huang, Guo, Peng & Porter, 2011).

Nanosensors can receive information and together with information and communications technology (ICT) they can analyze, record and report data. Advances in miniaturized instrumentation have also resulted in the development of biosensors capable of integrating bio-recognition and spectroscopy tools capable of addressing quality concerns in the food logistics (Li et al. 2010).

Bionanosensors have two basic principles which differ from chemical nanosensors: (1) the sensing elements are biological structures such as cells, enzymes, or nucleic acids; (2) the sensors are used to measure biological processes or physical changes (Malik et al. 2013).

The crucial step in the development of bionanosensors was the integration of biotechnology, nanotechnology and ICT, where industrial engineering and system science (studies of *systems*—from simple to complex) can develop systems that link all these technologies.

The sensing capacity of the detection systems in bionanosensors is being improved lastly by using variety of nanomaterials (metal or metal-oxide based magnetic nanoparticles, carbon based nanomaterials, quantum dots nanowires etc). The nanomaterials, used in electrical biosensors, should have a very high capacity for charge transfer, which makes them suitable to reach lower detection limits and higher sensitivity values. Gold nanomaterials are frequently reported as a nanomaterial of choice for bionanosensors due to their biocompatibility, conductivity, catalytic properties and high surface-to-volume ratio and high density (Li et al 2010).

Au-Nanoparticles (AuNPs)

Different *nanomaterials* can be used in nanosensors. The exact properties of nanoparticles and their correlation with their biological activity are poorly understood. Therefore, an extensive and complete characterization of nanomaterials is crucial (Dasgupta et al., 2015).

There are many kinds of nanoparticles used in biosensors. The major nanoparticle types widely applied in biosensor systems are metal nanoparticles (Pt, Ag, Au, Pd, Cu etc.). Nanosensor publications before 2009 revealed that gold (Au) nanoparticles (NPs) are the most frequently used among all the metal nanoparticles in biosensor systems (Huang, Peng, Guo, & Porter, 2010).

AuNPs, with the diameter of 1-100 nm, have a high surface-to-volume ratio and high surface energy to provide a stable immobilization of a large amount of biomolecules retaining their bioactivity. Moreover, AuNPs have an ability to permit fast and direct electron transfer between a wide range of electroactive species and electrode materials (Li, Schluesener, & Xu, 2010). They have excellent characteristics, such as biocompatibility, conductivity, catalytic properties, high surface-to-volume ratio. High-density AuNPs have attracted a lot of attention in (bio)nanosensors designing (Li et al., 2010).

Several methods have been described in the literature for the *synthesis* of AuNPs of various sizes and shapes. One limitation in using nanomaterial-based products is the limited possibility to synthesize in sufficient amounts and pure, i.e. industrial synthesis.

Control of Perishable Goods in Cold Logistic Chains

To introduce feedback control of perishable goods in Cold Supply Chains (CSC), gold nanoparticles (AuNPs) are very promising particles for nanosensors. The major challenge, and the reason why beneficial properties of AuNPs have not yet been introduced to the nanosensors, are the high-costs and large batch-to-batch variations in AuNP production by standard procedures, indicating that commercially available methods for the production of AuNPs are not sustainable for their large-scale application. Rudolf et al. (2014) have proven that Ultrasonic Spray Pyrolysis (USP) could provide an up-scaling production.

Depending on the intended use of the synthesized nanoparticles, they can be collected in an electrostatic field or suspended in a desired medium. USP as a nanoparticle production method is a relatively inexpensive and quite versatile technique for fine metallic, acidic and composite nanoparticles. Rudolf et al. (2014) have made an important step in the broad application of AuNPs.

NANOSENSORS IN THE INTERNET OF THINGS

Nanosensors for Controlling of Perishable Goods in CSC

In the agri-food sector, nanosensors could be designed to provide information in continuous time at critical control points in the cargo of agricultural product logistics over the period of time from the point when the food is produced or packaged to the time it is consumed. The latest developments have resulted in nanosensors being able to provide quality assurance by tracking microbes, toxins, and contaminants through the food processing chain by using data capture for automatic control functions and documentation. They can be attached to containers, pallets and individual items in cargo to function as active transport tracking devices. A variety of characteristic volatile compounds are produced by microorganisms. The most common causal organisms of food rotting are bacteria. Foul odor is a clear indication of food degradation which may be detected by visual and nasal sensation, but sometimes it may be too late, impractical and a further cause of poisoning. Therefore, it is more sensible to use rapid detection biosensors for the detection of these odors (Compagnone, McNeil, Athey, Dillio, & Guilbault, 1995).

Table 1. Estimation of storage life (in days) of some fruits and vegetables at different temperatures and unspecified humidity (based on Paul, 1999).

Fruit or Vegetable	Temperature °C									Optimum
	3	5	8	10	13	15	18	20	23	
Pears	60	40	25	19	13	12	11	10	9	<3
Avocado	12	20	40	24	16	15	11	10	9	8
Papaya	10	14	21	25	29	27	20	16	12	13
Banana	-	-	5	14	20	22	18	15	11	15
Sprouts	50	36	24	18	12	11	10	9	-	<3
Asparagus	19	17	14	13	10	9	7	6	-	<3
Lettuce	16	21	15	9	7	6	5	5	-	21
Sweet basil	4	5	8	10	12	13	12	11	9	13

Nanotechnology can be used also for preserving the quality and controlling the distribution of agricultural and other food products. Many agricultural products are either perishable or semi-perishable. These include fresh vegetables, fruits, meats, eggs, milk and dairy products, many processed foods, nutraceuticals, and pharmaceuticals (see the examples in Table 1).

Table 1 shows that the dynamics of deterioration depends on the temperature of cargo environment. There is an optimal temperature at which the life of perishable goods is the longest. But the optimal value depends also on humidity and physical influences. How the deterioration dynamics depends on temperature and other environmental factors, when they change the value, could be measured by nano-sensors and reported in the system of the feedback control of cargo developed in the environment of the Internet of Things (IoT). The improvement of shelf-life is one of the main areas for nanotechnology research to enhance the ability to preserve the freshness, quality and safety of food (see the paper by Dr. Q. Chaudhry in this special issue) (Chen & Yada, 2011).

Internet of Things (IoT)

The European Telecommunications Standards Institute (ETSI) and the EC, in the Cluster of European Research Projects (FP7), published the Strategic Research Roadmap on the Internet of Things - IoT (CERP-IoT, 2009). They announced that the IoT will shape the world and the society in the near future and therefore provided applicable recommendations, necessary to steer Europe on its course and make it beneficial for all citizens. On the list of research fields and the roadmap of future R&D, retail, logistics, Supply Chain Management (SCM), and Product Lifecycle Management (PLM) were highly ranked, while the use of the IoT in the transportation of goods, food traceability, different activities in agriculture in general as well as in breeding were also additionally exposed. The authors furthermore highlighted the challenges and the importance of the IoT in the insurance and reverse logistics industries, the recycling of items in a given supply chain.

The authors accepted the name "the Internet of Things" as first coined by Massachusetts Institute of Technology in the late 90s and referred to it as "devices or sensors connected world" where objects are connected, monitored, and optimized through either wired, wireless, or hybrid systems. As Greengard (2015) wrote, it is a "networked world of connected devices, objects, and people". Therefore CERP defined it as follows: "The Internet of Things is a dynamic global network infrastructure with self-configuring capabilities based on standard and interoperable communication protocols where physical and virtual things have identities, physical attributes, and virtual personalities, use intelligent interfaces, and are seamlessly integrated into the information network". The CERP-IoT group of researchers envisioned the Future Internet as comprising public/private infrastructures in Europe, which would be based on standard communication protocols and would merge computer networks, the Internet of Media (IoM), the Internet of Services (IoS), and the Internet of Things (IoT) into a common global IT platform of networks and networked things. The infrastructure would be extended by things as terminals of different kinds of ownership which the infrastructure would connect. This vision is perfectly described in Greengard's book (2015), where he offers a "guided tour" through this emerging world and describes how it will change the way we live and work.

From a supply chain perspective, the IoT may allow machine-enabled feedback control, where decision-making and interventions require minimum to zero human action. It deals with integrating and enabling information communication technologies, including RFID, wireless sensor networks, machine-to-machine systems, mobile apps, etc. The specific reference to "things" refers to the idea that crops, or manufactured products in general, will be part of the extended Internet, since they will be tagged and indexed by the producers/farmers during production/farming. Supply chain managers, or customers in general, can read tags via mobile applications and use the information connected to items in the supply chain to inform them of their purchases, a specific item's use, and disposal. Applications of the IoT are featured in sectors such as transportation, retail, and the food processing industry, as well as in product lifecycle management, recycling, and food traceability, which is particularly relevant to this chapter. According to CERP-IoT, the number of connected devices will reach up to 50 billion by 2020, and we can expect that food supply chains and other smart devices in post-harvest loss prevention will represent a good share in the structure of the different devices involved. The adoption of the IoT in post-harvest loss prevention has the potential to improve operational processes, reduce time lags, costs and risks, and consequently also the costs of insurance of the cargo, due to the IoT's transparency, traceability, adaptability, scalability, and flexibility, which is the topic of this chapter. Therefore, it is easy to forecast the IoT as an open system providing a new way of creating value to existing, in our day largely static, but now more dynamic and agile, information architectures, as adopted by a majority of supply chains. There is a need for data-driven business models and decision-making mechanisms to support this new environment, as stated in the Editorial of the Special issue of the International Journal of Production Economics: Supply chain management in the era of the internet of things (Zhou, Chong, Eric, 2015), where the authors intended to provide a forum to share some new and significant results pertaining to new infrastructure architectures, tools, techniques, and business models that can be applied in supply chains and open environments that are subject to greater uncertainty. Novel algorithms in system planning, control, and optimisation in an IoT environment, and some best practices in adopting the IoT in supply chain management have been presented in the aforementioned study. Using the IoT as described by Qiu, Luo, Xu, Zhong & Huang (2015) in their paper "Physical Assets and Service Sharing", we can create a Supply Hub in virtual cooperative supply chains, where scarce land resources and other capacities of individual members of a cooperative supply chain could improve their physical capacities by sharing the capacities of each partner. In IoT enabled cooperative supply chains, accurate real time information can track and trace physical assets and thus facilitate an automatic extension of the set of admissible solutions, which could improve the optimal solution of operation processes such as inventory management, warehousing management, production, delivery consolidation, truck loading, and transport route design. We would like to show how such cooperative supply chains can also benefit when a sudden change in the dynamics of deterioration of a crop detected by sensors in the supply chain immediately diverts the cargo to a more appropriate destination. Complementing the paper by Qiu et al. (2015), the paper "An RFID-based Intelligent Decision Support System Architecture for Production Monitoring and Scheduling in a Distributed Manufacturing Environment", written by Guo, Ngai, Yang & Liang (2015), proposes an architecture that integrates RFID and cloud technologies to capture real-time information from customer orders and the producer's (in our case the farmer's) plants, which optimizes decision-making on shipping and monitoring to achieve a more agile response and cost reduction. It emphasizes the importance of cloud technology for monitoring, scheduling and re-scheduling in time. As confirmed by Zhou et al. (2015), though their study focuses on the distributed labour-intensive manufacturing sector, the result can easily be extended, rescaled, reused, and customized – in our opinion also in the logistics of crops and other perishables. Based on the designs of Qiu et al. (2015), Guo et al. (2015), and Reaidy, Gunasekaran & Spalanzani (2015), as well as other authors featured in this special issue of the International Journal of production Economics, for instance papers by Fan, Tao, Deng and Li (2015), Yang and Liang (2015), Harris, Y. L. Wang and H. Y. Wang (2015), Ng, Scharf, Pogrebna & Maull (2015), Thiesse and Buckel

(2015), Wong, Lai, Chengnand Lun (2015), and Yu, Nachiappan, Ning and Edwards (2015), a rich set of decision-making tools is worked out to support various managers and end-users in fulfilling their daily tasks. These tools use "service-oriented models which are easy-to-deploy, flexible-to-access and simple-to-use through reconfigurable mechanisms and XML-based data" (Zhou et al., 2015) or other exchange principles. Here, however, we shall present a new approach based on the seminal work in the field of MRP theory, originating with Grubbström and his Linköping School and further developed together with researchers from Mediterranean Institute for Advanced Studies (the most important examples being: Bogataj, Vodopivec and Bogataj, 2005; Bogataj and Bogataj, 2007; Bogataj, Grubbström and Bogataj, 2011; Bogataj and Grubbström, 2012, 2013; Kovačić and Bogataj, 2013, 2015).

THEORETICAL BACKBONE TO DETERMINE ESSENTIAL DECISION VARIABLES IN FEEDBACK CONTROLLED LOGISTICS OF PERISHABLE GOODS

Net Present Value (NPV) Approach

As we have seen, Cold Supply Chain Management presently has at its disposal intelligent technology to improve post-harvest loss prevention. However, the current technology has to be improved to reduce lead times of detection, to improve the forecasting of deterioration trajectories, and to be able to make decisions regarding intervention with respect to the changed perishability dynamics, the final destination, and the reordering of items in an intelligent supply chain. Shorter lead times reduce perishability and therefore increase the added value of products, hence leading to an increase in the value of objective functions. This added value accumulates through the annuity stream of all net incomes in the chain. It can be achieved by developing a broader and more accurate environmental monitoring and feedback control based on improved information and communication technology. At the sensory level, the control of temperature, humidity, and mechanical interventions, as well as their impact on cargo and inventory in warehouses should be studied. For this purpose, new sensors should be developed, which would facilitate a direct insight into rotting or decaying during the logistics processes, enabling faster feedback control and therefore shorter lead times and less deteriorated items in cargo. How the lead time and other, even small, perturbations in timing influence the NPV in CSCs has been demonstrated by Bogataj et al. (2005). By adding constant monitoring of all these measured parameters and enabling real-time access to data through cloud computing, the CSC would allow real-time decision-making and management. Products exposed to risk need to have the opportunity to be sold in the nearest market and at a lower price, if necessary. The activation of cities for selling products exposed to risk locally with clear information about the process of deterioration is easier if the city is included in the CSC communication and cooperation system. In this chapter we discuss the possibility of measuring the impact of these technologies on the NPV and consequently on a well-informed management of CSCs, ranging from packaging and warehousing to distribution and waste disposal. We also supply a list of references providing information on how to improve post-harvest loss prevention and mitigate risks. An evaluation of the involvement of new technologies is also by possible through the NPV approach to extended material requirements planning procedures.

As concluded by Bogataj et al. (2015), cold supply chain is a science, a technology, and a process that combines applied bio-nanotechnology, innovations in the industrial engineering of sensor cooling and humidity, transportation, and applied mathematics. It is a science since it requires an understanding of the chemical and biological processes linked to perishability as well as a grasp of systems theory, which allows for the development of a theoretical framework for the control of systems with perturbed time lags. Secondly, it is a technology developed in engineering, which relies on physical means to ensure appropriate temperature conditions along the CSC. Thirdly, it is also a process since a series of tasks must be performed to prepare, store, transport, and monitor temperature- and humidity-sensitive cargo as well as have proper feedback control. Therefore, we have to merge the silos of separated knowledge in order to build an interdisciplinary and multidisciplinary science of post-harvest loss prevention. Considering the sensors as floating activity cells, as nodes in a graph of such a system, an extended MRP theory can be applied to determine optimal feedback control.

Early Writings on MRP Theory

The Material Requirements Planning (MRP) Theory was first developed for production planning purposes. Production was supposed to be running under one roof and there were no problems in transportation lead times. As presented in an overview in 1997 at Storlien (Grubbström, 1998a), the Linköping Approach was a line of research attempting to develop a theoretical background for Material Requirements Planning (MRP) for assembly, distribution and mixed systems. Although widely applied in practice, MRP was considered to be lacking a theoretical instrument for the determination of essential decision variables, such as order quantity, safety stock, and/or safety lead time. In planning decisions – as already criticized by Orlicky, who was the founder of the general MRP approach (Orlicky, 1975) – it often referred to the practical computer-based information systems applied in production processes. Continuing his previous considerations about the applications of the Laplace transform to certain economic problems, Grubbström (1967) defined the term more broadly to incorporate general production planning problems in which multilevel product structure and lead time have become important characteristics. Based on Leontief's "Input-Output Analysis" (1928, 1936), Koopman's "Activity Analysis" (1951), and Vazsonyi's (1955, 1958) approach to production and inventory control, the Linköping School published several papers on topics extending from the static and deterministic average cost approach, such as can be found in Grubbström's paper (1969) or in Grubbström and Lundquist (1989), to the NPV approach to one-level problems for continuous time dynamics using Laplace transforms, developed, for example, in the paper published by Grubbström and Lundquist (1977). Later, Thorstenson (1988) studied the discounted cash flow approach within the scope of economic and production research, and together with Grubbström (1993) published a theoretical analysis of the difference between the traditional and the annuity stream principles applied to inventory evaluation. An early survey of the development of this theory has been given by Grubbström and Jiang (1990). In the early nineties, Grubbström and Ovrin (1992) considered the Economic Order Quantity for multilevel cases in a z-transformed space of complex variables, still considering average costs and the annuity stream as criteria functions. This problem was later analyzed in the case of continuous time by Grubbström and Molinder (1994, 1996), also on a detailed coordination for only two levels, using the Laplace transform.

In multi-level production–inventory structures, the Input-Output approach has been used to enable a compact and distinct analysis. The theory of deterministic production-inventory models has therefore been developed, growing increasingly generalized until 1994, when Grubbström first analyzed stochastic MRP models (the Poisson and Gama distribution of events), in the beginning using only the one-level approach in the space of Laplace transformed, complex variables (Grubbström, 1994, 1996). Simultaneously, Bogataj and Horvat (1995, 1996) from the University of Ljubljana developed a stochastic approach to MRP assuming the Poisson-distributed demand. Two years later, Bogataj and Bogataj (1998a, 1998b) independently published a paper on the Compound distribution of demand in location – inventory problems. In that period, stochastic cases in general were also elaborated by Grubbström and Andersson (1994). Later on, Grubbström and Molinder (1994, 1996) developed the MRP stochastic continuous model, also focusing on safety stock. The question of safety stock is particularly important for modelling post-harvest loss prevention, although authors did not consider the perishability of goods in any of the aforementioned papers. To the developed models, Segerstedt (1995) added capacity requirements for deterministic cases, which were later generalized for stochastic L4L cases by Grubbström (1996c). While Segerstedt's approach considered average cost, Grubbström introduced the annuity stream as an objective function in the case of capacity constraints, after developing the annuity stream approach in a case without capacity constrains (Grubbström, 1996b).

An extensive list of papers followed the Storlien conference in 1997 (see Grubbström and Bogataj, 1998), which developed the missing ingredients of the theory, such as Grubbström (1998, 1999, 2012), Tang and Grubbström (2002, 2003, 2005, 2006), Grubbström and Tang (1999, 2000, 2006a, 2006b, 2012), Grubbström and Wang (2003), Tang, Grubbström and Zanoni (2004, 2007) and Zanoni and Grubbström (2004), Disney and Grubbström (2004), Zhou and Grubbström (2004), Grubbström and Huynh (2006, 2006b) and Naim, Wikner, and Grubbström (2007). Still, none of these papers considered the perishability of goods.

In the papers from the seventies and all up until the Storlien Conference, the cost approach was used in the majority of objective functions considering production-inventory problems. In the late nineties, the Linköping School and researchers from the University of Ljubljana continued with the study of problems, positing the annuity stream and its reverse transform to NPV as criteria functions. This approach is important because of the addition of risk evaluation as an objective function, such as the risk of an accelerated deterioration of perishables, provides additional clarity. The main sources of risks which influence the NPV in the MRP approach to production-inventory problems are described by D. Bogataj and M. Bogataj (2007), while perishable goods in MRP systems exposed to risk of deterioration have been elaborated in detail by M. Bogataj, L. Bogataj and Vodopivec (2005). In these papers, all links between two activity cells were modelled as a new node, so it was not possible to tackle the problem of timing in the detection of unexpected changes on the road in an appropriate way. The risk was uniformly distributed on a given link. Parallel to this paper, the controllability of a supply chain, controlled in a Laplace-transformed space, was studied by Bogataj, Bogataj and Ferbar -Tratar (2005). Furthermore, L. Bogataj and M. Bogataj showed how to evaluate additional investments in the capacities of MRP systems in 2007 (2007a) and a compact presentation of the lead-time perturbations in distribution networks was elaborated by M. Bogataj and L. Bogataj (2007b). 2007 saw the publication of the book *A compact representation of distribution and reverse logistics in the value chain*, which took into consideration the total supply chain, from production to distribution and reverse logistics (Grubbström, Bogataj and Bogataj, 2007). The realization dawned that for a global production, where production cells are no longer located under one roof, an extra transportation matrix should be introduced in the system, which was later, based on the ideas proposed by Bogataj and Bogataj (2004), elaborated by Bogataj, Grubbström and Bogataj (2011) when discussing the efficient location of industrial activity cells in a global supply chain, and by Bogataj and Grubbström (2012, 2013). However, in the study of dynamic properties elaborated in the lot sizing problem conducted by Grubbström, Bogataj and Bogataj (2010) the location and transportation matrices were not discussed.

Control of Perishable Goods in Cold Logistic Chains

After the introduction of transportation matrices, nothing prevents us from including in the system smart devices as floating activity cells in the Input-Output matrices of the Extended MRP Theory (EMRP) as described by Bogataj and Bogataj (2015) and applied to post-harvest loss prevention by Bogataj et al. (2015), while some practical experience on the transportation of perishable goods has already been achieved in the Spanish baby food industry (Kovačić et al., 2014).

Theoretical Backbone for Feedback-Controlled Logistics of Perishable Goods

Bogataj et al. (2015) based their idea on the following Grubbström's and Bogataj's formalization of the extended MRP model.

Early Writings

According to the basic MRP theory developed by Grubbström, the *j*th process is run on activity level (node) P_j, the volume of required inputs of item *i* at production unit *j* is $h_{ij}P_j$, and the volume of produced (transformed or uploaded, unloaded, conserved, detected as changed in quality) outputs of item *k* is $g_{kj}P_j$. The total of all inputs may then be collected into the column vector HP, and the total of all outputs into the column vector GP, from which the net production is determined as (G - H)P. In general, P (and thereby net production) will be a time-varying vector-valued function. In MRP systems, *lead times* are essential ingredients. A *lead time* is the latency between the initiation and completion of a process. If $P_j(t)$ is the rate of item *j*, planned to be completed at time *t*, then $h_{ij}P_j(t)$ of item *i* needs to be available for production (assembly, packaging…), and the lead time τ_j in advance of *t*, i.e. at time $(t - \tau_j)$. The volume $h_{ij}P_j$ of item *i*, previously having been a part of *available inventory* at time $(t - \tau_j)$, is marked for the specific production or manipulation $P_j(t)$, and thereby moved into *work-in-process or logistic process* (*activities in general*). At time *t*, when this activity is completed, the identity of the items type *i* disappears, and instead the newly produced items appear.

The Transportation Matrix with Timing, Including Sensors Installed in Transportation Units

While the item *i* is assumed to be previously located at location *i*, it will be available for activity *j* at location *j* before activity $P_j(t)$ starts, and it will need τ_{ij} to arrive there. We will introduce sensors attached to the items with known perishability β_{ij} in the flow between *i* and *j* as activity j', located on the link between two physical nodes. If change in perishability appears to have a higher intensity β'_{ij}, the sensor there reports on these changes in time delay $\tau_{ij} - \delta_{ij}$ before coming to the child node. If there are no perturbations, then delta is zero: $\delta_{ij} = 0$. In order to incorporate the lead times of transportation and manipulations in the nodes for the processes, we transform the relevant time functions into Laplace transforms in the frequency domain (as presented by Bogataj and Grubbström, 2012 and 2013). To simplify, Bogataj et al. considered a *linear logistic* system, for which the components of activity *j* need to be in place τ_j time units before completion the process (packaging, sorting, etc.), and sent from parent node i to j having a floating point of sensor in j' for additional time delay τ_{ij} in advance. By β they denoted the matrix of dynamics of rotting and decay from each parent node to each child node in the graph of the logistic chain and by β^* the matrix of perturbed dynamics. The values of coefficients in these two matrices were supposed to be improved by smart devices, dependent on temperature and humidity. They also denote the matrix of distances between the floating point and its destination node in case of perturbation by matrix δ, having coefficients in the row belonging to the index of the floating point and in the column of the index of the child node of each floating point. δ is the zero matrix if there are no perturbations in the system.

The matrix of perishability in CSC graph, denoted by $\tilde{\mathbf{H}}(s,\beta^*,\delta)$, which includes the floating points between each parent and child node of activity cells, as described in the matrix

$$\tilde{\mathbf{H}}(s,\beta^*,\delta) = \begin{bmatrix} 0 & 0 & \cdots & \cdots & \cdots & & 0 \\ h_{21}e^{s\beta^*_{21}\delta_{21}} & 0 & 0 & 0 & \cdots\cdots & & 0 \\ h_{31}e^{s\beta_{31}\tau_{31}} & h_{32}e^{s\beta_{32}(\tau_{31}-\delta_{21})} & 0 & 0 & \cdots\cdots & & 0 \\ & & h_{43}e^{s\beta^*_{43}\delta_{43}} & 0 & \cdots\cdots & & 0 \\ & & h_{53}e^{s\beta_{53}\delta_{53}} & h_{54}e^{\beta_{54}s(\tau_{53}-\delta_{43})} & \vdots & \cdots\cdots & \cdots\cdots \\ \cdots\cdots & \cdots\cdots & \cdots\cdots & \cdots\cdots & \vdots & \vdots & \vdots \\ 0 & 0 & 0 & 0 & \cdots & h_{n-1,n-2}e^{s\beta^*_{n-1,n-2}\delta_{n-1,n-2}} & 0 \\ 0 & 0 & 0 & 0 & \cdots & h_{n,n-2}e^{s\beta_{n,n-2}\delta_{n,n-2}} & h_{n,n-1}e^{s\beta_{n,n-1}(\tau_{n,n-2}-\delta_{n-1,n-2})} \end{bmatrix}, \quad (1)$$

and lead times in activity cells including delays of information in floating points by

$$\tau = \begin{bmatrix} e^{s\tau_1} & \cdots & 0 \\ \vdots & \ddots & \vdots \\ 0 & \cdots & e^{s\tau_n} \end{bmatrix} \quad (2)$$

By applying the time translation theorem in Laplace transforms, authors Bogataj and Bogataj (2015) have shown that input requirements as transforms can be written as

$$\tilde{\mathbf{H}}(s,\beta^*,\delta)\begin{bmatrix} e^{s\tau_1} & \cdots & 0 \\ \vdots & \ddots & \vdots \\ 0 & \cdots & e^{s\tau_n} \end{bmatrix}\tilde{\mathbf{P}}(s) = \breve{\mathbf{H}}(s,\beta^*,\delta)\tilde{\mathbf{P}}(s) \quad (3)$$

where $\breve{\mathbf{H}}(s,\beta^*,\delta)$ is the perturbed *generalized transportation- logistics activity input matrix* capturing the volumes of requirements as well as their advanced timing, and $\breve{\mathbf{H}}(s,\beta,[0])$ is its unperturbed case.

The vector $\mathbf{u}(s,\beta,[0]) = \breve{\mathbf{H}}(s,\beta,[0])\tilde{\mathbf{P}}(s)$ describes in a compact way all component volumes that need to be in place for the logistic plan $\tilde{P}(s)$ to be possible. The perturbations described by $\mathbf{u}(s,\beta^*,\delta) = \breve{\mathbf{H}}(s,\beta^*,\delta)\tilde{\mathbf{P}}(s)$ need some feedback control to get a solution close to the required one regarding legal constraints, technology and costs. If a component of the net production vector is negative, there is a need either for taking this amount from an available inventory or for importing this amount into the system, and if it is positive, it may be delivered (sold) to the environment of the system or added to an available inventory.

Similar to the fundamental equations derived by Grubbström in his early writings, using notation $\tilde{\mathbf{F}}(s)$ for the vector of deliveries from the system and $\tilde{\mathbf{F}}^*(s)$ as its perturbed case; $\tilde{P}(s)$ for given logistics plan, the available inventory $\tilde{\mathbf{R}}(s)$ would develop according to the following equation:

$$\tilde{\mathbf{R}}(s,\beta^*,\delta) = \frac{\mathbf{R}(0) + \left\{\mathbf{I} - \breve{\mathbf{H}}(s,\beta^*,\delta)\right\}\tilde{\mathbf{P}}(s) - \tilde{\mathbf{F}}^*(s)}{s}, \quad (4)$$

where $\mathbf{R}(0)$ collects initial available inventory levels. The division by s represents a time integration of the flows represented by other terms.

Like in previous writings of Grubbström, Bogataj and Bogataj (2015) also considered here *cyclical processes*, repeating themselves in constant time intervals T_j, $j = 1, 2, ..., m$, according to the plan $\tilde{P}(s)$ in the same way as in equation (12) of Grubbström (1998).

Introducing economic relationships, Bogataj et al. (2015a) and Bogataj and Bogataj (2015) assumed that perishables have unit economic values, which could be different in different nodes. If there is consumption planned only in the final node, the values of all other components are equal to 0. If the cargo suddenly becomes highly exposed to the risk of decay and smart devices recognize that, the system could report to a nearby city, which hosts a child node, so that a smart city could organize the transactions for such cargo locally at lower but acceptable prices, or could in the worst case organize a disposal of rotten goods.

Therefore, the authors have written the unperturbed price vector $\mathbf{p}$ as a row vector:

$$\mathbf{p} = \left[p_1, p_2, ..., p_n\right] \quad (5)$$

Usually at floating points it was assumed $p_2 = p_4 = ... = p_{n-1} = 0$, but in the case of perturbations the values could be lower than p_1 or even negative. Therefore, there exists a perturbed vector of lower or even negative prices:

$$\mathbf{p}^* = \left[p_1^*, p_2^*, ..., p_n^*\right] \quad (6)$$

Following the procedure of Bogataj and Grubbström (2012, 2013), the authors Bogataj and Bogataj (2015), and Bogataj et al. (2015b) expressed the NPV of unperturbed CSC and NPV* for the perturbed system as follows:

$$\mathrm{NPV} = \mathbf{p}\big(\mathbf{I} - \breve{\mathbf{H}}(\rho)\big)\tilde{\mathbf{P}}(\rho) - \mathbf{K}\tilde{\boldsymbol{\nu}}(\rho) \tag{7}$$

$$\mathrm{NPV}^* = \mathbf{p}^*\big(\mathbf{I} - \breve{\mathbf{H}}(\rho,\beta^*,\delta)\big)\tilde{\mathbf{P}}_1(\rho) - \mathbf{K}_1\tilde{\boldsymbol{\nu}}_1(\rho) \tag{8}$$

where $\mathbf{K}$ were ordering costs appearing at each node, also in floating points if necessary, collected into the row vector $\mathbf{K} = \big[K_1, K_2, \ldots, K_n\big]$, $\tilde{\boldsymbol{\nu}}(s) = \begin{bmatrix} \tilde{\nu}_1(s) \\ \vdots \\ \tilde{\nu}_m(s) \end{bmatrix}$ and ρ is a continuous interest rate, all similar as derived in Grubbström (1998).

As Explained by Bogataj and Bogataj (2015) and Bogataj et al. (2015b), at a given CSC the NPV^* depends on time delays in detection of deterioration and on proper forecasting of the development of deterioration, which all depend on the quality of smart devices and the proper and quick adaptation of temperature and humidity. All parameters are present in $\breve{\mathbf{H}}(\rho,\beta^*,\delta)$: the minimum delay in δ, the most accurate forecast of β^* and better communication of CSC management with cities which can be potential buyers. Therefore, supply chains and cities need smart devices to control, optimally decide and to communicate with agglomerations in the surroundings of CSC to quickly sell the products which are still of acceptable quality, in order to mitigate the consequences of risk realization. This is an example how such a skeleton of formalized cold supply chain could connect and formalize the connection of smart devices into a powerful feedback control of perishable goods.

FUTURE CHALLENGES, SOLUTIONS AND RECOMMENDATIONS

The number of patent applications in nanotechnology in general has increased for more than a tenfold during the last 20 years, demonstrating a great potential for commercial applications (Parisi et al., 2014). However, at present, most products of nanotechnology in the global food sector are in the R&D stage and only some are nearing the market. Reasons for this may be found in knowledge gaps which need further research as well as in safety issues which need to be addressed (Chen et al., 2013).

Millennium Development Goals (MEG) and Nanotechnologies in Agri-Food Sector

Cozzens et al. (2013) have investigated to what extent the research agendas of nanoscience and engineering have reflected the millennium development goals. These authors have taken into consideration the rank of nanotechnology application areas established in 2005 by Salamanca-Buentello et al. (2005), a Toronto group. The Toronto group identified specific areas where nanotechnologies could help developing countries achieve the millennium development goals (MEG). Among the highest ranking are agricultural productivity enhancement and food processing and storage. Cozzens et al. (2013) determined that among the technologies mentioned in the Toronto Group study, only a few are getting discernible attention. Among them are *bionanosensors* for soil quality and plant health monitoring and nanocomposites for food packaging. The application of bionanosensors in the agri-food sector is obviously worthwhile to be explored and broadened for other purposes as well. However, nanotechnologies are probably not enough to explore their full potential and must be seen as a part of a much larger collection of new technologies defined as 'converging technologies' (AZoNANO, 2005).

Nanotechnologies as Part of Converging Technologies and General Purpose Technologies

In both Europe and the United States of America, researchers and policy-makers have recognized the potential of converging technologies. However, Europe and USA are taking different approaches to their agenda.

The EC in its report on Converging Technologies, entitled *"Foresighting* the New Technology Wave", proposed Converging Technologies for the European Knowledge Society (CTEKS, 2010) and emphasized that it adopts a demand-driven approach in which converging technologies respond to societal needs and demands. The European approach urged to take the precautionary principle into account and made it "a priority to clarify the civil and societal benefits of this research to give them new legitimacy and to put them firmly in a context of positive social dynamics."

In the report of Committee for Scientific and Technological Policy (CSTP) within OECD (2014) the nanoscience and development of nanotechnology has been identified as a cornerstone of various visions of converging technologies. This report explores the characteristics of nanotechnology as it relates to technology convergence, examining four application areas in which nanotechnology plays a strong role (green packaging, food safety and security, pharmaceuticals and medical devices).

The US Government refers to convergence as NBIC (the integration of Nanotechnology, Biotechnology, Information Technology and Cognitive Science) with the goal of "improving human performance", both physically and cognitively. The US Governemtn .pushed the NBIC (nano, bio, info, cogno) approach to focus strongly on the enhancement of an individual human being.

Kreuchauff and Teichert (2014) provided evidence that nanotechnologies are evolving as general purpose technologies (GPT), as predicted by both scholars and practitioners. The application and convergence of GPT, biotechnologies, ICT, and nanotechnologies generates new paradigms and a huge expansion of innovation possibilities, advancing the best scientific and technological knowledge of different fields of research. The real future challenge is to exploit the great potential of their convergence.

Upscale Production, Repeatability, and Characterization of NPs

The potential of nanotechnology in agriculture is large, but a few issues are still to be addressed. The most important among them is how to increase the scale of production processes and lowering costs as well as the risk assessment issues.

The properties of NPs greatly depend on chemical composition and also on their surface characteristics. Unlike chemical compounds, where the chemical composition and purity is related to their activity, in the case of nanomaterials their activity depends of size, size distribution in suspensions, shape, surface area, surface chemistry, crystallinity, porosity, agglomeration state, surface charge, solubility, etc. It is to be highlighted that in most cases, the availability of facilities determines the type of characterization performed rather than the study design or experimental needs (Dhawan & Sharma, 2010). This needs to be changed in the future. Reference methods for nanomaterial characterization have to be established and harmonized on an international scale. Some of these issues are being addressed and the methods will be provided by the EU funded project NanoValid (http://www.nanovalid.eu/).

Regulation and Safety

Successful applications of nanotechnology to agri-food are limited due to many regulatory uncertainties of food companies. The agri-food sector is based on the experience with genetically modified food crops, therefore the technological changes in agriculture are perhaps cautious and balanced in their approach to nanotechnology's potential benefits.

We believe that the regulatory issue is a serious obstacle in the application of nanotechnology in the agri-food sector. But still there are many safer applications where nanomaterials are not coming in direct contact with the consumers or the environment and where their life cycle could be controlled.

Important challenge related to nanotechnology in agriculture and food industry refers to risk assessment issues Handford et al. (2014).

Nanotechnology is a rapidly emerging technology and its use raises many of the same safety issues as with any form of new technology. It needs to rely on the projections of safety, efficacy, and acceptable quality, with immediate revisions as new knowledge becomes available. In this technology, a continuous re-evaluation of risk is often needed.

Only the flexibility of regulation may foster innovation, while the improving use of pre-market and post-market information is the focus of attention by regulators, industry and NGOs alike. Some are particularly focused on developing more adaptive and discriminatory approaches to the management of risks and uncertainty during the development and the learning phase.

However, when revisions are embedded in regulatory and licensing frameworks, the industry may suffer from a climate of instability. The future challenge for regulates is to develop some form of certainty in regulation for the investors and operators.

How to Advance the Application of Nanotechnologies in General in Agri-Food Sector?

The agri-food and related sectors are facing same specific challenges to realize the potential of nanotechnologies. Some of them are common to all sectors while others are specific for the agri-food sector.

- In the case of the agri-food and related sectors the pre-market evaluation of nano-food products should be ensured as well as the labelling of nano-food products to inform the consumer. A global body for ensuring quality control and safety assessment of nano-food is needed.
- Among most important general future challenges in a successful application of nanotechnologies is the international research collaboration and networks that can address benefits and potential health as well as the environment risk. International and inter-sectoral collaboration could help develop clear and consistent international guidelines for risk assessment and a harmonized regulatory system on the global level.
- The promotion of industry best practices and self-regulation in the use of nanotechnologies for food and related applications would be of significant benefit.
- Focusing the attention and mobilizing resources to address Grand Challenges and solving the problems of "developing countries."; helping developing countries achieve the millennium development goals; identifying the areas, where nanotechnologies might actually improve the lives of the poor based on current analyses and research agendas.

Due to bibliometric analyses, Huang et al. (2010) suggested that a nanoparticle-based biosensor system is a highly *cross-disciplinary research area*. Therefore, to foster a cooperative research network and to facilitate a cross-field and cross-institutional research knowledge transfer, this field is ripe for stimulated research knowledge exchange.

CONCLUSION

The goal of the approach described here is the implementation of nanotechnologies in reducing the post-harvest loss which is in line with the millennium development goals.

Most food losses take place after the crop harvest, in the supply chains between the field and the final market. Around 1/3 of all crops and more than third of fruits and vegetables are lost in post-harvest logistics. The reasons for post-harvest loss vary widely, depending on crop type, the region of origin, the technology of harvesting and distribution, as well as the weather – especially temperature, humidity and other environmental conditions. There is a long list of causes of post-harvest loss, but there are three that need to be exposed: poor transportation and storage, lack of knowing new technologies and missing financial sources to invest in new technologies, and limited data. The Internet of things enables us to build databases needed for optimal control and to develop the sensors for feedback control of supply chains. Here, nanotechnology is of potential importance. We have explained how nanosensors could contribute to a better and quicker detection of required conditions of cargo and inventories of perishables in warehouses. Improved sensors can reduce lead time in decisions about changing regulations (temperature, humidity or even the direction towards the final destination of cargo). To improve decision-making in time and in quality, the formalization of such procedure is needed. One way to approach better management and control in post-harvest loss prevention is formalized by an extended MRP approach which helps improve decision-making in case of unexpected events in logistics of perishables and evaluate the impact of new technology which is improving the control, also the contribution of bionanosensors to NPV of all logistics activities in CSC. We can expect that the investigation of nano-materials in our laboratories will contribute to a higher NPV.

REFERENCES

Andersson, L. E., & Grubbström, R. W. (1994). *Asymptotic behavior of a stochastic multi-period inventory process with planned production*. Working paper WP – 210. Dept. of Production Economics, Linköping Institute of Technology.

Aulakh, J., & Regmi, A. (2013). *Post-harvest Food Losses Estimation - Development of Consistent Methodology*. Paper presented at the Agricultural & Applied Economics Associations 2013, AAEA & CAES Joint Annual Meeting, Washington, DC. Retrieved from http://worldfoodscience.com/article/postharvest-losses-and-food-waste-key-contributing-factors-african-food-insecurity-and#sthash.NMUXu0mB.dpuf

AZoNANO. (2005, Aug 15). *Converging Technologies - a Definition and an Overview of Some Different Viewpoints*. Retrieved August 24, 2015, from: http://www.azonano.com/article.aspx?ArticleID=1354

Bogataj, D., & Bogataj, M. (2007). Measuring the supply chain risk and vulnerability in frequency space. *International Journal of Production Economics, 108*(1-2), 291–301. doi:10.1016/j.ijpe.2006.12.017

Bogataj, D., & Bogataj, M. (2015). Floating points of a cold supply chain in an environment of the changing economic growth. In *Proc. Internat. Symp. on Operations Research SOR'15*. Bled: SDI-SOR.

Bogataj, D., Bogataj, M., Drobne, D., Ros-McDonnell, L., Rudolf, R., & Hudoklin, D. (2015a). Evaluation of investments in smart measurement devices controlling cold supply chains through nanotechnology, In *Proc. Internat. Symp. on Operations Research SOR'15*. Bled: SDI-SOR.

Bogataj, D., Drobne, D., Rudolf, R., Bogataj, M., & Hudoklin, D. (2015b). Evaluation of investments in smart measurement devices controlling Cold Supply Chains through nanotechnology where smart cities are hosting activity cells. In *Proceedings of 6th IESM Conference, I4e2*.

Bogataj, D., Vodopivec, R., & Bogataj, M. (2013). The extended MRP model for the evaluation and financing of superannuation schemes in a supply chain. *Technological and Economic Development of Economy, 19*(S1), S119-S133. doi:10.3846/20294913.2013.875957

Bogataj, L. (1995). Lead times and other delays in inventory systems. In L. Bogataj (Ed.), *Inventory Modelling - Lecture Notes of the International Postgraduate Summer School* (Vol. 1, pp. 23–25). Budapest: ISIR.

Bogataj, L., & Bogataj, M. (1998). *Input-output analysis applied to MRP models with compound distribution of total demand*. Lecture notes. New York: University of New York. Retrieved August 24, 2015, from: http://www.inforum.umd.edu/papers/ioconferences/1998/journali.pdf

Bogataj, L., & Bogataj, M. (2007). The study of optimal additional investments in capacities for reduction of delays in value chain. *International Journal of Production Economics, 108*(1-2), 281–290. doi:10.1016/j.ijpe.2006.12.016

Bogataj, L., Bogataj, M., & Ferbar Tratar, L. (2005) Controllability of supply chain, controlled in Laplace - transformed space. *International Journal of Applied Mathematics, 19*(3), 17-41.

Bogataj, L., & Grubbström, R. W. (Eds.). (1998). Input-output analysis and Laplace transforms in material requirements planning: Storlien. Portorož, Slovenia: University of ljubljana, FPP.

Bogataj, L., & Horvat, L. (1995). The application of game theory to MRP, input-output and multi-echelon inventory systems with exponentially distributed external demand. In *Proc. Internat. Symp. on Operations Research SOR'95* (pp. 25–36), Portoroz, Slovenia

Bogataj, L., & Horvat, L. (1996). Stochastic considerations of Grubbstrom-Molinder model of MRP, input-output and multiechelon inventory systems. *International Journal of Production Economics, 45*(1-3), 329–336. doi:10.1016/0925-5273(96)00050-3

Bogataj, M., & Bogataj, L. (1998). Compound distribution of demand in location - inventory problems. In S. Papachristos, & I. Ganas (Eds.), *Inventory Modelling in Production and Supply Chain* (pp. 15-46). Ioannina, Greece: University of Ioannina.

Bogataj, M., & Bogataj, L. (2004). On the compact presentation of the lead times perturbations in distribution networks. *International Journal of Production Economics*, *88*(2), 145–155. doi:10.1016/j.ijpe.2003.11.004

Bogataj, M., Bogataj, L., & Vodopivec, R. (2005). Stability of perishable goods in cold logistic chains. *International Journal of Production Economics*, *93/94*, 345–356. doi:10.1016/j.ijpe.2004.06.032

Bogataj, M., & Grubbström, R. W. (2012). On the representation of timing for different structures within MRP theory. *International Journal of Production Economics*, *140*(2), 749–755. doi:10.1016/j.ijpe.2011.04.016

Bogataj, M., & Grubbström, R. W. (2013). Transportation delays in reverse logistics. *International Journal of Production Economics*, *143*(2), 395–402. doi:10.1016/j.ijpe.2011.12.007

Bogataj, M., Grubbström, R. W., & Bogataj, L. (2011). Efficient location of industrial activity cells in a global supply chain. *International Journal of Production Economics*, *133*(1), 243–250. doi:10.1016/j.ijpe.2010.09.035

Bouwmeester, H., Dekkers, S., Noordam, M. Y., Hagens, W. I., Bulder, A. S., De Heer, C., & Sips, A. J. et al. (2009). Review of health safety aspects of nanotechnologies in food production. *Regulatory Toxicology and Pharmacology*, *53*(1), 52–62. doi:10.1016/j.yrtph.2008.10.008 PMID:19027049

Busch, L. (2008). Nanotechnologies, food, and agriculture: Next big thing or flash in the pan? *Agriculture and Human Values*, *25*(2), 215–218. doi:10.1007/s10460-008-9119-z

Buzby, J. C., & Hyman, J. (2012). Total and per capita value of food loss in the United States. *Food Policy*, *37*(5), 561–570. doi:10.1016/j.foodpol.2012.06.002

Buzby, J. C., & Jeffrey, H. (2011). Total and per capita value of food loss in the United States. *Food Policy*, *37*(5), 561–570. doi:10.1016/j.foodpol.2012.06.002

Cachon, G. P., & Fisher, M. (2000). Supply chain inventory management and the value of shared information. *Management Science*, *46*(8), 1032–1048. doi:10.1287/mnsc.46.8.1032.12029

CERP-IoT. (2009). *Internet of Things: Strategic Research Roadmap. Report from the Cluster of European Research Projects (FP7) on the Internet of Things (CERP-IoT)*. Retrieved August 24, 2015, from: http://www.internet-of-things-research.eu/pdf/IoT_Cluster_Strategic_Research_Agenda_2009.pdf

Chen, H., Roco, M. C., Son, J., Jiang, S., Larson, C. A., & Gao, Q. (2013). Global Global nanotechnology development from 1991 to 2012: Patents, scientific publications, and effect of NSF funding. *Journal of Nanoparticle Research*, *15*(9), 1–21. doi:10.1007/s11051-013-1951-4

Chen, H., & Yada, R. (2011). Nanotechnologies in agriculture: New tools for sustainable development. *Trends in Food Science & Technology*, *22*(11), 585–594. doi:10.1016/j.tifs.2011.09.004

Compagnone, D., McNeil, C. J., Athey, D., Di Ilio, C., & Guilbault, G. G. (1995). An amperometric NADH biosensor based on NADH oxidase from Thermus aquaticus. *Enzyme and Microbial Technology*, *17*(5), 472–476. doi:10.1016/0141-0229(94)00110-D

Cozzens, S., Cortes, R., Soumonni, O., & Woodson, T. (2013). Nanotechnology and the millennium development goals: Water, energy, and agri-food. *Journal of Nanoparticle Research, 15*(11), 10–23. doi:10.1007/s11051-013-2001-y

Dasgupta, N., Ranjan, S., Mundekkad, D., Ramalingam, C., Shanker, R., & Kumar, A. (2015). Nanotechnology in agro-food: From field to plate. *Food Research International, 69*, 381–400. doi:10.1016/j.foodres.2015.01.005

Development of research/innovation policies - Science and Technology Foresight. (2010, Jan 22). *Converging Technologies for a Diverse Europe*. Retrieved August 24, 2015, from: http://cordis.europa.eu/foresight/ntw-expert-group.htm

Dhawan, A., Sharma, V., & Parmar, D. (2009). Nanomaterials: A challenge for toxicologists. *Nanotoxicology, 3*(1), 1–9. doi:10.1080/17435390802578595

Disney, S. M., & Grubbström, R. W. (2004). Economic consequences of a production and inventory control policy. *International Journal of Production Research, 42*(17), 3419–3431. doi:10.1080/00207540410001727640

Fan, T. J., Tao, F., Deng, S., & Li, S. X. (2015). Impact of RFID technology on supply chain decisions with inventory inaccuracies. *International Journal of Production Economics, 159*, 117–125. doi:10.1016/j.ijpe.2014.10.004

Greengard, S. (2015). *The Internet of Things*. Cambridge, MA: MIT Press.

Grubbström, R. W. (1967). On the application of the Laplace transform to certain economic problems. *Management Science, 13*(7), 558–567. doi:10.1287/mnsc.13.7.558

Grubbström, R. W. (1969). *Produktionsekonomiska system- i statisk och dinamisk belysning [Production-economics systems- In static and dynamic Illumination]*. Stockholm, Sweden: Department of Industrial Economics and Management, Royal Institute of Technology.

Grubbström, R. W. (1991). The z-transform of tk. *Math. Scientist, 16*, 118–129.

Grubbström, R.W. (1994). *Stochastic relationship of a multi-period inventory process with planned production using transform methodology*. RR-126. Dept. of Production Economics. LiTH.

Grubbström, R. W. (1996). Material requirements planning and manufacturing resource planning. In M. Warner (Ed.), *International Encyclopedia of Business and Management*. London: Routledge.

Grubbström, R. W. (1996a). Stochastic properties of a production - inventory process with planned production using transform methodology. *International Journal of Production Economics, 45*(1-3), 407–419. doi:10.1016/0925-5273(96)00009-6

Grubbström, R. W. (1998). A net present value approach to safety stocks in planned production. *International Journal of Production Economics, 56/57*, 213–229. doi:10.1016/S0925-5273(97)00094-7

Grubbström, R. W. (1998a) The Linköping approach to Material Requirements Planning and Input-Output analysis, In L. Bogataj, R. W. Grubbström, (Eds.), Input-output analysis and Laplace transforms in Material Requirements Planning: Storlien (pp. 45-60). Portorož: University of Ljubljana, FPP.

Grubbström, R. W. (1999). A net present value approach to safety stocks in a multi-level MRP system. *International Journal of Production Economics, 59*(1/3), 361–375. doi:10.1016/S0925-5273(98)00016-4

Grubbström, R. W. (2012). The time-averaged L4L solution - a condition for long-run stability applying MRP theory. *International Journal of Production Research, 50*(21), 6099–6110. doi:10.1080/00207543.2011.653457

Grubbström, R. W., Bogataj, M., & Bogataj, L. (2007). *A compact representation of distribution and reverse logistics in the value chain, (Mathematical economics, operational reseach and logistics, no. 5).* Ljubljana, Slovenia: Faculty of Economics, KMOR.

Grubbström, R. W., Bogataj, M., & Bogataj, L. (2010). Optimal lotsizing within MRP theory. *Annual Reviews in Control, 34*(1), 89–100. doi:10.1016/j.arcontrol.2010.02.004

Grubbström, R. W., & Huynh, T. T. T. (2006). Multi-level, multi-stage capacity-constrained production-inventory systems in discrete time with non-zero lead times using MRP theory. *International Journal of Production Economics, 101*(1), 53–62. doi:10.1016/j.ijpe.2005.05.006

Grubbström, R. W., & Huynh, T. T. T. (2006b). Analysis of standard ordering policies within the framework of MRP theory. *International Journal of Production Research, 44*(18/19), 3759–3773. doi:10.1080/00207540600584839

Grubbström, R. W., & Jiang, Y. (1989). A survey and analysis of the application of the Laplace transform to present value problems. *Revista mat. scienze economiche sociale, 12*(1), 43–62.

Grubbström, R. W., & Lundquist, J. (1977). The Axsäter integrated production-inventory system interpreted in terms of the theory of relatively closed systems. *J. Cybernetics, 7*(1-2), 49–67. doi:10.1080/01969727708927548

Grubbström, R. W., & Lundquist, J. (1989). *Master scheduling, Input-Output analysis and production functions.* RR-121, Dept. of Production Economics, LiTH.

Grubbström, R. W., & Molinder, A. (1994). Further theoretical considerations on the relationship between MRP, input-outpt analysis and multi-echelon inventory systems. *International Journal of Production Economics, 35*(1-3), 299–311. doi:10.1016/0925-5273(94)90096-5

Grubbström, R. W., & Molinder, A. (1996). Safety production plans in MRP-systems using transform methodology. *International Journal of Production Economics, 46-47*, 297–309. doi:10.1016/0925-5273(95)00158-1

Grubbström, R. W., & Ovrin, P. (1992). Intertemporal generalization of the relationship between material requirements planning and input-output analysis. *International Journal of Production Economics, 26*(1-3), 311–318. doi:10.1016/0925-5273(92)90081-H

Grubbström, R. W., & Tang, O. (1999). Further developments on safety stocks in an MRP system applying Laplace transforms and input-output analysis. *International Journal of Production Economics, 60/61*, 381–387. doi:10.1016/S0925-5273(98)00141-8

Grubbström, R. W., & Tang, O. (2000). Modelling rescheduling activities in a multi-period production-inventory system. *International Journal of Production Economics, 68*(2), 123–135. doi:10.1016/S0925-5273(00)00050-5

Grubbström, R. W., & Tang, O. (2006). Optimal production opportunities in a remanufacturing system. *International Journal of Production Research, 44*(18/19), 3953–3966. doi:10.1080/00207540600806406

Grubbström, R. W., & Tang, O. (2006a). The moments and central moments of a compound distribution. *European Journal of Operational Research, 170*(1), 106–119. doi:10.1016/j.ejor.2004.06.012

Grubbström, R. W., & Tang, O. (2012). The space of solution alternatives in the optimal lotsizing problem for general assembly systems applying MRP theory. *International Journal of Production Economics, 140*(2), 765–777. doi:10.1016/j.ijpe.2011.01.012

Grubbström, R. W., & Wang, Z. (2003). A stochastic model of multi-level/multi-stage capacity-constrained production-inventory systems. *International Journal of Production Economics, 81/82*, 483–494. doi:10.1016/S0925-5273(02)00358-4

Gruère, G. P. (2012). Implications of nanotechnology growth in food and agriculture in OECD countries. *Food Policy, 37*(2), 191–198. doi:10.1016/j.foodpol.2012.01.001

65. Guo, Z. X., Ngai, E. W. T., Yang, C., & Liang, X. D. (2015). A RFID-based intelligent decision support system architecture for production monitoring and scheduling in a distributed manufacturing environment. *International Journal of Production Economics, 159*, 16–28. doi:10.1016/j.ijpe.2014.09.004

Handford, C. E., Dean, M., Henchion, M., Spence, M., Elliott, C. T., & Campbell, K. (2014). Implications of nanotechnology for the agri-food industry: Opportunities, benefits and risks. *Trends in Food Science & Technology, 40*(2), 226–241. doi:10.1016/j.tifs.2014.09.007

Harris, I., Wang, Y. L., & Wang, H. Y. (2015). ICT in multimodal transport and technological trends: Unleashing potential for the future. *International Journal of Production Economics, 159*, 88–103. doi:10.1016/j.ijpe.2014.09.005

Horvat, L., & Bogataj, L. (1996a). MRP, input-output analysis and multi-echelon inventory systems with exponentially distributed external demand. In *Proceedings of GLOCOSM Conference* (149–154). Bangalore, India.

Horvat, L., & Bogataj, L. (1996b). Game theory aspects of MRP, input-output and multi-echelon inventory systems. *9th Int Working Seminar on Production Economics, Preprints, 3*, 1–12.

Huang, L., Guo, Y., Peng, Z., & Porter, A. L. (2011). Characterising a technology development at the stage of early emerging applications: Nanomaterial enhanced biosensors. *Technology Analysis and Strategic Management, 23*(5), 527–544. doi:10.1080/09537325.2011.565666

Huang, L., Peng, Z., Guo, Y., & Porter, A. L. (2010). Identifying emerging nanoparticle roles in biosensors. *Journal of Business Chemistry, 7*(1), 15–30.

Islam, N., & Miyazaki, K. (2010). An empirical analysis of nanotechnology research domains. *Technovation, 30*(4), 229–237. doi:10.1016/j.technovation.2009.10.002

Jederman, R., Nicometo, M., Uysal, I., & Lang, W. (2014). Reducing food losses by intelligent food logistics. *Philosophical transactions. Series A, Mathematical, physical, and engineering sciences, 372*. 10.1098/rsta.2013.0302.

Kiaya, V. (2014). *Action Contre La Faim*. Retrieved August 24, 2015, from: http://www.actioncontrelafaim.org/sites/default/files/publications/fichiers/technical_paper_phl__.pdf

Koopmans, T. C. (Ed.). (1951). *Activity Analysis of Production amd Allocation*. New York, NY: Wiley.

Kovačić, D., & Bogataj, M. (2013). Reverse logistics facility location using cyclical model of extended MRP theory. *Central European Journal of Operations Research, 21*(1), 41–57. doi:10.1007/s10100-012-0251-x

Kovačić, D., Hontoria, E., Ros Mcdonnell, L., & Bogataj, M. (2015). Location and lead-time perturbations in multi-level assembly systems of perishable goods in Spanish baby food logistics. *Central European Journal of Operations Research, 23*(3), 607–623. doi:10.1007/s10100-014-0372-5

Kreuchauff, F., & Teichert, N. (2014). *Nanotechnology as general purpose technology*. Karlsruhe, Germany: Karlsruhe Institut für Technologie.

Leontief, W. W. (1928). *Die Wirtschaft als Kreislauf*. Thesis, Friedrich-Wilhelms-Universitat Berlin. Tübingen.

Leontief, W. W. (1936). Quantitative Input and Output Relations in the Economic System of the United States. *The Review of Economics and Statistics, 18*(3), 105–125. doi:10.2307/1927837

Li, Y., Schluesener, H. J., & Xu, S. (2010). Gold nanoparticle-based biosensors. *Gold Bulletin, 43*(1), 29–41. doi:10.1007/BF03214964

Lu, J., & Bowles, M. (2013). How Will Nanotechnology Affect Agricultural Supply Chains? *International Food and Agribusiness Management Review, 16*(2), 21–42.

Malik, P., Katyal, V., Malik, V., Asatkar, A., Inwati, G., & Mukherjee, T. K. (2013). Nanobiosensors: Concepts and Variations. *ISRN Nanomaterials. Article ID, 327435*, 9.

Molinder, A. (1995). *Material Requirements Planning Employing Input-Output Analysis and Laplace Transforms, PROFIL, 14*. Linköping: Production Economic Research.

Naim, M. M., Wikner, J., & Grubbström, R. W. (2007). A net present value assessment of make-to-order and make-to-stock manufacturing systems. *Omega, 35*(5), 524–532. doi:10.1016/j.omega.2005.09.006

National Institute of Food and Agriculture. (2015). *Nanotechnology Program*. Retrieved August 24, 2015, from: National Institute of Food and Agriculture: http://nifa.usda.gov/program/nanotechnology-program

Ng, I. C. L., Scharf, K., Pogrebna, G., & Maull, R. S. (2015). Contextual variety, internet-of- things and the choice of tailoring over platform: Mass customisation strategy in supply chain management. *International Journal of Production Economics, 159*, 76–87. doi:10.1016/j.ijpe.2014.09.007

Obayelu, A. E. (2014). *Postharvest Losses and Food Waste: The Key Contributing Factors to African Food Insecurity and Environmental Challenges*. Retrieved August 24, 2015, from: http://worldfoodscience.com/article/postharvest-losses-and-food-waste-key-contributing-factors-african-food-insecurity-and#sthash.NMUXu0mB.rzz4xZ5s.dpuf

Orlicky, J. A. (1975). *Material Requirements Planning*. New York, NY: McGrawHill.

Parfitt, J., Barthel, M., & Macnaughton, S. (2010). Food waste within food supply chains: quantification and potential for change to 2050. *Philosophical Transactions of the Royal Society B, 365*, 3065-3081. Retrieved from http://worldfoodscience.com/article/postharvest-losses-and-food-waste-key-contributing-factors-african-food-insecurity-and#sthash.NMUXu0mB.dpuf

Parisi, C., Vigani, M., & Rodríguez-Cerezo, E. (2014). *Proceedings of a workshop on "Nanotechnology for the agricultural sector: from research to the field"*. Luxembourg: Publications Office of the European Union.

Paull, R. E. (1999). Effects of temperature and relative humidity on fresh commodity quality. *Postharvest Biology and Technology, 15*(3), 263–277. doi:10.1016/S0925-5214(98)00090-8

Poudelet, E. (2013). *Assessment of nanomaterials in food, health and consumer products, Workshop on the Second Regulatory Review on Nanomaterials*. Retrieved from http://ec.europa.eu/enterprise/sectors/chemicals/files/reach/docs/events/nano-rev-ws-poudelet_en.pdf

Qiu, X., Luo, H., Xu, G. Y., Zhong, R. Y., & Huang, G. Q. (2015). Physical assets and service sharing for IoT-enabled Supply Hub in Industrial Park (SHIP). *International Journal of Production Economics, 159*, 4–15. doi:10.1016/j.ijpe.2014.09.001

Rai, A., Patnayakuni, R., & Seth, N. (2006). Firm performance impacts of digitally enabled supply chain integration capabilities. *Management Information Systems Quarterly, 30*(2), 225–246.

Ravichandran, R. (2010). Nanotechnology Applications in Food and Food Processing: Innovative Green Approaches, Opportunities and Uncertainties for Global Market. *International Journal of Green Nanotechnology: Physics and Chemistry, 1*(2), 72–96. doi:10.1080/19430871003684440

Reaidy, P., Gunasekaran, A., & Spalanzani, A. (2015). Bottom-up approach, internet of things, multi-agent system, RFID, ambient intelligence, collaborative ware-houses. *International Journal of Production Economics, 159*, 29–40. doi:10.1016/j.ijpe.2014.02.017

Rudolf, R., Tomic, S., Anzel, I., Zupančič Hartner, T., & Čolić, M. (2014). Microstructure and biocompatibility of gold-lanthanum strips. *Gold Bulletin, 47*(4), 263–273. doi:10.1007/s13404-014-0150-0

Sabourin, V., & Ayande, A. (2015). Commercial Opportunities and Market Demand for Nanotechnologies in Agribusiness Sector. *J. Technol. Manag. Innov., 10*(1), 40–51. doi:10.4067/S0718-27242015000100004

Salamanca-Buentello, F., Persad, D. L., Court, E. B., Martin, D. K., Daar, A. S., & Singer, P. A. (2005). Nanotechnology and the Developing World. *PLoS Medicine, 2*(5), e97. doi:10.1371/journal.pmed.0020097 PMID:15807631

Segerstedt, A. (1995). Multi-level production and inventory control problems: related to MRP and cover-time planning. Linköping: Production-Economic Research (Produktionsekonomisk forskning). (Profil; No. 13).

Tang, O., & Grubbström, R. W. (2002). Planning and replanning the master production schedule under demand uncertainty. *International Journal of Production Economics, 78*(3), 323–334. doi:10.1016/S0925-5273(00)00100-6

Tang, O., & Grubbström, R. W. (2003). The detailed coordination problem in a two-level assembly system with stochastic times. *International Journal of Production Economics, 81/82*(1), 415–429. doi:10.1016/S0925-5273(02)00296-7

Tang, O., & Grubbström, R. W. (2005). Considering stochastic lead times in a manufacturing/remanufacturing system with deterministic demands and returns. *International Journal of Production Economics, 93/94*, 285–300. doi:10.1016/j.ijpe.2004.06.027

Tang, O., & Grubbström, R. W. (2006). On using higher-order moments for stochastic inventory systems. *International Journal of Production Economics, 104*(2), 454–461. doi:10.1016/j.ijpe.2005.03.004

Tang, O., Grubbström, R. W., & Zanoni, S. (2004). Economic evaluation of disassembly processes in remanufacturing systems. *International Journal of Production Research, 42*(17), 3603–3617. doi:10.1080/00207540410001699435

Tang, O., Grubbström, R. W., & Zanoni, S. (2007). Planned lead time determination in a make-to-order remanufacturing system. *International Journal of Production Economics, 108*(1/2), 426–435. doi:10.1016/j.ijpe.2006.12.034

Thiesse, F., & Buckel, T. (2015). A comparison of RFID-based shelf replenishment policies in retail stores under suboptimal read rates. *International Journal of Production Economics, 159*, 126–136. doi:10.1016/j.ijpe.2014.09.002

Thomas, D. J., & Griffin, P. M. (1996). Coordinated supply chain management. *European Journal of Operational Research, 94*(1), 1–15. doi:10.1016/0377-2217(96)00098-7

Thorstenson, A. (1988). Capital Costs in Inventory Models — A Discounted Cash Flow Approach. Linköping:Production-Economic Research in Linköping, PROFIL.

Thorstenson, A., & Grubbström, R. W. (1993). Theoretical analysis of the difference between the traditional and the annuity stream principles applied to inventory evaluation. In R. Flavell (Ed.), Modelling Reality and Personal Modelling (pp. 296–326). Heidelberg: Physica-Verlag. doi:doi:10.1007/978-3-642-95900-4_19 doi:10.1007/978-3-642-95900-4_19

Vazsonyi, A. (1955). The use of mathematics in production and inventory control. *Management Science, 1*(3/4), 70–85.

Vickery, S. K., Jayaram, J., Droge, C., & Calantone, R. (2003). The effects of an integrative supply chain strategy on customer service and financial performance: An analysis of direct versus indirect relationships. *Journal of Operations Management, 21*(5), 523–539. doi:10.1016/j.jom.2003.02.002

Wong, C., Lai, K. H., Cheng, T. C. E., & Lun, V. Y. H. (2015). The role of IT-enabled collaborative decision making in inter-organizational information integration to improve customer service performance. *International Journal of Production Economics*, *159*, 56–65. doi:10.1016/j.ijpe.2014.02.019

Yu, J., Nachiappan, S. P., Ning, K., & Edwards, D. (2015). Product delivery service provider selection and customer satisfaction in the era of Internet Of Things: A Chinese e-retailers' perspective. *International Journal of Production Economics*, *159*, 104–116. doi:10.1016/j.ijpe.2014.09.031

Zanoni, S., & Grubbström, R. W. (2004). A note on an industrial strategy for stock management in supply chains: Modelling and performance evaluation. *International Journal of Production Research*, *42*(20), 4421–4426. doi:10.1080/00207540420002645 81

Zhou, L., Chong, A. Y. L., & Eric, W. T. (2015). Supply chain management in the era of the internet of things. *International Journal of Production Economics*, *159*, 1–3. doi:10.1016/j.ijpe.2014.11.014

Zhou, L., & Grubbström, R. W. (2004). Analysis of the effect of commonality in multi-level inventory systems applying MRP theory. *International Journal of Production Economics*, *90*(2), 251–263. doi:10.1016/S0925-5273(03)00208-1

ENDNOTE

[1]   (http://2020science.org/2010/08/24/value-added-nanotechnology/; http://www.azonano.com/article.aspx?ArticleID=3946)

Chapter 17
Understanding Toxicity of Nanomaterials in Biological Systems

Irshad Ahmad Wani
Jamia Millia Islamia, India

Tokeer Ahmad
Jamia Millia Islamia, India

ABSTRACT

Nanotechnology is a growing applied science having considerable global socioeconomic value. Nanoscale materials are casting their impact on almost all industries and all areas of society. A wide range of engineered nanoscale products has emerged with widespread applications in fields such as energy, medicine, electronics, plastics, energy and aerospace etc. While the market for nanotechnology products will have grown over one trillion US dollars by 2015, the presence of these material is likely to increase leading to increasing likelihood of exposure. The direct use of nanomaterials in humans for medical and cosmetic purposes dictates vigorous safety assessment of toxicity. Therefore this book chapter provides the detailed toxicity assessment of various types of nanomaterials.

INTRODUCTION

The field of nanotechnology is a fast-growing research niche (Ostiguy et al., 2006). Nanoparticles are particles that have at least one dimension in nanoregime, i.e., less than 100 nm. Nanotechnology is a comprensive term that involves the ability to work with materials at a nanometre scale. Nanotechnology has a potential applications in a range of sectors including energy (production, catalysis, storage), materials (lubricants, abrasives, paints, tires, and sportsware), electronics (chips and screens), optics, and remediation (pollution absorption, water filtering and disinfection), cosmetics (skin lotions and sun screens), and medicine (diagnostics and drug delivery) to food (additives and packaging) (Joner, Hartnik, & Amundsen, 2007; Rana & Kalaichelvan, 2013). The word "nano," is derived from the Greek "nanos" meaning "dwarf", and is becoming increasingly widespread in scientific literature. "Nano" word has

DOI: 10.4018/978-1-5225-0585-3.ch017

become so popular in modern science that many "nano-"derived words have recently find their places in dictionaries, including: nanometer, nanoscale, nanoscience, nanotechnology, nanostructure, nanotube, nanorod nanowire, and nanorobot and many more. Many new words like nanoelectronics, nanocrystal, nanovalve, nanoantenna, nanocavity, nanoscaffolds, nanofibers, nanomagnet, nanoporous, nanoarrays, nanolithography, nanopatterning, nanoencapsulation, etc. are used in respected and reputed publications, such as Science and Nature that are not yet widely recognized. The nanometer denotes one billionth of a meter or 10^{-9} meters and is a metric unit of length. The word 'nano' is popularly used as an adjective to describe objects, systems, or phenomena with characteristics arising from nanometer-scale structure and emphasize the atomic granularity that produces the unique phenomena observed in nanoscience. Most of the exciting properties begin to be apparent in systems smaller than 1000 nm, or 1 micrometer, 1 µm. Birth of the concept of nanotechnology is usually linked to a speech by Richard Feynman at the December 1959 meeting of the American Physical Society (Feyman, 1959) where he asked a very funny and strange question to the scientific audiences, "What would happen if we could arrange the atoms one by one the way we want them?"

Nanomaterials are materials possessing structural components smaller than 1 micrometer in at least one dimension. Many authors, however, limit the size of nanomaterials to 50 nm (Kittelson, 2001) or 100 nm (Borm et al., 2006), the choice of this upper limit being justified by the fact that some physical properties of nanoparticles approach those of bulk when their size reaches these values. However, a legitimate definition extends this upper size limit to 1 micron, the sub-micron range being classified as nano(Buzea, Pacheco, & Robbie, 2007b).

Nanoparticles are particles potentially as small as atomic and molecular length scales (~0.2 nm) or with at least one dimension falling in nanoregime. Nanoparticles can be amorphous or crystalline and their highly active surfaces can act as carriers for gases or liquid droplets. Nanoparticulate matter should be considered as a distinct state of apart from the solid, liquid, gaseous, and plasma states owing to their distinct properties like large surface area and quantum size effects.

Nanomaterials being versatile in nature has various applications in biology and medicine such as Fluorescent biological labels(Fu et al., 2007), drug and gene delivery(Yan & Chen, 2013)(Malmsten, 2013), bio detection of pathogens (Yang, Li, & Jiang, 2008), detection of proteins(Yi Zhang et al., 2013), probing of DNA structure (Mahtab & Murphy, 2005), Tissue engineering (Cipriano & Liu, 2013), tumour destruction via heating (hyperthermia) (Dutz & Hergt, 2014), MRI contrast enhancement (Na, Song, & Hyeon, 2009) etc

NANO VS. BULK MATERIALS AND NANOTOXICOLOGY PUBLICATION STATISTICS

Nanomaterials behave significantly different from bulk materials due to two primary factors which are surface effects and quantum effects (Roduner, 2006). These two factors affect the mechanical, electrical, optical and magnetic as well as chemical reactivity of the materials. Compared to the microparticles, nanoparticles have a very large surface area and high particle number per unit volume due to the high fraction of atoms at the surface. For example the ratio of surface area to volume (or mass) for a carbon particle with a diameter of 60 nm is 1000 times larger than a particle with a diameter of 60 µm. Since the material in nanoparticulate form presents a much large surface area, due to which chemical reactivity of the material is greatly enhanced roughly 1000 folds. The atoms situated at the surface has lower binding energy per atom due to the lesser no of neighbors than bulk atoms (Roduner, 2006). As a result, melting point reduces with reduction in particle size due to lesser binding energy at smaller particles sizes as followed from the Gibbs-Thomson equation. For example, the melting temperature of 3 nm gold nanoparticles is more than 300 degrees lower than the melting temperature of bulk gold (Roduner, 2006) (Figure 1).

There has been an exponential increase in the number of publications on the synthesis and properties of nanoparticles since 1090s (Figure 2) as indicated by the advanced search on Google Scholar Database. There is also an increase in number of publications discussing their toxicity for the past few years (Figure 2). However, the total no of papers on toxicity remains low in comparison to the total no of papers published on nanomaterials. Since Nanoscience and Nanotechnology encompasses a wide range of fields ranging from chemistry to physics, materials science and engineering, biology, medicine and electronics, which explains the reason of the vast no of papers published in this field. However, several reviews have been published addressing the nanotoxicology aspects of nanomaterials. Several are comparatively general(Borm et al., 2006; Nel, Xia, Mädler, & Li, 2006; X. Zhang, Hu, Li, Tao, & Wei, 2012) while others address selected aspects of nanoparticles toxicology such as occupational aspects of nanoparticle(Pfau, Sentissi, Weller, & Putnam, 2005); particle inhalation, retention, and clearance (Buzea, Pacheco, & Robbie, 2007a); pulmonary effects of inhaled particles (Oberdörster, 2000); inhalation and lung cancer(Borm, Schins, & Albrecht, 2004), oxidative mechanisms(Brown, Lockwood, & Sonawane, 2005; González-Flecha, 2004; Peters, 2005) and gastro-intestinal uptake of particles (Fubini & Hubbard, 2003).

Nanoparticle Synthesis

Various excellent review have been published on the synthesis and characterization of nanomaterials (Saber M Hussain et al., 2009; Rodriguez-Hernandez, Chécot, Gnanou, & Lecommandoux, 2005; Teow,

Figure 1. A plot of melting temperature of gold as a function of size

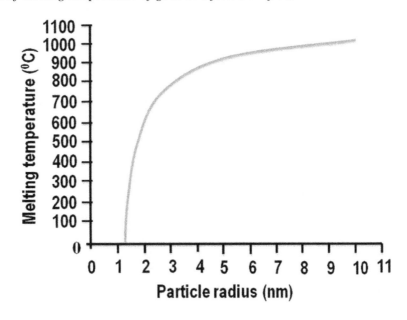

Figure 2. Evolution of no of papers on Synthesis, Applications and Toxicity on nanoparticles since 1990

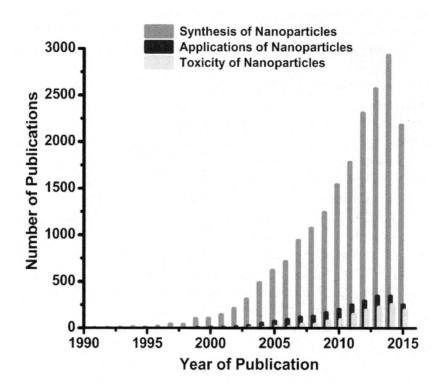

Asharani, Hande, & Valiyaveettil, 2011; Y. Wang & Cao, 2006). Two different but complementary approaches have been used for synthesizing nanomaterials (Figure 3). First one known as "top-down" approach where bulk materials are broken down into particles in nanorange by physical processes such as milling, grinding, etching and pyrolysis. In this approach particle size of the starting bulk material is reduced to nanorange using mechanical or chemical methods. Photolithography and silicon microfabrication are two processes often employed or prepared using the top-down approach (Mishra, Patel, & Tiwari, 2010). This approach has the advantage of the possibility of mass production for industrial purposes. However, this approach involves some disadvantages such as lengthy time consuming processes, generation of a broad feature size distribution, and imperfection or defects of the surface morphology generated. Since surface deformations can have a significant impact on the physical and chemical properties of the nanomaterials(Thakkar, Mhatre, & Parikh, 2010).e.g., pyrolysis involves the burning of organic precursor after forcing through an orfice at high pressures. The resulting ash produced is processed to recover the oxidized nanoparticles having wide particles size distribution. (Thakkar et al., 2010), thus consuming an enormous amount of energy so as to sustain high pressures. Bottom-up approach, involves creation of nano-objects by combining atomic scale materials. A variety of bottom up approaches exist involving comples coacervation of two oppositively charged polyelectrolyted, co-precipitation, salting out from aquous-organic mixture or solvent displacement, solvent emulsification, diffusion using oil-water microemulsions or using supercritical fluids. The process operated under ambient conditions leading to high energy saving. A wide variety of nanomaterials can be generated using this approach including nanoemulsions, liposomes, polymeric NPs, polymeric micelles, solid lipid NPs, protein NPs and carbohydrate such as chitosan, alginate, hyaluronic acid, dextran nanoparticles(Zonghua Liu, Jiao, Wang, Zhou, & Zhang, 2008; Szarpak et al., 2010). The most commonly employed method involves the chemical reduction of metal ions by a suitable reducing agent in the presence of a capping or stabilizing agent. The metal atoms obtained after reduction of the metal ions aggregate to form metal nanoparticles. The further growth of these metal nanoparticles is inhibited by the addition of the suitable capping agent which works either by conferring a suitable charge to the surface which repels neighboring naoparticles or by sterically preventing the agglomeration of two NPs (Thanh & Green, 2010). Examples of such capping agents include carboxylic acids or amines. Capping agent also serves important function such as controlling particle size, shape, charge and surface functionality which can lead to special properties. There has been a growing interest in synthesis of nanoparticles using biological approaches using bacteria (Reddy et al., 2010), yeast (Kowshik et al., 2003) fungi, plants and algae (Thakkar et al., 2010). This method offers advantage over conventional chemical methods as avoid the use of toxic organic solvents and reagents, generation of hazardous side-products, high cost and energy consumption. The nanoparticles produced by this approach relatively non-toxic and biocompatible. Nanoparticles of different materials ranging from from Ag, Au, CdS, magnetite and uranium were produced using biological approach (Nanda & Saravanan, 2009). A similar method has been used in the synthesis of Ag NPs using fungal cell filtrate (Gajbhiye, Kesharwani, Ingle, Gade, & Rai, 2009).

Toxicity of Different Nanoparticles

There has been tremendous expansion in research on potential application of nanotechnology in medicine ranging from exploiting nanomaterials as carriers of active pharmaceutical drugs in delivery and targeting applications to medical imaging purposes (Moghimi, Hunter, & Murray, 2005). These nanosized materials can affect biological functions by interaction with the biological tissues. Recent studies on potential nanotoxicity to human health and the eco-environment have drawn attention from both government agencies and the general public (Crist et al., 2013; Monteiro-Riviere, Inman, & Zhang, 2009). With the increasing commercialization and increasing human exposure of animals to nanomaterials, a safety assessment of these materials is needed. To facilitate such efforts, we herein present a brief survey the toxic effects of several organic and inorganic several nanocatalysts, mainly used in environmental remediation.

Carbon-Based Nanomaterials

There are three well-known carbon-based nanomaterials like fullerenes (Da Ros & Prato, 1999), carbon nanotubes (CNTs) (Ajayan & Zhou, 2001), and graphene (Sun et al., 2008). Regarding the cytotoxicity of C60 nanoparticles on various cell lines, controversial results have been published (Lewinski, Colvin, & Drezek, 2008). C60 nanoparticles have been found to be cytotoxic to some cell lines such as such as human dermal fibroblasts (HDFs)(Sayes et al., 2004, 2005) human liver carcinoma cells (HepG2) (Sayes et al., 2004, 2005), human epidermal keratinocytes (HEK)(Saathoff, Inman, Xia, Riviere, & Monteiro-Riviere, 2011), rat C6 glioma cells(Isakovic et al., 2006), and human U251 glioma cells(Isakovic et al., 2006), whereas no significant cytotoxic responses to pristine C60 nanoparticles were observed in murine macrophages(Fiorito, Serafino, Andreola, & Bernier, 2006), human monocyte-derived macrophages (Fiorito et al., 2006; Porter, Muller, Skepper, Midgley, & Welland, 2006), and Guinea pig alveolar macrophages(G. Jia et al., 2005). However it has been observed that surface functionalization

Figure 3. Two different approaches viz., Top down and Bottom up approaches for synthesis of nanomaterials

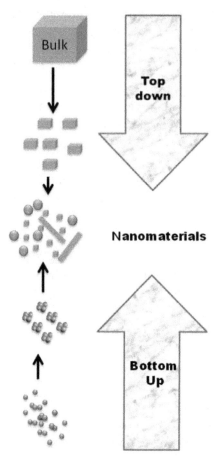

of C60 particles may partially alleviate their cytotoxicity and reduce reactive oxygen species (ROS) generation(Sayes et al., 2004), but not in all the cases, since polyhydroxylated fullerenes can still induce photooxidative stress in human cells(Vileno et al., 2006; Wielgus, Zhao, Chignell, Hu, & Roberts, 2010). In comparison to C60 nanoparticles, both single walled carbon nanotubes (SWCNTs) and multiwalled carbon nanotubes (MWCNTs) exhibit cytotoxic effects at high concentrations. Cytotoxicity of both types of carbon nanotubes has been observed in HEK(Monteiro-Riviere & Inman, 2006), human embryo kidney cells (HEK293)(Cui, Tian, Ozkan, Wang, & Gao, 2005), rat alveolar macrophage cells (NR8383)(Pulskamp, Diabaté, & Krug, 2007), human alveolar epithelial cells (A549) (Pulskamp et al., 2007), human skin fibroblasts (HSF42)(Ding et al., 2005), and human embryonic lung fibroblasts (IMR-90) (Ding et al., 2005). MWCNTs and SWCNTs induced pulmonary inflammatory responses and granuloma formation after intratracheal instillation in animal models (Chou et al., 2008; Lam, James, McCluskey, & Hunter, 2004; Muller et al., 2005). However literature reports have proven that SWCNTs are more toxic than MWCNTs and C60 nanoparticles (G. Jia et al., 2005; Pulskamp et al., 2007; Tian, Cui, Schwarz, Estrada, & Kobayashi, 2006). With surface functionalization the cytotoxicity effects of pristine SWCNTs can be lessened and the extent of cytotoxicity depends upon the degree of sidewall functionalization (Sayes et al., 2006).

Likewise graphene and its derivative like graphene oxide have been shown to have cytotoxicity effects in a dose dependant manner in various cells including A549, Henrietta Lacks cells (HeLa), National Institute of Health 3T3 mouse fibroblast cells (NIH-3T3), Sloan Kettering breast cancer cells (SKBR3), Michigan cancer foundation-7 breast cancer cells (MCF7), human neuroblastoma SH-SY5Y cells and even normal human lung cells (BEAS-2B)(Chang et al., 2011; Lv et al., 2012; Vallabani et al., 2011). Size and surface coating have been observed to influence the cytotoxicity of graphene and its derivatives. Reduced graphene oxide nanoplatelets (rGONPs) with average lateral dimensions (ALDs) of 11 ± 4 nm caused considerable cell destruction at the concentration of 1.0 µg/mL, while the rGONPs with ALDs of 3.8 ± 0.4 µm exhibited significant cytotoxic effects only at high concentrations of 100 µg/mL(Akhavan, Ghaderi, & Akhavan, 2012). In addition, surfaces coated with chitosan significantly reduce the haemolytic activity of GONP to human erythrocytes (RBCs) (Liao, Lin, Macosko, & Haynes, 2011).

Metal Nanoparticles

Although great progress have been made towards production of metal based nanoparticles worldwide both at the academic as well as at the industrial level, but still there is a serious lack of information about the impact of NPs on human health and environment, especially the potential for NP-induced toxicity(Canesi et al., 2008). Preliminary reports of the inherent toxicity of some NPs are available and indicate that they can affect biological behavior at the organ, tissue, cellular, subcellular, and protein levels (Schrand et al., 2010).

Aluminum Nanoparticles

Cellular interaction of aluminum oxide and aluminum nanoparticles including their effect on cell viability and cell phagocytosis, with reference to particle size and chemical composition has been investigated by Wagner et al (Wagner et al., 2007). Experiments were conducted to know the in vitro cellular effects of aluminium nanoparticles in rat alveolar macrophages (NR8383). Characterization techniques employed showed agglomeration in cell exposure media and Aluminium (Al) NP in the dose range of 100 to 250 µg/mL produced significantly reduced viability after 24 h of continuous exposure. No significant effect on the cellular viability was observed at 50, 80, or 120 nm Al NPs at 25 µg/mL for 24 h but cell phagocytotic ability was significantly hindered by exposure to same concentration (25 µg/mL) (Braydich-Stolle, Hussain, Schlager, & Hofmann, 2005).

Copper Nanoparticles

Copper nanoparticles (Cu NPs) have been recently investigated as a novel antimicrobial material, i.e. antiviral, antibacterial, antifouling, and antifungal and used in applications as biocides, antibiotic treatment alternatives, and nanocomposite coatings(Anyaogu, Fedorov, & Neckers, 2008; Cioffi et al., 2005). However, severe toxicity of Cu NPs have been demonstrated including heavy injuries in the kidney, liver, and spleen of mice after ingestion, as evident via histological analysis(Z. Chen et al., 2006). The Cu nanoparticles (23.5 nm) with oral LD50 of 413 mg/kg, is considered to be moderate toxic similar to Zn powder(Z. Chen et al., 2006; Suzuki, Toyooka, & Ibuki, 2007) in contrast to Cu microparticles (17 µm) which did not produce similar effects and were classified as nontoxic with LD50 values of >5000 mg/kg. Moreover, mice exposed to Cu NPs showed glomerulitis degeneration, and necrobiosis of renal tubules, but not in the mice exposed to Cu microparticles indicating that particle size and surface area are important material characteristics from a toxicological perspective. Meng et al gave the supporting evidence for the more efficient deposition of Cu NPs compared to micro-sized particles in renal tissues. They proposed that once inside the kidney, Cu NPs reacted with gastric juices and were converted to more toxic cupric ions (Meng et al., 2007).

Toxicity of Cu nanoparticles (80 nm) (1.5 mg/L; LC50) was also investigated and reported in zebrafish showing a decrease in gill Na^+/K^+ ATPase activity(Griffitt et al., 2007). Cu NP treatment also decreased blood urea nitrogen (BUN) levels and increased plasma alanine amino transferase (ALAT) levels. Additionally, dose-dependent damage of gill lamellae characterized by proliferation of epithelial cells, as well as edema of primary and secondary gill filaments, was observed after Cu NP treatment, indicating that the gill was the primary target organ for CuNP in zebrafish (Griffitt et al., 2007). In zebrafish exposed to Cu NPs, RT–PCR results showed higher levels of gene expression changes compared to $CuSO_4$ exposed fish. Furthermore, cluster analysis of these gene microarrays demonstrated that the transcriptional response induced by Cu NP was highly divergent (Griffitt et al., 2007).

Gold Nanoparticles

Several factors play important roles in determining the cytotoxicity of GNP, most important being the surface properties which have been shown to affect the biological interactions of GNPs (S. T. Kim, Saha, Kim, & Rotello, 2013). In one of the studies, polyethylenimine-modified GNPs has been shown to cause decrease in cell viability of in monkey kidney cells (COS-7) (Thomas & Klibanov, 2003). However, surface modified GNPs with different surface charges exhibited different cell uptake (Su et al., 2012) and toxic effects (Goodman, McCusker, Yilmaz, & Rotello, 2004) in cells. For instance, during cellular uptake, positively charged GNPs were reported to modulate the cell membrane potential and cause cell membrane perturbations. However, peptide functinalized GNPs were found to enter cells and targeted the nucleus without decreasing cell viability (Tkachenko A G, Xie H, Coleman D, Glomm W, Ryan J, Anderson M F, Frazen S, 2003). Apart from the surface properties, the particle size is another important physical parameter that determines the cytotoxicity of GNPs. Triphenylphosphine derivative-stabilized GNPs of 1.4 nm diameter exert most prominent cytotoxic effect when added to HeLa cells, SK-Mel-28 melanoma cells, L929 mouse fibroblasts, and mouse monocytic/macrophage cells (Pan Y, Neuss S, Leifert A, Fischer M, Wen F, Simon U, Schmid G, Brandau W, 2007). GNP-induced cytotoxicity was also found to depend on the cell type (Patra, Banerjee, Chaudhuri, Lahiri, & Dasgupta, 2007). In addition to cytotoxicity, GNPs cause an immunological response in cells. For instance, the surface hydrophobicity of GNPs dictates the immune response of splenocytes (Moyano et al., 2012).

The GNP dose is important for the toxicity of the nanomaterial *in vivo*. Intraperitoneal administration of 13 nm GNPs at the dose up to 0.4 mg/Kg/day was found to cause no acute or subacute physiological damage (Lasagna-Reeves et al., 2010), daily intraperitoneal administration of GNPs at the rate of 1.15 mg/Kg/day for fourteen days caused severe sickness in mice (Y. S. Chen, Hung, Liau, & Huang, 2009). Keeping the particle size of GNP in consideration as an important factor in vivo, GNPs of 3, 5, 50, or 100 nm at a dose of 8 mg/kg/week did not show harmful effects in mice. However, GNPs with a size ranging from 8 to 37 nm at the same dose caused a loss of structural integrity in the liver with an increase in Kupffer cells and the diffusion of white pulp in the spleen (Y. S. Chen et al., 2009). There are also some reports showing size dependant toxicity of gold nanoparticles with gold nanorods showing higher toxicity than gold nanospheres (Yinan Zhang, Xu, Li, Yu, & Chen, 2012)

The generation of ROS and reactive nitrogen species (RNS) is considered to be responsible for GNP-induced toxicity. Oxidative stress-related autophagy or necrosis in HeLa cells and MRC-5 human fetal lung fibroblasts were found to be induced by GNPs (J. J. Li, Hartono, Ong, Bay, & Yung, 2010; Pan et al., 2009). GNPs were also found to be responsible for the release of nitrogen oxide (NO) from endogenous S-nitroso adducts with thiol groups in blood serum (H. Y. Jia et al., 2008). GNPs also inhibited vascular endothelial growth factor (VEGF)-induced cell proliferation by affecting related signaling pathways (Bhattacharya et al., 2004; Kalishwaralal et al., 2011). Furthermore, respiratory activity of macrophages and the activity of macrophage mitochondrial enzymes (Staroverov et al., 2009) was found to be stimulated by both citrate-coated and antigen-conjugated GNPs (Staroverov et al., 2009). Figure 4 shows the schematic representation of the possible mechanistic pathway of nanoparticles on cell toxicity.

Silver Nanoparticles

Ag NPs have been previously seen wide use as antimicrobial agents and are have recently received considerable attention for use in possible defense and engineering applications (Arora, Jain, Rajwade, & Paknikar, 2008; Chadeau, Oulahal, Dubost, Favergeon, & Degraeve, 2010) Therefore, exposure to Ag nanoparticles underlines the underline the importance of assessing new uses of NPs for potential cytotoxicity. Such an evaluation was undertaken for Ag NP coated cotton fabric and the treated fabric caused no skin irritation and hence was found to be to be safe to guinea pigs (YeonáLee, KunáPark, MiáLee, & BumáPark, 2007). Ag NPs were has been reported to be safe to human cell carcinoma and human fibrosarcoma cell lines (Arora et al., 2008). However, Ag NPs have been found to agglomerate in the cytoplasm and in nuclei and induce oxidative stress-mediated cytotoxicity in human hepatoma cells Kim et al. (S. Kim et al., 2009). Interestingly, the mRNA level of oxidative stress-related genes was discovered to be regulated variably (S. Kim et al., 2009). Ag NPs due to their small size may also interact with the cellular genetic material but little is known about their genotoxicity. Surface chemistry based genotoxicty effects of Ag NPs on mouse embryonic fibroblasts, and on stem cells was reported

Figure 4. Possible mechanism of nanoparticle toxicity on cells

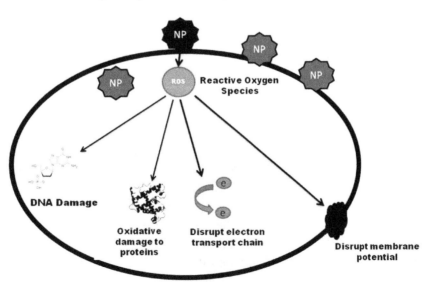

(Ahamed et al., 2008). It was found that polysaccharide functionalized Ag NPs exhibited more apoptosis than did uncoated Ag NPs (Ahamed et al., 2008). Ag NPs of 15 and 100 nm diameters reduced GSH levels, compromised cellular antioxidant defenses that led to ROS accumulation and induced toxicity in cells of a rat liver-derived BRL 3A cell line by generating oxidative stress. However, the mechanism of depletion of GSH by the action of Ag NPs is not yet known (S M Hussain, Hess, Gearhart, Geiss, & Schlager, 2005). Low and high concentrations of Ag NPs induced vasoconstriction and vasodilatation in isolated rat aortic rings, respectively when mediated by the endothelial cells of the aortic rings (Rosas-Hernández et al., 2009). Neurotoxic effects on rat hippocampal CA1 neurons were induced by Ag NPs, ranging from 32 to 380 nm in diameter. The resulting alterations in sodium current that probably led to the observed neuronal dysfunctioning were long lasting (Zhaowei Liu, Ren, Zhang, & Yang, 2009). Ag NPs (15 nm) though reported to be cytotoxic to a PC-12 Rattus norvegicus cell line and caused cell shrinkage but didn't affected cellular morphology (Saber M Hussain et al., 2006). Dose dependant cytotoxic effects were reported in some in vivo studies of Ag NPs (Sung et al., 2009). Free radical induced oxidative stress, alteration in gene expression, and apoptosis in brain tissue was reported due to the intraperitoneal administration of Ag NPs (Rahman et al., 2009).

METAL OXIDE NANOPARTICLES

Toxicity of some of the important metal oxide nanoparticles are briefly discussed below

Titanium Dioxide Nanoparticles

Both micro and nano structured Titanium dioxide (TiO_2) nanoparticles can break down environmental pollutants and hence have been widely used in environmental remediation due to their photocatalytic activity (Daghrir, Drogui, & El Khakani, 2013; Ingale et al., 2011; Tominaga, Kubo, & Hosoya, 2011). TiO_2 nanocatalysts combined with Ultraviolet A (UVA) radiations have been used in food processing industry to eliminate pathogenic microorganisms in food-contacting surfaces. Although considered to be non toxic, but with increased commercialization of TNP related products there are concerns about the possible toxicity which has motivated further research efforts in this direction (Griffitt, Hyndman, Denslow, & Barber, 2009).

Cytotoxicity assessments of TNPs have been performed in various cells, and the toxicity depends upon the cell type. TNPs up to the concentrations of 100 µg/mL have shown very little cytotoxicity in primary human peripheral blood mononuclear cells (PBMCs) (Andersson-Willman et al., 2012), monocyte-derived dendritic cells (MDDCs) (Andersson-Willman et al., 2012), rat peripheral blood neutrophils (Huang Tao, Nallathamby Prakash D, 2008) and primary cultures of human hematopoietic progenitor cells(Bregoli et al., 2009). However, in other cell lines a dose-dependent cytotoxicity was observed such as human skin fibroblasts (Dechsakulthorn, Hayes, Bakand, Joeng, & Winder, 2007), mouse fibroblasts (L929) (Jin, Zhu, Wang, & Lu, 2008), rabbit erythrocytes(S.-Q. Li et al., 2008), and human B-cell lymphoblastoid cells (J. J. Wang, Sanderson, & Wang, 2007). Moreover, genotoxicity studies confirms the DNA damage and increase the mutation frequency (Barillet et al., 2010; J. J. Wang et al., 2007; Wilhelmi et al., 2012) caused by TNP, but with a more serious DNA damage detected under ultraviolet radiation (Vevers & Jha, 2008). Therefore, the cell type appears to be a vital factor influencing

Understanding Toxicity of Nanomaterials in Biological Systems

the cytotoxicity of TNPs, which was also confirmed in a study of the responses of several eukaryotic cells to TNP exposure (Barillet et al., 2010).

The in vivo toxicity of TNPs has been investigated in various aquatic organisms and in animal models. A general growth inhibition was observed with variable EC50 values, ranging from 5.83(Aruoja, Dubourguier, Kasemets, & Kahru, 2009) to 241 mg/L (Vallabani et al., 2011) in freshwater algae, and in Pseudokirchneriella subcapitata. Biodistribution and toxicity of TNPs was also evaluated in rats and mice. It was observed that TNPs predominantly accumulated in the liver, in addition to the spleen, lung, and kidney after a single intravenous injection (5 mg/kg body weight) in wistar rats (Fabian et al., 2008). However no obvious toxic effect, immune response, or change in organ function was observed. TNP accumulation in CD-1 (Imprinting Control Region, ICR) mice after an abdominal cavity injection was found to be highest in Liver and lowest in heart with the order: liver > kidneys > spleen > lung > brain > heart(H. Liu et al., 2009).

Iron Oxide Nanoparticles

In vivo mouse studies have shown that tumor-targeted superparamagnetic iron oxide nanoparticles (SPION) may act as antitumor agents. Entrapment of these NPs, in the growing intravascular thrombi, led to cell death, hence, tumor-targeted SPION can be used to inhibit tumor growth (Simberg et al., 2009). Intratracheal administration of Fe_2O_3 NPs to male Sprague Dawley rats induced oxidative stress in the lungs (M.-T. Zhu et al., 2008). Such oxidative stress produced follicular hyperplasia, protein effusion, pulmonary capillary vessel hyperaemia, and alveolar lipoproteinosis.

SPIONS after entering into cells are probably degraded to iron ions within the lysosomes under the influence of hydrolytic enzymes. After crossing the nuclear and mitochondrial membrane these iron ions may generate reactive hydroxyl radicals via the Fenton reaction. These hydroxyl radicals then cause lipid peroxidation, DNA damage, and protein oxidation. In brief, cytotoxicity of SPIONP has been linked to cellular uptake and ROS generation. Although iron nanoparticles were shown to be biocompatible *in vivo* at a low concentration, these particles were found to be cytotoxic *in vitro*. Polyvinyl alcohol-coated superparamagnetic iron oxide NPs have caused shape and sizedependent toxicity to primary mouse connective tissue cells. Moreover, it has been observed that toxicity increases for NP forms in the following order: nanobeads > nanoworms > nanospheres. Moreover, with increase in hydrodynamic size toxicity decreased. Easy internalization of certain comparatively small particles into cells has been suggested as a possible reason for the size-dependent toxic effects (Mahmoudi, Simchi, Milani, & Stroeve, 2009).

Zinc Oxide (ZnO) Nanoparticles

Zinc oxide (ZnO) NPs have been widely used as ingredients in cosmetics, and in other dermatological preparations. Due to small size, these NPs always have some possibility to interact with DNA and cause damage. In fact, such cases have been reported on a human epidermal cell line (A431). Similarly, ZnO NPs have been reported to induce cytotoxicity in L2 cells (Hillegass et al., 2010). Lipid peroxidation (LPO) and oxidative stress mechanism are thought to be the causes of such genotoxic effects. Cytotoxicity has been found to be dependent on concentration and time of exposure. ZNO NPs showing higher cytotxicity at higher concentrations and higher exposure times (Sharma et al., 2009). Concentration and time-dependent increase in oxidative stress, intracellular [Ca^{2+}] levels, and cell membrane damage has been reported to occur in a cultured BEAS-2B cell line. In such cases alteration in gene expression from oxidative stress and apoptosis was considered to be the main reason behind the induced cytotoxicity (Huang, Aronstam, Chen, & Huang, 2010). ZnO NPs in 8–10 nm size range have induced more toxicity to human colon cancer cells (RKO) than has the micrometersized ZnO of less than 44 mm size. Depolarization-induced neuronal injury in rats by the activation of voltage-gated Na^+ channels, after exposure to ZnO NPs has been reported thereby giving a clue that these NPs can probably cause neuronal apoptosis (Zhao, Xu, Zhang, Ren, & Yang, 2009). In rats, the intratracheal instillation and inhalation exposure of ZnO NPs for 1-3 h induced metal-fume-fever-like responses in lungs that was characterized by short term lung inflammatory or cytotoxic responses(Warheit et al., 2007). Invivo pulmonary exposure of both nano and fine sized ZnO in rats demonstrated potent, but reversible, inflammation. (Hillegass et al., 2010). The effect of ZnO NPs has also been tested on aquatic species. Dose-dependent reduction in hatching rate and induction of pericardial edema in developing zebrafish embryos, and in larvae has been reported. The Zn^{2+} dependent mechanism was responsible for the embryonic toxicity that resulted from ROS generation (X. Zhu, Wang, Zhang, Chang, & Chen, 2009).

Cerium Oxide Nanoparticles

There has been an increased use of nanosized cerium oxide (CeO_2) in polishing and computer chip manufacturing (Limbach et al., 2005; Mädler, Stark, & Pratsinis, 2002) and as an additive to decrease diesel emissions (Jung, Kittelson, & Zachariah, 2005). However, CeO_2 nanoparticles of sizes in range of 15nm, to 45 nm NPs caused toxicity, ROS increase, GSH decrease, and induced oxidative stress-related genes such as heme oxygenase-1, catalase, glutathione-Stransferase, and thiorexoxin reductase in cultured human lung epithelial cells (BEAS-2B). It was found that increased ROS by CeO_2 NPs triggered the activation of cytosolic Caspase and chromatin condensation, causing toxicity via the apoptotic process (Park, Choi, Park, & Park, 2008). On the contrary, other reports showed high biocompatibility of ceria nanostructures, prevented renal degeneration induced by intracellular peroxidases and conferred radioprotection to normal cells compared to no protection for tumor cells(J. Chen, Patil, Seal, & McGinnis, 2006; Tarnuzzer, Colon, Patil, & Seal, 2005). This apparent incongruity might be due to the surface oxidation state of nanoceria to scavenge superoxides or act in a catalytic manner. on the other hand, no significant change in LDH leakage or cell morphology is observed in A549 lung cells exposed directly to flamespray synthesized CeO NPs for 10–30 min (Perez JM, Asati A, Nath S, 2008) but resulted in a decreases in the mean total lamellar body volume per cell, reduction of cell-cell contacts and a significant increase in 8-oxoguanine positive cells, however NP induced oxidative stress resulting in altered gene expression (Rothen-Rutishauser et al., 2009).

CONCLUSION

Nanotechnology is a fast-growing field of activity that will allow development of materials with brand-new properties. Various scientific groups are keen about this technology and are devoting themselves to the development of more, new, and better nanomaterials. In the near future, expectations are that no field will be left untouched by the magical benefits available through application of nanotechnology. However, in this pursuit the number of Quebec workers exposed to nanoparticles should increase over the next few years. Presently, there is only limited knowledge concerning the toxicological effects of NPs.

Before, NPs are commercially used it is most important that they be subjected to appropriate toxicity evaluation. Among the parameters of NPs that must be evaluated for their effect on toxicity are surface charges, types of coating material, and reactivity of NPs. Because of their very small size, these particles offer a large contact surface per mass unit.

Since Nanoparticles (NPs) have already found a wide range of applications around the world. In vitro and in vivo studies on NPs have revealed that most are toxic to animals. The types of effects that NPs have produced are those on the pulmonary, cardiac, reproductive, renal and cutaneous systems, as well as on various cell lines. After exposures, significant accumulations of NPs have been found in the lungs, brain, liver, spleen, and bones of test species. However, their toxic behavior varies with their size, shape, surface charge, type of coating material and reactivity. Dose, route of administration, and exposure are critical factors that affect the degree of toxicity produced by any particular type of NP. Since, toxicity of nanoparticles has been reviewed in several reports and the most severe problem is related to the carcinogenic potential of NPs that has been associated both to the chemistry as well as to the physical properties. In particular, chemistry was considered to be relevant for the oxidative DNA damage and the formation of radical oxygen species (ROS) involved *via* direct mechanisms, whereas size, morphology and surface seem to be more important in all the other indirect mechanisms that underlie of cancer. Nevertheless, the majority of studies on genotoxicity of NPs have been so far performed on cell cultures. Thus in order to obtain data more relevant to the knowledge on human exposure, more trials on animal and long-term exposure experiments should be carried out. We believe a careful and rigorous toxicity testing is necessary before any NP is declared to be safe for broad use. We also believe that an agreed upon testing system is needed that can be used to suitably, accurately, and economically assess the toxicity of NPs. Moreover, there is a timely need to pay attention on how to control the in vivo transport and toxicity of different type of nanoparticles as only few reports have been published in this direction (Y.-S. Chen et al., 2012)

ACKNOWLEDGMENT

The author thanks to Department of Chemistry, Jamia Millia Islamia, New Delhi, India, for providing excellent facilities and cordial research environment. The author also thanks to Dr. Tokeer Ahmad, Nanochemistry Laboratory, Jamia Millia Islamia, New Delhi and Dr. Mohd. Ashraf Shah, Electron Microscopy Centre, Department of Physics, National Institute of Technology, Hazratbal, Srinagar, for their encouragement and guidance.

REFERENCES

Ahamed, M., Karns, M., Goodson, M., Rowe, J., Hussain, S. M., Schlager, J. J., & Hong, Y. (2008). DNA damage response to different surface chemistry of silver nanoparticles in mammalian cells. *Toxicology and Applied Pharmacology*, 233(3), 404–410. doi:10.1016/j.taap.2008.09.015 PMID:18930072

Ajayan, P. M., & Zhou, O. Z. (2001). Applications of carbon nanotubes. In *Carbon nanotubes* (pp. 391–425). Springer. doi:10.1007/3-540-39947-X_14

Akhavan, O., Ghaderi, E., & Akhavan, A. (2012). Size-dependent genotoxicity of graphene nanoplatelets in human stem cells. *Biomaterials*, *33*(32), 8017–8025. doi:10.1016/j.biomaterials.2012.07.040 PMID:22863381

Andersson-Willman, B., Gehrmann, U., Cansu, Z., Buerki-Thurnherr, T., Krug, H. F., Gabrielsson, S., & Scheynius, A. (2012). Effects of subtoxic concentrations of TiO 2 and ZnO nanoparticles on human lymphocytes, dendritic cells and exosome production. *Toxicology and Applied Pharmacology*, *264*(1), 94–103. doi:10.1016/j.taap.2012.07.021 PMID:22842014

Anyaogu, K. C., Fedorov, A. V., & Neckers, D. C. (2008). Synthesis, characterization, and antifouling potential of functionalized copper nanoparticles. *Langmuir*, *24*(8), 4340–4346. doi:10.1021/la800102f PMID:18341370

Arora, S., Jain, J., Rajwade, J. M., & Paknikar, K. M. (2008). Cellular responses induced by silver nanoparticles: In vitro studies. *Toxicology Letters*, *179*(2), 93–100. doi:10.1016/j.toxlet.2008.04.009 PMID:18508209

Aruoja, V., Dubourguier, H.-C., Kasemets, K., & Kahru, A. (2009). Toxicity of nanoparticles of CuO, ZnO and TiO 2 to microalgae Pseudokirchneriella subcapitata. *The Science of the Total Environment*, *407*(4), 1461–1468. doi:10.1016/j.scitotenv.2008.10.053 PMID:19038417

Barillet, S., Simon-Deckers, A., Herlin-Boime, N., Mayne-L'Hermite, M., Reynaud, C., Cassio, D., & Carrière, M. et al. (2010). Toxicological consequences of TiO2, SiC nanoparticles and multi-walled carbon nanotubes exposure in several mammalian cell types: An in vitro study. *Journal of Nanoparticle Research*, *12*(1), 61–73. doi:10.1007/s11051-009-9694-y

Bhattacharya, R., Mukherjee, P., Xiong, Z., Atala, A., Soker, S., & Mukhopadhyay, D. (2004). Gold nanoparticles inhibit VEGF165-induced proliferation of HUVEC cells. *Nano Letters*, *4*(12), 2479–2481. doi:10.1021/nl0483789

Borm, P. J. A., Robbins, D., Haubold, S., Kuhlbusch, T., Fissan, H., Donaldson, K., & Lademann, J. et al. (2006). The potential risks of nanomaterials: A review carried out for ECETOC. *Particle and Fibre Toxicology*, *3*(1), 11. doi:10.1186/1743-8977-3-11 PMID:16907977

Borm, P. J. A., Schins, R. P. F., & Albrecht, C. (2004). Inhaled particles and lung cancer, part B: Paradigms and risk assessment. *International Journal of Cancer*, *110*(1), 3–14. doi:10.1002/ijc.20064 PMID:15054863

Braydich-Stolle, L., Hussain, S., Schlager, J. J., & Hofmann, M.-C. (2005). In vitro cytotoxicity of nanoparticles in mammalian germline stem cells. *Toxicological Sciences*, *88*(2), 412–419. doi:10.1093/toxsci/kfi256 PMID:16014736

Bregoli, L., Chiarini, F., Gambarelli, A., Sighinolfi, G., Gatti, A. M., Santi, P., & Cocco, L. et al. (2009). Toxicity of antimony trioxide nanoparticles on human hematopoietic progenitor cells and comparison to cell lines. *Toxicology*, *262*(2), 121–129. doi:10.1016/j.tox.2009.05.017 PMID:19482055

Brown, R. C., Lockwood, A. H., & Sonawane, B. R. (2005). Neurodegenerative diseases: An overview of environmental risk factors. *Environmental Health Perspectives*, *113*(9), 1250–1256. doi:10.1289/ehp.7567 PMID:16140637

Buzea, C., Pacheco, I. I., & Robbie, K. (2007a). Nanomaterials and nanoparticles: Sources and toxicity. *Biointerphases*, *2*(4), MR17–MR71. doi:10.1116/1.2815690 PMID:20419892

Buzea, C., Pacheco, I. I., & Robbie, K. (2007b). Nanomaterials and nanoparticles: Sources and toxicity. *Biointerphases*, *2*(4), MR17–R71. doi:10.1116/1.2815690 PMID:20419892

Canesi, L., Ciacci, C., Betti, M., Fabbri, R., Canonico, B., Fantinati, A., & Pojana, G. et al. (2008). Immunotoxicity of carbon black nanoparticles to blue mussel hemocytes. *Environment International*, *34*(8), 1114–1119. doi:10.1016/j.envint.2008.04.002 PMID:18486973

Chadeau, E., Oulahal, N., Dubost, L., Favergeon, F., & Degraeve, P. (2010). Anti-Listeria innocua activity of silver functionalised textile prepared with plasma technology. *Food Control*, *21*(4), 505–512. doi:10.1016/j.foodcont.2009.07.013

Chang, Y., Yang, S.-T., Liu, J.-H., Dong, E., Wang, Y., Cao, A., & Wang, H. et al. (2011). In vitro toxicity evaluation of graphene oxide on A549 cells. *Toxicology Letters*, *200*(3), 201–210. doi:10.1016/j.toxlet.2010.11.016 PMID:21130147

Chen, J., Patil, S., Seal, S., & McGinnis, J. F. (2006). Rare earth nanoparticles prevent retinal degeneration induced by intracellular peroxides. *Nature Nanotechnology*, *1*(2), 142–150. doi:10.1038/nnano.2006.91 PMID:18654167

Chen, Y.-S., Hung, Y.-C., Hong, M.-Y., Onischuk, A. A., Chiou, J. C., & Sorokina, I. V. (2012). ... Steve Huang, G. (2012). Control of In Vivo Transport and Toxicity of Nanoparticles by Tea Melanin. *Journal of Nanomaterials*, 1–11. doi:10.1155/2012/746960

Chen, Y. S., Hung, Y. C., Liau, I., & Huang, G. S. (2009). Assessment of the in vivo toxicity of gold nanoparticles. *Nanoscale Research Letters*, *4*(8), 858–864. doi:10.1007/s11671-009-9334-6 PMID:20596373

Chen, Z., Meng, H., Xing, G., Chen, C., Zhao, Y., Jia, G., & Zhao, F. et al. (2006). Acute toxicological effects of copper nanoparticles in vivo. *Toxicology Letters*, *163*(2), 109–120. doi:10.1016/j.toxlet.2005.10.003 PMID:16289865

Chou, C.-C., Hsiao, H.-Y., Hong, Q.-S., Chen, C.-H., Peng, Y.-W., Chen, H.-W., & Yang, P.-C. (2008). Single-walled carbon nanotubes can induce pulmonary injury in mouse model. *Nano Letters*, *8*(2), 437–445. doi:10.1021/nl0723634 PMID:18225938

Cioffi, N., Torsi, L., Ditaranto, N., Tantillo, G., Ghibelli, L., Sabbatini, L., & Traversa, E. et al. (2005). Copper nanoparticle/polymer composites with antifungal and bacteriostatic properties. *Chemistry of Materials*, *17*(21), 5255–5262. doi:10.1021/cm0505244

Cipriano, A. F. F., & Liu, H. (2013). *Nanomaterials in Tissue Engineering*. Nanomaterials in Tissue Engineering; doi:10.1533/9780857097231.2.334

Crist, R. M., Grossman, J. H., Patri, A. K., Stern, S. T., Dobrovolskaia, M. A., Adiseshaiah, P. P., & McNeil, S. E. et al. (2013). Common pitfalls in nanotechnology: Lessons learned from NCI's Nanotechnology Characterization Laboratory. *Integrative Biology*, *5*(1), 66–73. doi:10.1039/C2IB20117H PMID:22772974

Cui, D., Tian, F., Ozkan, C. S., Wang, M., & Gao, H. (2005). Effect of single wall carbon nanotubes on human HEK293 cells. *Toxicology Letters, 155*(1), 73–85. doi:10.1016/j.toxlet.2004.08.015 PMID:15585362

Da Ros, T., & Prato, M. (1999). Medicinal chemistry with fullerenes and fullerene derivatives. *Chemical Communications*, (8): 663–669. doi:10.1039/a809495k

Daghrir, R., Drogui, P., & El Khakani, M. A. (2013). Photoelectrocatalytic oxidation of chlortetracycline using Ti/TiO 2 photo-anode with simultaneous H 2 O 2 production. *Electrochimica Acta, 87*, 18–31. doi:10.1016/j.electacta.2012.09.020

Dechsakulthorn, F., Hayes, A., Bakand, S., Joeng, L., & Winder, C. (2007). In vitro cytotoxicity assessment of selected nanoparticles using human skin fibroblasts. *AATEX, 14*(Special Issue), 397–400.

Ding, L., Stilwell, J., Zhang, T., Elboudwarej, O., Jiang, H., Selegue, J. P., & Chen, F. F. et al. (2005). Molecular characterization of the cytotoxic mechanism of multiwall carbon nanotubes and nano-onions on human skin fibroblast. *Nano Letters, 5*(12), 2448–2464. doi:10.1021/nl051748o PMID:16351195

Dutz, S., & Hergt, R. (2014). Magnetic particle hyperthermia--a promising tumour therapy? *Nanotechnology, 25*(45), 452001. doi:10.1088/0957-4484/25/45/452001 PMID:25337919

Fabian, E., Landsiedel, R., Ma-Hock, L., Wiench, K., Wohlleben, W., & Van Ravenzwaay, B. (2008). Tissue distribution and toxicity of intravenously administered titanium dioxide nanoparticles in rats. *Archives of Toxicology, 82*(3), 151–157. doi:10.1007/s00204-007-0253-y PMID:18000654

Feyman, R. (1959). *There's Plenty of Room at the Bottom: An Invitation to Enter a New Field of Physics*. Retrieved from http://www.zyvex.com/nanotech/feynman.html

Fiorito, S., Serafino, A., Andreola, F., & Bernier, P. (2006). Effects of fullerenes and single-wall carbon nanotubes on murine and human macrophages. *Carbon, 44*(6), 1100–1105. doi:10.1016/j.carbon.2005.11.009

Fu, A., Gu, W., Boussert, B., Koski, K., Gerion, D., Manna, L., & Alivisatos, A. P. et al. (2007). Semiconductor quantum rods as single molecule fluorescent biological labels. *Nano Letters, 7*(1), 179–182. doi:10.1021/nl0626434 PMID:17212460

Fubini, B., & Hubbard, A. (2003). Reactive oxygen species (ROS) and reactive nitrogen species (RNS) generation by silica in inflammation and fibrosis. *Free Radical Biology & Medicine, 34*(12), 1507–1516. doi:10.1016/S0891-5849(03)00149-7 PMID:12788471

Gajbhiye, M., Kesharwani, J., Ingle, A., Gade, A., & Rai, M. (2009). Fungus-mediated synthesis of silver nanoparticles and their activity against pathogenic fungi in combination with fluconazole. *Nanomedicine; Nanotechnology, Biology, and Medicine, 5*(4), 382–386. doi:10.1016/j.nano.2009.06.005 PMID:19616127

González-Flecha, B. (2004). Oxidant mechanisms in response to ambient air particles. *Molecular Aspects of Medicine, 25*(1), 169–182. doi:10.1016/j.mam.2004.02.017 PMID:15051325

Goodman, C. M., McCusker, C. D., Yilmaz, T., & Rotello, V. M. (2004). Toxicity of gold nanoparticles functionalized with cationic and anionic side chains. *Bioconjugate Chemistry, 15*(4), 897–900. doi:10.1021/bc049951i PMID:15264879

Griffitt, R. J., Hyndman, K., Denslow, N. D., & Barber, D. S. (2009). Comparison of molecular and histological changes in zebrafish gills exposed to metallic nanoparticles. *Toxicological Sciences, 107*(2), 404–415. doi:10.1093/toxsci/kfn256 PMID:19073994

Griffitt, R. J., Weil, R., Hyndman, K. A., Denslow, N. D., Powers, K., Taylor, D., & Barber, D. S. (2007). Exposure to copper nanoparticles causes gill injury and acute lethality in zebrafish (Danio rerio). *Environmental Science & Technology, 41*(23), 8178–8186. doi:10.1021/es071235e PMID:18186356

Hillegass, J. M., Shukla, A., Lathrop, S. A., MacPherson, M. B., Fukagawa, N. K., & Mossman, B. T. (2010). Assessing nanotoxicity in cells in vitro. *Wiley Interdisciplinary Reviews: Nanomedicine and Nanobiotechnology*.

Huang, C.-C., Aronstam, R. S., Chen, D.-R., & Huang, Y.-W. (2010). Oxidative stress, calcium homeostasis, and altered gene expression in human lung epithelial cells exposed to ZnO nanoparticles. *Toxicology In Vitro, 24*(1), 45–55. doi:10.1016/j.tiv.2009.09.007 PMID:19755143

Hussain, S. M., Braydich-Stolle, L. K., Schrand, A. M., Murdock, R. C., Yu, K. O., Mattie, D. M., & Terrones, M. et al. (2009). Toxicity evaluation for safe use of nanomaterials: Recent achievements and technical challenges. *Advanced Materials, 21*(16), 1549–1559. doi:10.1002/adma.200801395

Hussain, S. M., Hess, K. L., Gearhart, J. M., Geiss, K. T., & Schlager, J. J. (2005). In vitro toxicity of nanoparticles in BRL 3A rat liver cells. *Toxicology In Vitro, 19*(7), 975–983. doi:10.1016/j.tiv.2005.06.034 PMID:16125895

Hussain, S. M., Javorina, A. K., Schrand, A. M., Duhart, H. M., Ali, S. F., & Schlager, J. J. (2006). The interaction of manganese nanoparticles with PC-12 cells induces dopamine depletion. *Toxicological Sciences, 92*(2), 456–463. doi:10.1093/toxsci/kfl020 PMID:16714391

Ingale, S. V., Wagh, P. B., Tripathi, A. K., Dudwadkar, A. S., Gamre, S. S., Rao, P. T., & Gupta, S. C. et al. (2011). Photo catalytic oxidation of TNT using TiO2-SiO2 nano-composite aerogel catalyst prepared using sol–gel process. *Journal of Sol-Gel Science and Technology, 58*(3), 682–688. doi:10.1007/s10971-011-2445-4

Isakovic, A., Markovic, Z., Todorovic-Markovic, B., Nikolic, N., Vranjes-Djuric, S., Mirkovic, M., & Nikolic, Z. et al. (2006). Distinct cytotoxic mechanisms of pristine versus hydroxylated fullerene. *Toxicological Sciences, 91*(1), 173–183. doi:10.1093/toxsci/kfj127 PMID:16476688

Jia, G., Wang, H., Yan, L., Wang, X., Pei, R., Yan, T., & Guo, X. et al. (2005). Cytotoxicity of carbon nanomaterials: Single-wall nanotube, multi-wall nanotube, and fullerene. *Environmental Science & Technology, 39*(5), 1378–1383. doi:10.1021/es0487291 PMID:15787380

Jia, H. Y., Liu, Y., Zhang, X. J., Han, L., Du, L. B., Tian, Q., & Xu, Y. C. (2008). Potential oxidative stress of gold nanoparticles by induced-NO releasing in serum. *Journal of the American Chemical Society, 131*(1), 40–41. doi:10.1021/ja808033w PMID:19072650

Jin, C.-Y., Zhu, B.-S., Wang, X.-F., & Lu, Q.-H. (2008). Cytotoxicity of titanium dioxide nanoparticles in mouse fibroblast cells. *Chemical Research in Toxicology, 21*(9), 1871–1877. doi:10.1021/tx800179f PMID:18680314

Joner, E. J., Hartnik, T., & Amundsen, C. E. (2007). Environmental fate and ecotoxicity of engineered nanoparticles. *Norwegian Pollution Control Authority Report No. TA, 2304,* 1–64.

Jung, H., Kittelson, D. B., & Zachariah, M. R. (2005). The influence of a cerium additive on ultrafine diesel particle emissions and kinetics of oxidation. *Combustion and Flame, 142*(3), 276–288. doi:10.1016/j.combustflame.2004.11.015

Kalishwaralal, K., Sheikpranbabu, S., BarathManiKanth, S., Haribalaganesh, R., Ramkumarpandian, S., & Gurunathan, S. (2011). RETRACTED ARTICLE: Gold nanoparticles inhibit vascular endothelial growth factor-induced angiogenesis and vascular permeability via Src dependent pathway in retinal endothelial cells. *Angiogenesis, 14*(1), 29–45. doi:10.1007/s10456-010-9193-x PMID:21061058

Kim, S., Choi, J. E., Choi, J., Chung, K.-H., Park, K., Yi, J., & Ryu, D.-Y. (2009). Oxidative stress-dependent toxicity of silver nanoparticles in human hepatoma cells. *Toxicology In Vitro, 23*(6), 1076–1084. doi:10.1016/j.tiv.2009.06.001 PMID:19508889

Kim, S. T., Saha, K., Kim, C., & Rotello, V. M. (2013). The role of surface functionality in determining nanoparticle cytotoxicity. *Accounts of Chemical Research, 46*(3), 681–691. doi:10.1021/ar3000647 PMID:23294365

Kittelson, D. B. (2001). *Recent measurements of nanoparticle emissions from engines.* Current Research on Diesel Exhaust Particles.

Kowshik, M., Ashtaputre, S., Kharrazi, S., Vogel, W., Urban, J., Kulkarni, S. K., & Paknikar, K. M. (2003). Extracellular synthesis of silver nanoparticles by a silver-tolerant yeast strain MKY3. *Nanotechnology, 14*(1), 95–100. doi:10.1088/0957-4484/14/1/321

Lam, C.-W., James, J. T., McCluskey, R., & Hunter, R. L. (2004). Pulmonary toxicity of single-wall carbon nanotubes in mice 7 and 90 days after intratracheal instillation. *Toxicological Sciences, 77*(1), 126–134. doi:10.1093/toxsci/kfg243 PMID:14514958

Lasagna-Reeves, C., Gonzalez-Romero, D., Barria, M. A., Olmedo, I., Clos, A., Ramanujam, V. M. S., & Soto, C. et al. (2010). Bioaccumulation and toxicity of gold nanoparticles after repeated administration in mice. *Biochemical and Biophysical Research Communications, 393*(4), 649–655. doi:10.1016/j.bbrc.2010.02.046 PMID:20153731

Lee, H. Y., Park, H. K., Lee, Y. M., Kim, K., & Park, S. B.YeonáLee. (2007). A practical procedure for producing silver nanocoated fabric and its antibacterial evaluation for biomedical applications. *Chemical Communications,* (28): 2959–2961. doi:10.1039/b703034g PMID:17622444

Lewinski, N., Colvin, V., & Drezek, R. (2008). Cytotoxicity of nanoparticles. *Small, 4*(1), 26–49. doi:10.1002/smll.200700595 PMID:18165959

Li, J. J., Hartono, D., Ong, C.-N., Bay, B.-H., & Yung, L.-Y. L. (2010). Autophagy and oxidative stress associated with gold nanoparticles. *Biomaterials, 31*(23), 5996–6003. doi:10.1016/j.biomaterials.2010.04.014 PMID:20466420

Li, S.-Q., Zhu, R.-R., Zhu, H., Xue, M., Sun, X.-Y., Yao, S.-D., & Wang, S.-L. (2008). Nanotoxicity of TiO 2 nanoparticles to erythrocyte in vitro. *Food and Chemical Toxicology, 46*(12), 3626–3631. doi:10.1016/j.fct.2008.09.012 PMID:18840495

Liao, K.-H., Lin, Y.-S., Macosko, C. W., & Haynes, C. L. (2011). Cytotoxicity of graphene oxide and graphene in human erythrocytes and skin fibroblasts. *ACS Applied Materials & Interfaces, 3*(7), 2607–2615. doi:10.1021/am200428v PMID:21650218

Limbach, L. K., Li, Y., Grass, R. N., Brunner, T. J., Hintermann, M. A., Muller, M., & Stark, W. J. et al. (2005). Oxide nanoparticle uptake in human lung fibroblasts: Effects of particle size, agglomeration, and diffusion at low concentrations. *Environmental Science & Technology, 39*(23), 9370–9376. doi:10.1021/es051043o PMID:16382966

Liu, H., Ma, L., Zhao, J., Liu, J., Yan, J., Ruan, J., & Hong, F. (2009). Biochemical toxicity of nano-anatase TiO2 particles in mice. *Biological Trace Element Research, 129*(1-3), 170–180. doi:10.1007/s12011-008-8285-6 PMID:19066734

Liu, Z., Jiao, Y., Wang, Y., Zhou, C., & Zhang, Z. (2008). Polysaccharides-based nanoparticles as drug delivery systems. *Advanced Drug Delivery Reviews, 60*(15), 1650–1662. doi:10.1016/j.addr.2008.09.001 PMID:18848591

Liu, Z., Ren, G., Zhang, T., & Yang, Z. (2009). Action potential changes associated with the inhibitory effects on voltage-gated sodium current of hippocampal CA1 neurons by silver nanoparticles. *Toxicology, 264*(3), 179–184. doi:10.1016/j.tox.2009.08.005 PMID:19683029

Lv, M., Zhang, Y., Liang, L., Wei, M., Hu, W., Li, X., & Huang, Q. (2012). Effect of graphene oxide on undifferentiated and retinoic acid-differentiated SH-SY5Y cells line. *Nanoscale, 4*(13), 3861–3866. doi:10.1039/c2nr30407d PMID:22653613

Mädler, L., Stark, W. J., & Pratsinis, S. E. (2002). Flame-made ceria nanoparticles. *Journal of Materials Research, 17*(06), 1356–1362. doi:10.1557/JMR.2002.0202

Mahmoudi, M., Simchi, A., Milani, A. S., & Stroeve, P. (2009). Cell toxicity of superparamagnetic iron oxide nanoparticles. *Journal of Colloid and Interface Science, 336*(2), 510–518. doi:10.1016/j.jcis.2009.04.046 PMID:19476952

Mahtab, R., & Murphy, C. J. (2005). Probing DNA structure with nanoparticles. *Methods in Molecular Biology (Clifton, N.J.), 303*, 179–190. doi:10.1385/1-59259-901-X:179 PMID:15923684

Malmsten, M. (2013). Inorganic nanomaterials as delivery systems for proteins, peptides, DNA, and siRNA. *Current Opinion in Colloid & Interface Science, 18*(5), 468–480. doi:10.1016/j.cocis.2013.06.002

Meng, H., Chen, Z., Xing, G., Yuan, H., Chen, C., Zhao, F., & Zhao, Y. et al. (2007). Ultrahigh reactivity and grave nanotoxicity of copper nanoparticles. *Journal of Radioanalytical and Nuclear Chemistry, 272*(3), 595–598. doi:10.1007/s10967-007-0630-2

Mishra, B., Patel, B. B., & Tiwari, S. (2010). Colloidal nanocarriers: A review on formulation technology, types and applications toward targeted drug delivery. *Nanomedicine; Nanotechnology, Biology, and Medicine, 6*(1), 9–24. doi:10.1016/j.nano.2009.04.008 PMID:19447208

Moghimi, S. M., Hunter, A. C., & Murray, J. C. (2005). Nanomedicine: Current status and future prospects. *The FASEB Journal*, *19*(3), 311–330. doi:10.1096/fj.04-2747rev PMID:15746175

Monteiro-Riviere, N. A., & Inman, A. O. (2006). Challenges for assessing carbon nanomaterial toxicity to the skin. *Carbon*, *44*(6), 1070–1078. doi:10.1016/j.carbon.2005.11.004

Monteiro-Riviere, N. A., Inman, A. O., & Zhang, L. W. (2009). Limitations and relative utility of screening assays to assess engineered nanoparticle toxicity in a human cell line. *Toxicology and Applied Pharmacology*, *234*(2), 222–235. doi:10.1016/j.taap.2008.09.030 PMID:18983864

Moyano, D. F., Goldsmith, M., Solfiell, D. J., Landesman-Milo, D., Miranda, O. R., Peer, D., & Rotello, V. M. (2012). Nanoparticle hydrophobicity dictates immune response. *Journal of the American Chemical Society*, *134*(9), 3965–3967. doi:10.1021/ja2108905 PMID:22339432

Muller, J., Huaux, F., Moreau, N., Misson, P., Heilier, J.-F., Delos, M., & Lison, D. et al. (2005). Respiratory toxicity of multi-wall carbon nanotubes. *Toxicology and Applied Pharmacology*, *207*(3), 221–231. doi:10.1016/j.taap.2005.01.008 PMID:16129115

Na, H., Song, I. C., & Hyeon, T. (2009). Inorganic nanoparticles for MRI contrast agents. *Advanced Materials*, *21*(21), 2133–2148. doi:10.1002/adma.200802366

Nanda, A., & Saravanan, M. (2009). Biosynthesis of silver nanoparticles from Staphylococcus aureus and its antimicrobial activity against MRSA and MRSE. *Nanomedicine; Nanotechnology, Biology, and Medicine*, *5*(4), 452–456. doi:10.1016/j.nano.2009.01.012 PMID:19523420

Nel, A., Xia, T., Mädler, L., & Li, N. (2006). Toxic potential of materials at the nanolevel. *Science*, *311*(5761), 622–627. doi:10.1126/science.1114397 PMID:16456071

Oberdörster, G. (2000). Pulmonary effects of inhaled ultrafine particles. *International Archives of Occupational and Environmental Health*, *74*(1), 1–8. doi:10.1007/s004200000185 PMID:11196075

Ostiguy, C., Lapointe, G., Ménard, L., Cloutier, Y., Trottier, M., Boutin, M., … Normand, C. (2006). *Nanoparticles: Current Knowledge about Occupational Health and Safety Risks and Prevention Measures*. Studies and Research, IRSST, Report R-470, Montreal. Retrieved from http://www.irsst.qc.ca/media/documents/pubirsst/r-470.pdf

Pan, Y., Leifert, A., Ruau, D., Neuss, S., Bornemann, J., Schmid, G., & Jahnen-Dechent, W. et al. (2009). Gold nanoparticles of diameter 1.4 nm trigger necrosis by oxidative stress and mitochondrial damage. *Small*, *5*(18), 2067–2076. doi:10.1002/smll.200900466 PMID:19642089

Pan, Y., Neuss, S., Leifert, A., Fischer, M., Wen, F., Simon, U., & Brandau, W. et al. (2007). Size-dependent cytotoxicity of gold nanoparticles. *Small*, *3*(11), 1941–1949. doi:10.1002/smll.200700378 PMID:17963284

Park, E.-J., Choi, J., Park, Y.-K., & Park, K. (2008). Oxidative stress induced by cerium oxide nanoparticles in cultured BEAS-2B cells. *Toxicology*, *245*(1), 90–100. doi:10.1016/j.tox.2007.12.022 PMID:18243471

Patra, H. K., Banerjee, S., Chaudhuri, U., Lahiri, P., & Dasgupta, A. K. (2007). Cell selective response to gold nanoparticles. *Nanomedicine; Nanotechnology, Biology, and Medicine*, *3*(2), 111–119. doi:10.1016/j.nano.2007.03.005 PMID:17572353

Perez, J. M., Asati, A., Nath, S. K. C., & Kaittanis, C. (2008). Synthesis of Biocompatible Detrax-coated Nanoceria with pH- dependent Antioxidant Properties. *Small, 4*(5), 552–556. doi:10.1002/smll.200700824 PMID:18433077

Peters, A. (2005). Particulate matter and heart disease: Evidence from epidemiological studies. *Toxicology and Applied Pharmacology, 207*(2), S477–S482. doi:10.1016/j.taap.2005.04.030 PMID:15990137

Pfau, J. C., Sentissi, J. J., Weller, G., & Putnam, E. A. (2005). Assessment of autoimmune responses associated with asbestos exposure in Libby, Montana, USA. *Environmental Health Perspectives*, 25–30. PMID:15626643

Porter, A. E., Muller, K., Skepper, J., Midgley, P., & Welland, M. (2006). Uptake of C 60 by human monocyte macrophages, its localization and implications for toxicity: Studied by high resolution electron microscopy and electron tomography. *Acta Biomaterialia, 2*(4), 409–419. doi:10.1016/j.actbio.2006.02.006 PMID:16765881

Pulskamp, K., Diabaté, S., & Krug, H. F. (2007). Carbon nanotubes show no sign of acute toxicity but induce intracellular reactive oxygen species in dependence on contaminants. *Toxicology Letters, 168*(1), 58–74. doi:10.1016/j.toxlet.2006.11.001 PMID:17141434

Rahman, M. F., Wang, J., Patterson, T. A., Saini, U. T., Robinson, B. L., Newport, G. D., & Ali, S. F. et al. (2009). Expression of genes related to oxidative stress in the mouse brain after exposure to silver-25 nanoparticles. *Toxicology Letters, 187*(1), 15–21. doi:10.1016/j.toxlet.2009.01.020 PMID:19429238

Rana, S., & Kalaichelvan, P. T. (2013). Ecotoxicity of nanoparticles. *ISRN Toxicology*, 574648. doi:10.1155/2013/574648 PMID:23724300

Reddy, A. S., Chen, C.-Y., Chen, C.-C., Jean, J.-S., Chen, H.-R., Tseng, M.-J., & Wang, J.-C. et al. (2010). Biological synthesis of gold and silver nanoparticles mediated by the bacteria Bacillus subtilis. *Journal of Nanoscience and Nanotechnology, 10*(10), 6567–6574. doi:10.1166/jnn.2010.2519 PMID:21137763

Rodriguez-Hernandez, J., Chécot, F., Gnanou, Y., & Lecommandoux, S. (2005). Toward "smart"nano-objects by self-assembly of block copolymers in solution. *Progress in Polymer Science, 30*(7), 691–724. doi:10.1016/j.progpolymsci.2005.04.002

Roduner, E. (2006). Size matters: Why nanomaterials are different. *Chemical Society Reviews, 35*(7), 583–592. doi:10.1039/b502142c PMID:16791330

Rosas-Hernández, H., Jiménez-Badillo, S., Martínez-Cuevas, P. P., Gracia-Espino, E., Terrones, H., Terrones, M., & González, C. et al. (2009). Effects of 45-nm silver nanoparticles on coronary endothelial cells and isolated rat aortic rings. *Toxicology Letters, 191*(2), 305–313. doi:10.1016/j.toxlet.2009.09.014 PMID:19800954

Rothen-Rutishauser, B., Grass, R. N., Blank, F., Limbach, L. K., Muhlfeld, C., Brandenberger, C., & Stark, W. J. et al. (2009). Direct combination of nanoparticle fabrication and exposure to lung cell cultures in a closed setup as a method to simulate accidental nanoparticle exposure of humans. *Environmental Science & Technology, 43*(7), 2634–2640. doi:10.1021/es8029347 PMID:19452928

Saathoff, J. G., Inman, A. O., Xia, X. R., Riviere, J. E., & Monteiro-Riviere, N. A. (2011). In vitro toxicity assessment of three hydroxylated fullerenes in human skin cells. *Toxicology In Vitro*, *25*(8), 2105–2112. doi:10.1016/j.tiv.2011.09.013 PMID:21964474

Sayes, C. M., Fortner, J. D., Guo, W., Lyon, D., Boyd, A. M., Ausman, K. D., & Hughes, J. B. et al. (2004). The differential cytotoxicity of water-soluble fullerenes. *Nano Letters*, *4*(10), 1881–1887. doi:10.1021/nl0489586

Sayes, C. M., Gobin, A. M., Ausman, K. D., Mendez, J., West, J. L., & Colvin, V. L. (2005). Nano-C 60 cytotoxicity is due to lipid peroxidation. *Biomaterials*, *26*(36), 7587–7595. doi:10.1016/j.biomaterials.2005.05.027 PMID:16005959

Sayes, C. M., Liang, F., Hudson, J. L., Mendez, J., Guo, W., Beach, J. M., & Billups, W. E. et al. (2006). Functionalization density dependence of single-walled carbon nanotubes cytotoxicity in vitro. *Toxicology Letters*, *161*(2), 135–142. doi:10.1016/j.toxlet.2005.08.011 PMID:16229976

Schrand, A. M., Rahman, M. F., Hussain, S. M., Schlager, J. J., Smith, D. a., & Syed, A. F. (2010). Metal-based nanoparticles and their toxicity assessment. *Wiley Interdisciplinary Reviews: Nanomedicine and Nanobiotechnology*, *2*(5), 544–568. doi:10.1002/wnan.103 PMID:20681021

Sharma, V., Shukla, R. K., Saxena, N., Parmar, D., Das, M., & Dhawan, A. (2009). DNA damaging potential of zinc oxide nanoparticles in human epidermal cells. *Toxicology Letters*, *185*(3), 211–218. doi:10.1016/j.toxlet.2009.01.008 PMID:19382294

Simberg, D., Zhang, W.-M., Merkulov, S., McCrae, K., Park, J.-H., Sailor, M. J., & Ruoslahti, E. (2009). Contact activation of kallikrein–kinin system by superparamagnetic iron oxide nanoparticles in vitro and in vivo. *Journal of Controlled Release*, *140*(3), 301–305. doi:10.1016/j.jconrel.2009.05.035 PMID:19508879

Staroverov, S. A., Aksinenko, N. M., Gabalov, K. P., Vasilenko, O. A., Vidyasheva, I. V., Shchyogolev, S. Y., & Dykman, L. A. (2009). Effect of gold nanoparticles on the respiratory activity of peritoneal macrophages. *Gold Bulletin*, *42*(2), 153–156. doi:10.1007/BF03214925

Su, G., Zhou, H., Mu, Q., Zhang, Y., Li, L., Jiao, P., & Yan, B. et al. (2012). Effective surface charge density determines the electrostatic attraction between nanoparticles and cells. *The Journal of Physical Chemistry C*, *116*(8), 4993–4998. doi:10.1021/jp211041m

Sun, X., Liu, Z., Welsher, K., Robinson, J. T., Goodwin, A., Zaric, S., & Dai, H. (2008). Nano-graphene oxide for cellular imaging and drug delivery. *Nano Research*, *1*(3), 203–212. doi:10.1007/s12274-008-8021-8 PMID:20216934

Sung, J. H., Ji, J. H., Park, J. D., Yoon, J. U., Kim, D. S., Jeon, K. S., & Han, J. H. et al. (2009). Subchronic inhalation toxicity of silver nanoparticles. *Toxicological Sciences*, *108*(2), 452–461. doi:10.1093/toxsci/kfn246 PMID:19033393

Suzuki, H., Toyooka, T., & Ibuki, Y. (2007). Simple and easy method to evaluate uptake potential of nanoparticles in mammalian cells using a flow cytometric light scatter analysis. *Environmental Science & Technology*, *41*(8), 3018–3024. doi:10.1021/es0625632 PMID:17533873

Szarpak, A., Cui, D., Dubreuil, F., De Geest, B. G., De Cock, L. J., Picart, C., & Auzély-Velty, R. (2010). Designing hyaluronic acid-based layer-by-layer capsules as a carrier for intracellular drug delivery. *Biomacromolecules*, *11*(3), 713–720. doi:10.1021/bm9012937 PMID:20108937

Tao, H., & Nallathamby, P. D. (2008). Photostable single-molecule nanoparticle optical biosensors for real-time sensing of single cytokine molecules and their binding reactions. *Journal of the American Chemical Society*, *130*(50), 17095–17105. doi:10.1021/ja8068853 PMID:19053435

Tarnuzzer, R. W., Colon, J., Patil, S., & Seal, S. (2005). Vacancy engineered ceria nanostructures for protection from radiation-induced cellular damage. *Nano Letters*, *5*(12), 2573–2577. doi:10.1021/nl052024f PMID:16351218

Teow, Y., Asharani, P. V., Hande, M. P., & Valiyaveettil, S. (2011). Health impact and safety of engineered nanomaterials. *Chemical Communications (Cambridge, England)*, *47*(25), 7025–7038. doi:10.1039/c0cc05271j PMID:21479319

Thakkar, K. N., Mhatre, S. S., & Parikh, R. Y. (2010). Biological synthesis of metallic nanoparticles. *Nanomedicine; Nanotechnology, Biology, and Medicine*, *6*(2), 257–262. doi:10.1016/j.nano.2009.07.002 PMID:19616126

Thanh, N. T. K., & Green, L. A. W. (2010). Functionalisation of nanoparticles for biomedical applications. *Nano Today*, *5*(3), 213–230. doi:10.1016/j.nantod.2010.05.003

Thomas, M., & Klibanov, A. M. (2003). Conjugation to gold nanoparticles enhances polyethylenimine's transfer of plasmid DNA into mammalian cells. *Proceedings of the National Academy of Sciences of the United States of America*, *100*(16), 9138–9143. doi:10.1073/pnas.1233634100 PMID:12886020

Tian, F., Cui, D., Schwarz, H., Estrada, G. G., & Kobayashi, H. (2006). Cytotoxicity of single-wall carbon nanotubes on human fibroblasts. *Toxicology In Vitro*, *20*(7), 1202–1212. doi:10.1016/j.tiv.2006.03.008 PMID:16697548

Tkachenko, A. G., Xie, H., Coleman, D., Glomm, W., Ryan, J., Anderson, M. F., & Frazen, S. (2003). Multifunctional gold nanoparticle-peptide complexes for nuclear targeting. *Journal of the American Chemical Society*, *125*(16), 4700–4701. doi:10.1021/ja0296935 PMID:12696875

Tominaga, Y., Kubo, T., & Hosoya, K. (2011). Surface modification of TiO 2 for selective photodegradation of toxic compounds. *Catalysis Communications*, *12*(9), 785–789. doi:10.1016/j.catcom.2011.01.021

Vallabani, N. V., Mittal, S., Shukla, R. K., Pandey, A. K., Dhakate, S. R., Pasricha, R., & Dhawan, A. (2011). Toxicity of graphene in normal human lung cells (BEAS-2B). *Journal of Biomedical Nanotechnology*, *7*(1), 106–107. doi:10.1166/jbn.2011.1224 PMID:21485826

Vevers, W. F., & Jha, A. N. (2008). Genotoxic and cytotoxic potential of titanium dioxide (TiO2) nanoparticles on fish cells in vitro. *Ecotoxicology (London, England)*, *17*(5), 410–420. doi:10.1007/s10646-008-0226-9 PMID:18491228

Vileno, B., Marcoux, P. R., Lekka, M., Sienkiewicz, A., Fehér, T., & Forró, L. (2006). Spectroscopic and photophysical properties of a highly derivatized C60 fullerol. *Advanced Functional Materials, 16*(1), 120–128. doi:10.1002/adfm.200500425

Wagner, A. J., Bleckmann, C. A., Murdock, R. C., Schrand, A. M., Schlager, J. J., & Hussain, S. M. (2007). Cellular interaction of different forms of aluminum nanoparticles in rat alveolar macrophages. *The Journal of Physical Chemistry B, 111*(25), 7353–7359. doi:10.1021/jp068938n PMID:17547441

Wang, J. J., Sanderson, B. J. S., & Wang, H. (2007). Cyto-and genotoxicity of ultrafine TiO 2 particles in cultured human lymphoblastoid cells. *Mutation Research/Genetic Toxicology and Environmental Mutagenesis, 628*(2), 99–106. doi:10.1016/j.mrgentox.2006.12.003 PMID:17223607

Wang, Y., & Cao, G. (2006). Synthesis and enhanced intercalation properties of nanostructured vanadium oxides. *Chemistry of Materials, 18*(12), 2787–2804. doi:10.1021/cm052765h

Warheit, D. B., Hoke, R. A., Finlay, C., Donner, E. M., Reed, K. L., & Sayes, C. M. (2007). Development of a base set of toxicity tests using ultrafine TiO 2 particles as a component of nanoparticle risk management. *Toxicology Letters, 171*(3), 99–110. doi:10.1016/j.toxlet.2007.04.008 PMID:17566673

Wielgus, A. R., Zhao, B., Chignell, C. F., Hu, D.-N., & Roberts, J. E. (2010). Phototoxicity and cytotoxicity of fullerol in human retinal pigment epithelial cells. *Toxicology and Applied Pharmacology, 242*(1), 79–90. doi:10.1016/j.taap.2009.09.021 PMID:19800903

Wilhelmi, V., Fischer, U., van Berlo, D., Schulze-Osthoff, K., Schins, R. P. F., & Albrecht, C. (2012). Evaluation of apoptosis induced by nanoparticles and fine particles in RAW 264.7 macrophages: Facts and artefacts. *Toxicology In Vitro, 26*(2), 323–334. doi:10.1016/j.tiv.2011.12.006 PMID:22198050

Yan, L., & Chen, X. (2013). *Nanomaterials for Drug Delivery. Nanocrystalline Materials: Their Synthesis-Structure-Property Relationships and Applications.* doi:10.1016/B978-0-12-407796-6.00007-5

Yang, H., Li, H., & Jiang, X. (2008). Detection of foodborne pathogens using bioconjugated nanomaterials. *Microfluidics and Nanofluidics, 5*(5), 571–583. doi:10.1007/s10404-008-0302-8

Zhang, X., Hu, W., Li, J., Tao, L., & Wei, Y. (2012). A comparative study of cellular uptake and cytotoxicity of multi-walled carbon nanotubes, graphene oxide, and nanodiamond. *Toxicological Reviews, 1*(1), 62–68.

Zhang, Y., Guo, Y., Xianyu, Y., Chen, W., Zhao, Y., & Jiang, X. (2013). Nanomaterials for ultrasensitive protein detection. *Advanced Materials, 25*(28), 3802–3819. doi:10.1002/adma.201301334 PMID:23740753

Zhang, Y., Xu, D., Li, W., Yu, J., & Chen, Y. (2012). Effect of size, shape, and surface modification on cytotoxicity of gold nanoparticles to human HEp-2 and canine MDCK cells. *Journal of Nanomaterials, 2012*, 7. doi:10.1155/2012/375496

Zhao, J., Xu, L., Zhang, T., Ren, G., & Yang, Z. (2009). Influences of nanoparticle zinc oxide on acutely isolated rat hippocampal CA3 pyramidal neurons. *Neurotoxicology, 30*(2), 220–230. doi:10.1016/j.neuro.2008.12.005 PMID:19146874

Zhu, M.-T., Feng, W.-Y., Wang, B., Wang, T.-C., Gu, Y.-Q., Wang, M., & Chai, Z.-F. et al. (2008). Comparative study of pulmonary responses to nano-and submicron-sized ferric oxide in rats. *Toxicology*, *247*(2), 102–111. doi:10.1016/j.tox.2008.02.011 PMID:18394769

Zhu, X., Wang, J., Zhang, X., Chang, Y., & Chen, Y. (2009). The impact of ZnO nanoparticle aggregates on the embryonic development of zebrafish (Danio rerio). *Nanotechnology*, *20*(19), 195103. doi:10.1088/0957-4484/20/19/195103 PMID:19420631

Compilation of References

Abdel-Khalek, A. A., Kadry, M. A. M., Badran, S. R., & Marie, M. A. S. (2015). Comparative toxicity of copper oxide bulk and nanoparticles in Nile Tilapia; *Oreochromis niloticus*: Biochemical and oxidative stress. *Journal of Basic & Applied Zoology*, *72*, 43–57. doi:10.1016/j.jobaz.2015.04.001

Abdelsayed, V., Aljarash, A., El-Shall, M. S., Al Othman, Z. A., & Alghamdi, A. H. (2009). Microwave synthesis of bimetallic nanoalloys and CO oxidation on ceria-supported nanoalloys. *Chemistry of Materials*, *21*(13), 2825–2834. doi:10.1021/cm9004486

Abebe, L. S., Smith, J. A., Narkiewicz, S., Oyanedel-Craver, V., Conaway, M., Singo, A., & Dillingham, R. et al. (2014). Ceramic water filters impregnated with silver nanoparticles as a point-of-use water-treatment intervention for HIV-positive individuals in Limpopo Province, South Africa: A pilot study of technological performance and human health benefits. *Journal of Water and Health*, *12*(2), 288–300. doi:10.2166/wh.2013.185 PMID:24937223

Abhijith, B. D., Ramesh, M., & Poopal, R. K. (2012). Sublethal toxicological evaluation of methyl parathion on some haematological and biochemical parameters in an Indian major carp *Catla catla*. *Comparative Clinical Pathology*, *21*(1), 55–61. doi:10.1007/s00580-010-1064-8

AbouZeid, K. M., Mohamed, M. B., & El-Shall, M. S. (2011). Hybrid Au–CdSe and Ag–CdSe Nanoflowers and Core–Shell Nanocrystals via One-Pot Heterogeneous Nucleation and Growth. *Small*, *7*(23), 3299–3307. doi:10.1002/smll.201100688 PMID:21994186

Achermann, M. (2010). Exciton–Plasmon Interactions in Metal– Semiconductor Nanostructures. *The Journal of Physical Chemistry Letters*, *1*(19), 2837–2843. doi:10.1021/jz101102e

Adam, N., Schmitt, C., Galceran, J., Companys, E., Vakurov, A., & Wallace, R. (2015). *The chronic toxicity of ZnO nanoparticles and ZnCl2 to Daphnia magna and the use of different methods to assess nanoparticle aggregation and dissolution. In Nanotoxicology*. Taylor & Francis.

Adam, N., Vakurov, A., Knapen, D., & Blust, R. (2015). The chronic toxicity of CuO nanoparticles and copper salt to Daphnia magna. *Journal of Hazardous Materials*, *283*, 416–422. doi:10.1016/j.jhazmat.2014.09.037 PMID:25464278

Adeleye, A. S., Conway, J. R., Garner, K., Huang, Y., Su, Y., & Keller, A. A. (2016). Engineered nanomaterials for water treatment and remediation: Costs, benefits, and applicability. *Chemical Engineering Journal*, *286*, 640–662. doi:10.1016/j.cej.2015.10.105

Adesina, A. A. (2004). Industrial exploitation of photocatalysis progress, perspectives and prospects. *Catalysis Surveys from Asia*, *8*(4), 265–273. doi:10.1007/s10563-004-9117-0

Compilation of References

Adhikari, S., Sarkar, B., Chatterjee, A., Mahapatra, C. T., & Ayyappan, S. (2004). Effects of cypermethrin and carbofuran haematological parameters and prediction of their recovery in a freshwater teleost, *Labeo rohita* (Hamilton). *Ecotoxicology and Environmental Safety*, *58*(2), 220–226. doi:10.1016/j.ecoenv.2003.12.003 PMID:15157576

Adriana, B., & Almodóva, A. N. M. (2007). Antimicrobial efficacy of *Curcuma zedoaria* extract as assessed by linear regression compared with commercial mouthrinses. *Brazilian Journal of Microbiology*, *38*(3), 440–445. doi:10.1590/S1517-83822007000300011

Agenson, K., Oh, J., Kikuta, T., & Urase, T. (2003). Rejection mechanisms of plastic additives and natural hormones in drinking water treated by nanofiltration. *Water Supply*, *3*(5), 311–319.

Agnihotri, S., Mukherji, S., & Mukherji, S. (2013). Immobilized silver nanoparticles enhance contact killing and show highest efficacy: Elucidation of the mechanism of bactericidal action of silver. *Nanoscale*, *5*(16), 7328–7340. doi:10.1039/c3nr00024a PMID:23821237

Agnihotri, S., Mukherji, S., & Mukherji, S. (2014). Size-controlled silver nanoparticles synthesized over the range 5–100 nm using the same protocol and their antibacterial efficacy. *RSC Advances*, *4*(8), 3974–3983. doi:10.1039/C3RA44507K

Ahamed, M., Akhtar, M. J., Alhadlaq, H. A., Khan, M. A. M., & Alrokayan, S. A. (2015). Comparative cytotoxic response of nickel ferrite nanoparticles in human liver HepG2 and breast MFC-7 cancer cells. *Chemosphere*, *135*, 278–288. doi:10.1016/j.chemosphere.2015.03.079 PMID:25966046

Ahamed, M., Karns, M., Goodson, M., Rowe, J., Hussain, S. M., Schlager, J. J., & Hong, Y. (2008). DNA damage response to different surface chemistry of silver nanoparticles in mammalian cells. *Toxicology and Applied Pharmacology*, *233*(3), 404–410. doi:10.1016/j.taap.2008.09.015 PMID:18930072

Ahmad, A., Mukherjee, P., Senapati, S., Mandal, D., Khan, M. I., Kumar, R., & Sastry, M. (2003). Extracellular biosynthesis of silver nanoparticles using the fungus Fusarium oxysporum. *Colloids and Surfaces. B, Biointerfaces*, *28*(4), 313–318. doi:10.1016/S0927-7765(02)00174-1

Ahmad, A., Tan, L., & Shukor, S. A. (2008). Dimethoate and atrazine retention from aqueous solution by nanofiltration membranes. *Journal of Hazardous Materials*, *151*(1), 71–77. doi:10.1016/j.jhazmat.2007.05.047 PMID:17587496

Ahmad, F., Liu, X., Zhou, Y., & Yao, H. (2015). An in vivo evaluation of acute toxicity of cobalt ferrite (CoFe2O4) nanoparticles in larval-embryo Zebrafish (Danio rerio). *Aquatic Toxicology (Amsterdam, Netherlands)*, *166*, 21–28. doi:10.1016/j.aquatox.2015.07.003 PMID:26197244

Ahn, C. H., Baek, Y., Lee, C., Kim, S. O., Kim, S., Lee, S., & Yoon, J. et al. (2012). Carbon nanotube-based membranes: Fabrication and application to desalination. *Journal of Industrial and Engineering Chemistry*, *18*(5), 1551–1559. doi:10.1016/j.jiec.2012.04.005

Airing Out Anthrax. (n.d.). Retrieved from http://ntrs.nasa.gov/archive/nasa/casi.ntrs.nasa.gov/20020080174.pdf

Ajayan, P. M., & Zhou, O. Z. (2001). Applications of carbon nanotubes. In *Carbon nanotubes* (pp. 391–425). Springer. doi:10.1007/3-540-39947-X_14

Akhavan, O., Ghaderi, E., & Akhavan, A. (2012). Size-dependent genotoxicity of graphene nanoplatelets in human stem cells. *Biomaterials*, *33*(32), 8017–8025. doi:10.1016/j.biomaterials.2012.07.040 PMID:22863381

Al-Asgah, N. A., Abdel-Warith, A. W. A., Younis, E. S. M., & Allam, H. Y. (2015). Haematological and biochemical parameters and tissue accumulations of cadmium in *Oreochromis niloticus* exposed to various concentrations of cadmium chloride. *Saudi Journal of Biological Sciences*, *22*(5), 543–550. doi:10.1016/j.sjbs.2015.01.002 PMID:26288556

Al-Bairuty, G. A., Shaw, B. J., Handy, R. D., & Henry, T. B. (2013). Histopathological effects of waterborne copper nanoparticles and copper sulphate on the organs of rainbow trout (*Oncorhynchus mykiss*). *Aquatic Toxicology (Amsterdam, Netherlands)*, *126*, 104–115. doi:10.1016/j.aquatox.2012.10.005 PMID:23174144

Albanese, A., Tang, P. S., & Chan, W. C. W. (2012). The Effect of Nanoparticle Size, Shape, and Surface Chemistry on Biological Systems. *Annual Review of Biomedical Engineering*, *14*(1), 1–16. doi:10.1146/annurev-bioeng-071811-150124 PMID:22524388

Albert, J., Luoto, J., & Levine, D. (2010). End-User Preferences for and Performance of Competing POU Water Treatment Technologies among the Rural Poor of Kenya. *Environmental Science & Technology*, *44*(12), 4426–4432. doi:10.1021/es1000566 PMID:20446726

Ali, M. A., Rehman, I., Iqbal, A., Din, S., Rao, A. Q., Latif, A., & Husnain, T. et al. (2014). Nanotechnology, a new frontier in Agriculture. *Advancements in Life Sciences*, *1*(3), 129–138.

Alinovi, R., Goldoni, M., Pinelli, S., Campanini, M., Aliatis, I., Bersani, D., & Mutti, A. et al. (2015). Oxidative and pro-inflammatory effects of cobalt and titanium oxide nanoparticles on aortic and venous endothelial cells. *Toxicology In Vitro*, *29*(3), 426–437. doi:10.1016/j.tiv.2014.12.007 PMID:25526690

Alkaladi, A., Nasr El-Deen, N. A. M., Afifi, M., & Abu Zinadah, O. A. (2015). Hematological and biochemical investigations on the effect of vitamin E and C on *Oreochromis niloticus* exposed to zinc oxide nanoparticles. *Saudi Journal of Biological Sciences*, *22*(5), 556–563. doi:10.1016/j.sjbs.2015.02.012 PMID:26288558

Allen, E. R., Hossner, L. R., Ming, D. W., & Henninger, D. L. (1993). Solubility and cation exchange in phosphate rock and saturated clinoptilolite mixtures. *Soil Science Society of America Journal*, *57*(5), 1368–1374. doi:10.2136/sssaj1993.03615995005700050034x PMID:11537990

Alves, J. O., Zhuo, C., Levendis, Y. A., & Tenório, J. A. (2011). Catalytic conversion of wastes from the bioethanol production into carbon nanomaterials. *Applied Catalysis B: Environmental*, *106*(3), 433–444. doi:10.1016/j.apcatb.2011.06.001

Alwan, S. F., Hadi, A. A., & Shokr, A. E. (2009). Alterations in haematological parameter of fresh water fish. *Tilapia zilli* exposed to Aluminium. *Journal of Science and Its Applications*, *3*(1), 12–19.

Amine, A., Mohammadi, H., Bourais, I., & Palleschi, G. (2006). Enzyme inhibition-based biosensors for food safety and environmental monitoring. *Biosensors & Bioelectronics*, *21*(8), 1405–1423. doi:10.1016/j.bios.2005.07.012 PMID:16125923

Amini, M., Jahanshahi, M., & Rahimpour, A. (2013). Synthesis of novel thin film nanocomposite (TFN) forward osmosis membranes using functionalized multi-walled carbon nanotubes. *Journal of Membrane Science*, *435*(0), 233–241. doi:10.1016/j.memsci.2013.01.041

Anandan, S., Grieser, F., & Ashokkumar, M. (2008). Sonochemical synthesis of Au− Ag core− shell bimetallic nanoparticles. *The Journal of Physical Chemistry C*, *112*(39), 15102–15105. doi:10.1021/jp806960r

An, B., Liang, Q. Q., & Zhao, D. Y. (2011). Removal of arsenic(V) from spent ion exchange brine using a new class of starch-bridged magnetite nanoparticles. *Water Research*, *45*(5), 1961–1972. doi:10.1016/j.watres.2011.01.004 PMID:21288549

An, B., & Zhao, D. Y. (2012). Immobilization of As(III) in soil and groundwater using a new class of polysaccharide stabilized Fe-Mn oxide nanoparticles. *Journal of Hazardous Materials*, *211*, 332–341. doi:10.1016/j.jhazmat.2011.10.062 PMID:22119304

Andersen, H., Siegrist, H., Halling-Sørensen, B., & Ternes, T. A. (2003). Fate of estrogens in a municipal sewage treatment plant. *Environmental Science & Technology*, *37*(18), 4021–4026. doi:10.1021/es026192a PMID:14524430

Compilation of References

Andersson, L. E., & Grubbström, R. W. (1994). *Asymptotic behavior of a stochastic multi-period inventory process with planned production.* Working paper WP – 210. Dept. of Production Economics, Linköping Institute of Technology.

Andersson, T., Nilsson, K., Sundahl, M., Westman, G., & Wennerstrom, O. (1992). C_{60} embedded in r-cyclodextrin - a water-soluble fullerene. *Journal of the Chemical Society. Chemical Communications*, (8): 604–606. doi:10.1039/C39920000604

Andersson-Willman, B., Gehrmann, U., Cansu, Z., Buerki-Thurnherr, T., Krug, H. F., Gabrielsson, S., & Scheynius, A. (2012). Effects of subtoxic concentrations of TiO 2 and ZnO nanoparticles on human lymphocytes, dendritic cells and exosome production. *Toxicology and Applied Pharmacology*, *264*(1), 94–103. doi:10.1016/j.taap.2012.07.021 PMID:22842014

Andreani, T., Kiill, C. P., Souza, A. L. R., Fangueiro, J. F., Fernandes, L., Doktorovová, S., & Silva, A. M. et al. (2014). Surface engineering of silica nanoparticles for oral insulin delivery: Characterization and cell toxicity studies. *Colloid Surf B*, *123*, 916–923. doi:10.1016/j.colsurfb.2014.10.047 PMID:25466464

Andrievsky, G. V., Kosevich, M. V., Vovk, O. M., Shelkovsky, V. S., & Vashchenko, L. A. (1995). On the production of an aqueous colloidal solution of fullerenes. *Journal of the Chemical Society. Chemical Communications*, (12): 1281–1282. doi:10.1039/c39950001281

Angel, B. M., Batley, G. E., Jarolimek, C. V., & Rogers, N. J. (2013). The impact of size on the fate and toxicity of nanoparticulate silver in aquatic systems. *Chemosphere*, *93*(2), 359–365. doi:10.1016/j.chemosphere.2013.04.096 PMID:23732009

Anonymous, N. (2009). Nanotechnology and nanoscience applicatios: revolution in India and beyond. *Strategic Appl. Integrating Nano Sciences.* Retrieved from www.sainsce.com

Antolin, M. C., Muro, I., & Sanchez-Diaz, M. (2010). Application of sewage sludge improves growth, photosynthesis and antioxidant activities of nodulated alfalfa plants under drought conditions. 68. *Environmental and Experimental Botany*, *68*(1), 75–82. doi:10.1016/j.envexpbot.2009.11.001

Anyaogu, K. C., Fedorov, A. V., & Neckers, D. C. (2008). Synthesis, characterization, and antifouling potential of functionalized copper nanoparticles. *Langmuir*, *24*(8), 4340–4346. doi:10.1021/la800102f PMID:18341370

Arbogast, J. W., Darmanyan, A. P., Foote, C. S., Rubin, Y., Diederich, F. N., Alvarez, M. M., & Whetten, R. L. et al. (1991). Photophysical properties of C_{60}. *Journal of Physical Chemistry*, *95*(1), 11–12. doi:10.1021/j100154a006

Arbogast, J. W., & Foote, C. S. (1991). Photophysical Properties of C_{70}. *Journal of the American Chemical Society*, *113*(23), 8886–8889. doi:10.1021/ja00023a041

Argonide Corporation Nano Ceramic Sterilization Filter. (n.d.). Retrieved from http://sbir.nasa.gov/SBIR/successes/ss/9-072text.html

Arif, M., Heo, K., Lee, B. Y., Lee, J., Seo, D. H., Seo, S., & Hong, S. et al. (2011). Metallic nanowire–graphene hybrid nanostructures for highly flexible field emission devices. *Nanotechnology*, *22*(35), 355709. doi:10.1088/0957-4484/22/35/355709 PMID:21828894

Armelles, G., Cebollada, A., García-Martín, A., & González, M. U. (2013). Magnetoplasmonics: Combining magnetic and plasmonic functionalities. *Advanced Optical Materials*, *1*(1), 10–35. doi:10.1002/adom.201200011

Armelles, G., González-Díaz, J. B., García-Martín, A., García-Martín, J. M., Cebollada, A., González, M. U., & Badenes, G. et al. (2008). Localized surface plasmon resonance effects on the magneto-optical activity of continuous Au/Co/Au trilayers. *Optics Express*, *16*(20), 16104–16112. doi:10.1364/OE.16.016104 PMID:18825249

Armendariz, V., Herrera, I., Jose-yacaman, M., Troiani, H., Santiago, P., & Gardea-Torresdey, J. L. (2004). Size controlled gold nanoparticle formation by Avena sativa biomass: Use of plants in nanobiotechnology. *Journal of Nanoparticle Research*, *6*(4), 377–382. doi:10.1007/s11051-004-0741-4

Arora, S., Jain, J., Rajwade, J. M., & Paknikar, K. M. (2008). Cellular responses induced by silver nanoparticles: In vitro studies. *Toxicology Letters*, *179*(2), 93–100. doi:10.1016/j.toxlet.2008.04.009 PMID:18508209

Arora, S., Rajwade, J. M., & Paknikar, K. M. (2012). Nanotoxicology and in vitro studies: The need of the hour. *Toxicology and Applied Pharmacology*, *258*(2), 151–165. doi:10.1016/j.taap.2011.11.010 PMID:22178382

Artuğ, G., & Hapke, J. (2006). Characterization of nanofiltration membranes by their morphology, charge and filtration performance parameters. *Desalination*, *200*(1), 178–180. doi:10.1016/j.desal.2006.03.287

Arunachalam, R., Dhanasingh, S., Kalimuthu, B., Uthirappan, M., Rose, C., & Mandal, A. B. (2012). Phytosynthesis of silver nanoparticles using Coccinia grandis leaf extract and its application in the photocatalytic degradation. *Colloids and Surfaces. B, Biointerfaces*, *94*, 226–230. doi:10.1016/j.colsurfb.2012.01.040 PMID:22348986

Aruoja, V., Dubourguier, H.-C., Kasemets, K., & Kahru, A. (2009). Toxicity of nanoparticles of CuO, ZnO and TiO_2 to microalgae Pseudokirchneriella subcapitata. *The Science of the Total Environment*, *407*(4), 1461–1468. doi:10.1016/j.scitotenv.2008.10.053 PMID:19038417

Aruoja, V., Pokhrel, S., Sihtmäe, M., Mortimer, M., Mädler, L., & Kahru, A. (2015). *Toxicity of 12 metal-based nanoparticles to algae, bacteria and protozoa. Environ. Sci.: Nano*. The Royal Society of Chemistry.

Asare, N., Instanes, C., Sandberg, W. J., Refsnes, M., Schwarze, P., Kruszewski, M., & Brunborg, G. (2012). Cytotoxic and genotoxic effects of silver nanoparticles in testicular cells. *Toxicology*, *291*(1–3), 65–72. doi:10.1016/j.tox.2011.10.022 PMID:22085606

Asghari, S., Johari, S. A., Lee, J. H., Kim, Y. S., Jeon, Y. B., & Choi, H. J. et al.. (2012). Toxicity of various silver nanoparticles compared to silver ions in Daphnia magna. *Journal of Nanobiotechnology*, *10*(14). doi:10.1186/1477-3155-10-14 PMID:22472056

Asharani, P. V., Lian Wu, Y., Gong, Z., & Valiyaveettil, S. (2008). Toxicity of silver nanoparticles in zebrafish models. *Nanotechnology*, *19*(25), 255102. doi:10.1088/0957-4484/19/25/255102 PMID:21828644

AshaRani, P. V., Low Kah Mun, G., Hande, M. P., & Valiyaveettil, S.AshaRani. (2009). Cytotoxicity and genotoxicity of silver nanoparticles in human cells. *ACS Nano*, *3*(2), 279–290. doi:10.1021/nn800596w PMID:19236062

Ates, M., Dugo, M. A., Demir, V., Arslan, Z., & Tchounwou, P. B. (2014). Effect of copper oxide nanoparticles to sheepshead minnow (*Cyprinodon variegatus*) at different salinities. *Digest Journal of Nanomaterials and Biostructures*, *9*(1), 369–377. PMID:25411584

Attia, Y. A., Buceta, D., Blanco-Varela, C., Mohamed, M. B., Barone, G., & López-Quintela, M. A. (2014). Structure-Directing and High-Efficiency Photocatalytic Hydrogen Production by Ag Clusters. *Journal of the American Chemical Society*, *136*(4), 1182–1185. doi:10.1021/ja410451m PMID:24410146

Aulakh, J., & Regmi, A. (2013). *Post-harvest Food Losses Estimation - Development of Consistent Methodology*. Paper presented at the Agricultural & Applied Economics Associations 2013, AAEA & CAES Joint Annual Meeting, Washington, DC. Retrieved from http://worldfoodscience.com/article/postharvest-losses-and-food-waste-key-contributing-factors-african-food-insecurity-and#sthash.NMUXu0mB.dpuf

Avanasi, R., Jackson, W. A., Sherwin, B., Mudge, J. F., & Anderson, T. A. (2014). C60 Fullerene Soil Sorption, Biodegradation, and Plant Uptake. *Environmental Science & Technology*, *48*(5), 2792–2797. doi:10.1021/es405306w

Compilation of References

AZoNANO. (2005, Aug 15). *Converging Technologies - a Definition and an Overview of Some Different Viewpoints.* Retrieved August 24, 2015, from: http://www.azonano.com/article.aspx?ArticleID=1354

Baalousha, M., Manciulea, A., Cumberland, S., Kendall, K., & Lead, J. R. (2008). Aggregation and surface properties of iron oxide nanoparticles: influence of pH and natural organic matter. *Environmental Toxicology and Chemistry / SETAC, 27*(9), 1875–1882.

Baalousha, M. (2009). Aggregation and disaggregation of iron oxide nanoparticles: Influence of particle concentration, pH and natural organic matter. *The Science of the Total Environment, 407*(6), 2093–2101. doi:10.1016/j.scitotenv.2008.11.022 PMID:19059631

Baalousha, M., & Lead, J. (2007). Characterization of natural aquatic colloids (< 5 nm) by flow-field flow fractionation and atomic force microscopy. *Environmental Science & Technology, 41*(4), 1111–1117. doi:10.1021/es061766n PMID:17593707

Baalousha, M., & Lead, J. (2012). Rationalizing nanomaterial sizes measured by atomic force microscopy, flow field-flow fractionation, and dynamic light scattering: Sample preparation, polydispersity, and particle structure. *Environmental Science & Technology, 46*(11), 6134–6142. doi:10.1021/es301167x PMID:22594655

Baalousha, M., Stolpe, B., & Lead, J. (2011). Flow field-flow fractionation for the analysis and characterization of natural colloids and manufactured nanoparticles in environmental systems: A critical review. *Journal of Chromatography. A, 1218*(27), 4078–4103. doi:10.1016/j.chroma.2011.04.063 PMID:21621214

Bacchetta, R., Santo, N., Fascio, U., Moschini, E., Freddi, S., & Chirico, G. (2011). *Nano-sized CuO, TiO2 and ZnO affect Xenopus laevis development. In Nanotoxicology.* Taylor & Francis.

Bacsik, Z., Atluri, R., Garcia-Bennett, A. E., & Hedin, N. (2010). Temperature-Induced Uptake of CO_2 and Formation of Carbamates in Mesocaged Silica Modified with n-Propylamines. *Langmuir, 26*(12), 10013–10024. doi:10.1021/la1001495

Badireddy, A. R., Hotze, E. M., Chellam, S., Alvarez, P., & Wiesner, M. R. (2007). Inactivation of Bacteriophages via photosensitization of fullerol nanoparticles. *Environmental Science & Technology, 41*(18), 6627–6632. doi:10.1021/es0708215

Baena-Nogueras, R. M., Pintado-Herrera, M. G., González-Mazo, E., & Lara-Martín, P. A. (2015). *Determination of Pharmaceuticals in Coastal Systems Using Solid Phase Extraction (SPE) Followed by Ultra Performance Liquid Chromatography–tandem Mass Spectrometry.* UPLC-MS/MS.

Baere, L. (2006). Will anaerobic digestion of solid waste survive in the future? *Water Science and Technology, 53*(8), 187–194. doi:10.2166/wst.2006.249 PMID:16784203

Bagrii, E. I., & Karaulove, E. N. (2001). New in fullerene chemistry (a review). *Petroleum Chemistry, 41*(5), 295–313.

Bain, R. E., Gundry, S. W., Wright, J. A., Yang, H., Pedley, S., & Bartram, J. K. (2012). Accounting for water quality in monitoring access to safe drinking-water as part of the Millennium Development Goals: Lessons from five countries. *Bulletin of the World Health Organization, 90*(3), 228–235. doi:10.2471/BLT.11.094284 PMID:22461718

Bai, X., Sandukas, S., Appleford, M., Ong, J. L., & Rabiei, A. (2012). Antibacterial effect and cytotoxicity of Ag-doped functionally graded hydroxyapatite coatings. *Journal of Biomedical Materials Research. Part B, Applied Biomaterials, 100*(2), 553–561. doi:10.1002/jbm.b.31985 PMID:22121007

Baker, T. J., Tyler, C. R., & Galloway, T. S. (2014). Impacts of metal and metal oxide nanoparticles on marine organisms. *Environmental Pollution, 186*, 257–271. doi:10.1016/j.envpol.2013.11.014 PMID:24359692

Balagna, C., Perero, S., Ferraris, S., Miola, M., Fucale, G., Manfredotti, C., & Ferraris, M. et al. (2012). Antibacterial coating on polymer for space application. *Materials Chemistry and Physics*, *135*(2-3), 714–722. doi:10.1016/j.matchemphys.2012.05.049

Bala, T., Singh, A., Sanyal, A., O'Sullivan, C., Laffir, F., Coughlan, C., & Ryan, K. M. (2013). Fabrication of noble metal-semiconductor hybrid nanostructures using phase transfer. *Nano Research*, *6*(2), 121–130. doi:10.1007/s12274-013-0287-9

Baldwin, D. H., Sandahl, J. F., Labenia, J. S., & Scholz, N. L. (2003). Sublethal effects of copper on coho salmon: Impacts on nonoverlapping receptor pathways in the peripheral olfactory nervous system. *Environmental Toxicology and Chemistry*, *22*(10), 2266–2274. doi:10.1897/02-428 PMID:14551988

Banerjee, S., Hemraj-Benny, T., & Wong, S. S. (2005). Covalent Surface Chemistry of Single-Walled Carbon Nanotubes. *Advanced Materials*, *17*(1), 17–29. doi:10.1002/adma.200401340

Bang, J. H., & Suslick, K. S. (2010). Applications of ultrasound to the synthesis of nanostructured materials. *Advanced Materials*, *22*(10), 1039–1059. doi:10.1002/adma.200904093 PMID:20401929

Banthí, J. C., Meneses-Rodríguez, D., García, F., González, M. U., García-Martín, A., Cebollada, A., & Armelles, G. (2012). High Magneto-Optical Activity and Low Optical Losses in Metal-Dielectric Au/Co/Au–SiO2 Magnetoplasmonic Nanodisks. *Advanced Materials*, *24*(10), OP36–OP41. PMID:22213149

Bao, F., Yao, J. L., & Gu, R. A. (2009). Synthesis of magnetic Fe2O3/Au core/shell nanoparticles for bioseparation and immunoassay based on surface-enhanced Raman spectroscopy. *Langmuir*, *25*(18), 10782–10787. doi:10.1021/la901337r PMID:19552373

Baraliya, J. D., & Joshi, H. H. (2014). Surface-engineered Core-Shell Nano-Size Ferrites and their Antimicrobial activity. *Solid State Physics: Proceedings of the 58th Dae Solid State Physics Symposium 2013, Pts a & B, 1591*, 429-431. doi:10.1063/1.4872627

Barani, H., Montazer, M., Samadi, N., Toliyat, T., Zadeh, M. K., & de Smeth, B. (2014). Application of Nano Silver/Lecithin on Wool through Various Methods: Antibacterial Properties and Cell Toxicity. *Journal of Engineered Fibers and Fabrics*, *9*(4), 126-134.

Barcaro, G., Sementa, L., Fortunelli, A., & Stener, M. (2015). Optical properties of nanoalloys. *Physical Chemistry Chemical Physics*, *17*(42), 27952–27967. doi:10.1039/C5CP00498E PMID:25875393

Barcellos, L. J. G., Kreutz, L. C., Quevedo, R. M., Fioreze, I., Soso, A. B., Cericato, L., & Ritter, F. et al. (2004). Nursery rearing of jundiá, *Rhamdia quelen* (Quoy and Gaimard) in cages: Cage type, stocking density and stress response to confinement. *Aquaculture (Amsterdam, Netherlands)*, *232*(1-4), 383–394. doi:10.1016/S0044-8486(03)00545-3

Barhate, R. S., Loong, C. K., & Ramakrishna, S. (2006). Preparation and characterization of nanofibrous filtering media. *Journal of Membrane Science*, *283*(1-2), 209–218. doi:10.1016/j.memsci.2006.06.030

Barillet, S., Simon-Deckers, A., Herlin-Boime, N., Mayne-L'Hermite, M., Reynaud, C., Cassio, D., & Carrière, M. et al. (2010). Toxicological consequences of TiO2, SiC nanoparticles and multi-walled carbon nanotubes exposure in several mammalian cell types: An in vitro study. *Journal of Nanoparticle Research*, *12*(1), 61–73. doi:10.1007/s11051-009-9694-y

Baronti, C., Curini, R., D'Ascenzo, G., Di Corcia, A., Gentili, A., & Samperi, R. (2000). Monitoring natural and synthetic estrogens at activated sludge sewage treatment plants and in a receiving river water. *Environmental Science & Technology*, *34*(24), 5059–5066. doi:10.1021/es001359q

Barrena, R., Casals, E., Colón, J., Font, X., Sánchez, A., & Puntes, V. (2009). Evaluation of the ecotoxicity of model nanoparticles. *Chemosphere*, *75*(7), 850–857. doi:10.1016/j.chemosphere.2009.01.078 PMID:19264345

Bastami, D. K., Moradlou, H. A., Zaragabadi, M. A., Mir, S. S. V., & Shakiba, M. M. (2009). Measurement of some haematological characteristics of the wild carp. *Comparative Clinical Pathology, 18*(3), 321–323. doi:10.1007/s00580-008-0802-7

Bastani, M., & Celik, N. (2015). Assessment of occupational safety risks in Floridian solid waste systems using Bayesian analysis. *Waste Management & Research*.

Batley, G. E., Kirby, J. K., & McLaughlin, M. J. (2012). Fate and risks of nanomaterials in aquatic and terrestrial environments. *Accounts of Chemical Research, 46*(3), 854–862. doi:10.1021/ar2003368 PMID:22759090

Batsmanova, L. M., Gonchar, L. M., Taran, N. Y., & Okanenko, A. A. (2013). Using a colloidal solution of metal nanoparticles as micronutrient fertiliser for cereals. *Proceedings of the International Conference on Nanomaterials: Applications and Properties*. Retrieved from http://nap.sumdu.edu.ua/index.php/nap/nap2013/paper/view/1097/504

Baughman, R. H., Zakhidov, A. A., & de Heer, W. A. (2002). Carbon Nanotubes--the Route Toward Applications. *Science, 297*(5582), 787–792. doi:10.1126/science.1060928 PMID:12161643

Baun, A., Hartmann, N. B., Grieger, K., & Kusk, K. O. (2008). Ecotoxicity of engineered nanoparticles to aquatic invertebrates: A brief review and recommendations for future toxicity testing. *Ecotoxicology (London, England), 17*(5), 387–395. doi:10.1007/s10646-008-0208-y PMID:18425578

Beeby, A., Eastoe, J., & Heenan, R. K. (1994). Solubilization of C_{60} in aqueous micellar solution. *Journal of the Chemical Society. Chemical Communications*, (2): 173–175. doi:10.1039/c39940000173

Beecroft, L. L., & Ober, C. K. (1997). Nanocomposite materials for optical applications. *Chemistry of Materials, 9*(6), 1302–1317. doi:10.1021/cm960441a

Beer, C., Foldbjerg, R., Hayashi, Y., Sutherland, D. S., & Autrup, H. (2012). Toxicity of silver nanoparticles—Nanoparticle or silver ion? *Toxicology Letters, 208*(3), 286–292. doi:10.1016/j.toxlet.2011.11.002 PMID:22101214

Begum, P., Ikhtiari, R., & Fugetsu, B. (2014). Potential impact of multi-walled carbon nanotubes exposure to the seedling stage of selected plant species. *Nanomaterials, 4*(2), 203–221. doi:10.3390/nano4020203

Behra, R., Sigg, L., Clift, M. J. D., Herzog, F., Minghetti, M., Johnston, B., & Rothen-Rutishauser, B. et al. (2013). Bioavailability of silver nanoparticles and ions: From a chemical and biochemical perspective. *Journal of the Royal Society, Interface, 10*(87), 20130396. doi:10.1098/rsif.2013.0396 PMID:23883950

Benhui, S. Y. S. (1996). Preparation and Application of Nanofiltration Membrane. *Petrochemical Design, 3*, 009.

Bennett, P., He, F., Zhao, D., Aiken, B., & Feldman, L. (2010). In situ testing of metallic iron nanoparticle mobility and reactivity in a shallow granular aquifer. *Journal of Contaminant Hydrology, 116*(1-4), 35–46. doi:10.1016/j.jconhyd.2010.05.006 PMID:20542350

Benn, T. M., & Westerhoff, P. (2008). Nanoparticle Silver Released into Water from Commercially Available Sock Fabrics. *Environmental Science & Technology, 42*(11), 41334139. doi:10.1021/es7032718 PMID:18589977

Bera, P., Kim, C. H., & Seok, S. I. (2010). High-yield synthesis of quantum-confined CdS nanorods using a new dimeric cadmium (II) complex of S-benzyldithiocarbazate as single-source molecular precursor. *Solid State Sciences, 12*(4), 532–535. doi:10.1016/j.solidstatesciences.2009.12.020

Berge, N. D., Ro, K. S., Mao, J., Flora, J. R., Chappell, M. A., & Bae, S. (2011). Hydrothermal carbonization of municipal waste streams. *Environmental Science & Technology, 45*(13), 5696–5703. doi:10.1021/es2004528 PMID:21671644

Berg, P., Hagmeyer, G., & Gimbel, R. (1997). Removal of pesticides and other micropollutants by nanofiltration. *Desalination*, *113*(2), 205–208. doi:10.1016/S0011-9164(97)00130-6

Berthelot, J. M. (1999). *Composite materials: mechanical behavior and structural analysis*. Springer-Verlag Berlin Heidelberg New York. doi:10.1007/978-1-4612-0527-2

Bessekhouad, Y., Robert, D., & Weber, J. V. (2004). Bi_2S_3/TiO_2 and CdS/TiO_2 Heterojunctions as an Available Configuration for Photocatalytic Degradation of Organic Pollutant. *Journal of Photochemistry and Photobiology A Chemistry*, *163*(3), 569–580. doi:10.1016/j.jphotochem.2004.02.006

Beyth, N., Houri-Haddad, Y., Domb, A., Khan, W., & Hazan, R. (2015). Alternative Antimicrobial Approach: Nano-Antimicrobial Materials. *Evidence-Based Complementary and Alternative Medicine*. doi:10.1155/2015/246012

Bhainsa, K. C., & D'souza, S. F. (2006). Extracellular biosynthesis of silver nanoparticles using the fungus Aspergillus fumigatus. *Colloids and Surfaces. B, Biointerfaces*, *47*(2), 160–164. doi:10.1016/j.colsurfb.2005.11.026 PMID:16420977

Bhakya, S., Muthukrishnan, S., Sukumaran, M., Muthukumar, M., Senthil Kumar, T., & Rao, M. V. (2015). Catalytic Degradation of Organic Dyes using Synthesized Silver Nanoparticles: A Green Approach. *Journal of Bioremediation & Biodegradation*, *6*(5), 1–9.

Bharde, A., Rautaray, D., Bansal, V., Ahmad, A., Sarkar, I., Yusuf, S. M., & Sastry, M. (2006). Extracellular biosynthesis of magnetite using fungi. *Small*, *2*(1), 135–141. doi:10.1002/smll.200500180 PMID:17193569

Bhattacharya, R., & Mukherjee, P. (2008). Biological properties of "naked" metal nanoparticles. *Advanced Drug Delivery Reviews*, *60*(11), 1289–1306. doi:10.1016/j.addr.2008.03.013 PMID:18501989

Bhattacharya, R., Mukherjee, P., Xiong, Z., Atala, A., Soker, S., & Mukhopadhyay, D. (2004). Gold nanoparticles inhibit VEGF165-induced proliferation of HUVEC cells. *Nano Letters*, *4*(12), 2479–2481. doi:10.1021/nl0483789

Bhattacharyya, A., Chandrasekar, R., Chandra, A. K., Epidi, T. T., & Prakasham, R. S. (2014). Application of Nanoparticles in sustainable Agriculture: Its Current Status. *Short Views on Insect Biochemistry and Molecular Biology*, *2*, 429–448.

Bhol, K. C., & Schechter, P. J. (2005). Topical nanocrystalline silver cream suppresses inflammatory cytokines and induces apoptosis of inflammatory cells in a murine model of allergic contact dermatitis. *The British Journal of Dermatology*, *152*(6), 1235–1242. doi:10.1111/j.1365-2133.2005.06575.x PMID:15948987

Bian, S.-W., Mudunkotuwa, I. A., Rupasinghe, T., & Grassian, V. H. (2011). Aggregation and dissolution of 4 nm ZnO nanoparticles in aqueous environments: Influence of pH, ionic strength, size, and adsorption of humic acid. *Langmuir*, *27*(10), 6059–6068. doi:10.1021/la200570n PMID:21500814

Binelli, A., Parolini, M., Cogni, D., Pedriali, A., & Provini, A. (2009). A multi-biomarker assessment of the impact of the antibacterial trimethoprim on the non-target organism zebra mussel (*Dreissena polymorpha*). *Comparative Biochemistry and Physiology*, (Part C): 150, 329–336. PMID:19481616

Blau, T. J., Taylor, J. S., Morris, K. E., & Mulford, L. (1992). DBP control by nanofiltration: Cost and performance. *Journal - American Water Works Association*, *84*(12), 104–116.

Blouina, M., Hodsonb, M. E., Delgadoc, E. A., Bakerd, G., Brussaarde, L., & Buttf, K. R. et al.. (2013). A review of earthworm impact on soil function and ecosystem services. *European Journal of Soil Science*, *64*(2), 161–182. doi:10.1111/ejss.12025

Bocquillon, G., Bogicevic, C., Fabre, C., & Rassat, A. (1993). C_{60} fullerene as carbon source for diamond synthesis. *Journal of Physical Chemistry*, *97*(49), 12924–12927. doi:10.1021/j100151a047

Bogataj, D., & Bogataj, M. (2015). Floating points of a cold supply chain in an environment of the changing economic growth. In *Proc. Internat. Symp. on Operations Research SOR'15*. Bled: SDI-SOR.

Bogataj, D., Bogataj, M., Drobne, D., Ros-McDonnell, L., Rudolf, R., & Hudoklin, D. (2015a). Evaluation of investments in smart measurement devices controlling cold supply chains through nanotechnology, In *Proc. Internat. Symp. on Operations Research SOR'15*. Bled: SDI-SOR.

Bogataj, D., Vodopivec, R., & Bogataj, M. (2013). The extended MRP model for the evaluation and financing of superannuation schemes in a supply chain. *Technological and Economic Development of Economy, 19*(S1), S119-S133. doi:10.3846/20294913.2013.875957

Bogataj, L., & Bogataj, M. (1998). *Input-output analysis applied to MRP models with compound distribution of total demand*. Lecture notes. New York: University of New York. Retrieved August 24, 2015, from: http://www.inforum.umd.edu/papers/ioconferences/1998/journali.pdf

Bogataj, L., & Grubbström, R. W. (Eds.). (1998). Input-output analysis and Laplace transforms in material requirements planning: Storlien. Portorož, Slovenia: University of ljubljana, FPP.

Bogataj, L., Bogataj, M., & Ferbar Tratar, L. (2005) Controllability of supply chain, controlled in Laplace - transformed space. *International Journal of Applied Mathematics, 19*(3), 17-41.

Bogataj, M., & Bogataj, L. (1998). Compound distribution of demand in location - inventory problems. In S. Papachristos, & I. Ganas (Eds.), *Inventory Modelling in Production and Supply Chain* (pp. 15-46). Ioannina, Greece: University of Ioannina.

Bogataj, D., & Bogataj, M. (2007). Measuring the supply chain risk and vulnerability in frequency space. *International Journal of Production Economics, 108*(1-2), 291–301. doi:10.1016/j.ijpe.2006.12.017

Bogataj, D., Drobne, D., Rudolf, R., Bogataj, M., & Hudoklin, D. (2015b). Evaluation of investments in smart measurement devices controlling Cold Supply Chains through nanotechnology where smart cities are hosting activity cells. In *Proceedings of 6th IESM Conference, I4e2*.

Bogataj, L. (1995). Lead times and other delays in inventory systems. In L. Bogataj (Ed.), *Inventory Modelling - Lecture Notes of the International Postgraduate Summer School* (Vol. 1, pp. 23–25). Budapest: ISIR.

Bogataj, L., & Bogataj, M. (2007). The study of optimal additional investments in capacities for reduction of delays in value chain. *International Journal of Production Economics, 108*(1-2), 281–290. doi:10.1016/j.ijpe.2006.12.016

Bogataj, L., & Horvat, L. (1995). The application of game theory to MRP, input-output and multi-echelon inventory systems with exponentially distributed external demand. In *Proc. Internat. Symp. on Operations Research SOR'95* (pp. 25–36), Portoroz, Slovenia

Bogataj, L., & Horvat, L. (1996). Stochastic considerations of Grubbstrom-Molinder model of MRP, input-output and multiechelon inventory systems. *International Journal of Production Economics, 45*(1-3), 329–336. doi:10.1016/0925-5273(96)00050-3

Bogataj, M., & Bogataj, L. (2004). On the compact presentation of the lead times perturbations in distribution networks. *International Journal of Production Economics, 88*(2), 145–155. doi:10.1016/j.ijpe.2003.11.004

Bogataj, M., Bogataj, L., & Vodopivec, R. (2005). Stability of perishable goods in cold logistic chains. *International Journal of Production Economics, 93/94*, 345–356. doi:10.1016/j.ijpe.2004.06.032

Bogataj, M., & Grubbström, R. W. (2012). On the representation of timing for different structures within MRP theory. *International Journal of Production Economics, 140*(2), 749–755. doi:10.1016/j.ijpe.2011.04.016

Bogataj, M., & Grubbström, R. W. (2013). Transportation delays in reverse logistics. *International Journal of Production Economics*, *143*(2), 395–402. doi:10.1016/j.ijpe.2011.12.007

Bogataj, M., Grubbström, R. W., & Bogataj, L. (2011). Efficient location of industrial activity cells in a global supply chain. *International Journal of Production Economics*, *133*(1), 243–250. doi:10.1016/j.ijpe.2010.09.035

Bogue, R. (2011). Nanocomposites: A review of technology and applications. *Assembly Automation*, *31*(2), 106–112. doi:10.1108/01445151111117683

Bohren, C. F., & Huffman, D. R. (1983). *Absorption and Scattering of Light by Small Particles*. New York: Wiley.

Bolea, E., Jiménez-Lamana, J., Laborda, F., & Castillo, J. (2011). Size characterization and quantification of silver nanoparticles by asymmetric flow field-flow fractionation coupled with inductively coupled plasma mass spectrometry. *Analytical and Bioanalytical Chemistry*, *401*(9), 2723–2732. doi:10.1007/s00216-011-5201-2 PMID:21750882

Bolong, N., Ismail, A., Salim, M. R., & Matsuura, T. (2009). A review of the effects of emerging contaminants in wastewater and options for their removal. *Desalination*, *239*(1), 229–246. doi:10.1016/j.desal.2008.03.020

Bonanni, V., Bonetti, S., Pakizeh, T., Pirzadeh, Z., Chen, J., Nogués, J., & Dmitriev, A. et al. (2011). Designer magnetoplasmonics with nickel nanoferromagnets. *Nano Letters*, *11*(12), 5333–5338. doi:10.1021/nl2028443 PMID:22029387

Bonfanti, P., Moschini, E., Saibene, M., Bacchetta, R., Rettighieri, L., Calabri, L., & Mantecca, P. et al. (2015). Do Nanoparticle Physico-Chemical Properties and Developmental Exposure Window Influence Nano ZnO Embryotoxicity in Xenopus laevis? *International Journal of Environmental Research and Public Health*, *12*(8), 8828–8848. doi:10.3390/ijerph120808828 PMID:26225989

Bong, S., Uhm, S., Kim, Y. R., Lee, J., & Kim, H. (2010). Graphene supported Pd electrocatalysts for formic acid oxidation. *Electrocatalysis*, *1*(2-3), 139–143. doi:10.1007/s12678-010-0021-2

Borden, R. C. (2007). Concurrent bioremediation of perchlorate and 1,1,1-trichloeoethane in an emulsified oil barrier. *Journal of Contaminant Hydrology*, *94*(1-2), 13–33. doi:10.1016/j.jconhyd.2007.06.002 PMID:17614158

Borden, R. C., Beckwith, W. J., Lieberman, M. T., Akladiss, N., & Hill, S. R. (2007). Enhanced anaerobic bioremediation of a TCE source at the Tarheel Army Missile Plant using EOS. *Remediation Journal*, *17*(3), 5–19. doi:10.1002/rem.20130

Borm, P. J. A., Robbins, D., Haubold, S., Kuhlbusch, T., Fissan, H., Donaldson, K., & Lademann, J. et al. (2006). The potential risks of nanomaterials: A review carried out for ECETOC. *Particle and Fibre Toxicology*, *3*(1), 11. doi:10.1186/1743-8977-3-11 PMID:16907977

Borm, P. J. A., Schins, R. P. F., & Albrecht, C. (2004). Inhaled particles and lung cancer, part B: Paradigms and risk assessment. *International Journal of Cancer*, *110*(1), 3–14. doi:10.1002/ijc.20064 PMID:15054863

Borovinskaya, O., Gschwind, S., Hattendorf, B., Tanner, M., & Gunther, D. (2014). Simultaneous mass quantification of nanoparticles of different composition in a mixture by microdroplet generator-ICPTOFMS. *Analytical Chemistry*, *86*(16), 8142–8148. doi:10.1021/ac501150c PMID:25014784

Bosetti, M., Masse, A., Tobin, E., & Cannas, M. (2002). Silver coated materials for external fixation devices: In vitro biocompatibility and genotoxicity. *Biomaterials*, *23*(3), 887–892. doi:10.1016/S0142-9612(01)00198-3 PMID:11771707

Bour, A., Mouchet, F., Verneuil, L., Evariste, L., Silvestre, J., Pinelli, E., & Gauthier, L. (2015). Toxicity of CeO_2 nanoparticles at different trophic levels – Effects on diatoms, chironomids and amphibians. *Chemosphere*, *120*, 230–236. doi:10.1016/j.chemosphere.2014.07.012 PMID:25086917

Compilation of References

Bouwmeester, H., Dekkers, S., Noordam, M. Y., Hagens, W. I., Bulder, A. S., De Heer, C., & Sips, A. J. et al. (2009). Review of health safety aspects of nanotechnologies in food production. *Regulatory Toxicology and Pharmacology*, *53*(1), 52–62. doi:10.1016/j.yrtph.2008.10.008 PMID:19027049

Boxall, A. B. A., Tiede, K., & Chaudhry, Q. (2007). Engineered nanomaterials in soils and water: How do they behave and could they pose a risk to human health? *Nanomedicine (London)*, *2*(6), 919–927. doi:10.2217/17435889.2.6.919 PMID:18095854

Boyer, P., Ménard, D., & Meunier, M. (2010). Nanoclustered Co− Au Particles Fabricated by Femtosecond Laser Fragmentation in Liquids. *The Journal of Physical Chemistry C*, *114*(32), 13497–13500. doi:10.1021/jp1037552

Boyle, D., Al-Bairuty, G. A., Ramsden, C. S., Sloman, K. A., Henry, T. B., & Handy, R. D. (2013). Subtle alterations in swimming speed distributions of rainbow trout exposed to titanium dioxide nanoparticles are associated with gill rather than brain injury. *Aquatic Toxicology (Amsterdam, Netherlands)*, *126*, 116–127. doi:10.1016/j.aquatox.2012.10.006 PMID:23178178

Bozich, J. S., Lohse, S. E., Torelli, M. D., Murphy, C. J., Hamers, R. J., & Klaper, R. D. (2014). Surface chemistry, charge and ligand type impact the toxicity of gold nanoparticles to Daphnia magna. *Environ. Sci.: Nano*, *1*(3), 260–270.

Braber, K. (1995). Anaerobic digestion of municipal solid waste: A modern waste disposal option on the verge of breakthrough. *Biomass and Bioenergy*, *9*(1), 365–376. doi:10.1016/0961-9534(95)00103-4

Braeken, L., Ramaekers, R., Zhang, Y., Maes, G., Van der Bruggen, B., & Vandecasteele, C. (2005). Influence of hydrophobicity on retention in nanofiltration of aqueous solutions containing organic compounds. *Journal of Membrane Science*, *252*(1), 195–203. doi:10.1016/j.memsci.2004.12.017

Brame, J., Long, M., Li, Q., & Alvarez, P. (2014). Trading oxidation power for efficiency: Differential inhibition of photo-generated hydroxyl radicals versus singlet oxygen. *Water Research*, *60*, 259–266. doi:10.1016/j.watres.2014.05.005

Brar, S. K., Verma, M., Tyagi, R. D., & Surampalli, R. Y. (2010). Engineered nanoparticles in wastewater and wastewater sludge evidence and impacts. *Waste Management (New York, N.Y.)*, *30*(3), 504–520. doi:10.1016/j.wasman.2009.10.012 PMID:19926463

Braydich-Stolle, L., Hussain, S., Schlager, J. J., & Hofmann, M. C. (2005). In vitro cytotoxicity of nanoparticles in mammalian germline stem cells. *Toxicological Sciences*, *88*(2), 412–419. doi:10.1093/toxsci/kfi256 PMID:16014736

Bregoli, L., Chiarini, F., Gambarelli, A., Sighinolfi, G., Gatti, A. M., Santi, P., & Cocco, L. et al. (2009). Toxicity of antimony trioxide nanoparticles on human hematopoietic progenitor cells and comparison to cell lines. *Toxicology*, *262*(2), 121–129. doi:10.1016/j.tox.2009.05.017 PMID:19482055

Brewer, A. K., & Striegel, A. M. (2011). Characterizing the size, shape, and compactness of a polydisperse prolate ellipsoidal particle via quadruple-detector hydrodynamic chromatography. *Analyst (London)*, *136*(3), 515–519. doi:10.1039/C0AN00738B PMID:21109889

British Columbia Ministry of Agriculture. Food and Fisheries. (1996). *Composting methods factsheet*. Retrieved August 2015, from http://www.al.gov.bc.ca/resmgmt/publist/300Series/382500-5.pdf

Brock, D. A., Douglas, T. E., Queller, D. C., & Strassmann, J. E. (2011). Primitive agriculture in a social amoeba. *Nature*, *469*(7330), 393–396. doi:10.1038/nature09668 PMID:21248849

Brotchie, A., Grieser, F., & Ashokkumar, M. (2008). Sonochemistry and sonoluminescence under dual-frequency ultrasound irradiation in the presence of water-soluble solutes. *The Journal of Physical Chemistry C*, *112*(27), 10247–10250. doi:10.1021/jp801763v

Brough, A., Hillman, D., & Perry, R. (1981). Capillary hydrodynamic chromatography-an investigation into operational characteristics. *Journal of Chromatography. A, 208*(2), 175–182. doi:10.1016/S0021-9673(00)81929-9

Brown, J., Chai, R., Wang, A., & Sobsey, M. D. (2012). Microbiological Effectiveness of Mineral Pot Filters in Cambodia. *Environmental Science & Technology, 46*(21), 12055–12061. doi:10.1021/es3027852 PMID:23030639

Brown, J., Proum, S., & Sobsey, M. D. (2009). Sustained use of a household-scale water filtration device in rural Cambodia. *Journal of Water and Health, 7*(3), 404–412. doi:10.2166/wh.2009.085 PMID:19491492

Brown, J., & Sobsey, M. D. (2010). Microbiological effectiveness of locally produced ceramic filters for drinking water treatment in Cambodia. *Journal of Water and Health, 8*(1), 1–11. doi:10.2166/wh.2009.007 PMID:20009242

Brown, J., Sobsey, M. D., & Loomis, D. (2008). Local drinking water filters reduce diarrheal disease in Cambodia: A randomized, controlled trial of the ceramic water purifier. *The American Journal of Tropical Medicine and Hygiene, 79*(3), 394–400. PMID:18784232

Brown, K. R., Walter, D. G., & Natan, M. J. (2000). Seeding of colloidal Au nanoparticle solutions. 2. Improved control of particle size and shape. *Chemistry of Materials, 12*(2), 306–313. doi:10.1021/cm980065p

Brown, R. C., Lockwood, A. H., & Sonawane, B. R. (2005). Neurodegenerative diseases: An overview of environmental risk factors. *Environmental Health Perspectives, 113*(9), 1250–1256. doi:10.1289/ehp.7567 PMID:16140637

Brullot, W., Strobbe, R., Bynens, M., Bloemen, M., Demeyer, P. J., Vanderlinden, W., & Verbiest, T. et al. (2014). Layer-by-Layer synthesis and tunable optical properties of hybrid magnetic–plasmonic nanocomposites using short bifunctional molecular linkers. *Materials Letters, 118*, 99–102. doi:10.1016/j.matlet.2013.12.057

Buch, P., Mohan, D. J., & Reddy, A. (2008). Preparation, characterization and chlorine stability of aromatic–cyclo-aliphatic polyamide thin film composite membranes. *Journal of Membrane Science, 309*(1), 36–44. doi:10.1016/j.memsci.2007.10.004

Buentello, S., Persad, D. L., Court, E. B., Martin, D. K., Daar, A. S., & Peter, A. (2005). Nanotechnology and the Developing World. *PLoS Medicine, 2*(5), e97. doi:10.1371/journal.pmed.0020097 PMID:15807631

Buffle, J., Wilkinson, K. J., Stoll, S., Filella, M., & Zhang, J. (1998). A generalized description of aquatic colloidal interactions: The three-colloidal component approach. *Environmental Science & Technology, 32*(19), 2887–2899. doi:10.1021/es980217h

Bui, N. N., & McCutcheon, J. R. (2013). Hydrophilic nanofibers as new supports for thin film composite membranes for engineered osmosis. *Environmental Science & Technology, 47*(3), 1761–1769. doi:10.1021/es304215g PMID:23234259

Bui, N. N., & McCutcheon, J. R. (2014). Nanofiber supported thin-film composite membrane for pressure-retarded osmosis. *Environmental Science & Technology, 48*(7), 4129–4136. doi:10.1021/es4037012 PMID:24387600

Buonomenna, M. G. (2013). Nano-enhanced reverse osmosis membranes. *Desalination, 314*, 73–88. doi:10.1016/j.desal.2013.01.006

Burd, A., Kwok, C. H., Hung, S. C., Chan, H. S., Gu, H., Lam, W. K., & Huang, L. (2007). A comparative study of the cytotoxicity of silver-based dressings in monolayer cell, tissue explant, and animal models. *Wound Repair and Regeneration, 15*(1), 94–104. doi:10.1111/j.1524-475X.2006.00190.x PMID:17244325

Burda, C., Chen, X., Narayanan, R., & El-Sayed, M. A. (2005). Chemistry and properties of nanocrystals of different shapes. *Chemical Reviews, 105*(4), 1025–1102. doi:10.1021/cr030063a PMID:15826010

Compilation of References

Burleson, D. J., Driessen, M. D., & Penn, R. L. (2005). On the characterization of environmental nanoparticles. *Journal of Environmental Science and Health. Part A, 39*(10), 2707–2753.

Bury, N. R., Walker, P. A., & Glover, C. N. (2003). Nutritive metal uptake in teleost fish. *The Journal of Experimental Biology, 206*(Pt 1), 11–23. doi:10.1242/jeb.00068 PMID:12456693

Busch, L. (2008). Nanotechnologies, food, and agriculture: Next big thing or flash in the pan? *Agriculture and Human Values, 25*(2), 215–218. doi:10.1007/s10460-008-9119-z

Bushell, G., & Amal, R. (2000). Measurement of fractal aggregates of polydisperse particles using small-angle light scattering. *Journal of Colloid and Interface Science, 221*(2), 186–194. doi:10.1006/jcis.1999.6532 PMID:10631019

Buyukhatipoglu, K., & Clyne, A. M. (2011). Superparamagnetic iron oxide nanoparticles change endothelial cell morphology and mechanics via reactive oxygen species formation. *Journal of Biomedical Materials Research. Part A, 96A*(1), 186–195. doi:10.1002/jbm.a.32972 PMID:21105167

Buzby, J. C., & Hyman, J. (2012). Total and per capita value of food loss in the United States. *Food Policy, 37*(5), 561–570. doi:10.1016/j.foodpol.2012.06.002

Buzea, C., Pacheco, I. I., & Robbie, K. (2007). Nanomaterials and nanoparticles: Sources and toxicity. *Biointerphases, 2*(4), 17–71. doi:10.1116/1.2815690 PMID:20419892

Byrne, M. T., & Gun'ko, Y. K. (2010). Recent Advances in Research on Carbon Nanotube–Polymer Composites. *Advanced Materials, 22*(15), 1672–1688. doi:10.1002/adma.200901545 PMID:20496401

Cabanas, R., Abarca, M. L., Bragulat, M. R., & Cabanes, F. J. (2009). In vitro activity of imazalil against *Penicillium expansum*: Comparison of the CLSI M38-A broth microdilution method with traditional techniques. *International Journal of Food Microbiology, 129*(1), 26–29. doi:10.1016/j.ijfoodmicro.2008.10.025 PMID:19059665

Caceci, M. S., & Billon, A. (1990). Evidence for large organic scatterers (50-200 nm diameter) in humic acid samples. *Organic Geochemistry, 15*(3), 335–350. doi:10.1016/0146-6380(90)90011-N

Cachon, G. P., & Fisher, M. (2000). Supply chain inventory management and the value of shared information. *Management Science, 46*(8), 1032–1048. doi:10.1287/mnsc.46.8.1032.12029

Callegari, A., Tonti, D., & Chergui, M. (2003). Photochemically grown silver nanoparticles with wavelength-controlled size and shape. *Nano Letters, 3*(11), 1565–1568. doi:10.1021/nl034757a

CalRecycle. (2012). *California's new goal: 75% recycling*. Retrieved June 2015, from http://www.calrecycle.ca.gov/75percent/Plan.pdf

Campos, E. V. R., Oliveira, J. L., & Fraceto, L. F. (2014). Applications of controlled release systems for fungicides, herbicides, acaricides, nutrients,and plant growth hormones: A review. *Adv. Sci. Eng. Med., 6*(4), 1–15. doi:10.1166/asem.2014.1538

Canadian Wildlife Service (CWS). (2002). *Pesticides and Wild Birds website*. Retrieved from http://www.cwsscf.ec.gc.ca/hww-fap/hww-fap.cfm?ID_species=90&lang=e

Cañas, J. E., Qi, B., Li, S., Maul, J. D., Cox, S. B., Das, S., et al. (2011). Acute and reproductive toxicity of nano-sized metal oxides (ZnO and TiO_2) to earthworms (Eisenia fetida). *J Environ Monit, 13*(12), 3351-3357. doi: 10.1039/c1em10497g

Canesi, L., Ciacci, C., Betti, M., Fabbri, R., Canonico, B., Fantinati, A., & Pojana, G. et al. (2008). Immunotoxicity of carbon black nanoparticles to blue mussel hemocytes. *Environment International, 34*(8), 1114–1119. doi:10.1016/j.envint.2008.04.002 PMID:18486973

Canesi, L., Ciacci, C., Fabbri, R., Marcomini, A., Pojana, G., & Gallo, G. (2012). Bivalve molluscs as a unique target group for nanoparticle toxicity. *Marine Environmental Research*, *76*, 16–21. doi:10.1016/j.marenvres.2011.06.005 PMID:21767873

Canesi, L., Fabbri, R., Gallo, G., Vallotto, D., Marcomini, A., & Pojana, G. (2010). Biomarkers in Mytilus galloprovincialis exposed to suspensions of selected nanoparticles (Nano carbon black, C60 fullerene, Nano-TiO2, Nano-SiO2). *Aquatic Toxicology (Amsterdam, Netherlands)*, *100*(2), 168–177. doi:10.1016/j.aquatox.2010.04.009 PMID:20444507

Cao, W. *Synthesis of nanomaterials by high energy ball milling*. Retrieved January 2016, from http://www.understandingnano.com/nanomaterial-synthesis-ball-milling.html

Cao, C., Kim, J. H., Yoon, D., Hwang, E. S., Kim, Y. J., & Baik, S. (2008). Optical Detection of DNA Hybridization Using Absorption Spectra of Single-Walled Carbon Nanotubes. *Materials Chemistry and Physics*, *112*(3), 738–741. doi:10.1016/j.matchemphys.2008.07.129

Cao, G., Meijernik, J., Brinkman, H., & Burggraaf, A. (1993). Permporometry study on the size distribution of active pores in porous ceramic membranes. *Journal of Membrane Science*, *83*(2), 221–235. doi:10.1016/0376-7388(93)85269-3

Cao, Y. C., Jin, R., & Mirkin, C. A. (2002). Nanoparticles with Raman spectroscopic fingerprints for DNA and RNA detection. *Science*, *297*(5586), 1536–1540. doi:10.1126/science.297.5586.1536 PMID:12202825

Cao, Y., Li, D., Jiang, F., Yang, Y., & Huang, Z. (2013). Engineering metal nanostructure for SERS application. *Journal of Nanomaterials*.

Carr, R., Hole, P., Malloy, A., Nelson, P., Wright, M., & Smith, J. (2009). Applications of nanoparticle tracking analysis in nanoparticle research-a mini-review. *European Journal of Parenteral & Pharmaceutical Sciences*, *14*(2), 45–50.

Caruso, R. A., Ashokkumar, M., & Grieser, F. (2002). Sonochemical formation of gold sols. *Langmuir*, *18*(21), 7831–7836. doi:10.1021/la020276f

Cascio, C., Geiss, O., Franchini, F., Ojea-Jimenez, I., Rossi, F., Gilliland, D., & Calzolai, L. (2015). Detection, quantification and derivation of number size distribution of silver nanoparticles in antimicrobial consumer products. *Journal of Analytical Atomic Spectrometry*, *30*(6), 1255–1265. doi:10.1039/C4JA00410H

Castiglione, M. R., Giorgetti, L., Geri, C., & Cremonini, R. (2011). The effects of nano-TiO_2 on seed germination, development and mitosis of root tip cells of *Vicia narbonensis* L. and *Zea mays* L. *Journal of Nanoparticle Research*, *13*(6), 2443–2449. doi:10.1007/s11051-010-0135-8

Cava, R. J. (1990). Structural chemistry and the local charge picture of copper oxide superconductors. *Science*, *247*(4943), 656–662. doi:10.1126/science.247.4943.656 PMID:17771881

Celik, E., Liu, L., & Choi, H. (2011). Protein fouling behavior of carbon nanotube/polyethersulfone composite membranes during water filtration. *Water Research*, *45*(16), 5287–5294. doi:10.1016/j.watres.2011.07.036 PMID:21862096

Celik, E., Park, H., Choi, H., & Choi, H. (2011). Carbon nanotube blended polyethersulfone membranes for fouling control in water treatment. *Water Research*, *45*(1), 274–282. doi:10.1016/j.watres.2010.07.060 PMID:20716459

CERP-IoT. (2009). *Internet of Things: Strategic Research Roadmap. Report from the Cluster of European Research Projects (FP7) on the Internet of Things (CERP-IoT)*. Retrieved August 24, 2015, from: http://www.internet-of-things-research.eu/pdf/IoT_Cluster_Strategic_Research_Agenda_2009.pdf

Chadeau, E., Oulahal, N., Dubost, L., Favergeon, F., & Degraeve, P. (2010). Anti-Listeria innocua activity of silver functionalised textile prepared with plasma technology. *Food Control*, *21*(4), 505–512. doi:10.1016/j.foodcont.2009.07.013

Chae, S.-R., Hotze, E. M., & Wiesner, M. R. (2009). Evaluation of the Oxidation of Organic Compounds by Aqueous Suspensions of Photosensitized Hydroxylated-C_{60} Fullerene Aggregates. *Environmental Science & Technology*, *43*(16), 6208–6213. doi:10.1021/es901165q

Chae, Y. J., Pham, C. H., Lee, J., Bae, E., Yi, J., & Gu, M. B. (2009). Evaluation of the toxic impact of silver nanoparticles on Japanese medaka (Oryzias latipes). *Aquatic Toxicology (Amsterdam, Netherlands)*, *94*(4), 320–327. doi:10.1016/j.aquatox.2009.07.019 PMID:19699002

Chaloupka, K., Malam, Y., & Seifalian, A. M. (2010). Nanosilver as a new generation of nanoproduct in biomedical applications. *Trends in Biotechnology*, *28*(11), 580–588. doi:10.1016/j.tibtech.2010.07.006 PMID:20724010

Chandran, S. P., Chaudhary, M., Pasricha, R., Ahmad, A., & Sastry, M. (2006). Synthesis of gold nanotriangles and silver nanoparticles using Aloevera plant extract. *Biotechnology Progress*, *22*(2), 577–583. doi:10.1021/bp0501423 PMID:16599579

Chandra, S., Bag, S., Bhar, R., & Pramanik, P. (2011). Sonochemical synthesis and application of rhodium–graphene nanocomposite. *Journal of Nanoparticle Research*, *13*(7), 2769–2777. doi:10.1007/s11051-010-0164-3

Chang, Y. N., Zhang, M., Xia, L., Zhang, J., & Xing, G. (2012). The toxic effects and mechanisms of CuO and ZnO nanoparticles. *Materials (Basel)*, *5*(12), 2850–2871. doi:10.3390/ma5122850

Chang, Y., Yang, S.-T., Liu, J.-H., Dong, E., Wang, Y., Cao, A., & Wang, H. et al. (2011). In vitro toxicity evaluation of graphene oxide on A549 cells. *Toxicology Letters*, *200*(3), 201–210. doi:10.1016/j.toxlet.2010.11.016 PMID:21130147

Chan, W.-F., Chen, H.-y., Surapathi, A., Taylor, M. G., Shao, X., Marand, E., & Johnson, J. K. (2013). Zwitterion Functionalized Carbon Nanotube/Polyamide Nanocomposite Membranes for Water Desalination. *ACS Nano*, *7*(6), 5308–5319. doi:10.1021/nn4011494 PMID:23705642

Chapman, P. M. (2006). Emerging Substances—Emerging Problems? *Environmental Toxicology and Chemistry*, *25*(6), 1445. doi:10.1897/06-025.1 PMID:16764460

Charpentier, P. A., Burgess, K., Wang, L., Chowdhury, R. R., Lotus, A. F., & Moula, G. (2012). Nano-TiO 2 /polyurethane composites for antibacterial and self-cleaning coatings. *Nanotechnology*, *23*(42), 425606. doi:10.1088/0957-4484/23/42/425606 PMID:23037881

Chase, B., Herron, N., & Holler, E. (1992). Vibrational spectroscopy of fullerenes (C60 and C70). Temperature dependant studies. *Journal of Physical Chemistry*, *96*(11), 4262–4266. doi:10.1021/j100190a029

Chehadi, Z., Alkees, N., Bruyant, A., Toufaily, J., Girardon, J. S., Capron, M., & Jradi, S. et al. (2016). Plasmonic enhanced photocatalytic activity of semiconductors for the degradation of organic pollutants under visible light. *Materials Science in Semiconductor Processing*, *42*, 81–84. doi:10.1016/j.mssp.2015.08.044

Chellam, S., Sharma, R. R., Shetty, G. R., & Wei, Y. (2008). Nanofiltration of pretreated Lake Houston water: Disinfection by-product speciation, relationships, and control. *Separation and Purification Technology*, *64*(2), 160–169. doi:10.1016/j.seppur.2008.09.007

Chen, C.-Y., & Chiang, C.-L. (2008). Preparation of cotton fibers with antibacterial silver nanoparticles. *Matterials Letters*, (62), 3607-3609.

Chenal, M., Rieger, J., Philippe, A., & Bouteiller, L. (2014). High Yield Preparation of All-Organic Raspberry-like Particles by Heterocoagulation via Hydrogen Bonding Interaction. *Polymer*, *55*(16), 3516–3524. doi:10.1016/j.polymer.2014.05.057

Chen, C., Unrine, J. M., Judy, J. D., Lewis, W. R., Guo, J., Mcnear, D. H. J. Jr, & Tsyusko, O. V. (2015). Toxicogenomic responses of the model legume Medicago truncatula to aged biosolids containing a mixture of nanomaterials (TiO2, Ag and ZnO) from a pilot wastewater treatment plant. *Environmental Science & Technology*, *49*(14), 8759–8768. doi:10.1021/acs.est.5b01211 PMID:26065335

Chen, H., Roco, M. C., Son, J., Jiang, S., Larson, C. A., & Gao, Q. (2013). Global Global nanotechnology development from 1991 to 2012: Patents, scientific publications, and effect of NSF funding. *Journal of Nanoparticle Research*, *15*(9), 1–21. doi:10.1007/s11051-013-1951-4

Chen, H., & Yada, R. (2011). Nanotechnologies in agriculture: New tools for sustainable development. *Trends in Food Science & Technology*, *22*(11), 585–594. doi:10.1016/j.tifs.2011.09.004

Chen, J., Albella, P., Pirzadeh, Z., Alonso-González, P., Huth, F., Bonetti, S., & Hillenbrand, R. (2011). Plasmonic nickel nanoantennas. *Small*, *7*(16), 2341–2347. doi:10.1002/smll.201100640 PMID:21678553

Chen, J.-L., Al-Abed, S. R., Ryan, J. A., & Li, Z. (2001). Effects of pH on dechlorination of trichloroethylene by zero-valent iron. *Journal of Hazardous Materials*, *B83*(3), 243–254. doi:10.1016/S0304-3894(01)00193-5 PMID:11348735

Chen, J., Patil, S., Seal, S., & McGinnis, J. F. (2006). Rare earth nanoparticles prevent retinal degeneration induced by intracellular peroxides. *Nature Nanotechnology*, *1*(2), 142–150. doi:10.1038/nnano.2006.91 PMID:18654167

Chen, K. L., & Bothun, G. D. (2014). Nanoparticles meet cell membranes: Probing nonspecific interactions using model membranes. *Environmental Science & Technology*, *48*(2), 873–880. doi:10.1021/es403864v PMID:24341906

Chen, K. L., Mylon, S. E., & Elimelech, M. (2007). Enhanced aggregation of alginate-coated iron oxide (Hematite) nanoparticles in the presence of calcium, strontium, and barium cations. *Langmuir*, *23*(11), 5920–5928. doi:10.1021/la063744k PMID:17469860

Chen, L. Q., Kang, B., & Ling, J. (2013). Cytotoxicity of cuprous oxide nanoparticles to fish blood cells: Hemolysis and internalization. *Journal of Nanoparticle Research*, *15*(3), 1507–1513. doi:10.1007/s11051-013-1507-7

Chen, P. J., Tan, S. W., & Wu, W. L. (2012). Stabilization or oxidation of nanoscale zero valent iron at environmentally relevant exposure changes bioavailability and toxicity in medaka fish. *Environmental Science & Technology*, *46*(15), 8431–8439. doi:10.1021/es3006783 PMID:22747062

Chen, R., Ni, H., Zhang, H., Yue, G., Zhan, W., & Xiong, P. (2013). A preliminary study on antibacterial mechanisms of silver ions implanted stainless steel. *Vacuum*, *89*, 249–253. doi:10.1016/j.vacuum.2012.05.025

Chen, S., Liu, J., & Zeng, H. (2005). Structure and antibacterial activity of silver-supporting activated carbon fibers. *Journal of Materials Science*, *40*(23), 6223–6231. doi:10.1007/s10853-005-3149-3

Chen, T. H., Lin, C. Y., & Tseng, M. C. (2011). Behavioral effects of titanium dioxide nanoparticles on larval zebrafish (*Danio rerio*). *Marine Pollution Bulletin*, *63*(5-12), 303–308. doi:10.1016/j.marpolbul.2011.04.017 PMID:21565364

Chen, W., Su, Y., Zhang, L., Shi, Q., Peng, J., & Jiang, Z. (2010). In situ generated silica nanoparticles as pore-forming agent for enhanced permeability of cellulose acetate membranes. *Journal of Membrane Science*, *348*(1–2), 75–83. doi:10.1016/j.memsci.2009.10.042

Chen, W., Zhao, J., Lee, J. Y., & Liu, Z. (2005). Microwave heated polyol synthesis of carbon nanotubes supported Pt nanoparticles for methanol electrooxidation. *Materials Chemistry and Physics*, *91*(1), 124–129. doi:10.1016/j.matchemphys.2004.11.003

Chen, X., Li, C., Grätzel, M., Kostecki, R., & Mao, S. S. (2012). Nanomaterials for renewable energy production and storage. *Chemical Society Reviews*, *41*(23), 7909–7937. doi:10.1039/c2cs35230c PMID:22990530

Chen, Y. S., Hung, Y. C., Liau, I., & Huang, G. S. (2009). Assessment of the in vivo toxicity of gold nanoparticles. *Nanoscale Research Letters*, *4*(8), 858–864. doi:10.1007/s11671-009-9334-6 PMID:20596373

Chen, Y.-S., Hung, Y.-C., Hong, M.-Y., Onischuk, A. A., Chiou, J. C., & Sorokina, I. V. (2012). ... Steve Huang, G. (2012). Control of In Vivo Transport and Toxicity of Nanoparticles by Tea Melanin. *Journal of Nanomaterials*, 1–11. doi:10.1155/2012/746960

Chen, Z., Meng, H., Xing, G., Chen, C., Zhao, Y., Jia, G., & Zhao, F. et al. (2006). Acute toxicological effects of copper nanoparticles in vivo. *Toxicology Letters*, *163*(2), 109–120. doi:10.1016/j.toxlet.2005.10.003 PMID:16289865

Chiang, I. W., Brinson, B. E., Smalley, R. E., Margrave, J. L., & Hauge, R. H. (2001). Purification and Characterization of Single-Wall Carbon Nanotubes. *The Journal of Physical Chemistry B*, *105*(6), 1157–1161. doi:10.1021/jp003453z

Childress, A. E., & Elimelech, M. (2000). Relating nanofiltration membrane performance to membrane charge (electrokinetic) characteristics. *Environmental Science & Technology*, *34*(17), 3710–3716. doi:10.1021/es0008620

Chinnamuthu, C. R., & Boopathi, P. M. (2009). Nanotechnology and Agroecosystem. *Madras Agriculture Journal*, *96*(1-6), 17–31.

Chinnamuthu, C. R., & Kokiladevi, E. (2007). Weed management through nanoherbicides. In C. R. Chinnamuthu, B. Chandrasekaran, & C. Ramasamy (Eds.), *Application of Nanotechnology in Agriculture*. Coimbatore, India: Tamil Nadu Agricultural University.

Chio, C. P., Chen, W. Y., Chou, W. C., Hsieh, N. H., Ling, M. P., & Liao, C. M. (2012). Assessing the potential risks to zebra fish posed by environmentally relevant copper and silver nanoparticles. *The Science of the Total Environment*, *420*, 111–118. doi:10.1016/j.scitotenv.2012.01.023 PMID:22326136

Chiou, C.-H., Wu, C.-Y., & Juang, R.-S. (2008). Influence of operating parameters on photocatalytic degradation of phenol in UV/TiO2 process. *Chemical Engineering Journal*, *139*(2), 322–329. doi:10.1016/j.cej.2007.08.002

Chiou, C.-S., & Shih, J.-S. (2000). Fullerene C_{60}-Cryptand Chromatographic Stationary Phase for Separations of Anions/Cations and Organic Molecules. *Analytica Chimica Acta*, *416*(2), 169–175. doi:10.1016/S0003-2670(00)00906-5

Choi, H.-, Son, M., Yoon, S. H., Celik, E., Kang, S., Park, H., & Choi, H et al.. (2015). Alginate fouling reduction of functionalized carbon nanotube blended cellulose acetate membrane in forward osmosis. *Chemosphere*, *136*, 204–210. doi:10.1016/j.chemosphere.2015.05.003 PMID:26022283

Choi, J. E., Kim, S., Ahn, J. H., Youn, P., Kang, J. S., Park, K., & Ryu, D.-Y. et al. (2010). Induction of oxidative stress and apoptosis by silver nanoparticles in the liver of adult zebrafish. *Aquatic Toxicology (Amsterdam, Netherlands)*, *100*(2), 151–159. doi:10.1016/j.aquatox.2009.12.012 PMID:20060176

Choi, S. M., Seo, M. H., Kim, H. J., & Kim, W. B. (2011). Synthesis and characterization of graphene-supported metal nanoparticles by impregnation method with heat treatment in H 2 atmosphere. *Synthetic Metals*, *161*(21), 2405–2411. doi:10.1016/j.synthmet.2011.09.008

Cho, M., Chung, H., Choi, W., & Yoon, J. (2005a). Different Inactivation Behaviors of MS-2 Phage and Escherichia coli in TiO_2 Photocatalytic Disinfection. *Applied and Environmental Microbiology*, *71*(1), 270–275. doi:10.1128/AEM.71.1.270-275.2005

Cho, M., Lee, J., Mackeyev, Y., Wilson, L. J., Alvarez, P. J. J., Hughes, J. B., & Kim, J. H. (2010). Visible Light Sensitized Inactivation of MS-2 Bacteriophage by a Cationic Amine-Functionalized C(60) Derivative. *Environmental Science & Technology*, *44*(17), 6685–6691. doi:10.1021/es1014967

Cho, S. J., Idrobo, J. C., Olamit, J., Liu, K., Browning, N. D., & Kauzlarich, S. M. (2005). Growth mechanisms and oxidation resistance of gold-coated iron nanoparticles. *Chemistry of Materials*, *17*(12), 3181–3186. doi:10.1021/cm0500713

Chou, C.-C., Hsiao, H.-Y., Hong, Q.-S., Chen, C.-H., Peng, Y.-W., Chen, H.-W., & Yang, P.-C. (2008). Single-walled carbon nanotubes can induce pulmonary injury in mouse model. *Nano Letters*, *8*(2), 437–445. doi:10.1021/nl0723634 PMID:18225938

Christen, V., Capelle, M., & Fent, K. (2013). Silver nanoparticles induce endoplasmatic reticulum stress response in zebrafish. *Toxicology and Applied Pharmacology*, *272*(2), 519–528. doi:10.1016/j.taap.2013.06.011 PMID:23800688

Chun, M. S., Park, O. O., & Kim, J. K. (1990). Flow and dynamic behavior of dilute polymer solutions in hydrodynamic chromatography. *Korean Journal of Chemical Engineering*, *7*(2), 126–137. doi:10.1007/BF02705057

Cioffi, N., Torsi, L., Ditaranto, N., Tantillo, G., Ghibelli, L., Sabbatini, L., & Traversa, E. et al. (2005). Copper nanoparticle/polymer composites with antifungal and bacteriostatic properties. *Chemistry of Materials*, *17*(21), 5255–5262. doi:10.1021/cm0505244

Cipriano, A. F. F., & Liu, H. (2013). *Nanomaterials in Tissue Engineering.* Nanomaterials in Tissue Engineering; doi:10.1533/9780857097231.2.334

Clark, I. A., Chaudhri, G., & Cowden, W. B. (1989). Some roles of free radicals in malaria. *Free Radical Biology & Medicine*, *6*(3), 315–321. doi:10.1016/0891-5849(89)90058-0 PMID:2663664

Clasen, T. F., Brown, J., Collin, S., Suntura, O., & Cairncross, S. (2004). Reducing diarrhea through the use of household-based ceramic water filters: A randomized, controlled trial in rural Bolivia. *The American Journal of Tropical Medicine and Hygiene*, *70*(6), 651–657. PMID:15211008

Clasen, T., & Boisson, S. (2006). Household-Based Ceramic Water Filters for the Treatment of Drinking Water in Disaster Response: An Assessment of a Pilot Programme in the Dominican Republic. *Water Practice & Technology*, *1*(2). doi:10.2166/wpt.2006031

Clasen, T., McLaughlin, C., Nayaar, N., Boisson, S., Gupta, R., Desai, D., & Shah, N. (2008). Microbiological Effectiveness and Cost of Disinfecting Water by Boiling in Semi-urban India. *The American Society of Tropical Medicine and Hygiene*, *79*(3), 407–413. PMID:18784234

Clasen, T., Nadakatti, S., & Menon, S. (2006). Microbiological performance of a water treatment unit designed for household use in developing countries. *Tropical Medicine & International Health*, *11*(9), 1399–1405. doi:10.1111/j.1365-3156.2006.01699.x PMID:16930262

Clasen, T., Parra, G. G., Boisson, S., & Collin, S. (2005). Household-Based Ceramic Water Filters for the Prevention of Diarrhea: A Randomized, Controlled Trial of a Pilot Program in Columbia. *The American Society of Tropical Medicine and Hygiene*, *73*(4), 790–795. PMID:16222027

Clemente, Z., Castro, V. L., Feitosa, L. O., Lima, R., Jonsson, C. M., Maia, A. H. N., & Fraceto, L. F. (2013). Fish exposure to nano-TiO2 under different experimental conditions: Methodological aspects for nanoecotoxicology investigations. *The Science of the Total Environment*, *463-464*, 647–656. doi:10.1016/j.scitotenv.2013.06.022 PMID:23845857

Clément, L., Hurel, C., & Marmier, N. (2013). Toxicity of TiO_2 nanoparticles to cladocerans, algae, rotifers and plants - Effects of size and crystalline structure. *Chemosphere*, *90*(3), 1083–1090. doi:10.1016/j.chemosphere.2012.09.013 PMID:23062945

Cockell, K. A., Hilton, J. W., & Bettger, W. J. (1992). Hepatobiliary and hematological effects of dietary di sodium arsenate heptahydrate in juvenile rainbow trout (*Oncorhynchus mykiss*). *Comparative Biochemistry and Physiology*, (Part C): 103, 453–458.

Colborn, T., vom Saal, F. S., & Soto, A. M. (1993). Developmental effects of endocrine-disrupting chemicals in wildlife and humans. *Environmental Health Perspectives*, *101*(5), 378–384. doi:10.1289/ehp.93101378 PMID:8080506

Coleman, J. N., Khan, U., Blau, W. J., & Gun'ko, Y. K. (2006). Small but strong: A review of the mechanical properties of carbon nanotube-polymer composites. *Carbon*, *44*(9), 1624–1652. doi:10.1016/j.carbon.2006.02.038

Collini, E., Fortunati, I., Scolaro, S., Signorini, R., Ferrante, C., Bozio, R., & Silvestrini, S. et al. (2010). A fullerene-distyrylbenzene photosensitizer for two-photon promoted singlet oxygen production. *Physical Chemistry Chemical Physics*, *12*(18), 4656–4666. doi:10.1039/b922740g

Colvin, V. L. (2003). The potential environmental impact of engineered nanomaterials. *Nat Biotech, 21*(10), 1166-1170.

Colvin, R. A., Fontaine, C. P., Laskowski, M., & Thomas, D. (2003). Zn^{2+} transporters and Zn^{2+} homeostasis in neurons. *European Journal of Pharmacology*, *479*(1-3), 171–185. doi:10.1016/j.ejphar.2003.08.067 PMID:14612148

Colvin, V. L., Schlamp, M. C., & Alivisatos, A. P. (1994). Light-emitting diodes made from cadmium selenide nanocrystals and a semiconducting polymer. *Nature*, *370*(6488), 354–357. doi:10.1038/370354a0

Colwell, R. R., Huq, A., Islam, M. S., Aziz, K. M. A., Yunus, M., Khan, N. H., & Russek-Cohen, E. et al. (2003). Reduction of cholera in Bangladeshi villages by simple filtration. *Proceedings of the National Academy of Sciences of the United States of America*, *100*(3), 1051–1055. doi:10.1073/pnas.0237386100 PMID:12529505

Compagnone, D., McNeil, C. J., Athey, D., Di Ilio, C., & Guilbault, G. G. (1995). An amperometric NADH biosensor based on NADH oxidase from Thermus aquaticus. *Enzyme and Microbial Technology*, *17*(5), 472–476. doi:10.1016/0141-0229(94)00110-D

Corma, A., & Serna, P. (2006). Chemoselective hydrogenation of nitro compounds with supported gold catalysts. *Science*, *313*(5785), 332–334. doi:10.1126/science.1128383 PMID:16857934

Cornell. (2003). *Common Pesticides in Groundwater*. Retrieved from http: //pmep.cce.cornell.edu/facts-slides-self/ slide-set/ gwater09.html

Corradini, E., de Moura, M. R., & Mattoso, L. H. C. (2010). A preliminary study of the incorparation of NPK fertilizer into chitosan nanoparticles. *Polymer Letters*, *4*(8), 509–515. doi:10.3144/expresspolymlett.2010.64

Corry, B. (2007). Designing Carbon Nanotube Membranes for Efficient Water Desalination. *The Journal of Physical Chemistry B*, *112*(5), 1427–1434. doi:10.1021/jp709845u PMID:18163610

Cosgrove, W. J., & Rijsberman, F. R. (2000). *World Water Vision: Making Water Everybody's Business*. London: Earthscan Publications.

Costi, R., Cohen, G., Salant, A., Rabani, E., & Banin, U. (2009). Electrostatic force microscopy study of single Au–CdSe hybrid nanodumbbells: Evidence for light-induced charge separation. *Nano Letters*, *9*(5), 2031–2039. doi:10.1021/nl900301v PMID:19435381

Costi, R., Saunders, A. E., Elmalem, E., Salant, A., & Banin, U. (2008). Visible light-induced charge retention and photocatalysis with hybrid CdSe-Au nanodumbbells. *Nano Letters*, *8*(2), 637–641. doi:10.1021/nl0730514 PMID:18197720

Couling, D. J., Bernot, R. J., Docherty, K. M., Dixon, J. K., & Maginn, E. J. (2006). Assessing the factors responsible for ionic liquid toxicity to aquatic organisms via quantitative structure–property relationship modeling. *Green Chemistry*, *8*(1), 82–90. doi:10.1039/B511333D

Cozzens, S., Cortes, R., Soumonni, O., & Woodson, T. (2013). Nanotechnology and the millennium development goals: Water, energy, and agri-food. *Journal of Nanoparticle Research*, *15*(11), 10–23. doi:10.1007/s11051-013-2001-y

Crane, R. A., & Scott, T. B. (2012). Nanoscale zero-valent iron: Future prospects for an emerging water treatment technology. *Journal of Hazardous Materials*, *211-212*, 112–125. doi:10.1016/j.jhazmat.2011.11.073 PMID:22305041

Creighton, J. A., & Eadon, D. G. (1991). Ultraviolet–visible absorption spectra of the colloidal metallic elements. *Journal of the Chemical Society, Faraday Transactions*, *87*(24), 3881–3891. doi:10.1039/FT9918703881

Crist, R. M., Grossman, J. H., Patri, A. K., Stern, S. T., Dobrovolskaia, M. A., Adiseshaiah, P. P., & McNeil, S. E. et al. (2013). Common pitfalls in nanotechnology: Lessons learned from NCI's Nanotechnology Characterization Laboratory. *Integrative Biology*, *5*(1), 66–73. doi:10.1039/C2IB20117H PMID:22772974

Cruz-Romero, M. C., Murphy, T., Morris, M., Cummins, E., & Kerry, J. P. (2013). Antimicrobial activity of chitosan, organic acids and nano-sized solubilisates for potential use in smart antimicrobially-active packaging for potential food applications. *Food Control*, *34*(2), 393–397. doi:10.1016/j.foodcont.2013.04.042

Cuenya, B. R. (2010). Synthesis and catalytic properties of metal nanoparticles: Size, shape, support, composition, and oxidation state effects. *Thin Solid Films*, *518*(12), 3127–3150. doi:10.1016/j.tsf.2010.01.018

Cui, D., Tian, F., Ozkan, C. S., Wang, M., & Gao, H. (2005). Effect of single wall carbon nanotubes on human HEK293 cells. *Toxicology Letters*, *155*(1), 73–85. doi:10.1016/j.toxlet.2004.08.015 PMID:15585362

Cushing, B. L., Kolesnichenko, V. L., & O'Connor, C. J. (2004). Recent advances in the liquid-phase syntheses of inorganic nanoparticles. *Chemical Reviews*, *104*(9), 3893–3946. doi:10.1021/cr030027b PMID:15352782

Da Ros, T., & Prato, M. (1999). Medicinal chemistry with fullerenes and fullerene derivatives. *Chemical Communications*, (8): 663–669. doi:10.1039/a809495k

Daghrir, R., Drogui, P., & El Khakani, M. A. (2013). Photoelectrocatalytic oxidation of chlortetracycline using Ti/TiO 2 photo-anode with simultaneous H 2 O 2 production. *Electrochimica Acta*, *87*, 18–31. doi:10.1016/j.electacta.2012.09.020

Dai, H. (2002). Carbon Nanotubes: Synthesis, Integration, and Properties. *Accounts of Chemical Research*, *35*(12), 1035–1044. doi:10.1021/ar0101640 PMID:12484791

Dai, L., Banta, G. T., Selck, H., & Forbes, V. E. (2015). Influence of copper oxide nanoparticle form and shape on toxicity and bioaccumulation in the deposit feeder, *Capitella teleta*. *Marine Environmental Research*, *111*, 99–106. doi:10.1016/j.marenvres.2015.06.010 PMID:26138270

Dameron, C. T., Reese, R. N., Mehra, R. K., Kortan, A. R., Carroll, P. J., Steigerwald, M. L., & Winge, D. R. et al. (1989). Biosynthesis of cadmium sulphide quantum semiconductor crystallites. *Nature*, *338*(6216), 596–597. doi:10.1038/338596a0

Danilovas, P. P., Navikaite, V., & Rutkaite, R. (2014). Preparation and Characterization of Potentially Antimicrobial Polymer Films Containing Starch Nano- and Microparticles. *Materials Science-Medziagotyra*, *20*(3), 283–288. doi:10.5755/j01.ms.20.3.5426

Dankovich, T. A. (2012). *Bactericidal Paper Containing Silver Nanoparticles for Water Treatment*. (Doctor of Philosophy Thesis). McGill University.

Compilation of References

Dankovich, T. A. (2014). Microwave-assisted incorporation of silver nanoparticles in paper for point-of-use water purification. *Royal Society of Chemistry*, *1*, 367–378. PMID:25400935

Dankovich, T. A., & Gray, D. G. (2011). Bactericidal Paper Impregnated with Silver Nanoparticles for Point-of-Use Water Treatment. *Environmental Science & Technology*, *45*(5), 1992–1998. doi:10.1021/es103302t PMID:21314116

Dankovich, T., Clinch, C., Weinronk, H., Dillingham, R., & Smith, J. A. (2014). (Manuscript submitted for publication). Inactivation of bacteria from contaminated streams in Limpopo, South Africa by silver- or copper-nanoparticle paper filters. *Water Research*.

Dasari, T. P., Pathakoti, K., & Hwang, H.-M. (2013). Determination of the mechanism of photoinduced toxicity of selected metal oxide nanoparticles (ZnO, CuO, Co3O4 and TiO2) to E. coli bacteria. *Journal of Environmental Sciences (China)*, *25*(5), 882–888. doi:10.1016/S1001-0742(12)60152-1 PMID:24218817

Das, G., Chakraborty, R., Gopalakrishnan, A., Baranov, D., Di Fabrizio, E., & Krahne, R. (2013). A new route to produce efficient surface-enhanced Raman spectroscopy substrates: Gold-decorated CdSe nanowires. *Journal of Nanoparticle Research*, *15*(5), 1–9. doi:10.1007/s11051-013-1596-3

Dasgupta, N., Ranjan, S., Mundekkad, D., Ramalingam, C., Shanker, R., & Kumar, A. (2015). Nanotechnology in agrofood: From field to plate. *Food Research International*, *69*, 381–400. doi:10.1016/j.foodres.2015.01.005

Das, R., Abd Hamid, S. B., Ali, M. E., Ismail, A. F., Annuar, M. S. M., & Ramakrishna, S. (2014). Multifunctional carbon nanotubes in water treatment: The present, past and future. *Desalination*, *354*, 160–179. doi:10.1016/j.desal.2014.09.032

Das, S., Gates, A. J., Abd, H. A., Rose, G. S., Picconatto, C. A., & Ellenbogen, J. C. (2007). Designs for Ultra-tiny, special-Purpose Nanoeleectronic Circuits. *IEEE Transactions on Circuits and Systems*, *154*(11), 2528–2540. doi:10.1109/TCSI.2007.907864

Dastjerdi, R., Montazer, M., & Shahsavan, S. (2009). A new method to stabilize nanoparticles on textile surfaces. *Colloids and Surfaces. A, Physicochemical and Engineering Aspects*, *345*(1-3), 202–210. doi:10.1016/j.colsurfa.2009.05.007

Dastjerdi, R., Montazer, M., & Shahsavan, S. (2010). A novel technique for producing durable multifunctional textiles using nanocomposite coating. *Colloids and Surfaces. B, Biointerfaces*, *81*(1), 32–41. doi:10.1016/j.colsurfb.2010.06.023 PMID:20675103

Davis, A. K. (2008). Ontogenetic changes in erythrocyte morphology in larval mole salamanders, *Ambystoma talpoideum*, measured with image analysis. *Comparative Clinical Pathology*, *17*(1), 23–28. doi:10.1007/s00580-007-0702-2

Dawson, A., & Kamat, P. V. (2001). Semiconductor-metal nanocomposites. Photoinduced fusion and photocatalysis of gold-capped TiO2 (TiO2/gold) nanoparticles. *The Journal of Physical Chemistry B*, *105*(5), 960–966. doi:10.1021/jp0033263

De Faria, A. F., Perreault, F., Shaulsky, E., Arias Chavez, L. H., & Elimelech, M. (2015). Antimicrobial Electrospun Biopolymer Nanofiber Mats Functionalized with Graphene Oxide-Silver Nanocomposites. *ACS Applied Materials & Interfaces*, *7*(23), 12751–12759. doi:10.1021/acsami.5b01639 PMID:25980639

De Gusseme, B., Hennebel, T., Christiaens, E., Saveyn, H., Verbeken, K., Fitts, J.P., Boon, N. & Verstraete, W., (2011). Virus disinfection in water by biogenic silver immobilized in polyvinylidene fluoride membranes. *Water Research*, *45*(4), 1856-1864.

De Gusseme, B., Sintubin, L., Hennebel, T., Boon, N., Verstraete, W., Baert, L., & Uyttendaele, M. (2010). Inactivation of viruses in water by biogenic silver: innovative and environmentally friendly disinfection technique. In *Bioinformatics and Biomedical Engineering (iCBBE), 2010 4th International Conference on* (pp. 1-5). IEEE. doi:10.1109/ICBBE.2010.5515631

De Gusseme, B., Sintubin, L., Baert, L., Thibo, E., Hennebel, T., Vermeulen, G., & Boon, N. et al. (2010). Biogenic silver for disinfection of water contaminated with viruses. *Applied and Environmental Microbiology*, *76*(4), 1082–1087. doi:10.1128/AEM.02433-09 PMID:20038697

De la Rubia, A., Rodríguez, M., León, V. M., & Prats, D. (2008). Removal of natural organic matter and THM formation potential by ultra-and nanofiltration of surface water. *Water Research*, *42*(3), 714–722. doi:10.1016/j.watres.2007.07.049 PMID:17765283

de Lannoy, C.-F., Jassby, D., Gloe, K., Gordon, A. D., & Wiesner, M. R. (2013). Aquatic Biofouling Prevention by Electrically Charged Nanocomposite Polymer Thin Film Membranes. *Environmental Science & Technology*, *47*(6), 2760–2768. doi:10.1021/es3045168 PMID:23413920

De Oliveira, J. L., Campos, E. V. R., Bakshi, M., Abhilash, P. C., & Fraceto, L. F. (2014). Application of nanotechnology for the encapsulation of botanicalinsecticides for sustainable agriculture: Prospects and promises. *Biotechnology Advances*, *32*(8), 1550–1561. doi:10.1016/j.biotechadv.2014.10.010 PMID:25447424

de Paiva, R., & Di Felice, R. (2008). Atomic and Electronic Structure at Au/CdSe Interfaces. *ACS Nano*, *2*(11), 2225–2236. doi:10.1021/nn8004608 PMID:19206387

De Rosa, M. C., Monreal, C., Schnitzer, M., Walsh, R., & Sultan, Y. (2010). Nanotechnology in fertilizers. *Nature Nanotechnology*, *5*(2), 91. doi:10.1038/nnano.2010.2 PMID:20130583

De, A., Bose, R., Kumar, A., & Mozumdar, S. (2014). Management of insect pests using nanotechnology: as modern approaches. In *Targeted delivery of pesticides using biodegradable polymeric nanoparticles* (pp. 29–33). New Delhi: Springer. doi:10.1007/978-81-322-1689-6_8

Deborde, M., & von Gunten, U. (2008). Reactions of chlorine with inorganic and organic compounds during water treatment-kinetics and mechanisums: A critical review. *Water Research*, *42*(1-2), 13–51. doi:10.1016/j.watres.2007.07.025 PMID:17915284

Dechsakulthorn, F., Hayes, A., Bakand, S., Joeng, L., & Winder, C. (2007). In vitro cytotoxicity assessment of selected nanoparticles using human skin fibroblasts. *AATEX*, *14*(Special Issue), 397–400.

Deguchi, S., Alargova, R. G., & Tsujii, K. (2001). Stable dispersions of fullerenes, C_{60} and C_{70}, in water. Preparation and characterization. *Langmuir*, *17*(19), 6013–6017. doi:10.1021/la010651o

Deng, X. L., Nikiforov, A., Vujosevic, D., Vuksanovic, V., Mugosa, B., Cvelbar, U., & Leys, C. et al. (2015). Antibacterial activity of nano-silver non-woven fabric prepared by atmospheric pressure plasma deposition. *Materials Letters*, *149*, 95–99. doi:10.1016/j.matlet.2015.02.112

Densilin, D. M., Srinivasan, S., Manju, P., & Sudha, S. (2011). Effect of Individual and Combined Application of Biofertilizers, Inorganic Fertilizer and Vermi compost on the Biochemical Constituents of Chilli (Ns - 1701). *Journal of Biofertilizers & Biopesticides*, *2*(02), 106. doi:10.4172/2155-6202.1000106

Department for Environment Food & Rural Affairs (DEFRA). (2007). *Advanced biological treatment of municipal solid waste.* Retrieved April 2015, from http://www.recycleforgloucestershire.com/CHttpHandler.ashx?id=59162&p=0

Department for Environment Food & Rural Affairs (DEFRA). (2013). *Advanced biological treatment of municipal solid waste*. Retrieved May 2015, from https://www.gov.uk/government/uploads/system/uploads/attachment_data/file/221037/pb13887-advanced-biological-treatment-waste.pdf

Department for Environment Food & Rural Affairs (DEFRA). (2013). *Mechanical biological treatment of municipal solid waste*. Retrieved May 2015, from https://www.gov.uk/government/uploads/system/uploads/attachment_data/file/221039/pb13890-treatment-solid-waste.pdf

Derfus, A. M., Chan, W. C., & Bhatia, S. N. (2004). Intracellular delivery of quantum dots for live cell labeling and organelle tracking. *Advanced Materials*, *16*(12), 961–966. doi:10.1002/adma.200306111

Destaye, A. G., Lin, C. K., & Lee, C. K. (2013). Glutaraldehyde vapor cross-linked nanofibrous PVA mat with in situ formed silver nanoparticles. *ACS Applied Materials & Interfaces*, *5*(11), 4745–4752. doi:10.1021/am401730x PMID:23668250

Development of research/innovation policies - Science and Technology Foresight. (2010, Jan 22). *Converging Technologies for a Diverse Europe*. Retrieved August 24, 2015, from: http://cordis.europa.eu/foresight/ntw-expert-group.htm

Devitt, E., Ducellier, F., Cote, P., & Wiesner, M. (1998). Effects of natural organic matter and the raw water matrix on the rejection of atrazine by pressure-driven membranes. *Water Research*, *32*(9), 2563–2568. doi:10.1016/S0043-1354(98)00043-8

Dhawan, A., Sharma, V., & Parmar, D. (2009). Nanomaterials: A challenge for toxicologists. *Nanotoxicology*, *3*(1), 1–9. doi:10.1080/17435390802578595

Diamanti-Kandarakis, E., Bourguignon, J.-P., Giudice, L. C., Hauser, R., Prins, G. S., Soto, A. M., & Gore, A. C. et al. (2009). Endocrine-disrupting chemicals: An Endocrine Society scientific statement. *Endocrine Reviews*, *30*(4), 293–342. doi:10.1210/er.2009-0002 PMID:19502515

Dias, C. R., Rosa, M. J., & de Pinho, M. N. (1998). Structure of water in asymmetric cellulose ester membranes—and ATR-FTIR study. *Journal of Membrane Science*, *138*(2), 259–267. doi:10.1016/S0376-7388(97)00226-3

Diederich, F., & Thilgen, C. (1996). Covalent fullerene chemistry. *Science*, *271*(5247), 317–323. doi:10.1126/science.271.5247.317

Dillingham, R., Pinkerton, R., Leger, P., Severe, P., Pape, J., & Fitzgerald, D. (2006). *Mortality in Haitian patients treated with antiretroviral therapy (ART) in a community setting*. Paper presented at the 13th Conference on Retroviruses and Opportunistic Infections, Denver, CO.

Dillingham, R., & Guerrant, R. L. (2004). Childhood stunting: Measuring and stemming the staggering costs of inadequate water and sanitation. *Lancet*, *363*(9403), 94–95. doi:10.1016/S0140-6736(03)15307-X PMID:14726158

DiMarzio, E., & Guttman, C. (1970). Separation by flow. *Macromolecules*, *3*(2), 131–146. doi:10.1021/ma60014a005

Dimitrijevic, N. M., & Kamat, P. V. (1993). Excited-state behavior and one-electron reduction of C_{60} in aqueous r-cyclodextrin solution. *Journal of Physical Chemistry*, *97*(29), 7623–7626. doi:10.1021/j100131a035

Dimkpa, C., Calder, A., & Gajjar, P. (2011). Interaction of silver nanoparticles with an environmentally beneficial bacterium, Pseudomonas chlororaphis. *Journal of Hazardous*.

Ding, L., Stilwell, J., Zhang, T., Elboudwarej, O., Jiang, H., Selegue, J. P., & Chen, F. F. et al. (2005). Molecular characterization of the cytotoxic mechanism of multiwall carbon nanotubes and nano-onions on human skin fibroblast. *Nano Letters*, *5*(12), 2448–2464. doi:10.1021/nl051748o PMID:16351195

Disney, S. M., & Grubbström, R. W. (2004). Economic consequences of a production and inventory control policy. *International Journal of Production Research*, *42*(17), 3419–3431. doi:10.1080/00207540410001727640

Doktorovova, S., Souto, E. B., & Silva, A. M. (2014). Nanotoxicology applied to solid lipid nanoparticles and nanostructured lipid carriers – A systematic review of in vitro data. *European Journal of Pharmaceutics and Biopharmaceutics*, *87*(1), 1–18. doi:10.1016/j.ejpb.2014.02.005 PMID:24530885

Domingos, R. F., Tufenkji, N., & Wilkinson, K. J. (2009). Aggregation of titanium dioxide nanoparticles: Role of a fulvic acid. *Environmental Science & Technology*, *43*(5), 1282–1286. doi:10.1021/es8023594 PMID:19350891

Dominguez, G. A., Lohse, S. E., Torelli, M. D., Murphy, C. J., Hamers, R. J., Orr, G., & Klaper, R. D. (2015). Effects of charge and surface ligand properties of nanoparticles on oxidative stress and gene expression within the gut of Daphnia magna. *Aquatic Toxicology (Amsterdam, Netherlands)*, *162*, 1–9. doi:10.1016/j.aquatox.2015.02.015 PMID:25734859

Donaldson, K., Stone, V., Tran, C. L., Kreyling, W., & Borm, P. J. A. (2004). Nanotoxicology. *Occupational and Environmental Medicine*, *61*(9), 727–728. doi:10.1136/oem.2004.013243 PMID:15317911

Dong, H., & Hinestroza, J. P. (2009). Metal Nanoparticles on Natural Cellulose Fibers: Electrostatic Assembly and In Situ Synthesis. *Applied Materials & Interfaces*, *1*(4), 797–803. doi:10.1021/am800225j PMID:20356004

Donlan, R. M., & Costerton, J. W. (2002). Biofilms: Survival mechanisms of clinically relevant microorganisms. *Clinical Microbiology Reviews*, *15*(2), 167–193. doi:10.1128/CMR.15.2.167-193.2002 PMID:11932229

Doshi, J., & Reneker, D. H. (1995). Electrospinning process and applications of electrospun fibers. *Journal of Electrostatics*, *35*(2-3), 151–160. doi:10.1016/0304-3886(95)00041-8

Dowling, A., Clift, R., Grobert, N., Hutton, D., Oliver, R., & O'neill, O. et al.. (2004). Nanoscience and nanotechnologies : Opportunities and uncertainties. *London The Royal Society The Royal Academy of Engineering Report*, *46*(July), 618–618.

Drabkin, D. L. (1946). Spectrometric studies, XIV: The crystallographic and optimal properties of the hemoglobin of man in comparison with those of other species. *The Journal of Biological Chemistry*, *164*, 703–723. PMID:21001166

Duan, J., Yu, Y., Li, Y., Yu, Y., Li, Y., Huang, P., & Sun, Z. et al. (2013). Developmental toxicity of CdTe QDs in zebrafish embryos and larvae. *Journal of Nanoparticle Research*, *15*(7), 1700. doi:10.1007/s11051-013-1700-8

Duan, J., Yu, Y., Shi, H., Tian, L., Guo, C., Huang, P., & Sun, Z. et al. (2013). Toxic effects of silica nanoparticles on zebrafish embryos and larvae. *PLoS ONE*, *8*(9), e74606. doi:10.1371/journal.pone.0074606 PMID:24058598

Dubascoux, S., Le Hecho, I., Hassellöv, M., Von Der Kammer, F., Potin Gautier, M., & Lespes, G. (2010). Field-flow fractionation and inductively coupled plasma mass spectrometer coupling: History, development and applications. *Journal of Analytical Atomic Spectrometry*, *25*(5), 613–623. doi:10.1039/b927500b

Dutz, S., & Hergt, R. (2014). Magnetic particle hyperthermia--a promising tumour therapy? *Nanotechnology*, *25*(45), 452001. doi:10.1088/0957-4484/25/45/452001 PMID:25337919

Du, W., Sun, Y., Ji, R., Zhu, J., Wu, J., & Guo, H. (2011). TiO_2 and ZnO nanoparticles negatively affect wheat growth and soil enzyme activities in agricultural soil. *Journal of Environmental Monitoring*, *13*(4), 822–828. doi:10.1039/c0em00611d PMID:21267473

Dwivedi, S., Siddiqui, M. A., Farshori, N. N., Ahamed, M., Musarrat, J., & Al-Khedhairy, A. A. (2014). Synthesis, characterization and toxicological evaluation of iron oxide nanoparticles in human lung alveolar epithelial cells. *Colloid Surf B*, *122*, 209–215. doi:10.1016/j.colsurfb.2014.06.064 PMID:25048357

EC. (2011). Recomendation on the definition of nanomaterial. *Off J Eur Communities, L274*, 1-40.

Echegoyen, Y., & Nerin, C. (2013). Nanoparticle release from nano-silver antimicrobial food containers. *Food and Chemical Toxicology*, *62*, 16–22. doi:10.1016/j.fct.2013.08.014 PMID:23954768

Eder, D. (2010). Carbon Nanotube–Inorganic Hybrids. *Chemical Reviews*, *110*(3), 1348–1385. doi:10.1021/cr800433k PMID:20108978

Efros, A. L., & Rosen, M. (2000). The Electronic Structure of Semiconductor Nanocrystals. *Annual Review of Materials Science*, *30*(1), 475–521. doi:10.1146/annurev.matsci.30.1.475

Ehdaie, B., Krause, C., & Smith, J. A. (2014). Porous Ceramic Tablet Embedded with Silver Nanopatches for Low- Cost Point-of-Use Water Purification. *Environmental Science & Technology*, *48*(23), 13901–13908. doi:10.1021/es503534c PMID:25387099

El Badawi, N., Ramadan, A. R., Esawi, A. M. K., & El-Morsi, M. (2014). Novel carbon nanotube–cellulose acetate nanocomposite membranes for water filtration applications. *Desalination*, *344*(0), 79–85. doi:10.1016/j.desal.2014.03.005

El Badawy, A. M., Luxton, T. P., Silva, R. G., Scheckel, K. G., Suidan, M. T., & Tolaymat, T. M. (2010). Impact of environmental conditions (pH, ionic strength, and electrolyte type) on the surface charge and aggregation of silver nanoparticles suspensions. *Environmental Science & Technology*, *44*(4), 1260–1266. doi:10.1021/es902240k PMID:20099802

Elgrabli, D., Floriani, M., Abella-Gallart, S., Meunier, L., Gamez, C., Delalain, P., & Lacroix, G. et al. (2008). Biodistribution and clearance of instilled carbon nanotubes in rat lung. *Particle and Fibre Toxicology*, *5*(1), 20. doi:10.1186/1743-8977-5-20 PMID:19068117

El-Kemary, M., El-Shamy, H., & El-Mehasseb, I. (2010). Photocatalytic degradation of ciprofloxacin drug in water using ZnO nanoparticles. *Journal of Luminescence*, *130*(12), 2327–2331. doi:10.1016/j.jlumin.2010.07.013

Elliott, D. W., & Zhang, W.-X. (2001). Field assessment of nanoscale bimetallic particles for groundwater treatment. *Environmental Science & Technology*, *35*(24), 4922–4926. doi:10.1021/es0108584 PMID:11775172

El-Nour, K. M. A., Eftaiha, A. A., Al-Warthan, A., & Ammar, R. A. (2010). Synthesis and applications of silver nanoparticles. *Arabian Journal of Chemistry, 3*(3), 135-140.

El-Temsah, Y. S., & Joner, E. J. (2012). Ecotoxicological effects on earthworms of fresh and aged nano-sized zero-valent iron (nZVI) in soil. *Chemosphere*, *89*(1), 76–82. doi:10.1016/j.chemosphere.2012.04.020 PMID:22595530

El-Temsah, Y. S., & Joner, E. J. (2012). Impact of Fe and Ag nanoparticles on seed germination and differences in bioavailability during exposure in aqueous suspension and soil. *Environmental Toxicology*, *27*(1), 42–49. doi:10.1002/tox.20610 PMID:20549639

El-Temsah, Y. S., & Joner, E. J. (2013). Effects of nano-sized zero-valent iron (nZVI) on DDT degradation in soil and its toxicity to collembola and ostracods. *Chemosphere*, *92*(1), 131–137. doi:10.1016/j.chemosphere.2013.02.039 PMID:23522781

Emadzadeh, D., Lau, W. J., & Ismail, A. F. (2013). Synthesis of thin film nanocomposite forward osmosis membrane with enhancement in water flux without sacrificing salt rejection. *Desalination*, *330*(0), 90–99. doi:10.1016/j.desal.2013.10.003

Emadzadeh, D., Lau, W. J., Matsuura, T., Ismail, A. F., & Rahbari-Sisakht, M. (2014). Synthesis and characterization of thin film nanocomposite forward osmosis membrane with hydrophilic nanocomposite support to reduce internal concentration polarization. *Journal of Membrane Science*, *449*(0), 74–85. doi:10.1016/j.memsci.2013.08.014

Emadzadeh, D., Lau, W. J., Matsuura, T., Rahbari-Sisakht, M., & Ismail, A. F. (2014). A novel thin film composite forward osmosis membrane prepared from PSf–TiO2 nanocomposite substrate for water desalination. *Chemical Engineering Journal*, *237*, 70–80. doi:10.1016/j.cej.2013.09.081

Emam, A. N., Mohamed, M. B., Girgis, E., & Rao, K. V. (2015). Hybrid magnetic–plasmonic nanocomposite: Embedding cobalt clusters in gold nanorods. *RSC Advances*, *5*(44), 34696–34703. doi:10.1039/C5RA01918D

Enrico Burello, E., & Wortha, A. P. (2011). A theoretical framework for predicting the oxidative stress potential of oxide nanoparticles. *Nanotoxicology*, *5*(2), 228–235. doi:10.3109/17435390.2010.502980 PMID:21609138

Envision New Zealand. (2003). *The road to zero waste*. Retrieved August 2015, from https://www.zerowaste.co.nz/assets/Reports/roadtozerowaste150dpi.pdf

Escalada, J. P., Pajares, A., Gianotti, J., Massad, W. A., Bertolotti, S., Amat-Guerri, F., & Garcia, N. A. (2006). Dye-sensitized photodegradation of the fungicide carbendazim and related benzimidazoles. *Chemosphere*, *65*(2), 237–244. doi:10.1016/j.chemosphere.2006.02.057

Esteban, P. P., Alves, D. R., Enright, M. C., Bean, J. E., Gaudion, A., Jenkins, A. T. A., & Arnot, T. C. et al. (2014). Enhancement of the Antimicrobial Properties of Bacteriophage-K via Stabilization using Oil-in-Water Nano-Emulsions. *Biotechnology Progress*, *30*(4), 932–944. doi:10.1002/btpr.1898 PMID:24616404

European Commission. (2010). *Being wise with waste: The EU's approach to waste management*. Retrieved August 2015, from http://ec.europa.eu/environment/waste/pdf/WASTE%20BROCHURE.pdf

European Environment Agency (EEA). (2013). *Managing municipal solid waste – A review of achievements in 32 European counties*. ISSN 1725-9177. Retrieved June 2015, from file:///C:/Users/Duygu/Downloads/Managing%20municipal%20solid%20waste.pdf

Eustis, S., & El-Sayed, M. A. (2006). Why gold nanoparticles are more precious than pretty gold: Noble metal surface plasmon resonance and its enhancement of the radiative and nonradiative properties of nanocrystals of different shapes. *Chemical Society Reviews*, *35*(3), 209–217. doi:10.1039/B514191E PMID:16505915

Fabian, E., Landsiedel, R., Ma-Hock, L., Wiench, K., Wohlleben, W., & Van Ravenzwaay, B. (2008). Tissue distribution and toxicity of intravenously administered titanium dioxide nanoparticles in rats. *Archives of Toxicology*, *82*(3), 151–157. doi:10.1007/s00204-007-0253-y PMID:18000654

Fabrega, J., Luoma, S. N., Tyler, C. R., Galloway, T. S., & Lead, J. R. (2011). Silver nanoparticles: Behaviour and effects in the aquatic environment. *Environment International*, *37*(2), 517–531. doi:10.1016/j.envint.2010.10.012 PMID:21159383

Fabricius, A.-L., Duester, L., Meermann, B., & Ternes, T. A. (2013). ICP-MS-based characterization of inorganic nanoparticles—sample preparation and off-line fractionation strategies. *Analytical and Bioanalytical Chemistry*, 1–13. PMID:24292431

Faisal, M., Abu Tariq, M., & Muneer, M. (2007). Photocatalysed degradation of two selected dyes in UV-irradiated aqueous suspensions of titania. *Dyes and Pigments*, *72*(2), 233–239. doi:10.1016/j.dyepig.2005.08.020

Faisal, M., Saquib, Q., Alatar, A. A., Al-Khedhairy, A. A., Hegazy, A. K., & Musarrat, J. (2013). Phytotoxic hazards of NiO-nanoparticles in tomato: A study on mechanism of cell death. *Journal of Hazardous Materials*, *250–251*, 318–332. doi:10.1016/j.jhazmat.2013.01.063 PMID:23474406

Fang, J., Cao, S. W., Wang, Z., Shahjamali, M. M., Loo, S. C. J., Barber, J., & Xue, C. (2012). Mesoporous plasmonic Au–TiO2 nanocomposites for efficient visible-light-driven photocatalytic water reduction. *International Journal of Hydrogen Energy*, *37*(23), 17853–17861. doi:10.1016/j.ijhydene.2012.09.023

Fan, R., Huang, Y. C., Grusak, M. A., Huang, C. P., & Sherrier, D. J. (2014). Effects of nano-TiO2 on the agronomically-relevant Rhizobium–legume symbiosis. *The Science of the Total Environment*, *466–467*, 503–512. doi:10.1016/j.scitotenv.2013.07.032 PMID:23933452

Fan, T. J., Tao, F., Deng, S., & Li, S. X. (2015). Impact of RFID technology on supply chain decisions with inventory inaccuracies. *International Journal of Production Economics*, *159*, 117–125. doi:10.1016/j.ijpe.2014.10.004

Fan, Z. L., Hu, J. Y., An, W., & Yang, M. (2013). Detection and occurrence of chlorinated byproducts of bisphenol A, nonylphenol, and estrogens in drinking water of China: Comparison to the parent compounds. *Environmental Science & Technology*, *47*(19), 10841–10850. doi:10.1021/es401504a PMID:24011124

Faraday, M. (1857). The Bakerian lecture: Experimental relations of gold (and other metals) to light. *Philosophical Transactions of the Royal Society of London*, *147*(0), 145–181. doi:10.1098/rstl.1857.0011

Farkas, A., Salanki, J., & Specziar, A. (2002). Relation between growth and the heavy metal concentration in organs of bream; *Abramis brama* L. populating Lake Balaton. *Archives of Environmental Contamination and Toxicology*, *43*(2), 236–243. doi:10.1007/s00244-002-1123-5 PMID:12115050

Federici, G., Shaw, B. J., & Handy, R. D. (2007). Toxicity of titanium dioxide nanoparticles to rainbow trout (*Oncorhynchus mykiss*): Gill injury, oxidative stress, and other physiological effects. *Aquatic Toxicology (Amsterdam, Netherlands)*, *84*(4), 415–430. doi:10.1016/j.aquatox.2007.07.009 PMID:17727975

Fedotov, P. S., Vanifatova, N. G., Shkinev, V. M., & Spivakov, B. Y. (2011). Fractionation and characterization of nano- and microparticles in liquid media. *Analytical and Bioanalytical Chemistry*, *400*(6), 1787–1804. doi:10.1007/s00216-011-4704-1 PMID:21318253

Fei, W., & Bart, H.-J. (2001). Predicting diffusivities in liquids by the group contribution method. *Chemical Engineering and Processing: Process Intensification*, *40*(6), 531–535. doi:10.1016/S0255-2701(00)00151-3

Fendorf, S. E., & Zasoski, R. J. (1992). Chromium(III) oxidation by delta-MnO_2. 1. Characterization. *Environmental Science & Technology*, *26*(1), 79–85. doi:10.1021/es00025a006

Fernández, A., Soriano, E., López-Carballo, G., Picouet, P., Lloret, E., Gavara, R., & Hernández-Muñoz, P. (2009). Preservation of aseptic conditions in absorbent pads by using silver nanotechnology. *Food Research International*, *42*(8), 1105–1112. doi:10.1016/j.foodres.2009.05.009

Fernández, J. G., Almeida, C. A., Fernández-Baldo, M. A., Felici, E., Raba, J., & Sanz, M. I. (2016). Development of nitrocellulose membrane filters impregnated with different biosynthesized silver nanoparticles applied to water purification. *Talanta*, *146*, 237–243. doi:10.1016/j.talanta.2015.08.060 PMID:26695258

Ferrando, R., Jellinek, J., & Johnston, R. L. (2008). Nanoalloys: From theory to applications of alloy clusters and nanoparticles. *Chemical Reviews*, *108*(3), 845–910. doi:10.1021/cr040090g PMID:18335972

Feyman, R. (1959). *There's Plenty of Room at the Bottom: An Invitation to Enter a New Field of Physics*. Retrieved from http://www.zyvex.com/nanotech/feynman.html

Filipe, V., Hawe, A., & Jiskoot, W. (2010). Critical evaluation of Nanoparticle Tracking Analysis (NTA) by NanoSight for the measurement of nanoparticles and protein aggregates. *Pharmaceutical Research*, *27*(5), 796–810. doi:10.1007/s11095-010-0073-2 PMID:20204471

Finlay-Moore, O., Hartel, P. G., & Cabrera, M. L. (2000). 17 beta-estradiol and testosterone in soil and runoff from grasslands amended with broiler litter. *Journal of Environmental Quality*, *29*(5), 1604–1611. doi:10.2134/jeq2000.00472425002900050030x

Finsy, R. (1994). Particle sizing by quasi-elastic light scattering. *Advances in Colloid and Interface Science*, *52*, 79–143. doi:10.1016/0001-8686(94)80041-3

Fiorito, S., Serafino, A., Andreola, F., & Bernier, P. (2006). Effects of fullerenes and single-wall carbon nanotubes on murine and human macrophages. *Carbon*, *44*(6), 1100–1105. doi:10.1016/j.carbon.2005.11.009

Fischer, H. C., & Chan, W. C. W. (2007). Nanotoxicity: The growing need for in vivo study. *Current Opinion in Biotechnology*, *18*(6), 565–571. doi:10.1016/j.copbio.2007.11.008 PMID:18160274

Fitzgerald, G. C., & Themelis, N. J. (2009). *Technical and economic analysis of pre-shredding municipal solid wastes prior to disposal*. (Unpublished MS thesis). Columbia University, New York, NY.

Fleming, D. A., & Williams, M. E. (2004). Size-controlled synthesis of gold nanoparticles via high-temperature reduction. *Langmuir*, *20*(8), 3021–3023. doi:10.1021/la0362829 PMID:15875823

Florence, A. T. (2005). Nanoparticle uptake by the oral route: Fulfilling its potential? *Drug Discovery Today. Technologies*, *2*(1), 75–81. doi:10.1016/j.ddtec.2005.05.019 PMID:24981758

Foldbjerg, R., Dang, D., & Autrup, H. (2011). Cytotoxicity and genotoxicity of silver nanoparticles in the human lung cancer cell line, A549. *Archives of Toxicology*, *85*(7), 743–750. doi:10.1007/s00204-010-0545-5 PMID:20428844

Foote, C. S., & Clennan, E. L. (1996). Properties & Reactions of Singlet Dioxygen. In C. S. Foote, J. S. Valentine, A. Greenburg, & J. F. Liebman (Eds.), *Active Oxygen in Chemistry* (pp. 105–140). London: Blackie Academic & Professional. doi:10.1007/978-94-007-0874-7

Fornasiero, F., Park, H. G., Holt, J. K., Stadermann, M., Grigoropoulos, C. P., Noy, A., & Bakajin, O. (2008). Ion exclusion by sub-2-nm carbon nanotube pores. *Proceedings of the National Academy of Sciences of the United States of America*, *105*(45), 17250–17255. doi:10.1073/pnas.0710437105 PMID:18539773

Fortner, J. D., Lyon, D. Y., Sayes, C. M., Boyd, A. M., Falkner, J. C., Hotze, E. M., & Hughes, J. B. et al. (2005). C_{60} in Water: Nanocrystal Formation and Microbial Response. *Environmental Science & Technology*, *39*(11), 4307–4316. doi:10.1021/es048099n

Foster, H. A., Ditta, I. B., Varghese, S., & Steele, A. (2011). Photocatalytic disinfection using titanium dioxide: Spectrum and mechanism of antimicrobial activity. *Applied Microbiology and Biotechnology*, *90*(6), 1847–1868. doi:10.1007/s00253-011-3213-7 PMID:21523480

Franci, G., Falanga, A., Galdiero, S., Palomba, L., Rai, M., Morelli, G., & Galdiero, M. (2015). Silver nanoparticles as potential antibacterial agents. *Molecules (Basel, Switzerland)*, *20*(5), 8856–8874. doi:10.3390/molecules20058856 PMID:25993417

Franco, R., Sánchez-Olea, R., Reyes-Reyes, E. M., & Panayiotidis, M. I. (2009). Environmental toxicity, oxidative stress and apoptosis: Ménage à Trois. *Mutation Research/Genetic Toxicology and Environmental Mutagenesis*, *674*(1–2), 3–22. doi:10.1016/j.mrgentox.2008.11.012 PMID:19114126

Franke, R., Rothe, J., Pollmann, J., Hormes, J., Bönnemann, H., Brijoux, W., & Hindenburg, T. (1996). A Study of the Electronic and Geometric Structure of Colloidal TiO⊙ 0.5 THF. *Journal of the American Chemical Society*, *118*(48), 12090–12097. doi:10.1021/ja953525d

Freeman, A. I., Halladay, L. J., & Cripps, P. (2012). The effect of silver impregnation of surgical scrub suits on surface bacterial contamination. *Veterinary Journal (London, England)*, *192*(3), 489–493. doi:10.1016/j.tvjl.2011.06.039 PMID:22015140

Fu, A., Gu, W., Boussert, B., Koski, K., Gerion, D., Manna, L., & Alivisatos, A. P. et al. (2007). Semiconductor quantum rods as single molecule fluorescent biological labels. *Nano Letters*, *7*(1), 179–182. doi:10.1021/nl0626434 PMID:17212460

Compilation of References

Fubini, B., & Hubbard, A. (2003). Reactive oxygen species (ROS) and reactive nitrogen species (RNS) generation by silica in inflammation and fibrosis. *Free Radical Biology & Medicine*, *34*(12), 1507–1516. doi:10.1016/S0891-5849(03)00149-7 PMID:12788471

Fu, F., Dionysiou, D. D., & Liu, H. (2014). The use of zero-valent iron for groundwater remediation and wastewater treatment: A review. *Journal of Hazardous Materials*, *267*, 194–205. doi:10.1016/j.jhazmat.2013.12.062 PMID:24457611

Fu, J., Ji, J., Fan, D., & Shen, J. (2006). Construction of antibacterial multilayer films containing nanosilver via layer-by-layer assembly of heparin and chitosan-silver ions complex. *Journal of Biomedical Materials Research. Part A*, *79A*(3), 665–674. doi:10.1002/jbm.a.30819 PMID:16832825

Fujiwara, K., Suematsu, H., Kiyomiya, E., Aoki, M., Sato, M., & Moritoki, N. (2008). Size-dependent toxicity of silica nano-particles to Chlorella kessleri. *Journal of Environmental Science and Health. Part A, Toxic/Hazardous Substances & Environmental Engineering*, *43*(10), 1167–1173. doi:10.1080/10934520802171675 PMID:18584432

Fu, P. P., Xia, Q., Hwang, H. M., Ray, P. C., & Yu, H. (2014). Mechanisms of nanotoxicity: Generation of reactive oxygen species. *Journal of Food and Drug Analysis*, *22*(1), 64–75. doi:10.1016/j.jfda.2014.01.005 PMID:24673904

Furno, F., Morley, K. S., Wong, B., Sharp, B. L., Arnold, P. L., & Howdle, S. M. et al.. (2004). Silver nanoparticles and polymeric medical devices: A new approach to prevention of infection? *The Journal of Antimicrobial Chemotherapy*, *54*(6), 1019–1024. doi:10.1093/jac/dkh478 PMID:15537697

Gabriel, U. U., Amakiri, E. U., & Ezeri, G. N. O. (2007). Haematology and gill pathology of *Clarias gariepinus* exposed to refined petroleum oil under laboratory conditions. *Journal of Animal and Veterinary Advances*, *6*, 461–465.

Gaiser, B. K., Fernandes, T. F., Jepson, M. A., Lead, J. R., Tyler, C. R., Baalousha, M., Biswas, A., et al. (2012). Inter-species comparisons on the uptake and toxicity of silver and cerium dioxide nanoparticles. *Environmental Toxicology and Chemistry / SETAC*, *31*(1), 144–54.

Gaiser, B. K., Fernandes, T. F., Jepson, M. A., Lead, J. R., Tyler, C. R., Baalousha, M., & Stone, V. et al. (2012). Inter-species comparisons on the uptake and toxicity of silver and cerium dioxide nanoparticles. *Environmental Toxicology and Chemistry*, *31*(1), 144–154. doi:10.1002/etc.703 PMID:22002553

Gajbhiye, M., Kesharwani, J., Ingle, A., Gade, A., & Rai, M. (2009). Fungus-mediated synthesis of silver nanoparticles and their activity against pathogenic fungi in combination with fluconazole. *Nanomedicine; Nanotechnology, Biology, and Medicine*, *5*(4), 382–386. doi:10.1016/j.nano.2009.06.005 PMID:19616127

Gangadharan, D., Harshvardan, K., Gnanasekar, G., Dixit, D., Popat, K. M., & Anand, P. S. (2010). Polymeric microspheres containing silver nanoparticles as a bactericidal agent for water disinfection. *Water Research*, *44*(18), 5481–5487. doi:10.1016/j.watres.2010.06.057 PMID:20673945

Gao, B., Lin, Y., Wei, S., Zeng, J., Liao, Y., Chen, L., & Hou, J. et al. (2012). Charge transfer and retention in directly coupled Au-CdSe nanohybrids. *Nano Research*, *5*(2), 88–98. doi:10.1007/s12274-011-0188-8

Gao, J., Sun, S. P., Zhu, W. P., & Chung, T. S. (2014). Polyethyleneimine (PEI) cross-linked P84 nanofiltration (NF) hollow fiber membranes for Pb2+ removal. *Journal of Membrane Science*, *452*, 300–310. doi:10.1016/j.memsci.2013.10.036

García, A., Delgado, L., Torà, J. A., Casals, E., González, E., Puntes, V., & Sánchez, A. et al. (2012). Effect of cerium dioxide, titanium dioxide, silver, and gold nanoparticles on the activity of microbial communities intended in wastewater treatment. *Journal of Hazardous Materials*, *199–200*, 64–72. doi:10.1016/j.jhazmat.2011.10.057 PMID:22088500

García, A., Espinosa, R., Delgado, L., Casals, E., González, E., Puntes, V., & Sánchez, A. et al. (2011). Acute toxicity of cerium oxide, titanium oxide and iron oxide nanoparticles using standardized tests. *Desalination, 269*(1–3), 136–141. doi:10.1016/j.desal.2010.10.052

García-Cambero, J. P., Herranz, A. L., Díaz, G. L., Cuadal, J. S., Castelltort, M. E. R., & Calvo, A. C. (2012). *Efectos letales y subletales de nanopartículas y material soluble de oro en el desarrollo de embriones de pez cebra*. Revista de Toxicología.

Gardea-Torresdey, J. L., Rico, C. M., & White, J. C. (2014). Trophic transfer, transformation, and impact of engineered nanomaterials in terrestrial environments. *Environmental Science & Technology, 48*(5), 2526–2540. doi:10.1021/es4050665 PMID:24499408

Gedda, G., Pandey, S., Lin, Y. C., & Wu, H. F. (2015). Antibacterial effect of calcium oxide nano-plates fabricated from shrimp shells. *Green Chemistry, 17*(6), 3276–3280. doi:10.1039/C5GC00615E

Geiger, C. L., Clausen, C. A., Brooks, K., Clausen, C., Huntley, C., Filipek, L., & Major, D. et al. (2003). Nanoscale and microscale iron emulsions for treating DNAPL. *ACS Symposium Series. American Chemical Society, 837*, 132–140. doi:10.1021/bk-2002-0837.ch009

Geiser, M., Rothen-Rutishauser, B., Kapp, N., Schürch, S., Kreyling, W., Schulz, H., & Gehr, P. et al. (2005). Ultrafine particles cross cellular membranes by nonphagocytic mechanisms in lungs and in cultured cells. *Environmental Health Perspectives, 113*(11), 1555–1560. doi:10.1289/ehp.8006 PMID:16263511

George, S., Gardner, H., Seng, E. K., Chang, H., Wang, C., Yu Fang, C. H., & Chan, W. K. et al. (2014). Differential effect of solar light in increasing the toxicity of silver and titanium dioxide nanoparticles to a ish cell line and Zebrafish embryos. *Environmental Science & Technology, 48*(11), 6374–6382. doi:10.1021/es405768n PMID:24811346

George, S., Lin, S., Ji, Z., Thomas, C. R., Li, L., Mecklenburg, M., & Nel, A. E. et al. (2012). Surface defects on plate-shaped silver nanoparticles contribute to its hazard potential in a fish gill cell line and zebrafish embryos. *ACS Nano, 6*(5), 3745–3759. doi:10.1021/nn204671v PMID:22482460

George, S., Pokhrel, S., Xia, T., Gilbert, B., Ji, Z., Schowalter, M., & Nel, A. E. et al. (2010). Use of a rapid cytotoxicity screening approach to engineer a safer zinc oxide nanoparticle through iron doping. *ACS Nano, 4*(1), 15–29. doi:10.1021/nn901503q PMID:20043640

Gerbec, J. A., Magana, D., Washington, A., & Strouse, G. F. (2005). Microwave-enhanced reaction rates for nanoparticle synthesis. *Journal of the American Chemical Society, 127*(45), 15791–15800. doi:10.1021/ja052463g PMID:16277522

Ghaemi, N., Madaeni, S. S., Alizadeh, A., Daraei, P., Badieh, M. M. S., Falsafi, M., & Vatanpour, V. (2012). Fabrication and modification of polysulfone nanofiltration membrane using organic acids: Morphology, characterization and performance in removal of xenobiotics. *Separation and Purification Technology, 96*, 214–228. doi:10.1016/j.seppur.2012.06.008

Ghaemi, N., Madaeni, S. S., Alizadeh, A., Daraei, P., Vatanpour, V., & Falsafi, M. (2012). Fabrication of cellulose acetate/sodium dodecyl sulfate nanofiltration membrane: Characterization and performance in rejection of pesticides. *Desalination, 290*, 99–106. doi:10.1016/j.desal.2012.01.013

Ghaemi, N., Madaeni, S. S., Alizadeh, A., Rajabi, H., Daraei, P., & Falsafi, M. (2012). Effect of fatty acids on the structure and performance of cellulose acetate nanofiltration membranes in retention of nitroaromatic pesticides. *Desalination, 301*, 26–41. doi:10.1016/j.desal.2012.06.008

Ghaffar, A., Ashraf, S., Hussain, R., Hussain, T., Shafique, M., Noreen, S., & Aslam, S. (2014). Clinico hematological disparities induced by triazophos (organophosphate) in Japanese quail. *Pakistan Veterinary Journal, 34*, 257–259.

Compilation of References

Ghasemzadeh, A. (2012). Global issues of food production. *Agrotechnol, 1*(2), 1–2. doi:10.4172/2168-9881.1000e102

Ghodake, G., Seo, Y. D., & Lee, D. S. (2011). Hazardous phytotoxic nature of cobalt and zinc oxide nanoparticles assessed using Allium cepa. *Journal of Hazardous Materials, 186*(1), 952–955. doi:10.1016/j.jhazmat.2010.11.018 PMID:21122986

Ghormade, V., Deshpande, M. V., & Paknikar, K. M. (2011). Perspectives for nano-biotechnology enabled protection and nutrition of plants. *Biotechnology Advances, 29*(6), 792–803. doi:10.1016/j.biotechadv.2011.06.007 PMID:21729746

Ghosh Chaudhuri, R., & Paria, S. (2011). Core/shell nanoparticles: Classes, properties, synthesis mechanisms, characterization, and applications. *Chemical Reviews, 112*(4), 2373–2433. doi:10.1021/cr100449n PMID:22204603

Gibbons, C., Rodriguez, R., Tallon, L., & Sobsey, M. (2010). Evaluation of positively charged alumina nanofibre cartridge filters for the primary concentration of noroviruses, adenoviruses and male-specific coliphages from seawater. *Journal of Applied Microbiology, 109*(2), 635–641. PMID:20202019

Giddings, J. C. (1993). Field-flow fractionation: analysis of macromolecular, colloidal, and particulate materials. *Science, 260*(5113), 1456–1465.

Gilbert, B., Lu, G., & Kim, C. S. (2007). Stable cluster formation in aqueous suspensions of iron oxyhydroxide nanoparticles. *Journal of Colloid and Interface Science, 313*(1), 152–159. doi:10.1016/j.jcis.2007.04.038 PMID:17511999

Gilbert, P., & Moore, L. E. (2005). Cationic antiseptics: Diversity of action under a common epithet. *Journal of Applied Microbiology, 99*(4), 703–715. doi:10.1111/j.1365-2672.2005.02664.x

Girgis, E., Khalil, W. K. B., Emam, A. N., Mohamed, M. B., & Rao, K. V. (2012). Nanotoxicity of gold and gold–cobalt nanoalloy. *Chemical Research in Toxicology, 25*(5), 1086–1098. doi:10.1021/tx300053h PMID:22486372

Giwa, A., Nkeonye, P. O., Bello, K. A., & Kolawole, K. A. (2012). Photocatalytic decolourization and degradation of C. I. basic blue 41 using TiO_2 nanoparticles. *J Environ Protect, 3*(9), 1063–1069. doi:10.4236/jep.2012.39124

Gluhoi, A. C., Lin, S. D., & Nieuwenhuys, B. E. (2004). The beneficial effect of the addition of base metal oxides to gold catalysts on reactions relevant to air pollution abatement. *Catalysis Today, 90*(3), 175–181. doi:10.1016/j.cattod.2004.04.025

Gogos, A., Knauer, K., & Bucheli, T. (2012). Nanomaterials in plant protection and fertilization: Current state, foreseen applications, and research priorities. *Journal of Agricultural and Food Chemistry, 60*(39), 9781–9792. doi:10.1021/jf302154y PMID:22963545

Gomes, T., Pereira, C. G., Cardoso, C., Pinheiro, J. P., Cancio, I., & Bebianno, M. J. (2012). Accumulation and toxicity of copper oxide nanoparticles in the digestive gland of *Mytilus galloprovincialis*. *Aquatic Toxicology, 118 – 119*, 72– 79.

Gomes, T., Pinheiro, J. P., Cancio, I., Pereira, C. G., Cardoso, C., & Bebianno, M. J. (2011). Effects of copper nanoparticles exposure in the mussel *Mytilus galloprovincialis*. *Environmental Science & Technology, 45*(21), 9356–9362. doi:10.1021/es200955s PMID:21950553

Gong, N., Shao, K., Feng, W., Lin, Z., Liang, C., & Sun, Y. (2011). Biotoxicity of nickel oxide nanoparticles and bioremediation by microalgae Chlorella vulgaris. *Chemosphere, 83*(4), 510–516. doi:10.1016/j.chemosphere.2010.12.059 PMID:21216429

González-Díaz, J. B., Sepúlveda, B., García-Martín, A., & Armelles, G. (2010). Cobalt dependence of the magneto-optical response in magnetoplasmonic nanodisks. *Applied Physics Letters, 97*(4), 043114. doi:10.1063/1.3474617

González-Flecha, B. (2004). Oxidant mechanisms in response to ambient air particles. *Molecular Aspects of Medicine, 25*(1), 169–182. doi:10.1016/j.mam.2004.02.017 PMID:15051325

Goodman, C. M., McCusker, C. D., Yilmaz, T., & Rotello, V. M. (2004). Toxicity of gold nanoparticles functionalized with cationic and anionic side chains. *Bioconjugate Chemistry*, *15*(4), 897–900. doi:10.1021/bc049951i PMID:15264879

Gopal, M., Kumar, R., & Goswami, A. (2012). Nano-pesticides - A recent approach for pest control. *The Journal of Plant Protection Sciences*, *4*(2), 1–7.

Gopal, R., Kaur, S., Ma, Z., Chan, C., Ramakrishna, S., & Matsuura, T. (2006). Electrospun nanofibrous filtration membrane. *Journal of Membrane Science*, *281*(1–2), 581–586. doi:10.1016/j.memsci.2006.04.026

Gottesman, R., Shukla, S., Perkas, N., Solovyov, L. A., Nitzan, Y., & Gedanken, A. (2010). Sonochemical Coating of Paper by Microbiocidal Silver Nanoparticles. *Langmuir*, *27*(2), 720–726. doi:10.1021/la103401z PMID:21155556

Gottschalk, F., & Nowack, B. (2011). The release of engineered nanomaterials to the environment. *Journal of environmental monitoring. JEM*, *13*(5), 1145–1155. PMID:21387066

Granot, E., Patolsky, F., & Willner, I. (2004). Electrochemical assembly of a CdS semiconductor nanoparticle monolayer on surfaces: Structural properties and photoelectrochemical applications. *The Journal of Physical Chemistry B*, *108*(19), 5875–5881. doi:10.1021/jp038004o

Gray, E. P., Bruton, T. A., Higgins, C. P., Halden, R. U., Westerhoff, P., & Ranville, J. F. (2012). Analysis of gold nanoparticle mixtures: A comparison of hydrodynamic chromatography (HDC) and asymmetrical flow field-flow fractionation (AF4) coupled to ICP-MS. *Journal of Analytical Atomic Spectrometry*, *27*(9), 1532–1539. doi:10.1039/c2ja30069a

Greengard, S. (2015). *The Internet of Things*. Cambridge, MA: MIT Press.

Gregory, T. R. (2001). The bigger the C-value, the larger the cell: Genome size and red blood cell size in vertebrates. *Blood Cells, Molecules & Diseases*, *27*(5), 830–843. doi:10.1006/bcmd.2001.0457 PMID:11783946

Griffin, L. P. (2012). *Anaerobic digestion of organic wastes: The impact of operating conditions on hydrolysis efficiency and microbial community composition.* (Doctoral dissertation). Colorado State University, Fort Collins, CO.

Griffitt, R. J., Hyndman, K., Denslow, N. D., & Barber, D. S. (2009). Comparison of molecular and histological changes in zebrafish gills exposed to metallic nanoparticles. *Toxicological Sciences : An Official Journal of the Society of Toxicology, 107*(2), 404–15.

Griffitt, R. J., Luo, J., Gao, J., Bonzongo, J.-C., & Barber, D. S. (2008). Effects of particle composition and species on toxicity of metallic nanomaterials in aquatic organisms. *Environmental Toxicology and Chemistry/SETAC, 27*(9), 1972–8.

Griffitt, R. J., Hyndman, K., Denslow, N. D., & Barber, D. S. (2009). Comparison of molecular and histological changes in zebrafish gills exposed to metallic nanoparticles. *Toxicological Sciences*, *107*(2), 404–415. doi:10.1093/toxsci/kfn256 PMID:19073994

Griffitt, R. J., Luo, J., Gao, J., Bonzongo, J.-C., & Barber, D. S. (2008). Effects of particle composition and species on toxicity of metallic nanomaterials in aquatic organisms. *Environmental Toxicology and Chemistry*, *27*(9), 1972–1978. doi:10.1897/08-002.1 PMID:18690762

Griffitt, R. J., Weil, R., Hyndman, K. A., Denslow, N. D., Powers, K., Taylor, D., & Barber, D. S. (2007). Exposure to Copper Nanoparticles Causes Gill Injury and Acute Lethality in Zebrafish (Danio rerio). *Environmental Science & Technology*, *41*(23), 8178–8186. doi:10.1021/es071235e PMID:18186356

Grillo, R., Rosa, A. H., & Fraceto, L. F. (2013). Poly(ε-caprolactone) nanocapsules carrying the herbicide atrazine: Effect of chitosan-coating agent on physico-chemical stability and herbicide release profile. *International Journal of Environmental Science and Technology*, *11*(6), 1691–1700. doi:10.1007/s13762-013-0358-1

Grubbström, R. W. (1998a) The Linköping approach to Material Requirements Planning and Input-Output analysis, In L. Bogataj, R. W. Grubbström, (Eds.), Input-output analysis and Laplace transforms in Material Requirements Planning: Storlien (pp. 45-60). Portorož: University of Ljubljana, FPP.

Grubbström, R. W., & Jiang, Y. (1989). A survey and analysis of the application of the Laplace transform to present value problems. *Revista mat. scienze economiche sociale, 12*(1), 43–62.

Grubbström, R. W., & Lundquist, J. (1989). *Master scheduling, Input-Output analysis and production functions*. RR-121, Dept. of Production Economics, LiTH.

Grubbström, R.W. (1994). *Stochastic relationship of a multi-period inventory process with planned production using transform methodology*. RR-126. Dept. of Production Economics. LiTH.

Grubbström, R. W. (1967). On the application of the Laplace transform to certain economic problems. *Management Science, 13*(7), 558–567. doi:10.1287/mnsc.13.7.558

Grubbström, R. W. (1969). *Produktionsekonomiska system- i statisk och dinamisk belysning [Production-economics systems- In static and dynamic Illumination]*. Stockholm, Sweden: Department of Industrial Economics and Management, Royal Institute of Technology.

Grubbström, R. W. (1991). The z-transform of tk. *Math. Scientist, 16*, 118–129.

Grubbström, R. W. (1996). Material requirements planning and manufacturing resource planning. In M. Warner (Ed.), *International Encyclopedia of Business and Management*. London: Routledge.

Grubbström, R. W. (1996a). Stochastic properties of a production - inventory process with planned production using transform methodology. *International Journal of Production Economics, 45*(1-3), 407–419. doi:10.1016/0925-5273(96)00009-6

Grubbström, R. W. (1998). A net present value approach to safety stocks in planned production. *International Journal of Production Economics, 56/57*, 213–229. doi:10.1016/S0925-5273(97)00094-7

Grubbström, R. W. (1999). A net present value approach to safety stocks in a multi-level MRP system. *International Journal of Production Economics, 59*(1/3), 361–375. doi:10.1016/S0925-5273(98)00016-4

Grubbström, R. W. (2012). The time-averaged L4L solution - a condition for long-run stability applying MRP theory. *International Journal of Production Research, 50*(21), 6099–6110. doi:10.1080/00207543.2011.653457

Grubbström, R. W., Bogataj, M., & Bogataj, L. (2007). *A compact representation of distribution and reverse logistics in the value chain, (Mathematical economics, operational reseach and logistics, no. 5)*. Ljubljana, Slovenia: Faculty of Economics, KMOR.

Grubbström, R. W., Bogataj, M., & Bogataj, L. (2010). Optimal lotsizing within MRP theory. *Annual Reviews in Control, 34*(1), 89–100. doi:10.1016/j.arcontrol.2010.02.004

Grubbström, R. W., & Huynh, T. T. T. (2006). Multi-level, multi-stage capacity-constrained production-inventory systems in discrete time with non-zero lead times using MRP theory. *International Journal of Production Economics, 101*(1), 53–62. doi:10.1016/j.ijpe.2005.05.006

Grubbström, R. W., & Huynh, T. T. T. (2006b). Analysis of standard ordering policies within the framework of MRP theory. *International Journal of Production Research, 44*(18/19), 3759–3773. doi:10.1080/00207540600584839

Grubbström, R. W., & Lundquist, J. (1977). The Axsäter integrated production-inventory system interpreted in terms of the theory of relatively closed systems. *J. Cybernetics, 7*(1-2), 49–67. doi:10.1080/01969727708927548

Grubbström, R. W., & Molinder, A. (1994). Further theoretical considerations on the relationship between MRP, input-outpt analysis and multi-echelon inventory systems. *International Journal of Production Economics*, *35*(1-3), 299–311. doi:10.1016/0925-5273(94)90096-5

Grubbström, R. W., & Molinder, A. (1996). Safety production plans in MRP-systems using transform methodology. *International Journal of Production Economics*, *46-47*, 297–309. doi:10.1016/0925-5273(95)00158-1

Grubbström, R. W., & Ovrin, P. (1992). Intertemporal generalization of the relationship between material requirements planning and input-output analysis. *International Journal of Production Economics*, *26*(1-3), 311–318. doi:10.1016/0925-5273(92)90081-H

Grubbström, R. W., & Tang, O. (1999). Further developments on safety stocks in an MRP system applying Laplace transforms and input-output analysis. *International Journal of Production Economics*, *60/61*, 381–387. doi:10.1016/S0925-5273(98)00141-8

Grubbström, R. W., & Tang, O. (2000). Modelling rescheduling activities in a multi-period production-inventory system. *International Journal of Production Economics*, *68*(2), 123–135. doi:10.1016/S0925-5273(00)00050-5

Grubbström, R. W., & Tang, O. (2006). Optimal production opportunities in a remanufacturing system. *International Journal of Production Research*, *44*(18/19), 3953–3966. doi:10.1080/00207540600806406

Grubbström, R. W., & Tang, O. (2006a). The moments and central moments of a compound distribution. *European Journal of Operational Research*, *170*(1), 106–119. doi:10.1016/j.ejor.2004.06.012

Grubbström, R. W., & Tang, O. (2012). The space of solution alternatives in the optimal lotsizing problem for general assembly systems applying MRP theory. *International Journal of Production Economics*, *140*(2), 765–777. doi:10.1016/j.ijpe.2011.01.012

Grubbström, R. W., & Wang, Z. (2003). A stochastic model of multi-level/multi-stage capacity-constrained production-inventory systems. *International Journal of Production Economics*, *81/82*, 483–494. doi:10.1016/S0925-5273(02)00358-4

Gruère, G. P. (2012). Implications of nanotechnology growth in food and agriculture in OECD countries. *Food Policy*, *37*(2), 191–198. doi:10.1016/j.foodpol.2012.01.001

Grun, M., Unger, K. K., Matsumoto, A., & Tstutsumi, K. (1999). Novel pathways for the preparation of mesoporous MCM-41 materials: Control of porosity and morphology. *Microporous and Mesoporous Materials*, *27*(2-3), 207–216. doi:10.1016/S1387-1811(98)00255-8

Gubbins, E. J., Batty, L. C., & Lead, J. R. (2011). Phytotoxicity of silver nanoparticles to Lemna minor L. *Environmental Pollution*, *159*(6), 1551–1559. doi:10.1016/j.envpol.2011.03.002 PMID:21450381

Guhr, K. I., Greaves, M. D., & Rotello, M. (1994). Reversible Covalent Attachment of C_{60} to a Polymer Support. *Journal of the American Chemical Society*, *116*(13), 5997–5998. doi:10.1021/ja00092a072

Guizard, C., Ayral, A., & Julbe, A. (2002). Potentiality of organic solvents filtration with ceramic membranes. A comparison with polymer membranes. *Desalination*, *147*(1), 275–280. doi:10.1016/S0011-9164(02)00552-0

Guldi, D. M., & Asmus, K. D. (1997). Photophysical properties of mono- and multiply-functionalized fullerene derivatives. *The Journal of Physical Chemistry A*, *101*(8), 1472–1481. doi:10.1021/jp9633557

Guldi, D. M., & Prato, M. (2000). Excited-State Properties of C_{60} Fullerene Derivatives. *Accounts of Chemical Research*, *33*(10), 695–703. doi:10.1021/ar990144m

Compilation of References

Gulrajani, M. L., Gupta, D., Periyasamy, S., & Muthu, S. G. (2008). Preparation and Application of Silver Nanoparticles on Silk for Imparting Antimicrobial Properties. *Journal of Applied Polymer Science*, *108*(1), 614–623. doi:10.1002/app.27584

Gunawan, P., Guan, C., Song, X., Zhang, Q., Leong, S. S. J., Tang, C., & Xu, R. et al. (2011). Hollow Fiber Membrane Decorated with Ag/MWNTs: Toward Effective Water Disinfection and Biofouling Control. *ACS Nano*, *5*(12), 10033–10040. doi:10.1021/nn2038725 PMID:22077241

Guo, D., Bi, H., Liu, B., Wu, Q., Wang, D., & Cui, Y. (2013). Reactive oxygen species-induced cytotoxic effects of zinc oxide nanoparticles in rat retinal ganglion cells. *Toxicology In Vitro*, *27*(2), 731–738. doi:10.1016/j.tiv.2012.12.001 PMID:23232460

Guo, J. (2004). Synchrotron radiation, soft X-ray spectroscopy and nanomaterials. *International Journal of Nanotechnology*, *1*(1), 193–225. doi:10.1504/IJNT.2004.003729

Guo, J. Z., Cui, H., Zhou, W., & Wang, W. (2008). Ag nanoparticle-catalyzed chemiluminescent reaction between luminol and hydrogen peroxide. *Journal of Photochemistry and Photobiology A Chemistry*, *193*(2), 89–96. doi:10.1016/j.jphotochem.2007.04.034

Guo, L., Yuan, W., Lu, Z., & Li, C. M. (2013). Polymer/nanosilver composite coatings for antibacterial applications. *Colloids and Surfaces. A, Physicochemical and Engineering Aspects*, *439*, 69–83. doi:10.1016/j.colsurfa.2012.12.029

Guo, R., Li, Y., Lan, J., Jiang, S., Liu, T., & Yan, W. (2013). Microwave-Assisted Synthesis of Silver Nanoparticles on Cotton Fabric Modified with 3-Aminopropyltrimethoxysilane. *Journal of Applied Polymer Science*. doi:10.1002/app.39636

Guo, S., & Wang, E. (2011). Noble metal nanomaterials: Controllable synthesis and application in fuel cells and analytical sensors. *Nano Today*, *6*(3), 240–264. doi:10.1016/j.nantod.2011.04.007

Guo, X., Zhao, Y., Qiu, Y., & Shi, X. (2015). Zero-valent iron nanoparticle-supported composite materials for environmental remediation applications. *Current Nanoscience*, *11*(6), 748–759. doi:10.2174/1573413711666150430223749

Gupta, B. K., Bhushan, B., Capp, C., & Coe, J. V. (1994). Materials characterization and effect of purity and ion implantation on the friction wear of sublimed fullerene films. *Journal of Materials Research*, *9*(11), 2823–2838. doi:10.1557/JMR.1994.2823

Gurr, J.-R., Wang, A. S. S., Chen, C.-H., & Jan, K.-Y. (2005). Ultrafine titanium dioxide particles in the absence of photoactivation can induce oxidative damage to human bronchial epithelial cells. *Toxicology*, *213*(1–2), 66–73. doi:10.1016/j.tox.2005.05.007 PMID:15970370

Gutierrez, M., Henglein, A., & Dohrmann, J. K. (1987). Hydrogen atom reactions in the sonolysis of aqueous solutions. *Journal of Physical Chemistry*, *91*(27), 6687–6690. doi:10.1021/j100311a026

Guzman, M., Dille, J., & Godet, S. (2012). Synthesis and antibacterial activity of silver nanoparticles against gram-positive and gram-negative bacteria. *Nanomedicine; Nanotechnology, Biology, and Medicine*, *8*(1), 37–45. doi:10.1016/j.nano.2011.05.007 PMID:21703988

Gwoździński, K., Roche, H., & Pérès, G. (1992). The comparison of the effects of heavy metal ions on antioxidant enzyme activities in human and fish *Dicentrarchus labrax* erythrocytes. *Comparative Biochemistry and Physiology*, *102*, 57–60. PMID:1358529

Gwynn, M. N., Portnoy, A., Rittenhouse, S. F., & Payne, D. J. (2010). Challenges of antibacterial discovery revisited. *Annals of the New York Academy of Sciences*, *1213*(1), 5–19. doi:10.1111/j.1749-6632.2010.05828.x PMID:21058956

Haag, W. R., & Hoigne, J. (1986). Singlet Oxygen in Surface Waters. 3. Photochemical Formation and Steady-State Concentrations in Various Types of Waters. *Environmental Science & Technology*, *20*(4), 341–348. doi:10.1021/es00146a005

Haag, W. R., Hoigne, J., Gassman, E., & Braun, A. M. (1984). Singlet oxygen in surface waters. 2. Quantum yields of its production by some natural humic materials as a function of wavelength. *Chemosphere*, *13*(5-6), 641–650. doi:10.1016/0045-6535(84)90200-5

Hackenberg, S., Scherzed, A., Kessler, M., Hummel, S., Technau, A., & Froelich, K. et al.. (2011). Silver nanoparticles: Evaluation of DNA damage, toxicity and functional impairment in human mesenchymal stem cells. *Toxicology Letters*, *201*(1), 27–33. doi:10.1016/j.toxlet.2010.12.001 PMID:21145381

Hackett, C., Durbin, T. D., Welch, W., & Pence, J. (2004). *Evaluation of conversion technology processes and products*. Retrieved April 2015, from http://energy.ucdavis.edu/files/05-06-2013-2004-evaluation-of-conversion-technology-processes-products.pdf

Haddon, R. C., Hebard, A. F., Rosseinsky, M. J., Murphy, D. W., Duclos, S. J., Lyons, K. B., & Thiel, F. A. et al. (1991). Conducting films of C_{60} and C_{70} by alkali-metal doping. *Nature*, *350*(6316), 320–322. doi:10.1038/350320a0

Hahm, J. I., & Lieber, C. M. (2004). Direct ultrasensitive electrical detection of DNA and DNA sequence variations using nanowire nanosensors. *Nano Letters*, *4*(1), 51–54. doi:10.1021/nl034853b

Haldar, K. K., Sinha, G., Lahtinen, J., & Patra, A. (2012). Hybrid colloidal Au-CdSe pentapod heterostructures synthesis and their photocatalytic properties. *ACS Applied Materials & Interfaces*, *4*(11), 6266–6272. doi:10.1021/am301859b PMID:23113704

Hall, H. K. Jr. (1957). Correlation of the Base Strengths of Amines. *Journal of the American Chemical Society*, *79*(20), 5441–5444. doi:10.1021/ja01577a030

Hall, J., Dravid, V. P., & Aslam, M. (2005). Templated Fabrication of Co@Au Core-Shell Nanorods Structure. *Nanoscape*, *2*(1), 67–71.

Hamano, T., Okuda, K., Mashino, T., Hirobe, M., Arakane, K., Ryu, A., & Nagano, T. et al. (1997). Singlet oxygen production from fullerene derivatives: Effect of sequential functionalization of the fullerene core. *Chemical Communications*, (1): 21–22. doi:10.1039/a606335g

Hamoda, M. F., Qdais, H. A., & Newham, J. (1998). Evaluation of municipal solid waste composting kinetics. *Resources, Conservation and Recycling*, *23*(4), 209–223. doi:10.1016/S0921-3449(98)00021-4

Han, B. (2015). *Degradation of estradiol in water and soil using a new class of stabilized manganese dioxide nanoparticles and hydrodechlorination of triclosan using supported palladium catalysts*. (Unpublished doctoral dissertation). Auburn University, Auburn, AL.

Han, B., Zhang, M., Zhao, D. Y., & Feng, Y. C. (2015). Degradation of aqueous and soil-sorbed estradiol using a new class of stabilized manganese oxide nanoparticles. *Water Research*, *70*, 288–299. PMID:25543239

Han, D.-W., Lee, M. S., Lee, M. H., Uzawa, M., & Park, J.-C. (2005). The use of silver-coated ceramic beads for sterilization of Sphingomonas sp. in drinking mineral water. *World Journal of Microbiology & Biotechnology*, *21*(6-7), 921–924. doi:10.1007/s11274-004-6721-0

Handford, C. E., Dean, M., Henchion, M., Spence, M., Elliott, C. T., & Campbell, K. (2014). Implications of nanotechnology for the agri-food industry: Opportunities, benefits and risks. *Trends in Food Science & Technology*, *40*(2), 226–241. doi:10.1016/j.tifs.2014.09.007

Handy, R. D., Kammer, F. V. D., Lead, J. R., Hassellóv, M., Owen, R., & Crane, M. (2008a). The ecotoxicology and chemistry of manufactured nanoparticles. *Ecotoxicology (London, England)*, *17*(4), 287–314. doi:10.1007/s10646-008-0199-8 PMID:18351458

Compilation of References

Handy, R. D., Owen, R., & Valsami-Jones, E. (2008). The ecotoxicology of nanoparticles and nanomaterials: Current status, knowledge gaps, challenges, and future needs. *Ecotoxicology (London, England), 17*(5), 315–325. doi:10.1007/s10646-008-0206-0 PMID:18408994

Hardman, R. (2006). A toxicologic review of quantum dots: Toxicity depends on physic chemical and environmental factors. *Environmental Health Perspectives, 114*(2), 165–172. doi:10.1289/ehp.8284 PMID:16451849

Hare, J. P., Kroto, H. W., & Taylor, R. (1991). Preparation and UV/Visible Spectra of Fullerenes C_{60} and C_{70}. *Chemical Physics Letters, 177*(4-5), 394-398.

Hariharan, C. (2006). Photocatalytic degradation of organic contaminants in water by ZnO nanoparticles: Revisited. *Appl Catalysis A, 304*, 55–61. doi:10.1016/j.apcata.2006.02.020

Harikrishnan, R., Balasundaram, C., & Heo, M. S. (2011). Impact of plant products on innate and adaptive immune system of cultured finfish and shellfish. *Aquaculture (Amsterdam, Netherlands), 317*(1-4), 1–15. doi:10.1016/j.aquaculture.2011.03.039

Harja, M., Bucur, D., Cimpeanu, S. M., Ciocinta, R. C., & Gurita, A. A. (2012). Conversion of ash on zeolites for soil application. *Journal of Food Agriculture and Environment, 10*(2), 1056–1059.

Har-Lavan, R., Ron, I., Thieblemont, F., & Cahen, D. (2008, April). Hybrid photovoltaic junctions: metal/molecular organic insulator/semiconductor MOIS solar cells. In *Photonics Europe* (pp. 70020M–70020M). International Society for Optics and Photonics. doi:10.1117/12.768652

Harris, I., Wang, Y. L., & Wang, H. Y. (2015). ICT in multimodal transport and technological trends: Unleashing potential for the future. *International Journal of Production Economics, 159*, 88–103. doi:10.1016/j.ijpe.2014.09.005

Hartmann, N. B., Von der Kammer, F., Hofmann, T., Baalousha, M., Ottofuelling, S., & Baun, A. (2010). Algal testing of titanium dioxide nanoparticles—Testing considerations, inhibitory effects and modification of cadmium bioavailability. *Toxicology, 269*(2-3), 190–197. doi:10.1016/j.tox.2009.08.008 PMID:19686796

Hasaneen, M. N. A., & Abdel-Aziz, H. M. M. (2014). Preparation of chitosan nanoparticles for loading with NPK fertilizer. *African Journal of Biotechnology, 13*(31), 3158–3164. doi:10.5897/AJB2014.13699

Hayat, K., Gondal, M. A., Khaled, M. M., & Ahmed, S. (2011). Effect of operational key parameters on photocatalytic degradation of phenol using nano nickel oxide synthesized by sol–gel method. *J Mol Catal A, 336*(1–2), 64–71. doi:10.1016/j.molcata.2010.12.011

Hebard, A. F. (1993). Buckminsterfullerene. *Annual Review of Materials Science, 23*(1), 159–191. doi:10.1146/annurev.ms.23.080193.001111

Heberer, T. (2002). Occurrence, fate, and removal of pharmaceutical residues in the aquatic environment: A review of recent research data. *Toxicology Letters, 131*(1), 5–17. doi:10.1016/S0378-4274(02)00041-3 PMID:11988354

Heckmann, L.-H., Hovgaard, M., Sutherland, D., Autrup, H., Besenbacher, F., & Scott-Fordsmand, J. (2011). Limit-test toxicity screening of selected inorganic nanoparticles to the earthworm Eisenia fetida. *Ecotoxicology (London, England), 20*(1), 226–233. doi:10.1007/s10646-010-0574-0 PMID:21120603

He, D., Zuo, C., Chen, S., Xiao, Z., & Ding, L. (2014). A highly efficient fullerene acceptor for polymer solar cells. *Physical Chemistry Chemical Physics, 16*(16), 7205–7208. doi:10.1039/c4cp00268g

Hedayati, A., Hoseini, S. M., & Hoseinifar, S. H. (2016). Response of plasma copper, ceruloplasmin, iron and ions in carp, *Cyprinus carpio* to waterborne copper ion and nanoparticle exposure. *Comparative Biochemistry and Physiology. Part C, Pharmacology, Toxicology & Endocrinology, 179*, 87–93. doi:10.1016/j.cbpc.2015.09.007 PMID:26408942

Hedberg, K., Hedberg, L., Buhl, M., Bethune, D. S., Brown, C. A., & Johnson, R. D. (1997). Molecular Structure of Free Molecules of the Fullerene C_{70} from Gas-Phase Electron Diffraction. *Journal of the American Chemical Society*, *119*(23), 5314–5320. doi:10.1021/ja970110e

He, F., Zhang, M., Qian, T. W., & Zhao, D. Y. (2009). Transport of carboxymethyl cellulose stabilized iron nanoparticles in porous media: Column experiments and modeling. *Journal of Colloid and Interface Science*, *334*(1), 96–102. doi:10.1016/j.jcis.2009.02.058 PMID:19383562

He, F., & Zhao, D. Y. (2007). Manipulating the size and dispersibility of zerovalent iron nanoparticles by use of carboxymethyl cellulose stabilizers. *Environmental Science & Technology*, *41*(17), 6216–6221. doi:10.1021/es0705543 PMID:17937305

He, F., & Zhao, D. Y. (2008). Comment on "Manipulating the size and dispersibility of zerovalent iron nanoparticles by use of carboxymethyl cellulose stabilizers" – Response. *Environmental Science & Technology*, *42*(9), 3480–3480. doi:10.1021/es8004255

He, F., Zhao, D., & Paul, C. (2010). Field assessment of carboxymethyl cellulose stabilized iron nanoparticles for in situ destruction of chlorinated solvents in source zones. *Water Research*, *44*(7), 2360–2370. doi:10.1016/j.watres.2009.12.041 PMID:20106501

Heinlaan, M., Ivask, A., Blinova, I., Dubourguier, H.-C., & Kahru, A. (2008). Toxicity of nanosized and bulk ZnO, CuO and TiO_2 to bacteria Vibrio fischeri and crustaceans Daphnia magna and Thamnocephalus platyurus. *Chemosphere*, *71*(7), 1308–1316. doi:10.1016/j.chemosphere.2007.11.047 PMID:18194809

He, J. H., Wan, Y. Q., & Yu, J. Y. (2005). Scaling law in electrospinning: Relationship between electric current and solution flow rate. *Polymer*, *46*(8), 2799–2801. doi:10.1016/j.polymer.2005.01.065

He, J., Ritalahti, K. M., Aiello, M. R., & Löffler, F. E. (2003a). Complete detoxification of vinyl chloride by an anaerobic enrichment culture and identification of the reductively dechlorinating population as a *Dehalococcoides* species. *Applied and Environmental Microbiology*, *69*(2), 996–100. doi:10.1128/AEM.69.2.996-1003.2003 PMID:12571022

He, J., Ritalahti, K. M., Yang, K., Koenigsberg, S. S., & Löffler, F. E. (2003b). Detoxification of vinyl chloride to ethene coupled to growth of an anaerobic bacterium. *Nature*, *424*(6944), 62–65. doi:10.1038/nature01717 PMID:12840758

He, M., Xiao, B., Hu, Z., Liu, S., Guo, X., & Luo, S. (2009). Syngas production from catalytic gasification of waste polyethylene: Influence of temperature on gas yield and composition. *International Journal of Hydrogen Energy*, *34*(3), 1342–1348. doi:10.1016/j.ijhydene.2008.12.023

Hendrickson, O. D., Zherdev, A. V., Gmoshinskii, I. V., & Dzantiev, B. B. (2014). Fullerenes: In vivo studies of biodistribution, toxicity, and biological action. *Nanotechnologies in Russia*, *9*(11-12), 601–617. doi:10.1134/S199507801406010X

Hendrix, K., & Vankelecom, I. F. (2013). Solvent-Resistant Nanofiltration Membranes. Encyclopedia of Membrane Science and Technology.

Hennebert, P., Avellan, A., Yan, J., & Aguerre-Chariol, O. (2013). Experimental evidence of colloids and nanoparticles presence from 25 waste leachates. *Waste Management (New York, N.Y.)*, *33*(9), 1870–1881. doi:10.1016/j.wasman.2013.04.014 PMID:23746986

Henn, K. W., & Waddill, D. W. (2006). Utilization of nanoscale zero-valent iron for source remediation - A case study. *Remediation Journal*, *16*(2), 57–77. doi:10.1002/rem.20081

Compilation of References

Henry, T. B., Menn, F.-M., Fleming, J. T., Wilgus, J., Compton, R. N., & Sayler, G. S. (2007). Attributing effects of aqueous C60 nano-aggregates to tetrahydrofuran decomposition products in larval zebrafish by assessment of gene expression. *Environmental Health Perspectives*, *115*(7), 1059–1065. doi:10.1289/ehp.9757 PMID:17637923

Heo, H. S., Park, H. J., Park, Y. K., Ryu, C., Suh, D. J., Suh, Y. W., & Kim, S. S. et al. (2010). Bio-oil production from fast pyrolysis of waste furniture sawdust in a fluidized bed. *Bioresource Technology*, *101*(1), S91–S96. doi:10.1016/j.biortech.2009.06.003 PMID:19560915

Hernández Battez, A., Viesca, J. L., González, R., Blanco, D., Asedegbega, E., & Osorio, A. (2010). Friction reduction properties of a CuO nano lubricant used as lubricant for a NiCrBSi coating. *Wear*, *268*(1-2), 325–328. doi:10.1016/j.wear.2009.08.018

Herring, N. P., AbouZeid, K., Mohamed, M. B., Pinsk, J., & El-Shall, M. S. (2011). Formation mechanisms of gold–zinc oxide hexagonal nanopyramids by heterogeneous nucleation using microwave synthesis. *Langmuir*, *27*(24), 15146–15154. doi:10.1021/la201698k PMID:21819068

Herrin, R. T., Andren, A. W., & Armstrong, D. E. (2001). Determination of Silver Speciation in Natural Waters. 1. Laboratory Tests of Chelex-100 Chelating Resin as a Competing Ligand. *Environmental Science & Technology*, *35*(10), 1953–1958. doi:10.1021/es001509x PMID:11393973

Herzig, J. P., Leclerc, D. M., & Legoff, P. (1970). Flow of suspensions through porous media - Application to deep filtration. *Industrial & Engineering Chemistry*, *62*(5), 8–35. doi:10.1021/ie50725a003

He, Y. T., Wilson, J. T., Su, C., & Wilkin, R. T. (2015). Review of abiotic degradation of chlorinated solvents by reactive iron minerals. *Ground Water Monitoring and Remediation*, *35*, 57–75.

Hillegass, J. M., Shukla, A., Lathrop, S. A., MacPherson, M. B., Fukagawa, N. K., & Mossman, B. T. (2010). Assessing nanotoxicity in cells in vitro. *Wiley Interdisciplinary Reviews: Nanomedicine and Nanobiotechnology*.

Hillie, T. (2007). *Nanocomputers and Swarm Intelligence* (p. 26). London: ISTE.

Hillie, T., & Hlophe, M. (2007). Nanotechnology and the challenge of clean water. *Nature Nanotechnology*, *2*(11), 663–664. doi:10.1038/nnano.2007.350 PMID:18654395

Hinds, B. J., Chopra, N., Rantell, T., Andrews, R., Gavalas, V., & Bachas, L. G. (2004). Aligned Multiwalled Carbon Nanotube Membranes. *Science*, *303*(5654), 62–65. doi:10.1126/science.1092048 PMID:14645855

Hino, T., Anzai, T., & Kuramoto, N. (2006). Visible-Light Induced Solvent-Free Photooxygenations of Organic Substrates by Using [C_{60}] Fullerene-Linked Silica Gels as Heterogeneous Catalysts and as Solid-Phase Reaction Fields. *Tetrahedron Letters*, *47*(9), 1429–1432. doi:10.1016/j.tetlet.2005.12.081

Hirai, H., Nakao, Y., & Toshima, N. (1978). Colloidal rhodium in poly (vinylpyrrolidone) as hydrogenation catalyst for internal olefins. *Chemistry Letters*, (5): 545–548. doi:10.1246/cl.1978.545

Hirai, H., Nakao, Y., & Toshima, N. (1979). Preparation of colloidal transition metals in polymers by reduction with alcohols or ethers. *Journal of Macromolecular Science. Chemistry*, *13*(6), 727–750. doi:10.1080/00222337908056685

Hirakawa, T., & Kamat, P. V. (2005). Charge separation and catalytic activity of Ag@ TiO2 core-shell composite clusters under UV-irradiation. *Journal of the American Chemical Society*, *127*(11), 3928–3934. doi:10.1021/ja042925a PMID:15771529

Hirsch, A., Li, Q., & Wudl, F. (1991). Globe-Trotting Hydrogens on the Surface of the Fullerene Compound $C_{60}H_6(N(CH_2CH_2)_2O)_6$. *Angewandte Chemie International Edition*, *30*(10), 1309–1310. doi:10.1002/anie.199113091

Hobbs, P. R., Sayre, K., & Gupta, R. (2008). The role of conservation agriculture in sustainable agriculture. *Philosophical Transactions of the Royal Society of London. Series B, Biological Sciences, 363*(1491), 543–555. doi:10.1098/rstb.2007.2169 PMID:17720669

Ho, C. L., Teo, S. S., Rahim, R. A., & Phang, S. M. (2010). Transcripts of *Gracilaria changii* that improve copper tolerance of *Escherichia coli*. *Asia-Pacific Journal of Molecular Biology and Biotechnology, 18*, 315–319.

Hohenblum, P., Gans, O., Moche, W., Scharf, S., & Lorbeer, G. (2004). Monitoring of selected estrogenic hormones and industrial chemicals in groundwaters and surface waters in Austria. *The Science of the Total Environment, 333*(1-3), 185–193. doi:10.1016/j.scitotenv.2004.05.009 PMID:15364528

Holt, K. B., & Bard, A. J. (2005). Interaction of Silver(I) Ions with the Respiratory Chain of Escherichia coli: An Electrochemical and Scanning Electrochemical Microscopy Study of the Antimicrobial Mechanism of Micromolar Ag+†. *Biochemistry, 44*(39), 13214–13223. doi:10.1021/bi0508542 PMID:16185089

Holtz, R. D., Lima, B. A., Souza Filho, A. G., Brocchi, M., & Alves, O. L. (2012). Nanostructured silver vanadate as a promising antibacterial additive to water-based paints. *Nanomedicine; Nanotechnology, Biology, and Medicine, 8*(6), 935–940. doi:10.1016/j.nano.2011.11.012 PMID:22197722

Holzinger, M., Vostrowsky, O., Hirsch, A., Hennrich, F., Kappes, M., Weiss, R., & Jellen, F. (2001). Sidewall Functionalization of Carbon Nanotubes. *Angewandte Chemie International Edition, 40*(21), 4002–4005. doi:10.1002/1521-3773(20011105)40:21<4002::AID-ANIE4002>3.0.CO;2-8 PMID:12404474

Hong, S., & Elimelech, M. (1997). Chemical and physical aspects of natural organic matter (NOM) fouling of nanofiltration membranes. *Journal of Membrane Science, 132*(2), 159–181. doi:10.1016/S0376-7388(97)00060-4

Hooper, H. L., Jurkschat, K., Morgan, A. J., Bailey, J., Lawlor, A. J., Spurgeon, D. J., & Svendsen, C. (2011). Comparative chronic toxicity of nanoparticulate and ionic zinc to the earthworm Eisenia veneta in a soil matrix. *Environment International, 37*(6), 1111–1117. doi:10.1016/j.envint.2011.02.019 PMID:21440301

Horvat, L., & Bogataj, L. (1996a). MRP, input-output analysis and multi-echelon inventory systems with exponentially distributed external demand. In *Proceedings of GLOCOSM Conference* (149–154). Bangalore, India.

Horvat, L., & Bogataj, L. (1996b). Game theory aspects of MRP, input-output and multi-echelon inventory systems. *9th Int Working Seminar on Production Economics, Preprints, 3*, 1–12.

Hossain, K. Z., Monreal, C. M., & Sayari, A. (2008). Adsorption of urease on PE-MCM-41 and its catalytic effect on hydrolysis of urea. *Colloid Surf. B., 62*(1), 42–50. doi:10.1016/j.colsurfb.2007.09.016 PMID:17961995

Hotze, E. M., Labille, J., Alvarez, P., & Wiesner, M. R. (2008). Mechanisms of photochemistry and reactive oxygen production by fullerene suspensions in water. *Environmental Science & Technology, 42*(11), 4175–4180. doi:10.1021/es702172w

Hotze, E. M., Phenrat, T., & Lowry, G. V. (2010). Nanoparticle aggregation: Challenges to understanding transport and reactivity in the environment. *Journal of Environmental Quality, 39*(6), 1909–1924. doi:10.2134/jeq2009.0462 PMID:21284288

Hristovski, K., Baumgardner, A., & Westerhoff, P. (2007). Selecting metal oxide nanomaterials for arsenic removal in fixed bed columns: From nanopowders to aggregated nanoparticle media. *Journal of Hazardous Materials, 147*(1–2), 265–274. doi:10.1016/j.jhazmat.2007.01.017 PMID:17254707

Hsieh, H. (1996). *Inorganic membranes for separation and reaction*. Elsevier.

Compilation of References

Hua, J., Vijver, M. G., Ahmad, F., Richardson, M. K., & Peijnenburgy, W. J. G. M. (2014). Toxicity of different-sized copper nano- and submicron particles and their shed copper ions to zebrafish embryos. *Environmental Toxicology and Chemistry, 33*(8), 1774–1782. doi:10.1002/etc.2615 PMID:24839162

Hua, M., Zhang, S., Pan, B., Zhang, W., Lv, L., & Zhang, Q. (2012). Heavy metal removal from water/wastewater by nanosized metal oxides: A review. *Journal of Hazardous Materials, 211–212*, 317–331. doi:10.1016/j.jhazmat.2011.10.016 PMID:22018872

Huang, C.-C., Aronstam, R. S., Chen, D.-R., & Huang, Y.-W. (2010). Oxidative stress, calcium homeostasis, and altered gene expression in human lung epithelial cells exposed to ZnO nanoparticles. *Toxicology In Vitro, 24*(1), 45–55. doi:10.1016/j.tiv.2009.09.007 PMID:19755143

Huang, H., Qu, X., Dong, H., Zhang, L., & Chen, H. (2013). Role of NaA zeolites in the interfacial polymerization process towards a polyamide nanocomposite reverse osmosis membrane. *RSC Advances, 3*(22), 8203–8207. doi:10.1039/c3ra40960k

Huang, H., Qu, X., Ji, X., Gao, X., Zhang, L., Chen, H., & Hou, L. (2013). Acid and multivalent ion resistance of thin film nanocomposite RO membranes loaded with silicalite-1 nanozeolites. *Journal of Materials Chemistry A, 1*(37), 11343–11349. doi:10.1039/c3ta12199b

Huang, H., & Yang, Y. (2008). Preparation of silver nanoparticles in inorganic clay suspensions. *Composites Science and Technology, 68*(14), 2948–2953. doi:10.1016/j.compscitech.2007.10.003

Huang, L., Guo, Y., Peng, Z., & Porter, A. L. (2011). Characterising a technology development at the stage of early emerging applications: Nanomaterial enhanced biosensors. *Technology Analysis and Strategic Management, 23*(5), 527–544. doi:10.1080/09537325.2011.565666

Huang, L., Manickam, S. S., & McCutcheon, J. R. (2013). Increasing strength of electrospun nanofiber membranes for water filtration using solvent vapor. *Journal of Membrane Science, 436*, 213–220. doi:10.1016/j.memsci.2012.12.037

Huang, L., & McCutcheon, J. R. (2014). Hydrophilic nylon 6,6 nanofibers supported thin film composite membranes for engineered osmosis. *Journal of Membrane Science, 457*, 162–169. doi:10.1016/j.memsci.2014.01.040

Huang, L., Peng, Z., Guo, Y., & Porter, A. L. (2010). Identifying emerging nanoparticle roles in biosensors. *Journal of Business Chemistry, 7*(1), 15–30.

Huang, L., Terakawa, M., Zhiyentayev, T., Huang, Y.-Y., Sawayama, Y., Jahnke, A., & Hamblin, M. R. et al. (2010). Innovative cationic fullerenes as broad-spectrum light-activated antimicrobials. *Nanomedicine; Nanotechnology, Biology, and Medicine, 6*(3), 442–452. doi:10.1016/j.nano.2009.10.005

Huang, L., Xuan, Y., Koide, Y., Zhiyentayev, T., Tanaka, M., & Hamblin, M. R. (2012). Type I and Type II mechanisms of antimicrobial photodynamic therapy: An in vitro study on gram-negative and gram-positive bacteria. *Lasers in Surgery and Medicine, 44*(6), 490–499. doi:10.1002/lsm.22045

Huang, X., Li, S., Wu, S., Huang, Y., Boey, F., Gan, C. L., & Zhang, H. (2012). Graphene Oxide-Templated Synthesis of Ultrathin or Tadpole-Shaped Au Nanowires with Alternating hcp and fcc Domains. *Advanced Materials, 24*(7), 979–983. doi:10.1002/adma.201104153 PMID:22252895

Huang, X., Tan, C., Yin, Z., & Zhang, H. (2014). 25th Anniversary Article: Hybrid Nanostructures Based on Two-Dimensional Nanomaterials. *Advanced Materials, 26*(14), 2185–2204. doi:10.1002/adma.201304964 PMID:24615947

Huang, X., Zhou, X., Wu, S., Wei, Y., Qi, X., Zhang, J., & Zhang, H. et al. (2010). Reduced Graphene Oxide-Templated Photochemical Synthesis and in situ Assembly of Au Nanodots to Orderly Patterned Au Nanodot Chains. *Small*, *6*(4), 513–516. doi:10.1002/smll.200902001 PMID:20077425

Huang, Y.-W., Wu, C.-, & Aronstam, R. S. (2010). Toxicity of Transition Metal Oxide Nanoparticles: Recent Insights from in vitro Studies. *Materials (Basel)*, *3*(10), 4842–4859. doi:10.3390/ma3104842

Huber, M. M., Canonica, S., Park, G. Y., & Von Gunten, U. (2003). Oxidation of pharmaceuticals during ozonation and advanced oxidation processes. *Environmental Science & Technology*, *37*(5), 1016–1024. doi:10.1021/es025896h PMID:12666935

Hu, C. W., Li, M., Cui, Y. B., Li, D. S., Chen, J., & Yang, L. Y. (2010). Toxicological effects of TiO2 and ZnO nanoparticles in soil on earthworm Eisenia fetida. *Soil Biology & Biochemistry*, *42*(4), 586–591. doi:10.1016/j.soilbio.2009.12.007

Hu, C., Peng, T., Hu, X., Nie, Y., Zhou, X., Qu, J., & He, H. (2009). Plasmon-induced photodegradation of toxic pollutants with Ag− AgI/Al2O3 under visible-light irradiation. *Journal of the American Chemical Society*, *132*(2), 857–862. doi:10.1021/ja907792d PMID:20028089

Huerta-García, E., Pérez-Arizti, J. A., Márquez-Ramírez, S. G., Delgado-Buenrostro, N. L., Chirino, Y. I., Iglesias, G. G., & López-Marure, R. (2014). Titanium dioxide nanoparticles induce strong oxidative stress and mitochondrial damage in glial cells. *Free Radical Biology & Medicine*, *73*, 84–94. doi:10.1016/j.freeradbiomed.2014.04.026 PMID:24824983

Hughes, M. D., Xu, Y. J., Jenkins, P., McMorn, P., Landon, P., Enache, D. I., & Stitt, E. H. et al. (2005). Tunable gold catalysts for selective hydrocarbon oxidation under mild conditions. *Nature*, *437*(7062), 1132–1135. doi:10.1038/nature04190 PMID:16237439

Hu, J., Wang, D., Wang, J., & Wang, J. (2012). Bioaccumulation of Fe_2O_3 (magnetic) nanoparticles in Ceriodaphnia dubia. *Environmental Pollution*, *162*, 216–222. doi:10.1016/j.envpol.2011.11.016 PMID:22243867

Hu, M., & Mi, B. (2013). Enabling graphene oxide nanosheets as water separation membranes. *Environmental Science & Technology*, *47*(8), 3715–3723. doi:10.1021/es400571g PMID:23488812

Hund-Rinke, K., & Simon, M. (2006). Ecotoxic effect of photocatalytic active nanoparticles (TiO_2) on algae and daphnids (8 pp). *Environ Sci Pollut Res*, *13*(4), 225–232. doi:10.1065/espr2006.06.311 PMID:16910119

Hunter, R. J. (2001). *Foundations of Colloid Science* (2nd ed.). Oxford University Press.

Huppmann, T., Yatsenko, S., Leonhardt, S., Krampe, E., Radovanovic, I., Bastian, M., & Wintermantel, E. (2014). Antimicrobial Polymers - the Antibacterial Effect of Photoactivated Nano Titanium Dioxide Polymer Composites. *Proceedings of Pps-29: The 29th International Conference of the Polymer - Conference Papers*, *1593*, 440-443. doi:10.1063/1.4873817

Huq, A., Xu, B., Chowdhury, M. A. R., Islam, M. S., Montilla, R., & Colwell, R. R. (1996). A Simple Filtration Method To Remove Plankton-Associated Vibrio cholerae in Raw Water Supplies in Developing Countries. *Applied and Environmental Microbiology*, *62*(7), 2508–2512. PMID:8779590

Huq, A., Yunus, M., Sohel, S. S., Bhuiya, A., Emch, M., Luby, S. P., & Colwell, R. R. et al. (2010). Simple Sari Cloth Filtration of Water Is Sustainable and Continues To Protect Villagers from Cholera in Matlab, Bangladesh. *mBio*, *1*(1), e00034-10. doi:10.1128/mBio.00034-10 PMID:20689750

Hussain, S. M., Braydich-Stolle, L. K., Schrand, A. M., Murdock, R. C., Yu, K. O., Mattie, D. M., & Terrones, M. et al. (2009). Toxicity evaluation for safe use of nanomaterials: Recent achievements and technical challenges. *Advanced Materials*, *21*(16), 1549–1559. doi:10.1002/adma.200801395

Hussain, S. M., Hess, K. L., Gearhart, J. M., Geiss, K. T., & Schlager, J. J. (2005). In vitro toxicity of nanoparticles in BRL 3A rat liver cells. *Toxicology In Vitro*, *19*(7), 975–983. doi:10.1016/j.tiv.2005.06.034 PMID:16125895

Hussain, S. M., Javorina, A. K., Schrand, A. M., Duhart, H. M., Ali, S. F., & Schlager, J. J. (2006). The interaction of manganese nanoparticles with PC-12 cells induces dopamine depletion. *Toxicological Sciences*, *92*(2), 456–463. doi:10.1093/toxsci/kfl020 PMID:16714391

Hussein, M. Z., Zainal, Z., Yahaya, A. H., & Foo, D. W. V. (2002). Controlled release of a plant growth regulator, 1-naphthaleneacetate from the lamella of Zn–Al-layered double hydroxide nanocomposite. *Journal of Controlled Release*, *82*(2-3), 417–427. doi:10.1016/S0168-3659(02)00172-4 PMID:12175754

Hutton, G., & Haller, L. (2004). *Evaluation of the costs and benefits of water and sanitation improvements at the global level*. Geneva: World Health Organization.

Hu, W., Peng, C., Luo, W., Lv, M., Li, X., Li, D., & Fan, C. et al. (2010). Graphene-Based Antibacterial Paper. *ACS Nano*, *4*(7), 4317–4323. doi:10.1021/nn101097v PMID:20593851

Hu, X., Cook, S., Wang, P., & Hwang, H.-. (2009). In vitro evaluation of cytotoxicity of engineered metal oxide nanoparticles. *The Science of the Total Environment*, *407*(8), 3070–3072. doi:10.1016/j.scitotenv.2009.01.033 PMID:19215968

Hu, X., Daun, J. K., & Scarth, R. (1994). Proportions of C18:1n−7 and C18:1n−9 fatty acids in canola seedcoat surface and internal lipids. *Journal of the American Oil Chemists' Society*, *71*(2), 221–222. doi:10.1007/BF02541560

Hu, Y. L., Qi, W., Han, F., Shao, J. Z., & Gao, J. Q. (2011). Toxicity evaluation of biodegradable chitosan nanoparticles using a zebrafish embryo model. *International Journal of Nanomedicine*, *6*, 3351–3359. PMID:22267920

Iarikov, D. D., Kargar, M., Sahari, A., Russel, L., Gause, K. T., Behkam, B., & Ducker, W. A. (2014). Antimicrobial surfaces using covalently bound polyallylamine. *Biomacromolecules*, *15*(1), 169–176. doi:10.1021/bm401440h PMID:24328284

Ibarra, L. E., Tarres, L., Bongiovanni, S., Barbero, C. A., Kogan, M. J., Rivarola, V. A., & Yslas, E. I. et al. (2015). Assessment of polyaniline nanoparticles toxicity and teratogenicity in aquatic environment using Rhinella arenarum model. *Ecotoxicology and Environmental Safety*, *114*, 84–92. doi:10.1016/j.ecoenv.2015.01.013 PMID:25617831

Ichikawa, Y., Kobayashi, M., Sasase, M., & Suemasu, T. (2008). Molecular beam epitaxy of semiconductor (BaSi2)/metal (CoSi2) hybrid structures on Si (111) substrates for photovoltaic application. *Applied Surface Science*, *254*(23), 7963–7967. doi:10.1016/j.apsusc.2008.04.011

IFC Advisory Services in Eastern Europe and Central Asia. (2012). *Municipal solid waste management: Opportunities for Russia, summary of key findings*. Retrieved July 2015, from http://www.ifc.org/wps/wcm/connect/a00336804bbed-60f8a5fef1be6561834/PublicationRussiaRREP-SolidWasteMngmt-2012-en.pdf?MOD=AJPERES

Iglesias-Silva, E., Rivas, J., Isidro, L. L., & Lopez-Quintela, M. A. (2007). Synthesis of silver-coated magnetite nanoparticles. *Journal of Non-Crystalline Solids*, *353*(8), 829–831. doi:10.1016/j.jnoncrysol.2006.12.050

Ilic, V., Šaponjić, Z., Vodnik, V., Lazović, S., Dimitrijević, S., Jovančić, P., & Radetić, M. et al. (2010). Bactericidal Efficiency of Silver Nanoparticles Deposited onto Radio Frequency Plasma Pretreated Polyester Fabrics. *Industrial & Engineering Chemistry Research*, *49*(16), 7287–7293. doi:10.1021/ie1001313

Ilic, V., Šaponjić, Z., Vodnik, V., Potkonjak, B., Jovančić, P., Nedeljković, J., & Radetić, M. (2009). The influence of silver content on antimicrobial activity and color of cotton fabrics functionalized with Ag nanoparticles. *Carbohydrate Polymers*, *78*(3), 564–569. doi:10.1016/j.carbpol.2009.05.015

Ingale, A. G., & Chaudhari, A. N. (2013). Biogenic synthesis of nanoparticles and potential applications: An eco-friendly approach. *Journal of Nanomedicine & Nanotechnology*, *4*, 165. doi:10.4172/2157-7439.1000165

Ingale, S. V., Wagh, P. B., Tripathi, A. K., Dudwadkar, A. S., Gamre, S. S., Rao, P. T., & Gupta, S. C. et al. (2011). Photo catalytic oxidation of TNT using TiO2-SiO2 nano-composite aerogel catalyst prepared using sol–gel process. *Journal of Sol-Gel Science and Technology, 58*(3), 682–688. doi:10.1007/s10971-011-2445-4

International Solid Waste Association (ISWA). (2013). *Alternative waste conversion technologies.* Retrieved April 2015, from file:///C:/Users/Duygu/Downloads/ISWA_WGER_WP_on__ATT_FINAL_2013_04_15%20(1).pdf

Iqbal, Z., Baughman, R. H., Ramakrishna, B. L., Khare, S., Murthy, N. S., Bornemann, H. J., & Morris, D. E. (1991). Superconductivity at 45-K in Rb/Tl codoped C_{60} and C_{60}/C_{70} mixtures. *Science, 254*(5033), 826–829. doi:10.1126/science.254.5033.826

Iram, M., Guo, C., Guan, Y., Ishfaq, A., & Liu, H. (2010). Adsorption and magnetic removal of neutral red dye from aqueous solution using Fe_3O_4 hollow nanospheres. *Journal of Hazardous Materials, 181*(1–3), 1039–1050. doi:10.1016/j.jhazmat.2010.05.119 PMID:20566240

Isakovic, A., Markovic, Z., Todorovic-Markovic, B., Nikolic, N., Vranjes-Djuric, S., Mirkovic, M., & Nikolic, Z. et al. (2006). Distinct cytotoxic mechanisms of pristine versus hydroxylated fullerene. *Toxicological Sciences, 91*(1), 173–183. doi:10.1093/toxsci/kfj127 PMID:16476688

Islam, N., & Miyazaki, K. (2010). An empirical analysis of nanotechnology research domains. *Technovation, 30*(4), 229–237. doi:10.1016/j.technovation.2009.10.002

Isquith, A. J., Abbott, E. A., & Walters, P. A. (1972). Surface-Bonded Antimicrobial Activity of an Organosilicon Quaternary Ammonium Chloride. *Applied Microbiology, 24*(6), 859-863. Retrieved from http://aem.asm.org/content/24/6/859.full.pdf

Ito, T., Sun, L., & Crooks, R. M. (2003). Simultaneous determination of the size and surface charge of individual nanoparticles using a carbon nanotube-based Coulter counter. *Analytical Chemistry, 75*(10), 2399–2406. doi:10.1021/ac034072v PMID:12918983

Ivask, A., Kurvet, I., Kasemets, K., Blinova, I., Aruoja, V., Suppi, S., & Kahru, A. et al. (2014). Size-dependent toxicity of silver nanoparticles to bacteria, yeast, algae, crustaceans and mammalian cells. *PLoS ONE, 9*(7), e102108. doi:10.1371/journal.pone.0102108 PMID:25048192

Jaberzadeh, A., Moaveni, P., Tohidi Moghadam, H. R., & Zahedi, H. (2013). Influence of bulk and nanoparticles titanium foliar application on some agronomic traits, seed gluten and starch contents of wheat subjected to water deficit stress. *Notulae Botanicae Horti Agrobotanici, 41*(1), 201–207.

Jadav, G. L., Aswal, V. K., & Singh, P. S. (2010). SANS study to probe nanoparticle dispersion in nanocomposite membranes of aromatic polyamide and functionalized silica nanoparticles. *Journal of Colloid and Interface Science, 351*(1), 304–314. doi:10.1016/j.jcis.2010.07.028 PMID:20701923

Jadav, G. L., & Singh, P. S. (2009). Synthesis of novel silica-polyamide nanocomposite membrane with enhanced properties. *Journal of Membrane Science, 328*(1–2), 257–267. doi:10.1016/j.memsci.2008.12.014

Jagadeesh, E., Khan, B., Chandran, P., & Khan, S. S. (2015). Toxic potential of iron oxide, CdS/Ag_2S composite, CdS and Ag_2S NPs on a fresh water alga Mougeotia sp. *Colloids and Surfaces. B, Biointerfaces, 125*, 284–290. doi:10.1016/j.colsurfb.2014.11.008 PMID:25465759

Jain, J., Arora, S., Rajwade, J. M., Omray, P., Khandelwal, S., & Paknikar, K. M. (2009). Silver nanoparticles in therapeutics: Development of an antimicrobial gel formulation for topical use. *Molecular Pharmaceutics, 6*(5), 1388–1401. doi:10.1021/mp900056g PMID:19473014

Jain, K. K. (2003). Nanodiagnostics: Application of nanotechnology in molecular diagnostics. *Expert Review of Molecular Diagnostics*, *3*(2), 153–161. doi:10.1586/14737159.3.2.153 PMID:12647993

Jain, P. K., & El-Sayed, M. A. (2007). Surface plasmon resonance sensitivity of metal nanostructures: Physical basis and universal scaling in metal nanoshells. *The Journal of Physical Chemistry C*, *111*(47), 17451–17454. doi:10.1021/jp0773177

Jain, P. K., & El-Sayed, M. A. (2007). Universal scaling of plasmon coupling in metal nanostructures: Extension from particle pairs to nanoshells. *Nano Letters*, *7*(9), 2854–2858. doi:10.1021/nl071496m PMID:17676810

Jain, P. K., Huang, X., El-Sayed, I. H., & El-Sayed, M. A. (2007). Review of some interesting surface plasmon resonance-enhanced properties of noble metal nanoparticles and their applications to biosystems. *Plasmonics*, *2*(3), 107–118. doi:10.1007/s11468-007-9031-1

Jain, P. K., Huang, X., El-Sayed, I. H., & El-Sayed, M. A. (2008). Noble metals on the nanoscale: Optical and photothermal properties and some applications in imaging, sensing, biology, and medicine. *Accounts of Chemical Research*, *41*(12), 1578–1586. doi:10.1021/ar7002804 PMID:18447366

Jain, P., & Pradeep, T. (2005). Potential of silver nanoparticle-coated polyurethane foam as an antibacterial water filter. *Biotechnology and Bioengineering*, *90*(1), 59–63. doi:10.1002/bit.20368 PMID:15723325

Jana, N. R., Gearheart, L., & Murphy, C. J. (2001). Wet chemical synthesis of high aspect ratio cylindrical gold nanorods. *The Journal of Physical Chemistry B*, *105*(19), 4065–4067. doi:10.1021/jp0107964

Jang, Y. J., Simer, C., & Ohm, T. (2006). Comparison of zinc oxide nanoparticles and its nano-crystalline particles on the photocatalytic degradation of methylene blue. *Materials Research Bulletin*, *41*(1), 67–77. doi:10.1016/j.materresbull.2005.07.038

Jarusutthirak, C., Mattaraj, S., & Jiraratananon, R. (2007). Factors affecting nanofiltration performances in natural organic matter rejection and flux decline. *Separation and Purification Technology*, *58*(1), 68–75. doi:10.1016/j.seppur.2007.07.010

Jarvie, H. P., Al-Obaidi, H., King, S. M., Bowes, M. J., Lawrence, M. J., Drake, A. F., & Dobson, P. J. et al. (2009). Fate of Silica Nanoparticles in Simulated Primary Wastewater Treatment. *Environmental Science & Technology*, *43*(22), 8622–8628. doi:10.1021/es901399q PMID:20028062

Jasuja, K., & Berry, V. (2009). Implantation and growth of dendritic gold nanostructures on graphene derivatives: Electrical property tailoring and Raman enhancement. *ACS Nano*, *3*(8), 2358–2366. doi:10.1021/nn900504v PMID:19702325

Jederman, R., Nicometo, M., Uysal, I., & Lang, W. (2014). Reducing food losses by intelligent food logistics. *Philosophical transactions. Series A, Mathematical, physical, and engineering sciences*, *372*. 10.1098/rsta.2013.0302.

Jemec, A., Drobne, D., Remskar, M., Sepcic, K., & Tisler, T. (2008). Effects of ingested nano-sized titanium dioxide on terrestrial isopods (Porcello scaber). *Environmental Toxicology and Chemistry*, *27*(9), 1904–1914. doi:10.1897/08-036.1 PMID:19086208

Jensen, A. W., & Daniels, C. (2003). Fullerene-Coated Beads as Reusable Catalysts. *The Journal of Organic Chemistry*, *68*(2), 207–210. doi:10.1021/jo025926z

Jeong, B.-H., Hoek, E. M. V., Yan, Y., Subramani, A., Huang, X., Hurwitz, G., & Jawor, A. et al. (2007). Interfacial polymerization of thin film nanocomposites: A new concept for reverse osmosis membranes. *Journal of Membrane Science*, *294*(1–2), 1–7. doi:10.1016/j.memsci.2007.02.025

Jia, G., Wang, H., Yan, L., Wang, X., Pei, R., Yan, T., & Guo, X. et al. (2005). Cytotoxicity of carbon nanomaterials: Single-wall nanotube, multi-wall nanotube, and fullerene. *Environmental Science & Technology*, *39*(5), 1378–1383. doi:10.1021/es048729l PMID:15787380

Jia, H. Y., Liu, Y., Zhang, X. J., Han, L., Du, L. B., Tian, Q., & Xu, Y. C. (2008). Potential oxidative stress of gold nanoparticles by induced-NO releasing in serum. *Journal of the American Chemical Society, 131*(1), 40–41. doi:10.1021/ja808033w PMID:19072650

Jiang, Y., Hua, Z., Zhao, Y., Liu, Q., Wang, F., & Zhang, Q. (2013). The Effect of Carbon Nanotubes on Rice Seed Germination and Root Growth. *Proceedings of the 2012 International Conference on Applied Biotechnology* (ICAB 2012).

Jiang, J., Oberdörster, G., & Biswas, P. (2008). Characterization of size, surface charge, and agglomeration state of nanoparticle dispersions for toxicological studies. *Journal of Nanoparticle Research, 11*(1), 77–89. doi:10.1007/s11051-008-9446-4

Jiang, J., Pang, S. Y., Ma, J., & Liu, H. (2012).Oxidation of phenolic endocrine disrupting chemicals by potassium permanganate in synthetic and real waters.[PubMed]. *Environmental Science & Technology, 46*(3), 1774–1781. doi:10.1021/es2035587

Jiang, L. Y., Huang, C., Chen, J. M., & Chen, X. (2009).Oxidative Transformation of 17 beta-estradiol by MnO2 in aqueous solution. *Archives of Environmental Contamination and Toxicology, 57*(2), 221–229.

Jia, X., Ma, X., Wei, D., Dong, J., & Qian, W. (2008). Direct formation of silver nanoparticles in cuttlebone-derived organic matrix for catalytic applications. *Colloids and Surfaces. A, Physicochemical and Engineering Aspects, 330*(2), 234–240. doi:10.1016/j.colsurfa.2008.08.016

Ji, J., Long, Z., & Lin, D. (2011). Toxicity of oxide nanoparticles to the green algae Chlorella sp. *Chemical Engineering Journal, 170*(2-3), 525–530. doi:10.1016/j.cej.2010.11.026

Jimenez-Hernandez, M. E., Manjon, F., Garcia-Fresnadillo, D., & Orellana, G. (2006). Solar water disinfection by singlet oxygen photogenerated with polymer-supported Ru(II) sensitizers. *Solar Energy, 80*(10), 1382–1387. doi:10.1016/j.solener.2005.04.027

Jimenez, M., Gomez, M., Bolea, E., Laborda, F., & Castillo, J. (2011). An approach to the natural and engineered nanoparticles analysis in the environment by inductively coupled plasma mass spectrometry. *International Journal of Mass Spectrometry, 307*(1), 99–104. doi:10.1016/j.ijms.2011.03.015

Jin, C.-Y., Zhu, B.-S., Wang, X.-F., & Lu, Q.-H. (2008). Cytotoxicity of titanium dioxide nanoparticles in mouse fibroblast cells. *Chemical Research in Toxicology, 21*(9), 1871–1877. doi:10.1021/tx800179f PMID:18680314

Jin, H., Jeng, E. S., Heller, D. A., Jena, P. V., Kirmse, R., Langowski, J., & Strano, M. S. (2007). Divalent Ion and Thermally Induced DNA Conformational Polymorphism on Single-walled Carbon Nanotubes. *Macromolecules, 40*(18), 6731–6739. doi:10.1021/ma070608t

Jin, R., Cao, Y. C., Hao, E., Métraux, G. S., Schatz, G. C., & Mirkin, C. A. (2003). Controlling anisotropic nanoparticle growth through plasmon excitation. *Nature, 425*(6957), 487–490. doi:10.1038/nature02020 PMID:14523440

Jin, R., Cao, Y., Mirkin, C. A., Kelly, K. L., Schatz, G. C., & Zheng, J. G. (2001). Photoinduced conversion of silver nanospheres to nanoprisms. *Science, 294*(5548), 1901–1903. doi:10.1126/science.1066541 PMID:11729310

Joner, E. J., Hartnik, T., & Amundsen, C. E. (2007). Environmental fate and ecotoxicity of engineered nanoparticles. *Norwegian Pollution Control Authority Report No. TA, 2304*, 1–64.

Jones, P. B. C. (2006). *A Nanotech Revolution in Agriculture and the Food Industry*. Retrieved from http://www.isb.vt.edu/articles/jun0605.htm

Jones, A. M., Garg, S., He, D., Pham, N., & Waite, T. D. (2011). Superoxide-mediated formation and charging of silver nanoparticles. *Environmental Science & Technology, 45*(4), 1428–1434. doi:10.1021/es103757c PMID:21265570

Jones, N., Ray, B., Ranjit, K. T., & Manna, A. C. (2008). Antibacterial activity of ZnO nanoparticle suspensions on a broad spectrum of microorganisms. *FEMS Microbiology Letters*, *279*(1), 71–76. doi:10.1111/j.1574-6968.2007.01012.x PMID:18081843

Jose Ruben, M., Jose Luis, E., Alejandra, C., Katherine, H., Juan, B. K., Jose Tapia, R., & Miguel Jose, Y. (2005). The bactericidal effect of silver nanoparticles. *Nanotechnology*, *16*(10), 2346–2353. doi:10.1088/0957-4484/16/10/059 PMID:20818017

Joseph, S., & Aluru, N. R. (2008). Why Are Carbon Nanotubes Fast Transporters of Water? *Nano Letters*, *8*(2), 452–458. doi:10.1021/nl072385q PMID:18189436

Jovanovi´c, B., & Pali´c, D. (2012). Immunotoxicology of non-functionalized engineered nanoparticles in aquatic organisms with special emphasis on fish—Review of current knowledge, gap identification, and call for further research. *Aquatic Toxicology*, *118– 119*, 141– 151.

Jung, H., Kittelson, D. B., & Zachariah, M. R. (2005). The influence of a cerium additive on ultrafine diesel particle emissions and kinetics of oxidation. *Combustion and Flame*, *142*(3), 276–288. doi:10.1016/j.combustflame.2004.11.015

Jun, P. H., Ho, K. S., Jung, K. H., & Ho, C. S. (2006). A New Composition of Nanosized Silica-Silver for Control of Various Plant Diseases. *The Plant Pathology Journal*, *22*(3), 295–302. doi:10.5423/PPJ.2006.22.3.295

Jun, X., Zhou, Z. H., & Hua, L. U. G. (2013). Effects of selected metal oxide nanoparticles on multiple biomarkers in *Carassius auratus*. *Biomedical and Environmental Sciences*, *26*(9), 742–749. PMID:24099608

Jurd, R. D. (1985). Specialization in teleost and anuran immune response: a comparative critique. In M. J. Manning & M. F. Tatner (Eds.), *Fish immunology* (pp. 9–28). London: Academic Press.

Kadam, D. M., Wang, C., Wang, S., Grewell, D., Lamsal, B., & Yu, C. (2014). Microstructure and Antimicrobial Functionality of Nano-Enhanced Protein-Based Biopolymers. *Transactions of the Asabe*, *57*(4), 1141-1150.

Kaegi, R., Voegelin, A., Sinnet, B., Zuleeg, S., Hagendorfer, H., Burkhardt, M., & Siegrist, H. (2011). Behavior of metallic silver nanoparticles in a pilot wastewater treatment plant. *Environmental Science & Technology*, *45*(9), 3902–3908. doi:10.1021/es1041892 PMID:21466186

Kah, M., Beulke, S., Tiede, K., & Hofmann, T. (2013). Nano-pesticides: State of knowledge, environmental fate and exposure modelling. *Critical Reviews in Environmental Science and Technology*, *43*(16), 1823–1867. doi:10.1080/10643389.2012.671750

Kah, M., & Hofmann, T. (2014). Nanopesticide research: Current trends and future priorities. *Environment International*, *63*, 224–235. doi:10.1016/j.envint.2013.11.015 PMID:24333990

Kah, M., Machinski, P., Koerner, P., Tiede, K., Grillo, R., Fraceto, L. F., & Hofmann, T. (2014). Analysing the fate of nanopesticides in soil and the applicability of regulatory protocols using a polymer-based nanoformulation of atrazine. *Environmental Science and Pollution Research International*, *21*(20), 11699–11707. doi:10.1007/s11356-014-2523-6 PMID:24474560

Kahru, A., Dubourguier, H. C., Blinova, I., Ivask, A., & Kasemets, K. (2008). Biotests and biosensors for ecotoxicology of metal oxide nanoparticles: A Minireview. *Sensors (Basel, Switzerland)*, *8*(8), 5153–5170. doi:10.3390/s8085153

Kalishwaralal, K., BarathManiKanth, S., Pandian, S. R. K., Deepak, V., & Gurunathan, S. (2010). Silver nanoparticles impede the biofilm formation by Pseudomonas aeruginosa and Staphylococcus epidermidis. *Colloids and Surfaces. B, Biointerfaces*, *79*(2), 340–344. doi:10.1016/j.colsurfb.2010.04.014 PMID:20493674

Kalishwaralal, K., Sheikpranbabu, S., BarathManiKanth, S., Haribalaganesh, R., Ramkumarpandian, S., & Gurunathan, S. (2011). RETRACTED ARTICLE: Gold nanoparticles inhibit vascular endothelial growth factor-induced angiogenesis and vascular permeability via Src dependent pathway in retinal endothelial cells. *Angiogenesis*, *14*(1), 29–45. doi:10.1007/s10456-010-9193-x PMID:21061058

Kallman, E. N., Vinka A. Oyanedel-Craver, A. M. A., & and James A. Smith, M. A. (2011). Ceramic Filters Impregnated with Silver Nanoparticles for Point-of-Use Water Treatment in Rural Guatemala. *Journal of Enviromental Engineering*, *137*, 407-415. doi: .000033010.1061/(ASCE)EE.1943-7870

Kampbell, D. H., & Vandegrift, S. A. (1998). Analysis of dissolved methane, ethane, and ethylene in ground water by a standard gas chromatographic technique. *Journal of Chromatographic Science*, *36*(5), 253–256. doi:10.1093/chromsci/36.5.253 PMID:9599433

Kanu, I., & Achi, O. K. (2011). Industrial effluents and their impact on water quality of receiving rivers in Nigeria. *J Appl Technol Environ Sanitat*, *1*(1), 75–86.

Karaulove, E. N., & Bagrii, E. I. (1999). Fullerenes: Functionalisation and prospects for the use of derivatives. *Russian Chemical Reviews*, *68*(11), 889–907. doi:10.1070/RC1999v068n11ABEH000499

Karel, F. B., Koparal, A. S., & Kaynak, E. (2015). Development of Silver Ion Doped Antibacterial Clays and Investigation of Their Antibacterial Activity. *Advances in Materials Science and Engineering*, *2015*, 1–6. doi:10.1155/2015/409078

Karimi, L., Yazdanshenas, M. E., Khajavi, R., Rashidi, A., & Mirjalili, M. (2014). Using graphene/TiO2 nanocomposite as a new route for preparation of electroconductive, self-cleaning, antibacterial and antifungal cotton fabric without toxicity. *Cellulose (London, England)*, *21*(5), 3813–3827. doi:10.1007/s10570-014-0385-1

Karlsson, H. L., Cronholm, P., Gustafsson, J., & Möller, L. (2008). Copper oxide nanoparticles are highly toxic: A comparison between metal oxide nanoparticles and carbon nanotubes. *Chemical Research in Toxicology*, *21*(9), 1726–1732. doi:10.1021/tx800064j PMID:18710264

Karlsson, U., Ringsberg, J. W., Johnson, E., Hoseini, M., & Ulfvarson, A. (2009). Experimental and numerical investigation of bulb impact with a ship side-shell structure. *Marine Technology*, *46*(1), 16–26.

Karn, B., Kuiken, T., & Otto, M. (2009). Nanotechnology and in situ remediation: A review of the benefits and potential risks. *Environmental Health Perspectives*, *117*(12), 1813. doi:10.1289/ehp.0900793 PMID:20049198

Karousis, N., Tagmatarchis, N., & Tasis, D. (2010). Current Progress on the Chemical Modification of Carbon Nanotubes. *Chemical Reviews*, *110*(9), 5366–5397. doi:10.1021/cr100018g PMID:20545303

Kar, S., Bindal, R. C., & Tewari, P. K. (2012). Carbon nanotube membranes for desalination and water purification: Challenges and opportunities. *Nano Today*, *7*(5), 385–389. doi:10.1016/j.nantod.2012.09.002

Kasuga, E., Kawakami, Y., Matsumoto, T., Hidaka, E., Oana, K., Ogiwara, N., & Honda, T. et al. (2011). Bactericidal activities of woven cotton and nonwoven polypropylene fabrics coated with hydroxyapatite-binding silver/titanium dioxide ceramic nanocomposite "Earth-plus". *International Journal of Nanomedicine*, *6*, 1937–1943. PMID:21931489

Kathad, U., & Gajera, H. P. (2014). Synthesis of copper nanoparticles by two different methods and size comparison. *International Journal of Pharma and Bio Sciences*, *5*(3), 533–540.

Kathiresan, K., Manivannan, S., Nabeel, M. A., & Dhivya, B. (2009). Studies on silver nanoparticles synthesized by a marine fungus, Penicillium fellutanum isolated from coastal mangrove sediment. *Colloids and Surfaces. B, Biointerfaces*, *71*(1), 133–137. doi:10.1016/j.colsurfb.2009.01.016 PMID:19269142

Kaur, S., Sundarrajan, S., Rana, D., Matsuura, T., & Ramakrishna, S. (2012). Influence of electrospun fiber size on the separation efficiency of thin film nanofiltration composite membrane. *Journal of Membrane Science, 392-393*, 101–111. doi:10.1016/j.memsci.2011.12.005

Kautsky, H. (1939). Quenching of luminescence by oxygen. *Transactions of the Faraday Society, 35*(1), 216-218.

Kavitha, C., Malarvizhi, A., Senthil Kumaran, S., & Ramesh, M. (2010). Toxicological effects of arsenate exposure on hematological, biochemical and liver transaminases activity in an Indian major carp, *Catla catla*. *Food and Chemical Toxicology, 48*(10), 2848–2854. doi:10.1016/j.fct.2010.07.017 PMID:20654677

Kawakami, H., Mikawa, M., & Nagaoka, S. (1996). Gas permeability and selectivity through asymmetric polyimide membranes. *Journal of Applied Polymer Science, 62*(7), 965–971. doi:10.1002/(SICI)1097-4628(19961114)62:7<965::AID-APP2>3.0.CO;2-Q

Kawata, K., Osawa, M., & Okabe, S. (2009). In vitro toxicity of silver nanoparticles at noncytotoxic doses to HepG2 human hepatoma cells. *Environmental Science & Technology, 43*(15), 6046–6051. doi:10.1021/es900754q PMID:19731716

Kearns, D. R. (1971). Physical and chemical properties of singlet molecular oxygen. *Chemical Reviews, 71*(4), 395–427. doi:10.1021/cr60272a004

Kedem, O., & Katchalsky, A. (1963). Permeability of composite membranes. Part 1.—Electric current, volume flow and flow of solute through membranes. *Transactions of the Faraday Society, 59*, 1918–1930. doi:10.1039/tf9635901918

Keenan, C. R., Goth-Goldstein, R., Lucas, D., & Sedlak, D. L. (2009). Oxidative stress induced by zero-valent iron nanoparticles and Fe(II) in human bronchial epithelial cells. *Environmental Science & Technology, 43*(12), 4555–4560. doi:10.1021/es9006383 PMID:19603676

Keller, A. A., Garner, K., Miller, R. J., & Lenihan, H. S. (2012). Toxicity of nano-zero valent iron to freshwater and marine organisms. *PLoS ONE, 7*(8), e43983. doi:10.1371/journal.pone.0043983 PMID:22952836

Kelly, F. M., & Johnston, J. H. (2011). Colored and Functional Silver Nanoparticle Wool Fiber Composites. *Applied Materials & Interfaces, 3*(4), 1083–1092. doi:10.1021/am101224v PMID:21381777

Kelly, K. L., Coronado, E., Zhao, L. L., & Schatz, G. C. (2003). The optical properties of metal nanoparticles: The influence of size, shape, and dielectric environment. *The Journal of Physical Chemistry B, 107*(3), 668–677. doi:10.1021/jp026731y

Kelty, S. P., Chen, C. C., & Lieber, C. M. (1991). Superconductivity at 30-K in Cesium-doped C_{60}. *Nature, 352*(6332), 223–225. doi:10.1038/352223a0

Kempener, R., & Neumann, F. (2014). *Salinity gradient energy*. IRENA.

Kerker, M. (1969). *The Scattering of Light and Other Electromagnetic Radiation*. New York: Academic.

Khabbazi, M., Harsij, M., Hedayati, S. A. A., Gholipoor, H., Gerami, M. H., & Farsani, H. G. (2015). Effect of CuO nanoparticles on some hematological indices of rainbow trout *Oncorhynchus mykiss* and their potential toxicity. *Nanomedicine (London), 2*(1), 67–73.

Khalavka, Y., & Sönnichsen, C. (2008). Growth of gold tips onto hyperbranched CdTe nanostructures. *Advanced Materials, 20*(3), 588–591. doi:10.1002/adma.200701518

Khalil-Abad, M. S., & Yazdanshenas, M. E. (2010). Superhydrophobic antibacterial cotton textiles. *Journal of Colloid and Interface Science, 351*(1), 293–298. doi:10.1016/j.jcis.2010.07.049 PMID:20709327

Khanna, P. K., Gaikwad, S., Adhyapak, P. V., Singh, N., & Marimuthu, R. (2007, October). Synthesis and characterization of copper nanoparticles. *Materials Letters*, *61*(25), 4711–4714. doi:10.1016/j.matlet.2007.03.014

Khattab, I. A., Ghaly, M. Y., Österlund, L., Ali, M. E. M., Farah, J. Y., Zaher, F. M., & Badawy, M. I. (2012). Photocatalytic degradation of azo dye Reactive Red 15 over synthesized titanium and zinc oxides photocatalysts: A comparative study. *Desalination and Water Treatment*, *48*(1-3), 120–129. doi:10.1080/19443994.2012.698803

Khlebtsov, N. G., & Dykman, L. A. (2010). Optical properties and biomedical applications of plasmonic nanoparticles. *Journal of Quantitative Spectroscopy & Radiative Transfer*, *111*(1), 1–35. doi:10.1016/j.jqsrt.2009.07.012

Khodakovskaya, M. V., De Silva, K., Biris, A. S., Dervishi, E., & Villagarcia, H. (2012). Carbon nanotubes induce growth enhancement of tobacco cells. *ACS Nano*, *6*(3), 2128–2135. doi:10.1021/nn204643g PMID:22360840

Khodakovskaya, M. V., De Silva, K., Nedosekin, D. A., Dervishi, E., Biris, A. S., Shashkov, E. V., & Zharov, V. P. et al. (2011). Complex genetic, photo thermal, and photo acoustic analysis of nanoparticle-plant interactions. *Proceedings of the National Academy of Sciences of the United States of America*, *108*(3), 1028–1033. doi:10.1073/pnas.1008856108 PMID:21189303

Khodakovskaya, M., Dervishi, E., Mahmood, M., Xu, Y., Li, Z., Watanabe, F., & Biris, A. S. (2009). Carbon nanotubes are able to penetrate plant seed coat and dramatically affect seed germination and plant growth. *ACS Nano*, *3*(10), 3221–3227. doi:10.1021/nn900887m PMID:19772305

Khot, L. R., Sankaran, S., Maja, J. M., Ehsani, R., & Schuster, E. (2012). Application of nanomaterials in agricultural production and crop protection: A review. *Crop Protection (Guildford, Surrey)*, *35*, 64–70. doi:10.1016/j.cropro.2012.01.007

Kiaune, L., & Singhasemanon, N. (2011). Pesticidal copper (I) oxide: Environmental fate and aquatic toxicity. *Reviews of Environmental Contamination and Toxicology*, *213*, 1–26. doi:10.1007/978-1-4419-9860-6_1 PMID:21541846

Kiaya, V. (2014). *Action Contre La Faim*. Retrieved August 24, 2015, from: http://www.actioncontrelafaim.org/sites/default/files/publications/fichiers/technical_paper_phl__.pdf

Kickelbick, G. (Ed.). (2007). *Hybrid materials: synthesis, characterization, and applications*. Germany: John Wiley & Sons.

Kim, K.-T., & Tanguay, R. L. (2014). The role of chorion on toxicity of silver nanoparticles in the embryonic zebrafish assay. *Environmental Health and Toxicology*, *29*, e2014021.

Kimbrough, D. E., & Suffet, I. (2002). Electrochemical removal of bromide and reduction of THM formation potential in drinking water. *Water Research*, *36*(19), 4902–4906. doi:10.1016/S0043-1354(02)00210-5 PMID:12448534

Kim, D., & Jon, S. (2012). Gold nanoparticles in image-guided cancer therapy. *Inorganica Chimica Acta*, *393*, 154–164. doi:10.1016/j.ica.2012.07.001

Kim, E., Kim, S.-H., Kim, H.-C., Lee, S., Lee, S., & Jeong, S. (2011). Growth inhibition of aquatic plant caused by silver and titanium oxide nanoparticles. *Toxicol Environ Health Sci*, *3*(1), 1–6. doi:10.1007/s13530-011-0071-8

Kim, E.-S., Hwang, G., Gamal El-Din, M., & Liu, Y. (2012). Development of nanosilver and multi-walled carbon nanotubes thin-film nanocomposite membrane for enhanced water treatment. *Journal of Membrane Science*, *394–395*(0), 37–48. doi:10.1016/j.memsci.2011.11.041

Kim, H. J., Baek, Y., Choi, K., Kim, D.-G., Kang, H., Choi, Y.-S., & Lee, J.-C. et al. (2014). The improvement of antibiofouling properties of a reverse osmosis membrane by oxidized CNTs. *RSC Advances*, *4*(62), 32802–32810. doi:10.1039/C4RA06489E

Kim, H. J., Choi, K., Baek, Y., Kim, D.-G., Shim, J., Yoon, J., & Lee, J.-C. (2014). High-Performance Reverse Osmosis CNT/Polyamide Nanocomposite Membrane by Controlled Interfacial Interactions. *ACS Applied Materials & Interfaces*, *6*(4), 2819–2829. doi:10.1021/am405398f PMID:24467487

Kim, H.-C., Choi, B. G., Noh, J., Song, K. G., Lee, S.-, & Maeng, S. K. (2014). Electrospun nanofibrous PVDF–PMMA MF membrane in laboratory and pilot-scale study treating wastewater from Seoul Zoo. *Desalination*, *346*, 107–114. doi:10.1016/j.desal.2014.05.005

Kim, J. K., Oh, B. R., Chun, Y. N., & Kim, S. W. (2006). Effects of temperature and hydraulic retention time on anaerobic digestion of food waste. *Journal of Bioscience and Bioengineering*, *102*(4), 328–332. doi:10.1263/jbb.102.328 PMID:17116580

Kim, J. S., Kuk, E., Yu, K. N., Kim, J.-H., Park, S. J., Lee, H. J., & Cho, M.-H. et al. (2007). Antimicrobial effects of silver nanoparticles. *Nanomedicine; Nanotechnology, Biology, and Medicine*, *3*(1), 95–101. doi:10.1016/j.nano.2006.12.001 PMID:17379174

Kim, J., Lee, C. W., & Choi, W. (2010). Platinized WO_3 as an Environmental Photocatalyst that Generates OH Radicals under Visible Light. *Environmental Science & Technology*, *44*(17), 6849–6854. doi:10.1021/es101981r

Kim, K. T., Klaine, S. J., Cho, J., Kim, S.-H., & Kim, S. D. (2010). Oxidative stress responses of Daphnia magna exposed to TiO_2 nanoparticles according to size fraction. *The Science of the Total Environment*, *408*(10), 2268–2272. doi:10.1016/j.scitotenv.2010.01.041 PMID:20153877

Kim, S. H., Kwak, S.-Y., Sohn, B.-H., & Park, T. H. (2003). Design of TiO 2 nanoparticle self-assembled aromatic polyamide thin-film-composite (TFC) membrane as an approach to solve biofouling problem. *Journal of Membrane Science*, *211*(1), 157–165. doi:10.1016/S0376-7388(02)00418-0

Kim, S. T., Saha, K., Kim, C., & Rotello, V. M. (2013). The role of surface functionality in determining nanoparticle cytotoxicity. *Accounts of Chemical Research*, *46*(3), 681–691. doi:10.1021/ar3000647 PMID:23294365

Kim, S. W., Kim, T., Kim, Y. S., Choi, H. S., Lim, H. J., Yang, S. J., & Park, C. R. (2012). Surface modifications for the effective dispersion of carbon nanotubes in solvents and polymers. *Carbon*, *50*(1), 3–33. doi:10.1016/j.carbon.2011.08.011

Kim, S., Choi, J. E., Choi, J., Chung, K.-H., Park, K., Yi, J., & Ryu, D.-Y. (2009). Oxidative stress-dependent toxicity of silver nanoparticles in human hepatoma cells. *Toxicology In Vitro*, *23*(6), 1076–1084. doi:10.1016/j.tiv.2009.06.001 PMID:19508889

Kim, W. B., Voitl, T., Rodriguez-Rivera, G. J., & Dumesic, J. A. (2004). Powering fuel cells with CO via aqueous polyoxometalates and gold catalysts. *Science*, *305*(5688), 1280–1283. doi:10.1126/science.1100860 PMID:15333837

King, S. M., & Jarvie, H. P. (2012). Exploring how organic matter controls structural transformations in natural aquatic nanocolloidal dispersions. *Environmental Science & Technology*, *46*(13), 6959–6967. doi:10.1021/es2034087 PMID:22260303

Kirchhoff, L. V., McClelland, K. E., Pinho, M. D. C., Araujo, J. G., De Sousa, M. A., & Guerrant, R. L. (1985). Feasibility and efficacy of in-home water chlorination in rural North-eastern Brazil. *The Journal of Hygiene*, *94*(2), 173–180. doi:10.1017/S0022172400061374 PMID:2985691

Kirschling, T. L., Gregory, K. B., Minkley, E. G. Jr, Lowry, G. V., & Tilton, R. D. (2010). Impact of nanoscale zero valent iron on geochemistry and microbial populations in trichloroethylene contaminated aquifer materials. *Environmental Science & Technology*, *44*(9), 3474–3480. doi:10.1021/es903744f PMID:20350000

Kiser, M. A., Ryu, H., Jang, H., Hristovski, K., & Westerhoff, P. (2010). Biosorption of nanoparticles to heterotrophic wastewater biomass. *Water Research*, *44*(14), 4105–4114. doi:10.1016/j.watres.2010.05.036 PMID:20547403

Kiso, Y., Sugiura, Y., Kitao, T., & Nishimura, K. (2001). Effects of hydrophobicity and molecular size on rejection of aromatic pesticides with nanofiltration membranes. *Journal of Membrane Science*, *192*(1), 1–10. doi:10.1016/S0376-7388(01)00411-2

Kittelson, D. B. (2001). *Recent measurements of nanoparticle emissions from engines*. Current Research on Diesel Exhaust Particles.

Kiwi, J., & Grätzel, M. (1979). Projection, size factors, and reaction dynamics of colloidal redox catalysts mediating light induced hydrogen evolution from water. *Journal of the American Chemical Society*, *101*(24), 7214–7217. doi:10.1021/ja00518a015

Klaine, S. J., Alvarez, P. J. J., Batley, G. E., Fernandes, T. F., Handy, R. D., Lyon, D. Y., & Lead, J. R. et al. (2008). Nanomaterials in the environment: Behavior, fate, bioavailability, and effects. *Environmental Toxicology and Chemistry*, *27*(9), 1825–1851. doi:10.1897/08-090.1 PMID:19086204

Klaper, R., Crago, J., Barr, J., Arndt, D., Setyowati, K., & Chen, J. (2009). Toxicity biomarker expression in daphinids exposed to manufactured nanoparticles: Changes in toxicity with functionalization. *Environmental Pollution*, *157*(4), 1152–1156. doi:10.1016/j.envpol.2008.11.010 PMID:19095335

Klaus, T., Joerger, R., Olsson, E., & Granqvist, C. G. (1999). Silver-based crystalline nanoparticles, microbially fabricated. *Proceedings of the National Academy of Sciences of the United States of America*, *96*(24), 13611–13614. doi:10.1073/pnas.96.24.13611 PMID:10570120

Klaysom, , Cath, , & Depuydt, , & Vankelecom, I. F. J. (2013). Forward and pressure retarded osmosis: Potential solutions for global challenges in energy and water supply. *Chemical Society Reviews*, *42*(6959). PMID:23778699

Klein, A. (2002). *Gasification: An alternative process for energy recovery and disposal of municipal solid wastes*. (Doctoral dissertation). Columbia University, New York, NY.

Kochuveedu, S. T., Jang, Y. H., & Kim, D. H. (2013). A study on the mechanism for the interaction of light with noble metal-metal oxide semiconductor nanostructures for various photophysical applications. *Chemical Society Reviews*, *42*(21), 8467–8493. doi:10.1039/c3cs60043b PMID:23925494

Köhler, J. M., Abahmane, L., Wagner, J., Albert, J., & Mayer, G. (2008). Preparation of metal nanoparticles with varied composition for catalytical applications in microreactors. *Chemical Engineering Science*, *63*(20), 5048–5055. doi:10.1016/j.ces.2007.11.038

Kohn, T., & Nelson, K. L. (2007). Sunlight-Mediated Inactivation of MS2 Coliphage via Exogenous Singlet Oxygen Produced by Sensitizers in Natural Waters. *Environmental Science & Technology*, *41*(1), 192–197. doi:10.1021/es061716i

Kolodziej, E. P., Harter, T., & Sedlak, D. L. (2004). Dairy wastewater, aquaculture, and spawning fish as sources of steroid hormones in the aquatic environment. *Environmental Science & Technology*, *38*(23), 6377–6384. doi:10.1021/es049585d PMID:15597895

Kolpin, D. W., Furlong, E. T., Meyer, M. T., Thurman, E. M., Zaugg, S. D., Barber, L. B., & Buxton, H. T. (2002). Pharmaceuticals, hormones, and other organic wastewater contaminants in US streams, 1999-2000: A national reconnaissance. *Environmental Science & Technology*, *36*(6), 1202–1211. doi:10.1021/es011055j PMID:11944670

Kong, L., Mukherjee, B., Chan, Y. F., & Zepp, R. G. (2013). Quenching and Sensitizing Fullerene Photoreactions by Natural Organic Matter. *Environmental Science & Technology*, *47*(12), 6189–6196. doi:10.1021/es304985w

Kong, X. Y., Ding, Y., & Wang, Z. L. (2004). Metal-semiconductor Zn-ZnO core-shell nanobelts and nanotubes. *The Journal of Physical Chemistry B*, *108*(2), 570–574. doi:10.1021/jp036993f

Kool, P. L., Ortiz, M. D., & van Gestel, C. A. M. (2011). Chronic toxicity of ZnO nanoparticles, non-nano ZnO and $ZnCl_2$ to Folsomia candida (Collembola) in relation to bioavailability in soil. *Environmental Pollution*, *159*(10), 2713–2719. doi:10.1016/j.envpol.2011.05.021 PMID:21724309

Koopmans, T. C. (Ed.). (1951). *Activity Analysis of Production amd Allocation*. New York, NY: Wiley.

Korobov, M. V., & Smith, A. L. (2000). Solubility of the Fullerenes. In K. M. Kadish & R. S. Ruoff (Eds.), *Fullerenes: Chemistry, Physics, and Technology* (pp. 54–60). New York: Wiley-Interscience.

Košutić, K., Furač, L., Sipos, L., & Kunst, B. (2005). Removal of arsenic and pesticides from drinking water by nanofiltration membranes. *Separation and Purification Technology*, *42*(2), 137–144. doi:10.1016/j.seppur.2004.07.003

Kovačić, D., & Bogataj, M. (2013). Reverse logistics facility location using cyclical model of extended MRP theory. *Central European Journal of Operations Research*, *21*(1), 41–57. doi:10.1007/s10100-012-0251-x

Kovačić, D., Hontoria, E., Ros Mcdonnell, L., & Bogataj, M. (2015). Location and lead-time perturbations in multi-level assembly systems of perishable goods in Spanish baby food logistics. *Central European Journal of Operations Research*, *23*(3), 607–623. doi:10.1007/s10100-014-0372-5

Kowshik, M., Ashtaputre, S., Kharrazi, S., Vogel, W., Urban, J., Kulkarni, S. K., & Paknikar, K. M. (2003). Extracellular synthesis of silver nanoparticles by a silver-tolerant yeast strain MKY3. *Nanotechnology*, *14*(1), 95–100. doi:10.1088/0957-4484/14/1/321

Koyuncu, I., Arikan, O. A., Wiesner, M. R., & Rice, C. (2008). Removal of hormones and antibiotics by nanofiltration membranes. *Journal of Membrane Science*, *309*(1), 94–101. doi:10.1016/j.memsci.2007.10.010

Kreibig, U., & Vollmer, M. (1995). *Optical properties of metal clusters*. Berlin, Germany: Springer. doi:10.1007/978-3-662-09109-8

Kreuchauff, F., & Teichert, N. (2014). *Nanotechnology as general purpose technology*. Karlsruhe, Germany: Karlsruhe Institut für Technologie.

Krishna Priya, K., Ramesh, M., Saravanan, M., & Ponpandian, N. (2015). Ecological risk assessment of silicon dioxide nanoparticles in a fresh-water fish *Labeo rohita*: Hematology, ionoregulation and gill Na^+/K^+ ATPase activity. *Ecotoxicology and Environmental Safety*, *120*, 295–302. doi:10.1016/j.ecoenv.2015.05.032 PMID:26094035

Kroto, H. W., Heath, J. R., O'Brien, S. C., Curl, R. F., & Smalley, R. E. (1985). C_{60}: Buckminsterfullerene. *Nature*, *318*(14), 162–163. doi:10.1038/318162a0

Kruk, M., Jaroniec, M., Kim, J. M., & Ryoo, R. (1999). Characterization of Highly Ordered MCM-41 Silicas Using X-ray Diffraction and Nitrogen Adsorption. *Langmuir*, *15*(16), 5279–5284. doi:10.1021/la990179v

Kumar, A., Vemula, P. K., Ajayan, P. M., & John, G. (2008). Silver-nanoparticle-embedded antimicrobial paints based on vegetable oil. *Nature Materials*, *7*(3), 236–241. doi:10.1038/nmat2099 PMID:18204453

Kumar, A., Zafaryab, M., Umar, A., Rizvi, M. M. A., Ansari, H. Z. A. F., & Ansari, S. G. (2015). Relief of oxidative stress using curcumin and glutathione functionalized ZnO nanoparticles in HEK-293 cell line. *J Biomed Nanotechnol*, *11*(11), 1913–1926. doi:10.1166/jbn.2015.2166 PMID:26554152

Kumari, M., Mukherjee, A., & Chandrasekaran, N. (2009). Genotoxicity of silver nanoparticles in Allium cepa. *The Science of the Total Environment*, *407*(19), 5243–5246. doi:10.1016/j.scitotenv.2009.06.024 PMID:19616276

Kundu, S., Peng, L., & Liang, H. (2008). A new route to obtain high-yield multiple-shaped gold nanoparticles in aqueous solution using microwave irradiation. *Inorganic Chemistry*, *47*(14), 6344–6352. doi:10.1021/ic8004135 PMID:18563880

Kuno, M. (2008). An overview of solution-based semiconductor nanowires: Synthesis and optical studies. *Physical Chemistry Chemical Physics*, *10*(5), 620–639. doi:10.1039/B708296G PMID:19791445

Kuzma, J. (2007). Moving forward responsibly: Oversight for the nanotechnology-biology interface. *Journal of Nanoparticle Research*, *9*(1), 165–182. doi:10.1007/s11051-006-9151-0

Kuzmanovic, D. A., Elashvili, I., Wick, C., O'Connell, C., & Krueger, S. (2003). Bacteriophage MS2: Molecular Weight and Spatial Distribution of the Protein and RNA Components by Small-Angle Neutron Scattering and Virus Counting. *Structure (London, England)*, *11*(11), 1339–1348. doi:10.1016/j.str.2003.09.021

Kwakman, P. H. S., & Zaat, S. A. J. (2012). Antibacterial components of honey. *IUBMB Life*, *64*(1), 48–55. doi:10.1002/iub.578 PMID:22095907

Kyriakopoulos, J., Tzirakis, M. D., Panagiotou, G. D., Alberti, M. N., Triantafyllidis, K. S., Giannakaki, S., & Lycourghiotis, A. et al. (2012). Highly active catalysts for the photooxidation of organic compounds by deposition of [60] fullerene onto the MCM-41 surface: A green approach for the synthesis of fine chemicals. *Applied Catalysis B: Environmental*, *117-118*, 36–48. doi:10.1016/j.apcatb.2011.12.024

Lacina, P., Dvorak, V., Vodickova, E., Barson, P., Kalivoda, J., & Goold, S. (2015). The Application of nano-sized zero-valent iron for in situ remediation of chlorinated ethylenes in groundwater: A field case study. *Water Environment Research*, *87*(4), 326–333. doi:10.2175/106143015X14212658613596 PMID:26462077

Lai, J. C. K., Lai, M. B., Jandhyam, S., Dukhande, V. V., Bhushan, A., & Daniels, C. K. et al.. (2008). Exposure to titanium dioxide and other metallic oxide nanoparticles induces cytotoxicity on human neural cells and fibroblasts. *International Journal of Nanomedicine*, *3*(4), 533–545. PMID:19337421

Lala, N. L., Ramaseshan, R., Bojun, L., Sundarrajan, S., Barhate, R. S., Ying-jun, L., & Ramakrishna, S. (2007). Fabrication of Nanofibers With Antimicrobial Functionality Used as Filters: Protection Against Bacterial Contaminants. *Biotechnology and Bioengineering*, *97*(6), 1357–1365. doi:10.1002/bit.21351 PMID:17274060

Lal, R. (2008). Promise and limitations of soils to minimize climate change. *Journal of Soil and Water Conservation*, *63*(4), 113A–118A. doi:10.2489/jswc.63.4.113A

Lam, C.-W., James, J. T., McCluskey, R., & Hunter, R. L. (2004). Pulmonary toxicity of single-wall carbon nanotubes in mice 7 and 90 days after intratracheal instillation. *Toxicological Sciences*, *77*(1), 126–134. doi:10.1093/toxsci/kfg243 PMID:14514958

Langenhoff, A. A. M., Brouwers-Ceiler, D. L., Engelberting, J. H. L., Quist, J. J., Wolkenfelt, J. G. P. N., Zehnder, A. J. B., & Schraa, G. (1997). Microbial reduction of manganese coupled to toluene oxidation. *FEMS Microbiology Ecology*, *22*(2), 119–127. doi:10.1111/j.1574-6941.1997.tb00363.x

Langhorst, M. A., Stanley, F. W. Jr, Cutie, S. S., Sugarman, J. H., Wilson, L. R., Hoagland, D. A., & Prud'homme, R. K. (1986). Determination of nonionic and partially hydrolyzed polyacrylamide molecular weight distributions using hydrodynamic chromatography. *Analytical Chemistry*, *58*(11), 2242–2247. doi:10.1021/ac00124a027

Lantagne, D. S., Quick, R., & Mintz, E. D. (2006). *Household Water Treatment and Safe Storage Options in Developing Coutries: A Review of Current Implementation Practices*. Academic Press.

Lantagne, D. S., & Clasen, T. F. (2001). Investigation of the Potters for Peace Colloidal Silver Impregnated Ceramic Filter Report 2: Field Investigations. *Environmental Science & Technology*, *46*(20), 11352–11360. doi:10.1021/es301842u PMID:22963031

Lantagne, D., Klarman, M., Mayer, A., Preston, K., Napotnik, J., & Jellison, K. (2010). Effect of production variables on microbiological removal in locally-produced ceramic filters for household water treatment. *International Journal of Environmental Health Research*, *20*(3), 171–187. doi:10.1080/09603120903440665 PMID:20162486

Larue, C., Laurette, J., Herlin-Boime, N., Khodja, H., Fayard, B., Flank, A.-M., & Carriere, M. et al. (2012). Accumulation, translocation and impact of TiO_2 nanoparticles in wheat (Triticum aestivum spp.): Influence of diameter and crystal phase. *The Science of the Total Environment*, *431*, 197–208. doi:10.1016/j.scitotenv.2012.04.073 PMID:22684121

Larue, C., Veronesi, G., Flank, A.-M., Surble, S., Herlin-Boime, N., & Carrière, M. (2012). Comparative uptake and impact of TiO_2 nanoparticles in wheat and rapeseed. *Journal of Toxicology and Environmental Health. Part A.*, *75*(13-15), 722–734. doi:10.1080/15287394.2012.689800 PMID:22788360

Lasagna-Reeves, C., Gonzalez-Romero, D., Barria, M. A., Olmedo, I., Clos, A., Ramanujam, V. M. S., & Soto, C. et al. (2010). Bioaccumulation and toxicity of gold nanoparticles after repeated administration in mice. *Biochemical and Biophysical Research Communications*, *393*(4), 649–655. doi:10.1016/j.bbrc.2010.02.046 PMID:20153731

Latassa, D., Enger, O., Thilgen, C., Habicher, T., Offermanns, H., & Diederich, F. (2002). Polysiloxane-Supported Fullerene Derivative as a New Heterogeneous Sensitiser for the Selective Photooxidation of Sulfides to Sulfoxides by $^1O_2$. *Journal of Materials Chemistry*, *12*(7), 1993–1995. doi:10.1039/b201141g

Latorre, B., Flores, V., Sara, A. M., & Roco, A. (1994). Dicarboximide resistant strains of *Botrytis cinerea* from table grapes in Chile: Survey and characterization. *Plant Disease*, *7*(10), 990–994. doi:10.1094/PD-78-0990

Lavanya, S., Ramesh, M., Kavitha, C., & Malarvizhi, A. (2011). Hematological, biochemical and ionoregulatory responses of Indian major carp *Catla catla* during chronic sublethal exposure to inorganic arsenic. *Chemosphere*, *82*(7), 977–985. doi:10.1016/j.chemosphere.2010.10.071 PMID:21094981

Leach, A. G., & Houk, K. N. (2002). Diels-Alder and Ene reactions of singlet oxygen, nitroso compounds and triazolinediones: Transition states and mechanisms from contemporary theory. *Chemical Communications*, (12): 1243–1255. doi:10.1039/b111251c

Lead, J. R., & Wilkinson, K. J. (2006). Aquatic colloids and nanoparticles: Current knowledge and future trends. *Environmental Chemistry*, *3*(3), 159–171. doi:10.1071/EN06025

Lead, J., Muirhead, D., & Gibson, C. (2005). Characterization of freshwater natural aquatic colloids by atomic force microscopy (AFM). *Environmental Science & Technology*, *39*(18), 6930–6936. doi:10.1021/es050386j PMID:16201613

Lee, B. I., Qi, L., & Copeland, T. (2005). Nanoparticles for materials design: Present & future. *Journal of Ceramic Processing & Research*, *6*(1), 31–40.

Lee, H. S., Im, S. J., Kim, J. H., Kim, H. J., Kim, J. P., & Min, B. R. (2008). Polyamide thin-film nanofiltration membranes containing TiO_2 nanoparticles. *Desalination*, *219*(1–3), 48–56. doi:10.1016/j.desal.2007.06.003

Lee, H. Y., Park, H. K., Lee, Y. M., Kim, K., & Park, S. B.YeonáLee. (2007). A practical procedure for producing silver nanocoated fabric and its antibacterial evaluation for biomedical applications. *Chemical Communications*, (28): 2959–2961. doi:10.1039/b703034g PMID:17622444

Lee, J., Chae, H.-R., Won, Y. J., Lee, K., Lee, C.-H., Lee, H. H., & Lee, J.- et al.. (2013). Graphene oxide nanoplatelets composite membrane with hydrophilic and antifouling properties for wastewater treatment. *Journal of Membrane Science, 448*, 223–230. doi:10.1016/j.memsci.2013.08.017

Lee, J., Fortner, J. D., Hughes, J. B., & Kim, J. H. (2007). Photochemical production of reactive oxygen species by C_{60} in the aqueous phase during UV irradiation. *Environmental Science & Technology, 41*(7), 2529–2535. doi:10.1021/es062066l

Lee, J., Hernandez, P., Lee, J., Govorov, A. O., & Kotov, N. A. (2007). Exciton–plasmon interactions in molecular spring assemblies of nanowires and wavelength-based protein detection. *Nature Materials, 6*(4), 291–295. doi:10.1038/nmat1869 PMID:17384635

Lee, J., Hong, S., Mackeyev, Y., Lee, C., Chung, E., Wilson, L. J., & Alvarez, P. J. J. et al. (2011). Photosensitized Oxidation of Emerging Organic Pollutants by Tetrakis C_{60} Aminofullerene-Derivatized Silica under Visible Light Irradiation. *Environmental Science & Technology, 45*(24), 10598–10604. doi:10.1021/es2029944

Lee, J., Kim, J., Shin, Y., Ryu, J., Eom, I., Lee, J. S., & Lee, B. et al. (2014). Serum and ultra structure responses of common carp (*Cyprinus carpio* L.) during long-term exposure to zinc oxide nanoparticles. *Ecotoxicology and Environmental Safety, 104*, 9–17. doi:10.1016/j.ecoenv.2014.01.040 PMID:24632117

Lee, J., Mackeyev, Y., Cho, M., Li, D., Kim, J.-H., Wilson, L. J., & Alvarez, P. J. J. (2009). Photochemical and Antimicrobial Properties of Novel C_{60} Derivatives in Aqueous Systems. *Environmental Science & Technology, 43*(17), 6604–6610. doi:10.1021/es901501k

Lee, J., Mackeyev, Y., Cho, M., Wilson, L. J., Kim, J.-H., & Alvarez, P. J. J. (2010). C_{60} Aminofullerene Immobilized on Silica as a Visible-Light-Activated Photocatalyst. *Environmental Science & Technology, 44*(24), 9488–9495. doi:10.1021/es1028475

Lee, J., Yamakoshi, Y., Hughes, J. B., & Kim, J. H. (2008). Mechanism of C_{60} photoreactivity in water: Fate of triplet state and radical anion and production of reactive oxygen species. *Environmental Science & Technology, 42*(9), 3459–3464. doi:10.1021/es702905g

Lee, K. J., Browning, L. M., Nallathamby, P. D., Desai, T., Cherukuri, P. K., & Xu, X.-H. N. (2012). In vivo quantitative study of sized-dependent transport and toxicity of single silver nanoparticles using zebrafish embryos. *Chemical Research in Toxicology, 25*(5), 1029–1046. doi:10.1021/tx300021u PMID:22486336

Lee, K. J., Browning, L. M., Nallathamby, P. D., Osgood, C. J., & Xu, X.-H. N. (2013). Silver nanoparticles induce developmental stage-specific embryonic phenotypes in zebrafish. *Nanoscale, 5*(23), 11625–11636. doi:10.1039/c3nr03210h PMID:24056877

Lee, K. J., Browning, L. M., Nallathamby, P. D., & Xu, X.-H. N. (2013). Study of charge-dependent transport and toxicity of peptide-functionalized silver nanoparticles using zebrafish embryos and single nanoparticle plasmonic spectroscopy. *Chemical Research in Toxicology, 26*(6), 904–917. doi:10.1021/tx400087d PMID:23621491

Lee, K. J., Nallathamby, P. D., Browning, L. M., Desai, T., Cherukuri, P. K., & Xu, X.-H. N. (2012). Single nanoparticle spectroscopy for real-time in vivo quantitative analysis of transport and toxicity of single nanoparticles in single embryos. *Analyst (London), 137*(13), 2973–2986. doi:10.1039/c2an35293a PMID:22563577

Lee, K. J., Nallathamby, P. D., Browning, L. M., Osgood, C. J., & Xu, X.-H. N. (2007). In vivo imaging of transport and biocompatibility of single silver nanoparticles in early development of zebrafish embryos. *ACS Nano, 1*(2), 133–143. doi:10.1021/nn700048y PMID:19122772

Lee, K. P., Arnot, T. C., & Mattia, D. (2011). A review of reverse osmosis membrane materials for desalination—Development to date and future potential. *Journal of Membrane Science, 370*(1-2), 1–22. doi:10.1016/j.memsci.2010.12.036

Lee, K. S., & El-Sayed, M. A. (2006). Gold and silver nanoparticles in sensing and imaging: Sensitivity of plasmon response to size, shape, and metal composition. *The Journal of Physical Chemistry B*, *110*(39), 19220–19225. doi:10.1021/jp062536y PMID:17004772

Lee, L. S., Strock, T. J., Sarmah, A. K., & Rao, P. S. C. (2003). Sorption and dissipation of testosterone, estrogens, and their primary transformation products in soils and sediment. *Environmental Science & Technology*, *37*(18), 4098–4105. doi:10.1021/es020998t PMID:14524441

Lee, S. S., Bai, H., Liu, Z., & Sun, D. D. (2013). Novel-structured electrospun TiO2/CuO composite nanofibers for high efficient photocatalytic cogeneration of clean water and energy from dye wastewater. *Water Research*, *47*(12), 4059–4073. doi:10.1016/j.watres.2012.12.044 PMID:23541306

Lee, S. S., Li, W. L., Kim, C., Cho, M. J., Lafferty, B. J., & Fortner, J. D. (2015). Surface functionalized manganese ferrite nanocrystals for enhanced uranium sorption and separation in water. *Journal of Materials Chemistry A*, *3*(43), 21930–21939. doi:10.1039/C5TA04406E

Lee, S., Bi, X., Reed, R. B., Ranville, J. F., Herckes, P., & Westerhoff, P. (2014). Nanoparticle size detection limits by single particle ICP-MS for 40 elements. *Environmental Science & Technology*, *48*(17), 10291–10300. doi:10.1021/es502422v PMID:25122540

Lee, S., & Lee, C.-H. (2007). Effect of membrane properties and pretreatment on flux and NOM rejection in surface water nanofiltration. *Separation and Purification Technology*, *56*(1), 1–8. doi:10.1016/j.seppur.2007.01.007

Lee, S.-W., Kim, S.-M., & Choi, J. (2009). Genotoxicity and ecotoxicity assays using the freshwater crustacean Daphnia magna and the larva of the aquatic midge Chironomus riparius to screen the ecological risks of nanoparticle exposure. *Environmental Toxicology and Pharmacology*, *28*(1), 86–91. doi:10.1016/j.etap.2009.03.001 PMID:21783986

Lee, W.-M., Kwak, J. I., & An, Y.-J. (2012). Effect of silver nanoparticles in crop plants Phaseolus radiatus and Sorghum bicolor: Media effect on phytotoxicity. *Chemosphere*, *86*(5), 491–499. doi:10.1016/j.chemosphere.2011.10.013 PMID:22075051

Lee, Y., Garcia, M. A., Frey Huls, N. A., & Sun, S. (2010). Synthetic tuning of the catalytic properties of Au-Fe3O4 nanoparticles. *Angewandte Chemie*, *122*(7), 1293–1296. doi:10.1002/ange.200906130 PMID:20077449

Lee, Y.-H., Cheng, F.-Y., Chiu, H.-W., Tsai, J.-C., Fang, C.-Y., Chen, C.-W., & Wang, Y.-J. (2014). Cytotoxicity, oxidative stress, apoptosis and the autophagic effects of silver nanoparticles in mouse embryonic fibroblasts. *Biomaterials*, *35*(16), 4706–4715. doi:10.1016/j.biomaterials.2014.02.021 PMID:24630838

Lee, Y., Yoon, J., & von Gunten, U. (2005). Kinetics of the oxidation of phenols and phenolic endocrine disruptors during water treatment with ferrate (Fe(VI)). *Environmental Science & Technology*, *39*(22), 8978–8984. doi:10.1021/es051198w PMID:16323803

Leggo, P. J. (2000). An investigation of plant growth in an organo–zeolitic substrate and its ecological significance. *Plant and Soil*, *219*(1/2), 135–146. doi:10.1023/A:1004744612234

Lengke, M. F., Fleet, M. E., & Southam, G. (2006). Morphology of gold nanoparticles synthesized by filamentous cyanobacteria from gold (I)-thiosulfate and gold (III)-chloride complexes. *Langmuir*, *22*(6), 2780–2787. doi:10.1021/la052652c PMID:16519482

Leontief, W. W. (1928). *Die Wirtschaft als Kreislauf*. Thesis, Friedrich-Wilhelms-Universitat Berlin. Tübingen.

Leontief, W. W. (1936). Quantitative Input and Output Relations in the Economic System of the United States. *The Review of Economics and Statistics*, *18*(3), 105–125. doi:10.2307/1927837

LeOuay, B., & Stellacci, F. (2015). Antibacterial activity of silver nanoparticles: A surface science insight. *Nano Today*, *10*(3), 339–354. doi:10.1016/j.nantod.2015.04.002

LePape, H., Solano-Serena, F., Contini, P., Devillers, C., Maftah, A., & Leprat, P. (2002). Evaluation of the anti-microbial properties of an activated carbon fibre supporting silver using a dynamic method. *Carbon*, *40*(15), 2947–2954. doi:10.1016/S0008-6223(02)00246-4

LePape, H., Solano-Serena, F., Contini, P., Devillers, C., Maftah, A., & Leprat, P. (2004). Involvement of reactive oxygen species in the bactericidal activity of activated carbon fibre supporting silver: Bactericidal activity of ACF(Ag) mediated by ROS. *Journal of Inorganic Biochemistry*, *98*(6), 1054–1060. doi:10.1016/j.jinorgbio.2004.02.025 PMID:15149815

Lespes, G., & Gigault, J. (2011). Hyphenated analytical techniques for multidimensional characterisation of submicron particles: A review. *Analytica Chimica Acta*, *692*(1), 26–41. doi:10.1016/j.aca.2011.02.052 PMID:21501709

Lewinski, N., Colvin, V., & Drezek, R. (2008). Cytotoxicity of nanoparticles. *Small*, *4*(1), 26–49. doi:10.1002/smll.200700595 PMID:18165959

Li, K.-E., Chang, Z.-Y., Shen, C.-X., Yaon, N., Siddiqui, M. H., . . . (Eds.). (2015). Toxicity of Nanomaterials to Plants. In M. H. Siddiqui, M. H. Al-Whaibi & F. Mohammad (Eds.), Nanotechnology and Plant Sciences (pp. 101-123). Springer International Publishing. doi:10.1007/978-3-319-14502-0_6

Liang, J., Deng, Z., Jiang, X., Li, F., & Li, Y. (2002). Photoluminescence of tetragonal ZrO2 nanoparticles synthesized by microwave irradiation. *Inorganic Chemistry*, *41*(14), 3602–3604. doi:10.1021/ic025532q PMID:12099861

Liang, Q. Q., & Zhao, D. Y. (2014). Immobilization of arsenate in a sandy loam soil using starch-stabilized magnetite nanoparticles. *Journal of Hazardous Materials*, *271*, 16–23. doi:10.1016/j.jhazmat.2014.01.055 PMID:24584068

Liang, Q. Q., Zhao, D. Y., Qian, T. W., Freeland, K., & Feng, Y. C. (2011). Effects of stabilizers and water chemistry on arsenate sorption by polysaccharide-stabilized magnetite nanoparticles. *Industrial & Engineering Chemistry Research*, *51*(5), 2407–2418. doi:10.1021/ie201801d

Liang, S., Liu, X. L., Yang, Y. Z., Wang, Y. L., Wang, J. H., Yang, Z. J., & Zhang, Z. et al. (2012). Symmetric and Asymmetric Au–AgCdSe Hybrid Nanorods. *Nano Letters*, *12*(10), 5281–5286. doi:10.1021/nl3025505 PMID:22947073

Liao, K.-H., Lin, Y.-S., Macosko, C. W., & Haynes, C. L. (2011). Cytotoxicity of graphene oxide and graphene in human erythrocytes and skin fibroblasts. *ACS Applied Materials & Interfaces*, *3*(7), 2607–2615. doi:10.1021/am200428v PMID:21650218

Lica, G. C., Zelakiewicz, B. S., Constantinescu, M., & Tong, Y. (2004). Charge dependence of surface plasma resonance on 2 nm octanethiol-protected Au nanoparticles: Evidence of a free-electron system. *The Journal of Physical Chemistry B*, *108*(52), 19896–19900. doi:10.1021/jp045302s

Li, F., Liang, Z., Zheng, X., Zhao, W., Wu, M., & Wang, Z. (2015). Toxicity of nano-TiO2 on algae and the site of reactive oxygen species production. *Aquatic Toxicology (Amsterdam, Netherlands)*, *158*, 1–13. doi:10.1016/j.aquatox.2014.10.014 PMID:25461740

Li, H., Zhang, J., Wang, T., Luo, W., Zhou, Q., & Jiang, G. (2008). Elemental selenium particles at nano-size (Nano-Se) are more toxic to Medaka (Oryzias latipes) as a consequence of hyper-accumulation of selenium: A comparison with sodium selenite. *Aquatic Toxicology (Amsterdam, Netherlands)*, *89*(4), 251–256. doi:10.1016/j.aquatox.2008.07.008 PMID:18768225

Li, J. J., Hartono, D., Ong, C.-N., Bay, B.-H., & Yung, L.-Y. L. (2010). Autophagy and oxidative stress associated with gold nanoparticles. *Biomaterials*, *31*(23), 5996–6003. doi:10.1016/j.biomaterials.2010.04.014 PMID:20466420

Li, J. W., & Liang, W. J. (2007). Loss of Characteristic Absorption Bands of C_{60} Conjugation Systems in the Addition with Aliphatic Amines. *Spectrochimica Acta Part A*, *67*(5), 1346–1350. doi:10.1016/j.saa.2006.10.022

Li, J., & Liu, C. Y. (2010). Ag/graphene heterostructures: Synthesis, characterization and optical properties. *European Journal of Inorganic Chemistry*, *2010*(8), 1244–1248. doi:10.1002/ejic.200901048

Li, J., Zhao, T., Chen, T., Liu, Y., Ong, C. N., & Xie, J. (2015). Engineering noble metal nanomaterials for environmental applications. *Nanoscale*, *7*(17), 7502–7519. doi:10.1039/C5NR00857C PMID:25866322

Li, L. S., Hu, J., Yang, W., & Alivisatos, A. P. (2001). Band gap variation of size-and shape-controlled colloidal CdSe quantum rods. *Nano Letters*, *1*(7), 349–351. doi:10.1021/nl015559r

Li, L., Daou, T. J., Texier, I., Kim Chi, T. T., Liem, N. Q., & Reiss, P. (2009). Highly luminescent CuInS2/ZnS core/shell nanocrystals: Cadmium-free quantum dots for in vivo imaging. *Chemistry of Materials*, *21*(12), 2422–2429. doi:10.1021/cm900103b

Li, M., Schnablegger, H., & Mann, S. (1999). Coupled synthesis and self-assembly of nanoparticles to give structures with controlled organization. *Nature*, *402*(6760), 393–395. doi:10.1038/46509

Lima, R., Seabra, A. B., & Durán, N. (2012). Silver nanoparticles: A brief review of cytotoxicity and genotoxicity of chemically and biogenically synthesized nanoparticles. *Journal of Applied Toxicology*, *32*(11), 867–879. doi:10.1002/jat.2780 PMID:22696476

Limbach, L. K., Li, Y., Grass, R. N., Brunner, T. J., Hintermann, M. A., Muller, M., & Stark, W. J. et al. (2005). Oxide nanoparticle uptake in human lung fibroblasts: Effects of particle size, agglomeration, and diffusion at low concentrations. *Environmental Science & Technology*, *39*(23), 9370–9376. doi:10.1021/es051043o PMID:16382966

Lim, J., Eggeman, A., Lanni, F., Tilton, R. D., & Majetich, S. A. (2008). Synthesis and Single-Particle Optical Detection of Low-Polydispersity Plasmonic-Superparamagnetic Nanoparticles. *Advanced Materials*, *20*(9), 1721–1726. doi:10.1002/adma.200702196

Lin, S., Zhao, Y., Nel, A. E., & Lin, S. (2013). Zebrafish: an in vivo model for nano EHS studies. Small (Weinheim an der Bergstrasse, Germany), 9(9-10), 1608–18.

Li, N., Sioutas, C., Cho, A., Schmitz, D., Misra, C., Sempf, J., & Nel, A. et al. (2003). Ultrafine particulate pollutants induce oxidative stress and mitochondrial damage. *Environmental Health Perspectives*, *111*(4), 455–460. doi:10.1289/ehp.6000 PMID:12676598

Lin, D., & Xing, B. (2007). Phytotoxicity of nanoparticles: Inhibition of seed germination and root growth. *Environmental Pollution*, *150*(2), 243–250. doi:10.1016/j.envpol.2007.01.016 PMID:17374428

Lin, D., & Xing, B. (2008). Root uptake and phytotoxicity of ZnO nanoparticles. *Environmental Science & Technology*, *42*(15), 5580–5585. doi:10.1021/es800422x PMID:18754479

Lind, M. L., Eumine Suk, D., Nguyen, T.-V., & Hoek, E. M. V. (2010). Tailoring the Structure of Thin Film Nanocomposite Membranes to Achieve Seawater RO Membrane Performance. *Environmental Science & Technology*, *44*(21), 8230–8235. doi:10.1021/es101569p PMID:20942398

Lind, M. L., Jeong, B.-H., Subramani, A., Huang, X., & Hoek, E. M. V. (2009). Effect of mobile cation on zeolite-polyamide thin film nanocomposite membranes. *Journal of Materials Research*, *24*(5), 1624–1631. doi:10.1557/jmr.2009.0189

Lin, G., Tian, M., Lu, Y.-L., Zhang, X.-J., & Zhang, L.-Q. (2006). Morphology, Antimicrobial and Mechanical Properties of Nano-TiO2/Rubber Composites Prepared by Direct Blending. *Polymer Journal*, *38*(5), 498–502. doi:10.1295/polymj.38.498

Linhua, H., Zhenyu, W., & Baoshan, X. (2009). Effect of sub acute exposure to TiO2 nanoparticles on oxidative stress and histopathological changes in Juvenile Carp (*Cyprinus carpio*). *Journal of Environmental Sciences (China)*, *21*(10), 1459–1466. doi:10.1016/S1001-0742(08)62440-7 PMID:20000003

Linic, S., Christopher, P., & Ingram, D. B. (2011). Plasmonic-metal nanostructures for efficient conversion of solar to chemical energy. *Nature Materials*, *10*(12), 911–921. doi:10.1038/nmat3151 PMID:22109608

Lin, J. M., Lin, H. Y., Cheng, C. L., & Chen, Y. F. (2006). Giant enhancement of bandgap emission of ZnO nanorods by platinum nanoparticles. *Nanotechnology*, *17*(17), 4391–4394. doi:10.1088/0957-4484/17/17/017

Lin, K. D., Liu, W. P., & Gan, J. (2009b). Reaction of tetrabromobisphenol A (TBBPA) with manganese dioxide: Kinetics, products, and pathways. *Environmental Science & Technology*, *43*(12), 4480–4486. doi:10.1021/es803622t PMID:19603665

Lin, K., Liu, W., & Gan, J. (2009a). Oxidative removal of bisphenol A by manganese dioxide: Efficacy, products, and pathways. *Environmental Science & Technology*, *43*(10), 3860–3864. doi:10.1021/es900235f PMID:19544899

Link, S., & El-Sayed, M. A. (2000). Shape and size dependence of radiative, non-radiative and photothermal properties of gold nanocrystals. *International Reviews in Physical Chemistry*, *19*(3), 409–453. doi:10.1080/01442350050034180

Link, S., & El-Sayed, M. A. (2003). Optical properties and ultrafast dynamics of metallic nanocrystals. *Annual Review of Physical Chemistry*, *54*(1), 331–366. doi:10.1146/annurev.physchem.54.011002.103759 PMID:12626731

Link, S., Wang, Z. L., & El-Sayed, M. A. (1999). Alloy formation of gold-silver nanoparticles and the dependence of the plasmon absorption on their composition. *The Journal of Physical Chemistry B*, *103*(18), 3529–3533. doi:10.1021/jp990387w

Lin, M., Lindsay, H., Weitz, D., Klein, R., Ball, R., & Meakin, P. (1990). Universal diffusion-limited colloid aggregation. *Journal of Physics Condensed Matter*, *2*(13), 3093–3113. doi:10.1088/0953-8984/2/13/019

Lin, S. H., & Wiesner, M. R. (2012). Theoretical Investigation on the Steric Interaction in Colloidal Deposition. *Langmuir*, *28*(43), 15233–15245. doi:10.1021/la302201g PMID:22978750

Lin, S., Huang, R., Cheng, Y., Liu, J., Lau, B. L. T., & Wiesner, M. R. (2013). Silver nanoparticle-alginate composite beads for point-of-use drinking water disinfection. *Water Research*, *47*(12), 3959–3965. doi:10.1016/j.watres.2012.09.005 PMID:23036278

Lin, S., Reppert, J., Hu, Q., Hudson, J. S., Reid, M. L., & Ratnikova, T. A. et al.. (2009). Uptake, translocation, and transmission of carbon nanomaterials in rice plants. *Small*, *5*(10), 1128–1132. doi:10.1002/smll.200801556 PMID:19235197

Lin, Y.-L., Chiang, P.-C., & Chang, E.-E. (2007). Removal of small trihalomethane precursors from aqueous solution by nanofiltration. *Journal of Hazardous Materials*, *146*(1), 20–29. doi:10.1016/j.jhazmat.2006.11.050 PMID:17212977

Li, Q., Mahendra, S., Lyon, D. Y., Brunet, L., Liga, M. V., Li, D., & Alvarez, P. J. J. (2008). Antimicrobial nanomaterials for water disinfection and microbial control: Potential applications and implications. *Water Research*, *42*(18), 4591–4602. doi:10.1016/j.watres.2008.08.015 PMID:18804836

Li, S.-Q., Zhu, R.-R., Zhu, H., Xue, M., Sun, X.-Y., Yao, S.-D., & Wang, S.-L. (2008). Nanotoxicity of TiO 2 nanoparticles to erythrocyte in vitro. *Food and Chemical Toxicology*, *46*(12), 3626–3631. doi:10.1016/j.fct.2008.09.012 PMID:18840495

Li, S., Wallis, L. K., Diamond, S. A., Ma, H., & Hoff, D. J. (2014). Species sensitivity and dependence on exposure conditions impacting the phototoxicity of TiO_2 nanoparticles to benthic organisms. *Environmental Toxicology and Chemistry*, *33*(7), 1563–1569. doi:10.1002/etc.2583 PMID:24846372

Compilation of References

Li, S., Wallis, L. K., Ma, H., & Diamond, S. A. (2014). Phototoxicity of TiO_2 nanoparticles to a freshwater benthic amphipod: Are benthic systems at risk? *The Science of the Total Environment*, *466-467*, 800–808. doi:10.1016/j.scitotenv.2013.07.059 PMID:23973546

Liu, H., Lee, Y.-Y., Norsten, T. B., & Chong, K. (2013). In situ formation of anti-bacterial silver nanoparticles on cotton textiles. *Journal of Industrial Textiles*. doi:10.1177/1528083713481833

Liu, F., Wen, L. X., Li, Z. Z., Yu, W., Sun, H. Y., & Chen, J. F. (2006). Porous hollow silica nanoparticles as controlled delivery system for water soluble pesticide. *Materials Research Bulletin*, *41*(12), 2268–2275. doi:10.1016/j.materresbull.2006.04.014

Liu, H. H., & Cohen, Y. (2014). Multimedia environmental distribution of engineered nanomaterials. *Environmental Science & Technology*, *48*(6), 3281–3292. doi:10.1021/es405132z PMID:24548277

Liu, H., Ma, L., Zhao, J., Liu, J., Yan, J., Ruan, J., & Hong, F. (2009). Biochemical toxicity of nano-anatase $TiO2$ particles in mice. *Biological Trace Element Research*, *129*(1-3), 170–180. doi:10.1007/s12011-008-8285-6 PMID:19066734

Liu, J., Rinzler, A. G., Dai, H., Hafner, J. H., Bradley, R. K., Boul, P. J., & Smalley, R. E. et al. (1998). Fullerene Pipes. *Science*, *280*(5367), 1253–1256. doi:10.1126/science.280.5367.1253 PMID:9596576

Liu, L., Son, M., Park, H., Celik, E., Bhattacharjee, C., & Choi, H. (2014). Efficacy of CNT-bound polyelectrolyte membrane by spray-assisted layer-by-layer (LbL) technique on water purification. *RSC Advances*, *4*(62), 32858–32865. doi:10.1039/C4RA05272B

Liu, L., Veerappan, V., Pu, Q., Cheng, C., Wang, X., Lu, L., & Guo, G. et al. (2013). High-Resolution Hydrodynamic Chromatographic Separation of Large DNA Using Narrow, Bare Open Capillaries: A Rapid and Economical Alternative Technology to Pulsed-Field Gel Electrophoresis? *Analytical Chemistry*, *86*(1), 729–736. doi:10.1021/ac403190a PMID:24274685

Liu, M., & Guyot-Sionnest, P. (2005). Mechanism of silver (I)-assisted growth of gold nanorods and bipyramids. *The Journal of Physical Chemistry B*, *109*(47), 22192–22200. doi:10.1021/jp054808n PMID:16853888

Liu, P., & Zhao, M. (2009). Silver nanoparticle supported on halloysite nanotubes catalyzed reduction of 4-nitrophenol (4-NP). *Applied Surface Science*, *255*(7), 3989–3993. doi:10.1016/j.apsusc.2008.10.094

Liu, R. P., Xu, W., He, Z., Lan, H. C., Liu, H. J., & Prasai, T. et al. (2015). Adsorption of antimony(V) onto Mn(II)-enriched surfaces of manganese-oxide and Fe-Mn binary oxide. *Chemosphere*, *138*, 616–624. PMID:26218341

Liu, X. T., Mu, X. Y., Wu, X. L., Meng, L. X., Guan, W. B., & Ma, Y. Q. et al. (2014). Toxicity of multi-walled carbon nanotubes, graphene oxide, and reduced graphene oxide to zebrafish embryos. *Biomedical and environmental sciences. BES*, *27*(9), 676–683. PMID:25256857

Liu, X., Feng, Z., Zhang, S., Zhang, J., Xiao, Q., & Wang, Y. (2006). Preparation and testing of cementing nano-subnano composites of slow or controlled release of fertilizers. *Scientia Agricultura Sinica*, *39*, 1598–1604.

Liu, Y. Y., Su, G. X., Zhang, B., Jiang, G. B., & Yan, B. (2011). Nanoparticle-based strategies for detection and remediation of environmental pollutants. *Analyst (London)*, *136*(5), 872–877. doi:10.1039/c0an00905a PMID:21258678

Liu, Y., Wang, R., Ma, H., Hsiao, B. S., & Chu, B. (2013). High-flux microfiltration filters based on electrospun polyvinylalcohol nanofibrous membranes. *Polymer (United Kingdom)*, *54*(2), 548–556. doi:10.1016/j.polymer.2012.11.064

Liu, Z. P., Jenkins, S. J., & King, D. A. (2005). Origin and activity of oxidized gold in water-gas-shift catalysis. *Physical Review Letters*, *94*(19), 196102. doi:10.1103/PhysRevLett.94.196102 PMID:16090190

Liu, Z., Jiao, Y., Wang, Y., Zhou, C., & Zhang, Z. (2008). Polysaccharides-based nanoparticles as drug delivery systems. *Advanced Drug Delivery Reviews*, *60*(15), 1650–1662. doi:10.1016/j.addr.2008.09.001 PMID:18848591

Liu, Z., Ren, G., Zhang, T., & Yang, Z. (2009). Action potential changes associated with the inhibitory effects on voltage-gated sodium current of hippocampal CA1 neurons by silver nanoparticles. *Toxicology*, *264*(3), 179–184. doi:10.1016/j.tox.2009.08.005 PMID:19683029

Liu, Z., Shi, L., Shi, Z., Liu, X. H., Zi, J., Zhou, S. M., & Xia, Y. J. et al. (2009). Magneto-optical Kerr effect in perpendicularly magnetized Co/Pt films on two-dimensional colloidal crystals. *Applied Physics Letters*, *95*(3), 032502. doi:10.1063/1.3182689

Liu, Z., Yan, J., Miao, Y. E., Huang, Y., & Liu, T. (2015). Catalytic and antibacterial activities of green-synthesized silver nanoparticles on electrospun polystyrene nanofiber membranes using tea polyphenols. *Composites. Part B, Engineering*, *79*, 217–223. doi:10.1016/j.compositesb.2015.04.037

Li, X. L., Lu, Z. S., & Li, Q. (2013). Multilayered films incorporating CdTe quantum dots with tunable optical properties for antibacterial application. *Thin Solid Films*, *548*, 336–342. doi:10.1016/j.tsf.2013.09.088

Li, X., Lian, J., Lin, M., & Chan, Y. (2010). Light-induced selective deposition of metals on gold-tipped CdSe-seeded CdS nanorods. *Journal of the American Chemical Society*, *133*(4), 672–675. doi:10.1021/ja1076603 PMID:21174430

Li, X.-Q., Elliott, D. W., & Zhang, W.-X. (2006). Zero-valent iron nanoparticles for abatement of environmental pollutants: Materials and engineering aspects. *Critical Reviews in Solid State and Material Sciences*, *31*(4), 111–122. doi:10.1080/10408430601057611

Li, Y., Schluesener, H. J., & Xu, S. (2010). Gold nanoparticle-based biosensors. *Gold Bulletin*, *43*(1), 29–41. doi:10.1007/BF03214964

Li, Z. (2003). Use of surfactant-modified zeolite as fertilizer carriers to control nitrate release. *Microporous and Mesoporous Materials*, *61*(1-3), 181–188. doi:10.1016/S1387-1811(03)00366-4

Lkhagvajav, N., Koizhaiganova, M., Yasa, I., Celik, E., & Sari, O. (2015). Characterization and antimicrobial performance of nano silver coatings on leather materials. *Brazilian Journal of Microbiology*, *46*(1), 41–48. doi:10.1590/S1517-838220130446 PMID:26221087

Llacuna, S., Gorriz, A., Riera, M., & Nadal, J. (1996). Effects of air pollution on hematological parameters in passerine birds. *Archives of Environmental Contamination and Toxicology*, *31*(1), 148–152. doi:10.1007/BF00203919 PMID:8688002

Loeb, S., Titelman, L., Korngold, E., & Freiman, J. (1997). Effect of porous support fabric on osmosis through a Loeb-Sourirajan type asymmetric membrane. *Journal of Membrane Science*, *129*(2), 243–249. doi:10.1016/S0376-7388(96)00354-7

London Borough of Lambeth. (2010). *Municipal solid waste management strategy 2011-2031*. Retrieved August 2015, from https://www.lambeth.gov.uk/sites/default/files/rr-lambeth-waste-strategy-waste-prevention-plan.pdf

Lonnen, J., Kilvington, S., Kehoe, S. C., Al-Touati, F., & McGuigan, K. G. (2005). Solar and Photocatalytic Disinfection of Protozoan, Fungal and Bacterial Microbes in Drinking Water. *Water Research*, *39*(5), 877–883. doi:10.1016/j.watres.2004.11.023

Lo, P. K., Karam, P., & Sleiman, H. F. (2010). Loading and selective release of Cargo in DNA nano tubes with logtitudinal variation. *Nature Chemistry*, *2*(4), 319–328. doi:10.1038/nchem.575 PMID:21124515

López-Serrano, A., Olivas, R. M., Landaluze, J. S., & Cámara, C. (2014). Nanoparticles: A global vision. Characterization, separation, and quantification methods. Potential environmental and health impact. *Analytical Methods*, *6*(1), 38–56. doi:10.1039/C3AY40517F

Louie, S. M., Tilton, R. D., & Lowry, G. V. (2013). Effects of molecular weight distribution and chemical properties of natural organic matter on gold nanoparticle aggregation. *Environmental Science & Technology*. Retrieved from http://www.ncbi.nlm.nih.gov/pubmed/23550560

Loutfy, R. O., Lowe, T. P., Moravsky, A. P., & Katagiri, S. (2002). Commercial Production of Fullerenes and Carbon Nanotubes. In E. Ōsawa (Ed.), *Perspectives of Fullerene Nanotechnology* (pp. 35–46). Dordrecht: Springer Netherlands. doi:10.1007/0-306-47621-5_4

Lovern, S. B., Owen, H. A., & Klaper, R. (2008). Electron microscopy of gold nanoparticle intake in the gut of Daphnia magna. *Nanotoxicology*, *2*(1), 43–48. doi:10.1080/17435390801935960

Lovern, S. B., Strickler, J. R., & Klaper, R. (2007). Behavioral and Physiological Changes in Daphnia magna when Exposed to Nanoparticle Suspensions (Titanium Dioxide, Nano-C60, and C60HxC70Hx). *Environmental Science & Technology*, *41*(12), 4465–4470. doi:10.1021/es062146p

Lowry, G. V., Gregory, K. B., Apte, S. C., & Lead, J. R. (2012). Transformations of nanomaterials in the environment. *Environmental Science & Technology*, *46*(13), 6893–6899. doi:10.1021/es300839e PMID:22582927

Lu, G. H., Chen, W., & Li, Y. (2011). Effects of PAHs on biotransformation enzymatic activities in fish. *Chemical Research in Chinese Universities*, *27*, 413–416.

Lu, G., Li, H., Liusman, C., Yin, Z., Wu, S., & Zhang, H. (2011). Surface enhanced Raman scattering of Ag or Au nanoparticle-decorated reduced graphene oxide for detection of aromatic molecules. *Chemical Science*, *2*(9), 1817–1821. doi:10.1039/c1sc00254f

Lu, G., Mao, S., Park, S., Ruoff, R. S., & Chen, J. (2009). Facile, noncovalent decoration of graphene oxide sheets with nanocrystals. *Nano Research*, *2*(3), 192–200. doi:10.1007/s12274-009-9017-8

Lu, J., & Bowles, M. (2013). How Will Nanotechnology Affect Agricultural Supply Chains? *International Food and Agribusiness Management Review*, *16*(2), 21–42.

Lukhele, L. P., Mamba, B. B., Musee, N., & Wepener, V. (2015). *Acute Toxicity of Double-Walled Carbon Nanotubes to Three Aquatic Organisms*. Academic Press.

Lu, S.-y., Wu, D., Wang, Q.-, Yan, J., Buekens, A. G., & Cen, K.-. (2011). Photocatalytic decomposition on nano-TiO2: Destruction of chloroaromatic compounds. *Chemosphere*, *82*(9), 1215–1224. doi:10.1016/j.chemosphere.2010.12.034 PMID:21220149

Lu, W., Wang, B., Zeng, J., Wang, X., Zhang, S., & Hou, J. G. (2005). Synthesis of core/shell nanoparticles of Au/CdSe via Au-Cd bialloy precursor. *Langmuir*, *21*(8), 3684–3687. doi:10.1021/la0469250 PMID:15807621

Lu, X., Jordan, B., & Berge, N. D. (2012). Thermal conversion of municipal solid waste via hydrothermal carbonization: Comparison of carbonization products to products from current waste management techniques. *Waste Management (New York, N.Y.)*, *32*(7), 1353–1365. doi:10.1016/j.wasman.2012.02.012 PMID:22516099

Lu, Y. H., Zhou, M., Zhang, C., & Feng, Y. P. (2009). Metal-embedded graphene: A possible catalyst with high activity. *The Journal of Physical Chemistry C*, *113*(47), 20156–20160. doi:10.1021/jp908829m

Lu, Z. J., Lin, K. D., & Gan, J. (2011). Oxidation of bisphenol F (BPF) by manganese dioxide. *Environmental Pollution*, *159*(10), 2546–2551. doi:10.1016/j.envpol.2011.06.016 PMID:21741139

Lu, Z. S., Dai, T. H., Huang, L. Y., Kurup, D. B., Tegos, G. P., Jahnke, A., & Hamblin, M. R. et al. (2010). Photodynamic therapy with a cationic functionalized fullerene rescues mice from fatal wound infections. *Nanomedicine (London)*, *5*(10), 1525–1533. doi:10.2217/nnm.10.98

Lu, Z., Li, C. M., Bao, H., Qiao, Y., & Bao, Q. (2009). Photophysical mechanism for quantum dots-induced bacterial growth inhibition. *Journal of Nanoscience and Nanotechnology*, *9*(5), 3252–3255. doi:10.1166/jnn.2009.022 PMID:19452999

Lv, M., Zhang, Y., Liang, L., Wei, M., Hu, W., Li, X., & Huang, Q. (2012). Effect of graphene oxide on undifferentiated and retinoic acid-differentiated SH-SY5Y cells line. *Nanoscale*, *4*(13), 3861–3866. doi:10.1039/c2nr30407d PMID:22653613

Lv, Y., Liu, H., Wang, Z., Liu, S., Hao, L., Sang, Y., & Boughton, R. I. et al. (2009). Silver nanoparticle-decorated porous ceramic composite for water treatment. *Journal of Membrane Science*, *331*(1), 50–56. doi:10.1016/j.memsci.2009.01.007

Lyon, D. Y., Adams, L. K., Falkner, J. C., & Alvarez, P. J. J. (2006). Antibacterial Activity of Fullerene Water Suspensions: Effects of Preparation Method and Particle Size. *Environmental Science & Technology*, *40*(14), 4360–4366. doi:10.1021/es0603655

Lyon, J. L., Fleming, D. A., Stone, M. B., Schiffer, P., & Williams, M. E. (2004). Synthesis of Fe oxide core/Au shell nanoparticles by iterative hydroxylamine seeding. *Nano Letters*, *4*(4), 719–723. doi:10.1021/nl035253f

Maartens, A., Swart, P., & Jacobs, E. (1999). Feed-water pretreatment: Methods to reduce membrane fouling by natural organic matter. *Journal of Membrane Science*, *163*(1), 51–62. doi:10.1016/S0376-7388(99)00155-6

Maceda-Veiga, A., Figuerola, J., Martínez-Silvestre, A., Viscor, G., Ferrari, N., & Pacheco, M. (2015). Inside the Redbox: Applications of haematology in wildlife monitoring and ecosystem health assessment. *The Science of the Total Environment*, *514*, 322–332. PMID:25668285

Machado, A. B. M., Drummond, G. M., & Paglia, A. P. (2008). *Livro vermelho da fauna brasileira ameaçada de extinção* (Vol. II). Brasília, Belo Horizonte: MMA / Fundação Biodiversitas.

Mädler, L., Stark, W. J., & Pratsinis, S. E. (2002). Flame-made ceria nanoparticles. *Journal of Materials Research*, *17*(06), 1356–1362. doi:10.1557/JMR.2002.0202

Madsen, H. T., & Søgaard, E. G. (2014). Applicability and modelling of nanofiltration and reverse osmosis for remediation of groundwater polluted with pesticides and pesticide transformation products. *Separation and Purification Technology*, *125*, 111–119. doi:10.1016/j.seppur.2014.01.038

Ma, G. H., He, J., Rajiv, K., Tang, S. H., Yang, Y., & Nogami, M. (2004). Observation of resonant energy transfer in Au: CdS nanocomposite. *Applied Physics Letters*, *84*(23), 4684–4686. doi:10.1063/1.1760220

Maggini, M., Scorrano, G., Prato, M., Brusatin, G., Innocenzi, P., Guglielmi, M., & Bozio, R. et al. (1995). C_{60} derivatives embedded in sol-gel silica films. *Advanced Materials*, *7*(4), 404–406. doi:10.1002/adma.19950070414

Ma, H., Bertsch, P. M., Glenn, T. C., Kabengi, N. J., & Williams, P. L. (2009). Toxicity of manufactured zinc oxide nanoparticles in the nematode Caenorhabditis elegans. *Environmental Toxicology and Chemistry*, *28*(6), 1324–1330. doi:10.1897/08-262.1 PMID:19192952

Ma, H., Brennan, A., & Diamond, S. A. (2012a). Photocatalytic reactive oxygen species production and phototoxicity of titanium dioxide nanoparticles are dependent on the solar ultraviolet radiation spectrum. *Environmental Toxicology and Chemistry*, *31*(9), 2099–2107. doi:10.1002/etc.1916 PMID:22707245

Ma, H., Brennan, A., & Diamond, S. A. (2012b). Phototoxicity of TiO_2 nanoparticles under solar radiation to two aquatic species: Daphnia magna and Japanese medaka. *Environmental Toxicology and Chemistry*, *31*(7), 1621–1629. doi:10.1002/etc.1858 PMID:22544710

Compilation of References

Ma, H., Wallis, L. K., Diamond, S., Li, S., Canas-Carrell, J., & Parra, A. (2014). Impact of solar UV radiation on toxicity of ZnO nanoparticles through photocatalytic reactive oxygen species (ROS) generation and photo-induced dissolution. *Environmental Pollution, 193*, 165–172. doi:10.1016/j.envpol.2014.06.027 PMID:25033018

Ma, H., Williams, P. L., & Diamond, S. A. (2013). Ecotoxicity of manufactured ZnO nanoparticles – A review. *Environmental Pollution, 172*, 76–85. doi:10.1016/j.envpol.2012.08.011 PMID:22995930

Mahamallik, P., Saha, S., & Pal, A. (2015). Tetracycline degradation in aquatic environment by highly porous MnO2 nanosheet assembly. *Chemical Engineering Journal, 276*, 155–165. doi:10.1016/j.cej.2015.04.064

Mahdavi, S., Jalali, M., & Afkhami, A. (2012). Heavy metals removal from aqueous solutions using TiO2, MgO, and Al_2O_3 nanoparticles. *Chemical Engineering Communications, 200*(3), 448–470. doi:10.1080/00986445.2012.686939

Mahmoodi, N. M., Arami, M., Limaee, N. Y., & Gharanjig, K. (2007). Photocatalytic degradation of agricultural N-heterocyclic organic pollutants using immobilized nanoparticles of titania. *Journal of Hazardous Materials, 145*(1–2), 65–71. doi:10.1016/j.jhazmat.2006.10.089 PMID:17145132

Mahmoudi, M., Simchi, A., Milani, A. S., & Stroeve, P. (2009). Cell toxicity of superparamagnetic iron oxide nanoparticles. *Journal of Colloid and Interface Science, 336*(2), 510–518. doi:10.1016/j.jcis.2009.04.046 PMID:19476952

Mahtab, R., & Murphy, C. J. (2005). Probing DNA structure with nanoparticles. *Methods in Molecular Biology (Clifton, N.J.), 303*, 179–190. doi:10.1385/1-59259-901-X:179 PMID:15923684

Majumder, M., & Ajayan, P. M. (2010). Carbon Nanotube Membranes: A New Frontier in Membrane Science. In D. Enrico & G. Lidietta (Eds.), *Comprehensive Membrane Science and Engineering* (pp. 291–310). Oxford: Elsevier. doi:10.1016/B978-0-08-093250-7.00038-4

Malik, P., Katyal, V., Malik, V., Asatkar, A., Inwati, G., & Mukherjee, T. K. (2013). Nanobiosensors: Concepts and Variations. *ISRN Nanomaterials. Article ID, 327435*, 9.

Malmsten, M. (2013). Inorganic nanomaterials as delivery systems for proteins, peptides, DNA, and siRNA. *Current Opinion in Colloid & Interface Science, 18*(5), 468–480. doi:10.1016/j.cocis.2013.06.002

Malvindi, M. A., De Matteis, V., Galeone, A., Brunetti, V., Anyfantis, G. C., Athanassiou, A., & Pompa, P. P. et al. (2014). Toxicity assessment of silica coated iron oxide nanoparticles and biocompatibility improvement by surface engineering. *PLoS ONE, 9*(1), e85835. doi:10.1371/journal.pone.0085835 PMID:24465736

Maneerung, T., Tokura, S., & Rujiravanit, R. (2007). (in press). Impregnation of silver nanoparticles into bacterial cellulose for antimicrobial wound dressing. *Carbohydrate Polymers*.

Manimegalai, G. S., Shanthakumar, S., & Sharma, C. (2014). Silver nanoparticles: Synthesis and application in mineralization of pesticides using membrane support. *Int Nano Lett., 4*(105), 1–5.

Manning, B. A., Fendorf, S. E., Bostick, B., & Suarez, D. L. (2002). Arsenic(III) oxidation and arsenic(V) adsorption reactions on synthetic birnessite. *Environmental Science & Technology, 36*(5), 976–981. doi:10.1021/es0110170 PMID:11918029

Manno, D., Filippo, E., Di Giulio, M., & Serra, A. (2008). Synthesis and characterization of starch-stabilized Ag nanostructures for sensors applications. *Journal of Non-Crystalline Solids, 354*(52), 5515–5520. doi:10.1016/j.jnoncrysol.2008.04.059

Mansell, B. L., & Drewes, J. E. (2004). Fate of steroidal hormones during soil-aquifer treatment. *Ground Water Monitoring and Remediation, 24*(2), 94–101. doi:10.1111/j.1745-6592.2004.tb00717.x

Mansour, A.S., Emam, A.N., Mohamed, M.B. (2016). *Remarkable Enhancement of Optical and Photocatalytic Properties of Au-CdSe Tetrapods upon Loading into Graphene Oxide.* (Unpublished to be submitted in Preparation).

Manzo, S., Miglietta, M. L., Rametta, G., Buono, S., & Di Francia, G. (2013). Toxic effects of ZnO nanoparticles towards marine algae Dunaliella tertiolecta. *The Science of the Total Environment, 445*, 371–376. doi:10.1016/j.scitotenv.2012.12.051 PMID:23361041

Marquardt, D., Vollmer, C., Thomann, R., Steurer, P., Mülhaupt, R., Redel, E., & Janiak, C. (2011). The use of microwave irradiation for the easy synthesis of graphene-supported transition metal nanoparticles in ionic liquids. *Carbon, 49*(4), 1326–1332. doi:10.1016/j.carbon.2010.09.066

Maryan, A. S., Montazer, M., & Harifi, T. (2015). Synthesis of nano silver on cellulosic denim fabric producing yellow colored garment with antibacterial properties. *Carbohydrate Polymers, 115*, 568–574. doi:10.1016/j.carbpol.2014.08.100 PMID:25439933

Mashino, T., Nishikawa, D., Takahashi, K., Usui, N., Yamori, T., Seki, M., & Mochizuki, M. et al. (2003). Antibacterial and antiproliferative activity of cationic fullerene derivatives. *Bioorganic & Medicinal Chemistry Letters, 13*(24), 4395–4397. doi:10.1016/j.bmcl.2003.09.040

Mason, T. J. (2009). Sonoelectrochemical synthesis of nanoparticles. *Molecules (Basel, Switzerland), 14*(10), 4284–4299. doi:10.3390/molecules14104284 PMID:19924064

Matsuura, T. (2001). Progress in membrane science and technology for seawater desalination—a review. *Desalination, 134*(1), 47–54. doi:10.1016/S0011-9164(01)00114-X

Matthies, K., Bitter, H., Deobald, N., Heinle, M., Diedel, R., Obstaand, U., & Brenner-Weissa, G. (2015). Morphology, composition and performance of a ceramic filter for household water treatment in Indonesia. *Water Practice & Technology, 10*(2), 361–370. doi:10.2166/wpt.2015.044

Maurer-Jones, M., Gunsolus, I. L., Murphy, C. J., & Haynes, C. L. (2013). Toxicity of engineered nanoparticles in the environment. *Analytical Chemistry, 85*(6), 3036–3049. doi:10.1021/ac303636s PMID:23427995

Maurette, M. T., Oliveros, E., Infelta, P. P., Ramsteiner, K., & Braun, A. M. (1983). Singlet Oxygen and Superoxide - Experimental Differentiation and Analysis. *Helvetica Chimica Acta, 66*(2), 722–733. doi:10.1002/hlca.19830660236

Maurice, P., & Namjesnik-Dejanovic, K. (1999). Aggregate structures of sorbed humic substances observed in aqueous solution. *Environmental Science & Technology, 33*(9), 1538–1541. doi:10.1021/es981113+

Mauter, M. S., & Elimelech, M. (2008). Environmental Applications of Carbon-Based Nanomaterials. *Environmental Science & Technology, 42*(16), 5843–5859. doi:10.1021/es8006904 PMID:18767635

Ma, X., & Bouchard, D. (2009). Formation of Aqueous Suspensions of Fullerenes. *Environmental Science & Technology, 43*(2), 330–336. doi:10.1021/es801833p

Ma, X., Gurung, A., & Deng, Y. (2013). Phytotoxicity and uptake of nanoscale zero-valent iron (nZVI) by two plant species. *The Science of the Total Environment, 443*, 844–849. doi:10.1016/j.scitotenv.2012.11.073 PMID:23247287

Ma, Y., Kuang, L., He, X., Bai, W., Ding, Y., Zhang, Z., & Chai, Z. et al. (2010). Effects of rare earth oxide nanoparticles on root elongation of plants. *Chemosphere, 78*(3), 273–279. doi:10.1016/j.chemosphere.2009.10.050 PMID:19897228

Maymo-Gatell, X., Chien, Y., Gossett, J. M., & Zender, S. H. (1997). Isolation of a bacterium that reductively dechlorinates tetrachloroethene to ethene. *Science, 276*(5318), 1568–1571. doi:10.1126/science.276.5318.1568 PMID:9171062

Maysinger, D. (2007). Nanoparticles and cells: Good companions and doomed partnerships. *Organic & Biomolecular Chemistry*, *5*(15), 2335–2342. doi:10.1039/b704275b PMID:17637950

Ma, Z., Kotaki, M., Inai, R., & Ramakrishna, S. (2005). Potential of nanofiber matrix as tissue-engineering scaffolds. *Tissue Engineering*, *11*(1-2), 101–109. doi:10.1089/ten.2005.11.101 PMID:15738665

Mazon, A. F., & Fernandes, M. N. (1999). Toxicity and differential tissue accumulation of copper in the tropical freshwater fish, *Prochilodus scrofa* (Prochilodontidae). *Bulletin of Environmental Contamination and Toxicology*, *63*(6), 797–804. doi:10.1007/s001289901049 PMID:10594155

McCarty, P. L. (1997). Breathing with chlorinated solvents. *Science*, *276*(5318), 1521–1522. doi:10.1126/science.276.5318.1521 PMID:9190688

McDonnell, A. M. P., Beving, D., Wang, A. J., Chen, W., & Yan, Y. S. (2005). Hydrophilic and antimicrobial zeolite coatings for gravity-independent water separation. *Advanced Functional Materials*, *15*(2), 336–340. doi:10.1002/adfm.200400183

McGlynn, W. (n.d.). *Guidelines for the Use of Chlorine Bleach as a Sanitizer in Food Processing Operations*. Retrieved from http://ucfoodsafety.ucdavis.edu/files/26437.pdf

McHugh, A. J., & Brenner, H. (1984). Particle size measurement using chromatography. *Critical Reviews in Analytical Chemistry*, *15*(1), 63–117. doi:10.1080/10408348408085429

McLamore, E. S., Diggs, A., Calvo Marzal, P., Shi, J., Blakeslee, J. J., Peer, W. A., & Porterfield, D. M. et al. (2010). Noninvasive quantification of endogenous root auxin transport using an integrated flux microsensor technique. *The Plant Journal*, *63*(6), 1004–1016. doi:10.1111/j.1365-313X.2010.04300.x PMID:20626658

McLaren, A., Valdes-Solis, T., Li, G., & Tsang, S. C. (2009). Shape and size effects of ZnO nanocrystals on photocatalytic activity. *Journal of the American Chemical Society*, *131*(35), 12540–12541. doi:10.1021/ja9052703 PMID:19685892

McPhillips, J., Murphy, A., Jonsson, M. P., Hendren, W. R., Atkinson, R., Hook, F., & Pollard, R. J. et al. (2010). High-performance biosensing using arrays of plasmonic nanotubes. *ACS Nano*, *4*(4), 2210–2216. doi:10.1021/nn9015828 PMID:20218668

Mecha, C. A., & Pillay, V. L. (2014). Development and evaluation of woven fabric microfiltration membranes impregnated with silver nanoparticles for potable water treatment. *Journal of Membrane Science*, *458*, 149–156. doi:10.1016/j.memsci.2014.02.001

Meireles, M., Bessieres, A., Rogissart, I., Aimar, P., & Sanchez, V. (1995). An appropriate molecular size parameter for porous membranes calibration. *Journal of Membrane Science*, *103*(1), 105–115. doi:10.1016/0376-7388(94)00311-L

Melechko, A. V., Merkulov, V. I., McKnight, T. E., Guillorn, M. A., Klein, K. L., Lowndes, D. H., & Simpson, M. L. (2005). Vertically aligned carbon nanofibers and related structures: Controlled synthesis and directed assembly. *Journal of Applied Physics*, *97*(4), 041301. doi:10.1063/1.1857591

Melle, S., Menéndez, J. L., Armelles, G., Navas, D., Vázquez, M., Nielsch, K., & Gösele, U. et al. (2003). Magneto-optical properties of nickel nanowire arrays. *Applied Physics Letters*, *83*(22), 4547–4549. doi:10.1063/1.1630840

Mellor, J. E., Smith, J. A., Learmonth, G. P., Netshandama, V. O., & Dillingham, R. A. (2012). Modeling the Complexities of Water, Hygiene, and Health in Limpopo Province, South Africa. *Environmental Science & Technology*, *46*(24), 13512–13520. doi:10.1021/es3038966 PMID:23186073

Mellor, J., Abebe, L., Ehdaie, B., Dillingham, R., & Smith, J. (2014). Modeling the sustainability of a ceramic water filter intervention. *Water Research*, *49*, 286–299. doi:10.1016/j.watres.2013.11.035 PMID:24355289

Menard, A., Drobne, D., & Jemec, A. (2011). Ecotoxicity of nanosized TiO_2. Review of in vivo data. *Environmental Pollution*, *159*(3), 677–684. doi:10.1016/j.envpol.2010.11.027 PMID:21186069

Meng, H., Chen, Z., Xing, G., Yuan, H., Chen, C., Zhao, F., & Zhao, Y. et al. (2007). Ultrahigh reactivity and grave nanotoxicity of copper nanoparticles. *Journal of Radioanalytical and Nuclear Chemistry*, *272*(3), 595–598. doi:10.1007/s10967-007-0630-2

Menon, S. K., Modi, N. R., Pandya, A., & Lodha, A. (2013). Ultrasensitive and specific detection of dimethoate using a p-sulphonato-calix[4]resorcinarene functionalized silver nanoprobe in aqueous solution. *RSC Advances*, *3*(27), 10623–10627. doi:10.1039/c3ra40762d

Merhari, L. (Ed.). (2009). *Hybrid nanocomposites for nanotechnology*. Springer, US. doi:10.1007/978-0-387-30428-1

Metreveli, G., Philippe, A., & Schaumann, G. E. (2014). Disaggregation of silver nanoparticle homoaggregates in a river water matrix. *The Science of the Total Environment*. PMID:25433382

Miao, A.-J., Luo, Z., Chen, C.-S., Chin, W.-C., Santschi, P. H., & Quigg, A. (2010). Intracellular uptake: A possible mechanism for silver engineered nanoparticle toxicity to a freshwater alga Ochromonas danica. *PLoS ONE*, *5*(12), e15196. doi:10.1371/journal.pone.0015196 PMID:21203552

Miao, A.-J., Schwehr, K. A., Xu, C., Zhang, S.-J., Luo, Z., Quigg, A., & Santschi, P. H. (2009). The algal toxicity of silver engineered nanoparticles and detoxification by exopolymeric substances. *Environmental Pollution*, *157*(11), 3034–3041. doi:10.1016/j.envpol.2009.05.047 PMID:19560243

Mikhaylova, M., Kim, D. K., Bobrysheva, N., Osmolowsky, M., Semenov, V., Tsakalakos, T., & Muhammed, M. (2004). Superparamagnetism of magnetite nanoparticles: Dependence on surface modification. *Langmuir*, *20*(6), 2472–2477. doi:10.1021/la035648e PMID:15835712

Millán, G., Agosto, F., & Vázquez, M. (2008). Use of clinoptilolite as a carrier for nitrogen fertilizers in soils of the Pampean regions of Argentina. *Ciencia e investigación agraria. Agr*, *35*(3), 293–302.

Miller, G. P. (2006). Reactions Between Aliphatic Amines and $[C_{60}]$ Fullerene: A Review. *Comptes Rendus. Chimie*, *9*(7-8), 952–959. doi:10.1016/j.crci.2005.11.020

Miller, J. S. (2005). Rose bengal-sensitized photooxidation of 2-chlorophenol in water using solar simulated. *Water Research*, *39*(2-3), 412–422. doi:10.1016/j.watres.2004.09.019

Miller, R. J., Bennett, S., Keller, A. A., Pease, S., & Lenihan, H. S. (2012). TiO_2 nanoparticles are phototoxic to marine phytoplankton. *PLoS ONE*, *7*(1), e30321. doi:10.1371/journal.pone.0030321 PMID:22276179

Millstone, J. E., Park, S., Shuford, K. L., Qin, L., Schatz, G. C., & Mirkin, C. A. (2005). Observation of a quadrupole plasmon mode for a colloidal solution of gold nanoprisms. *Journal of the American Chemical Society*, *127*(15), 5312–5313. doi:10.1021/ja043245a PMID:15826156

Miralles, P., Church, T. L., & Harris, A. T. (2012). Toxicity, uptake, and translocation of engineered nanomaterials in vascular plants. *Environmental Science & Technology*, *46*(17), 9224–9239. doi:10.1021/es202995d PMID:22892035

Misdan, N., Lau, W. J., & Ismail, A. F. (2012). Seawater Reverse Osmosis (SWRO) desalination by thin-film composite membrane—Current development, challenges and future prospects. *Desalination*, *287*(0), 228–237. doi:10.1016/j.desal.2011.11.001

Mishra, B., Patel, B. B., & Tiwari, S. (2010). Colloidal nanocarriers: A review on formulation technology, types and applications toward targeted drug delivery. *Nanomedicine; Nanotechnology, Biology, and Medicine*, *6*(1), 9–24. doi:10.1016/j.nano.2009.04.008 PMID:19447208

Mishra, S., Singh, B. R., Singh, A., Keswani, C., Naqvi, A. H., & Singh, H. B. (2014). Biofabricated Silver Nanoparticles Act as a Strong Fungicide against *Bipolaris sorokiniana* Causing Spot Blotch Disease in Wheat. *PLoS ONE, 9*(5), e97881. doi:10.1371/journal.pone.0097881 PMID:24840186

Mitrano, D. M., Lesher, E. K., Bednar, A., Monserud, J., Higgins, C. P., & Ranville, J. F. (2011). Detecting nanoparticulate silver using single-particle inductively coupled plasma-mass spectrometry. *Environmental Toxicology and Chemistry, 31*(1), 115–121. doi:10.1002/etc.719 PMID:22012920

Mizukoshi, Y., Oshima, R., Maeda, Y., & Nagata, Y. (1999). Preparation of platinum nanoparticles by sonochemical reduction of the Pt (II) ion. *Langmuir, 15*(8), 2733–2737. doi:10.1021/la9812121

Mody, V. V., Siwale, R., Singh, A., & Mody, H. R. (2010). Introduction to metallic nanoparticles. *Journal of Pharmacy and Bioallied Sciences, 2*(4), 282. doi:10.4103/0975-7406.72127 PMID:21180459

Moghimi, S. M., Hunter, A. C., & Murray, J. C. (2005). Nanomedicine: Current status and future prospects. *The FASEB Journal, 19*(3), 311–330. doi:10.1096/fj.04-2747rev PMID:15746175

Mohamed, M. B., AbouZeid, K. M., Abdelsayed, V., Aljarash, A. A., & El-Shall, M. S. (2010). Growth mechanism of anisotropic gold nanocrystals via microwave synthesis: Formation of dioleamide by gold nanocatalysis. *ACS Nano, 4*(5), 2766–2772. doi:10.1021/nn9016179 PMID:20392051

Mohd Omar, F., Abdul Aziz, H., & Stoll, S. (2014). Aggregation and disaggregation of ZnO nanoparticles: Influence of pH and adsorption of Suwannee River humic acid. *The Science of the Total Environment, 468-469*, 195–201. PMID:24029691

Mohmood, I., Lopes, C., Lopes, I., Ahmad, I., Duarte, A., & Pereira, E. (2013). Nanoscale materials and their use in water contaminants removal—a review. *Environ Sci Pollut Res, 20*(3), 1239–1260. doi:10.1007/s11356-012-1415-x PMID:23292223

Mokari, T., Rothenberg, E., Popov, I., Costi, R., & Banin, U. (2004). Selective growth of metal tips onto semiconductor quantum rods and tetrapods. *Science, 304*(5678), 1787–1790. doi:10.1126/science.1097830 PMID:15205530

Mokari, T., Sztrum, C. G., Salant, A., Rabani, E., & Banin, U. (2005). Formation of asymmetric one-sided metal-tipped semiconductor nanocrystal dots and rods. *Nature Materials, 4*(11), 855–863. doi:10.1038/nmat1505

Molinder, A. (1995). *Material Requirements Planning Employing Input-Output Analysis and Laplace Transforms, PROFIL, 14*. Linköping: Production Economic Research.

Montaño, M. D., Badiei, H. R., Bazargan, S. & Ranville, J. (2014). Improvements in the detection and characterization of engineered nanoparticles using spICP-MS with microsecond dwell times. *Environmental Science: Nano*.

Montazer, M., Alimohammadi, F., Shamei, A., & Rahimi, M. K. (2012). Durable antibacterial and cross-linking cotton with colloidal silver nanoparticles and butane tetracarboxylic acid without yellowing. *Colloids and Surfaces. B, Biointerfaces, 89*, 196–202. doi:10.1016/j.colsurfb.2011.09.015 PMID:21978552

Montazer, M., Alimohammadi, F., Shamei, A., & Rahimi, M. K. (2012). In situ synthesis of nano silver on cotton using Tollens' reagent. *Carbohydrate Polymers, 87*(2), 1706–1712. doi:10.1016/j.carbpol.2011.09.079

Montazer, M., Shamei, A., & Alimohammadi, F. (2012). Stabilized nanosilver loaded nylon knitted fabric using BTCA without yellowing. *Progress in Organic Coatings, 74*(1), 270–276. doi:10.1016/j.porgcoat.2012.01.003

Monteiro-Riviere, N. A., & Inman, A. O. (2006). Challenges for assessing carbon nanomaterial toxicity to the skin. *Carbon, 44*(6), 1070–1078. doi:10.1016/j.carbon.2005.11.004

Monteiro-Riviere, N. A., Inman, A. O., & Zhang, L. W. (2009). Limitations and relative utility of screening assays to assess engineered nanoparticle toxicity in a human cell line. *Toxicology and Applied Pharmacology*, *234*(2), 222–235. doi:10.1016/j.taap.2008.09.030 PMID:18983864

Montgomery, J. H. (2000). *Groundwater Chemicals* (3rd ed.). Boca Raton, FL: Lewis Publishers and CRC Press.

Moore, M. N. (2006). Do nanoparticles present ecotoxicological risks for the health of the aquatic environment? *Environment International*, *32*(8), 967–976. doi:10.1016/j.envint.2006.06.014 PMID:16859745

Moor, K. J., Snow, S. D., & Kim, J.-H. (2015). Differential Photoactivity of Aqueous [C60] and [C70] Fullerene Aggregates. *Environmental Science & Technology*, *49*(10), 5990–5998. doi:10.1021/acs.est.5b00100

Moor, K. J., Valle, D., Li, C., & Kim, J.-H. (2015). *Improving the Visible Light Photoactivity of Supported Fullerene Photocatalysts Through the Use of* [C70]. Fullerene.

Moor, K., & Kim, J.-H. (2014). Simple Synthetic Method Toward Solid Supported C_{60} Visible LightActivated Photocatalysts. *Environmental Science & Technology*, *48*(5), 2785–2791. doi:10.1021/es405283w

Morales-Flores, N., Pal, U., & Sánchez Mora, E. (2011). Photocatalytic behavior of ZnO and Pt-incorporated ZnO nanoparticles in phenol degradation. *Appl Catalysis A*, *394*(1–2), 269–275. doi:10.1016/j.apcata.2011.01.011

Morrison, F. Jr, & Osterle, J. (1965). Electrokinetic energy conversion in ultrafine capillaries. *The Journal of Chemical Physics*, *43*(6), 2111–2115. doi:10.1063/1.1697081

Mortimer, J. C., Coxon, K. M., Laohavisit, A., & Davies, J. M. (2009). Hemi independent soluble and membrane associated peroxidase activity of a Zea mays annexin preparations. *Plant Signaling & Behavior*, *4*(5), 428–430. doi:10.4161/psb.4.5.8297 PMID:19816107

Mortimer, M., Kasemets, K., & Kahru, A. (2010). Toxicity of ZnO and CuO nanoparticles to ciliated protozoa Tetrahymena thermophila. *Toxicology*, *269*(2-3), 182–189. doi:10.1016/j.tox.2009.07.007 PMID:19622384

Mort, J., Ziolo, R., Machonkin, M., Huffman, D. R., & Ferguson, M. I. (1991). Electrical conductivity studies of undoped solid films of $C_{60/70}$. *Chemical Physics Letters*, *186*(2-3), 284–286. doi:10.1016/S0009-2614(91)85142-J

Moura, J. C. V. P., Oliveira Campos, A. M. F., & Griffiths, J. (1997). Synthesis and evaluation of phenothiazine singlet oxygen sensitising dyes for application in cancer phototherapy. *Phosphorus, Sulfur, and Silicon and the Related Elements*, *120*(1), 459–460. doi:10.1080/10426509708545597

Moyano, D. F., Goldsmith, M., Solfiell, D. J., Landesman-Milo, D., Miranda, O. R., Peer, D., & Rotello, V. M. (2012). Nanoparticle hydrophobicity dictates immune response. *Journal of the American Chemical Society*, *134*(9), 3965–3967. doi:10.1021/ja2108905 PMID:22339432

Mpenyana-Monyatsi, L., Mthombeni, N. H., Onyango, M. S., & Momba, M. N. (2012). Cost-effective filter materials coated with silver nanoparticles for the removal of pathogenic bacteria in groundwater. *International Journal of Environmental Research and Public Health*, *9*(1), 244–271. doi:10.3390/ijerph9010244 PMID:22470290

Mubarak Ali, D., Thajuddin, N., Jeganathan, K., & Gunasekaran, M. (2011). Plant extract mediated synthesis of silver and gold nanoparticles and its antibacterial activity against clinically isolated pathogens. *Colloids and Surfaces. B, Biointerfaces*, *85*(2), 360–365. doi:10.1016/j.colsurfb.2011.03.009 PMID:21466948

Mueller, N. C., Bruns, B. J., Černík, M., Rissing, P., Rickerby, D., & Nowack, B. (2012). Application of nanoscale zero valent iron (NZVI) for groundwater remediation in Europe. *Environmental Science and Pollution Research International*, *19*(2), 550–558. doi:10.1007/s11356-011-0576-3 PMID:21850484

Compilation of References

Mukherjee, P., Ahmad, A., Mandal, D., Senapati, S., Sainkar, S. R., Khan, M. I., & Sastry, M. (2001). Fungus-mediated synthesis of silver nanoparticles and their immobilization in the mycelial matrix: A novel biological approach to nanoparticle synthesis. *Nano Letters*, *1*(10), 515–519. doi:10.1021/nl0155274

Mukherjee, S. G., O'Claonadh, N., Casey, A., & Chambers, G. (2012). Comparative in vitro cytotoxicity study of silver nanoparticle on two mammalian cell lines. *Toxicology In Vitro*, *26*(2), 238–251. doi:10.1016/j.tiv.2011.12.004 PMID:22198051

Mukhopadhyay, S. S. (2014). Nanotechnology in agriculture: Prospects and constraints. *Nanotechnology, Science and Applications*, *7*, 63–71. doi:10.2147/NSA.S39409 PMID:25187699

Mulder, M. (1996). *Basic principles of membrane technology*. Springer Science & Business Media. doi:10.1007/978-94-009-1766-8

Muller, J., Huaux, F., Moreau, N., Misson, P., Heilier, J.-F., Delos, M., & Lison, D. et al. (2005). Respiratory toxicity of multi-wall carbon nanotubes. *Toxicology and Applied Pharmacology*, *207*(3), 221–231. doi:10.1016/j.taap.2005.01.008 PMID:16129115

Mu, R., Xu, Z., Li, L., Shao, Y., Wan, H., & Zheng, S. (2010). On the photocatalytic properties of elongated TiO_2 nanoparticles for phenol degradation and Cr(VI) reduction. *Journal of Hazardous Materials*, *176*(1–3), 495–502. doi:10.1016/j.jhazmat.2009.11.057 PMID:19969418

Muranyi, P., Schraml, C., & Wunderlich, J. (2010). Antimicrobial efficiency of titanium dioxide-coated surfaces. *Journal of Applied Microbiology*, *108*(6), 1966–1973. doi:10.1111/j.1365-2672.2009.04594.x PMID:19886892

Murphy, C. J. (2002). Nanocubes and nanoboxes. *Science*, *298*(5601), 2139–2141. doi:10.1126/science.1080007 PMID:12481122

Murphy, J. D., & McKeogh, E. (2004). Technical, economic and environmental analysis of energy production from municipal solid waste. *Renewable Energy*, *29*(7), 1043–1057. doi:10.1016/j.renene.2003.12.002

Murray, C., Norris, D. J., & Bawendi, M. G. (1993). Synthesis and characterization of nearly monodisperse CdE (E= sulfur, selenium, tellurium) semiconductor nanocrystallites. *Journal of the American Chemical Society*, *115*(19), 8706–8715. doi:10.1021/ja00072a025

Murray, J. W. (1974). Surface chemistry of hydrous manganese-dioxide. *Journal of Colloid and Interface Science*, *46*(3), 357–371. doi:10.1016/0021-9797(74)90045-9

Murray, J. W. (1975). The interaction of metal ions at the manganese dioxide-solution interface. *Geochimica et Cosmochimica Acta*, *39*(4), 505–519. doi:10.1016/0016-7037(75)90103-9

Murray, W. A., & Barnes, W. L. (2007). Plasmonic materials. *Advanced Materials*, *19*(22), 3771–3782. doi:10.1002/adma.200700678

Musale, D. A., & Kumar, A. (2000). Solvent and pH resistance of surface crosslinked chitosan/poly (acrylonitrile) composite nanofiltration membranes. *Journal of Applied Polymer Science*, *77*(8), 1782–1793. doi:10.1002/1097-4628(20000822)77:8<1782::AID-APP15>3.0.CO;2-5

Naderi, M. R., & Danesh-Shahraki, A. (2013). Nanofertilizers and their roles in sustainable agriculture. *International Journal of Agriculture and Crop Sciences*, *5*(19), 2229–2232.

Naderi, M. R., & Danesh-Shahraki, A. (2013). Nanofertilizers and their roles in sustainable agriculture. *Intl J Agri Crop Sci*, *5*(19), 2229–2232.

Na, H., Song, I. C., & Hyeon, T. (2009). Inorganic nanoparticles for MRI contrast agents. *Advanced Materials*, *21*(21), 2133–2148. doi:10.1002/adma.200802366

Naim, M. M., Wikner, J., & Grubbström, R. W. (2007). A net present value assessment of make-to-order and make-to-stock manufacturing systems. *Omega*, *35*(5), 524–532. doi:10.1016/j.omega.2005.09.006

Nair, B., & Pradeep, T. (2002). Coalescence of nanoclusters and formation of submicron crystallites assisted by Lactobacillus strains. *Crystal Growth & Design*, *2*(4), 293–298. doi:10.1021/cg0255164

Nair, P. M. G., Park, S. Y., & Choi, J. (2013). Evaluation of the effect of silver nanoparticles and silver ions using stress responsive gene expression in Chironomus riparius. *Chemosphere*, *92*(5), 592–599. doi:10.1016/j.chemosphere.2013.03.060 PMID:23664472

Nakamura, E., & Isobe, H. (2003). Functionalized fullerenes in water. The first 10 years of their chemistry, biology, and nanoscience. *Accounts of Chemical Research*, *36*(11), 807–815. doi:10.1021/ar030027y

Nalawade, P., Mukherjee, T., & Kapoor, S. (2012). High-yield synthesis of multispiked gold nanoparticles: Characterization and catalytic reactions. *Colloids and Surfaces. A, Physicochemical and Engineering Aspects*, *396*, 336–340. doi:10.1016/j.colsurfa.2012.01.018

Nalbandian, M. J., Zhang, M., Sanchez, J., Choa, Y.-H., Cwiertny, D. M., & Myung, N. V. (2015). Synthesis and optimization of BiVO4 and co-catalyzed BiVO4 nanofibers for visible light-activated photocatalytic degradation of aquatic micropollutants. *Journal of Molecular Catalysis A Chemical*, *404–405*, 18–26. doi:10.1016/j.molcata.2015.04.003

Nalbandian, M. J., Zhang, M., Sanchez, J., Kim, S., Choa, Y. H., Cwiertny, D. M., & Myung, N. V. (2015). Synthesis and optimization of Ag-TiO$_2$ composite nanofibers for photocatalytic treatment of impaired water sources. *Journal of Hazardous Materials*, *299*, 141–148. doi:10.1016/j.jhazmat.2015.05.053 PMID:26101968

Nanda, A., & Saravanan, M. (2009). Biosynthesis of silver nanoparticles from Staphylococcus aureus and its antimicrobial activity against MRSA and MRSE. *Nanomedicine; Nanotechnology, Biology, and Medicine*, *5*(4), 452–456. doi:10.1016/j.nano.2009.01.012 PMID:19523420

Naqvi, S., Samim, M., Abdin, M. Z., Ahmed, F. J., Maitra, A. N., & Prashant, C. K. et al.. (2010). Concentration-dependent toxicity of iron oxide nanoparticles mediated by increased oxidative stress. *International Journal of Nanomedicine*, *5*, 983–989. doi:10.2147/IJN.S13244 PMID:21187917

National Institute of Food and Agriculture. (2015). *Nanotechnology Program*. Retrieved August 24, 2015, from: National Institute of Food and Agriculture: http://nifa.usda.gov/program/nanotechnology-program

Nations, S. L. (2009, May 1). *Acute and developmental toxicity of metal oxide nanoparticles (ZnO, TiO2, Fe2O3, and CuO) in Xenopus laevis*. Retrieved August 22, 2015, from https://repositories.tdl.org/ttu-ir/handle/2346/16511

Nations, S., Wages, M., Cañas, J. E., Maul, J., Theodorakis, C., & Cobb, G. P. (2011). Acute effects of Fe2O3, TiO2, ZnO and CuO nanomaterials on Xenopus laevis. *Chemosphere*.

Nations, S., Long, M., Wages, M., Maul, J. D., Theodorakis, C. W., & Cobb, G. P. (2015). Sub chronic and chronic developmental effects of copper oxide (CuO) nanoparticles on *Xenopus laevis*. *Chemosphere*, *135*, 166–174. PMID:25950410

Nations, S., Long, M., Wages, M., Maul, J. D., Theodorakis, C. W., & Cobb, G. P. (2015). Subchronic and chronic developmental effects of copper oxide (CuO) nanoparticles on Xenopus laevis. *Chemosphere*, *135*, 166–174. doi:10.1016/j.chemosphere.2015.03.078 PMID:25950410

Nations, S., Wages, M., Cañas, J. E., Maul, J., Theodorakis, C., & Cobb, G. P. (2011). Acute effects of Fe_2O_3, TiO_2, ZnO and CuO nanomaterials on Xenopus laevis. *Chemosphere*, *83*(8), 1053–1061. doi:10.1016/j.chemosphere.2011.01.061 PMID:21345480

Navabpour, P., Ostovarpour, S., Hampshire, J., Kelly, P., Verran, J., & Cooke, K. (2014). The effect of process parameters on the structure, photocatalytic and self-cleaning properties of TiO2 and Ag-TiO2 coatings deposited using reactive magnetron sputtering. *Thin Solid Films*, *571*(Part 1), 75–83. doi:10.1016/j.tsf.2014.10.040

Navarro, D. A., Kookana, R. S., McLaughlin, M. J., & Kirby, J. K. (2016). Fullerol as a Potential Pathway for Mineralization of Fullerene Nanoparticles in Biosolid-Amended Soils. *Environmental Science & Technology Letters*, *3*(1), 7–12. doi:10.1021/acs.estlett.5b00292

Navarro, E., Baun, A., Behra, R., Hartmann, N., Filser, J., Miao, A.-J., & Sigg, L. et al. (2008). Environmental behavior and ecotoxicity of engineered nanoparticles to algae, plants, and fungi. *Ecotoxicology (London, England)*, *17*(5), 372–386. doi:10.1007/s10646-008-0214-0 PMID:18461442

Nazarenus, M., Zhang, Q., Soliman, M. G., del Pino, P., Pelaz, B., Carregal-Romero, S., & Parak, W. J. et al. (2014). In vitro interaction of colloidal nanoparticles with mammalian cells: What have we learned thus far? *Beilstein Journal of Nanotechnology*, *5*, 1477–1490. doi:10.3762/bjnano.5.161 PMID:25247131

Nel, A. E., Mädler, L., Velegol, D., Xia, T., Hoek, E. M. V., Somasundaran, P., & Thompson, M. et al. (2009). Understanding biophysicochemical interactions at the nano-bio interface. *Nature Materials*, *8*(7), 543–557. doi:10.1038/nmat2442 PMID:19525947

Nel, A., Xia, T., Mädler, L., & Li, N. (2006). Toxic potential of materials at the nanolevel. *Science*, *311*(5761), 622–627. doi:10.1126/science.1114397 PMID:16456071

Nelson, D. A., & Morris, M. W. (1989). Basic methodology. Hematology and coagulation. In D. A. Nelson, & J. B. Henry (Eds.), Clinical Diagnosis and Management by Laboratory Methods (pp. 578 –625). Saunder Company.

Nelson, S. M., Mahmoud, T., Beaux, M. II, Shapiro, P., McIlroy, D. N., & Stenkamp, D. L. (2010). Toxic and teratogenic silica nanowires in developing vertebrate embryos. *Nanomedicine; Nanotechnology, Biology, and Medicine*, *6*(1), 93–102. doi:10.1016/j.nano.2009.05.003 PMID:19447201

Nemamcha, A., Rehspringer, J. L., & Khatmi, D. (2006). Synthesis of palladium nanoparticles by sonochemical reduction of palladium (II) nitrate in aqueous solution. *The Journal of Physical Chemistry B*, *110*(1), 383–387. doi:10.1021/jp0535801 PMID:16471546

Nepal, D., Balasubramanian, S., Simonian, A. L., & Davis, V. A. (2008). Strong antimicrobial coatings: Single-walled carbon nanotubes armored with biopolymers. *Nano Letters*, *8*(7), 1896–1901. doi:10.1021/nl080522t PMID:18507479

Newton, K. M., Puppala, H. L., Kitchens, C. L., Colvin, V. L., & Klaine, S. J. (2013). Silver nanoparticle toxicity to Daphnia magna is a function of dissolved silver concentration. *Environmental Toxicology and Chemistry / SETAC*, *32*(10), 2356–64.

Nghiem, D., Schäfer, A., & Waite, T. (2002). *Adsorption of estrone on nanofiltration and reverse osmosis membranes in water and wastewater treatment*. Academic Press.

Nghiem, L. D., Schäfer, A. I., & Elimelech, M. (2004). Removal of natural hormones by nanofiltration membranes: Measurement, modeling, and mechanisms. *Environmental Science & Technology*, *38*(6), 1888–1896. doi:10.1021/es034952r PMID:15074703

Nghiem, L., Manis, A., Soldenhoff, K., & Schäfer, A. (2004). Estrogenic hormone removal from wastewater using NF/RO membranes. *Journal of Membrane Science*, *242*(1), 37–45. doi:10.1016/j.memsci.2003.12.034

Ng, I. C. L., Scharf, K., Pogrebna, G., & Maull, R. S. (2015). Contextual variety, internet-of- things and the choice of tailoring over platform: Mass customisation strategy in supply chain management. *International Journal of Production Economics*, *159*, 76–87. doi:10.1016/j.ijpe.2014.09.007

Ng, K. W., Khoo, S. P. K., Heng, B. C., Setyawati, M. I., Tan, E. C., Zhao, X., & Loo, J. S. C. et al. (2011). The role of the tumor suppressor p53 pathway in the cellular DNA damage response to zinc oxide nanoparticles. *Biomaterials*, *32*(32), 8218–8225. doi:10.1016/j.biomaterials.2011.07.036 PMID:21807406

Ngo, C., Van, D., & Voorde, M. H. (2014). Nanotechnologies in agriculture and food. In *Nanotechnology in a nutshell* (pp. 233–247). New York: Springer. doi:10.2991/978-94-6239-012-6_13

Ngo, Y. H., Li, D., Simon, G. P., & Garnier, G. (2011). Paper surfaces functionalized by nanoparticles. *Advances in Colloid and Interface Science*, *163*(1), 23–38. doi:10.1016/j.cis.2011.01.004 PMID:21324427

Nikoobakht, B., & El-Sayed, M. A. (2003). Preparation and growth mechanism of gold nanorods (NRs) using seed-mediated growth method. *Chemistry of Materials*, *15*(10), 1957–1962. doi:10.1021/cm020732l

Nischwitz, V., & Goenaga-Infante, H. (2012). Improved sample preparation and quality control for the characterisation of titanium dioxide nanoparticles in sunscreens using flow field flow fractionation on-line with inductively coupled plasma mass spectrometry. *Journal of Analytical Atomic Spectrometry*, *27*(7), 1084–1092. doi:10.1039/c2ja10387g

Nogueira, P. F. M., Nakabayashi, D., & Zucolotto, V. (2015). The effects of graphene oxide on green algae Raphidocelis subcapitata. *Aquatic Toxicology (Amsterdam, Netherlands)*, *166*, 29–35. doi:10.1016/j.aquatox.2015.07.001 PMID:26204245

Nogueira, V., Lopes, I., Rocha-Santos, T. A. P., Rasteiro, M. G., Abrantes, N., Gonçalves, F., & Pereira, R. et al. (2015). Assessing the ecotoxicity of metal nano-oxides with potential for wastewater treatment. *Environ Sci Pollut Res*, *22*(17), 13212–13224. doi:10.1007/s11356-015-4581-9 PMID:25940480

No, H. K., Meyers, S. P., Prinyawiwatkul, W., & Xu, Z. (2007). Applications of chitosan for improvement of quality and shelf life of foods: A review. *Journal of Food Science*, *72*(5), 87–100. doi:10.1111/j.1750-3841.2007.00383.x PMID:17995743

Noureen, A., & Jabeen, F. (2015). The toxicity, ways of exposure and effects of Cu nanoparticles and Cu bulk salts on different organisms. *International Journal of Biosciences*, *6*(2), 147–156. doi:10.12692/ijb/6.2.147-156

Nover, D. M., McKenzie, E. R., Joshi, G., & Fleenor, W. E. (2013). Assessment of colloidal silver impregnated ceramic bricks for small-scale drinking water treatment applications. *International Journal for Service Learning*, *8*(1), 18–35.

Nowack, B., Ranville, J. F., Diamond, S., Gallego-Urrea, J. A., Metcalfe, C., Rose, J., & Klaine, S. J. et al. (2012). Potential scenarios for nanomaterial release and subsequent alteration in the environment. *Environmental Toxicology and Chemistry*, *31*(1), 50–59. doi:10.1002/etc.726 PMID:22038832

Noy, A., Park, H. G., Fornasiero, F., Holt, J. K., Grigoropoulos, C. P., & Bakajin, O. (2007). Nanofluidics in carbon nanotubes. *Nano Today*, *2*(6), 22–29. doi:10.1016/S1748-0132(07)70170-6

Nussey, G., Van Vuren, J., & du Preez, H. H. (1995). Effect of copper on the haematology and osmoregulation of the Mozambique tilapia, *Oreochromis mossambicus* (Cichlidae). *Comparative Biochemistry and Physiology. Part C, Pharmacology, Toxicology & Endocrinology*, *111*(3), 369–380. doi:10.1016/0742-8413(95)00063-1

O'Carroll, D., Sleep, B., Krol, M., Boparai, H., & Kocur, C. (2013). Nanoscale zero valent iron and bimetallic particles for contaminated site remediation. *Advances in Water Resources*, *51*, 104–122. doi:10.1016/j.advwatres.2012.02.005

Compilation of References

O'Hara, S., Krug, T., Quinn, J., Clausen, C., & Geiger, C. (2006). Field and laboratory evaluation of the treatment of DNAPL source zones using emulsified zero-valent iron. *Remediation Journal*, *16*(2), 35–56. doi:10.1002/rem.20080

Obayelu, A. E. (2014). *Postharvest Losses and Food Waste: The Key Contributing Factors to African Food Insecurity and Environmental Challenges*. Retrieved August 24, 2015, from: http://worldfoodscience.com/article/postharvest-losses-and-food-waste-key-contributing-factors-african-food-insecurity-and#sthash.NMUXu0mB.rzz4xZ5s.dpuf

Oberdörster, G. (2000). Pulmonary effects of inhaled ultrafine particles. *International Archives of Occupational and Environmental Health*, *74*(1), 1–8. doi:10.1007/s004200000185 PMID:11196075

Oberdörster, G., Maynard, A., Donaldson, K., Castranova, V., Fitzpatrick, J., Ausman, K., & Olin, S. et al. (2005). Principles for characterizing the potential human health effects from exposure to nanomaterials: Elements of a screening strategy. *Particle and Fibre Toxicology*, *2*(1), 1. doi:10.1186/1743-8977-2-8 PMID:16209704

Oberdörster, G., Oberdörster, E., & Oberdörster, J. (2005). Nanotoxicology: An emerging discipline evolving from studies of ultrafine particles. *Environmental Health Perspectives*, *113*(7), 823–839. doi:10.1289/ehp.7339 PMID:16002369

Oberdörster, G., Sharp, Z., Atudorei, V., Elder, A., Gelein, R., Lunts, A., & Cox, C. et al. (2002). Extrapulmonary translocation of ultrafine carbon particles following whole-body inhalation exposure of rats. *Journal of Toxicology and Environmental Health. Part A.*, *65*(20), 1531–1543. doi:10.1080/00984100290071658 PMID:12396867

Oberholster, P. J., Musee, N., Botha, M., Chelule, P. K., Focke, W. W., & Ashton, P. J. (2011). Assessment of the effect of nanomaterials on sediment-dwelling invertebrate Chironomus tentans larvae. *Ecotoxicology and Environmental Safety*, *74*(3), 416–423. doi:10.1016/j.ecoenv.2010.12.012 PMID:21216008

Ogawa, K., Dy, J. T., Kobuke, Y., Ogura, S. I., & Okura, I. (2007). Singlet oxygen generation and photocytotoxicity against tumor cell by two-photon absorption. *Molecular Crystals and Liquid Crystals*, *471*(1), 61–67. doi:10.1080/15421400701545270

Ogden, L. E. (2013). Nanoparticles in the Environment: Tiny Size, Large Consequences?. *Bioscience*, *63*(3), 236–236. doi:10.1525/bio.2013.63.3.17

O'Halloran, K. (2006). Toxicological considerations of contaminants in the terrestrial environment for ecological risk assessment. *Human and Ecological Risk Assessment*, *12*(1), 74–83. doi:10.1080/10807030500428603

Ohko, Y., Iuchi, K. I., Niwa, C., Tatsuma, T., Nakashima, T., Iguchi, T., & Fujishima, A. et al. (2002). 17 beta-estrodial degradation by TiO_2 photocatalysis as means of reducing estrogenic activity. *Environmental Science & Technology*, *36*(19), 4175–4181. doi:10.1021/es011500a PMID:12380092

Okasha, A. M., Mohamed, M. B., Abdallah, T., Basily, A. B., Negm, S., & Talaat, H. (2010, March). Characterization of core/shell (Ag/CdSe) nanostructure using photoacoustic spectroscopy.[). IOP Publishing.]. *Journal of Physics: Conference Series*, *214*(1), 012131. doi:10.1088/1742-6596/214/1/012131

Okitsu, K. (2011). Sonochemical synthesis of metal nanoparticles. In *Theoretical and Experimental Sonochemistry Involving Inorganic Systems* (pp. 131–150). Springer Netherlands.

Okitsu, K., Ashokkumar, M., & Grieser, F. (2005). Sonochemical synthesis of gold nanoparticles: Effects of ultrasound frequency. *The Journal of Physical Chemistry B*, *109*(44), 20673–20675. doi:10.1021/jp0549374 PMID:16853678

Okitsu, K., Sharyo, K., & Nishimura, R. (2009). One-pot synthesis of gold nanorods by ultrasonic irradiation: The effect of pH on the shape of the gold nanorods and nanoparticles. *Langmuir*, *25*(14), 7786–7790. doi:10.1021/la9017739 PMID:19545140

Oladimeji, A. A., Quadri, S. U., & DeFreitas, A. S. W. (1984). Long-term effects of arsenic accumulation in rainbow trout, *Salmo gairdneri*. *Bulletin of Environmental Contamination and Toxicology*, *32*(1), 732–741. doi:10.1007/BF01607564 PMID:6743863

Oldenburg, S. J., Averitt, R. D., Westcott, S. L., & Halas, N. J. (1998). Nanoengineering of optical resonances. *Chemical Physics Letters*, *288*(2), 243–247. doi:10.1016/S0009-2614(98)00277-2

Ong, K. J., MacCormack, T. J., Clark, R. J., Ede, J. D., Ortega, V. A., Felix, L. C., & Goss, G. G. et al. (2014). Widespread nanoparticle-assay interference: Implications for nanotoxicity testing. *PLoS ONE*, *9*(3), e90650. doi:10.1371/journal.pone.0090650 PMID:24618833

Opara, L. (2004). *Emerging Technological Innovation Triad for Smart Agriculture in the 21st Century. Part I. Prospects and Impacts of Nanotechnology in Agriculture. In Agricultural Engineering International: the CIGR Journal of Scientific Research and Development* (Vol. 6). Invited Overview Paper.

Orlicky, J. A. (1975). *Material Requirements Planning*. New York, NY: McGrawHill.

Osborne, O. J., Johnston, B. D., Moger, J., Balousha, M., Lead, J. R., Kudoh, T., & Tyler, C. R. (2012). Effects of particle size and coating on nanoscale Ag and TiO2 exposure in zebrafish (Danio rerio) embryos. In *Nanotoxicology*. Taylor & Francis. Retrieved January 22, 2016, from http://www.tandfonline.com/doi/abs/10.3109/17435390.2012.737484?journalCode=inan20

Oshode, O. A., Bakare, A. A., Adeogun, A. O., Efuntoye, M. O., & Sowunmi, A. A. (2008). Ecotoxicological assessment using *Clarias gariepius* and microbial characterization of leachate from municipal solid waste landfill. *International Journal of Environmental of Research*, *2*(4), 391–400.

Ostaszewska, T., Chojnacki, M., Kamaszewski, M., & Chwalibóg, E. S. (2016). Histopathological effects of silver and copper nanoparticles on the epidermis, gills, and liver of Siberian sturgeon. *Environmental Science and Pollution Research International*, *23*(2), 1621–1633. doi:10.1007/s11356-015-5391-9 PMID:26381783

Ostiguy, C., Lapointe, G., Ménard, L., Cloutier, Y., Trottier, M., Boutin, M., … Normand, C. (2006). *Nanoparticles: Current Knowledge about Occupational Health and Safety Risks and Prevention Measures*. Studies and Research, IRSST, Report R-470, Montreal. Retrieved from http://www.irsst.qc.ca/media/documents/pubirsst/r-470. pdf

Otero, J., Mazarrasa, O., Villasante, J., Silva, V., Prádanos, P., Calvo, J., & Hernández, A. (2008). Three independent ways to obtain information on pore size distributions of nanofiltration membranes. *Journal of Membrane Science*, *309*(1), 17–27. doi:10.1016/j.memsci.2007.09.065

Otles, S., & Yalcin, B. (2010). Nano-biosensors as new tool for detection of food quality and safety. *LogForum*, *6*(4), 67–70.

Oukarroum, A., Bras, S., Perreault, F., & Popovic, R. (2012). Inhibitory effects of silver nanoparticles in two green algae, Chlorella vulgaris and Dunaliella tertiolecta. *Ecotoxicology and Environmental Safety*, *78*, 80–85. doi:10.1016/j.ecoenv.2011.11.012 PMID:22138148

Owolade, O. F., Ogunleti, D. O., & Adenekan, M. O. (2008). Titanium dioxide affects disease development and yield of edible cowpea. *Electronic Journal of Environmental Agricultural and Food Chemistry*, *7*(50), 2942–2947.

Oyanedel-Craver, V., & Smith, J. A. (2008). Sustainable colloidal-silver-impregnated ceramic filter for point-of-use water treatment. *Environmental Science & Technology*, *42*(3), 927–933. doi:10.1021/es071268u PMID:18323124

Ozel, R. E., Wallace, K. N., & Andreescu, S. (2014). Alterations of intestinal serotonin following nanoparticle exposure in embryonic zebrafish. *Environmental Science. Nano*, *2014*(1), 27–36.

Özkar, S., & Finke, R. G. (2002). Nanocluster formation and stabilization fundamental studies. 2. Proton sponge as an effective H+ scavenger and expansion of the anion stabilization ability series. *Langmuir, 18*(20), 7653–7662. doi:10.1021/la020225i

Pace, H. E., Rogers, N. J., Jarolimek, C., Coleman, V. A., Gray, E. P., Higgins, C. P., & Ranville, J. F. (2012). Single Particle Inductively Coupled Plasma-Mass Spectrometry: A Performance Evaluation and Method Comparison in the Determination of Nanoparticle Size. *Environmental Science & Technology, 46*(22), 12272–12280. doi:10.1021/es301787d PMID:22780106

Page, K., Wilson, M., & Parkin, I. P. (2009). Antimicrobial surfaces and their potential in reducing the role of the inanimate environment in the incidence of hospital-acquired infections. *Journal of Materials Chemistry, 19*(23), 3819. doi:10.1039/b818698g

Pal, A., Gin, K. Y. H., Lin, A. Y. C., & Reinhard, M. (2010). Impacts of emerging organic contaminants on freshwater resources: Review of recent occurrences, sources, fate and effects. *The Science of the Total Environment, 408*(24), 6062–6069. doi:10.1016/j.scitotenv.2010.09.026 PMID:20934204

Panda, A. B., Glaspell, G., & El-Shall, M. S. (2007). Microwave synthesis and optical properties of uniform nanorods and nanoplates of rare earth oxides. *The Journal of Physical Chemistry C, 111*(5), 1861–1864. doi:10.1021/jp0670283

Panda, A. B., Glaspell, G., & El-Shall, M. S. (2008). Microwave Synthesis of Highly Aligned Ultra Narrow Semiconductor Rods and Wires[J. Am. Chem. Soc. 2006, 128, 2790-2791]. *Journal of the American Chemical Society, 130*(12), 4203–4203. doi:10.1021/ja8003717 PMID:16506744

Pandya, S. (n.d.). *Nanocomposites and its Applications – Review*. Retrieved July 20, 2015 from https://www.academia.edu/3038972/Nanocomposites_and_its_Applications-Review

Pang, C., Selck, H., Misra, S. K., Berhanu, D., Dybowska, A., Valsami-Jones, E., & Forbes, V. E. (2012). Effects of sediment-associated copper to the deposit-feeding snail, *Potamopyrgus antipodarum*: A comparison of Cu added in aqueous form or as nano- and micro-CuO particles. *Aquatic Toxicology, 106 - 107*, 114 - 122.

Pant, H. R., Kim, H. J., Joshi, M. K., Pant, B., Park, C. H., Kim, J. I., & Kim, C. S. et al. (2014). One-step fabrication of multifunctional composite polyurethane spider-web-like nanofibrous membrane for water purification. *Journal of Hazardous Materials, 264*, 25–33. doi:10.1016/j.jhazmat.2013.10.066 PMID:24269971

Pan, Y., Leifert, A., Ruau, D., Neuss, S., Bornemann, J., Schmid, G., & Jahnen-Dechent, W. et al. (2009). Gold nanoparticles of diameter 1.4 nm trigger necrosis by oxidative stress and mitochondrial damage. *Small, 5*(18), 2067–2076. doi:10.1002/smll.200900466 PMID:19642089

Pan, Y., Neuss, S., Leifert, A., Fischer, M., Wen, F., Simon, U., & Brandau, W. et al. (2007). Size-dependent cytotoxicity of gold nanoparticles. *Small, 3*(11), 1941–1949. doi:10.1002/smll.200700378 PMID:17963284

Papavassiliou, G. C. (1979). Optical properties of small inorganic and organic metal particles. *Progress in Solid State Chemistry, 12*(3), 185–271. doi:10.1016/0079-6786(79)90001-3

Parfitt, J., Barthel, M., & Macnaughton, S. (2010). Food waste within food supply chains: quantification and potential for change to 2050. *Philosophical Transactions of the Royal Society B, 365*, 3065-3081. Retrieved from http://worldfoodscience.com/article/postharvest-losses-and-food-waste-key-contributing-factors-african-food-insecurity-and#sthash.NMUXu0mB.dpuf

Parham, H., Zargar, B., & Rezazadeh, M. (2012). Removal, preconcentration and spectrophotometric determination of picric acid in water samples using modified magnetic iron oxide nanoparticles as an efficient adsorbent. *Materials Science and Engineering C, 32*(7), 2109–2114. doi:10.1016/j.msec.2012.05.044

Parikh, D. V., Fink, T., Rajasekharan, K., Sachinvala, N. D., Sawhney, A. P. S., Calamari, T. A., & Parikh, A. D. (2005). Antimicrobial Silver/Sodium Carboxymethyl Cotton Dressings for Burn Wounds. *Textile Research Journal, 75*(2), 134–138. doi:10.1177/004051750507500208

Parisi, C., Vigani, M., & Rodríguez-Cerezo, E. (2014). *Proceedings of a workshop on "Nanotechnology for the agricultural sector: from research to the field"*. Luxembourg: Publications Office of the European Union.

Park, E.-J., Choi, J., Park, Y.-K., & Park, K. (2008). Oxidative stress induced by cerium oxide nanoparticles in cultured BEAS-2B cells. *Toxicology, 245*(1), 90–100. doi:10.1016/j.tox.2007.12.022 PMID:18243471

Park, E.-J., Yi, J., Chung, K.-H., Ryu, D.-Y., Choi, J., & Park, K. (2008). Oxidative stress and apoptosis induced by titanium dioxide nanoparticles in cultured BEAS-2B cells. *Toxicology Letters, 180*(3), 222–229. doi:10.1016/j.toxlet.2008.06.869 PMID:18662754

Park, E.-J., Yi, J., Kim, Y., Choi, K., & Park, K. (2010). Silver nanoparticles induce cytotoxicity by a Trojan-horse type mechanism. *Toxicology In Vitro, 24*(3), 872–878. doi:10.1016/j.tiv.2009.12.001 PMID:19969064

Park, J., Choi, W., Kim, S. H., Chun, B. H., Bang, J., & Lee, K. B. (2010). Enhancement of Chlorine Resistance in Carbon Nanotube Based Nanocomposite Reverse Osmosis Membranes. *Desalination and Water Treatment, 15*(1-3), 198–204. doi:10.5004/dwt.2010.1686

Park, M. V. D. Z., Neigh, A. M., Vermeulen, J. P., de la Fonteyne, L. J. J., Verharen, H. W., Briedé, J. J., & de Jong, W. H. et al. (2011). The effect of particle size on the cytotoxicity, inflammation, developmental toxicity and genotoxicity of silver nanoparticles. *Biomaterials, 32*(36), 9810–9817. doi:10.1016/j.biomaterials.2011.08.085 PMID:21944826

Park, S. J., Cheedrala, R. K., Diallo, M. S., Kim, C., Kim, I. S., & Goddard, W. A. (2012). Nanofiltration membranes based on polyvinylidene fluoride nanofibrous scaffolds and crosslinked polyethyleneimine networks. *Journal of Nanoparticle Research, 14*(7), 884. doi:10.1007/s11051-012-0884-7

Park, S.-J., & Jang, Y.-S. (2003). Preparation and Characterization of Activated Carbon Fibers Supported with Silver Metal for Antibacterial Behavior. *Journal of Colloid and Interface Science, 261*(2), 238–243. doi:10.1016/S0021-9797(03)00083-3 PMID:16256528

Pasricha, R., Gupta, S., & Srivastava, A. K. (2009). A Facile and Novel Synthesis of Ag–Graphene-Based Nanocomposites. *Small, 5*(20), 2253–2259. doi:10.1002/smll.200900726 PMID:19582730

Pastoriza-Santos, I., & Liz-Marzán, L. M. (2002). Formation of PVP-protected metal nanoparticles in DMF. *Langmuir, 18*(7), 2888–2894. doi:10.1021/la015578g

Patra, H. K., Banerjee, S., Chaudhuri, U., Lahiri, P., & Dasgupta, A. K. (2007). Cell selective response to gold nanoparticles. *Nanomedicine; Nanotechnology, Biology, and Medicine, 3*(2), 111–119. doi:10.1016/j.nano.2007.03.005 PMID:17572353

Paul, K. T., Satpathy, S. K., Manna, I., Chakraborty, K. K., & Nando, G. B. (2007). Preparation and characterization of nano structured materials from fly ash: A waste from thermal power stations, by high energy ball milling. *Nanoscale Research Letters, 2*(8), 397–404. doi:10.1007/s11671-007-9074-4

Paull, R. E. (1999). Effects of temperature and relative humidity on fresh commodity quality. *Postharvest Biology and Technology, 15*(3), 263–277. doi:10.1016/S0925-5214(98)00090-8

Paul, S., Jamie, R., Harrison, R. M., Jones, I. P., & Stoll, S. (2005). Characterization of humic substances by environmental scanning electron microscopy. *Environmental Science & Technology, 39*(7), 1962–1966. doi:10.1021/es0489543 PMID:15871224

Paulsen, F. G., Shojaie, S. S., & Krantz, W. B. (1994). Effect of evaporation step on macrovoid formation in wet-cast polymeric membranes. *Journal of Membrane Science, 91*(3), 265–282. doi:10.1016/0376-7388(94)80088-X

Payne, D. J., Gwynn, M. N., Holmes, D. J., & Pompliano, D. L. (2007). Drugs for bad bugs: confronting the challenges of antibacterial discovery. *Nat Rev Drug Discov, 6*(1), 29-40. Retrieved from http://www.nature.com/nrd/journal/v6/n1/pdf/nrd2201.pdf10.1038/nrd2201

Pelaez, M., Nolan, N. T., Pillai, S. C., Seery, M. K., Falaras, P., Kontos, A. G., & Dionysiou, S. D. et al. (2012). A review on the visible light active titanium dioxide photocatalysts for environmental applications. *Applied Catalysis B: Environmental, 125*, 331–349. doi:10.1016/j.apcatb.2012.05.036

Pelley, A. J., & Tufenkji, N. (2008).Effect of particle size and natural organic matter on the migration of nano- and microscale latex particles in saturated porous media. *Journal of Colloid and Interface Science, 321*(1), 74–83.

Pena, D. J., Mbindyo, J. K., Carado, A. J., Mallouk, T. E., Keating, C. D., Razavi, B., & Mayer, T. S. (2002). Template growth of photoconductive metal-CdSe-metal nanowires. *The Journal of Physical Chemistry B, 106*(30), 7458–7462. doi:10.1021/jp0256591

Pena, M. E., Korfiatis, G. P., Patel, M., Lippincott, L., & Meng, X. (2005). Adsorption of As(V) and As(III) by nanocrystalline titanium dioxide. *Water Research, 39*(11), 2327–2337. doi:10.1016/j.watres.2005.04.006 PMID:15896821

Pendergast, M. M., & Hoek, E. M. V. (2011). A review of water treatment membrane nanotechnologies. *Energy & Environmental Science, 4*.

Pendergast, M. T. M., Nygaard, J. M., Ghosh, A. K., & Hoek, E. M. V. (2010). Using nanocomposite materials technology to understand and control reverse osmosis membrane compaction. *Desalination, 261*(3), 255–263. doi:10.1016/j.desal.2010.06.008

Peng, X., Palma, S., Fisher, N. S., & Wong, S. S. (2011). Effect of morphology of ZnO nanostructures on their toxicity to marine algae. *Aquatic Toxicology (Amsterdam, Netherlands), 102*(3-4), 186–196. doi:10.1016/j.aquatox.2011.01.014 PMID:21356181

Peng, Z. A., & Peng, X. (2001). Formation of high-quality CdTe, CdSe, and CdS nanocrystals using CdO as precursor. *Journal of the American Chemical Society, 123*(1), 183–184. doi:10.1021/ja003633m PMID:11273619

Penlidis, A., Hamielec, A., & MacGregor, J. (1983). Hydrodynamic and Size Exclusion Chromatography of Particle Suspensions-An Update. *Journal of Liquid Chromatography, 6*(sup002S2), 179–217. doi:10.1080/01483918308062874

Pepper, D. (1988). RO-fractionation membranes. *Desalination, 70*(1), 89–93. doi:10.1016/0011-9164(88)85046-X

Peralta-Videa, J. R., Zhao, L., Lopez-Moreno, M. L., de la Rosa, G., Hong, J., & Gardea-Torresdey, J. L. (2011). Nanomaterials and the environment: A review for the biennium 2008–2010. *Journal of Hazardous Materials, 186*(1), 1–15. doi:10.1016/j.jhazmat.2010.11.020 PMID:21134718

Pereira, V. J., Galinha, J., Crespo, M. T. B., Matos, C. T., & Crespo, J. G. (2012). Integration of nanofiltration, UV photolysis, and advanced oxidation processes for the removal of hormones from surface water sources. *Separation and Purification Technology, 95*, 89–96. doi:10.1016/j.seppur.2012.04.013

Perelshtein, I., Lipovsky, A., Perkas, N., Gedanken, A., Moschini, E., & Mantecca, P. (2015). The influence of the crystalline nature of nano-metal oxides on their antibacterial and toxicity properties. *Nano Research, 8*(2), 695–707. doi:10.1007/s12274-014-0553-5

Perez-de-Luque, A., & Rubiales, D. (2009). Nanotechnology for parasitic plant control. *Pest Management Science, 65*(5), 540–545. doi:10.1002/ps.1732 PMID:19255973

Perez, J. M., Asati, A., Nath, S. K. C., & Kaittanis, C. (2008). Synthesis of Biocompatible Detrax-coated Nanoceria with pH- dependent Antioxidant Properties. *Small*, *4*(5), 552–556. doi:10.1002/smll.200700824 PMID:18433077

Perez-Juste, J., Liz-Marzan, L. M., Carnie, S., Chan, D. Y., & Mulvaney, P. (2004). Electric-field-directed growth of gold nanorods in aqueous surfactant solutions. *Advanced Functional Materials*, *14*(6), 571–579. doi:10.1002/adfm.200305068

Pergantis, S. A., Jones-Lepp, T. L., & Heithmar, E. M. (2012). Hydrodynamic chromatography online with single particle-inductively coupled plasma mass spectrometry for ultratrace detection of metal-containing nanoparticles. *Analytical Chemistry*, *84*(15), 6454–6462. doi:10.1021/ac300302j PMID:22804728

Perreault, F., Oukarroum, A., Melegari, S. P., Matias, W. G., & Popovic, R. (2012). Polymer coating of copper oxide nanoparticles increases nanoparticles uptake and toxicity in the green alga *Chlamydomonas reinhardtii*. *Chemosphere*, *87*(11), 1388–1394. doi:10.1016/j.chemosphere.2012.02.046 PMID:22445953

Perreault, F., Popovic, R., & Dewez, D. (2014). Different toxicity mechanisms between bare and polymer-coated copper oxide nanoparticles in *Lemna gibba*. *Environmental Pollution*, *185*, 219–227. doi:10.1016/j.envpol.2013.10.027 PMID:24286697

Perret, D., Leppard, G. G., Müller, M., Belzile, N., De Vitre, R., & Buffle, J. (1991). Electron microscopy of aquatic colloids: Non-perturbing preparation of specimens in the field. *Water Research*, *25*(11), 1333–1343. doi:10.1016/0043-1354(91)90111-3

Perry, M., & Linder, C. (1989). Intermediate reverse osmosis ultrafiltration (RO UF) membranes for concentration and desalting of low molecular weight organic solutes. *Desalination*, *71*(3), 233–245. doi:10.1016/0011-9164(89)85026-X

Peters, A. (2005). Particulate matter and heart disease: Evidence from epidemiological studies. *Toxicology and Applied Pharmacology*, *207*(2), S477–S482. doi:10.1016/j.taap.2005.04.030 PMID:15990137

Petersen, R. J. (1993). Composite reverse osmosis and nanofiltration membranes. *Journal of Membrane Science*, *83*(1), 81–150. doi:10.1016/0376-7388(93)80014-O

Peters, K., Unger, R., Kirkpatrick, C. J., Gatti, A., & Monari, E. (2004). Effects of nano-scaled particles on endothelial cell function in vitro: Studies on viability, proliferation and inflammation. *Journal of Materials Science. Materials in Medicine*, *15*(4), 321–325. doi:10.1023/B:JMSM.0000021095.36878.1b PMID:15332593

Peter-Varbanets, M., Zurbrugg, C., Swartz, C., & Pronk, W. (2009). Decentralized systems for potable water and the potential of membrane technology. *Water Research*, *43*(2), 245–265. doi:10.1016/j.watres.2008.10.030 PMID:19010511

Petri, D., Glover, C. N., Ylving, S., Kolas, K., Fremmersvik, G., Waagbo, R., & Berntssen, M. H. G. (2006). Sensitivity of Atlantic salmon (*Salmo salar*) to dietary endosulfan as assessed by haematology, blood biochemistry, and growth parameters. *Aquatic Toxicology (Amsterdam, Netherlands)*, *80*(3), 207–216. doi:10.1016/j.aquatox.2006.07.019 PMID:17081631

Petrović, M., Gonzalez, S., & Barceló, D. (2003). Analysis and removal of emerging contaminants in wastewater and drinking water. *TrAC Trends in Analytical Chemistry*, *22*(10), 685–696. doi:10.1016/S0165-9936(03)01105-1

Pfau, J. C., Sentissi, J. J., Weller, G., & Putnam, E. A. (2005). Assessment of autoimmune responses associated with asbestos exposure in Libby, Montana, USA. *Environmental Health Perspectives*, 25–30. PMID:15626643

Pfuetzner, S., Meiss, J., Petrich, A., Riede, M., & Leo, K. (2009). Improved bulk heterojunction organic solar cells employing C_{70} fullerenes. *Applied Physics Letters*, *94*(22), 223307. doi:10.1063/1.3148664

Phenrat, T., Cihan, A., Kim, H. J., Mital, M., Illangasekare, T., & Lowry, G. V. (2010). Transport and deposition of polymer-modified Fe-0 nanoparticles in 2-D heterogeneous porous media: Effects of particle concentration, Fe-0 content, and coatings. *Environmental Science & Technology*, *44*(23), 9086–9093.

Compilation of References

Philippe, A. (2015). *Hydrodynamic Chromatography for Studying Interactions between Colloids and Dissolved Organic Matter in the Environment*. Academic Press.

Philippe, A., Gangloff, M., Rakcheev, D., & Schaumann, G. (2014). Evaluation of hydrodynamic chromatography coupled with inductively coupled plasma mass spectrometry detector for analysis of colloids in environmental media-effects of colloid composition, coating and shape. *Analytical Methods*, *6*(21), 8722–8728. doi:10.1039/C4AY01567C

Philippe, A., & Schaumann, G. E. (2014a). Evaluation of Hydrodynamic Chromatography Coupled with UV-Visible, Fluorescence and Inductively Coupled Plasma Mass Spectrometry Detectors for Sizing and Quantifying Colloids in Environmental Media. *PLoS ONE*, *9*(2), e90559. doi:10.1371/journal.pone.0090559 PMID:24587393

Philippe, A., & Schaumann, G. E. (2014b). Interactions of dissolved organic matter with natural and engineered inorganic colloids: A review. *Environmental Science & Technology*, *48*(16), 8946–8962. doi:10.1021/es502342r PMID:25082801

Pirela, S. V., Miousse, I. R., Lu, X., Castranova, V., Thomas, T., Qian, Y., & Demokritou, P. et al. (2015). Effects of Laser Printer-Emitted Engineered Nanoparticles on Cytotoxicity, Chemokine Expression, Reactive Oxygen Species, DNA Methylation, and DNA Damage: A Comprehensive Analysis in Human Small Airway Epithelial Cells, Macrophages, and Lymphoblasts. *Environmental Health Perspectives*. doi:10.1289/ehp.1409582 PMID:26080392

Piumetti, M., Hussain, M., Fino, D., & Russo, N. (2015). Mesoporous silica supported Rh catalysts for high concentration N_2O decomposition. *Applied Catalysis B: Environmental*, *165*(0), 158–168. doi:10.1016/j.apcatb.2014.10.008

Plakas, K., Karabelas, A., Wintgens, T., & Melin, T. (2006). A study of selected herbicides retention by nanofiltration membranes—the role of organic fouling. *Journal of Membrane Science*, *284*(1), 291–300. doi:10.1016/j.memsci.2006.07.054

Plaschke, M., Römer, J., & Kim, J. I. (2002). Characterization of Gorleben groundwater colloids by atomic force microscopy. *Environmental Science & Technology*, *36*(21), 4483–4488. doi:10.1021/es0255148 PMID:12433155

Plata, D. L., Gschwend, P. M., & Reddy, C. M. (2008). Industrially synthesized single-walled carbon nanotubes: Compositional data for users, environmental risk assessments, and source apportionment. *Nanotechnology*, *19*(18), 185706. doi:10.1088/0957-4484/19/18/185706 PMID:21825702

Podila, R., & Brown, J. M. (2013). Toxicity of engineered nanomaterials: A physicochemical perspective. *Journal of Biochemical and Molecular Toxicology*, *27*(1), 50–55. doi:10.1002/jbt.21442 PMID:23129019

Polonini, H. C., Brandão, H. M., Raposo, N. R. B., Brandão, M. A. F., Mouton, L., Couté, A., & Brayner, R. et al. (2015). Size-dependent ecotoxicity of barium titanate particles: The case of Chlorella vulgaris green algae. *Ecotoxicology (London, England)*, *24*(4), 938–948. doi:10.1007/s10646-015-1436-6 PMID:25763523

Poole, C. P., & Owens, F. J. (2003). *Introduction to Nanotechnology*. New York: Wiley-Interscience.

Porter, A. E., Gass, M., Muller, K., Skepper, J. N., Midgley, P., & Welland, M. (2007). Visualizing the uptake of C60 to the cytoplasm and nucleus of human monocyte-derived macrophage cells using energy-filtered transmission electron microscopy and electron tomography. *Environmental Science & Technology*, *41*(8), 3012–3017. doi:10.1021/es062541f PMID:17533872

Porter, A. E., Muller, K., Skepper, J., Midgley, P., & Welland, M. (2006). Uptake of C 60 by human monocyte macrophages, its localization and implications for toxicity: Studied by high resolution electron microscopy and electron tomography. *Acta Biomaterialia*, *2*(4), 409–419. doi:10.1016/j.actbio.2006.02.006 PMID:16765881

Post, J. E. (1999). Manganese oxide minerals: Crystal structures and economic and environmental significance. *Proceedings of the National Academy of Sciences of the United States of America*, *96*(7), 3447–3454. doi:10.1073/pnas.96.7.3447 PMID:10097056

Poudelet, E. (2013). *Assessment of nanomaterials in food, health and consumer products, Workshop on the Second Regulatory Review on Nanomaterials*. Retrieved from http://ec.europa.eu/enterprise/sectors/chemicals/files/reach/docs/events/nano-rev-ws-poudelet_en.pdf

Powell, R. M., Puls, R. W., Blowes, D. W., Vogan, J. L., Gillham, R. W., Powell, P. D. ... Landis, R. (1998). Permeable reactive barrier technologies for contaminant remediation. EPA/600/R-98/125, US EPA, Washington DC.

Poynton, H. C., Lazorchak, J. M., Impellitteri, C. A., Blalock, B. J., Rogers, K., Allen, H. J., & Govindasmawy, S. et al. (2012). Toxicogenomic responses of nanotoxicity in Daphnia magna exposed to silver nitrate and coated silver nanoparticles. *Environmental Science & Technology*, *46*(11), 6288–6296. doi:10.1021/es3001618 PMID:22545559

Pradeep, T., & Anshup, . (2009). Noble metal nanoparticles for water purification: A critical review. *Thin Solid Films*, *517*(24), 6441–6478. doi:10.1016/j.tsf.2009.03.195

Prasad, R., Bagde, U. S., & Varma, A. (2012). Intellectual property rights and agricultural biotechnology: An overview. *African Journal of Biotechnology*, *11*(73), 13746–13752. doi:10.5897/AJB12.262

Prasad, R., Kumar, V., & Prasad, K. S. (2014). Nanotechnology in sustainable agriculture: Present Concerns and future aspects. *African Journal of Biotechnology*, *13*(6), 705–713. doi:10.5897/AJBX2013.13554

Prasad, T. N. V. K. V., Sudhakar, P., Sreenivasulu, Y., Latha, P., Munaswamy, V., Reddy, K. R., & Pradeep, T. et al. (2012). Effect of nanoscale zinc oxide particles on the germination, growth and yield of peanut. *Journal of Plant Nutrition*, *35*(6), 905–927. doi:10.1080/01904167.2012.663443

Prat, F., Stackow, R., Bernstein, R., Qian, W. Y., Rubin, Y., & Foote, C. S. (1999). Triplet-state properties and singlet oxygen generation in a homologous series of functionalized fullerene derivatives. *The Journal of Physical Chemistry A*, *103*(36), 7230–7235. doi:10.1021/jp991237o

Praveena, S. M., & Aris, A. Z. (2015). *Application of Low-Cost Materials Coated with Silver Nanoparticle as Water Filter in Escherichia coli Removal*. Water Qual Expo Health; doi:10.1007/s12403-015-0167-5

Premanathan, M., Karthikeyan, K., Jeyasubramanian, K., & Manivannan, G. (2011). Selective toxicity of ZnO nanoparticles toward Gram-positive bacteria and cancer cells by apoptosis through lipid peroxidation. *Nanomedicine (London)*, *7*(2), 184–192. doi:10.1016/j.nano.2010.10.001 PMID:21034861

Prodan, E., Radloff, C., Halas, N. J., & Nordlander, P. (2003). A hybridization model for the plasmon response of complex nanostructures. *Science*, *302*(5644), 419–422. doi:10.1126/science.1089171 PMID:14564001

Proulx, K., & Wilkinson, K. (2014). Separation, detection and characterization of engineered nanoparticles in natural waters using hydrodynamic chromatography and multi-method detection (light scattering, analytical ultracentrifugation and single particle ICP-MS). *Environmental Chemistry*, *11*(4), 392. doi:10.1071/EN13232

Pulskamp, K., Diabaté, S., & Krug, H. F. (2007). Carbon nanotubes show no sign of acute toxicity but induce intracellular reactive oxygen species in dependence on contaminants. *Toxicology Letters*, *168*(1), 58–74. doi:10.1016/j.toxlet.2006.11.001 PMID:17141434

Puntes, V. F., Krishnan, K. M., & Alivisatos, A. P. (2001). Colloidal nanocrystal shape and size control: The case of cobalt. *Science*, *291*(5511), 2115–2117. doi:10.1126/science.1057553 PMID:11251109

Puntes, V. F., Krishnan, K. M., & Alivisatos, P. (2001). Synthesis, self-assembly, and magnetic behavior of a two-dimensional superlattice of single-crystal ε-Co nanoparticles. *Applied Physics Letters*, *78*(15), 2187–2189. doi:10.1063/1.1362333

Puntes, V. F., Krishnan, K., & Alivisatos, A. P. (2002). Synthesis of colloidal cobalt nanoparticles with controlled size and shapes. *Topics in Catalysis*, *19*(2), 145–148. doi:10.1023/A:1015252904412

Compilation of References

Puntes, V. F., Zanchet, D., Erdonmez, C. K., & Alivisatos, A. P. (2002). Synthesis of hcp-Co nanodisks. *Journal of the American Chemical Society*, *124*(43), 12874–12880. doi:10.1021/ja027262g PMID:12392435

Puthussery, J., Kosel, T. H., & Kuno, M. (2009). Facile Synthesis and Size Control of II–VI Nanowires Using Bismuth Salts. *Small*, *5*(10), 1112–1116. doi:10.1002/smll.200801838 PMID:19334010

Qian, H., Peng, X., Han, X., Ren, J., Sun, L., & Fu, Z. (2013). Comparison of the toxicity of silver nanoparticles and silver ions on the growth of terrestrial plant model Arabidopsis thaliana. *Journal of Environmental Sciences (China)*, *25*(9), 1947–1955. doi:10.1016/S1001-0742(12)60301-5 PMID:24520739

Qian, Y. G., Yao, J., Russel, M., Chen, K., & Wang, X. Y. (2015). Characterization of green synthesized nano-formulation (ZnO-A. vera) and their antibacterial activity against pathogens. *Environmental Toxicology and Pharmacology*, *39*(2), 736–746. doi:10.1016/j.etap.2015.01.015 PMID:25723342

Qin, W., & Li, X. (2010). A theoretical study on the catalytic synergetic effects of Pt/graphene nanocomposites. *The Journal of Physical Chemistry C*, *114*(44), 19009–19015. doi:10.1021/jp1072523

Qiuli, W., Abdelli, N., Yiping, L., Min, Z., & Wei, W. (2013). Comparison of toxicities from three metal oxide nanoparticles at environmental relevant concentrations in nematode *Caenorhabditis elegans*. *Chemosphere*, *90*(3), 1123–1131. doi:10.1016/j.chemosphere.2012.09.019 PMID:23062833

Qiu, X., Luo, H., Xu, G. Y., Zhong, R. Y., & Huang, G. Q. (2015). Physical assets and service sharing for IoT-enabled Supply Hub in Industrial Park (SHIP). *International Journal of Production Economics*, *159*, 4–15. doi:10.1016/j.ijpe.2014.09.001

Quang, D. V., Sarawade, P. B., Jeon, S. J., Kim, S. H., Kim, J.-K., Chai, Y. G., & Kim, H. T. (2013). Effective water disinfection using silver nanoparticle containing silica beads. *Applied Surface Science*, *266*, 280–287. doi:10.1016/j.apsusc.2012.11.168

Qu, F., Li, N. B., & Luo, H. Q. (2013). Transition from nanoparticles to nanoclusters: Microscopic and spectroscopic investigation of size-dependent physicochemical properties of polyamine-functionalized silver nanoclusters. *The Journal of Physical Chemistry C*, *117*(7), 3548–3555. doi:10.1021/jp3091792

Quinn, J., Geiger, C., Clausen, C., Brooks, K., Coon, C., O'Hara, S., & Holdsworth, T. et al. (2005). Field demonstration of DNAPL dehalogenation using emulsified zero-valent iron. *Environmental Science & Technology*, *39*(5), 1309–1318. doi:10.1021/es0490018 PMID:15787371

Quirós, J., Borges, J. P., Boltes, K., Rodea-Palomares, I., & Rosal, R. (2015). Antimicrobial electrospun silver-, copper- and zinc-doped polyvinylpyrrolidone nanofibers. *Journal of Hazardous Materials*, *299*, 298–305. doi:10.1016/j.jhazmat.2015.06.028 PMID:26142159

Qu, L., Peng, Z. A., & Peng, X. (2001). Alternative routes toward high quality CdSe nanocrystals. *Nano Letters*, *1*(6), 333–337. doi:10.1021/nl0155532

Qu, X., Alvarez, P. J. J., & Li, Q. (2013). Applications of nanotechnology in water and wastewater treatment. *Water Research*, *47*(12), 3931–3946. doi:10.1016/j.watres.2012.09.058 PMID:23571110

Qu, X., Brame, J., Li, Q., & Alvarez, P. J. J. (2012). Nanotechnology for a Safe and Sustainable Water Supply: Enabling Integrated Water Treatment and Reuse. *Accounts of Chemical Research*, *46*(3), 834–843. doi:10.1021/ar300029v

Radetic, M. (2013). Functionalization of textile materials with silver nanoparticles. *Journal of Materials Science*, *48*(1), 95–107. doi:10.1007/s10853-012-6677-7

Radjenović, J., Petrović, M., Ventura, F., & Barceló, D. (2008). Rejection of pharmaceuticals in nanofiltration and reverse osmosis membrane drinking water treatment. *Water Research*, *42*(14), 3601–3610. doi:10.1016/j.watres.2008.05.020 PMID:18656225

Rady, H. S., Emam, A. N., Mohamed, M. B., & El-shall, M. S. (2016). (Manuscript submitted for publication). Graphene Interface Enhances the Photochemical Synthesis, Stability and Photothermal Effect of Plasmonic Nanostructures. *Carbon*.

Rahale, S. (2010). *Nutrient release pattern of nano – fertilizer formulations*. (Thesis, Ph.D). TNAU, Tamil Nadu.

Rahman, M. F., Wang, J., Patterson, T. A., Saini, U. T., Robinson, B. L., Newport, G. D., & Ali, S. F. et al. (2009). Expression of genes related to oxidative stress in the mouse brain after exposure to silver-25 nanoparticles. *Toxicology Letters*, *187*(1), 15–21. doi:10.1016/j.toxlet.2009.01.020 PMID:19429238

Rahman, Q. I., Ahmad, M., Misra, S., & Lohani, M. (2013). Effective photocatalytic degradation of rhodamine B dye by ZnO nanoparticles. *Materials Letters*, *91*, 170–174. doi:10.1016/j.matlet.2012.09.044

Rahman, Q., Lohani, M., Dopp, E., Pemsel, H., Jonas, L., Weiss, D. G., & Schiffmann, D. (2002). Evidence that ultrafine titanium dioxide induces micronuclei and apoptosis in syrian hamster embryo fibroblasts. *Environmental Health Perspectives*, *110*(8), 797–800. doi:10.1289/ehp.02110797 PMID:12153761

Rai, A., Patnayakuni, R., & Seth, N. (2006). Firm performance impacts of digitally enabled supply chain integration capabilities. *Management Information Systems Quarterly*, *30*(2), 225–246.

Rai, M., Yadav, A., & Gade, A. (2009). Silver nanoparticles as a new generation of antimicrobials. *Biotechnology Advances*, *27*(1), 76–83. doi:10.1016/j.biotechadv.2008.09.002 PMID:18854209

Rai, S., Kookana, R. S., Boxall, A. B. A., Reeves, P. T., Ashauer, R., Beulke, S., & Van den Brink, P. J. et al. (2014). Nanopesticides: Guiding Principles for Regulatory Evaluation of Environmental Risks. *Journal of Agricultural and Food Chemistry*, *62*(19), 4227–4240. doi:10.1021/jf500232f PMID:24754346

Rai, V., Acharya, S., & Dey, N. (2012). Implications of nanobiosensors in agriculture. *Journal of Biomaterials and Nanobiotechnology*, *3*(02), 315–324. doi:10.4236/jbnb.2012.322039

Rajaeian, B., Rahimpour, A., Tade, M. O., & Liu, S. (2013). Fabrication and characterization of polyamide thin film nanocomposite (TFN) nanofiltration membrane impregnated with TiO2 nanoparticles. *Desalination*, *313*(0), 176–188. doi:10.1016/j.desal.2012.12.012

Rajathi, K., & Sridhar, S. (2013). Green Synthesized Silver Nanoparticles From The Medicinal Plant Wrightia Tinctoria and Its Antimicrobial Potential. *International Journal of ChemTech Research*, *5*(4), 1707–1713.

Raj, T., Billing, B. K., Kaur, N., & Singh, N. (2015). Design, synthesis and antimicrobial evaluation of dihydropyrimidone based organic-inorganic nano-hybrids. *RSC Advances*, *5*(58), 46654–46661. doi:10.1039/C5RA08765A

Rakcheev, D., Philippe, A., & Schaumann, G. E. (2013). Hydrodynamic chromatography coupled with single particle-inductively coupled plasma mass spectrometry for investigating nanoparticles agglomerates. *Analytical Chemistry*, *85*(22), 10643–10647. doi:10.1021/ac4019395 PMID:24156639

Raliya, R., & Tarafdar, J. C. (2013). ZnO nanoparticle biosynthesis and its effect on phosphorous-mobilizing enzyme secretion and gum contents in Clusterbean (*Cyamopsis tetragonoloba* L.). *Agriculture Research*, *2*(1), 48–57. doi:10.1007/s40003-012-0049-z

Ramakrishna, S., Fujihara, K., Teo, W. E., Yong, T., Ma, Z., & Ramaseshan, R. (2006). Electrospun nanofibers: Solving global issues. *Materials Today*, *9*(3), 40–50. doi:10.1016/S1369-7021(06)71389-X

Rameshaiah, G. N., Pallavi, J., & Shabnam, S. (2015). Nanofertilizers and nanosensors – an attempt for developing smart agriculture. *International Journal of Engineering Research and General Science, 3*(1), 314–320.

Ramesh, M., Sankaran, M., Veera-Gowtham, V., & Poopal, R. K. (2014). Hematological, biochemical and enzymological responses in an Indian major carp *Labeo rohita* induced by sublethal concentration of waterborne selenite exposure. *Chemico-Biological Interactions, 207*, 67–73. doi:10.1016/j.cbi.2013.10.018 PMID:24183823

Ramesh, M., Srinivasan, R., & Saravanan, M. (2009). Effect of atrazine (Herbicide) on blood parameters of common carp *Cyprinus carpio* (Actinopterygii: Cypriniformes). *African Journal of Environmental Science and Technology, 3*(12), 453–458.

Ramsden, C. S., Smith, T. J., Shaw, B. J., & Handy, R. D. (2009). Dietary exposure to titanium dioxide nanoparticles in rainbow trout (*Oncorhynchus mykiss*): No effect on growth, but subtle biochemical disturbances in the brain. *Ecotoxicology (London, England), 18*(7), 939–951. doi:10.1007/s10646-009-0357-7 PMID:19590957

Ramsden, J. J. (2005). What is nanotechnology? *Nanotechnology Perceptions., 1*(1), 3–17. doi:10.4024/N03RA05/01.01

Ramskov, T., Croteau, M. N., Forbes, V. E., & Selck, H. (2015). Biokinetics of different-shaped copper oxide nanoparticles in the freshwater gastropod, *Potamopyrgus antipodarum*. *Aquatic Toxicology (Amsterdam, Netherlands), 163*, 71–80. doi:10.1016/j.aquatox.2015.03.020 PMID:25863028

Rana, S., & Kalaichelvan, P. T. (2013). Ecotoxicity of nanoparticles. *ISRN Toxicology, 574648*. doi:10.1155/2013/574648 PMID:23724300

Rang, M., Jones, A. C., Zhou, F., Li, Z. Y., Wiley, B. J., Xia, Y., & Raschke, M. B. (2008). Optical near-field mapping of plasmonic nanoprisms. *Nano Letters, 8*(10), 3357–3363. doi:10.1021/nl801808b PMID:18788789

Rao, G. P., Lu, C., & Su, F. (2007). Sorption of divalent metal ions from aqueous solution by carbon nanotubes: A review. *Separation and Purification Technology, 58*(1), 224–231. doi:10.1016/j.seppur.2006.12.006

Rao, S., & Shekhawat, G. S. (2014). Toxicity of ZnO engineered nanoparticles and evaluation of their effect on growth, metabolism and tissue specific accumulation in Brassica juncea. *J Environ Chem Eng, 2*(1), 105–114. doi:10.1016/j.jece.2013.11.029

Raskar, S. V., & Laware, S. L. (2014). Effect of zinc oxide nanoparticles on cytology and seed germination in onion. *Int.J.Curr.Microbiol.App.Sci, 3*(2), 467–473.

Ravichandran, R. (2010). Nanotechnology Applications in Food and Food Processing: Innovative Green Approaches, Opportunities and Uncertainties for Global Market. *International Journal of Green Nanotechnology: Physics and Chemistry, 1*(2), 72–96. doi:10.1080/19430871003684440

Ravindra, S., Mohan, Y. M., Reddy, N. N., & Raju, K. M. (2010). Fabrication of antibacterial cotton fibres loaded with silver nanoparticles via "Green Approach". *Colloids and Surfaces. A, Physicochemical and Engineering Aspects, 367*(1-3), 31–40. doi:10.1016/j.colsurfa.2010.06.013

Rayner, J. (2009). *Current Practices in Manufacturing of Ceramic Pot Filters for Water Treatment. (Master of Science)*. Loughborough University.

Rayner, J., Zhang, H., Schubert, J., Lennon, P., Lantagne, D., & Oyanedel-Craver, V. (2013). Laboratory Investigation into the Effect of Silver Application on the Bacterial Removal Efficacy of Filter Material for Use on Locally Produced Ceramic Water Filters for Household Drinking Water Treatment. *ACS Sustainabe Chemical Engineering, 1*, 737–745. doi:10.1021/sc400068p

Razeeb, K. M., Podporska-Carroll, J., Jamal, M., Hasan, M., Nolan, M., McCormack, D. E., & Pillai, S. C. et al. (2014). Antimicrobial properties of vertically aligned nano-tubular copper. *Materials Letters*, *128*, 60–63. doi:10.1016/j.matlet.2014.04.130

Reaidy, P., Gunasekaran, A., & Spalanzani, A. (2015). Bottom-up approach, internet of things, multi-agent system, RFID, ambient intelligence, collaborative ware-houses. *International Journal of Production Economics*, *159*, 29–40. doi:10.1016/j.ijpe.2014.02.017

Recillas, S., García, A., González, E., Casals, E., Puntes, V., Sánchez, A., & Font, X. (2011). Use of CeO_2, TiO_2 and Fe_3O_4 nanoparticles for the removal of lead from water: Toxicity of nanoparticles and derived compounds. *Desalination*, *277*(1–3), 213–220. doi:10.1016/j.desal.2011.04.036

Reda, M. (2011). In situ production of silver nanoparticle on cotton fabric and its antimicrobial evaluation. *Cellulose (London, England)*, *18*(1), 75–82. doi:10.1007/s10570-010-9455-1

Reddy, A. S., Chen, C.-Y., Chen, C.-C., Jean, J.-S., Chen, H.-R., Tseng, M.-J., & Wang, J.-C. et al. (2010). Biological synthesis of gold and silver nanoparticles mediated by the bacteria Bacillus subtilis. *Journal of Nanoscience and Nanotechnology*, *10*(10), 6567–6574. doi:10.1166/jnn.2010.2519 PMID:21137763

Redmond, R. W., & Kochevar, I. E. (2006). Spatially Resolved Cellular Responses to Singlet Oxygen. *Photochemistry and Photobiology*, *82*(5), 1178–1186. doi:10.1562/2006-04-14-IR-874

Reetz, M. T., & Lohmer, G. (1996). Propylene carbonate stabilized nanostructured palladium clusters as catalysts in Heck reactions. *Chemical Communications*, (16): 1921–1922. doi:10.1039/cc9960001921

Reetz, M. T., & Maase, M. (1999). Redox-Controlled Size-Selective Fabrication of Nanostructured Transition Metal Colloids. *Advanced Materials*, *11*(9), 773–777. doi:10.1002/(SICI)1521-4095(199906)11:9<773::AID-ADMA773>3.0.CO;2-1

Regoli, F., & Principato, G. (1995). Glutathione, glutathione-dependent and antioxidant enzymes in mussel, *Mytilus galloprovincialis*, exposed to metals under field and laboratory conditions: Implications for the use of biochemical biomarkers. *Aquatic Toxicology (Amsterdam, Netherlands)*, *31*(2), 143–164. doi:10.1016/0166-445X(94)00064-W

Regueiro, M. N., Monceau, P., & Hodeau, J. L. (1992). Crushing C_{60} to diamond at room temperature. *Nature*, *355*(6357), 237–239. doi:10.1038/355237a0

Rehman, S., Ullah, R., Butt, A. M., & Gohar, N. D. (2009). Strategies of Making TiO_2 and ZnO Visible Light Active. *Journal of Hazardous Materials*, *170*(2-3), 560–569. doi:10.1016/j.jhazmat.2009.05.064

Remédios, C., Rosário, F., & Bastos, V. (2012). Environmental nanoparticles interactions with plants: Morphological, physiological, and genotoxic aspects. *Le Journal de Botanique*, *2012*. doi:10.1155/2012/751686

Remucal, C. K., & Ginder-Vogel, M. (2014). A critical review of the reactivity of manganese oxides with organic contaminants. *Environmental Science Processes & Impacts*, *16*(6), 1247–1266. doi:10.1039/c3em00703k PMID:24791271

Remya, A. S., Ramesh, M., Saravanan, M., Poopal, R. K., Bharathi, S., & Nataraj, D. (2015). Iron oxide nanoparticles to an Indian major carp, *Labeo rohita*: Impacts on hematology, ionoregulation and gill Na^+/K^+ATPase activity. *Journal of King Saud University – Science*, *27*, 151 – 160.

Remyla, S. R., Ramesh, M., Sajwan, K. S., & Senthil Kumar, K. (2008). Influence of zinc on cadmium induced haematological and biochemical responses in a freshwater teleost fish *Catla catla*. *Fish Physiology and Biochemistry*, *34*(2), 169–174. doi:10.1007/s10695-007-9157-2 PMID:18649034

Ren, D., Colosi, L. M., & Smith, J. A. (2013). Evaluating the Sustainability of Ceramic Filters for Point-of-Use Drinking Water Treatment. *Environmental Science & Technology*, *47*(19), 11206–11213. doi:10.1021/es4026084 PMID:23991752

Compilation of References

Ren, D., & Smith, J. A. (2013). Retention and Transport of Silver Nanoparticles in a Ceramic Porous Medium Used for Point-of-Use Water Treatment. *Environmental Science & Technology*, *47*(8), 3825–3832. doi:10.1021/es4000752 PMID:23496137

Ren, G., Hu, D., Cheng, E. W., Vargas-Reus, M. A., Reip, P., & Allaker, R. P. (2009). Characterisation of copper oxide nanoparticles for antimicrobial applications. *International Journal of Antimicrobial Agents*, *33*(6), 587–590. doi:10.1016/j.ijantimicag.2008.12.004 PMID:19195845

Rengifo-Herrera, J. A., Sanabria, J., Machuca, F., Dierolf, C. F., Pulgarin, C., & Orellana, G. (2007). A comparison of solar photocatalytic inactivation of waterborne E-coli using tris (2,2 '-bipyridine)-ruthenium(II), rose bengal, and TiO_2. *Journal of Solar Energy Engineering-Transactions of the Asme*, *129*(1), 135–140. doi:10.1115/1.2391319

Ren, W., Fang, Y., & Wang, E. (2011). A binary functional substrate for enrichment and ultrasensitive SERS spectroscopic detection of folic acid using graphene oxide/Ag nanoparticle hybrids. *ACS Nano*, *5*(8), 6425–6433. doi:10.1021/nn201606r PMID:21721545

Richardson, S. D., & Postigo, C. (2012). *Drinking water disinfection by-products. In Emerging organic contaminants and human health* (pp. 93–137). Springer.

Richardson, S. D., & Ternes, T. A. (2005). Water analysis: Emerging contaminants and current issues. *Analytical Chemistry*, *77*(12), 3807–3838. doi:10.1021/ac058022x PMID:15952758

Rico, C. M., Majumdar, S., Duarte-Gardea, M., Peralta-Videa, J. R., & Gardea-Torresdey, J. L. (2011). Interaction of nanoparticles with edible plants and their possible implications in the food chain. *Journal of Agricultural and Food Chemistry*, *59*(8), 3485–3498. doi:10.1021/jf104517j PMID:21405020

Ríos, J. L., & Recio, M. C. (2005). Medicinal plants and antimicrobial activity. *Journal of Ethnopharmacology*, *100*(1–2), 80–84. doi:10.1016/j.jep.2005.04.025 PMID:15964727

Rivera-Gil, P., Jimenez De Aberasturi, D., Wulf, V., Pelaz, B., Del Pino, P., Zhao, Y., & Parak, W. J. et al. (2013). The Challenge To Relate the Physicochemical Properties of Colloidal Nanoparticles to Their Cytotoxicity. *Accounts of Chemical Research*, *46*(3), 743–749. doi:10.1021/ar300039j PMID:22786674

Rivera-Utrilla, J., Sanchez-Polo, M., Ferro-Garcia, M. A., Prados-Joya, G., & Ocampo-Perez, R. (2013). Pharmaceuticals as emerging contaminants and their removal from water. A review. *Chemosphere*, *93*(7), 1268–1287. doi:10.1016/j.chemosphere.2013.07.059 PMID:24025536

Robichaud, C. O., Uyar, A. E., Darby, M. R., Zucker, L. G., & Wiesner, M. R. (2009). Estimates of upper bounds and trends in nano-TiO_2 production as a basis for exposure assessment. *Environmental Science & Technology*, *43*(12), 4227–4233. doi:10.1021/es8032549 PMID:19603627

Roco, M. C. (2005). International perspective on government nanotechnology funding in 2005. *Journal of Nanoparticle Research*, *7*(6), 707–712. doi:10.1007/s11051-005-3141-5

Rodriguez-Hernandez, J., Chécot, F., Gnanou, Y., & Lecommandoux, S. (2005). Toward "smart"nano-objects by self-assembly of block copolymers in solution. *Progress in Polymer Science*, *30*(7), 691–724. doi:10.1016/j.progpolymsci.2005.04.002

Roduner, E. (2006). Size matters: Why nanomaterials are different. *Chemical Society Reviews*, *35*(7), 583–592. doi:10.1039/b502142c PMID:16791330

Roe, D., Karandikar, B., Bonn-Savage, N., Gibbins, B., & Roullet, J. B. (2008). Antimicrobial surface functionalization of plastic catheters by silver nanoparticles. *The Journal of Antimicrobial Chemotherapy, 61*(4), 869–876. doi:10.1093/jac/dkn034 PMID:18305203

Rogers, N. J., Franklin, N. M., Apte, S. C., & Batley, G. E. (2007). The importance of physical and chemical characterization in nanoparticle toxicity studies. *Integrated Environmental Assessment and Management, 3*(2), 303–304. doi:10.1002/ieam.5630030219 PMID:17477301

Roh, J.-y., Sim, S. J., Yi, J., Park, K., Chung, K. H., Ryu, D.-y., & Choi, J. (2009). Ecotoxicity of silver nanoparticles on the soil nematode Caenorhabditis elegans using functional ecotoxicogenomics. *Environmental Science & Technology, 43*(10), 3933–3940. doi:10.1021/es803477u PMID:19544910

Romani, R., Antognelli, C., Baldracchini, F., Santis, A., Isani, G., Giovannini, E., & Rosi, G. (2003). Increased acetylcholinesterase activities in specimens of *Sparus auratus* exposed to sublethal concentrations. *Chemico-Biological Interactions, 145*(3), 321–329. doi:10.1016/S0009-2797(03)00058-9 PMID:12732458

Roman, M., Rigo, C., Castillo-Michel, H., Munivrana, I., Vindigni, V., Mivceti'c, I., & Cairns, W. R. et al. (2015). Hydrodynamic chromatography coupled to single-particle ICP-MS for the simultaneous characterization of AgNPs and determination of dissolved Ag in plasma and blood of burn patients. *Analytical and Bioanalytical Chemistry*, 1–16. PMID:26396079

Römer, I., White, T. A., Baalousha, M., Chipman, K., Viant, M. R., & Lead, J. R. (2011). Aggregation and dispersion of silver nanoparticles in exposure media for aquatic toxicity tests. *Journal of Chromatography. A, 1218*(27), 4226–4233. doi:10.1016/j.chroma.2011.03.034 PMID:21529813

Rosas-Hernández, H., Jiménez-Badillo, S., Martínez-Cuevas, P. P., Gracia-Espino, E., Terrones, H., Terrones, M., & González, C. et al. (2009). Effects of 45-nm silver nanoparticles on coronary endothelial cells and isolated rat aortic rings. *Toxicology Letters, 191*(2), 305–313. doi:10.1016/j.toxlet.2009.09.014 PMID:19800954

Rosenfeldt, E. J., & Linden, K. G. (2004). Degradation of endocrine disrupting chemicals bisphenol A, ethinyl estradiol, and estradiol during UV photolysis and advanced oxidation processes. *Environmental Science & Technology, 38*(20), 5476–5483. doi:10.1021/es035413p PMID:15543754

Rothen-Rutishauser, B., Grass, R. N., Blank, F., Limbach, L. K., Muhlfeld, C., Brandenberger, C., & Stark, W. J. et al. (2009). Direct combination of nanoparticle fabrication and exposure to lung cell cultures in a closed setup as a method to simulate accidental nanoparticle exposure of humans. *Environmental Science & Technology, 43*(7), 2634–2640. doi:10.1021/es8029347 PMID:19452928

Rubert, K. F., & Pedersen, J. A. (2006). Kinetics of oxytetracycline reaction with a hydrous manganese oxide. *Environmental Science & Technology, 40*(23), 7216–7221. doi:10.1021/es060357o PMID:17180969

Rudolf, R., Tomic, S., Anzel, I., Zupančič Hartner, T., & Čolić, M. (2014). Microstructure and biocompatibility of gold-lanthanum strips. *Gold Bulletin, 47*(4), 263–273. doi:10.1007/s13404-014-0150-0

Ruiz, P., Katsumiti, A., Nieto, J. A., Bori, J., Romero, A. J., Reip, P., & Cajaraville, M. P. et al. (2015). Short-term effects on antioxidant enzymes and long-term genotoxic and carcinogenic potential of CuO nanoparticles compared to bulk CuO and ionic copper in mussels *Mytilus galloprovincialis*. *Marine Environmental Research, 111*, 107–120. doi:10.1016/j.marenvres.2015.07.018 PMID:26297043

Rule, K. L., Ebbett, V. R., & Vikesland, P. J. (2005). Formation of chloroform and chlorinated organics by free-chlorine-mediated oxidation of triclosan. *Environmental Science & Technology, 39*(9), 3176–3185. doi:10.1021/es048943+ PMID:15926568

Compilation of References

Ruoff, R. S., Tse, D. S., Malhotra, R., & Lorents, D. C. (1993). Solubility of C_{60} in a variety of solvents. *Journal of Physical Chemistry*, *97*(13), 3379–3383. doi:10.1021/j100115a049

Ruparelia, J., Chatterjee, A., Duttagupta, S., & Mukherji, S. (2008). Strain specificity in antimicrobial activity of silver and copper nanoparticles. *Acta Biomaterialia*, *4*(3), 707–716. doi:10.1016/j.actbio.2007.11.006 PMID:18248860

Rusia, V., & Sood, S. K. (1992). Routine hematological studies. In L. Kanai (Ed.), *Medical Laboratory Technology* (pp. 252–258). Tata McGraw Hill Publishing.

Saari, E. E. (1998). Leaf blight diseases and associated soilborne fungal pathogens of wheat in south and southeast Asia. In E. Duveiller, H. J. Dubin, J. Reeves, & A. McNab (Eds.), *Helminthosporium Blights of Wheat: Spot Blotch and Tan Spot* (pp. 37–51). CIMMYT.

Saathoff, J. G., Inman, A. O., Xia, X. R., Riviere, J. E., & Monteiro-Riviere, N. A. (2011). In vitro toxicity assessment of three hydroxylated fullerenes in human skin cells. *Toxicology In Vitro*, *25*(8), 2105–2112. doi:10.1016/j.tiv.2011.09.013 PMID:21964474

Saaty, T. L. (1980). *The analytic hierarchy process*. New York, NY: McGraw-Hill.

Sabourin, V., & Ayande, A. (2015). Commercial Opportunities and Market Demand for Nanotechnologies in Agribusiness Sector. *J. Technol. Manag. Innov.*, *10*(1), 40–51. doi:10.4067/S0718-27242015000100004

Saharan, V., Mehrotra, A., Khatik, R., Rawal, P., Sharma, S. S., & Pal, A. (2013). Synthesis of chitosan based nanoparticles and their in vitro evaluation against phytopathogenic fungi. *International Journal of Biological Macromolecules*, *62*, 677–683. doi:10.1016/j.ijbiomac.2013.10.012 PMID:24141067

Sahoo, N. G., Rana, S., Cho, J. W., Li, L., & Chan, S. H. (2010). Polymer nanocomposites based on functionalized carbon nanotubes. *Progress in Polymer Science*, *35*(7), 837–867. doi:10.1016/j.progpolymsci.2010.03.002

Saidani-Scott, H., Tierney, M., & Sánchez-Silva, F. (2009). Experimental Study of Water Filtering Using Textiles as in Traditional Methods. *Applied Mechanics and Materials*, *15*, 15–20. doi:10.4028/www.scientific.net/AMM.15.15

Sakthivadivel, R. (1966). *Theory and mechanism of filtration of non-colloidal fines through a porous medium. Hydraulic Engineering Laboratory*. Berkeley: University of Califonia.

Sakthivadivel, R. (1969). *Clogging of a granular porous medium by sediment. Hydraulic Engineering Laboratory*. Berkeley: University of Califonia.

Sakthivel, S., Neppolian, B., Shankar, M. V., Arabindoo, B., Palanichamy, M., & Murugesan, V. (2003). Solar photocatalytic degradation of azo dye: Comparison of photocatalytic efficiency of ZnO and TiO2. *Solar Energy Materials and Solar Cells*, *77*(1), 65–82. doi:10.1016/S0927-0248(02)00255-6

Salant, A., Amitay-Sadovsky, E., & Banin, U. (2006). Directed self-assembly of gold-tipped CdSe nanorods. *Journal of the American Chemical Society*, *128*(31), 10006–10007. doi:10.1021/ja063192s PMID:16881617

Sampaio, F. G., Boijink, C. L., Bichara dos Santos, L. R., Tie Oba, E., Kalinin, A. L., Barreto Luiz, A. J., & Rantin, F. T. (2012). Antioxidant defenses and biochemical changes in the neotropical fish pacu, *Piaractus mesopotamicus*: Responses to single and combined copper and hypercarbia exposure. *Comparative Biochemistry and Physiology*, (Part C): 156, 178–186. PMID:22796211

Sanches, S., Galinha, C., Crespo, M. B., Pereira, V., & Crespo, J. (2013). Assessment of phenomena underlying the removal of micropollutants during water treatment by nanofiltration using multivariate statistical analysis. *Separation and Purification Technology*, *118*, 377–386. doi:10.1016/j.seppur.2013.07.020

Sanches, S., Penetra, A., Rodrigues, A., Ferreira, E., Cardoso, V. V., Benoliel, M. J., & Crespo, J. G. et al. (2012). Nanofiltration of hormones and pesticides in different real drinking water sources. *Separation and Purification Technology, 94*, 44–53. doi:10.1016/j.seppur.2012.04.003

Sánchez, A., Recillas, S., Font, X., Casals, E., González, E., & Puntes, V. (2011). Ecotoxicity of, and remediation with, engineered inorganic nanoparticles in the environment. *TrAC Trends Analytical Chem, 30*(3), 507–516. doi:10.1016/j.trac.2010.11.011

Sanchez, D. A., Consiglio, A., & Richaud, Y. (2012). Efficient generation of A9 midbrain dopaminergic neurons by lentiviral delivery of LMX1A in human embryonic stem cells and induced pluripotent stem cells. *Human Gene Therapy, 23*(1), 56–69. doi:10.1089/hum.2011.054 PMID:21877920

Sancho, E., Cerón, J. J., & Ferrando, M. D. (2000). Cholinesterase activity and hematological parameters as biomarkers of sublethal molinate exposure in *Anguilla anguilla*. *Ecotoxicology and Environmental Safety, 46*(1), 81–86. doi:10.1006/eesa.1999.1888 PMID:10805997

Sangpour, P., Hashemi, F., & Moshfegh, A. Z. (2010). Photoenhanced degradation of methylene blue on cosputtered M: TiO_2 (M= Au, Ag, Cu) nanocomposite systems: a comparative study. *The Journal of Physical Chemistry C, 114*(33), 13955–13961. doi:10.1021/jp910454r

Sang, W., Morales, V. L., Zhang, W., Stoof, C. R., Gao, B., Schatz, A. L., & Steenhuis, T. S. et al. (2013). Quantification of colloid retention and release by straining and energy minima in variably saturated porous media. *Environmental Science & Technology, 47*(15), 8256–8264. PMID:23805840

Santo, C. E., Quaranta, D., & Grass, G. (2012). Antimicrobial metallic copper surfaces kill *Staphylococcus haemolyticus* via membrane damage. *Microbiology Open, 1*(1), 46–52.

Saravanan, M., Karthika, S., Malarvizhi, A., & Ramesh, M. (2011b). Ecotoxicological impacts of clofibric acid and diclofenac in common carp (*Cyprinus carpio*) fingerlings: Hematological, biochemical, ionoregulatory and enzymological responses. *Journal of Hazardous Materials, 195*, 188–194. doi:10.1016/j.jhazmat.2011.08.029 PMID:21885190

Saravanan, M., Prabhu Kumar, K., & Ramesh, M. (2011a). Haematological and biochemical responses of freshwater teleost fish *Cyprinus carpio* (Actinopterygii: Crypriniformes) during acute and chronic sublethal exposure to lindane. *Pesticide Biochemistry and Physiology, 100*(3), 206–211. doi:10.1016/j.pestbp.2011.04.002

Saravanan, M., Usha Devi, K., Malarvizhi, A., & Ramesh, M. (2012). Effects of Ibuprofen on hematological, biochemical and enzymological parameters of blood in an Indian major carp, *Cirrhinus mrigala*. *Environmental Toxicology and Pharmacology, 34*(1), 14–22. doi:10.1016/j.etap.2012.02.005 PMID:22418069

Sastry, R. K., Rashmi, H. B., Rao, N. H., & Ilyas, S. M. (2010). Integrating nanotechnology into agri-food systems research in India: A conceptual framework. *Technol. Forecast. Soc., 77*(4), 639–648. doi:10.1016/j.techfore.2009.11.008

Satapanajaru, T., Anurakpongsatorn, P., Pengthamkeerati, P., & Boparai, H. (2008). Remediation of atrazine-contaminated soil and water by nano zerovalent iron. *Water, Air, and Soil Pollution, 192*(1-4), 349–359. doi:10.1007/s11270-008-9661-8

Sathishkumar, M., Sneha, K., & Yun, Y.-S. (2010). Immobilization of silver nanoparticles synthesized using Curcuma longa tuber powder and extract on cotton cloth for bactericidal activity. *Bioresource Technology, 101*(20), 7958–7965. doi:10.1016/j.biortech.2010.05.051 PMID:20541399

Saunders, A. E., Popov, I., & Banin, U. (2006). Synthesis of hybrid CdS-Au colloidal nanostructures. *The Journal of Physical Chemistry B, 110*(50), 25421–25429. doi:10.1021/jp065594s PMID:17165989

Sau, T. K., & Murphy, C. J. (2004). Room temperature, high-yield synthesis of multiple shapes of gold nanoparticles in aqueous solution. *Journal of the American Chemical Society, 126*(28), 8648–8649. doi:10.1021/ja047846d PMID:15250706

Savić, R., Luo, L., Eisenberg, A., & Maysinger, D. (2003). Micellar nanocontainers distribute to defined cytoplasmic organelles. *Science, 300*(5619), 615–618. doi:10.1126/science.1078192 PMID:12714738

Sayes, C. M., Fortner, J. D., Guo, W., Lyon, D., Boyd, A. M., Ausman, K. D., & Hughes, J. B. et al. (2004). The differential cytotoxicity of water-soluble fullerenes. *Nano Letters, 4*(10), 1881–1887. doi:10.1021/nl0489586

Sayes, C. M., Gobin, A. M., Ausman, K. D., Mendez, J., West, J. L., & Colvin, V. L. (2005). Nano-C 60 cytotoxicity is due to lipid peroxidation. *Biomaterials, 26*(36), 7587–7595. doi:10.1016/j.biomaterials.2005.05.027 PMID:16005959

Sayes, C. M., Liang, F., Hudson, J. L., Mendez, J., Guo, W., Beach, J. M., & Billups, W. E. et al. (2006). Functionalization density dependence of single-walled carbon nanotubes cytotoxicity in vitro. *Toxicology Letters, 161*(2), 135–142. doi:10.1016/j.toxlet.2005.08.011 PMID:16229976

Schaep, J., Vandecasteele, C., Mohammad, A. W., & Bowen, W. R. (2001). Modelling the retention of ionic components for different nanofiltration membranes. *Separation and Purification Technology, 22*(1-2), 169–179. doi:10.1016/S1383-5866(00)00163-5

Schäfer, A. I., Fane, A. G., & Waite, T. D. (2005). *Nanofiltration: principles and applications*. Elsevier.

Schimid, G. (1994). *Clusters and colloids: from theory to application*. New York: VCH. doi:10.1002/9783527616077

Schirg, P., & Widmer, F. (1992). Characterisation of nanofiltration membranes for the separation of aqueous dye-salt solutions. *Desalination, 89*(1), 89–107. doi:10.1016/0011-9164(92)80154-2

Schoen, D. T., Schoen, A. P., Hu, L., Kim, H. S., Heilshorn, S. C., & Cui, Y. (2010). High Speed Water Sterilization Using One-Dimensional Nanostructures. *Nano Letters, 10*(9), 3628–3632. doi:10.1021/nl101944e PMID:20726518

Schouteden, K., & Van Haesendonck, C. (2010). Narrow Au (111) terraces decorated by self-organized Co nanowires: A low-temperature STM/STS investigation. *Journal of Physics Condensed Matter, 22*(25), 255504. doi:10.1088/0953-8984/22/25/255504 PMID:21393803

Schrand, A. M., Rahman, M. F., Hussain, S. M., Schlager, J. J., Smith, D. a., & Syed, A. F. (2010). Metal-based nanoparticles and their toxicity assessment. *Wiley Interdisciplinary Reviews: Nanomedicine and Nanobiotechnology, 2*(5), 544–568. doi:10.1002/wnan.103 PMID:20681021

Schuch, H., & Wohleben, W. (2010). Measurement of Particle Size Distribution of Polymer Latexes (L. M. Gugliotta & J. R. Veda, Eds.). Academic Press.

Schull, C. (1948). The determination of pore size distribution from gas adsorption data. *Journal of the American Chemical Society, 70*(4), 1405–1410. doi:10.1021/ja01184a034

Schwab, F., Zhai, G., Kern, M., Turner, A., Schnoor, J. L., & Wiesner, M. R. (2015). Barriers, pathways and processes for uptake, translocation and accumulation of nanomaterials in plants – Critical review. *Nanotoxicology*, 1–22. doi:10.3109/17435390.2015.1048326 PMID:26067571

Scott, N. R. (2007). Nanotechnology opportunities in agriculture and food systems. Biological and environmental engineering, Cornell university NSF nanoscale science and engineering grantees conference, Arlington, VA.

Scott, N., & Chen, H. (2002). *Nanoscale science and engineering for agriculture and food systems*. National Planning Workshop, Washington, DC. Available from: http://www.nseafs.cornell.edu/web.roadmap.pdf

Scott-Fordsmand, J. J., Krogh, P. H., Schaefer, M., & Johansen, A. (2008). The toxicity testing of double-walled nanotubes-contaminated food to Eisenia veneta earthworms. *Ecotoxicology and Environmental Safety, 71*(3), 616–619. doi:10.1016/j.ecoenv.2008.04.011 PMID:18514310

Scown, T. M., van Aerle, R., & Tyler, C. R. (2010). Do engineered nanoparticles pose a significant threat to the aquatic environment? *Critical Reviews in Toxicology, 40*(7), 653–670. doi:10.3109/10408444.2010.494174 PMID:20662713

Scrinis, G., & Lyons, K. (2007). The emerging nano-corporate paradigm: Nanotechnology and the transformation of nature, food and agri-food systems. *International Journal of Sociology of Agriculture and Food, 15*, 22–44.

Scrivens, W. A., Cassell, A. M., North, B. L., & Tour, J. M. (1994). Single Column Purification of Gram Quantities of C_{70}. *Journal of the American Chemical Society, 116*(15), 6939–6940. doi:10.1021/ja00094a060

Sedki, M., Mohamed, M. B., Fawzy, M., Abdelrehim, D. A., & Abdel-Mottaleb, M. M. (2015). Phytosynthesis of silver–reduced graphene oxide (Ag–RGO) nanocomposite with an enhanced antibacterial effect using Potamogeton pectinatus extract. *RSC Advances, 5*(22), 17358–17365.

Segerstedt, A. (1995). Multi-level production and inventory control problems: related to MRP and cover-time planning. Linköping: Production-Economic Research (Produktionsekonomisk forskning). (Profil; No. 13).

Sekhon, B. S. (2014). Nanotechnology in agri-food production: An overview. *Nanotechnology, Science and Applications, 7*, 31–53. doi:10.2147/NSA.S39406 PMID:24966671

Selli, E., Bianchi, C. L., Pirola, C., Cappelletti, G., & Ragaini, V. (2008). Efficiency of 1,4-dichlorobenzene degradation in water under photolysis, photocatalysis on TiO_2 and sonolysis. *Journal of Hazardous Materials, 153*(3), 1136–1141. doi:10.1016/j.jhazmat.2007.09.071 PMID:17976904

Semião, A. J., & Schäfer, A. I. (2013). Removal of adsorbing estrogenic micropollutants by nanofiltration membranes. Part A—Experimental evidence. *Journal of Membrane Science, 431*, 244–256. doi:10.1016/j.memsci.2012.11.080

Sergeev, G. B. (2003). Cryochemistry of metal nanoparticles. *Journal of Nanoparticle Research, 5*(5-6), 529–537. doi:10.1023/B:NANO.0000006153.65107.42

Sergeev, G. B., & Shabatina, T. I. (2008). Cryochemistry of nanometals. *Colloids and Surfaces. A, Physicochemical and Engineering Aspects, 313*, 18–22. doi:10.1016/j.colsurfa.2007.04.064

Serio, M. A., Kroo, E., Bassilakis, R., Wójtowicz, M. A., & Suuberg, E. M. (2001). *A prototype pyrolyzer for solid waste resource recovery in space* (No. 2001-01-2349). SAE Technical Paper.

Serio, M. A., Chen, Y., Wójtowicz, M. A., & Suuberg, E. M. (2000). Pyrolysis processing of mixed solid waste streams. *ACS Div. of Fuel Chem. Prepr, 45*(3), 466–474.

Servin, A. D., Castillo-Michel, H., Hernandez-Viezcas, J. A., Diaz, B. C., Peralta-Videa, J. R., & Gardea-Torresdey, J. L. (2012). Synchrotron micro-XRF and micro-XANES confirmation of the uptake and translocation of TiO_2 nanoparticles in cucumber (Cucumis sativus) plants. *Environmental Science & Technology, 46*(14), 7637–7643. doi:10.1021/es300955b PMID:22715806

Servin, A. D., Morales, M. I., Castillo-Michel, H., Hernandez-Viezcas, J. A., Munoz, B., Zhao, L., & Gardea-Torresdey, J. L. et al. (2013). Synchrotron Verification of TiO_2 accumulation in cucumber fruit: A possible pathway of TiO_2 nanoparticle transfer from soil into the food chain. *Environmental Science & Technology, 47*(20), 11592–11598. doi:10.1021/es403368j PMID:24040965

Seshadri, R., Govindaraj, A., Nagarajan, R., Pradeep, T., & Rao, C. N. R. (1992). Addition of Amines and Halogens to Fullerenes C_{60} and C_{70}. *Tetrahedron Letters, 33*(15), 2069–2070. doi:10.1016/0040-4039(92)88144-T

Compilation of References

Shahid, M., McDonagh, A., Kim, H., & Shon, H. K. (2015). Magnetised titanium dioxide (TiO_2) for water purification: Preparation, characterisation and application. *Desalination and Water Treatment*, *54*(4-5), 979–1002. doi:10.1080/19443994.2014.911119

Shankar, S. S., Rai, A., Ahmad, A., & Sastry, M. (2004). Biosynthesis of silver and gold nanoparticles from extracts of different parts of the geranium plant. *Applications in Nanotechnology*, *1*, 69–77.

Shankar, S. S., Rai, A., Ahmad, A., & Sastry, M. (2004). Rapid synthesis of Au, Ag, and bimetallic Au core–Ag shell nanoparticles using Neem (Azadirachta indica) leaf broth. *Journal of Colloid and Interface Science*, *275*(2), 496–502. doi:10.1016/j.jcis.2004.03.003 PMID:15178278

Sharma, V. K., Li, X. Z., Graham, N., & Doong, R. A. (2008). Ferrate(VI) oxidation of endocrine disruptors and antimicrobials in water. *Journal of Water Supply: Research & Technology - Aqua*, *57*(6), 419–426. doi:10.2166/aqua.2008.077

Sharma, V. K., Yngard, R. A., & Lin, Y. (2009). Silver nanoparticles: Green synthesis and their antimicrobial activities. *Advances in Colloid and Interface Science*, *145*(1), 83–96. doi:10.1016/j.cis.2008.09.002 PMID:18945421

Sharma, V., Anderson, D., & Dhawan, A. (2012). Zinc oxide nanoparticles induce oxidative DNA damage and ROS-triggered mitochondria mediated apoptosis in human liver cells (HepG2). *Apoptosis*, *17*(8), 852–870. doi:10.1007/s10495-012-0705-6 PMID:22395444

Sharma, V., Shukla, R. K., Saxena, N., Parmar, D., Das, M., & Dhawan, A. (2009). DNA damaging potential of zinc oxide nanoparticles in human epidermal cells. *Toxicology Letters*, *185*(3), 211–218. doi:10.1016/j.toxlet.2009.01.008 PMID:19382294

Sharqawy, M. H., Zubair, S. M., & Lienhard, V. J. H. (2011). Second law analysis of reverse osmosis desalination plants: An alternative design using pressure retarded osmosis. *Energy*, *36*(11), 6617–6626. doi:10.1016/j.energy.2011.08.056

Shaviv, E., & Banin, U. (2010). Synergistic Effects on Second Harmonic Generation of Hybrid CdSe– Au Nanoparticles. *ACS Nano*, *4*(3), 1529–1538. doi:10.1021/nn901778k PMID:20192238

Shaw, B. J., Al-Bairuty, G. A., & Handy, R. D., 2012. Effects of waterborne copper nano-particles and copper sulphate on rainbow trout (*Oncorhynchus mykiss*): physiology and accumulation. *Aquatic Toxicology*, *116 – 117*, 90 – 101.

Shaw, B. J., & Handy, R. D. (2011). Physiological effects of nanoparticles on fish: A comparison of nanometals versus metal ions. *Environment International*, *37*(6), 1083–1097. doi:10.1016/j.envint.2011.03.009 PMID:21474182

Shelimov, K. B., Esenaliev, R. O., Rinzler, A. G., Huffman, C. B., & Smalley, R. E. (1998). Purification of single-wall carbon nanotubes by ultrasonically assisted filtration. *Chemical Physics Letters*, *282*(5–6), 429–434. doi:10.1016/S0009-2614(97)01265-7

Shen, C., Shen, Y., Wen, Y., Wang, H., & Liu, W. (2011). Fast and highly efficient removal of dyes under alkaline conditions using magnetic chitosan-Fe(III) hydrogel. *Water Research*, *45*, 5200-5210

Shen, Y.F., Tang, J., Nie, Z.H., Wang, Y.D., Ren, Y., & Zuo, L. (2009). Tailoring size and structural distortion of Fe3O4 nanoparticles for the purification of contaminated water. *Bioresource Technology*, *100*, 4139-4146.

Sheng, Z., & Liu, Y. (2011). Effects of silver nanoparticles on wastewater biofilms. *Water Research*, *45*(18), 6039–6050. doi:10.1016/j.watres.2011.08.065 PMID:21940033

Sherry, L. J., Jin, R., Mirkin, C. A., Schatz, G. C., & Van Duyne, R. P. (2006). Localized surface plasmon resonance spectroscopy of single silver triangular nanoprisms. *Nano Letters*, *6*(9), 2060–2065. doi:10.1021/nl061286u PMID:16968025

Shevchenko, E. V., Ringler, M., Schwemer, A., Talapin, D. V., Klar, T. A., Rogach, A. L., & Alivisatos, A. P. et al. (2008). Self-assembled binary superlattices of CdSe and Au nanocrystals and their fluorescence properties. *Journal of the American Chemical Society*, *130*(11), 3274–3275. doi:10.1021/ja710619s PMID:18293987

Shipway, A. N., Lahav, M., Gabai, R., & Willner, I. (2000). Investigations into the electrostatically induced aggregation of Au nanoparticles. *Langmuir*, *16*(23), 8789–8795. doi:10.1021/la000316k

Shi, W., Zeng, H., Sahoo, Y., Ohulchanskyy, T. Y., Ding, Y., Wang, Z. L., & Prasad, P. N. et al. (2006). A general approach to binary and ternary hybrid nanocrystals. *Nano Letters*, *6*(4), 875–881. doi:10.1021/nl0600833 PMID:16608302

Shi, X., Wang, S., Meshinchi, S., Van Antwerp, M. E., Bi, X., Lee, I., & Baker, J. R. (2007). Dendrimer – entrapped gold nanoparticles as a platform for cancer cell targeting and imaging. *Small*, *3*(7), 1245–1252. doi:10.1002/smll.200700054 PMID:17523182

Shi, Z., Lian, Y., Liao, F., Zhou, X., Gu, Z., Zhang, Y., & Iijima, S. (1999). Purification of single-wall carbon nanotubes. *Solid State Communications*, *112*(1), 35–37. doi:10.1016/S0038-1098(99)00278-1

Shore, L. S., & Shemesh, M. (2003). Naturally produced steroid hormones and their release into the environment. *Pure and Applied Chemistry*, *75*(11-12), 1859–1871. doi:10.1351/pac200375111859

Shu, H.-Y., Chang, M.-C., & Chang, C.-C. (2009). Integration of nanosized zero-valent iron particles addition with UV/H_2O_2 process for purification of azo dye Acid Black 24 solution. *Journal of Hazardous Materials*, *167*(1–3), 1178–1184. doi:10.1016/j.jhazmat.2009.01.106 PMID:19250743

Shukla, R. K., Sharma, V., Pandey, A. K., Singh, S., Sultana, S., & Dhawan, A. (2011). ROS-mediated genotoxicity induced by titanium dioxide nanoparticles in human epidermal cells. *Toxicology In Vitro*, *25*(1), 231–241. doi:10.1016/j.tiv.2010.11.008 PMID:21092754

Siddiqui, M. H., & Al-Whaibi, M. H. (2014). Role of nano-SiO2 in germination of tomato (*Lycopersicum esculentum* seeds Mill.). *Saudi Journal of Biological Sciences*, *21*(1), 13–17. doi:10.1016/j.sjbs.2013.04.005 PMID:24596495

Siddiqui, S., Goddard, R. H., & Bielmyer-Fraser, G. K. (2015). Comparative effects of dissolved copper and copper oxide nanoparticle exposure to the sea anemone, *Exaiptasia pallid*. *Aquatic Toxicology (Amsterdam, Netherlands)*, *160*, 205–213. doi:10.1016/j.aquatox.2015.01.007 PMID:25661886

Simberg, D., Zhang, W.-M., Merkulov, S., McCrae, K., Park, J.-H., Sailor, M. J., & Ruoslahti, E. (2009). Contact activation of kallikrein–kinin system by superparamagnetic iron oxide nanoparticles in vitro and in vivo. *Journal of Controlled Release*, *140*(3), 301–305. doi:10.1016/j.jconrel.2009.05.035 PMID:19508879

Simonet, B. M., & Valcárcel, M. (2009). Monitoring nanoparticles in the environment. *Analytical and Bioanalytical Chemistry*, *393*(1), 17–21. doi:10.1007/s00216-008-2484-z PMID:18974979

Singh, G., Stephan, C., Westerhoff, P., Carlander, D., & Duncan, T. V. (2014). Measurement methods to detect, characterize, and quantify engineered nanomaterials in foods. *Comprehensive Reviews in Food Science and Food Safety*, *13*(4), 693–704. doi:10.1111/1541-4337.12078

Singh, H., & Reddy, T. (1990). Effect of copper sulfate on hematology, blood chemistry, and hepato-somatic index of an Indian catfish, *Heteropneustes fossilis* (Bloch), and its recovery. *Ecotoxicology and Environmental Safety*, *20*(1), 30–35. doi:10.1016/0147-6513(90)90043-5 PMID:2226241

Singh, M., Singh, S., Prasad, S., & Gambhir, I. S. (2008). Nanotechnology in medicine and antibacterial effect of silver nanoparticles. *Digest Journal of Nanomaterials and Biostructures*, *3*(3), 115–122.

Compilation of References

Sintubin, L., Awoke, A. A., Wang, Y., Van der Ha, D., & Verstraete, W. (2012). Enhanced disinfection efficiencies of solar irradiation by biogenic silver. *Annals of Microbiology*, *62*(1), 187–191. doi:10.1007/s13213-011-0245-2

Sinzig, J., de Jongh, L. J., Bönnemann, H., Brijoux, W., & Köppler, R. (1998). Antiferromagnetism of colloidal [MnO· 0.3 THF] x. *Applied Organometallic Chemistry*, *12*(5), 387–391. doi:10.1002/(SICI)1099-0739(199805)12:5<387::AID-AOC743>3.0.CO;2-S

Small, H. (1974). Hydrodynamic chromatography a technique for size analysis of colloidal particles. *Journal of Colloid and Interface Science*, *48*(1), 147–161. doi:10.1016/0021-9797(74)90337-3

Smidt, H., & de Vos, W. M. (2004). Anaerobic microbial dehalogenation. *Annual Review of Microbiology*, *58*(1), 43–73. doi:10.1146/annurev.micro.58.030603.123600 PMID:15487929

Smith, C. J., Shaw, B. J., & Handy, R. D. (2007). Toxicity of single walled carbon nanotubes on rainbow trout, (*Onchorhynchus mykiss*): Respiratory toxicity, organ pathologies and other physiological effects. *Aquatic Toxicology (Amsterdam, Netherlands)*, *82*(2), 94–109. doi:10.1016/j.aquatox.2007.02.003 PMID:17343929

Snow, S. D., Lee, J., & Kim, J. H. (2012). Photochemical and photophysical properties of sequentially functionalized fullerenes in the aqueous phase. *Environmental Science & Technology*, *46*(24), 13227–13234. doi:10.1021/es303237v

Snow, S. D., Park, K., & Kim, J.-. (2014). Cationic Fullerene Aggregates with Unprecedented Virus Photoinactivation Efficiencies in Water. *Environmental Science & Technology Letters*, *1*(6), 290–294. doi:10.1021/ez5001269

Sobsey, M. D., Stauber, C. E., Casanova, L. M., Brown, J. M., & Elliott, M. A. (2008). Point of Use Household Drinking Water Filtration: A Practical, Effective Solution for Providing Sustained Access to Safe Drinking Water in the Developing World. *Environmental Science & Technology*, *42*(12), 4261–4267. doi:10.1021/es702746n PMID:18605542

Soenen, S. J., Parak, W. J., Rejman, J., & Manshian, B. (2015). (Intra)Cellular Stability of Inorganic Nanoparticles: Effects on Cytotoxicity, Particle Functionality, and Biomedical Applications. *Chemical Reviews*, *115*(5), 2109–2135. doi:10.1021/cr400714j PMID:25757742

Sohn, E. K., Chung, Y. S., Johari, S. A., Kim, T. G., Kim, J. K., Lee, J. H., Lee, Y. H., et al. (2015). *Acute Toxicity Comparison of Single-Walled Carbon Nanotubes in Various Freshwater Organisms*. Academic Press.

Solanki, P., Bhargava, A., Chhipa, H., Jain, N., & Panwar, J. (2015). In M. Rai et al. (Eds.), *Nano-fertilizers and Their Smart Delivery System* (pp. 81–101). Nanotechnologies in Food and Agriculture, Springer International Publishing Switzerland; doi:10.1007/978-3-319-14024-7_4

Sondi, I., & Salopek-Sondi, B. (2004). Silver nanoparticles as antimicrobial agent: A case study on E. coli as a model for Gram-negative bacteria. *Journal of Colloid and Interface Science*, *275*(1), 177–182. doi:10.1016/j.jcis.2004.02.012 PMID:15158396

Song, G., Gao, Y., Wu, H., Hou, W., Zhang, C., & Ma, H. (2012). Physiological effect of anatase TiO_2 nanoparticles on Lemna minor. *Environmental Toxicology and Chemistry*, *31*(9), 2147–2152. doi:10.1002/etc.1933 PMID:22760594

Song, K. L., Gao, A. Q., Cheng, X., & Xie, K. L. (2015). Preparation of the superhydrophobic nano-hybrid membrane containing carbon nanotube based on chitosan and its antibacterial activity. *Carbohydrate Polymers*, *130*, 381–387. doi:10.1016/j.carbpol.2015.05.023 PMID:26076639

Song, L., Connolly, M., Fernández-Cruz, M. L., Vijver, M. G., Fernández, M., Conde, E., & Navas, J. M. et al. (2013). Species-specific toxicity of copper nanoparticles among mammalian and piscine cell lines. *Nanotoxicology*, 1–11. PMID:23600739

Song, L., Vijver, M. G., Peijnenburg, W. J. G. M., Galloway, T. S., & Tyle, C. R. (2015). A comparative analysis on the in vivo toxicity of copper nanoparticles in three species of freshwater fish. *Chemosphere*, *139*, 181–189. doi:10.1016/j.chemosphere.2015.06.021 PMID:26121603

Song, U., Jun, H., Waldman, B., Roh, J., Kim, Y., Yi, J., & Lee, E. J. (2013). Functional analyses of nanoparticle toxicity: A comparative study of the effects of TiO_2 and Ag on tomatoes (Lycopersicon esculentum). *Ecotoxicology and Environmental Safety*, *93*, 60–67. doi:10.1016/j.ecoenv.2013.03.033 PMID:23651654

Song, X., Liu, Z., & Sun, D. D. (2013). Energy recovery from concentrated seawater brine by thin-film nanofiber composite pressure retarded osmosis membranes with high power density. *Energy & Environmental Science*, *6*(4), 1199–1210. doi:10.1039/c3ee23349a

Song, Y., Liu, F., & Sun, B. (2005). Preparation, characterization, and application of thin film composite nanofiltration membranes. *Journal of Applied Polymer Science*, *95*(5), 1251–1261. doi:10.1002/app.21338

Soni, D., Naoghare, P. K., Saravanadevi, S., & Pandey, R. A. (2015). Release, transport and toxicity of engineered nanoparticles. In D. M. Whitacre (Ed.), *Reviews of environmental contamination and toxicology* (Vol. 234). Switzerland: Springer. doi:10.1007/978-3-319-10638-0_1

Son, M., Hyeon-gyu, C., Liu, L., Celik, E., & Park, H. (2015). Efficacy of carbon nanotube positioning in the polyethersulfone support layer on the performance of thin-film composite membrane for desalination. *Chemical Engineering Journal*, *266*, 376–384. doi:10.1016/j.cej.2014.12.108

Son, M., Park, H., Liu, L., Choi, H., Kim, J. H., & Choi, H. (2016). Thin-film nanocomposite membrane with CNT positioning in support layer for energy harvesting from saline water. *Chemical Engineering Journal*, *284*, 68–77. doi:10.1016/j.cej.2015.08.134

Son, S. U., Jang, Y., Yoon, K. Y., Kang, E., & Hyeon, T. (2004). Facile synthesis of various phosphine-stabilized monodisperse palladium nanoparticles through the understanding of coordination chemistry of the nanoparticles. *Nano Letters*, *4*(6), 1147–1151. doi:10.1021/nl049519+

Sorribas, S., Gorgojo, P., Téllez, C., Coronas, J., & Livingston, A. G. (2013). High flux thin film nanocomposite membranes based on metal–organic frameworks for organic solvent nanofiltration. *Journal of the American Chemical Society*, *135*(40), 15201–15208. doi:10.1021/ja407665w PMID:24044635

Spadaro, D., Garibaldi, A., & Martines, G. F. (2004). Control of *Penicillium expansum* and *Botrytis cinerea* on apple combining a biocontrol agent with hot water dipping and acibenzolar-S-methyl, baking soda, or ethanol application. *Postharvest Biology and Technology*, *33*(2), 141–151. doi:10.1016/j.postharvbio.2004.02.002

Spasova, M., Salgueiriño-Maceira, V., Schlachter, A., Hilgendorff, M., Giersig, M., Liz-Marzán, L. M., & Farle, M. (2005). Magnetic and optical tunable microspheres with a magnetite/gold nanoparticle shell. *Journal of Materials Chemistry*, *15*(21), 2095–2098. doi:10.1039/b502065d

Speth, T. F., Gusses, A. M., & Summers, R. S. (2000). Evaluation of nanofiltration pretreatments for flux loss control. *Desalination*, *130*(1), 31–44. doi:10.1016/S0011-9164(00)00072-2

Spiegler, K., & Kedem, O. (1966). Thermodynamics of hyperfiltration (reverse osmosis): Criteria for efficient membranes. *Desalination*, *1*(4), 311–326. doi:10.1016/S0011-9164(00)80018-1

Spitalsky, Z., Tasis, D., Papagelis, K., & Galiotis, C. (2010). Carbon nanotube-polymer composites: Chemistry, processing, mechanical and electrical properties. *Progress in Polymer Science*, *35*(3), 357–401. doi:10.1016/j.progpolymsci.2009.09.003

Stamatialis, D. F., Dias, C. R., & de Pinho, M. N. (1999). Atomic force microscopy of dense and asymmetric cellulose-based membranes. *Journal of Membrane Science*, *160*(2), 235–242. doi:10.1016/S0376-7388(99)00089-7

Stampoulis, D., Sinha, S. K., & White, J. C. (2009). Assay-dependent phytotoxicity of nanoparticles to plants. *Environmental Science & Technology*, *43*(24), 9473–9479. doi:10.1021/es901695c PMID:19924897

Staroverov, S. A., Aksinenko, N. M., Gabalov, K. P., Vasilenko, O. A., Vidyasheva, I. V., Shchyogolev, S. Y., & Dykman, L. A. (2009). Effect of gold nanoparticles on the respiratory activity of peritoneal macrophages. *Gold Bulletin*, *42*(2), 153–156. doi:10.1007/BF03214925

Steffens, W. (1989). *Principles of Fish Nutrition* (p. 384). Chichester, UK: Ellis Harwood.

Stephenson, L. M. (1980). Mechanism of the singlet oxygen Ene reaction. *Tetrahedron Letters*, *21*(11), 1005–1008. doi:10.1016/S0040-4039(00)78824-1

Stoisits, R. F., Poehlein, G. W., & Vanderhoff, J. W. (1976). Mathematical modeling of hydrodynamic chromatography. *Journal of Colloid and Interface Science*, *57*(2), 337–344. doi:10.1016/0021-9797(76)90208-3

Stone, A. T. (1987). Reductive dissolution of manganese (III/IV) oxides by substituted phenols. *Environmental Science & Technology*, *21*(10), 979–988. doi:10.1021/es50001a011 PMID:19994996

Stone, V., Nowack, B., Baun, A., van den Brink, N., von der Kammer, F., Dusinska, M., & Fernandes, T. F. et al. (2010). Nanomaterials for environmental studies: Classification, reference material issues, and strategies for physico-chemical characterisation. *The Science of the Total Environment*, *408*(7), 1745–1754. doi:10.1016/j.scitotenv.2009.10.035 PMID:19903569

Straatsma, J., Bargeman, G., Van der Horst, H., & Wesselingh, J. (2002). Can nanofiltration be fully predicted by a model? *Journal of Membrane Science*, *198*(2), 273–284. doi:10.1016/S0376-7388(01)00669-X

Strathmann, H., Scheible, P., & Baker, R. (1971). A rationale for the preparation of Loeb-Sourirajan-type cellulose acetate membranes. *Journal of Applied Polymer Science*, *15*(4), 811–828. doi:10.1002/app.1971.070150404

Striegel, A. M. (2012). Hydrodynamic chromatography: Packed columns, multiple detectors, and microcapillaries. *Analytical and Bioanalytical Chemistry*, *402*(1), 1–5. doi:10.1007/s00216-011-5334-3 PMID:21901463

Striegel, A. M., & Brewer, A. K. (2012). Hydrodynamic Chromatography. *Annual Review of Analytical Chemistry*, *5*(1), 15–34. doi:10.1146/annurev-anchem-062011-143107 PMID:22708902

Striolo, A. (2006). The Mechanism of Water Diffusion in Narrow Carbon Nanotubes. *Nano Letters*, *6*(4), 633–639. doi:10.1021/nl052254u PMID:16608257

Stuart, S. (2003). *Development of Resistance in Pest Populations*. Retrieved from Http://Www.Nd.Edu/Chem191/E2.Html

Subramanian, K. S., Manikandan, A., Thirunavukkarasu, M., & Rahale, C. S. (2015). Nano-fertilizers for Balanced Crop Nutrition in Nanotechnologies in Food and Agriculture. Springer international Publishing.

Subramanian, K. S., Paulraj, C., & Natarajan, S. (2008). Nanotechnological approaches in nutrient management. In C. R. Chinnamuthu, B. Chandrasekaran, & C. Ramasamy (Eds.), *Nanotechnology applications in agriculture, TNAU technical bulletin* (pp. 37–42). Coimbatore: TNAU.

Subramanian, V., Wolf, E. E., & Kamat, P. V. (2003). Influence of metal/metal ion concentration on the photocatalytic activity of TiO2-Au composite nanoparticles. *Langmuir*, *19*(2), 469–474. doi:10.1021/la026478t

Subramanian, V., Wolf, E. E., & Kamat, P. V. (2004). Catalysis with TiO2/gold nanocomposites. Effect of metal particle size on the Fermi level equilibration. *Journal of the American Chemical Society*, *126*(15), 4943–4950. doi:10.1021/ja0315199 PMID:15080700

Su, C. H., Wu, P. L., & Yeh, C. S. (2003). Sonochemical synthesis of well-dispersed gold nanoparticles at the ice temperature. *The Journal of Physical Chemistry B*, *107*(51), 14240–14243. doi:10.1021/jp035451v

Su, C., Puls, R. W., Krug, T. A., Watling, M. T., O'Hara, S. K., Quinn, J. W., & Ruiz, N. E. (2012). A two and half-year-performance evaluation of a field test on treatment of source zone tetrachloroethene and its chlorinated daughter products using emulsified zero valent iron nanoparticles. *Water Research*, *46*(16), 5071–5084. doi:10.1016/j.watres.2012.06.051 PMID:22868086

Su, C., Puls, R. W., Krug, T. A., Watling, M. T., O'Hara, S. K., Quinn, J. W., & Ruiz, N. E. (2013). Travel distance and transformation of injected emulsified zero valent iron nanoparticles in the subsurface during two and half years. *Water Research*, *47*(12), 4095–4106. doi:10.1016/j.watres.2012.12.042 PMID:23562563

Su, G., Zhou, H., Mu, Q., Zhang, Y., Li, L., Jiao, P., & Yan, B. et al. (2012). Effective surface charge density determines the electrostatic attraction between nanoparticles and cells. *The Journal of Physical Chemistry C*, *116*(8), 4993–4998. doi:10.1021/jp211041m

Sugunan, A., & Dutta, J. (2008). Nanotechnology: Environmental Aspects (vol. 2). Wiley-VCH.

Suman, P. R., Jain, V. K., & Varma, A. (2010). Role of nanomaterials in symbiotic fungus growth enhancement. *Current Science*, *99*, 1189–1191.

Sun, A. H., Xiong, Z. G., & Xu, Y. M. (2008). Removal of malodorous organic sulfides with molecular oxygen and visible light over metal phthalocyanine. *Journal of Hazardous Materials*, *152*(1), 191–195. doi:10.1016/j.jhazmat.2007.06.105

Sung, J. H., Ji, J. H., Park, J. D., Yoon, J. U., Kim, D. S., Jeon, K. S., & Han, J. H. et al. (2009). Subchronic inhalation toxicity of silver nanoparticles. *Toxicological Sciences*, *108*(2), 452–461. doi:10.1093/toxsci/kfn246 PMID:19033393

Sun, L., Wu, W., Tian, Q., Lei, M., Liu, J., Xiao, X., & Jiang, C. et al. (2015). *In situ Oxidation and Self-Assembly Synthesis of Dumbbell-Like α-Fe2O3/Ag/AgX (X= Cl, Br, I)*. Heterostructures with Enhanced Photocatalytic Properties. ACS Sustainable Chemistry & Engineering.

Sun, X., Liu, Z., Welsher, K., Robinson, J. T., Goodwin, A., Zaric, S., & Dai, H. (2008). Nano-graphene oxide for cellular imaging and drug delivery. *Nano Research*, *1*(3), 203–212. doi:10.1007/s12274-008-8021-8 PMID:20216934

Sun, Y. P., Guduru, R., Lawson, G. E., Mullins, J. E., Guo, Z. X., Quinlan, J., & Gord, J. R. et al. (2000). Photophysical and electron-transfer properties of mono- and multiple-functionalized fullerene derivatives. *The Journal of Physical Chemistry B*, *104*(19), 4625–4632. doi:10.1021/jp0000329

Sun, Y.-P., Fu, K., Lin, Y., & Huang, W. (2002). Functionalized Carbon Nanotubes: Properties and Applications. *Accounts of Chemical Research*, *35*(12), 1096–1104. doi:10.1021/ar010160v PMID:12484798

Suppan, S. (2013). *Nanomaterials in soil: our future food chain?*. Academic Press.

Suresh, A. K., Pelletier, D. A., & Doktycz, M. J. (2013). Relating nanomaterial properties and microbial toxicity. *Nanoscale*, *5*(2), 463–474. doi:10.1039/C2NR32447D PMID:23203029

Suslick, K. S., Fang, M. M., & Hyeon, T. (1996). Sonochemical synthesis of iron colloids. *Journal of the American Chemical Society*, *118*(47), 11960–11961. doi:10.1021/ja961807n

Sutter, E., Wang, B., Albrecht, P., Lahiri, J., Bocquet, M. L., & Sutter, P. (2012). Templating of arrays of Ru nanoclusters by monolayer graphene/Ru Moirés with different periodicities. *Journal of Physics Condensed Matter*, *24*(31), 314201. doi:10.1088/0953-8984/24/31/314201 PMID:22820349

Suvetha, L., Ramesh, M., & Saravanan, M. (2010). Influence of cypermethrin toxicity on ionic regulation and gill Na^+/K^+-ATPase activity of a freshwater teleost fish *Cyprinus carpio*. *Environmental Toxicology and Pharmacology*, *29*(1), 44–49. doi:10.1016/j.etap.2009.09.005 PMID:21787581

Suvith, V. S., & Philip, D. (2014). Catalytic degradation of methylene blue using biosynthesized gold and silver nanoparticles. *Spectrochimica Acta. Part A: Molecular and Biomolecular Spectroscopy*, *118*, 526–532. doi:10.1016/j.saa.2013.09.016 PMID:24091344

Suzuki, H., Toyooka, T., & Ibuki, Y. (2007). Simple and easy method to evaluate uptake potential of nanoparticles in mammalian cells using a flow cytometric light scatter analysis. *Environmental Science & Technology*, *41*(8), 3018–3024. doi:10.1021/es0625632 PMID:17533873

Sweeney, R. Y., Mao, C., Gao, X., Burt, J. L., Belcher, A. M., Georgiou, G., & Iverson, B. L. (2004). Bacterial biosynthesis of cadmium sulfide nanocrystals. *Chemistry & Biology*, *11*(11), 1553–1559. doi:10.1016/j.chembiol.2004.08.022 PMID:15556006

Szarpak, A., Cui, D., Dubreuil, F., De Geest, B. G., De Cock, L. J., Picart, C., & Auzély-Velty, R. (2010). Designing hyaluronic acid-based layer-by-layer capsules as a carrier for intracellular drug delivery. *Biomacromolecules*, *11*(3), 713–720. doi:10.1021/bm9012937 PMID:20108937

Tabata, A., Kashiwada, S., Ohnishi, Y., Ishikawa, H., Miyamoto, N., Itoh, M., & Magara, Y. (2001). Estrogenic influences of estradiol-17 beta, p-nonylphenol and bis-phenol-A on japanese medaka (Oryzias latipes) at detected environmental concentrations. *Water Science and Technology*, *43*(2), 109–116. PMID:11380168

Talapin, D. V., Lee, J. S., Kovalenko, M. V., & Shevchenko, E. V. (2009). Prospects of colloidal nanocrystals for electronic and optoelectronic applications. *Chemical Reviews*, *110*(1), 389–458. doi:10.1021/cr900137k PMID:19958036

Tammisetti, R. (2010). *Research on the Effectiveness of Using Cloth as a Filter to Remove Turbidity from Water*. Academic Press.

Tang, J., Birkedal, H., McFarland, E. W., & Stucky, G. D. (2003). Self-assembly of CdSe/CdS quantum dots by hydrogen bonding on Au surfaces for photoreception. *Chemical Communications (Cambridge)*, (18): 2278–2279. doi:10.1039/b306888a PMID:14518873

Tang, O., & Grubbström, R. W. (2002). Planning and replanning the master production schedule under demand uncertainty. *International Journal of Production Economics*, *78*(3), 323–334. doi:10.1016/S0925-5273(00)00100-6

Tang, O., & Grubbström, R. W. (2003). The detailed coordination problem in a two-level assembly system with stochastic times. *International Journal of Production Economics*, *81/82*(1), 415–429. doi:10.1016/S0925-5273(02)00296-7

Tang, O., & Grubbström, R. W. (2005). Considering stochastic lead times in a manufacturing/remanufacturing system with deterministic demands and returns. *International Journal of Production Economics*, *93/94*, 285–300. doi:10.1016/j.ijpe.2004.06.027

Tang, O., & Grubbström, R. W. (2006). On using higher-order moments for stochastic inventory systems. *International Journal of Production Economics*, *104*(2), 454–461. doi:10.1016/j.ijpe.2005.03.004

Tang, O., Grubbström, R. W., & Zanoni, S. (2004). Economic evaluation of disassembly processes in remanufacturing systems. *International Journal of Production Research*, *42*(17), 3603–3617. doi:10.1080/00207540410001699435

Tang, O., Grubbström, R. W., & Zanoni, S. (2007). Planned lead time determination in a make-to-order remanufacturing system. *International Journal of Production Economics, 108*(1/2), 426–435. doi:10.1016/j.ijpe.2006.12.034

Tang, S. C. N., & Lo, I. M. C. (2013). Magnetic nanoparticles: Essential factors for sustainable environmental applications. *Water Research, 47*(8), 2613–2632. doi:10.1016/j.watres.2013.02.039 PMID:23515106

Tang, S., Meng, X., Lu, H., & Zhu, S. (2009). PVP-assisted sonoelectrochemical growth of silver nanostructures with various shapes. *Materials Chemistry and Physics, 116*(2-3), 464–468. doi:10.1016/j.matchemphys.2009.04.004

Tanori, J., & Pileni, M. P. (1997). Control of the shape of copper metallic particles by using a colloidal system as template. *Langmuir, 13*(4), 639–646. doi:10.1021/la9606097

Tano, T., Esumi, K., & Meguro, K. (1989). Preparation of organopalladium sols by thermal decomposition of palladium acetate. *Journal of Colloid and Interface Science, 133*(2), 530–533. doi:10.1016/S0021-9797(89)80069-4

Tao, H., & Nallathamby, P. D. (2008). Photostable single-molecule nanoparticle optical biosensors for real-time sensing of single cytokine molecules and their binding reactions. *Journal of the American Chemical Society, 130*(50), 17095–17105. doi:10.1021/ja8068853 PMID:19053435

Tarafdar, J. C., Raliya, R., Mahawar, H., & Rathore, I. (2014). Development of zinc nanofertilizer to enhance crop production in pearl millet (*Pennisetum americanum*). *Agriculture Research, 3*(3), 257–262. doi:10.1007/s40003-014-0113-y

Tarafdar, J. C., Sharma, S., & Raliya, R. (2013). Nanotechnology: Interdisciplinary science of applications. *African Journal of Biotechnology, 12*(3), 219–226.

Tarnuzzer, R. W., Colon, J., Patil, S., & Seal, S. (2005). Vacancy engineered ceria nanostructures for protection from radiation-induced cellular damage. *Nano Letters, 5*(12), 2573–2577. doi:10.1021/nl052024f PMID:16351218

Tavares, K. P., Caloto-Oliveira, A., Vicentini, D. S., Melegari, S. P., Matias, W. G., Barbosa, S., & Kummrow, F. (2014). Acute toxicity of copper and chromium oxide nanoparticles to *Daphnia similis*. *Ecotoxicology and Environmental Contamination, 9*(1), 43–50. doi:10.5132/eec.2014.01.006

Tedesco, S., Doyle, H., Blasco, J., Redmond, G., & Sheehan, D. (2010). Oxidative stress and toxicity of gold nanoparticles in Mytilus edulis. *Aquatic Toxicology (Amsterdam, Netherlands), 100*(2), 178–186. doi:10.1016/j.aquatox.2010.03.001 PMID:20382436

Tegart, G. (2009). Energy and nanotechnologies: Priority areas for Australia's future. *Technological Forecasting and Social Change, 76*(9), 1240–1246. doi:10.1016/j.techfore.2009.06.010

Tegos, G. P., Demidova, T. N., Arcila-Lopez, D., Lee, H., Wharton, T., Gali, H., & Hamblin, M. R. (2005). Cationic fullerenes are effective and selective antimicrobial photosensitizers. *Chemistry & Biology, 12*(10), 1127–1135. doi:10.1016/j.chembiol.2005.08.014

Tejamaya, M., Römer, I., Merrifield, R. C., & Lead, J. R. (2012). Stability of citrate, PVP, and PEG coated silver nanoparticles in ecotoxicology media. *Environmental Science & Technology, 46*(13), 7011–7017. doi:10.1021/es2038596 PMID:22432856

Tellez-Banuelos, M. C., Santerre, A., Casas-Solis, J., Bravo-Cuellar, A., & Zaitseva, G. (2009). Oxidative stress in macrophages from spleen of Nile tilapia (*Oreochromis niloticus*) exposed to sublethal concentration of endosulfan. *Fish & Shellfish Immunology, 27*(2), 105–111. doi:10.1016/j.fsi.2008.11.002 PMID:19049881

Temnov, V. V., Armelles, G., Woggon, U., Guzatov, D., Cebollada, A., Garcia-Martin, A., & Bratschitsch, R. et al. (2010). Active magneto-plasmonics in hybrid metal–ferromagnet structures. *Nature Photonics, 4*(2), 107–111. doi:10.1038/nphoton.2009.265

Compilation of References

Templeton, A. C., Pietron, J. J., Murray, R. W., & Mulvaney, P. (2000). Solvent refractive index and core charge influences on the surface plasmon absorbance of alkanethiolate monolayer-protected gold clusters. *The Journal of Physical Chemistry B, 104*(3), 564–570. doi:10.1021/jp991889c

Teng, X., Feygenson, M., Wang, Q., He, J., Du, W., Frenkel, A. I., & Aronson, M. et al. (2009). Electronic and magnetic properties of ultrathin Au/Pt nanowires. *Nano Letters, 9*(9), 3177–3184. doi:10.1021/nl9013716 PMID:19645434

Teng, X., Han, W., Wang, Q., Li, L., Frenkel, A. I., & Yang, J. C. (2008). Hybrid Pt/Au nanowires: Synthesis and electronic structure. *The Journal of Physical Chemistry C, 112*(38), 14696–14701. doi:10.1021/jp8054685

Teorell, T. (1953). Transport processes and electrical phenomena in ionic membranes. *Progress in Biophysics and Biophysical Chemistry, 3*, 305–369.

Teow, Y., Asharani, P. V., Hande, M. P., & Valiyaveettil, S. (2011). Health impact and safety of engineered nanomaterials. *Chemical Communications (Cambridge, England), 47*(25), 7025–7038. doi:10.1039/c0cc05271j PMID:21479319

Ternes, T. A., Stumpf, M., Mueller, J., Haberer, K., Wilken, R. D., & Servos, M. (1999). Behavior and occurrence of estrogens in municipal sewage treatment plants - I. Investigations in Germany, Canada and Brazil. *The Science of the Total Environment, 225*(1-2), 81–90. doi:10.1016/S0048-9697(98)00334-9 PMID:10028705

Thakkar, K. N., Mhatre, S. S., & Parikh, R. Y. (2010). Biological synthesis of metallic nanoparticles. *Nanomedicine; Nanotechnology, Biology, and Medicine, 6*(2), 257–262. doi:10.1016/j.nano.2009.07.002 PMID:19616126

Thakkar, M. N., Mhatre, S., & Parikh, R. Y. (2010). Biological synthesis of metallic nanoparticles. *Nanotecho.l Biol. Med., 6*, 257–262.

Thanh, N. T. K., & Green, L. A. W. (2010). Functionalisation of nanoparticles for biomedical applications. *Nano Today, 5*(3), 213–230. doi:10.1016/j.nantod.2010.05.003

Theron, J., Walker, J. A., & Cloete, T. E. (2008). Nanotechnology and water treatment: Applications and emerging opportunities. *Critical Reviews in Microbiology, 34*(1), 43–69. doi:10.1080/10408410701710442 PMID:18259980

Thess, A., Lee, R., Nikolaev, P., Dai, H., Petit, P., Robert, J., & Smalley, R. E. et al. (1996). Crystalline Ropes of Metallic Carbon Nanotubes. *Science, 273*(5274), 483–487. doi:10.1126/science.273.5274.483 PMID:8662534

Thiesse, F., & Buckel, T. (2015). A comparison of RFID-based shelf replenishment policies in retail stores under suboptimal read rates. *International Journal of Production Economics, 159*, 126–136. doi:10.1016/j.ijpe.2014.09.002

Thilgen, C., Herrmann, A., & Diederich, F. (1997). The Covalent Chemistry of Higher Fullerenes: C_{70} and Beyond. *Angewandte Chemie International Edition, 36*(21), 2268–2280. doi:10.1002/anie.199722681

Thomas, D. J., & Griffin, P. M. (1996). Coordinated supply chain management. *European Journal of Operational Research, 94*(1), 1–15. doi:10.1016/0377-2217(96)00098-7

Thomas, K. G., & Kamat, P. V. (2003). Chromophore-functionalized gold nanoparticles. *Accounts of Chemical Research, 36*(12), 888–898. doi:10.1021/ar030030h PMID:14674780

Thomas, M., & Klibanov, A. M. (2003). Conjugation to gold nanoparticles enhances polyethylenimine's transfer of plasmid DNA into mammalian cells. *Proceedings of the National Academy of Sciences of the United States of America, 100*(16), 9138–9143. doi:10.1073/pnas.1233634100 PMID:12886020

Thorstenson, A. (1988). Capital Costs in Inventory Models — A Discounted Cash Flow Approach. Linköping:Production-Economic Research in Linköping, PROFIL.

Thorstenson, A., & Grubbström, R. W. (1993). Theoretical analysis of the difference between the traditional and the annuity stream principles applied to inventory evaluation. In R. Flavell (Ed.), Modelling Reality and Personal Modelling (pp. 296–326). Heidelberg: Physica-Verlag. doi:doi:10.1007/978-3-642-95900-4_19 doi:10.1007/978-3-642-95900-4_19

Tian, D. S., Dong, Q., Pan, D. J., He, Y., Yu, Z. Y., Xie, M. J., & Wang, W. (2007). Attenuation of astrogliosis by suppressing of microglial proliferation with the cell cycle inhibitor olomoucine in rat spinal cord injury model. *Brain Research*, *1154*, 206–214. doi:10.1016/j.brainres.2007.04.005 PMID:17482149

Tian, E. L., Zhou, H., Ren, Y. W., mirza, Z., Wang, X. Z., & Xiong, S. W. (2014). Novel design of hydrophobic/hydrophilic interpenetrating network composite nanofibers for the support layer of forward osmosis membrane. *Desalination*, *347*, 207–214. doi:10.1016/j.desal.2014.05.043

Tian, F., Cui, D., Schwarz, H., Estrada, G. G., & Kobayashi, H. (2006). Cytotoxicity of single-wall carbon nanotubes on human fibroblasts. *Toxicology In Vitro*, *20*(7), 1202–1212. doi:10.1016/j.tiv.2006.03.008 PMID:16697548

Tian, M., Qiu, C., Liao, Y., Chou, S., & Wang, R. (2013). Preparation of polyamide thin film composite forward osmosis membranes using electrospun polyvinylidene fluoride (PVDF) nanofibers as substrates. *Separation and Purification Technology*, *118*, 727–736. doi:10.1016/j.seppur.2013.08.021

Tian, M., Wang, R., Goh, K., Liao, Y., & Fane, A. G. (2015). Synthesis and characterization of high-performance novel thin film nanocomposite PRO membranes with tiered nanofiber support reinforced by functionalized carbon nanotubes. *Journal of Membrane Science*, *486*, 151–160. doi:10.1016/j.memsci.2015.03.054

Tiede, K., Boxall, A. B., Tear, S. P., Lewis, J., David, H., & Hassellov, M. (2008). Detection and characterization of engineered nanoparticles in food and the environment. *Food Additives & Contaminants Part A Chemistry, Analysis, Control. Exposure & Risk Assessment.*, *25*(7), 795–821.

Tiede, K., Boxall, A. B., Tiede, D., Tear, S. P., David, H., & Lewis, J. (2009). A robust size-characterisation methodology for studying nanoparticle behaviour in "real" environmental samples, using hydrodynamic chromatography coupled to ICP-MS. *Journal of Analytical Atomic Spectrometry*, *24*(7), 964–972. doi:10.1039/b822409a

Tiede, K., Boxall, A. B., Wang, X., Gore, D., Tiede, D., Baxter, M., & Lewis, J. et al. (2010). Application of hydrodynamic chromatography-ICP-MS to investigate the fate of silver nanoparticles in activated sludge. *Journal of Analytical Atomic Spectrometry*, *25*(7), 1149–1154. doi:10.1039/b926029c

Tiede, K., Boxall, A., Tear, S., Lewis, J., David, H., & Hassellov, M. (2008). Detection and characterization of engineered nanoparticles in food and the environment-a review. *Food Additives and Contaminants*, *25*(07), 795–821. doi:10.1080/02652030802007553 PMID:18569000

Tijssen, R., Bos, J., & Van Kreveld, M. E. (1986). Hydrodynamic chromatography of macromolecules in open microcapillary tubes. *Analytical Chemistry*, *58*(14), 3036–3044. doi:10.1021/ac00127a030

Tillman, D., Cassman, K. G., Matson, P. A., Naylor, R., & Polasky, S. (2002). Agricultural sustainability and intensive production practices. *Nature*, *418*(6898), 671–677. doi:10.1038/nature01014 PMID:12167873

Tipping, E. (2002). *Cation binding by humic substances*. Cambridge University Press. doi:10.1017/CBO9780511535598

Tiraferri, A., Vecitis, C. D., & Elimelech, M. (2011). Covalent Binding of Single-Walled Carbon Nanotubes to Polyamide Membranes for Antimicrobial Surface Properties. *ACS Applied Materials & Interfaces*, *3*(8), 2869–2877. doi:10.1021/am200536p PMID:21714565

Tiwari, D. K., Behari, J., & Sen, P. (2008). Application of nanoparticles in waste water treatment. *World Appl Sci J*, *3*(3), 417–433.

Tkachenko, A. G., Xie, H., Coleman, D., Glomm, W., Ryan, J., Anderson, M. F., & Frazen, S. (2003). Multifunctional gold nanoparticle-peptide complexes for nuclear targeting. *Journal of the American Chemical Society, 125*(16), 4700–4701. doi:10.1021/ja0296935 PMID:12696875

Toh, Y. S., Lim, F., & Livingston, A. (2007). Polymeric membranes for nanofiltration in polar aprotic solvents. *Journal of Membrane Science, 301*(1), 3–10.

Tominaga, Y., Kubo, T., & Hosoya, K. (2011). Surface modification of TiO 2 for selective photodegradation of toxic compounds. *Catalysis Communications, 12*(9), 785–789. doi:10.1016/j.catcom.2011.01.021

Tongying, P., Plashnitsa, V. V., Petchsang, N., Vietmeyer, F., Ferraudi, G. J., Krylova, G., & Kuno, M. (2012). Photocatalytic hydrogen generation efficiencies in one-dimensional CdSe heterostructures. *The Journal of Physical Chemistry Letters, 3*(21), 3234–3240. doi:10.1021/jz301628b PMID:26296035

Torney, F., Trewyn, B. G., Lin, V. S., & Wang, K. (2007). Mesoporous silica nanoparticles deliver DNA and chemicals into plants. *Nature Nanotechnology, 2*(5), 295–300. doi:10.1038/nnano.2007.108 PMID:18654287

Torrado, J. F., González-Díaz, J. B., González, M. U., García-Martín, A., & Armelles, G. (2010). Magneto-optical effects in interacting localized and propagating surface plasmon modes. *Optics Express, 18*(15), 15635–15642. doi:10.1364/OE.18.015635 PMID:20720945

Tourinho, P. S., van Gestel, C. A. M., Jurkschat, K., Soares, A. M. V. M., & Loureiro, S. (2015). Effects of soil and dietary exposures to Ag nanoparticles and $AgNO_3$ in the terrestrial isopod Porcellionides pruinosus. *Environmental Pollution, 205*, 170–177. doi:10.1016/j.envpol.2015.05.044 PMID:26071943

Tran, Q. H., & Le, A. T. (2013). Silver nanoparticles: Synthesis, properties, toxicology, applications and perspectives. *Advances in Natural Sciences: Nanoscience and Nanotechnology, 4*(3), 033001.

Tranquada, J. M., Sternlieb, B. J., Axe, J. D., Nakamura, Y., & Uchida, S. (1995). Evidence for stripe correlations of spins and holes in copper oxide superconductors. *Nature, 375*(6532), 561–563. doi:10.1038/375561a0

Triboulet, S., Aude-Garcia, C., Armand, L., Gerdil, A., Diemer, H., Proamer, F., & Rabilloud, T. et al. (2014). Analysis of cellular responses of macrophages to zinc ions and zinc oxide nanoparticles: A combined targeted and proteomic approach. *Nanoscale, 6*(11), 6102–6114. doi:10.1039/c4nr00319e PMID:24788578

Trujillo-Reyes, J., Majumdar, S., Botez, C. E., Peralta-Videa, J. R., & Gardea-Torresdey, J. L. (2014). Exposure studies of core-shell Fe/Fe_3O_4 and Cu/CuO NPs to lettuce (Lactuca sativa) plants: Are they a potential physiological and nutritional hazard? *Journal of Hazardous Materials, 267*, 255–263. doi:10.1016/j.jhazmat.2013.11.067 PMID:24462971

Tufenkji, N., & Elimelech, M. (2004). Correlation equation for predicting single-collector efficiency in physicochemical filtration in saturated porous media. *Environmental Science & Technology, 38*(2), 529–536. doi:10.1021/es034049r PMID:14750730

Tufenkji, N., Miller, G. F., Ryan, J. N., Harvey, R. W., & Elimelech, M. (2004). Transport of cryptosporidium oocysts in porous media: Role of straining and physicochemical filtration. *Environmental Science & Technology, 38*(22), 5932–5938. doi:10.1021/es049789u PMID:15573591

Tuoriniemi, J., Cornelis, G., & Hassellov, M. (2012). Size discrimination and detection capabilities of single-particle ICPMS for environmental analysis of silver nanoparticles. *Analytical Chemistry, 84*(9), 3965–3972. doi:10.1021/ac203005r PMID:22483433

Turkevich, J. (1985). Colloidal gold. Part II. *Gold Bulletin, 18*(4), 125–131. doi:10.1007/BF03214694

Turkevich, J., Garton, G., & Stevenson, P. C. (1954). The color of colloidal gold. *Journal of Colloid Science, 9*, 26–35. doi:10.1016/0095-8522(54)90070-7

Turkevich, J., & Kim, G. (1970). Palladium: Preparation and catalytic properties of particles of uniform size. *Science, 169*(3948), 873–879. doi:10.1126/science.169.3948.873 PMID:17750062

Ukrainczyk, L., & McBride, M. B. (1992). Oxidation of phenol in acidic aqueous suspensions of manganese oxides. *Archive of Clays and Clay Minerals, 40*(2), 157–166. doi:10.1346/CCMN.1992.0400204

Ullah, R., Zuberi, A., Naeem, M., & Ullah, S. (2015). Toxicity to hematology of liver, brain and gills during acute exposure of Mahseer (*Tor putitora*) to cypermethrin. *Pakistan Journal of Agriculture & Biology, 12* - 18.

Umut, E., Pineider, F., Arosio, P., Sangregorio, C., Corti, M., Tabak, F., & Ghigna, P. et al. (2012). Magnetic, optical and relaxometric properties of organically coated gold–magnetite (Au–Fe_3O_4) hybrid nanoparticles for potential use in biomedical applications. *Journal of Magnetism and Magnetic Materials, 324*(15), 2373–2379. doi:10.1016/j.jmmm.2012.03.005

United States Environmental Protection Agency (EPA). (2012). *Greening EPA: Federal waste diversion and recycling requirements*. Retrieved June 2015, from http://www.epa.gov/greeningepa/waste/requirements.htm

United States Environmental Protection Agency (EPA). (2015). *Advancing sustainable materials management: Assessing trends in material generation, recycling and disposal in the United States*. Retrieved June 2015, from http://www.epa.gov/wastes/nonhaz/municipal/pubs/2013_advncng_smm_fs.pdf

United States Environmental Protection Agency (EPA). (2015). *Municipal solid waste*. Retrieved July 2015, from http://www.epa.gov/epawaste/nonhaz/municipal/

URS Corporation. (2005). *Evaluation of alternative solid waste processing technologies*. Retrieved June 2015, from http://lacitysan.org/solid_resources/strategic_programs/alternative_tech/PDF/final_report.pdf

Urynowicz, M. A. (2008). In situ chemical oxidation with permanganate: Assessing the competitive interactions between target and nontarge compounds. *Soil & Sediment Contamination, 17*(1), 53–62. doi:10.1080/15320380701741412

Usenko, C. Y., Harper, S. L., & Tanguay, R. L. (2007). In vivo evaluation of carbon fullerene toxicity using embryonic zebrafish. *Carbon, 45*(9), 1891–1898. doi:10.1016/j.carbon.2007.04.021 PMID:18670586

Usenko, C. Y., Harper, S. L., & Tanguay, R. L. (2008). Fullerene C60 exposure elicits an oxidative stress response in embryonic zebrafish. *Toxicology and Applied Pharmacology, 229*(1), 44–55. doi:10.1016/j.taap.2007.12.030 PMID:18299140

Uyak, V., Koyuncu, I., Oktem, I., Cakmakci, M., & Toroz, I. (2008). Removal of trihalomethanes from drinking water by nanofiltration membranes. *Journal of Hazardous Materials, 152*(2), 789–794. doi:10.1016/j.jhazmat.2007.07.082 PMID:17768007

Valhondo, C., Carrera, J., Ayora, C., Tubau, I., Martinez-Landa, L., Nodler, K., & Licha, T. (2015). Characterizing redox conditions and monitoring attenuation of selected pharmaceuticals during artificial recharge through a reactive layer. *The Science of the Total Environment, 512*, 240–250. doi:10.1016/j.scitotenv.2015.01.030 PMID:25625636

Vallabani, N. V., Mittal, S., Shukla, R. K., Pandey, A. K., Dhakate, S. R., Pasricha, R., & Dhawan, A. (2011). Toxicity of graphene in normal human lung cells (BEAS-2B). *Journal of Biomedical Nanotechnology, 7*(1), 106–107. doi:10.1166/jbn.2011.1224 PMID:21485826

Van Aerle, R., Lange, A., Moorhouse, A., Paszkiewicz, K., Ball, K., Johnston, B. D., & Santos, E. M. et al. (2013). Molecular mechanisms of toxicity of silver nanoparticles in zebrafish embryos. *Environmental Science & Technology, 47*(14), 8005–8014. doi:10.1021/es401758d PMID:23758687

Compilation of References

Van der Bruggen, B., Schaep, J., Maes, W., Wilms, D., & Vandecasteele, C. (1998). Nanofiltration as a treatment method for the removal of pesticides from ground waters. *Desalination*, *117*(1), 139–147. doi:10.1016/S0011-9164(98)00081-2

Van Wyk, H. (2015). Antibiotic resistance[review]. *SA Pharmaceutical Journal*, *82*(3), 20–23. PMID:26415379

Vazsonyi, A. (1955). The use of mathematics in production and inventory control. *Management Science*, *1*(3/4), 70–85.

Velasco-Santamaría, Y. M., Handy, R. D., & Sloman, K. A. (2011). Endosulfan affects health variables in adult zebrafish (*Danio rerio*) and induces alterations in larvae development. *Comparative Biochemistry and Physiology Part C Toxicology & Pharmacology*, *153*(4), 372–380. doi:10.1016/j.cbpc.2011.01.001 PMID:21262389

Velisek, J., Svobodova, Z., & Piackova, V. (2009). Effects of acute exposure to bifenthrin on some haematological, biochemical and histopathological parameters of rainbow trout (*Oncorhynchus mykiss*). *Veterinarni Medicina*, *54*(3), 131–137.

Verma, S. (2002). *Anaerobic digestion of biodegradable organics in municipal solid wastes.* (Master's thesis). Columbia University, New York, NY.

Verma, A., Uzun, O., Hu, Y., Hu, Y., Han, H. S., Watson, N., & Stellacci, F. et al. (2008). Surface structure regulated cell membrane penetration by monolayer protected nanoparticles. *Nature Materials*, *7*(7), 588–595. doi:10.1038/nmat2202 PMID:18500347

Verweij, H., Schillo, M. C., & Li, J. (2007). Fast Mass Transport Through Carbon Nanotube Membranes. *Small*, *3*(12), 1996–2004. doi:10.1002/smll.200700368 PMID:18022891

Vevers, W. F., & Jha, A. N. (2008). Genotoxic and cytotoxic potential of titanium dioxide (TiO2) nanoparticles on fish cells in vitro. *Ecotoxicology (London, England)*, *17*(5), 410–420. doi:10.1007/s10646-008-0226-9 PMID:18491228

Vicario-Parés, U., Castañaga, L., Lacave, J. M., Oron, M., Reip, P., Berhanu, D., & Orbea, A. et al. (2014). Comparative toxicity of metal oxide nanoparticles (CuO, ZnO and TiO_2) to developing zebrafish embryos. *Journal of Nanoparticle Research*, *16*(8), 2550. doi:10.1007/s11051-014-2550-8

Vickery, S. K., Jayaram, J., Droge, C., & Calantone, R. (2003). The effects of an integrative supply chain strategy on customer service and financial performance: An analysis of direct versus indirect relationships. *Journal of Operations Management*, *21*(5), 523–539. doi:10.1016/j.jom.2003.02.002

Vidoni, O., Philippot, K., Amiens, C., Chaudret, B., Balmes, O., Malm, J. O., & Casanove, M. J. et al. (1999). Novel, spongelike ruthenium particles of controllable size stabilized only by organic solvents. *Angewandte Chemie International Edition*, *38*(24), 3736–3738. doi:10.1002/(SICI)1521-3773(19991216)38:24<3736::AID-ANIE3736>3.0.CO;2-E PMID:10649342

Vileno, B., Marcoux, P. R., Lekka, M., Sienkiewicz, A., Fehér, T., & Forró, L. (2006). Spectroscopic and photophysical properties of a highly derivatized C60 fullerol. *Advanced Functional Materials*, *16*(1), 120–128. doi:10.1002/adfm.200500425

Vinodgopal, K., He, Y., Ashokkumar, M., & Grieser, F. (2006). Sonochemically prepared platinum-ruthenium bimetallic nanoparticles. *The Journal of Physical Chemistry B*, *110*(9), 3849–3852. doi:10.1021/jp060203v PMID:16509663

Vinodgopal, K., Neppolian, B., Lightcap, I. V., Grieser, F., Ashokkumar, M., & Kamat, P. V. (2010). Sonolytic design of graphene−Au nanocomposites. simultaneous and sequential reduction of graphene oxide and Au (III). *The Journal of Physical Chemistry Letters*, *1*(13), 1987–1993. doi:10.1021/jz1006093

Vishnupriya, S., Chaudhari, K., Jagannathan, R., & Pradeep, T. (2013). Single-Cell Investigations of Silver Nanoparticle-Bacteria Interactions. *Particle & Particle Systems Characterization*, *30*(12), 1056–1062. doi:10.1002/ppsc.201300165

Visnapuu, M., Joost, U., Juganson, K., Kunnis-Beres, K., Kahru, A., Kisand, V., & Ivask, A. (2013). Dissolution of Silver Nanowires and Nanospheres Dictates Their Toxicity to Escherichia coli. *Biomed Res Int*. doi:10.1155/2013/819252

Vithiya, K., & Sen, S. (2011). Biosynthesis of nanoparticles. *Int J Pharm Sci Res, 2*, 2781–2785.

Volland, M., Hampel, M., Martos-Sitcha, J. A., Trombini, C., Martínez-Rodríguez, G., & Blasco, J. (2015). Citrate gold nanoparticle exposure in the marine bivalve Ruditapes philippinarum: Uptake, elimination and oxidative stress response. *Environmental Science and Pollution Research International, 22*(22), 17414–17424. doi:10.1007/s11356-015-4718-x PMID:25994271

Von der Kammer, F., Baborowski, M., & Friese, K. (2005). Field-flow fractionation coupled to multi-angle laser light scattering detectors: Applicability and analytical benefits for the analysis of environmental colloids. *Analytica Chimica Acta, 552*(1), 166–174. doi:10.1016/j.aca.2005.07.049

Von der Kammer, F., Ferguson, P. L., Holden, P. A., Masion, A., Rogers, K. R., Klaine, S. J., & Unrine, J. M. et al. (2012). Analysis of engineered nanomaterials in complex matrices (environment and biota): General considerations and conceptual case studies. *Environmental Toxicology and Chemistry, 31*(1), 32–49. doi:10.1002/etc.723 PMID:22021021

Vos, J. G., Dybing, E., Greim, H. A., Ladefoged, O., Lambré, C., Tarazona, J. V., & Vethaak, A. D. et al. (2000). Health effects of endocrine-disrupting chemicals on wildlife, with special reference to the European situation. *CRC Critical Reviews in Toxicology, 30*(1), 71–133. doi:10.1080/10408440091159176 PMID:10680769

Vroblesky, D. A., Petkewich, M. D., Lowery, M. A., & Landmeyer, J. E. (2011). Sewer as a source and sink of chlorinated-solvent groundwater contamination, Marine Corps Recruit Depot, Parris Island, South Carolina. *Ground Water Monitoring and Remediation, 31*(4), 63–69. doi:10.1111/j.1745-6592.2011.01349.x

Vulliet, E., Wiest, L., Baudot, R., & Grenier-Loustalot, M.-F. (2008). Multi-residue analysis of steroids at sub-ng/L levels in surface and ground-waters using liquid chromatography coupled to tandem mass spectrometry. *Journal of Chromatography. A, 1210*(1), 84–91. doi:10.1016/j.chroma.2008.09.034 PMID:18823894

Wagner, A. J., Bleckmann, C. A., Murdock, R. C., Schrand, A. M., Schlager, J. J., & Hussain, S. M. (2007). Cellular interaction of different forms of aluminum nanoparticles in rat alveolar macrophages. *The Journal of Physical Chemistry B, 111*(25), 7353–7359. doi:10.1021/jp068938n PMID:17547441

Waller, K., Swan, S. H., DeLorenze, G., & Hopkins, B. (1998). Trihalomethanes in drinking water and spontaneous abortion. *Epidemiology (Cambridge, Mass.), 9*(2), 134–140. doi:10.1097/00001648-199803000-00006 PMID:9504280

Wallis, L. K., Diamond, S. A., Ma, H., Hoff, D. J., Al-Abed, S. R., & Li, S. (2014). Chronic TiO_2 nanoparticle exposure to a benthic organism, Hyalella azteca: Impact of solar UV radiation and material surface coatings on toxicity. *The Science of the Total Environment, 499*, 356–362. doi:10.1016/j.scitotenv.2014.08.068 PMID:25203828

Walters, C. R., Pool, E. J., & Somerset, V. S. (2014). Ecotoxicity of silver nanomaterials in the aquatic environment: A review of literature and gaps in nano-toxicological research. *Journal of Environmental Science and Health. Part A, Toxic/Hazardous Substances & Environmental Engineering, 49*(13), 1588–1601. doi:10.1080/10934529.2014.938536 PMID:25137546

Wang, H., Fan, W., Xue, F., Wang, X., Li, X., & Guo, L. (2015). Chronic effects of six micro/nano-Cu_2O crystals with different structures and shapes on Daphnia magna. *Environmental Pollution (Barking, Essex : 1987), 203*, 60–8. R

Wang, Y.-J., He, Z.-Z., Fang, Y.-W., Xu, Y., Chen, Y.-N., Wang, G.-Q., Yang, Y.-Q., et al. (2014). Effect of titanium dioxide nanoparticles on zebrafish embryos and developing retina. *International Journal of Ophthalmology, 7*(6), 917–23.

Wang, C., Huang, X. B., Deng, W. L., Chang, C. L., Hang, R. Q., & Tang, B. (2014). A nano-silver composite based on the ion-exchange response for the intelligent antibacterial applications. *Materials Science & Engineering C-Materials for Biological Applications*, *41*, 134–141. doi:10.1016/j.msec.2014.04.044 PMID:24907746

Wang, C., Yin, H., Chan, R., Peng, S., Dai, S., & Sun, S. (2009). One-pot synthesis of oleylamine coated AuAg alloy NPs and their catalysis for CO oxidation. *Chemistry of Materials*, *21*(3), 433–435. doi:10.1021/cm802753j

Wang, G., Van Hove, M. A., Ross, P. N., & Baskes, M. I. (2005). Monte Carlo simulations of segregation in Pt-Ni catalyst nanoparticles. *The Journal of Chemical Physics*, *122*(2), 024706. doi:10.1063/1.1828033 PMID:15638613

Wang, H. Y., Gao, B., Wang, S. S., Fang, J., Xue, Y. W., & Yang, K. (2015). Removal of Pb(II), Cu(II), and Cd(II) from aqueous solutions by biochar derived from $KMnO_4$ treated hickory wood. *Bioresource Technology*, *197*, 356–362. doi:10.1016/j.biortech.2015.08.132 PMID:26344243

Wang, H., & Huang, Y. (2011). Prussian-blue-modified iron oxide magnetic nanoparticles as effective peroxidase-like catalysts to degrade methylene blue with H_2O_2. *Journal of Hazardous Materials*, *191*(1-3), 163–169. doi:10.1016/j.jhazmat.2011.04.057 PMID:21570769

Wang, H., Li, L., Zhang, X., & Zhang, S. (2010). Polyamide thin-film composite membranes prepared from a novel triamine 3, 5-diamino-N-(4-aminophenyl)-benzamide monomer and m-phenylenediamine. *Journal of Membrane Science*, *353*(1), 78–84. doi:10.1016/j.memsci.2010.02.033

Wang, H., Su, Y., Zhao, H., Yu, H., Chen, S., Zhang, Y., & Quan, X. (2014). Photocatalytic Oxidation of Aqueous Ammonia Using Atomic Single Layer Graphitic-C3N4. *Environmental Science & Technology*, *48*(20), 11984–11990. doi:10.1021/es503073z

Wang, H., Wick, R. L., & Xing, B. (2009). Toxicity of nanoparticulate and bulk ZnO, Al_2O_3 and TiO_2 to the nematode Caenorhabditis elegans. *Environmental Pollution*, *157*(4), 1171–1177. doi:10.1016/j.envpol.2008.11.004 PMID:19081167

Wang, H., Xie, C., Zhang, W., Cai, S., Yang, Z., & Gui, Y. (2007). Comparison of dye degradation efficiency using ZnO powders with various size scales. *Journal of Hazardous Materials*, *141*(3), 645–652. doi:10.1016/j.jhazmat.2006.07.021 PMID:16930825

Wang, J. J., Sanderson, B. J. S., & Wang, H. (2007). Cyto-and genotoxicity of ultrafine TiO 2 particles in cultured human lymphoblastoid cells. *Mutation Research/Genetic Toxicology and Environmental Mutagenesis*, *628*(2), 99–106. doi:10.1016/j.mrgentox.2006.12.003 PMID:17223607

Wang, J., Zhu, X., Zhang, X., Zhao, Z., Liu, H., George, R., & Chen, Y. et al. (2011). Disruption of zebrafish (Danio rerio) reproduction upon chronic exposure to TiO_2 nanoparticles. *Chemosphere*, *83*(4), 461–467. doi:10.1016/j.chemosphere.2010.12.069 PMID:21239038

Wang, L. J., Guo, Z. M., Li, T. J., & Li, M. (2001). The nano structure SiO_2 in the plants. *Chinese Science Bulletin*, *46*, 625–631.

Wang, L. M., Ding, Y., Shen, Y., Cai, Z. S., Zhang, H. F., & Xu, L. H. (2014). Study on properties of modified nano-TiO2 and its application on antibacterial finishing of textiles. *Journal of Industrial Textiles*, *44*(3), 351–372. doi:10.1177/1528083713487758

Wang, L. X., He, S., Wu, X. M., Liang, S. S., Mu, Z. L., Wei, J., & Wei, S. C. et al. (2014). Polyetheretherketone/nano-fluorohydroxyapatite composite with antimicrobial activity and osseointegration properties. *Biomaterials*, *35*(25), 6758–6775. doi:10.1016/j.biomaterials.2014.04.085 PMID:24835045

Wang, M., Wang, C., Young, K. L., Hao, L., Medved, M., Rajh, T., & Stamenkovic, V. R. et al. (2012). Cross-linked heterogeneous nanoparticles as bifunctional probe. *Chemistry of Materials*, *24*(13), 2423–2425. doi:10.1021/cm300381f

Wang, P., Huang, B., Dai, Y., & Whangbo, M. H. (2012). Plasmonic photocatalysts: Harvesting visible light with noble metal nanoparticles. *Physical Chemistry Chemical Physics*, *14*(28), 9813–9825. doi:10.1039/c2cp40823f PMID:22710311

Wang, R., Liu, Y., Li, B., Hsiao, B. S., & Chu, B. (2012). Electrospun nanofibrous membranes for high flux microfiltration. *Journal of Membrane Science*, *392-393*, 167–174. doi:10.1016/j.memsci.2011.12.019

Wang, R.-M., Wang, B.-Y., He, Y.-F., Lv, W.-H., & Wang, J.-F. (2010). Preparation of composited Nano-TiO2 and its application on antimicrobial and self-cleaning coatings. *Polymers for Advanced Technologies*, *21*(5), 331–336. doi:10.1002/pat.1432

Wang, T., Long, X., Cheng, Y., Liu, Z., & Yan, S. (2014). The potential toxicity of copper nanoparticles and copper sulphate on juvenile *Epinephelus coioides*. *Aquatic Toxicology (Amsterdam, Netherlands)*, *152*, 96–104. doi:10.1016/j.aquatox.2014.03.023 PMID:24742820

Wang, T., Long, X., Liu, Z., Cheng, Y., & Yan, S. (2015). Effect of copper nanoparticles and copper sulphate on oxidation stress, cell apoptosis and immune responses in the intestines of juvenile *Epinephelus coioides*. *Fish & Shellfish Immunology*, *44*, 674–682. PMID:25839971

Wang, W., Mai, K., Zhang, W., Ai, Q., Yao, C., Li, H., & Liufu, Z. (2009). Effects of dietary copper on survival, growth and immune response of juvenile abalone, *Haliotis discus hannai* Ino. *Aquaculture (Amsterdam, Netherlands)*, *297*(1–4), 122–127. doi:10.1016/j.aquaculture.2009.09.006

Wang, X. G., Qiao, J. W., Yuan, F. X., Hang, R. Q., Huang, X. B., & Tang, B. (2014). In situ growth of self-organized Cu-containing nano-tubes and nano-pores on Ti-90 (-) (x)Cu(10)Alx (x=0,45) alloys by one-pot anodization and evaluation of their antimicrobial activity and cytotoxicity. *Surface and Coatings Technology*, *240*, 167–178. doi:10.1016/j.surfcoat.2013.12.036

Wang, X., Chen, X., Yoon, K., Fang, D., Hsiao, B. S., & Chu, B. (2005). High flux filtration medium based on nanofibrous substrate with hydrophilic nanocomposite coating. *Environmental Science & Technology*, *39*(19), 7684–7691. doi:10.1021/es050512j PMID:16245845

Wang, X.-L., Tsuru, T., Nakao, S.-i., & Kimura, S. (1995). Electrolyte transport through nanofiltration membranes by the space-charge model and the comparison with Teorell-Meyer-Sievers model. *Journal of Membrane Science*, *103*(1), 117–133. doi:10.1016/0376-7388(94)00317-R

Wang, Y. (1992). Photoconductivity of fullerene doped polymers. *Nature*, *356*(6370), 585–587. doi:10.1038/356585a0

Wang, Y., & Cao, G. (2006). Synthesis and enhanced intercalation properties of nanostructured vanadium oxides. *Chemistry of Materials*, *18*(12), 2787–2804. doi:10.1021/cm052765h

Wang, Y., Li, Y., Fortner, J. D., Hughes, J. B., Abriola, L. M., & Pennell, K. D. (2008). Transport and Retention of Nanoscale C_{60} Aggregates in Water-Saturated Porous Media. *Environmental Science & Technology*, *42*(10), 3588–3594. doi:10.1021/es800128m

Wang, Z. G., Zhou, R., Jiang, D., Song, J. E., Xu, Q., & Si, J. (2015). Toxicity of Graphene Quantum Dots in Zebrafish Embryo. *Biomedical and Environmental Sciences*, *28*(5), 341–351. PMID:26055561

Ward, H. C., & Dutta, J. (2005). Nanotechnology for agriculture and food systems-A view. *Proc. Of the 2nd International Conference on Innovations in Food Processing Technology and Engineering*.

Compilation of References

Warheit, D. B., Hoke, R. A., Finlay, C., Donner, E. M., Reed, K. L., & Sayes, C. M. (2007). Development of a base set of toxicity tests using ultrafine TiO 2 particles as a component of nanoparticle risk management. *Toxicology Letters*, *171*(3), 99–110. doi:10.1016/j.toxlet.2007.04.008 PMID:17566673

Watanabe, T., Fujimoto, R., Sawai, J., Kikuchi, M., Yahata, S., & Satoh, S. (2014). Antibacterial Characteristics of Heated Scallop-Shell Nano-Particles. *Biocontrol Science, 19*(2), 93-97. Retrieved from https://www.jstage.jst.go.jp/article/bio/19/2/19_93/_pdf

Watanabe, T., Nakajima, A., Wang, R., Minabe, M., Koizumi, S., Fujishima, A., & Hashimoto, K. (1999). Photocatalytic activity and photoinduced hydrophilicity of titanium dioxide coated glass. *Thin Solid Films*, *351*(1–2), 260–263. doi:10.1016/S0040-6090(99)00205-9

Watson, G. S., Green, D. W., Schwarzkopf, L., Li, X., Cribb, B. W., Myhra, S., & Watson, J. A. (2015). A gecko skin micro/nano structure - A low adhesion, superhydrophobic, anti-wetting, self-cleaning, biocompatible, antibacterial surface. *Acta Biomaterialia*, *21*, 109–122. doi:10.1016/j.actbio.2015.03.007 PMID:25772496

Watson, J. T., & Sparkman, O. D. (2007). *Introduction to mass spectrometry: instrumentation, applications, and strategies for data interpretation*. John Wiley & Sons. doi:10.1002/9780470516898

Weber, S., Gallenkemper, M., Melin, T., Dott, W., & Hollender, J. (2004). Efficiency of nanofiltration for the elimination of steroids from water. *Water Science and Technology*, *50*(5), 9–14. PMID:15497823

Weber, W. J. Jr. (2001). *Environmental Systems and Processes: Principles, Modeling, and Design*. New York: Wiley-Interscience.

Wei, Y.-T., Wu, S.-C., Chou, C.-M., Che, C.-H., Tsai, S.-M., & Lien, H.-L. (2010). Influence of nanoscale zero-valent iron on geochemical properties of groundwater and vinyl chloride degradation: A field case study. *Water Research*, *44*(1), 131–140. doi:10.1016/j.watres.2009.09.012 PMID:19800096

Wendelaar Bonga, S. E. (1997). The stress response in fish. *Physiological Reviews*, *77*, 591–625. PMID:9234959

Wen, Y., Xing, F., He, S., Song, S., Wang, L., Long, Y., & Fan, C. et al. (2010). A graphene-based fluorescent nanoprobe for silver (I) ions detection by using graphene oxide and a silver-specific oligonucleotide. *Chemical Communications*, *46*(15), 2596–2598. doi:10.1039/b924832c PMID:20449319

Wepener, V., van-Vuren, J. H. J., & Du-Preez, H. H. (1992). The effect of iron and manganese at an acidic pH on the hematology of the banded Tilapia (*Tilapia sparrmanii*, Smith). *Bulletin of Environmental Contamination and Toxicology*, *49*(4), 613–619. doi:10.1007/BF00196307 PMID:1421857

Westerhoff, P., Song, G., Hristovski, K., & Kiser, M. (2011). Occurrence and removal of titanium at full scale wastewater treatment plants: Implications for TiO_2 nanomaterials. *Journal of Environmental Monitoring*, *13*(5), 1195–1203. doi:10.1039/c1em10017c PMID:21494702

Wetz, F., Soulantica, K., Falqui, A., Respaud, M., Snoeck, E., & Chaudret, B. (2007). Hybrid Co–Au nanorods: Controlling Au nucleation and location. *Angewandte Chemie*, *119*(37), 7209–7211. doi:10.1002/ange.200702017 PMID:17685370

Wharton, T., Kini, V. U., Mortis, R. A., & Wilson, L. J. (2001). New non-ionic, highly water soluble derivatives of C_{60} designed for biological compatibility. *Tetrahedron Letters*, *42*(31), 5159–5162. doi:10.1016/S0040-4039(01)00956-X

Whitby, M., & Quirke, N. (2007). Fluid flow in carbon nanotubes and nanopipes. *Nat Nano*, *2*(2), 87–94. doi:10.1038/nnano.2006.175 PMID:18654225

Wielgus, A. R., Zhao, B., Chignell, C. F., Hu, D.-N., & Roberts, J. E. (2010). Phototoxicity and cytotoxicity of fullerol in human retinal pigment epithelial cells. *Toxicology and Applied Pharmacology*, *242*(1), 79–90. doi:10.1016/j.taap.2009.09.021 PMID:19800903

Wiench, K., Wohlleben, W., Hisgen, V., Radke, K., Salinas, E., Zok, S., & Landsiedel, R. (2009). Acute and chronic effects of nano- and non-nano-scale TiO(2) and ZnO particles on mobility and reproduction of the freshwater invertebrate Daphnia magna. *Chemosphere*, *76*(10), 1356–1365. doi:10.1016/j.chemosphere.2009.06.025 PMID:19580988

Wienk, M. M., Kroon, J. M., Verhees, W. J. H., Knol, J., Hummelen, J. C., Van Hal, P. A., & Janssen, R. A. J. (2003). Efficient Methano[70]fullerene/MDMO-PPV Bulk Heterojunction Photovoltaic Cells. *Angewandte Chemie International Edition*, *42*(29), 3371–3375. doi:10.1002/anie.200351647

Wiesner, M. R., Lowry, G. V., Alvarez, P., Dionysiou, D., & Biswas, P. (2006). Assessing the risks of manufactured nanomaterials. *Environmental Science & Technology*, *40*(14), 4336–4345. doi:10.1021/es062726m PMID:16903268

Wigginton, N. S., Haus, K. L., & Hochella, M. F. Jr. (2007). Aquatic environmental nanoparticles. *Journal of Environmental Monitoring*, *9*(12), 1306–1316. doi:10.1039/b712709j PMID:18049768

Wilhelmi, V., Fischer, U., van Berlo, D., Schulze-Osthoff, K., Schins, R. P. F., & Albrecht, C. (2012). Evaluation of apoptosis induced by nanoparticles and fine particles in RAW 264.7 macrophages: Facts and artefacts. *Toxicology In Vitro*, *26*(2), 323–334. doi:10.1016/j.tiv.2011.12.006 PMID:22198050

Wilkin, R. T., Lee, T. R., McNeil, M. S., Su, C., & Adair, C. (2014). Fourteen-year assessment of a permeable reactive barrier for treatment of hexavalent chromium and trichloroethylene. In R. Naidu & V. Birke (Eds.), *Permeable Reactive Barrier: Sustainable Groundwater Remediation* (pp. 99–107). CRC Press.

Wilkinson, F., & Brummer, J. G. (1981). Rate constants for the decay and reactions of the lowest electronically excited singlet-state of molecular-oxygen in solution. *Journal of Physical and Chemical Reference Data*, *10*(4), 809-1000.

Wilkinson, F., Helman, W. P., & Ross, A. B. (1995). Rate Constants for the Decay and Reactions of the Lowest Electronically Excited Singlet State of Molecular Oxygen in Solution. An Expanded and Revised Compilation. *Journal of Physical and Chemical Reference Data*, *24*(2), 663-677. doi:10.1063/1.555965

Wilkinson, K. J., Balnois, E., Leppard, G. G., & Buffle, J. (1999). Characteristic features of the major components of freshwater colloidal organic matter revealed by transmission electron and atomic force microscopy. *Colloids and Surfaces. A, Physicochemical and Engineering Aspects*, *155*(2-3), 287–310. doi:10.1016/S0927-7757(98)00874-7

Williams, R. B., Jenkins, B. M., & Nguyen, D. (2003). Solid waste conversion: A review and database of current and emerging technologies. Report prepared for California Integrated Waste Management Board, Davis, CA.

Williams, D. (2008). The relationship between biomaterials and nanotechnology. *Biomaterials*, *29*(12), 1737–1738. doi:10.1016/j.biomaterials.2008.01.003 PMID:18249439

Willis, K. P., Osada, S., & Willerton, K. L. (2010). Plasma gasification: lessons learned at Eco-Valley WTE facility. In *18th Annual North American Waste-to-Energy Conference* (pp. 133-140). New York, NY: American Society of Mechanical Engineers. doi:10.1115/NAWTEC18-3515

Wilson, B., Williams, N., Liss, B., & Wilson, B. (2013). *A comparative assessment of commercial technologies for conversion of solid waste to energy*. Retrieved June 2015, from http://www.itigroup.co/uploads/files/59_Comparative_WTE-Technologies-Mar-_2014.pdf

Wohlleben, W. (2012). Validity range of centrifuges for the regulation of nanomaterials: From classification to as-tested coronas. *Journal of Nanoparticle Research*, *14*(12), 1–18. doi:10.1007/s11051-012-1300-z PMID:23239934

Compilation of References

Wong, C., Lai, K. H., Cheng, T. C. E., & Lun, V. Y. H. (2015). The role of IT-enabled collaborative decision making in inter-organizational information integration to improve customer service performance. *International Journal of Production Economics*, *159*, 56–65. doi:10.1016/j.ijpe.2014.02.019

World Bank Group. (2014). Results-based financing for municipal solid waste (91861 v2). Washington, DC: Author.

World Energy Council. (2013). *World Energy Council issues official statement ahead of 22nd World Energy Congress*. Retrieved August 2015, from https://www.worldenergy.org/news-and-media/news/world-energy-council-issues-official-statement-ahead-of-22nd-world-energy-congress/

Wu, C., Zhang, S., Yang, D., & Jian, X. (2009). Preparation, characterization and application of a novel thermal stable composite nanofiltration membrane. *Journal of Membrane Science*, *326*(2), 429–434. doi:10.1016/j.memsci.2008.10.033

Wudl, F. (2002). Fullerene materials. *Journal of Materials Chemistry*, *12*(7), 1959–1963. doi:10.1039/b201196d

Wu, H., Tang, B., & Wu, P. (2010). MWNTs/polyester thin film nanocomposite membrane: An approach to overcome the trade-off effect between permeability and selectivity. *The Journal of Physical Chemistry C*, *114*(39), 16395–16400. doi:10.1021/jp107280m

Wu, J., Hou, Y., & Gao, S. (2011). Controlled synthesis and multifunctional properties of FePt-Au heterostructures. *Nano Research*, *4*(9), 836–848. doi:10.1007/s12274-011-0140-y

Wu, S. G., Huang, L., Head, J., Chen, D.-R., Kong, I.-C., & Tang, Y. J. (2012). Phytotoxicity of metal oxide nanoparticles is related to both dissolved metals ions and adsorption of particles on seed surfaces. *J Petrol Environ Biotechnol*, *3*(4).

Wu, S., He, Q., Zhou, C., Qi, X., Huang, X., Yin, Z., & Zhang, H. et al. (2012). Synthesis of Fe 3 O 4 and Pt nanoparticles on reduced graphene oxide and their use as a recyclable catalyst. *Nanoscale*, *4*(7), 2478–2483. doi:10.1039/c2nr11992g PMID:22388949

Wu, W., He, Q., Chen, H., Tang, J., & Nie, L. (2007). Sonochemical synthesis, structure and magnetic properties of air-stable Fe3O4/Au nanoparticles. *Nanotechnology*, *18*(14), 145609. doi:10.1088/0957-4484/18/14/145609

Wu, Z. C., Zhang, Y., Tao, T. X., Zhang, L., & Fong, H. (2010). Silver nanoparticles on amidoxime fibers for photo-catalytic degradation of organic dyes in waste water. *Applied Surface Science*, *257*(3), 1092–1097. doi:10.1016/j.apsusc.2010.08.022

Xia, G., Liu, T., Wang, Z., Hou, Y., Dong, L., Zhu, J., & Qi, J. (2015). The effect of silver nanoparticles on zebrafish embryonic development and toxicology. *Artificial Cells, Nanomedicine, and Biotechnology*, 1–6.

Xia, T., Kovochich, M., Liong, M., Mädler, L., Gilbert, B., Shi, H., & Nel, A. E. et al. (2008). Comparison of the mechanism of toxicity of zinc oxide and cerium oxide nanoparticles based on dissolution and oxidative stress properties. *ACS Nano*, *2*(10), 2121–2134. doi:10.1021/nn800511k PMID:19206459

Xie, X.-L., Mai, Y.-W., & Zhou, X.-P. (2005). Dispersion and alignment of carbon nanotubes in polymer matrix: A review. *Materials Science and Engineering R Reports*, *49*(4), 89–112. doi:10.1016/j.mser.2005.04.002

Xin, Q., Rotchell, J. M., Cheng, J., Yi, J., & Zhang, Q. (2015). Silver nanoparticles affect the neural development of zebrafish embryos. *Journal of applied Toxicology*. Retrieved August 17, 2015, from http://www.ncbi.nlm.nih.gov/pubmed/25976698

Xiong, Z., He, F., Zhao, D. Y., & Barnett, M. O. (2009). Immobilization of mercury in sediment using stabilized iron sulfide nanoparticles. *Water Research*, *43*(20), 5171–5179. doi:10.1016/j.watres.2009.08.018 PMID:19748651

Xue, C. H., Chen, J., Yin, W., Jia, S. T., & Ma, J. Z. (2012). Superhydrophobic conductive textiles with antibacterial property by coating fibers with silver nanoparticles. *Applied Surface Science, 258*(7), 2468–2472. doi:10.1016/j.apsusc.2011.10.074

Xu, J. F., Ji, W., Shen, Z. X., Tang, S. H., Ye, X. R., Jia, D. Z., & Xin, X. Q. (1999). Preparation and characterization of CuO nanocrystals. *Journal of Solid State Chemistry, 147*(2), 516–519. doi:10.1006/jssc.1999.8409

Xu, J., Wang, Y., & Zhu, Y. (2013). Nanoporous Graphitic Carbon Nitride with Enhanced Photocatalytic Performance. *Langmuir, 29*(33), 10566–10572. doi:10.1021/la402268u

Xu, L., Xu, C., Zhao, M. R., Qiu, Y. P., & Sheng, G. D. (2008). Oxidative removal of aqueous steroid estrogens by manganese oxides. *Water Research, 42*(20), 5038–5044. doi:10.1016/j.watres.2008.09.016 PMID:18929389

Xu, P., Zeng, G. M., Huang, D. L., Feng, C. L., Hu, S., Zhao, M. H., & Liu, Z. F. et al. (2012). Use of iron oxide nanomaterials in wastewater treatment: A review. *The Science of the Total Environment, 424*, 1–10. doi:10.1016/j.scitotenv.2012.02.023 PMID:22391097

Xu, Z., Hou, Y., & Sun, S. (2007). Magnetic core/shell Fe3O4/Au and Fe3O4/Au/Ag nanoparticles with tunable plasmonic properties. *Journal of the American Chemical Society, 129*(28), 8698–8699. doi:10.1021/ja073057v PMID:17590000

Yakub, I., & Soboyejo, W. O. (2012). Adhesion of E. coli to silver-or copper-coated porous clay ceramic surfaces. *Journal of Applied Physics, 111*(12), 124324. doi:10.1063/1.4722326

Yamakoshi, Y. N., Yagami, T., Fukuhara, K., Sueyoshi, S., & Miyata, N. (1994). Solubilization of fullerenes into water with polyvinylpyrrolidone applicable to biological tests. *Journal of the Chemical Society. Chemical Communications*, (4): 517–518. doi:10.1039/c39940000517

Yamakoshi, Y., Umezawa, N., Ryu, A., Arakane, K., Miyata, N., Goda, Y., & Nagano, T. et al. (2003). Active oxygen species generated from photoexcited fullerene (C_{60}) as potential medicines: O_2^- versus $^1O_2$. *Journal of the American Chemical Society, 125*(42), 12803–12809. doi:10.1021/ja0355574

Yan, L., & Chen, X. (2013). *Nanomaterials for Drug Delivery. Nanocrystalline Materials: Their Synthesis-Structure-Property Relationships and Applications.* doi:10.1016/B978-0-12-407796-6.00007-5

Yang, B., Ying, G. G., Zhao, J. L., Zhang, L. J., Fang, Y. X., & Nghiem, L. D. (2010). Oxidation of triclosan by ferrate: Reaction kinetics, products identification and toxicity evaluation. *Journal of Hazardous Materials, 186*(1), 227–235. doi:10.1016/j.jhazmat.2010.10.106 PMID:21093982

Yang, H., Li, H., & Jiang, X. (2008). Detection of foodborne pathogens using bioconjugated nanomaterials. *Microfluidics and Nanofluidics, 5*(5), 571–583. doi:10.1007/s10404-008-0302-8

Yang, H., Liu, C., Yang, D., Zhang, H., & Xi, Z. (2009). Comparative study of cytotoxicity, oxidative stress and genotoxicity induced by four typical nanomaterials: The role of particle size, shape and composition. *Journal of applied toxicology. JAT, 29*(1), 69–78. PMID:18756589

Yang, J., Tian, C., Wang, L., Tan, T., Yin, J., Wang, B., & Fu, H. (2012). In situ reduction, oxygen etching, and reduction using formic acid: An effective strategy for controllable growth of monodisperse palladium nanoparticles on graphene. *ChemPlusChem, 77*(4), 301–307. doi:10.1002/cplu.201100058

Yang, L., & Watts, D. J. (2005). Particle surface characteristics may play an important role in phytotoxicity of alumina nanoparticles. *Toxicology Letters, 158*(2), 122–132. doi:10.1016/j.toxlet.2005.03.003 PMID:16039401

Yang, P., Gai, S., & Lin, J. (2012). Functionalized mesoporous silica materials for controlled drug delivery. *Chemical Society Reviews, 41*(9), 3679–3698. doi:10.1039/c2cs15308d

Compilation of References

Yang, Y., Nogami, M., Shi, J., Chen, H., Ma, G., & Tang, S. (2005). Enhancement of third-order optical nonlinearities in 3-dimensional films of dielectric shell capped Au composite nanoparticles. *The Journal of Physical Chemistry B*, *109*(11), 4865–4871. doi:10.1021/jp045854a PMID:16863140

Yang, Z., Chu, H., Jin, Z., Zhou, W., & Li, Y. (2010). Preparation and properties of CdS/Au composite nanorods and hollow Au tubes. *Chinese Science Bulletin*, *55*(10), 921–926. doi:10.1007/s11434-010-0061-2

Yantasee, W., Warner, C.L., Sangvanich, T., Addleman, R.S., Carter, T.G., Wiacek, R.J., Fryxell, G.E., Timchalk, C., & Warner, M.G. (2007). Removal of heavy metals from aqueous systems with thiol functionalized superparamagnetic nanoparticles. *Environmental Science and Technology, 41*, 5114-5119.

Yan, W., Lien, H.-L., Koel, B. E., & Zhang, W.-X. (2013). Iron nanoparticles for environmental clean-up: Recent developments and future outlook. *Environmental Science: Processes & Impacts*, *15*, 63–77. PMID:24592428

Yan, X. F., Jie, Z. Q., Zhao, L. H., Yang, H., Yang, S. P., & Liang, J. (2014). High-efficacy antibacterial polymeric micro/nano particles with N-halamine functional groups. *Chemical Engineering Journal*, *254*, 30–38. doi:10.1016/j.cej.2014.05.114

Yarin, A. L., Koombhongse, S., & Reneker, D. H. (2001). Taylor cone and jetting from liquid droplets in electrospinning of nanofibers. *Journal of Applied Physics*, *90*(9), 4836–4846. doi:10.1063/1.1408260

Yasar, D., Celik, N., & Sharit, J. (2016). Evaluation of advanced thermal solid waste management technologies for sustainability in Florida. *International Journal of Performability Engineering*, *12*(01), 63–78.

Yeow, M., Liu, Y., & Li, K. (2004). Morphological study of poly (vinylidene fluoride) asymmetric membranes: Effects of the solvent, additive, and dope temperature. *Journal of Applied Polymer Science*, *92*(3), 1782–1789. doi:10.1002/app.20141

Ye, R., Yu, X., Yang, S., Yuan, J., & Yang, X. (2013). Effects of Silica Dioxide Nanoparticles on the Embryonic Development of Zebrafish. *Integrated Ferroelectrics*, *147*(1), 166–174. doi:10.1080/10584587.2013.792625

Yin, J., Kim, E.-S., Yang, J., & Deng, B. (2012). Fabrication of a novel thin-film nanocomposite (TFN) membrane containing MCM-41 silica nanoparticles (NPs) for water purification. *Journal of Membrane Science*, *423–424*(0), 238–246. doi:10.1016/j.memsci.2012.08.020

Yin, T., Walker, H. W., Chen, D., & Yang, Q. (2014). Influence of pH and ionic strength on the deposition of silver nanoparticles on microfiltration membranes. *Journal of Membrane Science*, *449*, 9–14. doi:10.1016/j.memsci.2013.08.020

Yin, Z., He, Q., Huang, X., Zhang, J., Wu, S., Chen, P., & Zhang, H. et al. (2012). Real-time DNA detection using Pt nanoparticle-decorated reduced graphene oxide field-effect transistors. *Nanoscale*, *4*(1), 293–297. doi:10.1039/C1NR11149C PMID:22089471

Yip, N. Y., Tiraferri, A., Phillip, W. A., Schiffman, J. D., Hoover, L. A., Kim, Y. C., & Elimelech, M.Ngai Yin Yip. (2011). Thin-Film Composite Pressure Retarded Osmosis Membranes for Sustainable Power Generation from Salinity Gradients. *Environmental Science & Technology*, *45*(10), 4360–4369. doi:10.1021/es104325z PMID:21491936

Yokel, R. A., & MacPhail, R. C. (2011). Engineered nanomaterials: Exposures, hazards, and risk prevention. *Journal of Occupational Medicine and Toxicology (London, England)*, *6*(1), 7. doi:10.1186/1745-6673-6-7 PMID:21418643

Yoon, B., Häkkinen, H., Landman, U., Wörz, A. S., Antonietti, J. M., Abbet, S., & Heiz, U. et al. (2005). Charging effects on bonding and catalyzed oxidation of CO on Au8 clusters on MgO. *Science*, *307*(5708), 403–407. doi:10.1126/science.1104168 PMID:15662008

Yoon, K. Y., Byeon, J. H., Park, C. W., & Hwang, J. (2008). Antimicrobial Effect of Silver Particles on Bacterial Contamination of Activated Carbon Fibers. *Environmental Science & Technology, 42*(4), 1251–1255. doi:10.1021/es0720199 PMID:18351101

Yoon, K. Y., Byeon, J. H., Park, J. H., & Hwang, J. (2007). Susceptibility constants of *Escherichia coli* and *Bacillus subtilis* to silver and copper nanoparticles. *The Science of the Total Environment, 373*(2-3), 572–575. doi:10.1016/j.scitotenv.2006.11.007 PMID:17173953

Yoon, K., Hsiao, B. S., & Chu, B. (2009). High flux nanofiltration membranes based on interfacially polymerized polyamide barrier layer on polyacrylonitrile nanofibrous scaffolds. *Journal of Membrane Science, 326*(2), 484–492. doi:10.1016/j.memsci.2008.10.023

Yslas, E. I., Ibarra, L. E., Peralta, D. O., Barbero, C. A., Rivarola, V. A., & Bertuzzi, M. L. (2012). Polyaniline nanofibers: Acute toxicity and teratogenic effect on Rhinella arenarum embryos. *Chemosphere, 87*(11), 1374–1380. doi:10.1016/j.chemosphere.2012.02.033 PMID:22386461

Yuan, Z., Li, J., Cui, L., Xu, B., Zhang, H., & Yu, C. P. (2013). Interaction of silver nanoparticles with pure nitrifying bacteria. *Chemosphere, 90*(4), 1404–1411. doi:10.1016/j.chemosphere.2012.08.032 PMID:22985593

Yu, C., & Wang, H. (2010). Light-Induced Bipolar-Resistance Effect Based on Metal–Oxide–Semiconductor Structures of Ti/SiO2/Si. *Advanced Materials, 22*(9), 966–970. doi:10.1002/adma.200903070 PMID:20217821

Yuhas, B. D., Habas, S. E., Fakra, S. C., & Mokari, T. (2009). Probing compositional variation within hybrid nanostructures. *ACS Nano, 3*(11), 3369–3376. doi:10.1021/nn901107p PMID:19813744

Yu, J., Nachiappan, S. P., Ning, K., & Edwards, D. (2015). Product delivery service provider selection and customer satisfaction in the era of Internet Of Things: A Chinese e-retailers' perspective. *International Journal of Production Economics, 159*, 104–116. doi:10.1016/j.ijpe.2014.09.031

Yu, J., Zhang, W. Y., Li, Y., Wang, G., Yang, L. D., Jin, J. F., & Huang, M. H. et al. (2015). Synthesis, characterization, antimicrobial activity and mechanism of a novel hydroxyapatite whisker/nano zinc oxide biomaterial. *Biomedical Materials, 10*(1). doi. *Artn, 015001.* doi:10.1088/1748-6041/10/1/015001

Yun, J., Jin, D., Lee, Y. S., & Kim, H. I. (2010). Photocatalytic treatment of acidic waste water by electrospun composite nanofibers of pH-sensitive hydrogel and TiO2. *Materials Letters, 64*(22), 2431–2434. doi:10.1016/j.matlet.2010.08.001

Yu, S. J., Yin, Y. G., & Liu, J. F. (2013). Silver nanoparticles in the environment. *Environmental Science: Processes & Impacts, 15*(1), 78–92. PMID:24592429

Zafar, S. (2009). *Gasification of municipal solid waste*. Retrieved May 2015, from http://www.altenergymag.com/content.php?issue_number=09.06.01&article=zafar

Zanoni, S., & Grubbström, R. W. (2004). A note on an industrial strategy for stock management in supply chains: Modelling and performance evaluation. *International Journal of Production Research, 42*(20), 4421–4426. doi:10.1080/00207540420000264581

Zasloff, M. (2002). Antimicrobial peptides of multicellular organisms. *Nature, 415*(6870), 389-395. Retrieved from http://www.nature.com/nature/journal/v415/n6870/pdf/415389a.pdf

Zazouli, M. A., Susanto, H., Nasseri, S., & Ulbricht, M. (2009). Influences of solution chemistry and polymeric natural organic matter on the removal of aquatic pharmaceutical residuals by nanofiltration. *Water Research, 43*(13), 3270–3280. doi:10.1016/j.watres.2009.04.038 PMID:19520413

Compilation of References

Zeng, F., Hou, C., Wu, S., Liu, X., Tong, Z., & Yu, S. (2007). Silver nanoparticles directly formed on natural macroporous matrix and their anti-microbial activities. *Nanotechnology, 18*(5), 055605. doi:10.1088/0957-4484/18/5/055605

Zeng, H., Cai, W., Liu, P., Xu, X., Zhou, H., Klingshirn, C., & Kalt, H. (2008). ZnO-based hollow nanoparticles by selective etching: Elimination and reconstruction of metal– semiconductor interface, improvement of blue emission and photocatalysis. *ACS Nano, 2*(8), 1661–1670. doi:10.1021/nn800353q PMID:19206370

Zhai, Y., Han, L., Wang, P., Li, G., Ren, W., Liu, L., & Dong, S. et al. (2011). Superparamagnetic plasmonic nanohybrids: Shape-controlled synthesis, TEM-induced structure evolution, and efficient sunlight-driven inactivation of bacteria. *ACS Nano, 5*(11), 8562–8570. doi:10.1021/nn201875k PMID:21951020

Zhang, Z. D. (2004). Nanocapsules Encyclopedia of Nanoscience and Nanotechnology. Academic Press.

Zhang, H. C., & Huang, C. H. (2003). Oxidative transformation of triclosan and chlorophene by manganese oxides. *Environmental Science & Technology, 37*(11), 2421–2430. doi:10.1021/es026190q PMID:12831027

Zhang, J., Du, J., Han, B., Liu, Z., Jiang, T., & Zhang, Z. (2006). Sonochemical formation of single-crystalline gold nanobelts. *Angewandte Chemie, 118*(7), 1134–1137. doi:10.1002/ange.200503762 PMID:16389606

Zhang, J., Post, M., Veres, T., Jakubek, Z. J., Guan, J., Wang, D., & Simard, B. et al. (2006). Laser-assisted synthesis of superparamagnetic Fe@Au core-shell nanoparticles. *The Journal of Physical Chemistry B, 110*(14), 7122–7128. doi:10.1021/jp0560967 PMID:16599475

Zhang, J., Sun, B., Xiong, X. M., Gao, N. Y., Song, W. H., Du, E. D., & Zhou, G. M. et al. (2014). Removal of emerging pollutants by Ru/TiO$_2$-catalyzed permanganate oxidation. *Water Research, 63*, 262–270. doi:10.1016/j.watres.2014.06.028 PMID:25016299

Zhang, J., Zhang, Y., Chen, Y., Du, L., Zhang, B., Zhang, H., & Wang, K. et al. (2012). Preparation and characterization of novel polyethersulfone hybrid ultrafiltration membranes bending with modified halloysite nanotubes loaded with silver nanoparticles. *Industrial & Engineering Chemistry Research, 51*(7), 3081–3090. doi:10.1021/ie202473u

Zhang, L., & Fang, M. (2010). Nanomaterials in pollution trace detection and environmental improvement. *Nano Today, 5*(2), 128–142. doi:10.1016/j.nantod.2010.03.002

Zhang, L., Shi, G.-Z., Qiu, S., Cheng, L.-H., & Chen, H.-L. (2011). Preparation of high-flux thin film nanocomposite reverse osmosis membranes by incorporating functionalized multi-walled carbon nanotubes. *Desalination and Water Treatment, 34*(1-3), 19–24. doi:10.5004/dwt.2011.2801

Zhang, Q., Xie, J., Yu, Y., Yang, J., & Lee, J. Y. (2010). Tuning the crystallinity of Au nanoparticles. *Small, 6*(4), 523–527. doi:10.1002/smll.200902033 PMID:20108243

Zhang, S., Shao, T., Bekaroglu, S. S. K., & Karanfil, T. (2009). The impacts of aggregation and surface chemistry of carbon nanotubes on the adsorption of synthetic organic compounds. *Environmental Science & Technology, 43*(15), 5719–5725. doi:10.1021/es900453e PMID:19731668

Zhang, W.-X. (2003). Nanoscale iron particles for environmental remediation: An overview. *Journal of Nanoparticle Research, 5*(3/4), 323–332. doi:10.1023/A:1025520116015

Zhang, W., Yang, T., Huang, D. M., & Jiao, K. (2008). Electro-chemical Sensing of DNA Immobilization and Hybridization Based on Carbon Nanotubes/Nano Zinc Oxide/ Chitosan Composite Film. *Chinese Chemical Letters, 19*(5), 589–591. doi:10.1016/j.cclet.2008.03.012

Zhang, X., Hu, W., Li, J., Tao, L., & Wei, Y. (2012). A comparative study of cellular uptake and cytotoxicity of multi-walled carbon nanotubes, graphene oxide, and nanodiamond. *Toxicological Reviews, 1*(1), 62–68.

Zhang, X., Niu, H., Yan, J., & Cai, Y. (2011). Immobilizing silver nanoparticles onto the surface of magnetic silica composite to prepare magnetic disinfectant with enhanced stability and antibacterial activity. *Colloids and Surfaces. A, Physicochemical and Engineering Aspects*, *375*(1-3), 186–192. doi:10.1016/j.colsurfa.2010.12.009

Zhang, Y., Guo, Y., Xianyu, Y., Chen, W., Zhao, Y., & Jiang, X. (2013). Nanomaterials for ultrasensitive protein detection. *Advanced Materials*, *25*(28), 3802–3819. doi:10.1002/adma.201301334 PMID:23740753

Zhang, Y., Van der Bruggen, B., Chen, G., Braeken, L., & Vandecasteele, C. (2004). Removal of pesticides by nanofiltration: Effect of the water matrix. *Separation and Purification Technology*, *38*(2), 163–172. doi:10.1016/j.seppur.2003.11.003

Zhang, Y., Xu, D., Li, W., Yu, J., & Chen, Y. (2012). Effect of size, shape, and surface modification on cytotoxicity of gold nanoparticles to human HEp-2 and canine MDCK cells. *Journal of Nanomaterials*, *2012*, 7. doi:10.1155/2012/375496

Zhan, J., Kolesnichenko, I., Sunkara, B., He, J., McPherson, G. L., Piringer, G., & John, V. T. (2011). Multifunctional iron – carbon nanocomposits through an aerosol-based process for the in situ remediation of chlorinated hydrocarbons. *Environmental Science & Technology*, *45*(5), 1949–1954. doi:10.1021/es103493e PMID:21299241

Zhao, X., Wang, S., Wu, Y., You, H., & Lv, L. (2013). Acute ZnO nanoparticles exposure induces developmental toxicity, oxidative stress and DNA damage in embryo-larval zebrafish. *Aquatic Toxicology*, *136 – 137*, 49 – 59.

Zhao, H., Qiu, S., Wu, L., Zhang, L., Chen, H., & Gao, C. (2014). Improving the performance of polyamide reverse osmosis membrane by incorporation of modified multi-walled carbon nanotubes. *Journal of Membrane Science*, *450*(0), 249–256. doi:10.1016/j.memsci.2013.09.014

Zhao, J., Wang, Z., Liu, X., Xie, X., Zhang, K., & Xing, B. (2011). Existing research on nanotoxicity has concentrated on empirical evaluation of the toxicity of various nanoparticles. *Journal of Hazardous Materials*, *197*, 304–310. doi:10.1016/j.jhazmat.2011.09.094 PMID:22014442

Zhao, J., Xu, L., Zhang, T., Ren, G., & Yang, Z. (2009). Influences of nanoparticle zinc oxide on acutely isolated rat hippocampal CA3 pyramidal neurons. *Neurotoxicology*, *30*(2), 220–230. doi:10.1016/j.neuro.2008.12.005 PMID:19146874

Zhao, L., Chu, P. K., Zhang, Y., & Wu, Z. (2009). Antibacterial coatings on titanium implants. *Journal of Biomedical Materials Research. Part B, Applied Biomaterials*, *91B*(1), 470–480. doi:10.1002/jbm.b.31463 PMID:19637369

Zhe, J., Jagtiani, A., Dutta, P., Hu, J., & Carletta, J. (2007). A micromachined high throughput Coulter counter for bioparticle detection and counting. *Journal of Micromechanics and Microengineering*, *17*(2), 304–313. doi:10.1088/0960-1317/17/2/017

Zheng, L., Hong, F. S., Lu, S. P., & Liu, C. (2005). Effect of nano-TiO_2 on strength of naturally and growth aged seeds of spinach. *Biological Trace Element Research*, *104*(1), 83–91. doi:10.1385/BTER:104:1:083 PMID:15851835

Zhou, H., Zou, F., Koh, K., & Lee, J. (2014). Multifunctional magnetoplasmonic nanomaterials and their biomedical applications. *Journal of Biomedical Nanotechnology, 10*(10), 2921-2949.

Zhou, B., Hermans, S., & Gabor, A. (2004). *Nanotechnology in catalysis*. doi:10.1007/978-1-4419-9048-8

Zhou, H., Lee, J., Park, T. J., Lee, S. J., Park, J. Y., & Lee, J. (2012). Ultrasensitive DNA monitoring by Au–Fe 3 O 4 nanocomplex. *Sensors and Actuators. B, Chemical*, *163*(1), 224–232. doi:10.1016/j.snb.2012.01.040

Zhou, K., Wang, R., Xu, B., & Li, Y. (2006). Synthesis, characterization and catalytic properties of CuO nanocrystals with various shapes. *Nanotechnology*, *17*(15), 3939–3943. doi:10.1088/0957-4484/17/15/055

Zhou, L., Chong, A. Y. L., & Eric, W. T. (2015). Supply chain management in the era of the internet of things. *International Journal of Production Economics*, *159*, 1–3. doi:10.1016/j.ijpe.2014.11.014

Zhou, L., & Grubbström, R. W. (2004). Analysis of the effect of commonality in multi-level inventory systems applying MRP theory. *International Journal of Production Economics*, *90*(2), 251–263. doi:10.1016/S0925-5273(03)00208-1

Zhou, M., Li, J., Ye, Z., Ma, C., Wang, H., Huo, P., & Yan, Y. et al. (2015). Transferring charge and energy of Ag@CdSe QDs-rGO core-shell plasmonic photocatalyst for enhanced visible light photocatalytic activity. *ACS Applied Materials & Interfaces*, *7*(51), 28231–28243. doi:10.1021/acsami.5b06997 PMID:26669327

Zhou, X., Huang, X., Qi, X., Wu, S., Xue, C., Boey, F. Y., & Zhang, H. et al. (2009). In situ synthesis of metal nanoparticles on single-layer graphene oxide and reduced graphene oxide surfaces. *The Journal of Physical Chemistry C*, *113*(25), 10842–10846. doi:10.1021/jp903821n

Zhou, Z., Son, J., Harper, B., Zhou, Z., & Harper, S. (2015). Influence of surface chemical properties on the toxicity of engineered zinc oxide nanoparticles to embryonic zebrafish. *Beilstein Journal of Nanotechnology*, *6*(1), 1568–1579. doi:10.3762/bjnano.6.160 PMID:26425408

Zhu, H., Han, J., Xiao, J. Q., & Jin, Y. (2008). Uptake, translocation, and accumulation of manufactured iron oxide nanoparticles by pumpkin plants. *J Environ Monit*, *10*(6), 713-717. doi: 10.1039/b805998e

Zhuang, J., Qi, J., & Jin, Y. (2005). Retention and transport of amphiphilic colloids under unsaturated flow conditions: Effect of particle size and surface property. *Environmental Science & Technology*, *39*(20), 7853–7859. doi:10.1021/es050265j PMID:16295847

Zhuang, P., & Pavlostathis, S. G. (1995). Effect of temperature, pH, and electron donor on the microbial reductive dechlorination of chloroalkenes. *Chemosphere*, *31*(6), 3537–3548. doi:10.1016/0045-6535(95)00204-L

Zhu, B., Liu, G.-L., Ling, F., Song, L.-S., & Wang, G.-X. (2015). Development toxicity of functionalized single-walled carbon nanotubes on rare minnow embryos and larvae. *Nanotoxicology*, *9*(5), 579–590. doi:10.3109/17435390.2014.957253 PMID:25211547

Zhu, C., Guo, S., & Dong, S. (2012). PdM (M= Pt, Au) Bimetallic Alloy Nanowires with Enhanced Electrocatalytic Activity for Electro-oxidation of Small Molecules. *Advanced Materials*, *24*(17), 2326–2331. doi:10.1002/adma.201104951 PMID:22473584

Zhu, H. N., Lu, Q. X., & Abdollahi, K. (2005). Seed coat structure of Pinus koraiensis. *Microscopy and Microanalysis*, *11*(S02), 1158–1159. doi:10.1017/S1431927605503635

Zhu, J., Palchik, O., Chen, S., & Gedanken, A. (2000). Microwave assisted preparation of CdSe, PbSe, and Cu2-x Se nanoparticles. *The Journal of Physical Chemistry B*, *104*(31), 7344–7347. doi:10.1021/jp001488t

Zhu, M. T., Feng, W. Y., Wang, B., Wang, T. C., Gu, Y. Q., Wang, M., & Chai, Z. F. et al. (2008). Comparative study of pulmonary responses to nano and submicron-sized ferric oxide in rats. *Toxicology*, *247*(2-3), 102–111. doi:10.1016/j.tox.2008.02.011 PMID:18394769

Zhu, M., Chen, P., & Liu, M. (2011). Graphene oxide enwrapped Ag/AgX (X= Br, Cl) nanocomposite as a highly efficient visible-light plasmonic photocatalyst. *ACS Nano*, *5*(6), 4529–4536. doi:10.1021/nn200088x PMID:21524132

Zhuo, C., & Levendis, Y. A. (2014). Upcycling waste plastics into carbon nanomaterials: A review. *Journal of Applied Polymer Science*, *131*(4), n/a. doi:10.1002/app.39931

Zhu, S., Xu, T., Fu, H., Zhao, J., & Zhu, Y. (2007). Synergetic Effect of Bi_2WO_6 Photocatalyst with C_{60} and Enhanced Photoactivity under Visible Irradiation. *Environmental Science & Technology*, *41*(17), 6234–6239. doi:10.1021/es070953y

Zhu, W. P., Sun, S. P., Gao, J., Fu, F. J., & Chung, T. S. (2014). Dual-layer polybenzimidazole/polyethersulfone (pbi/pes) nanofiltration (nf) hollow fiber membranes for heavy metals removal from wastewater. *Journal of Membrane Science*, *456*, 117–127. doi:10.1016/j.memsci.2014.01.001

Zhu, X., Chang, Y., & Chen, Y. (2010). Toxicity and bioaccumulation of TiO_2 nanoparticle aggregates in Daphnia magna. *Chemosphere*, *78*(3), 209–215. doi:10.1016/j.chemosphere.2009.11.013 PMID:19963236

Zhu, X., Tian, S., & Cai, Z. (2012). Toxicity assessment of iron oxide nanoparticles in Zebrafish (Danio rerio) early life stages. *PLoS ONE*, *7*(9), e46286. doi:10.1371/journal.pone.0046286 PMID:23029464

Zhu, X., Wang, J., Zhang, X., Chang, Y., & Chen, Y. (2009). The impact of ZnO nanoparticle aggregates on the embryonic development of zebrafish (Danio rerio). *Nanotechnology*, *20*(19), 195103. doi:10.1088/0957-4484/20/19/195103 PMID:19420631

Zhu, Y., Wang, D., Jiang, L., & Jin, J. (2014). Recent progress in developing advanced membranes for emulsified oil/water separation. *NPG Asia Mater*, *6*(5), e101. doi:10.1038/am.2014.23

Zimmermann, R. (1966). Condensation Polymers: By Interfacial and Solution Methods. Von PW Morgan. John Wiley & Sons, New York London-Sydney 1965. 1. Aufl., XVIII, 561 S., zahlr. Abb., mehrere Tab., geb.£ 9.10.–. *Angewandte Chemie*, *78*(16), 787–787. doi:10.1002/ange.19660781632

About the Contributors

Sung Hee Joo is an Assistant Professor of Environmental Engineering at the University of Miami and an affiliate of the Dr. John T. McDonald Biomedical Nanotechnology Institute. Dr. Joo received PhD in environmental engineering at the University of New South Wales. Following her time in Australia, she conducted research on the formation and pathways of nitrogenous disinfection by-products during chlorine and chloramine disinfection at Yale University. She and her colleagues expanded research involving the development and applications of stabilized bimetallic nanomaterials for in situ remediation of chlorinated hydrocarbons. Dr. Joo has expertise in the field of advanced treatment technologies for emerging environmental contaminants, environmental nanotechnology, chemical nanoscience, the innovative processes of water/wastewater treatment, the application of membrane technology in wastewater, and the fate & transport of contaminants in the environment. She is a recipient of the Provost Research Award, USEPA's STAA, NRC, and YCC award of ACS.

* * *

Tatiana Andreani is graduated in Industrial Pharmacy at the University of Maringá (UEM), Brazil. She received her Ph.D. degree in Chemical and Biological Sciences at the University of Tras-os-Montes and Alto Douro, Portugal (UTAD) in 2014. Her research lines focus on the design and characterization of new and functional nanomaterials for drug delivery and targeting with high efficacy and low toxicity.

Jiyeol Bae, M.S. 2011 Environmental Science and Engineering, Gwangju Institute of Science and Technology (GIST) B.S. 2008 Environmental Engineering, University of Seoul.

César A Barbero, Full Professor (UNRC) Superior Researcher (CONICET) 141 publications, h index 31 (2015).

David Bogataj, Assist. Prof. DDr., actuary, PhD: Law and Real Estate management and PhD Quality management is Assisted Professor for Quantitative methods. His background is actuarial science and has more than 20 years expertise in statistical methods, actuarial models and risk theory, information systems and general systems theory in insurance and supply chain management. In the last 4 years he is working on the problems of supply systems with perturbed delays (recently at Universidad Politécnica de Cartagena, Spain), BigData and reproducible research.

Sirine Bouguerra is a Tunisian doctoral researcher with particular interests in Engineering biology, Environmental biotechnology and ecotoxicological risk assessment. Prior to enrolling doctoral studies, she Hold's her master degree in Environmental Biotechnology from University of Sfax, Tunisia. She collaborated in international scientific project: Sfp project (N° 983311) entitled - Remediation Processes in Uranium and other Mining Explorations, funded by the North Atlantic Treaty Organization (NATO). She have opportunity to get 9 months investigator fellowship from (UNESCO/ Japan Young Researchers' Fellowship Program) (Ref: ERI/NCS/FLP/LZF/11.236) for working in University of Aveiro, Portugal, Through years as researcher in Engineering school of Sfax in University of Tunisia; Biology Department in University of Aveiro, Portugal and in Biology department in Faculty of science, University of Porto, Portugal. She collaborated in National Portuguese project (PTDC/AAC-AMB/120697/2010) untitled: ReaLISE – Derivation of Risk LImits for the Protection of Soil Ecosystems from emerging compounds, within an investigation fellowship from Fundacão para a Ciência a Tecnologia (Portugal). She was author and co-author in several scientific papers and international communication deal in, waste water treatment, impact of mining pollution on soil biota and microbial community and impact of manufactured nanomaterials on environmental ecosystem. Now, Sirine's research interests lie in identifying risks of emerged nanomaterials in environment, particularly to terrestrial ecosystem.

Emrah Celik received PhD degree at University of Arizona in the field of Mechanical Engineering. Dr. Celik then continued postdoctoral training on bio-nanotechnology area at University of Miami at the department of Physiology and Biophysics. Dr. Celik continues to serve as an assistant professor at University of Miami and performs research on microfluidics, single molecule detection, cancer screening and cell mechanics.

Nurcin Celik is an Assistant Professor in the Department of Industrial Engineering at the University of Miami. She received her M.S. and Ph.D. degrees in Systems and Industrial Engineering from the University of Arizona. Her research interests lie in the areas of integrated modeling and decision making for large-scale, complex and dynamic systems with applications to electric power grids, semiconductor manufacturing, and solid waste systems. She has received several awards, including 2014 Eliahu I. Jury Early Career Research Award, 2013 AFOSR Young Investigator Award, 2011 IAMOT Outstanding Research Project Award, 2009 IIE Outstanding Graduate Research Award, the 2009 IERC Best Ph.D. Scientific Poster Award, and 2007 Diversity in Science and Engineering Award from Women in Science and Engineering Program.

Heechul Choi received PhD degree at Texas A&M University in 1995. He is professor at Gwangju Institute of Science and Technology (GIST) in Republic of Korea. His research interest is focused on application and implication of environmental nanotechnology, i.e., nano-material incorporated membrane for fresh water and energy generation, nano-architectured adsorbent for water purification and desalination, nano-fibrous membrane and filter for water and air using electrospinning technique, remediation for contaminated soil and groundwater using nanomaterials, water reuse and reclamation by natural purification and contaminant transport and modeling through porous media and nanomaterials.

About the Contributors

Hyeon-gyu Choi received bachelor's degree at Pusan National University in Republic of Korea in 2008 and MS and PhD degrees at Gwangju Institute of Science and Technology (GIST) in 2016. He is senior research scientist at Korea Research Institute of Chemical Technology (KRICT). He is interested in environmental nanotechnology based on separation and adsorption processes and conventional membrane treatment for water and energy harvesting.

Damjana Drobne, Prof. Dr., head of the group for Nanobiology and Nanotoxicology at the University of Ljubljana and professor for (environmental) toxicology and for zoology. Her background is in general biology (zoology) and has more than 20 years expertise in experimental toxicity. She has established own research group of Nanobiology and Nanotoxicology at the University of Ljubljana in 2008. Her research group has developed a series of methods for testing effects of metals or nanomaterials on cells, tissue and organisms as well as how to follow internal distribution of metals once entering the organisms. Up to now she has published together with co-authors more than 100 original scientific publications. Prof. Dr. Damjana Drobne is lecturing at undergraduate as well as postgraduate study programs at University of Ljubljana courses on zoology, toxicology, nanobiology and nanotoxicology. She has supervised or co-supervised 17 PhD students and numerous undergraduate students.

Antara Dutta Borah has completed her B.Sc (agriculture) and M.Sc (agriculture) in Agronomy from Assam Agriculture university, India. Presently working as research fellow in Rain Forest Research Institute, Jorhat, Assam.

Ahmed Nabile Emam earned his B.Sc and M.Sc degrees in medical biophysics in 2006 from Helwan University, and 2010 from Cairo University, respectively. In October 2006, he joined the Center of Excellence for Advanced Sciences "CEAS" (Nobel Project) in National Research Center "NRC", where he was engaged in Advanced Materials & Nanotechnology Lab as a temporary contract. Following his graduate work in M.Sc thesis, he promoted as an Assistant Researcher in Biomaterials Department, and CEAS. His research work in master thesis focused on the biophysical study of semiconductor nanoparticles doped with different magnetic materials (i.e. Co2+ ions) to be used in biomedical applications, under supervision of Dr. Mona Bakr Mohamed, who was the former postgraduate student of Prof. Mostafa A. El-Sayed. From December 2011 till now, Mr. Emam was a PhD student in Laser Sciences and Nanotechnology at the National Institute of Laser Enhanced Sciences "NILES" – Cairo University. In Sep. 2012, he joined to Ultrasound Imaging and Therapeutics Research Laboratory, Biomedical Engineering Department, University of Texas at Austin – USA as a Visiting Researcher till Jan. 2013. In this period; his research work focused on engineering of nanomaterials to be used as Photo-acoustic Tomography Imaging (PAT). Mr. Emam's research interests are in the areas of Nanotechnology including synthesis; characterization of different nanoparticles includes Magnetic, Plasmonic, Semiconductor and Diluted Magnetic Semiconductors NPs with different sizes and shapes. In addition to that studying the toxicological evaluation, and using of them in biomedical applications especially in biomedical imaging and diagnosis of diseases in early stages. Mr. Emam has published 2 articles publications and submits another 3 articles to international scientific journals. Distinctions & Awards: The Best M.Sc thesis 2010 Award, Faculty of Science, Cairo University; Partnership & Ownership Initiative (ParOwn) Travel Grant for Young Researcher up to 3 Months, funded by Egyptian Government (Cultural & Missions Section).

Scott Alan Floerke is a senior undergraduate student in the Mechanical and Aerospace Engineering Department at the University of Miami. Mr. Floerke has major research interests in the areas of atomic force microscopy, cell mechanics and nanotechnology. He hails from Tulsa, Oklahoma.

Ana Gavina is graduated in Biology and Geology by University of Aveiro in 2009, completed master's degree in Applied Biology – Toxicology and Ecotoxicology by University of Aveiro, in 2011. Currently is doing a PhD in Biology in Faculty of Science of University of Porto. She has been dedicated to ecotoxicology, mostly terrestrial, and trying to understand the effects of nanomaterials and their interaction with other contaminants (such as pesticides and metals) to terrestrial ecosystem. Furthermore, she is trying to determine risk limits for the nanomaterials tested.

Krishna Giri is Scientist B at the Rain Forest Research Institute Jorhat, Assam (Indian Council of Forestry Research and Education, Dehradun, Ministry of Environment, Forests & Climate Change, GoI, India) and has acquired M. Sc. and Ph. D. degrees in Environmental Sciences from G.B. Pant University of Agriculture and Technology, Pantnagar in the year 2010 and 2013 respectively. He has published about 10 research papers in peer reviewed scientific journals of National and International repute and also contributed papers to National and International seminars, symposiums and workshops. He is actively engaged in teaching and research activities in the field of carbon sequestration and shifting cultivation in Northeast India.

Kaliappan Krishnapriya has completed her graduate programme in Zoology and at present she is doing Ph.D. programme in the field of Zoology. Her Ph.D. topic is toxicity of nano particles on aquatic organisms.

Thomas A. Krug is an environmental engineer with Geosyntec Consultants Inc., Canada.

Wen Liu got his Bachelor Degree in Nankai University in 2009, and then got the PhD of Environmental Engineering in Peking University in 2014. Afterwards, he moved to Auburn University to start the postdoctoral program in Dongye Zhao's group at Civil Engineering Department. His research area includes environmental nanotechnology, material synthesis and application, technologies for remediation of heavy-metal contaminated waters, transport of nanomaterials and contaminates.

Isabel Lopes is a Principal Researcher at the Department of Biology and CESAM, University of Aveiro, where she develops research in the scientific areas of ecological risk assessment, aquatic ecotoxicological and microevolution due to chemical pollution, by using amphibians and other aquatic species as model organisms.

Ahmed Sadek Mansour gained a B.Sc. in Chemistry on 2003, Faculty of Science – Helwan University. In 2011, he earned a M.Sc. degree in Photochemistry and Nanotechnology, National Institute of Laser Enhanced Science (NILES) – Cairo University. Now he is a Ph.D. student at NILES. Research Focus His research work basically is focusing on synthesis of Nanomaterials include Semiconductor nanomaterials (i.e. CdSe, CdTe, CdS, ZnSe, etc…), Plasmonic nanomaterials (i.e. Ag, Au, Pt and etc….), Metal oxides (i.e. TiO_2, ZnO, and etc…), Magnetic nanomaterials (i.e. Fe_3O_4), and its nanocomposites

About the Contributors

such as (Metal\Metal, Metal\Semiconductor, Semiconductor\Semiconductor) and carbon nanomaterials such as Carbon dots and Graphene for different applications such as solar cells, Light Emitting Diode (LED) and Water Desalination.

Gaurav Mishra has completed his M. Sc. and Ph. D degree in Soil Science from G. B. Pant University of Agriculture and Technology, Pantnagar Presently he is working as a scientist B at the Rain Forest Research Institute, Jorhat, Assam. He has published more than 10 research papers in National and International reputed journals.

Mona Bakr Mohamed gained her B.Sc. 1991 and her MSc, in Chemistry in 1994. She received her PhD in 2002 from Georgia Tech, Atlanta, USA under supervision of Prof. Mostafa El-Sayed, then post-doctoral at University of Lausanne and senior research scientist in EPFL Switzerland. Dr. Mona Bakr Mohamed is Currently Associate Prof. at National Institute of Laser Enhanced Science, Cairo University. Research Focus Her research interest is focused on synthesis and characterization of metallic, magnetic and semiconductor nanocrystals of different shape and size, as well as their optical properties and ultrafast dynamics using different laser techniques. Her research group 22 graduate students are working on constructing novel nanomaterials for solar cell applications, photoelectronic devices, biomedical imaging, cancer therapy, nanocatalysis, and water treatment. Publications & Awards: She has 66 publications published in international journals such as Small, ACS Nano, Advanced Functional Mater., J. Phys. Chem, Physical Rev., Nano letters, Chemical phys. Letters, RSC Advances she supervised many students and formed the scientific school of more than 35 MSc and PhD students in various universities all over Egypt Dr. Mohamed receives three national grants from STDF and IMC to support her work on nanomaterials for solar cell applications and for biomedical imaging. Finally; for her work in the area of nanomaterials, Dr. Mohamed was elected by the academy of scientific research (ASRT) to State Encouragement Award in Advanced Technological Science "Basic Science" in 2010, Misr El-Khir for highly cited scientists and Cairo University for Scientific Publishing in 2010, 2011, 2012, 2013, and 2014, respectively.

Kyle Moor is a graduate student working on light-activated materials for disinfection.

Verónica Inês Jesus Oliveira Nogueiraa is a post-doctoral researcher at the Faculty of Sciences, University of Porto, Portugal. She has a PhD in Biology and a MSc degree in Toxicology and Ecotoxicology, from the University of Aveiro. Her research has been mainly focused on the safety application of nanomaterials for water and wastewater treatment. Her main research area has been ecotoxicology, mostly aquatic, of nanomaterials and the use of these nanomaterials as catalysts in the treatment of real effluents, namely the olive oil mill effluent.

Kathryn Gwenyth Nunnelley is a PhD. candidate at the University of Virginia in the Civil and Environmental Engineering department studying the use of metal nanoparticles in porous media for point-of-use water treatment. Received an M.E. from the University of Virginia in Civil and Environmental Engineering in 2015 and B.S. from Auburn University in 2013 in Polymer and Fiber Engineering.

Suzanne K. O'Hara is an environmental engineer with Geosyntec Consultants Inc., Canada.

Shailesh Pandey is working as a Scientist at Rain Forest Research Institute, Jorhat, Assam. He did his Masters and Ph.D. in Plant Pathology from G.B.P.U.A and T. Pantnagar. He is actively engaged in teaching, research and extension activities in the field of plant pathology. Dr. Pandey has over 10 scientific publications to his credit in national and international journals of repute. His area of interest is population biology of plant pathogens, biological control and plant disease management.

Ruth Pereira has a PhD in Biology, she is assistant professor with habilitation at the Faculty of Sciences of the University of Porto (UP), and Researcher at the Interdisciplinary Center of Marine and Environmental Research (CIIMAR-UP), where she coordinates the research group in Risk Assessment: Soil and Water Interactions. She is author/co-author of about 100 WOS papers (70% Q1), 6 books and 13 book chapters. Ruth Pereira has coordinated and participated as researcher in several national and international funded projects. She is/was supervisor of 14 PhD Students, 5 post-doc researchers and 32 MSc students. She is member of the external panel of advisors in risk assessment of the European Commission (SANCO).

Allan Philippe, after studying and working in the field of organic chemistry, completed his PhD degree at the University Koblenz-Landau, Germany, on the development of hydrodynamic chromatography coupled to ICP-MS and the interactions between natural organic matter and engineered nanoparticles. He is currently working as a scientific assistant at the University Koblenz-Landau on the development of new detection techniques for colloids in natural media. His interests cover also, among others, fate of engineered colloids in the environment, single particle ICP-MS, nanoparticle tracking analysis and agglomeration kinetics.

Robert W. Puls is a soil scientist (retired) from USEPA.

Jacqueline W. Quinn is an environmental engineer with NASA Kennedy Space Center, USA.

Mathan Ramesh has completed his graduate and Ph.D. programme in the field of Toxicology. At present he is working as a professor at Department of Zoology, Bharathiar University, Coimbatore, India. His field of specialization is Aquatic Toxicology and to his credit he has published more than 125 research articles in referred journals.

Teresa Rocha-Santos is graduated in Analytical Chemistry (1996) and PhD in Chemistry (2000). Since 2014, she has been a principal researcher at Centre for Environmental and Marine Studies (CESAM), University of Aveiro. She has more than 100 publications in journals from Science Citation Index. She has been the leader of 4 RTD projects (3 funded by Portuguese Science Foundation (FCT) and 1 bilateral cooperation FCT/CNRST) and has been a participant (researcher) in 7 RTD projects funded by FCT, 1 project funded by Calouste Gulbenkian Foundation, 1 RTD project funded by NATO Science and 1 RTD project funded by European Commission. Her research interests are the development of novel methods for environmental, food and heath care applications (fit for purpose); and the study of organic contaminants' fate and behaviour in the environment and during wastewater treatment.

About the Contributors

Nicolas Augustus Rongione has recently graduated from the Aerospace and Mechanical Engineering Department at the University of Miami. Mr. Rongione continues his PhD education at the University of Miami as a graduate research assistant. His major research interests are nanotechnology and microfluidics.

Nancy E. Ruiz is an environmental engineer with US Navy, USA.

James A. Smith is the Henry L. Kinnier Professor of Environmental Engineering in the Department of Civil and Environmental Engineering at the University of Virginia. He received his B.S. and M.S. degrees in Civil Engineering from Virginia Tech in 1983 and 1984, respectively. He received his Ph.D. in Civil Engineering from Princeton University in 1992. He has worked as a research hydrologist with the U.S. Geological Survey from 1985 to 1992. In 1992, he accepted his current position as a faculty member in the Civil and Environmental Engineering Department at the University of Virginia. Mr. Smith has served as the UPS Foundation Visiting Professor of Environmental Engineering at Stanford University (1998-99) and as the William R. Kenan Visiting Professor for Distinguished Teaching at Princeton University (2004-05). At the University of Virginia, he has been the recipient of the Alumni Board of Trustees Teaching Award (1997) and has held the Cavalier's Distinguished Teaching Chair (2000-02). He was selected to receive the AEESP/McGraw Hill Outstanding Teaching Award in 2002. Mr. Smith is a Fellow of the American Society of Civil Engineers and is the founder of PureMadi, a not-for-profit organization working to solve global water and health problems by working at the interface of water, societal, and human health disciplines. He is also a member of the Water and Health in Limpopo (WHIL) program in collaboration with the University of Venda in Thohoyandou, South Africa. His research interests include sustainable point-of-use water treatment technologies for the developing world and their impact on human health, the disinfection properties of zero-valent nano-silver and nano-copper particles, organic vapor transport in the vadose zone, low-impact development (LID) technologies for stormwater runoff, the fate and transport of emerging environmental pollutants, the engineering properties of organoclays, phytoremediation, and bacterial chemotaxis in porous media.

Moon Son has received B.S. from Yonsei University, Republic of Korea in 2010 as environmental engineering, and now PhD student in Gwangju Institute of Science and Technology, GIST. He is doing research on nano-enhanced membrane for fresh water and energy harvesting. He is undertaking research in not only performance evaluation of nano-enhanced membrane, but also synthesis of it.

Chunming Su is a Soil Scientist in the Subsurface Remediation Branch, Ground Water and Ecosystems Restoration Division, National Risk Management Research Laboratory, Office of Research and Development, United States Environmental Protection Agency. Dr. Su holds a Ph.D. in Soil Science from Washington State University. Before joining the EPA, he worked as a project scientist with ManTech Environmental Research Services Corporation, Ada, Oklahoma; as a National Research Council Resident Research Associate, Ada, Oklahoma; and a term soil scientist for USDA, Riverside, California. He serves as a National Remedy Review Board member for Superfund sites cleanup. He is an author of more than 100 publications and a co-recipient of a U.S. patent. He has served as a reviewer for 70 scientific journals and a reviewer of proposals to the Department of Commerce, EPA, USDA, and USGS. He has received several EPA Scientific and Technological Achievement Awards and EPA ORD Honor Awards. He has served as a mentor to dozens of National Research Council Resident Research Associates (regular and senior university faculty fellows), McNair Scholars, and Environmental Research Apprentice Program students. Dr. Su's research focus is on applications and implications of environmental nanotechnology with respect to the fate and transport of nanomaterials in the subsurface, and in situ treatment of organic (chlorinated solvents) and inorganic (chromate, arsenic, nitrate, etc.) contaminants in groundwater and soils using permeable reactive barrier technologies and monitored natural attenuation approaches.

Irshad Ahmad Wani was born and raised in state, Jammu and Kashmir, India. He received his M.Sc in Physical chemistry from Jamia Millia Islamia, New Delhi in 2007. He received his Ph.D. in 2012 from the same university under the tender supervision of Dr. Tokeer Ahmad. His Ph.D. mainly focused on synthesis of noble metal nanoparticles and their oxides by employing various chemical-biological routes and studying their application in biomedical area. He has been able to publish a good number of research papers in various esteemed international journals. His research work has been cited by a number of scientific journals. Currently he is working as lecturer in chemistry serving J&K govt. education services. He is also a peer reviewer of various international journals and a member of various scientific societies. His current research interests include Nanoparticle-Cancer, Nanoparticle-Protein and carbohydrate interactions, conjugation of inorganic nanoparticles with various biologically significant molecules and role of nanoparticles in sensor applications.

Mark T. Watling is an environmental engineer with Geosyntec Consultants Inc., Canada.

Duygu Yasar is a Ph.D. student in the Department of Industrial Engineering at the University of Miami. She earned her B.Sc. degree in the field of Industrial Engineering from the Yildiz Technical University, Turkey. She graduated in the top 5 percent of her class. Her research interests evolve around sustainable solid waste treatment, evaluation of solid waste management technologies, multi-criteria decision analysis, advanced thermal treatment of municipal waste, and modeling of integrated large-scale and complex systems.

Edith Inés Yslas was born in Argentina, in 1972. She graduated as a Microbiologist at the Universidad Nacional de Río Cuarto in 2000. She got the doctorate in Biology Science at the Universidad Nacional de Río Cuarto in 2005. Actually, she is a permanent research of the CONICET and Professor at the Universidad Nacional de Río Cuarto.

Index

A

Abiotic Dechlorination 93, 102, 105, 111
Aggregation 7, 14, 117, 119, 137-138, 141-143, 147, 193, 221, 237, 280, 330, 334, 339, 362-363
Agri-food 377-378, 380-381, 390, 392
Agrochemicals 174, 184
Analytic Hierarchy Process 204
Anisotropic Structures 245, 247, 255
Antibacterial 37-38, 40-41, 59, 185-187, 190, 192-196, 279, 300, 339, 353, 409
Antimicrobial 10, 27, 30-31, 38-40, 54, 56, 138, 142, 186-196, 202-203, 210, 277-278, 285, 295, 303, 353, 409, 411
Aquatic compartment 300
Aquatic Environment 19, 72, 81, 140, 296, 299, 330-331, 336-337, 339, 341, 353-354, 360-361
Ash 206, 209, 214, 230, 406

B

Biofilm 56, 186-187, 202, 280
Biogas 204, 206, 218-220, 225, 230
Biomarkers 361
Biomaterial 193-194, 202
Biopolymer 193, 203
Biotic Dechlorination 111
Bulk Copper 352, 354, 356-363

C

Carboxymethyl Cellulose (CMC) 112, 117-129, 135, 172
Ceramics 27, 41, 75, 353
Chlorinated Volatile Organic Compounds 94, 111
Coatings 40, 187-188, 190, 193, 278, 334, 353, 409
Cold Supply Chain 376, 384, 390
Combustion 205, 207, 213-214, 225, 230
Comparative Study 282, 352
Cooperative Supply Chain 383
Core Diameter 5, 18, 26
Cu NPs 352-363, 409-410
Cytotoxicity 192-193, 196, 203, 279, 296, 310-312, 407-411, 413-414

D

Dense Non-Aqueous Phase Liquid 92, 94, 111
Disinfection 28-32, 35, 80-81, 83, 114, 137-138, 140, 142, 153, 185-188, 278, 280, 403

E

Ecotoxicity 94, 294, 296, 300, 302, 304, 330-332, 341, 363
Effective Diameter 5, 10, 16-18, 26
Electrospinning 57-58, 71
Electrospun nanofiber 50-51, 57-61
Emerging Contaminants 72-74, 79, 81-82, 84, 94, 112
Emerging Organic Contaminants (EOCs) 112-115, 135
Endocrine Disrupting Chemicals (EDCs) 73, 81-82, 113-114, 116, 129, 135-136
Environmental Nanotechnology 92, 94, 111
Estradiol 81, 125, 127-128
Exposure 42, 79-80, 146, 172, 191, 277, 298-300, 302-304, 309, 311-312, 330-332, 336, 339-341, 356, 361-362, 403, 407, 409, 411, 413-415

F

Fiber 28, 37-41, 57-59, 71, 311
Filtration 3-5, 9, 13, 28, 31-35, 38-39, 41, 51, 57-59, 61, 71-72, 74, 78, 80, 83-84, 126, 149-150, 153, 166, 280
Floating Point 388
Forward Osmosis (FO) 51, 60-61, 71
Fractal Agglomerates 17, 26
Fuel Cells 283
Fullerene 137-143, 145-154

G

Groundwater 28-29, 80, 92-98, 102-105, 107, 111-118, 129, 135, 300
Groundwater Remediation 94, 111, 115
Gyration Diameter 5, 26

H

Hematology 362
Hybrid Nanocomposites 231-232, 248, 251-252, 254, 277, 280, 282, 285
Hydrodynamic Chromatography 2, 5, 10, 26
Hydrodynamic Diameter 5, 7-8, 10, 26, 119-121
hydrophilic functionality 137

I

ICP-MS 1-3, 5, 8-10, 13, 16, 18-19, 26
Inactivation of MS2 Bacteriophage 137, 147-148
In-Situ Remediation 112, 114, 116-118, 129, 136
Internet of Things 381-383, 393
Iron Oxides 2, 6, 107, 299

L

Labeo rohita 352, 354-355, 362-363
Landfill 116, 205-206, 210, 212, 225, 230
Laplace Transforms 385, 387-388
Logistics 376, 378-384, 386-389, 393

M

Magneto-Plasmonic 246-247, 249, 280
Manganese Dioxide 118
Mass Flux 93, 95, 97, 104, 107, 111
Maximum Contaminant Levels 94, 111
Mechanism of Action 187, 189, 193-194, 203, 278, 330, 339
Metal Nanoparticles 39, 232-233, 235, 244, 251, 258, 340, 353, 380, 407, 409
Metal oxide 119, 190-191, 280, 282-283, 300, 303-304, 336, 353, 412
Methane 93, 97, 99, 101, 218-219, 230
Microfiltration (MF) 41, 51, 58, 71, 74, 153
MRP Theory 384-385, 387
Municipal Solid Waste 204-205, 215

N

Nano Risk 174, 184
Nanoalloys 231-232, 237-239, 242, 244, 247, 249
Nano-Enhanced Membrane (NeM) 51, 54, 71
Nanofertilizers 164-167, 170, 174
Nanofilms 189-190, 192-193, 195
Nanofiltration 59, 72-84, 173
Nanoformulations 165, 171
Nanomaterials 3, 19, 30, 42, 51-54, 57, 59, 61, 71, 92, 94, 111, 153, 166, 169-170, 172, 174, 188-190, 193-196, 205, 209-210, 226, 231-233, 242, 249, 258-259, 277, 282, 294-296, 300, 302, 330-337, 339, 341, 354, 360, 377-378, 380, 391-392, 403-408, 414
Nanoparticle 5, 35-36, 39, 41, 112, 123, 126, 128-129, 189, 191, 193, 232, 234, 236, 246, 251, 253, 331-332, 334-336, 339-340, 352, 361, 380-381, 405, 411
Nanoparticle Synthesis 193, 246, 405
Nanoparticles 1, 3, 8-10, 16-19, 30-31, 35-41, 51, 71, 112, 116-118, 120-123, 126, 128-129, 135-136, 153, 165-167, 169-174, 184, 189-195, 205, 232-235, 237-242, 244-246, 248-254, 257, 259, 276-280, 282-283, 285, 304, 311, 330-336, 339-340, 352-354, 360-363, 376, 378, 380-381, 403-415
Nanopesticides 164, 171-172
Nanoremediation 92-94, 111, 166, 173
Nanosensors 164-165, 167, 172-173, 184, 378-382, 393
Nanotechnology 27, 50-51, 54, 57, 61, 92, 94, 111, 164-168, 170-174, 184-185, 188-189, 194-196, 205-206, 208-210, 225-226, 294-295, 300, 312, 330-331, 353, 360, 376-380, 382, 390-393, 403-405, 407, 414
nanotoxicity 240, 330-332, 339, 407
Nanowires 41, 189, 205, 249, 251, 257, 340, 380
Nanozeolites 184
National Priorities List (NPL) 111
Non-Aqueous Phase Liquids 111

O

Oxidation 30, 59, 114-116, 120-123, 126-127, 137, 139-140, 151, 153, 244, 280-281, 296, 335, 413-414
Oxidative Stress 279, 303, 308-311, 335, 340-341, 360, 412-414
Oxide Nanoparticles 171, 191, 311, 336, 362, 412-414

P

Paper 31, 35-38, 41, 193, 206, 208-213, 340, 382-383, 385-386
Perishable Goods 376-377, 381-382, 384, 386-387, 390
Pesticides 79-80, 164-167, 173, 184, 282, 295-296, 304
Photodegradation 276, 282-283
Physicochemical Characteristic 330

Plasma 5, 8, 16, 26, 40, 206-208, 213-214, 230, 404, 410
Pressure-Retarded Osmosis (PRO) 51, 53-54, 56, 60-61, 71
Pretreatment 74, 83-84, 212-213, 215

Q

Quantum Dots 138, 187, 193, 196, 251-252, 285, 353, 380

R

Reactive Oxygen Species 137, 187, 192-193, 279, 295, 335, 340, 408
Recycling 145, 166, 204, 207, 210-212, 220, 225-226, 230, 382-383
Reductive Dechlorination 93, 98, 102, 105, 117
Retention 10, 15, 17-18, 30-31, 77, 80-84, 173, 217, 219, 405
Reverse Osmosis (RO) 51, 54, 56, 71, 74, 78, 80, 83-84

S

Semiconductor 138, 166, 242, 250-251, 253-254, 259, 282-283, 299
Sensors 247, 250, 280, 285, 331, 376-378, 380, 382-385, 387, 393
Silver 10, 16-17, 27, 30-32, 34-42, 51-52, 54, 59, 153, 171, 174, 187, 189-191, 193-194, 239-242, 245, 251, 257, 277-279, 285, 295, 300, 303-304, 309, 311, 332, 336-337, 339, 411
Size Exclusion 4-5, 10, 41, 58, 80, 82-83
SP-ICP-MS 5, 8-10, 13-14, 16, 18, 26

Stabilized Nanoparticles 112, 117-119, 121, 129, 135-136
Steroidal Estrogens 113, 136
Superfund Site 92-93, 95, 107, 111
Sustainability 107, 164, 196, 204, 207, 226, 302
Sustainable Agriculture 165-167, 184
Syngas 213-215, 230
Synthetic Biology 172, 184

T

Tetrachloroethene 92, 94, 111
Textiles 27, 31, 38-39, 166, 331, 353
Toxicity 81, 153-154, 187, 190-191, 203, 210, 276, 279, 294-295, 298-300, 302-304, 309-312, 330-337, 340-341, 352-355, 360-363, 403, 405-407, 409-415
Transport 32, 55-58, 71, 78, 94, 107, 116, 128-129, 154, 193, 277, 302, 334, 339, 376-377, 381, 383, 385, 415

U

Ultrafiltration (UF) 51, 71, 74, 84

W

Wastewater Treatment 16, 72-73, 76, 79, 84, 113-114, 192, 209-210, 232, 294-296, 298-300, 310-311
Water-Energy Nexus 50-51, 61
Windrow 215, 217-218, 221, 223, 230

Receive Free Lifetime E-Access or Free Hardcover*

Purchase a print book or e-book through the IGI Global Online Bookstore and receive the alternate version for free! Shipping fees apply.

www.igi-global.com

Recommended Reference Books

Take **20% Off** through the IGI Global Online Bookstore¨

ISBN: 978-1-4666-2038-4
© 2013; 2,102 pp.
Take 20% Off:* $1,560

ISBN: 978-1-4666-4852-4
© 2014; 1,872 pp.
Take 20% Off:* $1,476

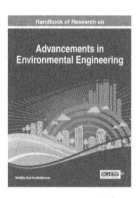

ISBN: 978-1-4666-7336-6
© 2015; 660 pp.
Take 20% Off:* $276

ISBN: 978-1-4666-2824-3
© 2013; 313 pp.
Take 20% Off:* $144

ISBN: 978-1-4666-4317-8
© 2014; 338 pp.
Take 20% Off:* $156

ISBN: 978-1-4666-4995-8
© 2014; 324 pp.
Take 20% Off:* $156

*IGI Global now offers the option to purchase hardcover, e-access, or hardcover + e-access for one price! You choose the format that best suits your needs. This offer is only valid on purchases made directly through IGI Global's Online Bookstore and not intended for use by book distributors or wholesalers. Shipping fees will be applied for hardcover purchases during checkout if this option is selected. E-Access will remain available for the lifetime of the publication and will only allow access to the edition purchased. Free Lifetime E-Access is only available to individuals and single institutions that purchase printed publications directly through IGI Global. Sharing the Free Lifetime E-Access is prohibited and will result in the termination of e-access. **20% discount cannot be combined with any other offer. Only valid on purchases made directly through IGI Global's Online Bookstore and not intended for use by book distributors or wholesalers. Not applicable on databases.

Publishing Progressive Information Science and Technology Research Since 1988

www.igi-global.com   Sign up at www.igi-global.com/newsletters   facebook.com/igiglobal   twitter.com/igiglobal

CPSIA information can be obtained
at www.ICGtesting.com
Printed in the USA
BVHW010830250620
R10897400001B/R108974PG581832BVX27B/19